Calcium:
A Matter of Life or Death

Edited by

Joachim Krebs

NMR Based Structural Biology,
Max Planck Institute for Biophysical Chemistry, Am Fassberg 11, D-37077 Göttingen,
Germany and Institute of Biochemistry, HPM1, Swiss Federal Institute of Technology,
Schafmattstrasse 18, CH-8093 Zürich, Switzerland

Marek Michalak

Department of Biochemistry,
3-56 Medical Sciences Building, University of Alberta,
Edmonton, Alberta, Canada T6G 2H7

ELSEVIER

Amsterdam · Boston · Heidelberg · London · New York · Oxford
Paris · San Diego · San Francisco · Singapore · Sydney · Tokyo

Elsevier
Radarweg 29, PO Box 211, 1000 AE Amsterdam, The Netherlands
The Boulevard, Langford Lane, Kidlington, Oxford OX5 1GB, UK
First edition 2007

Library of Congress Cataloging-in-Publication Data
A catalog record for this book is available from the Library of Congress

British Library Cataloguing in Publication Data
Calcium: a matter of life or death. — (New comprehensive biochemistry; v. 41)
1. Calcium – Physiological effect
2. Calcium in the body I. Krebs, J. (Joachim), 1940– II. Michalak, Marek 572.5′16

ISBN: 978-0-444-52805-6
ISBN: 978-0-444-80303-0 (Series)
ISSN: 0167-7306

For information on all Elsevier publications
visit our website at books.elsevier.com

Printed and bound in the United Kingdom

Transferred to Digital Print 2010

Contents

Preface

Calcium is a versatile carrier of signals regulating many aspects of cellular activity such as fertilization to create a new life and programmed cell death to end it. Calcium homeostasis is strictly controlled by channels, pumps and exchangers functioning as gates for calcium entry and release. Therefore, calcium message might be alike the two faces of Janus, the God of beginnings and endings, gates and doors. Just as the safety of a home may be breached, the dysregulation of calcium homeostasis may lead to many severe diseases.

Calcium controls virtually all cellular functions including energy metabolism, protein phosphorylation and de-phosphorylation, muscle contraction and relaxation, embryogenesis and subsequent development, cell differentiation and proliferation, gene expression, secretion, learning and memory, membrane excitability, cell-cycle progression and apoptosis. Calcium fulfills these functions because there is (i) a steep and tightly controlled concentration gradient of ionized Ca^{2+} across cellular membranes with typically 100–200 nM intracellular calcium concentration in resting cells and millimolar concentrations of calcium in the extracellular space and within intracellular organelles and (ii) a highly specific interaction of calcium with calcium-binding proteins resulting in modulations of many cellular functions. Given that calcium is such a versatile messenger, the field of calcium signaling is continuously and rapidly expanding. In this book, we review the most recent developments in calcium signaling by leading experts in the field. This volume is a state-of-the-art summary of our present knowledge in the quickly growing field of calcium signaling. It provides insight into the impressive progress made in many areas of calcium signaling and reminds us of how much remains to be learned.

We are grateful to all contributors for their enthusiasm and support of this exciting project. We are indebted to Elsevier Science and to Giorgio Bernardi, the general editor of this series, for the opportunity to create this book. We hope it provides a stimulating guide to workers in this research area and to a broader scientific community with a general interest in the fascinating field of calcium signaling. Our sincere thanks also go to Adriaan Klinkenberg, Tari Broderick and Anne Russum from Elsevier who kindly helped in all aspects of editing this volume. Last but not least, we are very thankful to our wives, Eva Krebs-Roubicek and Hanna Michalak, for their patience and understanding during the process of editing this book.

Joachim Krebs
Switzerland

Marek Michalak
Canada

Other volumes in the series

Volume 1. *Membrane Structure* (1982)
J.B. Finean and R.H. Michell (Eds.)

Volume 2. *Membrane Transport* (1982)
S.L. Bonting and J.J.H.H.M. de Pont (Eds.)

Volume 3. *Stereochemistry* (1982)
C. Tamm (Ed.)

Volume 4. *Phospholipids* (1982)
J.N. Hawthorne and G.B. Ansell (Eds.)

Volume 5. *Prostaglandins and Related Substances* (1983)
C. Pace-Asciak and E. Granstrom (Eds.)

Volume 6. *The Chemistry of Enzyme Action* (1982)
M.I. Page (Ed.)

Volume 7. *Fatty Acid Metabolism and its Regulation* (1982)
S. Numa (Ed.)

Volume 8. *Separation Methods* (1984)
Z. Deyl (Ed.)

Volume 9. *Bioenergetics* (1985)
L. Ernster (Ed.)

Volume 10. *Glycolipids* (1985)
H. Wiegandt (Ed.)

Volume 11a. *Modern Physical Methods in Biochemistry, Part A* (1985)
A. Neuberger and L.L.M. van Deenen (Eds.)

Volume 11b. *Modern Physical Methods in Biochemistry, Part B* (1988)
A. Neuberger and L.L.M. van Deenen (Eds.)

Volume 12. *Sterols and Bile Acids* (1985)
H. Danielsson and J. Sjovall (Eds.)

Volume 13. *Blood Coagulation* (1986)
R.F.A. Zwaal and H.C. Hemker (Eds.)

Volume 14. *Plasma Lipoproteins* (1987)
A.M. Gotto Jr. (Ed.)

Volume 16. *Hydrolytic Enzymes* (1987)
A. Neuberger and K. Brocklehurst (Eds.)

Volume 17. *Molecular Genetics of Immunoglobulin* (1987)
F. Calabi and M.S. Neuberger (Eds.)

Volume 18a. *Hormones and Their Actions, Part 1* (1988)
B.A. Cooke, R.J.B. King and H.J. van der Molen (Eds.)

Volume 18b. *Hormones and Their Actions, Part 2 – Specific Action of Protein Hormones* (1988)
B.A. Cooke, R.J.B. King and H.J. van der Molen (Eds.)

Volume 19. *Biosynthesis of Tetrapyrroles* (1991)
P.M. Jordan (Ed.)

Volume 20. *Biochemistry of Lipids, Lipoproteins and Membranes* (1991)
D.E. Vance and J. Vance (Eds.) – Please see Vol. 31 – revised edition

Volume 21. *Molecular Aspects of Transfer Proteins* (1992)
J.J. de Pont (Ed.)

Volume 22. *Membrane Biogenesis and Protein Targeting* (1992)
W. Neupert and R. Lill (Eds.)

Volume 23. *Molecular Mechanisms in Bioenergetics* (1992)
L. Ernster (Ed.)

Volume 24. *Neurotransmitter Receptors* (1993)
F. Hucho (Ed.)

Volume 25. *Protein Lipid Interactions* (1993)
A. Watts (Ed.)

Volume 26. *The Biochemistry of Archaea* (1993)
M. Kates, D. Kushner and A. Matheson (Eds.)

Volume 27. *Bacterial Cell Wall* (1994)
J. Ghuysen and R. Hakenbeck (Eds.)

List of contributors*

James B. Ames 95
Chemistry Department, University of California, Davis, CA 95616, USA

Curtis F. Barrett 127
Department of Molecular and Cellular Physiology, Stanford University Medical Center, Stanford, CA 94305-5345, USA

Michael John Berridge 485
The Babraham Institute, Babraham, Cambridge CB2 4AT, UK

Jacob Brenner 403
Department of Neurobiology, Stanford University School of Medicine, 299 Campus Drive, Fairchild Building Rm. 227, Stanford, CA 94305, USA

Yu-Qing Cao 127
Washington University Pain Center, Basic Research Division, Department of Anesthesiology, Washington University School of Medicine, St. Louis, MO 63110, USA

Ernesto Carafoli 3, 179
Venetian Institute of Molecular Medicine and Department of Biochemistry, University of Padova, Viale Colombo 3, 35121 Padova, Italy

Felice A. Chow 345
Department of Cell Biology, FibroGen, Inc., South San Francisco, CA 94080, USA; Department of Pharmacology and Cancer Biology, Duke University Medical Center, Box 3813, Durham, NC 27710-3813, USA

Jong Tai Chun 425
Cell Signaling Laboratory, Stazione Zoologica "Anton Dohrn" Villa Comunale, I-80121, Naples, Italy

Leonard Dode 229
Laboratorium voor Fysiologie, K.U. Leuven, Campus Gasthuisberg O/N, Herestraat 49 Bus 802, B3000 Leuven, Belgium

*Authors' names are followed by the starting page number(s) of their contribution(s).

Ricardo Dolmetsch 403
Department of Neurobiology, Stanford University School of Medicine, 299 Campus Drive, Fairchild Building Rm. 227, Stanford, CA 94305, USA

Stefan Feske 365
Harvard Medical School and the CBR Institute for Biomedical Research, Harvard Medical School, 200 Longwood Avenue, Boston, MA 02115, USA

Natalia Gomez-Ospina 403
Department of Neurobiology, Stanford University School of Medicine, 299 Campus Drive, Fairchild Building Rm. 227, Stanford, CA 94305, USA

Claus W. Heizmann 51
Division of Clinical Chemistry and Biochemistry, Department of Pediatrics, University of Zürich, Steinwiesstrasse 75, CH-8032 Zürich, Switzerland

Patrick G. Hogan 365
Harvard Medical School and the CBR Institute for Biomedical Research, Harvard Medical School, 200 Longwood Avenue, Boston, MA 02115, USA

Mitsuhiko Ikura 95
Division of Signaling Biology, Ontario Cancer Institute and Department of Medical Biophysics, University of Toronto, Toronto, Ontario, Canada M5G 1L7; Toronto Medical Discovery Tower, MaRS Centre, 4th Floor, Room 4-804, 101 College Street, Toronto, Ontario, Canada M5G 1L7

Allison Kraus 199
Department of Biochemistry, University of Alberta, Edmonton, Alberta, Canada T6G 2H7

Joachim Krebs 51, 167
NMR Based Structural Biology, MPI for Biophysical Chemistry, Am Fassberg 11, D-37077 Göttingen, Germany; Institute of Biochemistry, HPM1, Swiss Federal Institute of Technology, ETH-Hönggerberg, Schafmattstrasse 18, CH-8093 Zürich, Switzerland

F. Anthony Lai 287
Department of Cardiology, Wales Heart Research Institute, School of Medicine, Cardiff University, Cardiff CF14 4XN, UK

Anthony R. Means 345
Department of Pharmacology and Cancer Biology, Duke University Medical Center, Box 3813, Durham, NC 27710-3813, USA

Marek Michalak 199
Department of Biochemistry, University of Alberta, Edmonton, Alberta, Canada T6G 2H7

Katsuhiko Mikoshiba 267
The Institute of Medical Science, The University of Tokyo, and RIKEN, Brain Science Institute, Calcium Oscillation Project, ICORP-SORST, JST, 4-6-1 Shirokanedai, Minato-ku, Tokyo 108-8639, Japan

Andrew L. Miller 445
Department of Biology, The Hong Kong University of Science and Technology, Clear Water Bay, Hong Kong, People's Republic of China

Ludwig Missiaen 229
Laboratorium voor Fysiologie, K.U. Leuven, Campus Gasthuisberg O/N, Herestraat 49 Bus 802, B3000 Leuven, Belgium

Claudia Ortega 179
Venetian Institute of Molecular Medicine and Department of Biochemistry, University of Padova, Viale Colombo 3, 35121 Padova, Italy

Saida Ortolano 179
Venetian Institute of Molecular Medicine and Department of Biochemistry, University of Padova, Viale Colombo 3, 35121 Padova, Italy

Erika S. Piedras-Rentería 127
Department of Physiology, Loyola University Medical Center, Stritch School of Medicine, Maywood, IL 60156-5500, USA

Tullio Pozzan 471
Department of Biomedical Sciences and CNR Institute of Neuroscience, University of Padova, Viale Colombo 3, 35121 Padova, Italy; Venetian Institute of Molecular Medicine and Interdipartimental Centre for the Study of Cell Signals, University of Padova, Via Orus 2, 1-35121 Padova, Italy

Luc Raeymaekers 229
Laboratorium voor Fysiologie, K.U. Leuven, Campus Gasthuisberg O/N, Herestraat 49 Bus 802, B3000 Leuven, Belgium

Anjana Rao 365
Harvard Medical School and the CBR Institute for Biomedical Research, Harvard Medical School, 200 Longwood Avenue, Boston, MA 02115, USA

Rosario Rizzuto 471
Department of Experimental and Diagnostic Medicine, Section of General Pathology, ER-GenTech laboratory and Interdisciplinary Centre for the Study of Inflammation (ICSI), University of Ferrara, Via Borsari 46, 44100 Ferrara, Italy

Luigia Santella 425
Cell Signaling Laboratory, Stazione Zoologica "Anton Dohrn" Villa Comunale, I-80121, Naples, Italy

Peter B. Stathopulos 95
Division of Signaling Biology, Ontario Cancer Institute and Department of Medical Biophysics, University of Toronto, Toronto, Ontario, Canada M5G 1L7

Chikashi Toyoshima 219
Institute of Molecular and Cellular Biosciences, The University of Tokyo, 1-1-1 Yayoi, Bunkyo-ku, Tokyo 113-0032, Japan

Richard W. Tsien 127
Department of Molecular and Cellular Physiology, Stanford University Medical Center, Stanford, CA 94305-5345, USA

Jo Vanoevelen 229
Laboratorium voor Fysiologie, K.U. Leuven, Campus Gasthuisberg O/N, Herestraat 49 Bus 802, B3000 Leuven, Belgium

Sarah E. Webb 445
Department of Biology, The Hong Kong University of Science and Technology, Clear Water Bay, Hong Kong, People's Republic of China

Robert J. P. Williams 23
Inorganic Chemistry Laboratory, University of Oxford, South Parks Road, Oxford OX1 3QR, UK

Frank Wuytack 229
Laboratorium voor Fysiologie, K.U. Leuven, Campus Gasthuisberg O/N, Herestraat 49 Bus 802, B3000 Leuven, Belgium

Spyros Zissimopoulos 287
Department of Cardiology, Wales Heart Research Institute, School of Medicine, Cardiff University, Cardiff CF14 4XN, UK

History and Evolution of Calcium Biochemistry

J. Krebs and M. Michalak (Editors)
Calcium: A Matter of Life or Death

ISSN: 0167-7306/DOI: 10.1016/S0167-7306(06)41001-2

CHAPTER 1

The unusual history and unique properties of the calcium signal

Ernesto Carafoli

Venetian Institute of Molecular Medicine and Department of Biochemistry, University of Padova, Viale Colombo 3, 35121 Padova, Italy, Tel.: +39 049 7923 240; Fax: +39 049 8276 125; E-mail: ernesto.carafoli@unipd.it

Abstract

Calcium (Ca^{2+}) is the most universal carrier of biological signals: it modulates cell life from its origin at fertilization to its end in the apoptotic process. The signaling function of Ca^{2+} has had an unusual history. Discovered serendipitously at the end of the nineteenth century, it received almost no attention for decades. After its rediscovery in the 1960s, it grew in popularity and importance at an exponential pace. As progress advanced, it was recognized that the signaling function of Ca^{2+} had a number of unique properties. One is the ability of Ca^{2+} to act both as a first and as a second messenger. Ca^{2+} may recognize a canonical seven-transmembrane domain receptor at the external side of the plasma membrane, initiating an internal signaling cascade that may even involve Ca^{2+} itself. Another distinctive property of the Ca^{2+} signal is autoregulation. It occurs at the transcriptional and post-translational levels, as the expression and activity of a number of proteins involved in the transport of Ca^{2+} and in the processing of its signal are regulated by Ca^{2+} itself. Most importantly, the Ca^{2+} signal shows ambivalence. Cells need Ca^{2+} to correctly carry out most of their important functions. To this aim, they have developed a sophisticated array of means to carefully control its concentration and movements. But damage of various degrees, up to cell death, invariably follows the failure of the cell systems to properly control Ca^{2+}.

Keywords: calcium and disease, calcium channels, calcium sensors, calcium transporters, signal autoregulation

1. Preamble

The evolutionary transition from unicellular to multicellular life brought with it the division of labor among cells. This in turn generated the necessity of exchanging signals among the cells that formed the new organisms: as is well known, as a rule monocellular organisms do not depend on the exchange of signals, their mutual interactions being largely confined to the exchange of nutrients. Cell signaling is thus essentially a hallmark of pluricellular organisms, in which a multiplicity of external signals, termed first messengers, is translated inside cells into specific responses mediated by a less numerous group of newly produced chemicals, the second messengers. External signals, e.g., hormones, had long been known, but second messengers have only arrived on the scientific scene more recently, with the discovery of cyclicAMP in the 1950s. One of the second messengers, however, had become known much earlier: it was calcium (Ca^{2+}), which was the main actor in a landmark, even if serendipitous, experiment published by Ringer in 1883 [1]. At the time of the experiment, Ca^{2+} was already recognized as an important structural element in bone and teeth, but Ringer showed that it also had a completely unexpected additional function: it carried the signal that promoted the contraction of isolated hearts. Ringer's experiment, which is shown in Fig. 1, is now justly famous. Peculiarly, however, it remained essentially forgotten for decades, until a new set of discoveries brought it back to center stage, initiating the extraordinarily popular area of Ca^{2+} signaling. The advancement of knowledge rapidly became all-encompassing: as large amounts of new information were gathered, it gradually became clear that Ca^{2+} had a number of properties that made it unique among all other carriers of biological information. Three of these properties are particularly striking and will be reviewed in some detail in this contribution. The first is the ability of Ca^{2+} to act both as a first and as a second messenger. The second is the autoregulative property of the Ca^{2+} signal, i.e., Ca^{2+} itself may control the generation and the regulation of the informxation it carries. Finally, and perhaps most importantly, the Ca^{2+} signal has a clear ambivalent character: although essential to the

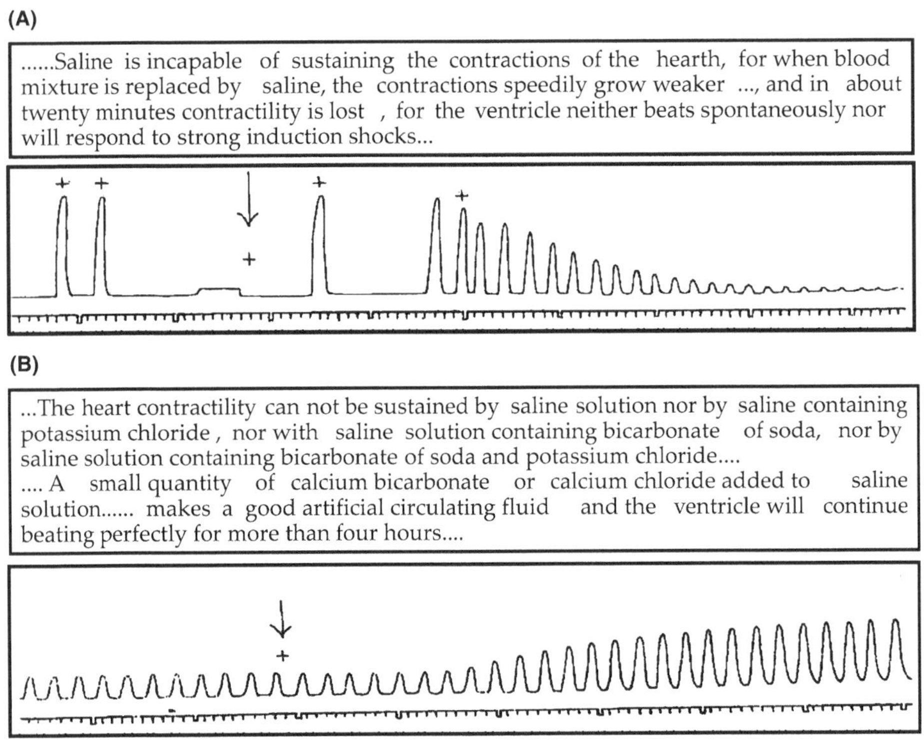

Fig. 1. The experiment of Ringer on the contraction of isolated frog ventricles [1]. **(A)** Traces in which the blood mixture was replaced by saline at the point indicated by the arrow. **(B)** The effect of addition of 3.5 ml of calcium chloride to the medium in which the heart was suspended. The figure also shows the comments by Ringer in the text.

correct functioning of cells, it can easily become a messenger of death. Prior to discussing these special properties, some general concepts of Ca^{2+} signaling will succinctly discussed to provide some common ground of understanding. For a detailed discussion of these principles, a number of comprehensive recent reviews should be consulted [2–5].

2. *General principles of Ca^{2+} signaling*

Eukaryotic cells are surrounded by media containing free Ca^{2+} concentrations that exceed 1 mM, but manage to maintain a free Ca^{2+} concentration in the cytoplasm that is four orders of magnitude lower. This very low internal concentration is dictated by the necessity to avoid the precipitation of Ca^{2+}, given the poor solubility of Ca^{2+}-phosphate salts. It also prevents the prohibitive expenditures of energy that would otherwise be necessary to significantly change the background concentration of Ca^{2+} in the vicinity of the targets that must be regulated. If the resting free Ca^{2+} in the cytosol were much higher than nM, considerably larger amounts of energy would have to be invested to change it substantially. The total Ca^{2+} concentration within cells is naturally much higher than nM but is reduced to the sub-μM ionic range first by binding to membrane (acidic) phospholipids, to low molecular weight metabolites, to inorganic ions like phosphate, and then by complexation to specific proteins. These belong to several classes: one comprises the membrane intrinsic proteins that operate as Ca^{2+} transporters in the plasma membrane and in the membranes of the organelles. These proteins play the most important role in the maintenance of the cellular Ca^{2+} homeostasis as they move Ca^{2+} back and forth between the cytosol, where most of the targets of the Ca^{2+} signal are located, and the extracellular spaces or the lumenal spaces of the organelles. Depending on whether large amounts of Ca^{2+} must be moved, or whether only the fine-tuning of Ca^{2+} down to very low concentration levels is required, low or high Ca^{2+} affinity transporters operate. The high-affinity transport mode demands ATPases, which are located in the plasma membrane (plasma membrane Ca^{2+} pumps, PMCA pumps), in the endo(sarco)plasmic reticulum (endo(sarco) plasmic reticulum Ca^{2+} pumps, SERCA pumps), and in the Golgi membranes (secretory pathway Ca^{2+} pumps, SPCA pumps). Low-affinity transport relies instead on Na/Ca exchangers (NCXs) in the plasma membrane and in the inner mitochondrial membrane and on the electrophoretic Ca^{2+} uptake uniporter of the inner membrane of mitochondria. Proteins also form channels for the specific traffic of Ca^{2+} through the plasma membrane and the membranes of some organelles. In the plasma membrane, these channels belong to several groups, depending essentially on the gating mechanisms: the voltage-gated channels are tetrameric structures that are particularly active in excitable cells, the ligand-gated channels are operated by a number of chemicals of which the neurotransmitters are the best characterized examples, and the store-operated channels (SOCs), which comprise several members of the now very popular of the transient receptor potential (trp) channels, are gated by a still poorly understood mechanism triggered by the emptying of the cellular Ca^{2+} stores. Ligand-gated channels also operate in some organelles, essentially the endo(sarco) plasmic reticulum and the Golgi system. The most important ligand that gates these internal channels, inositol 1,4,5-tris-phosphate (InsP3) [6], is generated in response to the interaction of extracellular stimuli (first messengers) with plasma membrane receptors, e.g., purinergic or adrenergic receptors. Other ligands for intracellular Ca^{2+} channels that have become popular more recently are cyclic-ADP ribose (cADPr) and the nicotinic acid analogue of NADP (NAADP): unfortunately, the information available on the mechanism(s) that connects their generation to the interaction of first messengers with plasma membrane receptors is still very limited. Fig. 2 offers an overall panorama of the control of Ca^{2+} in eukaryotic cells.

Another class of proteins specifically able to bind Ca^{2+} is that of soluble proteins (or of proteins organized in non-membranous structures, e.g., the myofibrils or the cytoskeleton). A minority of these proteins merely operate as Ca^{2+} buffers, lowering its free concentration in the cytoplasm to the sub-μM level. The majority, instead, in addition to contributing to the lowering of the free Ca^{2+} concentration, perform the additional important function of processing its signal: as they complex Ca^{2+}, they

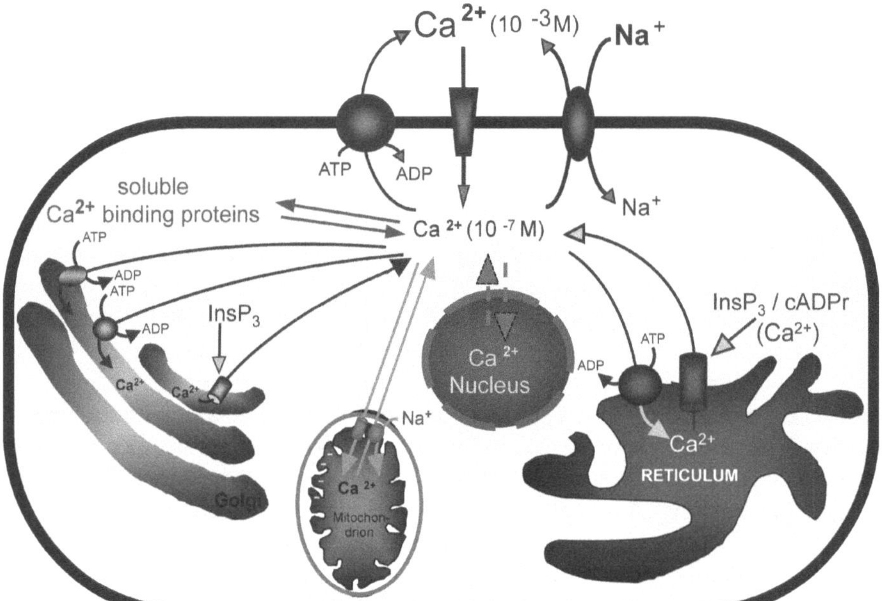

Fig. 2. A global view of cellular Ca^{2+} homeostasis. The figure shows the systems that transport Ca^{2+} across the plasma membrane and the membranes of the organelles. The transport of Ca^{2+} in and out of the nucleus is represented by dashed arrows to indicate the present controversy over whether the traffic of Ca^{2+} across the nuclear envelope occurs continuously and passively through the pores or whether the pores are somehow gated. For simplicity, only one Ca^{2+} channel type is indicated in the plasma membrane. Ca^{2+} enters the reticulum and the Golgi through ATPases: the SERCA pump in the reticulum, and the SERCA pump plus the SPCA pump in the Golgi. It exits from these two compartments through channels activated by ligands: InsP3 and cADPr in the reticulum, InsP3 in the Golgi. The novel Ca^{2+}-linked messenger NAADP is not indicated in the figure as the membrane system on which it acts has net yet been conclusively identified. Ca^{2+} enters mitochondria through the electrophoretic uniporter, which is energized by the negative inside membrane potential maintained by the respiratory chain. It leaves them through exchangers, of which the most important and best characterized is a NCX. The figure also shows the two systems that export Ca^{2+} from cells, a high-affinity low-capacity PMCA pump and a low-affinity large-capacity NCX. Finally, it shows soluble proteins that buffer Ca^{2+} and may also decode its message (the Ca^{2+} sensor proteins) (See Color Plate 1, p. 503).

decode its information before conveying it to (enzyme) targets with which they interact. These proteins are thus appropriately defined as Ca^{2+} sensors. They belong to several classes: the annexins, gelsolin, proteins containing C-2 domains, and the EF-hand proteins. Most of them, e.g., the annexins and gelsolin, process the Ca^{2+} signal for the benefit of a single target function. The most important of these Ca^{2+}-binding proteins, the EF-hand proteins [7], are instead a very large family of related proteins, now estimated to number more than 600 members [8]. They modulate a vast number of targets, each one of them, as a rule, processing the Ca^{2+} signal for the benefit of a single enzyme/protein. The most important of them, calmodulin, transmits instead Ca^{2+} information to scores of targets. The EF-hand proteins bind Ca^{2+} to a variable number of helix-loop-helix motifs in which two approximately orthogonal helical domains flank a non-helical 12-amino-acid loop in which Ca^{2+} coordinates to side chain and carbonyl oxygens of five invariant residues and in some cases to a water oxygen [7]. This binding mode, which was first described in parvalbumin about 40 years ago [7], has remained essentially valid as a structural paradigm for all EF-hand proteins. Since the mid 1990s, however, variants of this basic Ca^{2+}-binding mode have been described, including the contribution of coordinating oxygens from domains outside the helix-loop-helix motif and even from neighboring proteins.

The fine structural details of the decoding of the Ca^{2+} signal are understood down to the atomic level only for the case of calmodulin [9]. This is a dumbbell-shaped protein that contains two helix-loop-helix Ca^{2+}-binding motifs in each of the two terminal globular lobes. It undergoes a first conformational change upon binding Ca^{2+} that exposes hydrophobic methionine residues, of which calmodulin is very rich, predisposing the proteins for the interaction with specific binding domains in target enzymes. As this occurs, a second conformational change collapses the elongated protein along its central helix to wrap it, hairpin style, around the binding domain of target enzymes [10], completing the transmission of the Ca^{2+} signal. As mentioned, calmodulin is the only Ca^{2+} sensor protein for which such a sophisticated level of molecular understanding is presently available. Possibly, some basic molecular features of the processing of the Ca^{2+} signal by calmodulin may be valid for other EF-hand proteins as well. However, the dumbbell shape, which is an essential ingredient in the processing of the Ca^{2+} signal by calmodulin, seems to be the exception rather than the rule in the family of EF-hand proteins. As for the fine structural details of the processing of the Ca^{2+} signal by proteins different from those of the EF-hand family, only very scarce information is currently available.

Many EF-hand proteins, calmodulin being the most obvious example, are regulatory subunits that become temporarily associated with target proteins. In some cases, however, the association may be irreversible and persist even in the absence of Ca^{2+}. Again, this is the case for some of the targets of calmodulin regulation, e.g., phosphorylase kinase. In other cases, calmodulin may become associated in the canonical reversible way but to enzymes that already possess their own calmodulin like EF-hand subunit. This is, for example, the case for the Ca^{2+}-activated protein phosphatase calcineurin, which is thus the target of dual Ca^{2+} regulation. Other cases of dual Ca^{2+} regulation exist, even if slightly at variance with that of calcineurin. The protease calpain is one striking example: it contains a calmodulin like Ca^{2+}-binding domain covalently bound within the catalytic subunit. Then, a separate EF-hand subunit also interacts with the catalytic subunit. Finally, numerous proteins, enzymes or otherwise, contain specific binding sites for Ca^{2+}, which thus regulates their activity directly, without the intermediation of proteins like calmodulin. These proteins, of which one important example is protein kinase C, can thus also be considered as bona fide Ca^{2+} sensors.

3. Calcium is both a first and a second messenger

Signaling operations typically involve the interaction of messenger chemicals (the first messengers) with the surface of target cells, which then process the incoming message and initiate an internal signaling chain which is normally mediated by soluble signaling molecules (the second messengers, see above). This chain of events is the most common way of exchanging information between cells but is by no means unique. Cells can also communicate with each other through direct contacts in the form of gap junctions or through surface proteins that recognize protein counterparts on the surface of other cells. As for first messengers, some may bypass the plasma membrane and penetrate directly into target cells. Once there, they interact with receptors in the cytoplasm and/or the nucleus and act without the intermediation of second messengers. These alternative possibilities are all interesting and important: however, the classical way of exchanging information remains that based on the interaction of first messengers with plasma membrane receptors and on the generation of diffusible second messengers inside cells.

In considering Ca^{2+} based on the points above, a number of peculiarities emerge that are difficult to reconcile with the canonical second messenger concept. Ca^{2+} certainly behaves as a diffusible second messenger generated inside cells in response to the interaction of a number of first messengers with plasma membrane receptors. However, the increase of cell Ca^{2+} is not directly linked to the processing of first messenger signals, as is the case, for instance, for cAMP. It occurs one step further down the signaling chain and follows the generation of another bona fide second messenger, e.g., InsP3, which will then liberate Ca^{2+} from intracellular stores. Strictly speaking, then, Ca^{2+} should be defined as a third messenger. But Ca^{2+} can also penetrate directly into cells from the external spaces to initiate the intracellular signaling chain. This, however, is not unique to Ca^{2+}: other second messengers, e.g., the NO radical, may also have this dual origin (interstitial spaces and intracellular ambient).

The peculiarities above may be considered minor and essentially formal. Another distinctive property of the Ca^{2+} signal, however, is more substantial. Ca^{2+} has the ability, now being documented in a growing number of cell types, to recognize specific G-protein-linked plasma membrane receptors. Following the interaction, a G-protein-mediated chain of events is initiated that may even end in the modulation of the release of Ca^{2+} itself from intracellular stores.

The finding that Ca^{2+} acted on a plasma membrane receptor as any other canonical first messenger came as a surprise [11]: the concentration of first messengers is very low and necessarily fluctuates around cells, whereas that of Ca^{2+} is high and remains remarkably constant. Thus, the possibility that extracellular Ca^{2+}

would act as a first messenger had not been generally considered. However, it had long been known that cells secreting the calciotropic hormones that regulate the homeostasis of Ca^{2+} in the organism, e.g., the cells of the parathyroid gland, modify the release of the hormones in response to perturbations of external Ca^{2+}. Indications that the mechanism by which these cells sensed the changes in external Ca^{2+} may have involved a plasma membrane receptor came much more recently. The first clear findings in this direction were the discoveries that raising the level of external Ca^{2+} induced a transient and then a sustained elevation of Ca^{2+} in parathyroid cells [12], which was linked to the activation of phospholipase C [13]. This type of response to extracellular Ca^{2+} was then documented in a number of other cell types, particularly those of the kidney (see Ref. 11 for a review). Eventually, the plasma membrane Ca^{2+} receptor was cloned, first from a bovine parathyroid cDNA library [14] and then from a number of other cells. Among them are those that are involved in the regulation of organismic Ca^{2+} homeostasis, i.e., the C-cells of the thyroid and kidney cells. The Ca^{2+} receptor, which is routinely referred to as the Ca^{2+} sensor, is predicted to be topographically organized into three domains (Fig. 3): a 600-residue

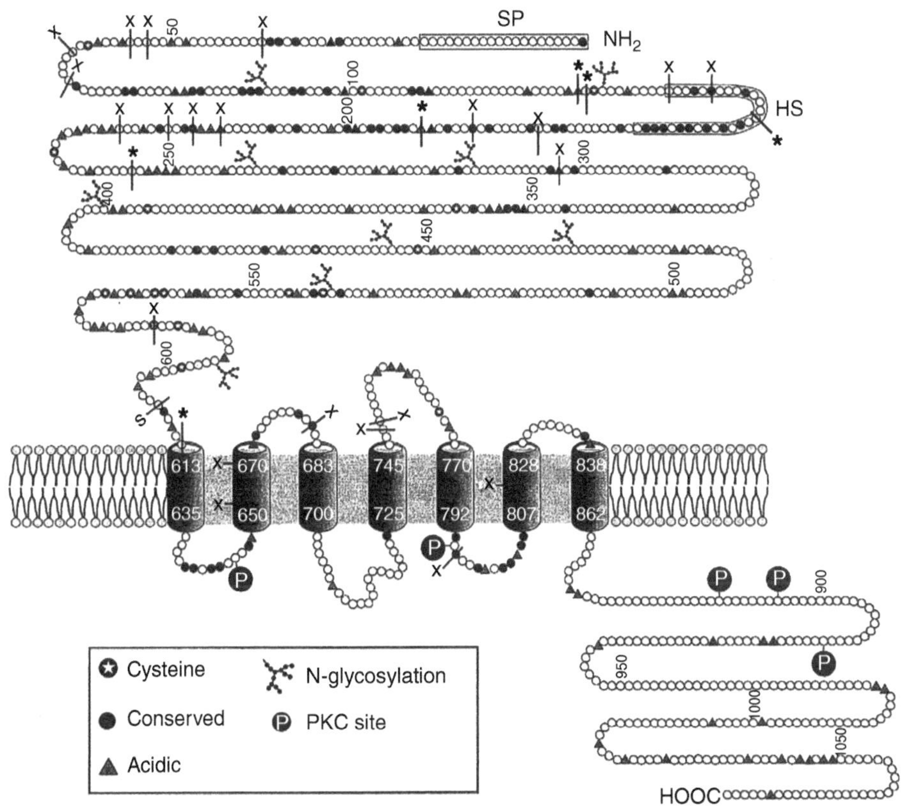

Fig. 3. The plasma membrane Ca^{2+} receptor (the Ca^{2+} sensor). Adapted from Ref. 11. *, activating mutations; **X**, inactivating mutations; SP, signal sequence; HS, a hydrophobic domain.

N-terminal extracellular domain, a mid domain containing the canonical seven-transmembrane helices of G-protein-linked receptors, and a 200-residue intracellular C-terminal domain. The extracellular portion contains several acidic regions, similar to those of other low-affinity Ca^{2+}-binding proteins, that could be the binding site(s) for external Ca^{2+}. Analysis of the transcripts of the receptor have identified them in several tissues, including some, e.g., brain, that are not known to play a role in the regulation of general Ca^{2+} metabolism. As for the function of the Ca^{2+} sensor, its interaction with external Ca^{2+} depresses the release of parathormone by parathyroid cells [11] and activates that of calcitonin by the C-cells of the thyroid [15].

Clearly, then, Ca^{2+}, universally considered as an intracellular carrier of information, is also a bona fide extracellular carrier of information, conveying signals to cells involved in the production of calciotropic hormones and possibly to other cell types as well. This has led to the novel concept of Ca^{2+} as a "calciotropic" hormone, analogous to vitamin D, calcitonin, and parathormone [11].

4. The Ca^{2+} signal has autoregulatory properties

The autoregulatory property of the Ca^{2+} signal has gained prominence more recently. Autoregulation occurs both at the transcriptional and post-transcriptional levels. The earliest findings at the transcriptional level were those on the genes of some Ca^{2+} transporters of the plasma membrane (the PMCA pumps and the NCXs) and of the intracellular membranes (the InsP3 receptor). Cerebellar granule neurons [16–19] express all four basic isoforms of the PMCA pump, all three basic isoforms of the NCX, and InsP3 receptor type 1. When prepared from neonatal rats and cultured in vitro, they mature in a few days but succumb apoptotically unless their plasma membrane is partially depolarized to permit a modestly increased influx of Ca^{2+} through voltage-gated (L type) channels. The resulting modest increase of cytosolic Ca^{2+}, which is evidently necessary for the long-term survival of the cultured neurons, completely alters the pattern of expression of membrane Ca^{2+} transporters. Isoforms 2 and 3 of the PMCA pumps are slowly upregulated, whereas the expression of PMCA 4 is rapidly and dramatically downregulated: this isoform disappears nearly completely after less than 1 h of culture under conditions of partially depolarized plasma membrane. The expression of PMCA 1, by contrast, remains quantitatively constant, but the isoform experiences a splicing switch which privileges a C-terminally truncated, presumably less active variant. Of the three basic NCX isoforms, under these depolarizing conditions, NCX 1 remains expressed at a constant level for days, NCX 3 is slowly upregulated, whereas NCX 2 disappears nearly completely in less than 30 min. The autoregulatory aspect of the process is further underlined by the fact that the downregulation of the expression of PMCA 4 and NCX 2 is mediated by calcineurin, the protein phosphatase that is the target of dual regulation by Ca^{2+}. The expression of the InsP3 receptor becomes upregulated during the prolonged survival of the cultured neurons promoted by the membrane-depolarizing conditions: also in this case, the process is mediated by calcineurin. The end result of these complex Ca^{2+}-induced changes in the pattern of expression of Ca^{2+} transporters is the threefold to fourfold increase of cytosolic Ca^{2+}, which somehow permits the long-term survival of cerebellar granule neurons in culture. That such a modest increase should demand such a dramatic change in the expression pattern of so many Ca^{2+} transporters may at a first glance seem amazing. However, the phenomenon underlines very clearly the importance of regulating cell Ca^{2+} with absolute precision: to promote the survival of the neurons, cytosolic Ca^{2+} must increase, but only to the relatively modest level that is necessary and no more. The reasons why cultured neurons should demand a new set-point of cytosolic Ca^{2+} are not understood at the moment. But it is clear that to achieve this aim with the required precision, a battery of membrane transporters must be reprogrammed.

An important recent development in the transcriptional autoregulation of the Ca^{2+} message is that linked to the function of the downstream regulatory element antagonistic modulator (DREAM), an EF-hand protein that acts as a silencer of an increasing number of genes [20]. Ca^{2+}-free DREAM binds to specific DNA sites (DRE sites) in the promoter of genes, repressing transcription. When Ca^{2+} becomes bound to DREAM, presumably as a result of its increase in the environment, the DRE sites release DREAM and transcription resumes. Fig. 4 illustrates pictorially the function of DREAM. Very recently, DREAM has been shown to control the transcription of the gene for NCX 3, an isoform of the exchanger that is particularly

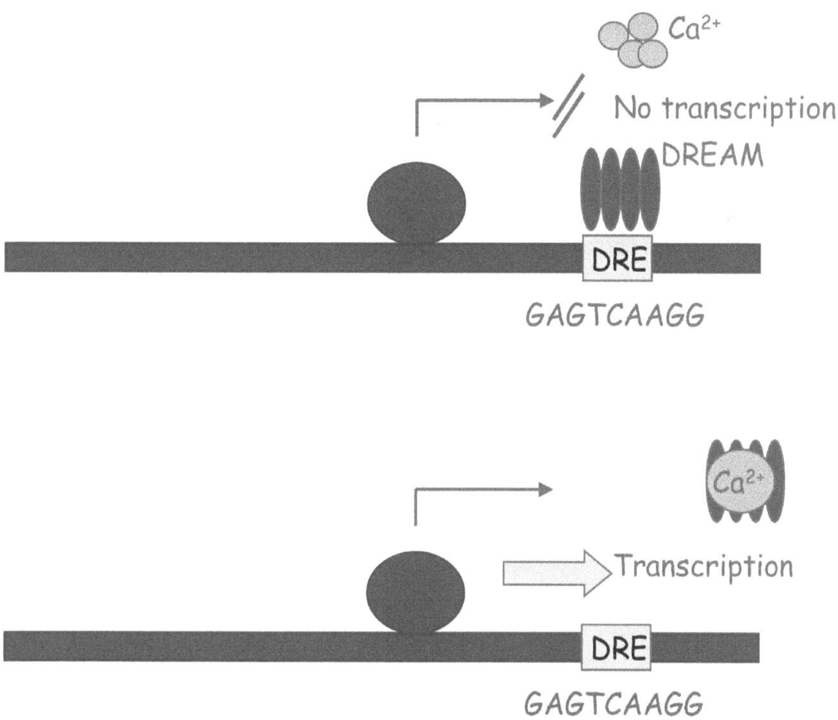

Fig. 4. Effect of the downstream regulatory element antagonistic modulator (DREAM) on gene transcription. The DREAM tetramer binds to DNA sites (DRE sites GAGT) ideally located between the TATA box and the TS site. Each DREAM monomer contains 4 EF hands, of which only three are operational (See Color Plate 2, p. 504).

important to Ca^{2+} homeostasis in neurons [21]. The effect of DREAM is specific for NCX 3: if DREAM is prevented from binding Ca^{2+}, cultured neurons lose the ability to efficiently export it. As a result, they experience cytosolic Ca^{2+} overload and become more liable to damaging procedures.

Cases of autoregulation of the Ca^{2+} message at the post-transcriptional level have been known for a longer time and are more numerous. An interesting discussion may begin with the well-known regulation of the PMCA ATPase by calmodulin. The regulation by calmodulin is reversible, but the pump may also be activated, in this case irreversibly, by the endogenous protease calpain, which is also controlled by Ca^{2+} (see Section 2). The matter of the PMCA pump is of interest from another angle as well, as the enzyme has recently been suggested to be a modulator of Ca^{2+}-calmodulin-dependent enzymes [22]. One of the Ca^{2+}-calmodulin-dependent enzymes that the PMCA pump would modulate, in this case by inhibiting it, is calcineurin [23]. Calcineurin, in turn, inhibits the activity of NCX 1, the cardiac isoform of the NCX [24]. Another example of the autoregulatory character of the Ca^{2+} signal that is worth quoting is the gating by Ca^{2+} of the intracellular Ca^{2+} channels in the endo(sarco)plasmic reticulum.

5. *The ambivalent nature of the Ca^{2+} signal*

The information gathered in decades on the Ca^{2+} signal shows very clearly that no cell would function properly without the messages that Ca^{2+} carries to it. However, as repeatedly emphasized in the sections above, it is vital that the messages be delivered, and processed in the cell interior, in a carefully controlled way. An increase of free Ca^{2+} much above the optimal cytosolic concentration of 100–200 nM can be tolerated, and may even be necessary, but only for a short time. That is, it must occur in the form of repetitive transients: rapid Ca^{2+} oscillations indeed are a convenient device to which cells resort to modulate functions that may require free Ca^{2+} concentrations significantly exceeding those of the normal cytosol at rest. The issue is essentially of time: if the Ca^{2+} increase would become sustained, all Ca^{2+} controlled functions would become permanently activated (the Ca^{2+} control of cell functions is most frequently activatory) including those that are potentially detrimental to cell life, e.g., proteases, phospholipases, and nucleases. Their persistent activation would lead to various degrees of cell damage until, in the absence of successful rescue attempts, cell death would eventually ensue. In the end, this is again an issue of time: cells can cope with situations of Ca^{2+} overload, provided that the duration of the emergency is not excessively protracted. This is so because the mitochondria can accumulate very large amounts of the Ca^{2+} that has inundated the cytoplasm and can do so as they also accumulate inorganic phosphate to store excess Ca^{2+} in the matrix as an insoluble phosphate salt (hydroxyapatite). Mitochondria thus buy precious time for cells, enabling them to survive Ca^{2+} storms. Clearly, however, mitochondria are only temporary safety devices, as the energy they transform can be either used to synthesize ATP or to take up Ca^{2+}: as long as they actively take up Ca^{2+}, they do not synthesize ATP. If they are forced to continuously take up Ca^{2+}, the cessation of ATP synthesis will deprive cells of the ATP necessary to clear Ca^{2+} out of the cytoplasm through the pumps. This will initiate a vicious circle that will eventually end with cell death.

Naturally, Ca^{2+} catastrophe is an extreme case that results from the global, massive, and protracted dysregulation of Ca^{2+} homeostasis. But even in the cases of Ca^{2+}-promoted cell death, one must distinguish between necrotic death, which is clearly an unwanted negative phenomenon, and the process of programmed cell death (apoptosis). Even if it involves the death of cells, the latter still is one of the meaningful ways in which cells process the Ca^{2+} signal, in this case to control processes essential to the life of organisms like organ modeling or tissue renewal. Even if Ca^{2+} may be used as a final biochemical tool to execute both types of cell death, it is thus important to regard apoptosis essentially as a positive phenomenon. Apoptotic cell death is mediated by a family of proteases, the caspases, that are not Ca^{2+} dependent but may process downstream enzymes that may generate Ca^{2+} overload, eventually activating Ca^{2+} stimulated hydrolytic activities that may in the end mediate cell demise. An interesting development in this context is the recent finding [25] that activated caspases cleave the PMCA pump, inactivating it and generating the Ca^{2+} overload situation that activates the killer enzymes above.

The massive dysfunctions of Ca^{2+} signaling that terminate with cell death underline in a striking way the ambivalent nature of the Ca^{2+} signal. To use a metaphor, it is as if the decision to choose Ca^{2+} as a determinant for function would force cells to continuously walk along the edge of a precipice, the risk of a faux pas that would lead to the fatal fall to the abyss being ever present. However, a number of less dramatic conditions also exist in which the Ca^{2+} signal is not deranged globally and dramatically but in subtler ways that only destabilize individual participants in the signaling operation. Most of these conditions, certainly the most interesting among them, are genetic, and affect one or more actors in the complex chain of processes involved in the generation, processing, and control of the Ca^{2+} signal. Diseases that originate from defects of Ca^{2+} sensor proteins, including enzymes that are directly modulated by Ca^{2+}, or of Ca^{2+} transporters and channels have now become numerous. Non-genetic conditions may also be characterized by phenotypes in which the derangement of the Ca^{2+} signal is prominent, e.g., some cancer types. However, whether the Ca^{2+} defect is at the origin of these conditions, or merely a consequence of them, is frequently unclear.

As a comprehensive discussion of all Ca^{2+}-signaling dysfunctions would be beyond the scope and space limits of this contribution, a selection will be made of those that are best understood and that have aspects that are of particular interest. It has also been decided not to discuss the now already large literature on the genetic manipulations (gene knockouts, various transgenics) of Ca^{2+} regulators (i.e., transporters) and Ca^{2+} sensor proteins.

6. *Diseases originating from defects of Ca^{2+} sensor proteins*

As mentioned, Ca^{2+} sensor proteins transmit the processed Ca^{2+} message to targets placed downstream in the signaling chain, but enzymes also exist that decode directly the Ca^{2+} message for their own benefit. Various degrees of complexity in the operation have been alluded to in the preceding sections, e.g., Ca^{2+}-dependent enzymes even exist that process directly the Ca^{2+} signal while also accepting information from separate Ca^{2+} sensor proteins: the cases of calcineurin and of the calpains are striking examples. EF-hand proteins are the most important Ca^{2+} sensor proteins, and calmodulin is the most important among them. Because calmodulin distributes Ca^{2+} information to scores of targets, many of which are essential to cell life, inactivating defects of calmodulin would thus in all likelihood be lethal. It is thus easy to understand why three separate genes should code for calmodulin in mammals [26] and why no genetic conditions based on specific calmodulin defects have been described. By contrast, and as could have been anticipated, numerous diseases originating from defects of committed EF-hand Ca^{2+} sensor proteins have been described.

Non-EF-hand proteins have been studied less intensively than EF-hand proteins and are thus less well understood functionally. However, it is interesting that clues on the function of some of them have in some cases emerged from the discovery that molecularly well understood diseases could be traced back to their defects.

6.1. *A disease involving gelsolin*

Gelsolin is an actin-binding protein regulated by Ca^{2+} and phosphoinositides, which exists as a cytosolic protein of 80 kDa and as a slightly larger secreted protein [27–29]. The ability to bind actin is key to the function of gelsolin, which regulates the cytoskeleton inside cells and acts externally as a scavenger for actin secreted by injured cells to regulate blood viscosity. Both forms of gelsolin are composed of six domains, each one containing a Ca^{2+}-binding site in the actin-free state. The binding of Ca^{2+} promotes the binding of actin, and the latter creates two additional Ca^{2+}-binding sites in domains 1 and 4. Dissociation of bound actin is promoted by phosphoinositides. Both gelsolin forms are substrates of furin (α-gelsolinase), which is one of the Ca^{2+}-dependent proteases known as proprotein convertases. These proteases cycle between the trans-Golgi network and the cell surface, activating precursor proteins in the secretory pathway by removing pro-domains and signal sequences [30]. Although a typical consensus sequence for furin attack exists in gelsolin, no cleavage of the protein has been reported in the secretory pathway because Ca^{2+}, which is very concentrated in the pathway, stabilizes gelsolin against furin cleavage [31]. The connection of gelsolin with disease has been discovered when a mutated form of the protein was identified in an amyloid disease known as familial amyloidosis of Finnish type (FAF). As all amyloidoses, FAF is characterized by aggregates of misfolded peptides in the extracellular spaces, a process that in this case originates from the inability of mutated gelsolin to bind Ca^{2+}. In the mutated protein, a single aspartic acid residue (D 187) is replaced by an asparagine or a tyrosine, a change that removes the ability of domain 2 to bind Ca^{2+} and is sufficient to destabilize conformationally the protein in a way that exposes the furin cleavage site. As a result, a smaller cleavage product is generated (68 kDa) which is secreted and used as a substrate of a still unknown β-gelsolinase which in successive cleavages processes the protein to small molecular weight amyloidogenic peptides.

6.2. *Annexinopathies*

Annexins are a family of Ca^{2+}- and membrane-binding proteins, of which six basic isoforms have been recognized [32]. They bind to acidic phospholipids in a Ca^{2+}-dependent way through a consensus sequence termed the endonexin fold in each of four repeats of about 70 residues. Annexin VI, however, contains eight such repeats, probably due to gene duplication. In three dimensions, annexins have a slightly curved shape with the convex side facing the membrane. They self-associate to form trimers. The Ca^{2+}-binding sites are located on the convex side of the protein and are formed by a 17-residue consensus sequence that coordinates Ca^{2+} to the carbonyl oxygens of its N-terminal portion, to a carboxylic side chain about 40 residues downstream, and to water molecules. Annexins have been associated to a number of functions, of which that related to the inflammatory process is considered particularly important. In this context, the inhibition of phospholipase A2 has received the largest attention. Other functions of annexins are the aggregation and fusion of membrane vesicles, the inhibition of blood coagulation, and ion channel activity. Defects of some annexin isoforms have been variously associated to diseases, the best documented cases being those of annexin II and annexin VI. Annexin II, and its binding partner S-100 A10, which is another Ca^{2+}-binding protein, have been reported to be secreted onto the surface of monocytes and macrophages to regulate fibrinolysis by binding plasminogen and tissue plasminogen activator, and promoting the activation and release of surface plasmin [33]. Annexin II has also been reported to be upregulated in leukemic cells of patients with promyelocytic leukemia, which is a hemorrhagic condition. Interestingly, the expression of annexin II becomes downregulated by the treatment of these cells with agents that block the hemorrhagic complications of the disease [34]. Dysfunction of the clot-dissolving function of annexin II may lead to atherothrombotic disease [35]. As for annexin VI, its main function has been claimed to be the regulation of a number of processes that control the intracellular homeostasis of Ca^{2+}, and hence the general functioning, of cardiomyocytes [36]. Overexpression of annexin VI in transgenic mouse hearts causes heart hypertrophy [37], whereas that of a dominant-negative variant has been claimed to hamper the removal of Ca^{2+} from the cytoplasm into the sarcoplasmic reticulum. Thus, annexin VI appears to play a role in heart which is analogous to that of phospholamban (PLN) [36].

6.3. *Calpainopathies*

Calpains are a family of intracellular proteases, of which a dozen members have so far been described [38]. The best known are heterodimers, in which a catalytic subunit binds Ca^{2+} to several C-terminal EF-hand motifs as well as to a couple of non-EF-hand motifs located closer to the active site [39]. The smaller regulatory subunit is a canonical EF-hand protein with homology to calmodulin. As mentioned in Section 2, calpains are thus a striking example of enzymes that are under multiple control of the Ca^{2+} signal. With respect to the case of calcineurin, which has also been emphasized in Section 2, Ca^{2+} regulation of calpains has an even greater degree of complexity, as the protease binds

Ca^{2+} to covalently bound EF-hand motifs, to a separate EF-hand subunit, but also to two Ca^{2+}-binding sites of novel type located close to the catalytic center. The two dimeric calpains designated as m and μ are ubiquitous. Other tissue-specific calpains, by contrast, are monomeric and may not contain the endogenous EF-hand motifs. However, as they still contain the non-EF-hand Ca^{2+}-binding sites, they are still likely to be controlled by Ca^{2+}. Calpains are regulatory proteases that do not demolish protein substrates completely; rather, they remove from them limited peripheral portions, frequently leading to their activation. The two ubiquitous dimeric calpains have been involved in a long list of cell dysfunction states related to disturbances of Ca^{2+} homeostasis which could have lead to abnormal calpain activation. Alzheimer's disease, cataract, and multiple sclerosis have frequently been mentioned: a particularly interesting and well documented case is that of excitotoxic neuronal death [25]. Whereas not all described cases of m or μ calpain involvement in pathology have been convincingly shown to have in vivo significance, this is instead clearly the case for two tissue-specific atypical calpains, calpain III (p94) and calpain X. Calpain II is monomeric and muscle-specific: inactivating mutations of its gene are responsible for a mild form of limb girdle muscular dystrophy (type 2A) [40]. The problem with calpain III is that only one plausible substrate in vivo has so far been described: IkB, the inhibitor of the NF-kB, which is the inhibitor of the Rel transcription factor family [41]. Considering that the absence of calpain III leads to an event as catastrophic as cell dystrophy, it would be logical to expect that the protease has additional in vivo substrates.

The genetic defect of calpain X, another monomeric calpain that does not contain EF-hand Ca^{2+}-binding motifs but could still be Ca^{2+} sensitive as it still contains the non-EF-hand Ca^{2+}-binding sites, has been shown to be linked to the increased susceptibility to type 2 diabetes [42].

6.4. Dysfunctions of neuronal Ca^{2+} sensor proteins

These proteins (NCS) are expressed in the central nervous system and in retinal photoreceptors [43]. They all possess four EF-hand Ca^{2+}-binding motifs (some of which may be not operational) and, as a rule, are N-terminally myrystoylated. Myrystoylation may be a means to target NCSs to membranes following the Ca^{2+}-promoted exposure of the myrystoyl group. Members of the NCSs family include the recoverins and the guanylate cyclase-activating proteins (GCAPs), which are primarily expressed in retinal photoreceptors where they play a crucial role in the regulation of phototransduction, the frequenins, the visinin-like proteins (VILIPs), and the K+ channel interacting proteins (KChIPs), which are all widely expressed in the central nervous system where they play a number of roles: among them, the regulation of neurotransmitter release and of the activity of A-type K+ channels, of which they are constitutive subunits. Interestingly, the gene that codes for KchIP also encodes calsenilin, a cytosolic EF-hand protein that interacts with presenilin, a protein that is genetically linked to Alzheimer's disease. The same gene also encodes DREAM, the previously mentioned nuclear protein that acts as a Ca^{2+}-binding transcriptional repressor [20]. The case of DREAM as been emphasized in Section 4 as a particularly striking example of the autoregulation of the Ca+ signal, as one of the genes that are controlled by DREAM encodes one of the plasma membrane transporters of Ca^{2+} (NCX 3). A number of genetic conditions interfere with the Ca^{2+}-binding ability of NCSs and generate pathological phenotypes that involve primarily the retina. A severe autosomal-dominant form of retinal cone dystrophy is produced by a Tyr 99-Cys mutation in CGPA1 [44], whereas the aberrant expression of recoverin outside the retina in non-neuronal cancer cells initiates an autoimmunal disorder termed cancer-associated retinopathy [45]. In the disorder, recoverin becomes an auto-antigen that triggers the degeneration of the photoreceptors.

6.5. *Calcium channelopathies*

Defects have been described both in the plasma membrane and the intracellular Ca^{2+} channels. The InsP3 receptor has recently been linked to Huntington's disease [46], and as a result of the increasing interest in trp channels, reports on their involvement in Ca^{2+}-linked disease states are now beginning to appear [47,48]. But the best documented and most intensively studied internal Ca^{2+} channelopathies are those involving the sarcoplasmic reticulum ryanodine receptor and the plasma membrane voltage-gated channels.

6.5.1. *The ryanodine receptor*

Genetic defects of the ryanodine receptor have been described in skeletal muscle and in heart. The two skeletal muscle diseases are malignant hyperthermia (MH) [49] and central core disease (CCD) [50]. Both are caused by mutations occurring mostly in two regions of the ryanodine receptor. MH is a contracturing condition caused by the flooding of muscle cytoplasm with Ca^{2+} massively released from the sarcoplasmic reticulum under the precipitating influence of local anesthetics. CCD is caused by the continuous abnormal leakiness of the ryanodine receptor, presumably forcing mitochondria in the central portion of the fiber, where Ca^{2+} would remain higher due to the distance from the plasma membrane, to continuously accumulate Ca^{2+} until they disintegrate leaving behind the necrosis foci.

Defects of the cardiac ryanodine receptor are more complex, as the receptor in this case contains four molecules of an immunophilin (FKBP12) that stabilizes the receptor preventing the unwanted activity of its Ca^{2+} channel at rest. The presence of the immunophilin explains why the immunosuppressant FK506, which targets FKBP12, increases heart contractility, resulting eventually in cardiac hypertrophy and, in extreme cases, in heart failure. The increased heart contractility, which may trigger fatal arrhythmias, in patients with catecholaminergic polymorphic ventricular tachycardia (CPVT) recognizes a similar pathogenic origin. Mutations in the gene for the ryanodine receptor reduce the affinity for FKBP12, resulting in enhanced channel activity and in the dysregulation of Ca^{2+} homeostasis [51]. Arrhythmias also characterize arrhythmogenic right ventricular dysplasia type 2 (ARVD2), causing partial degeneration of the left ventricle, electrical instability, and sudden death [52]. In ARVD2 patients, mutations have been identified in the region of the ryanodine receptor where the MH and CCD mutations also cluster.

6.5.2. The plasma membrane voltage-gated Ca^{2+} channel

All voltage-gated Ca^{2+} channels are tetrameric structures that contain a pore-forming α1-subunit and a number of accessory subunits that regulate channel properties and may also have other functions: for example, they may influence the expression of the pore-forming subunit [53]. Of the 10 α1 genes so far cloned, four encode L-type α1-subunits and three encode P/Q, R, and N-type α1-subunits. All these channel types are gated by high voltage. The last three cloned genes encode α1-subunits of T-type channels, which are instead gated by low voltage. A number of human disease conditions caused by dysfunctions of Ca^{2+} channels linked to mutations in the L- and P/Q-type pore-forming subunits have been described, mostly defining phenotypes of migraine, epilepsy, and ataxia. Several similar defects of voltage-gated channels, which could be models for human diseases, have also been described in mice. The α1-subunit mutations are of various type, ranging from missense mutations to large deletions [54]. The predominant expression of the P/Q-type channels in the nervous tissue explains the frequency of neurological symptoms in patients that bear P/Q-type channel mutations. So far, only one form of human disease, a form of childhood absence epilepsy, has been traced back to mutations in the pore-forming subunit of T-type channels [55].

Plasma membrane Ca^{2+} channelopathies now define a large and expanding topic. To supplement the necessarily succinct discussion presented here, a cartoon depicting all the mutations that have been described so far is offered in Fig. 5.

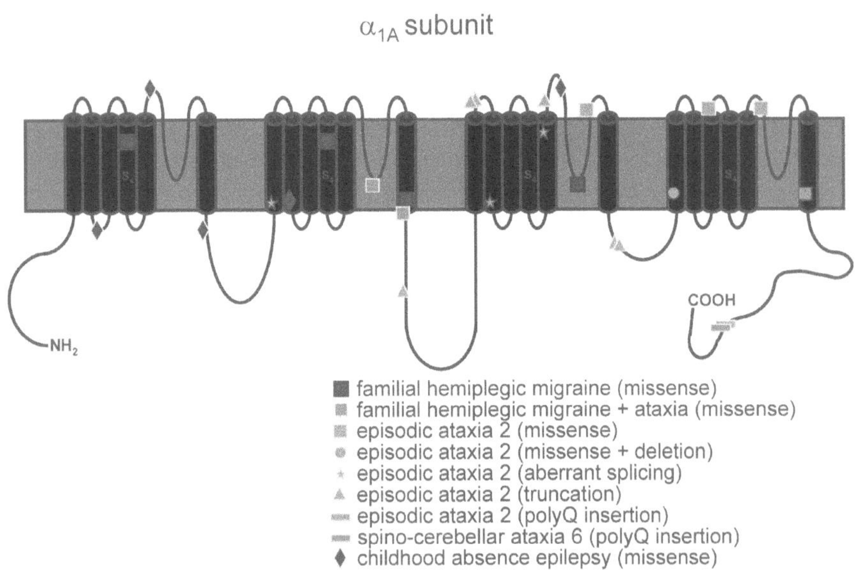

Fig. 5. Mutations in the pore-forming subunit of voltage-gated Ca^{2+} channels. The subunit is a fourfold repeat of regions consisting of six transmembrane domains, in which the loop connecting transmembrane domains 5 and 6 on the external side of the membrane fold in to form the channel. The figure shows the disease phenotypes described so far. Adapted from Ref. 56 (See Color Plate 3, p. 504).

6.6. Ca^{2+} pump defects

Of the three Ca^{2+} pump types known in mammals, two, that of the endo(sarco)plasmic reticulum (SERCA pump) and that of the Golgi system (SPCA pump), have been unambiguously linked to human diseases. The third pump (the plasma membrane pump, PMCA) had originally been linked to mouse disease phenotypes with impaired hearing. In humans, similar phenotypes have also been recently described, but the affected individuals, in addition to the PMCA defect, also had a defect in an unrelated gene [57–58] (see below). In addition to genetic defects, acquired disease states, e.g., cancer, may also affect Ca^{2+} pump function through changes in their expression [59], while other pathological states, e.g., neurotoxicity, may lead to Ca^{2+} pump inactivation.

Mutations in the gene encoding the SERCA pump isoform of skeletal muscle lead to its truncation and cause Brody's disease, a muscle condition in which the loss of SERCA pump function generates a phenotype of exercise-induced impairment of relaxation, which is clearly linked to the inability to clear Ca^{2+} out of the muscle cytoplasm [60]. Null mutations in the gene encoding SERCA 2, which is the isoform of the pump abundantly expressed in keratinocytes, cause Darier–White's disease, a skin condition complicated by neurological and psychiatric disturbances [61]. The abnormal Ca^{2+} elevation in the cytoplasm of keratinocytes leads to the fragmentation of the desmosome-keratin filaments that connect suprabasal epidermal cells, and impairs their adhesion. SERCA 2 is also abundantly expressed in cardiomyocytes, where it is functionally regulated by the reversible association with the small membrane intrinsic protein phospholamban (PLN). This protein interacts with both the intramembrane and the cytosolic portions of the pump, maintaining it inhibited by associating with a binding site close to the pump catalytic center. The removal of PLN from the binding site through phosphorylation of its Ser-16 by the cAMP-dependent protein kinase reactivates the pump. The disease designated as dilating cardiomyopathy has been linked, first experimentally in mice by ablating the PLN gene [62] and later in humans by the identification of two naturally occurring PLN mutations, to loss of PLN function. This could be related to the decreased level of PLN expression or of PLN phosphorylation [63].

The secretory pathway SPCA pump, which is one of the two resident Ca^{2+} pumps of the Golgi membranes (the other is a bona fide SERCA pump), is also highly expressed in human keratinocytes. A number of mutations in its gene, including missense mutations, frameshift insertions, deletions, splice-site mutations, and nonsense mutations, have been detected in humans suffering from Hailey–Hailey's disease, a skin disorder in which, as was the case for Darier–White's disease, Ca^{2+} increases in the keratinocytes, impairing the adhesion of epidermal cells [64–66].

An interesting phenotype with loss of balance and hearing defects has been described in several mice strains bearing inactivating mutations in the gene of isoform 2 of the PMCA pump, which is prominently expressed in the outer hair cell layer of the organ of Corti in the inner ear. The variant of the pump present in the stereocilia of the outer hair cells, which are crucial to the auditory function, is an alternatively spliced version truncated C-terminally, bearing an insert further upstream in the sequence next to a domain that binds activatory acidic phospholipids. Preliminary evidence suggests that this variant of the pump is less efficient than any other variant tested so far [66,67]. It is plausible to suggest that the stereocilia of the outer hair cells, which are important to the regulation of the Ca^{2+} level in the endolymph, may demand a pump type with a limited ability to export Ca^{2+}.

7. Some final comments

Calcium signaling has now become a favorite topic for review writers: one could quote at least a dozen that have appeared in recent issues of prestigious journals. Several important factors have played roles in shaping the popularity of Ca^2, but the unusual properties if its signaling function cannot be underrated. This contribution has singled out some of the properties that had seemed particularly striking: there was, unavoidably, an element of subjectivity in the selection; other aspects could have also been considered, e.g., the fact that Ca^{2+} signaling presides over both the beginning and the end of cell life (Ca^{2+} regulates the creation of new life at fertilization and its end at the conclusion of the vital cycle of cells), or the fact that the choice of Ca^{2+} as a cellular carrier of signals was probably not an evolutionary option, but an evolutionary necessity. Once the decision was made to use phosphate as the energetic currency, it obviously became vital to develop ways and means to maintain Ca^{2+} inside cells at a very low levels to avoid the precipitation of Ca-phosphate salts, an accomplishment that created ideal conditions for Ca^{2+} to be used as a signaling agent.

This contribution has placed special emphasis on one distinguishing aspect of the signaling function of Ca^{2+} that is likely to become a hot topic in the future: the pathological deviations of the signaling function. The times when this topic was only timidly alluded to Ca^{2+} aficionados are likely to soon become distant past. The disease connection can be safely predicted to be on its way to become very important not only to cell biochemistry and physiology but to medicine as well.

References

1. Ringer, S. (1883) *J. Physiol.* 4, 29–43.
2. Carafoli, E. (1987) *Annu. Rev. Biochem.* 56, 395–433.
3. Williams, R.J.P. (1999) in E. Carafoli and C.B. Klee (Eds) *Calcium as a Cellular Regulator*, Oxford University Press, pp. 3–27.
4. Carafoli, E., Santella, L., Branca, D. and Brini, M. (2001) *Crit. Rev. Biochem. Mol. Biol.* 36, 107–260.
5. Carafoli, E. (2004) *Proc. Natl. Acad. Sci. USA* 99, 1115–1122.
6. Berridge, M.J. (1993) *Nature* 361, 315–325.
7. Kretsinger, R.H. and Noockholds, C.E. (1973) *J. Biol. Chem.* 248, 3313–3326.
8. Nakayama, S., Kawasaki, H., and Kretsinger, R. (2000) in E. Carafoli and J. Krebs (Eds) *Calcium Homeostasis*, Springer, Berlin, pp. 30–58.
9. Klee, C.B. and Vanaman, T.C. (1982) *Adv. Protein Chem.* 35, 213–321.
10. Ikura, M., Clore, G.M., Gronenborn, A.M., Zhu, G., Klee, C.B. and Bax, A. (1992) *Science* 256, 632–638.
11. Brown, E.M., Quinn, S.M. and Vassilev, P.M. (1999) in E. Carafoli and C.B. Klee (Eds) *Calcium as a Cellular Regulator*, Oxford University Press, pp. 295–310.

12. Nemeth, E.F. and Scarpa, A. (1986) *FEBS Lett.* 203, 15–19.
13. Kifor, O., Moore, F.D., Jr., Wang, P., Goldstein, M., Vassilev, P., Kifor, I., Hebert, S.C. and Brown, E.M. (1996) *J. Clin. Endocrinol. Metab.* 81, 1598–1606.
14. Brown, E.M., Gamba, G., Riccardi, D., Lombardi, M., Butters, R., Kifor, O., Sun, A., Hediger, M.A., Lytton, J., and Hebert, S.C. (1993) *Nature* 366, 575–580.
15. McGehee, D.S., Aldersberg, M., Liu, K.P., Hsuing, S., Heath, M.J. and Tamir, H. (1997) *J. Physiol.* 502, 31–44.
16. Guerini, D., Garcia-Martin, E., Gerber, A., Volbracht, C., Leist, M., Merino, C.G. and Carafoli, E. (1999) *J. Biol. Chem.* 274, 1667–1676.
17. Guerini, D., Wang, X., Li, L., Genazzani, A. and Carafoli, E. (2000) *J. Biol. Chem.* 275, 3706–3712.
18. Li, L., Guerini, D. and Carafoli, E. (2000) *J. Biol. Chem.* 275, 20903–20910.
19. Geanzzani, A., Carafoli, E. and Guerini, D. (1999) *Proc. Natl. Acad. Sci. USA* 96, 5797–5801.
20. Carrion, A.M., Link, W.A., Ledo, F., Mellström, B. and Naranjo, J.R. (1999) *Nature* 398, 80–84.
21. Gomez-Villafuertes, R., Torres, B., Barrio, J., Savignac, M., Gabellini, N., Rizzato, F., Pintado, B., Gutierrez-Adan, A., Mellstrom, B., Carafoli, E. and Naranjo, J.R. (2005) *J. Neurosci.* 25, 10822–10830.
22. Schuh, K., Uldrijan, S., Telkamp, M., Rothlein, N. and Neyses, L. (2001) *J. Cell. Biol.* 155, 201–205.
23. Buch, M.H., Pickard, A., Rodriguez, A., Gillies, S., Maass, A.H., Emerson, M., Cartwright, E.J., Williams, J.C., Oceandy, D., Redondo, J.M., Neyses, L. and Armesilla, A.L. (2005) *J. Biol. Chem.* 280, 29479–29487.
24. Katanosaka, Y., Iwata, Y., Kobayashi, Y., Shibasaki, F., Wakabayashi, S. and Shigekawa, M. (2005) *J. Biol. Chem.* 280, 5764–5772.
25. Bano, D., Young, K.W., Guerin, C.J., Lefeuvre, R., Rothwell, N.J., Naldini, L., Rizzuto, R., Carafoli, E. and Nicotera, P. (2005) *Cell* 120, 275–285.
26. Fisher, R., Koller, M., Flura, M., Mathews, S., Strehler-Page, M.A., Krebs, J., Penniston, J.T., Carafoli, E. and Strehler, E.E. (1988) *J. Biol. Chem.* 263, 17055–17062.
27. Yin, H.L. (1987) *Bioassays* 7, 176–179.
28. Yin, H.L., Kwiakowski, D.J., Mole, J.E. and Cole, F.S. (1984) *J. Biol. Chem.* 259, 5271–5276.
29. Lee, W.M. and Galbraith, R.M. (1992) *N. Engl. J. Med.* 326, 1335–1341.
30. Molloy, S.S., Anderson, E.D., Jean, F. and Thomas, G. (1999) *Trends Cell. Biol.* 9, 28–35.
31. Chen, C.D., Huff, M.E., Matteson, J., Page, L., Phillips, R., Kelly, J.W. and Balch, W.E. (2001) *EMBO J.* 20, 6277–6287.
32. Hayes, M.J. and Moss, S.E. (2004) *Biochem. Biophys. Res. Commun.* 322, 1166–1170.
33. Brownstein, C., Deora, A.B., Jacovina, A.T., Weintraub, R., Gertler, M., Khan, K.M., Falcone, D.J. and Hajjar, K.A. (2004) *Blood* 103, 317–324.
34. Menell, J.S., Cesarman, G.M., Jacovina, A.T., McLaughlin, M.A., Lev, E.A. and Hajjar, K.A. (1999) *N. Engl. J. Med.* 340, 994–1004.
35. Hajjar, K.A. and Krishnan, S. (1999) *Trends Cardiovasc. Med.* 9, 128–138.
36. Kaetzel, M.A. and Dedman, J.R. (2004) *Biochem. Biophys. Res. Commun.* 322, 1171–1177.
37. Gunteski-Hamblin, A.M., Song, G., Walsh, R.A., Frenzke, M., Boivin, G.P., Dorn, G.W., 2nd, Kaetzel, M.A., Horseman, N.D. and Dedman, J.R. (1996) *Am. J. Physiol.* 270, H1091–H1100.
38. Sorimachi, H., Ishiura, S. and Suzuki, K. (1997) *Biochem. J.* 328, 721–732.
39. Moldoveanu, T., Jia, Z. and Davies, P.L. (2004) *J. Biol. Chem.* 279, 6106–6114.
40. Richard, I., Broux, O., Allamand, V., Fougerousse, F., Chinnikulchai, N., Bourg, N., Brenguier, L., Devaud, C., Pasturaud, P., Roudaut, C., Hillaire, D., Passos-Bueno, M.-R., Zatz, M., Tischfield, J.A., Fardeau, M., Jackson, C.E., Cohen, D. and Beckmann, J.S. (1995) *Cell* 81, 27–40.
41. Baghdiguian, S., Martin, M., Richard, I., Pons, F., Astier, C., Bourg, N., Hay, R.T., Chemaly, R., Halaby, G., Loiselet, J., Anderson, L.V., Lopez de Munain, A., Fardeau, M., Mangeat, P., Beckmann, J.S. and Lefranc, G. (1999) *Nat. Med.* 5, 503–511.
42. Horikawa, Y., Oda, N., Cox, N.J., Li, X., Orho-Melander, M., Hara, M., Hinokio, Y., Lindner, T.H., Mashima, H., Schwarz, P.E., del Bosque-Plata, L., Horikawa, Y., Oda, Y., Yoshiuchi, I., Colilla, S.,

Polonsky, K.S., Wei, S., Concannon, P., Iwasaki, N., Schulze, J., Baier, L.J., Bogardus, C., Groop, L., Boerwinkle, E., Hanis, C.L. and Bell, G.I. (2000) *Nat. Genet.* 26, 163–175.
43. Burgoyne, R.D. and Weiss, J.L. (2001) *Biochem. J.* 353, 1–12.
44. Sokal, I., Li, N., Surgucheva, I., Warren, M.J., Payne, A.M., Bhattacharya, S.S., Baehr, W. and Palczewski, K. (1998) *Mol. Cell* 2, 129–133.
45. Polans, A.S., Witkowska, D., Haley, T.L., Amundson, D., Baizer, L. and Adamus, G. (1995) *Proc. Natl. Acad. Sci. USA* 92, 9176–9180.
46. Tang, T.S., Tu, H., Chan, E.Y., Maximov, A., Wang, Z., Wellington, C.L., Hayden, M.R. and Bezprozvanny, I. (2003) *Neuron* 39, 227–239.
47. Wissenbach, U., Niemeyer, B., Himmerkus, N., Fixemer, T., Bonkhoff, H. and Flockerzi, V. (2004) *Biochem. Biophys. Res. Commun.* 322, 1359–1363.
48. Anyatonwu, G.I. and Ehrlich, B.E. (2004) *Biochem. Biophys. Res. Commun.* 322, 1364–1373.
49. MacLennan, D.H. and Phillips, M.S. (1992) *Science* 256, 789–794.
50. Quane, K.A., Healy, J.M., Keating, K.E., Manning, B.M., Couch, F.J., Palmucci, L.M., Doriguzzi, C., Fagerlund, T.H., Berg, K., Ording, H., Bendixen, D., Mortier, W., Linz, U., Muller, C.R. and McCarthy, T.V. (1993) *Nat. Genet.* 5, 51–55.
51. Wehrens, X.H., Lehnart, S.E., Huang, F., Vest, J.A., Reiken, S.R., Mohler, P.J., Sun, J., Guatimosim, S., Song, L.S., Rosemblit, N., D'Armiento, J.M., Napolitano, C., Memmi, M., Priori, S.G., Lederer, W.J. and Marks, A.R. (2003) *Cell* 113, 829–840.
52. Tiso, N., Stephan, D.A., Nava, A., Bagattin, A., Devaney, J.M., Stanchi, F., Larderet, G., Brahmbhatt, B., Brown, K., Bauce, B., Muriago, M., Basso, C., Thiene, G., Danieli, G.A. and Rampazzo, A. (2001) *Hum. Mol. Genet.* 10, 189–194.
53. Ertel, E.A., Campbell, K.P., Harpold, M.M., Hofmann, F., Mori, Y., Perez-Reyes, E., Schwartz, A., Snutch, T.P., Tanabe, T., Birnbaumer, L., Tsien, R.W. and Catterall, W.A. (2000) *Neuron* 25, 533–535.
54. Meir, A. and Dolphin, A.C. (2002) *Pflugers Arch.* 444, 263–275.
55. Khosravani, H., Altier, C., Simms, B., Hamming, K.S., Snutch, T.P., Mezeyova, J., McRory, J.E. and Zamponi, G.W. (2004) *J. Biol. Chem.* 279, 9681–9684.
56. Carafoli, E. (2004) *Trends Biochem. Sci.* 29, 371–379.
57. Schultz, J.M., Yang, Y., Caride, A.J., Filoteo, A.G., Penheiter, A.R., Lagziel, A., Morell, R.J., Mohiddin, S.A., Fananapazir, L., Madeo, A.C., Penniston, J.T. and Griffith, A.J. (2005) *N. Engl. J. Med.* 352, 1557–1564.
58. Legrand, G., Humez, S., Slomianny, C., Dewailly, E., Vanden Abeele, F., Mariot, P., Wuytack, F. and Prevarskaya, N. (2001) *J. Biol. Chem.* 276, 47608–47614.
59. Brody, I.A. (1969) *N. Engl. J. Med.* 281, 187–192.
60. Sakuntabhai, A., Ruiz-Perez, V., Carter, S., Jacobsen, N., Burge, S., Monk, S., Smith, M., Munro, C.S., O'Donovan, M., Craddock, N., Kucherlapati, R., Rees, J.L., Owen, M., Lathrop, G.M., Monaco, A.P., Strachan, T. and Hovnanian, A. (1999) *Nat. Genet.* 21, 271–277.
61. Luo, W., Grupp, I.L., Harrer, J., Ponniah, S., Grupp, G., Duffy, J.J., Doetschman, T. and Kranias, E.G. (1994) *Circ. Res.* 75, 401–409.
62. Haghighi, K., Gregory, K.N. and Kranias, E.G. (2004) *Biochem. Biophys. Res. Commun.* 322, 1214–1222.
63. Hailey, H. and Hailey, H. (1939) *Arch. Dermatol. Siphilol.* 39, 679–685.
64. Hu, Z., Bonifas, J.M., Beech, J., Bench, G., Shigihara, T., Ogawa, H., Ikeda, S., Mauro, T. and Epstein, E.H., Jr. (2000) *Nat. Genet.* 24, 61–65.
65. Sudbrak, R., Brown, J., Dobson-Stone, C., Carter, S., Ramser, J., White, J., Healy, E., Dissanayake, M., Larregue, M., Perrussel, M., Lehrach, H., Munro, C.S., Strachan, T., Burge, S., Hovnanian, A. and Monaco, A.P. (2000) *Hum. Mol. Genet.* 9, 1131–1140.
66. Wood, J.D., Muchinsky, S.J., Filoteo, A.G., Penniston, J.T. and Tempel, B.L. (2004) *J. Assoc. Res. Otolaryngol.* 5, 99–110.
67. Brini, M., Ortolano, S., Di Leva, F. and Carafoli, E. (2006) manuscript in preparation.

J. Krebs and M. Michalak (Editors)
Calcium: A Matter of Life or Death

ISSN: 0167-7306/DOI: 10.1016/S0167-7306(06)41002-4

CHAPTER 2

The evolution of the biochemistry of calcium

Robert J. P. Williams

Inorganic Chemistry Laboratory, University of Oxford, South Parks Road, Oxford OX1 3QR, UK, Tel.: +44 (0) 1865 272621; Fax: +44 (0) 1865 272690; E-mail: bob.williams@chem.ox.ac.uk

Abstract

The chemistry of the calcium (Ca) ion does not appear of great interest, but it has been an essential element in biological evolution. Initially, procaryotic cells treated it as an intracellular poison rejecting it together with sodium and chloride ions. The resulting functions of Ca apart from being pumped out of cells were on the cell exterior, assisting such activities as wall stability and extra-cytoplasmic digestion. These functions have continued in a modified form to this day. The great significance of Ca arose with the very beginnings of eucaryote multi-compartment cells. Here communication between compartments is essential, and additionally a great advantage to cell survival, now of longer life, arose from knowledge of the extracellular environment. Here the large adventitious outside cytoplasmic Ca gradient provided an immediate solution for message transmission. It was further aided by ion storage in many internal vesicles. The development needed new channel and response proteins. As multi-compartment eucaryotes evolved further to multicellular organisms and then animals with brains, a succession of functions for Ca signals developed coupled to new organic messengers so that Ca became essential to the whole homeostatic integration of organisms and their activities. There is an inevitability about this progression seen in the multiplicity of old and new Ca proteins. Slowly too the complexity of Ca minerals in shells and bones arose. The latest place in the evolution of the functional value of Ca is mankind's application of it, outside biological evolution, in materials science, cements and plasters for example.

Keywords: calcium channels, calcium messengers, calcium minerals, calcium proteins, calcium pumps, calcium triggers

1. *Introduction*

The evolution of the biochemistry of calcium (Ca) is constrained by the availability of the element in Earth's waters and its mundane inorganic chemistry. Hence, its biochemistry is not of great variety, but despite this poverty of variation, it is of huge significance. The reason lies in the fact that initially cells had to reduce the cytoplasmic concentration of Ca ions to say 10^{-6} M from $>10^{-3}$ M in the sea. As a consequence, a considerable gradient of Ca ions was created across the cytoplasmic membrane. While in the primitive cell with an inflexible wall and a single compartment only external Ca had structural and some catalytic functions, it was the evolution of a flexible exterior and a compartmentalized interior that naturally led to the utilization of the Ca^{2+} gradient in messages. The diversity of this use of Ca^{2+} as an information carrier started when oxidative chemistry increased in importance some two billion years ago. During the more recent eukaryote periods, as diversity of compartments increased, the Ca^{2+} gradient was pressed into messenger service in a multitude of ways. There was an ever-increasing requirement for novel organic messengers and proteins as receptors linked to calcium signals. Ca functions outside cells increased too with the development of extracellular fluids and matrices. Finally, humans have exploited Ca in many structural units, concretes, plasters, etc., materials outside the potential of biological activity and not linked to genes. In this chapter, we provide the steps of the above evolution indicating how Ca became unavoidably more or more evolved in living systems.

2. *Inorganic chemistry of Ca*

This account of the evolution of the functions of the Ca^{2+} is unavoidably speculative in part. Any history has the underlying weakness that it is not testable by experiment but is suggested by inference from a picture of events, back in time. There are some well-recognized events in the history of life which we accept nowadays, but there are always lingering doubts over their full veracity and interpretation. These general limitations apply to a description of the evolution of Ca functions but not to the general chemistry of the element. The beginning of the story of Ca and its connection with cells has to be the big bang that led in time to the giant stars and the formation of the elements of the periodic table (Fig. 1) and their abundances [1]. Our interest lies in the fact that Ca is an abundant element, and from what we know of its chemical properties today, it is not surprising that it is concentrated in various minerals, mainly oxides and silicates of the deep mantle of Earth whereas

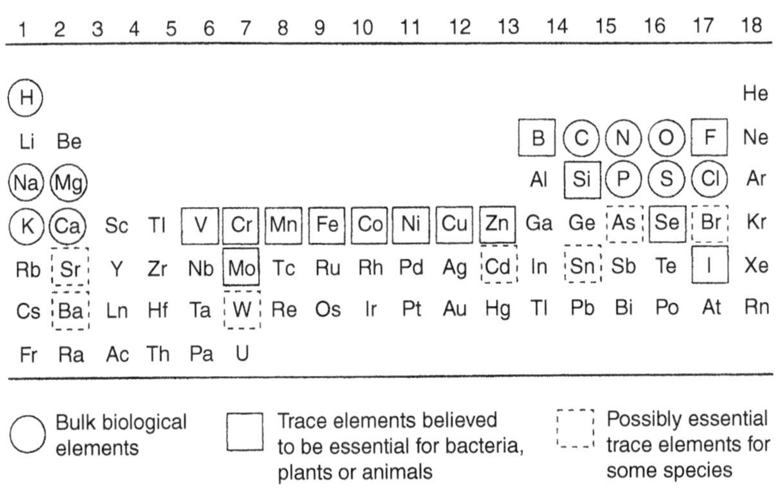

Fig. 1. The periodic table indicating essential elements.

in the surface rocks it is mainly present in carbonates. It has always been available as an ion in the sea at roughly 10 mM, the equilibrium concentration with mineral carbonates at the pH of the sea (around 7.5). We need to see these properties in relationship to those of magnesium (Mg) and several transition metal ions as they interact with Ca functions in cells to some degree. Mg is more abundant than Ca, whereas the transition metals except iron (Fe), which is of greater abundance, are less abundant. Hence, the main silicates of the underlying mantle are of Mg and Fe. The seawater solubility of Mg has been roughly fixed over time and is again related to its surface minerals, including Ca/Mg carbonate, at around 10–50 mM. The transition metal ions are harder to discuss because they changed in their combination with time, being found in sulphides first and then oxides, and hence in availability. We have described elsewhere these relative affinities and the effect of changing combination on the sea contents over time [1]. In essence, when there was H_2S in the sea, no transition metal ion except perhaps manganese ion (Mn^{2+}) had a presence above that of Fe^{2+} at say 10^{-7} M, and most had concentrations far below this. The only other metal ion then present that I shall concern myself with here is strontium ion, Sr^{2+}, as this has a carbonate of solubility similar to that of Ca though it also has an insoluble sulfate. In the developing oxidizing atmosphere and sea, we know that while the concentrations of Mg^{2+}, Ca^{2+} and Sr^{2+} have been maintained throughout evolution from the anaerobic period, those of Fe, especially, and Mn have dropped dramatically through oxidation and precipitation of Fe_2O_3 and MnO_2. We shall have little need to refer to other transition metal ions, such as those of Ni, Co and Mo here, but there are very important links between cellular Ca-chemistry and those of Mg, Mn and Fe, and some links to those of zinc (Zn) and copper (Cu) in more recent times as they were liberated from their sulphides by oxidation.

2.1. Model complex ion chemistry

Apart from the importance of elements in minerals, the complexes with organic chemicals of these metal ions, Mn, Mg, Fe and Ca, are important. The formation of chelates of all of them is well studied with a variety of donor non-metal atoms involved in direct binding [2]. Only Fe is seriously bound to organic sulphides and only Mn and Fe are seriously bound to nitrogen donors, either aliphatic or aromatic. (We shall not need to discuss Mg trapped kinetically in chlorophyll.) Mg and Ca ions more than those of the other two ions are bound only to oxygen donors of ethers, alcohols, carboxylates and phosphates, but the bindings to these last ligand atoms only becomes significant in chelates. Intrinsically, the bare ions bind to single non-metal donor with strength in the order $Fe^{2+} > Mn^{2+} > Mg^{2+} > Ca^{2+}$ due to their electron affinities. A secondary reason for the inability of Ca^{2+} to bind is that the pK_a values of sulphides and amines, mostly, are several units higher than the pH of the sea (7–8) and limit binding. When we turn to neutral O-donors or O-anions their pK_a values are lower but the binding constants to single RCO_2^- and RPO_4^{2-} remain of little consequence in

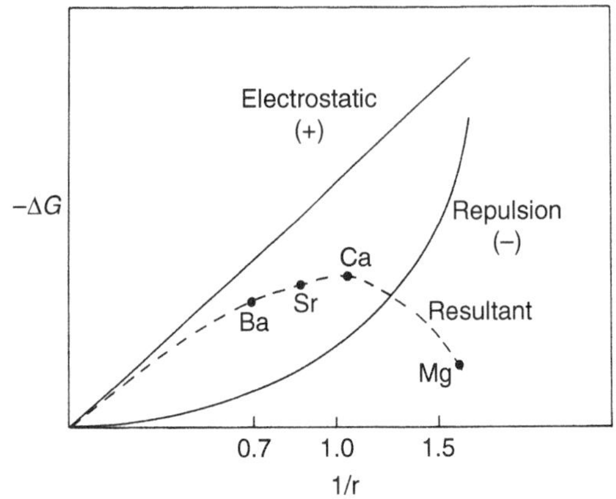

Fig. 2. A schematic plot of the free energy of binding of cations of different radius, *r*, to a chelate ligand here providing a best hole which fits a radius of 1.0 Å that is of Ca^{2+}.

complexes. It is to chelates we must turn to observe binding, but as the number of donors of the structure of the chelating agent increases, a factor other than intrinsic binding strength (electron affinity of ions) becomes important, i.e. ion size (Fig. 2).

Now the structure of the chelate as well as the number of binding groups of it is important opposite ion size especially because the ion sizes are so different – Sr^{2+} (1.15 Å), Ca^{2+} (1.00 Å), Mn^{2+} (0.75 Å), Mg^{2+} and Fe^{2+} (0.60 Å). If a ligand chelate forms a constructed cavity, a ring or a tight horse-shoe, such that it fits one of these ions but not the others, then binding is controlled by fitting in the presence of competition with H_2O.

$$M^{2+}(H_2O)_n + L \rightleftarrows ML(H_2O)m + (n-m)H_2O$$

The hydration of ions in water (55 M) is of very adaptable structure and is then competitive with poorly fitting chelates. A simple example is that Fe^{2+} and Mg^{2+} can fit inside porphyrin (chlorophyll) rings, losing H_2O, whereas Sr^{2+}, Ca^{2+} and Mn^{2+} cannot. (Note that the only other cations ever found in porphyrin type rings are also those of around 0.6 Å in radius, Co, Ni and rarely Cu.) It is not necessary to form such a ring as a fitting constraint. A large open horse-shoe chelate can bind effectively depending on the ability of the ion to coordinate at good bond distances; for example, Ca^{2+} can bind better than Mg^{2+} to EGTA (contrast EDTA), which is therefore the selective organic reagent for Ca^{2+} (Fig. 3). The solubility of sulphates and carbonates as opposed to fluorides and hydroxides is based on the better packing in larger holes created by the large anions by larger cations. Thus, we observe preferential insolubility of $SrSO_4$ and $CaCO_3$ as opposed to MgF_2 and $Mg(OH)_2$. The relative stability of Mg^{2+} and Ca^{2+} complexes is well understood, and a particular problem for cell chemistry is that Ca^{2+} and Mg^{2+} bind equally

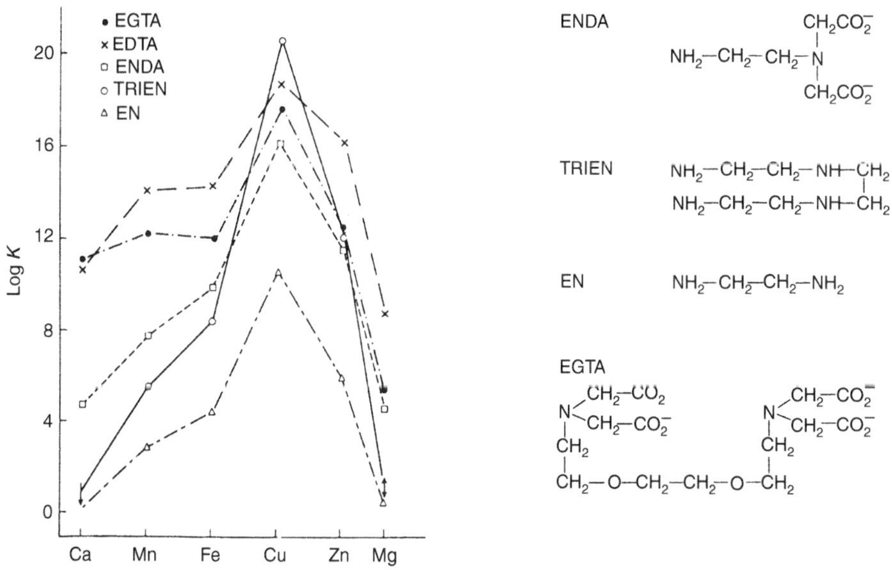

Fig. 3. The logarithm of the stability constants of a series of ligands, formulae shown, with divalent ions. Note the drop of log*K* for Mg^{2+} for EGTA, which is not seen for Ca^{2+}.

strongly to many shapeless, chelating phosphates and carboxylates common in cells, e.g. ATP, DNA and RNA. Note that Mn^{2+} is a relatively large ion, and frequently it is located with Ca^{2+} outside the cytoplasm.

2.2. *Exchange rates*

In this account, we shall assume that most Ca^{2+} complexes exchange at equilibrium with the observed free Ca^{2+} concentration permitted in a given vessel or cell compartment at a given time. This is largely true too of Mg^{2+}, Mn^{2+} and Fe^{2+}, but they are sometimes held in kinetic traps such as porphyrins. However, Mn^{3+} and Fe^{3+} do not exchange at all rapidly when their binding is strong [3]. The binding constants described above, Section 2.1, and on/off rates are for isolated binding sites. If the binding site is cooperative with other bindings say to Ca or other metal ions such as Zn, or if there follows further events after binding such as the interaction of the binding protein with other biopolymers, then in both cases the binding and retention time of Ca may be increased. This allows the period of Ca action to be extended from the fast cytoplasm times for off rates of a millisecond to a much longer time in principle. Many other chapters of this book deal with events due to long retention times. For example to be effective in transcription, which is considered to be a property of one known protein dependent on Ca, downstream regulatory element antagonistic modulator (DREAM), retention has to be considerable. The main concern of this chapter however is homeostasis associated mainly with fast off-reactions.

3. Primitive earth conditions

We can only infer from geological records the nature of early waters in which life could have begun [4]. Let us assume that the temperature was around 20°C, the contents of the sea in the presence of sulphide were about 10^{-7} M Fe^{2+}, 10^{-7} M Mn^{2+}, 10^{-2} M Ca^{2+} and 5×10^{-2} M Mg^{2+}. The concentration of especially Ca^{2+} is not only sufficient to bind many organic chelating molecules, e.g. multi-carboxylates and phosphates, but it tends to precipitate them while Mg^{2+}, Mn^{2+} and Fe^{2+} are more likely to form more soluble complexes. There was no free oxygen present.

4. Procaryote cellular chemistry: concentrations in the cytoplasm

We can now turn to procaryote cell chemistry noting immediately that cells can act in ways not foreseeable directly from model chemistry. Cells are energized flow systems and not just static, and features of them are time dependent. This means that the interior of cells is restricted by functional needs and rules, and certain rules are independent of genetic information and operated before there was DNA, although all the systems used today are DNA dependent for expression. The DNA independent rules are as follows:

(1) All ions relevant here are in fast exchange.
(2) As Rule 1 applies, all ionic interactions inside cells are at equilibrium.
(3) In order for DNA/RNA/protein systems to begin to function, the concentrations of certain ions, e.g. Na^+, Ca^{2+} and perhaps Mn^{2+}, were limited by outward pumps whereas those of others, e.g. K^+ and Mg^{2+}, entered. In particular Ca^{2+} precipitates many phosphates. The ion gradients across cell membranes must have been established very early in evolution essentially before there was a code. (Amongst anions, chloride is pumped out and phosphate is pumped in.)

Observed free ion concentrations in the most primitive and all subsequent cell cytoplasm are found to be restricted as follows: K^+ 10^{-1} M, Na^+ $<10^{-2}$ M, Mg^{2+} 10^{-3} M, $Ca^{2+} < 10^{-6}$ M, Fe^{2+} and Mn^{2+} approximately 10^{-7} M, and all other transition metal ions $<10^{-9}$ M (Fig. 4). Binding constants follow the general rules observed outside cells as seen in model complexes but described in detail elsewhere [3]. It follows that any Ca^{2+} in the cytoplasm of procaryote cells is unlikely to interfere with the binding of other cations and Ca^{2+} cannot compete with Mg^{2+} for most O-donors in the cytoplasm. Do not forget however that we deduce the concentrations of ions in the earliest cells from our knowledge of existing anaerobic bacteria and archaea.

Calcium proteins that undoubtedly arose early are then those functional in the Ca pumps for outward rejection of Ca^{2+} (Fig. 5). Their binding constants on the outside of the cell membrane are of the order of 10^3 M^{-1}, whereas the pumps have internal binding constants of 10^6 M^{-1} or greater. Their present-day DNA sequences suggest that they arose early in evolution at the same time as outward pumps for any poisonous heavy metal ion, Zn^{2+} and Cu^{2+} [5]. The concentrations of these heavy

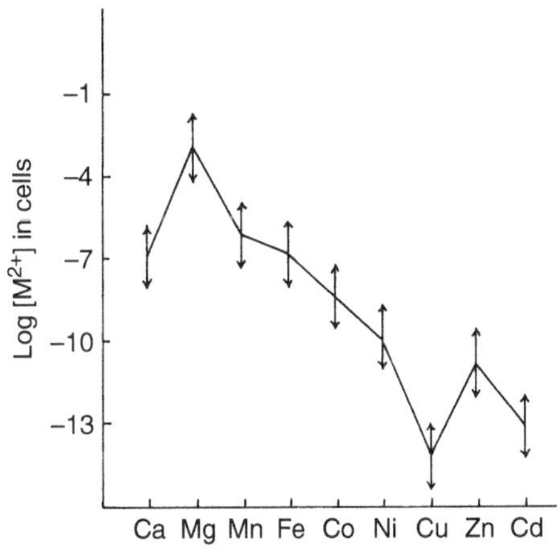

Fig. 4. The generalized plot of free divalent ion concentrations in the cytoplasm of all cells.

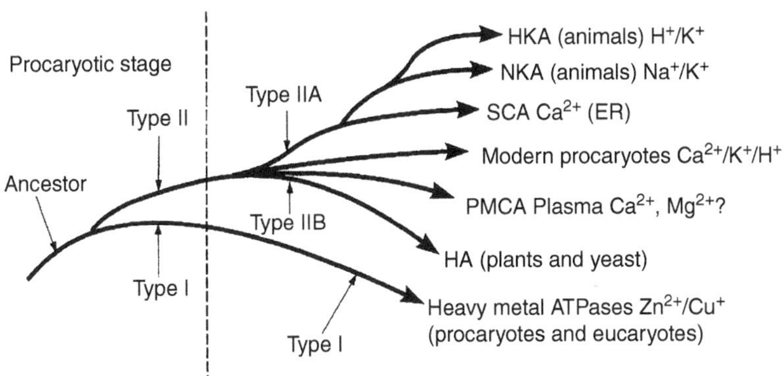

Fig. 5. The evolution of the ATPase pumps. Notice how the Type II pumps expand, especially after the broken line, when eucaryotes appear.

elements in the environment arose much later and hence rejection of free ions to avoid poisoning is increasingly necessary. Note the binding of Ca^{2+} here is often to four or five groups such as neutral O-donors and one or two carboxylates plus three or four water molecules making a 7- or 8-coordinate structure. The only other available ion forming even a nearly similar structure is Mn^{2+}, e.g. with EDTA.

4.1. The first functions of Ca

Given the above data concerning environmental and cell concentrations in Section 4, Ca was clearly expelled from ‘cells’ as a very early necessity to avoid precipitations, for example, of RNA and DNA, and competition with the more effective Lewis acid, Mg^{2+}, in the cytoplasm, the only initial internal compartment. Outside the cell and on external surfaces, Ca could assume roles in catalysis and cross-linking of any protective coat of the cell. Today, we see its protective property in the stabilization of the walls of bacteria and this use becomes general through almost all life. Ca is generally more effective in cross-linking than any other available cation due to its availability, large size and its ability to coordinate in a variety of geometries. Its catalytic ability though weak was useful in hydrolysis of proteins, lipids, saccharides and nucleotides outside cells, where the catalysts act as an aid to providing small molecule nutrients for procaryotes. In some cases, external Ca is a trigger of catalysis by forcing a conformation change on a protein, a pro-enzyme rejected from a cell. All these activities remain common today in many classes of organisms (see Sections 5 and 6), but their uses have developed.

A very different activity of Ca arose a little later in photosystem II. This is the membrane bound photo-system for the capture of hydrogen from water with release of dioxygen. The key unit in the process is a cluster of Mn_4Ca, which may also bind chloride, from which O_2 is released. The structure is relatively confidently given [6] now, but the question arises as to how such a cluster arose. The problem is very different from that of the clusters of all other metal ion clusters formed from Fe^{2+}/Fe^{3+} [with or without Ni^{2+}, Mo(VI) or V(V)]. As stated, these three ions can bind quite strongly to sulphide, and on doing so, some of the Fe^{2+} become Fe^{3+} when these clusters contain a stable Fe^{2+}, Fe^{3+} core. It is not possible to make manganese clusters of this kind as Mn^{2+} has too weak an affinity for sulphide or thiolates. In fact, the Mn^{2+} ions in the oxygen-generating site are bound to O-donor ligands such as carboxylate with but one histidine in the cluster. The Mn sites are then very like those often found as Ca^{2+}-binding sites. Taken together with the fact that Ca^{2+} is still found at the site in this cluster suggests that the evolution of this site developed from the possibility that all the O-donor ligands bound Ca^{2+} at first but that by chance two such sites came close to one another. While Mn^{2+} at 10^{-6} M is likely to bind but only with limited affinity to these single Ca^{2+} sites if two such Mn^{2+} ions are bound at closely adjacent such sites, they readily become $Mn^{3+}.O^{2-}.Mn^{3+}$ even at low redox potential. They are then quite stable at pH 7 and do not exchange readily. My suggestion is then that the external surface cluster has come about accidentally through the coming together of several sites for Ca^{2+} which became a four Mn^{3+} cluster plus one Ca^{2+}. It is no surprise that one Ca remains and we must ask the functional significance of it. The simplest explanation is that it is the binding of one Ca^{2+} that helps to form the proteins' geometry for taking up linked Mn^{2+} ions. Today, Ca^{2+} ion is still regulatory for oxygen release, and the suggestion that Ca^{2+} initiates synthesis of the cluster on the outside of the membrane is very important as the whole of this protein is replaced once every half hour!

5. *The initial single cell eucaryotes*

(It is useful at this stage of this chapter to have general references on Calcium biochemistry [7–10].) The first eucaryotes were single cells with organelles (chloroplasts and mitochondria), an endoplasmic reticulum and many vesicles. Their evolution is dependent on the increase in the level of oxygen and its environmental consequences

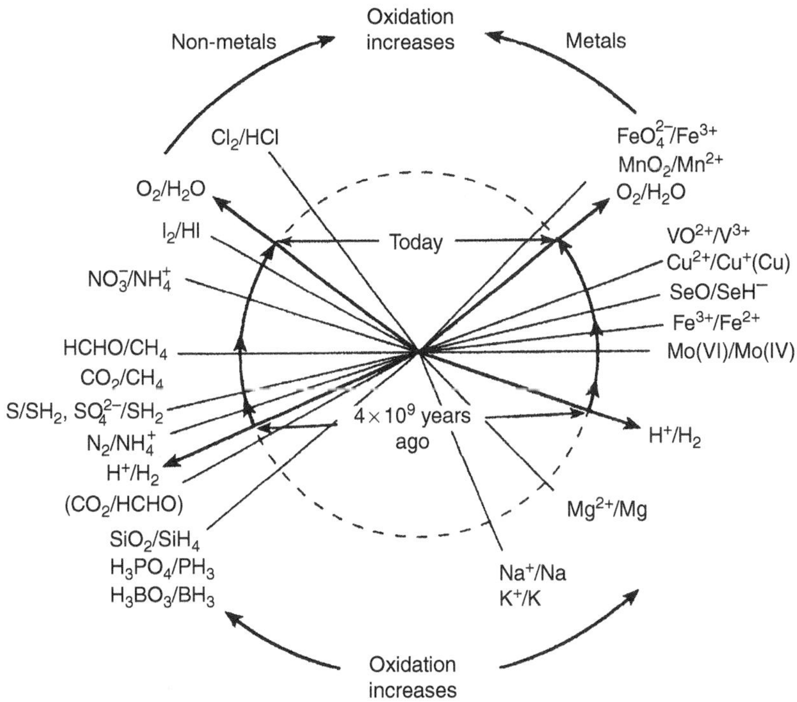

Fig. 6. The changes of oxidation state of elements in the environment as oxygen increased in the atmosphere from the H^+/H_2 potential to the O_2/H_2O potential.

(Fig. 6) [1]. How they came about is unknown for there are too many seemingly simultaneous advances in metabolism and structure. In this chapter, I am concerned only with the roles of Ca in these novel larger organisms. The obvious feature is that a cell of many compartments needs a communication network (Fig. 7). Moreover, such a larger cell has of necessity a longer life and must be more aware of the threats and opportunities presented by its environment to give it survival value, again needing an alerting mechanism. Both these internal and environmental communication systems are found to be dependent on stimulated entry of Ca. It is Ca which passes through the outer membranes when a cell meets novel circumstances and then activates, with the assistance of phosphorylation and Mg and with frequent release from the inside of the membrane of such molecules as IP_3. This required a special Ca^{2+} entry channel protein. The Ca interacts internally but at a distance with many kinases activating them through conformation changes of the novel receptor protein, calmodulin. The Ca signal can also be amplified by Ca^{2+}-induced release of Ca from various Ca stores in vesicles and in the reticulum. Thus, Ca can create the passage of repetitive flood waves of Ca concentration across a cell starting from a point. The stimulation can be due to contact with surfaces or with food particles when the food, including bacteria, is ingested often in vesicles. Thus, a series of both physical and chemical events are set in train to take advantage of or to avoid environmental changes.

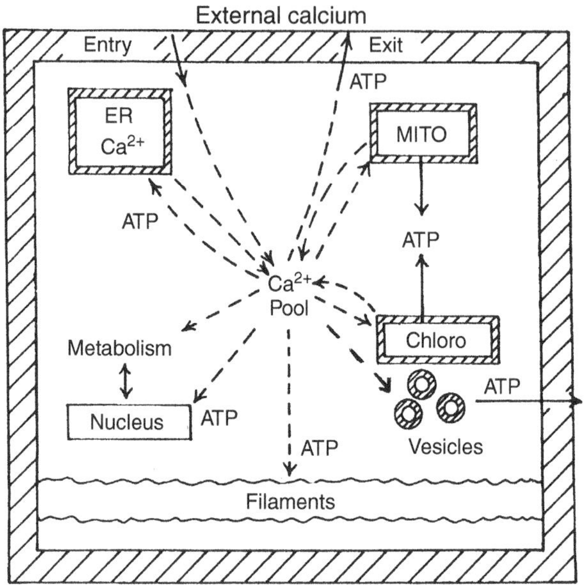

Fig. 7. A compartmentalized eucaryote single cell indicating the way in which Ca^{2+} signals to the endoplasmic reticulum (ER), mitochondria (MITO), chloroplasts (CHLORO), the filaments, the nucleus and some vesicles.

Such activities require energy of ATP, so that ATP synthesis must be stimulated by the internal Ca communication system. We observe this stimulation in the mitochondria, where dehydrogenases are activated, and in the chloroplast, where Ca has several roles, as well as on the O_2 release mentioned before (Fig. 7).

Now a eucaryotic cell also reacts physically to the environment extending binding molecules to hold to surfaces or engulfing whole procaryote cells. The physical response is activated by Ca entry and binding to contractile filaments. The cell must recover quickly from any invasive Ca pulse because higher levels of Ca are dangerous and the cell must be made ready for the next external event. Recovery is achieved rapidly through pumps or exchangers rejecting Ca^{2+} into the extensive weaving endoplasmic reticula and the equally extensive weaving mitochondria and chloroplasts in addition to outward pumping by the ATPase pumps of the outer membrane. The vesicles have pumps similar to those of the outer membrane directed into their structures, i.e. in the opposite direction to the Ca channels for release of Ca from the cell cytoplasm. The organelles also take in Ca quickly by Na/Ca exchangers but then release it again slowly. This complex set of operations ensures quick recovery without great loss of Ca to the extracellular environment. The overall result is a massive temporary cell response which is not connected to genes although all the novel proteins are so connected. The major separation of slow gene from fast environmental responses was a novel achievement of single cell eucaryotes. However, the Ca^{2+} of the reticula, the Golgi and the nucleus is involved in slower processes too.

Table 1
Some classes of calcium proteins

Protein	Location and function
Calmodulin[a]	Cytoplasm, trigger of kinases, etc.
Calcineurin[a]	Cytoplasm, trigger of phosphatases
Annexins	Internal, associated with lipids, trigger
C-2 domains	Part of several membrane-link enzymes
S-100[a]	Internal and external: buffer, messenger, trigger
EGF domains	External growth factor but general protein assembly control, e.g. fibrillin
GLA domains	External, associated with bone
Cadherins	Cell–cell adhesion
Calsequestrin	Calcium store in reticula
ATPases	Calcium pumps

EGF, epidermal growth factor.
[a]EF-hand proteins.

Now it is clear that this huge communication network required several novel Ca-binding proteins with binding constants to match the elevation of Ca concentration on stimulation. The list of some of these cytoplasmic proteins is included in Table 1. They all have binding constants of $>10^6\ M^{-1}$, e.g. calmodulin. A further series of proteins was required in the vesicles to store Ca in the presence of the considerable free Ca^{2+} concentration of 10^{-3} M and to release it quickly. The proteins are named calsequestrin and calreticulin. They have binding constants of around $10^3\ M^{-1}$ and may bind as many as 40 Ca^{2+} ions loosely, i.e. ready for release. Of course, the output Ca channel proteins of vesicles are new too and all must be selective for Ca^{2+} over Mg^{2+} and Mn^{2+}. Some vesicles have quite separate but similar pumps for Mn^{2+} uptake, e.g. see the Golgi, and can have novel pumps for Ca^{2+}.

Another function worthy of note is a novel activity of eucaryotes in their reticulum, where both protein folding and glycosylation of proteins for export occurs. These activities are managed in the reticulum under the control of Ca chaperones. The chaperone function is to aid protein folding and quite different from the S-100 Ca carrier proteins, sometimes also labelled (wrongly) as 'chaperones'. In the Golgi, the activity of Ca^{2+} can be sustained – see also Mn^{2+}.

We shall look at the novel protein structures which have evolved after we have described the function of Ca^{2+} in multicellular organisms.

5.1. Early mineralization [11]

Mineralization was probably accidental in the earliest organisms. The chert of the most ancient deposits shows little organization. The first organized mineral deposits appear to be those of corals. Here, either calcite or aragonite is formed based on external filaments. The Ca isotope composition suggests that it is likely to be closely

Table 2
The main inorganic calcium solids in biological systems [11]

Anion	Formula	Occurrence	function
Carbonate	$CaCO_3$	Widespread in animals and plants	Exoskeleton Gravity, Ca store Eye lens
Phosphate	$Ca_{10}(PO_4)_6(OH)_2$	Shells, some bacteria, bones and teeth	Skeletal, Ca store (piezoelectric)
Oxalate	$Ca(COO)_2{\cdot}H_2O$	Insect eggs Vertebrate stones	Deterrent Cytoskeleton
	$Ca(COO)_2{\cdot}2H_2O$	Abundant in plants	Ca store
Sulphate	$CaSO_4{\cdot}2H_2O$	Coelenterate statocysts	Gravity S store, Ca store

in equilibrium with environmental Ca, but note the solubility product is not that of purely inorganic mineral and there will be some small isotope fractionation due to external growth on the preferred axis linked to the external matrix [12]. The next level of complexity is shown in the formaniferae, which have calcite shells partly formed away from external Ca [12]. A complete separation is seen in the coccolith, where the calcite shell is completed piece by piece internally before export to form the shell. There are also organisms that can use $CaCO_3$ in gravity sensors, though in some $BaSO_4$ is used. There are further organisms, Acantharia, which build an internal vesicular shell of great complexity of $SrSO_4$. We return to minerals later when control over Ca is whole body control in multicellular eucaryotes (Table 2).

6. *Multicellular eucaryotes*

There is little doubt that it is the rise of oxygen which forces the evolution of organization (Fig. 8) [1]. It is not a random search for any kind of life of high survival value but a necessary search for ways of surviving a poisonous chemical, oxygen, and its oxidation products, not so much with outright advantage but by utilizing both the changing environment and all earlier changes of organisms. Survival utilizing very diverse chemistry such as that of reduction in the cytoplasm and oxidation demands compartmental separation as seen in the periplasm of bacteria and the vesicles of single cell eucaryotes. Much though these three forms of life may appear competitive on the small scale of similar species of each form, overall life is a highly cooperative unity, including symbiotic procaryotes (organelles), that has evolved in one system of changing environments. We have described this broad feature of evolution as one of cooperative chemotypes driven in an effort to optimalize energy capture in the environment/organism ecosystem [3]. Now the next step from single cell eucaryotes was to multicellular eucaryotes which have two major novel features – extracellular matrices and cellular differentiation leading on to a third novelty, extracellular fluids separate from the environment. Advantage lies in the use of separation of function, now in differentiated cells in organs, leading to improved activities but at the cost of a longer

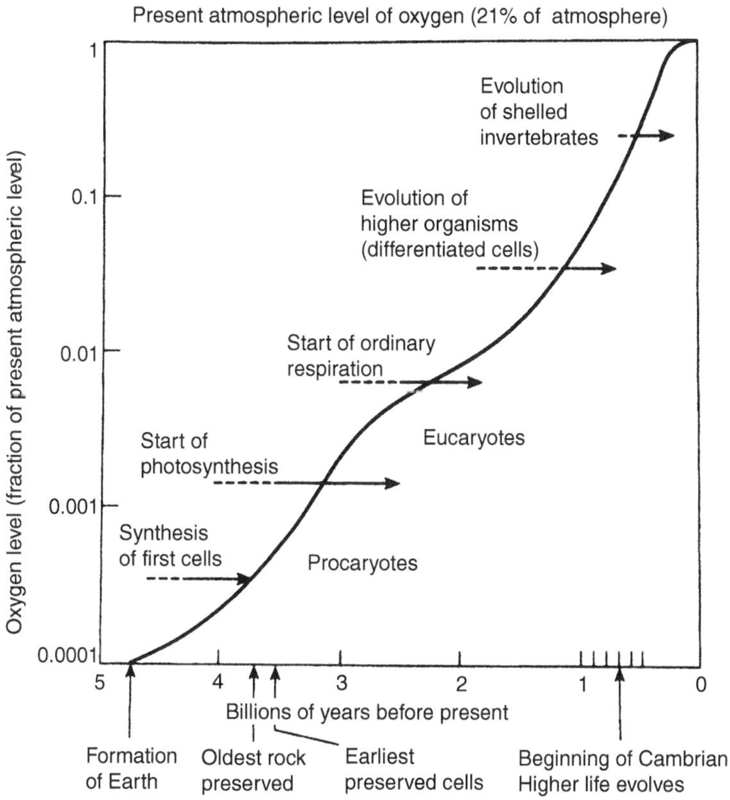

Fig. 8. The relationship between the rise in oxygen in the atmosphere, and therefore of the oxidation potential of many elements (see Fig. 7), and the complexity of organisms.

life and a need for greater protection. All such development demands greater and greater refinement of signalling for whole body integration and environmental recognition (Table 3). The general signalling messengers of the cytoplasm and immediately across cell membranes must not be confused however with cell–cell selective communication. Even so each signalling system has to talk to the ones operative in earlier cell organization so that the whole of a multicellular eucaryote becomes a unity. Remember it is the increase of oxygen and products of oxidation in the environment which makes this evolution possible, but the oxygen arose from the primitive demand to reduce C/N compounds of the environment to forms of use in cytoplasmic synthesis. This reductive synthesis has to be kept as the major polymers of cells must be made.

6.1. Calcium in multicellular eucaryotes

For convenience, we divide these eucaryotes into two groups – those with no nerves and those with nerves, higher animals [3]. We shall place the higher plants with the animals without nerves. This is a little arbitrary of course but we are examining

Table 3
Calcium message-dependent effects in cells today

Effect (system)	Proteins
Muscle contraction	Troponin
Energy availability	Mitochondrial dehydrogenases
Oxygen formation	Chloroplast Mn protein
Membrane activities	IP_3-receptor
	Certain kinases (calmodulin)
	C-2 domains
Apoptosis	Calpain
Water-soluble hormone release transmitter release	Vesicle-trafficking proteins
Flow of vesicles	Tubulin-related proteins
Fertilization	Many
Morphogenesis	Many
Differentiation	Many
Cell division	?
Genetic connection	Phosphatases (calcineurin) kinases
Nerves and brain	Many
General metabolism	c-AMP, c-GMP cyclases
Nitroxide synthetase	NO generation, muscle relaxation

purely the use of Ca, and therefore we are concerned with chemical properties of this ion and related ions, i.e. chemotypes not genotypes. The major new features of the multicellular organisms without nerves are as follows:

(1) cell differentiation;
(2) cell positioning held by connective tissue;
(3) cell–cell communication now by organic messengers (Fig. 9).

In differentiated cells, we know rather little about the way Ca is distributed in different cells of soft organs. The cellular distribution may be very similar to that in single cell eucaryotes. Turning to cell position and connective tissue, we observe that the extracellular proteins with long extensions often have Ca^{2+}-binding centres. Generally speaking, this is an extension of the early use of Ca^{2+} in the maintenance of cell wall structures as seen in procaryotes. The Ca can be bound to peptide repeats in proteins such as epidermal growth factor (EGF) domains or to saccharides, the second being more common in plants. Thus, Ca in evolution is assisting the increase of the use of space. New use of space demands an increase in communication, and here we see the greatest exploitation of the functional value of Ca.

In fact, the communication network expanded greatly through the use of oxidized organic molecules (Tables 3 and 4). These molecules appear as hormones and transmitters, e.g. hydroxylated sterols and adrenaline. Some such as the sterols pass through membranes and circulate at low levels at all times, only being adjusted slowly (Fig. 9). They do not connect to fast Ca message systems but interact with slow Zn proteins [1]. All such hormonal responses are very slow – often on the scale

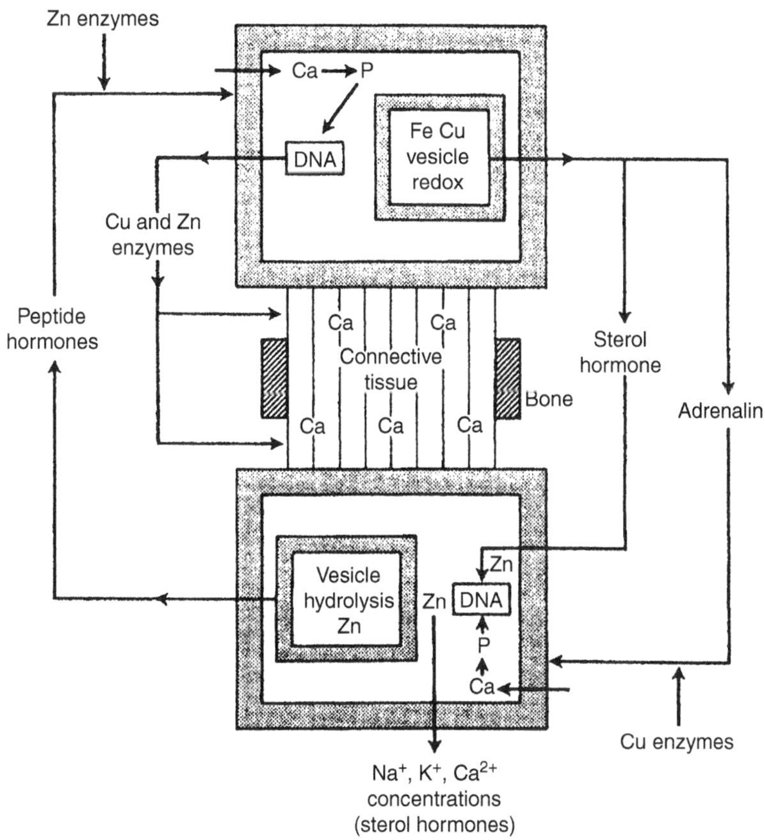

Fig. 9. The role of calcium (Ca) in multicellular organisms became involved with the formation of the extracellular matrix (later including bone) and the required communication system based on fast organic messengers going between cells, see adrenaline. Ca is not involved so much with hormonal messages, see sterol hormones. There is now a gross interplay between Ca^{2+} chemistry and those of Cu and Zn.

Table 4

Organic messengers produced by oxidation

Messenger	Production	Reception	Destruction
NO	arginine oxidation (haem)	G-protein	? Haem oxidation
Steroids	Cholesterol oxidation (haem)	Zn-fingers	Haem oxidation
Amidated peptides	Cu oxides	(Ca^{2+} release)	Zn peptidases
Adrenaline	Fe/Cu oxidase	(Ca^{2+} release)	Cu enzyme?
δ-OH tryptamine	Fe/Cu oxidase	(Ca^{2+} release)	Cu enzyme?
Thyroxine	Haem (Fe) peroxidase	Zn finger?[a]	Se enzyme
Retinoic acid	Retinol (vitamin A) oxidation	Zn finger?[a]	?

[a]In the nuclear receptor super family of transcriptional receptors.

of days or years. Organic charged molecules, such as adrenaline and peptides, do not pass through membranes but act on the outer cell membrane causing Ca channel opening. Thus, Ca movement now stimulates the fast internal cooperation of cells and organs due initially to organic charged transmitters. Its action in the cytoplasm is closely similar to that we have described in single cell eucaryotes.

Some eucaryote cells in these organisms are large, and free diffusion of Ca across them is slow due to the rejection of the ion from the cytoplasm to 10^{-7} M free Ca^{2+}. These large cells frequently have S-100 proteins to carry Ca across them, e.g. epithial cells.

Now the cells may also have digestive proteins in their vesicles which take in large macromolecules. These proteins may be Ca^{2+}-activated.

6.2. The extracellular matrix

Particularly interesting are the developments of special new sites for Ca^{2+} binding in extracellular fluids due to oxidation of protein side chains. One very interesting oxidation is the hydroxylation of aspartate in the EGF domains. This allows novel cross-linking of EGF repeats especially in proteins of the fibrin group. I do not think they are found in plants, but the EGF domain is very common in animals. Another example is the oxidation of glutamate to γ-carboxy-glutamate or GLA. This amino acid is found in several blood-clotting proteins.

There are other oxidations of amino acids in proteins of extracellular matrices including hydroxyproline and hydroxylysine of collagen, but I do not know of Ca^{2+} involvement. Ca binds to several glycoproteins, lectins, of plants. Of course, it also binds to the phosphoproteins of bone and there it is a probable nucleation point for the function of apatite.

The management of Ca^{2+} proteins is linked to two new vitamins, K, for GLA production, and vitamin D, which controls several proteins for Ca^{2+} uptake, and the process is curiously light dependent.

6.3. The extracellular fluids and later the nerves

To refine signalling in multicellular organisms, it is clearly advantageous to control the local extracellular fluid within a single body. Such a fluid can have contents managed in part by the central DNA code. We see closer and closer control of ions including Ca^{2+} ions until the elaboration of the body fluids of higher animals where equilibrium is established with apatite, Ca hydroxy phosphate. These fluids now carry response proteins to alert the body to biological as well as chemical threats.

Now these fluids also contain well-controlled concentrations of Na^+ and K^+ (Table 5). These ions are not used in organisms to bring about chemical change as they bind too weakly. The ions with chloride became useful in the physical transmission of nerves and then in the brain, but their effects at nerve termini are very much Ca^{2+}-dependent. Note how the weak or modestly binding ions rejected by primitive

Table 5
Extracellular fluids (mM/l)

	Na^+	K^+	Ca^{2+}	Mg^{2+}	Cl^-
Sea water	470	10	10.2	54	550
Typical fresh water	0.6	0.01	2.0	0.2	0.5
Marine invertebrates	450	10	10–20	10–50	530
Freshwater invertebrates	10–100	0.5–4	10	0.5–5	10–150
Terrestrial animals	10–150	5–30	4–15	1–40	20–150

cells, Na^+, Cl^- and Ca^{2+} (and Mn^{2+}), respectively, are a throw-out or waste turned into a use. This is also probably the case for Cu and Zn as they became available, being of much lower value for procaryotes than the ever-increasingly sophisticated eucaryotes.

6.4. Animals with brains

The last step in biological evolution which we see is that of the development of animals with brains. The brain was dependent on the earlier evolution of nerves as a very fast connection 'wire' between organs often linking motion to sense organs – sight, touch and hearing. The relay of information along the nerve is through the Na^+/K^+ potential, but Na^+ and K^+ although physically active are chemically relatively inert because they do not bind organic molecules. It is observed that the nerve message at its terminal can only be effective if the current-carrying potential switch of Na^+/K^+ ions opens a channel to allow Ca^{2+} entry at the nerve terminal. The Ca^{2+} then interacts with the internal fibres (muscles) or with internal vesicles storing organic transmitters (nerve relays at synapses). Sensing is also directly related to Ca even down to the level of the response of eyes to light (Fig. 10). We see an ever-increasing and correlated use of both organic and Ca signals and at the same time a vast increase in the use of Ca binding proteins (Table 6). The brain probably has the most sophisticated set of signals of any organ (Table 7).

7. Minerals in higher organisms [11]

The Cambrian explosion generated a variety of external mineralization of Ca carbonates, examples of many of which are with us today (Table 2), and of several other salts. The analysis of these shells shows small crystals of the carbonate embedded in a glycoprotein matrix. The exact growth pattern of a shell appears to be based on the simple growth of a cone in a variety of twisted forms [13]. The cone patterns fit a logarithmic growth curve in many cases. The implication is that the mineral form has a mixed origin due to control of the body of the organism and simple mathematical constraint. We are all familiar with the very different, later, internal mineral apaptite

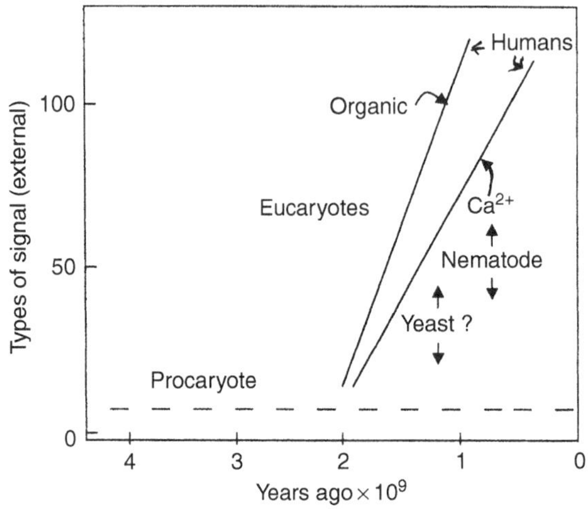

Fig. 10. The growth of signalling of two kinds, organic (left line) and calcium (right line) with time.

Table 6
Distribution of different Ca^{2+}-binding protein motifs in organisms [14]

	Binding proteins					
	Excalibur	EF-hand	C-2	Annexins	Calrecticulum	S-100
Archaea	–	6*	–	–	–	–
Bacteria	17	68*	–	–	–	–
Yeasts	–	38	27	1	4	–
Fungi	–	116	51	4	6	–
Plants	–	499	242	45	40	–
Animals	–	2540	762	160	69	107

The table is based on DNA sequences available in 2004. The activities of the proteins are unknown in most cases.
[a]Single hands only see Chapters 3 and 5 (this volume).

of bones of multicellular higher animals. In effect, bone is not just a static structural support for our body but a flexible living tissue, unlike a shell, and an aid to Ca^{2+} homeostasis in the extracellular fluids. It is piezoelectric responding to stress locally and therefore dependent on life style. Bone is also part of the Ca store for Ca^{2+} in circulating fluids. Very intriguing, it is still connected to two vitamins in these fluids. The homeostasis of Ca itself is linked to epithial uptake and kidney rejection. In epithial cells, Ca^{2+} uptake is connected to the S-100 transport protein, the synthesis of which is connected to vitamin D and sunlight exposure. Hormonal controls of Ca metabolism are of great importance in old animals. Now the Ca transport in the blood involves the g-carboxy-glutamate residue due to vitamin K activity. The complex network linking light and oxygen with these vitamins and

Table 7

Messenger compounds between cells in the brain and connected to calcium

Fast signals
Excitatory (+)/inhibitory (−)
Glutamate (+)
Glycine (−)
GABA (γ-aminobutyric acid) (−)
Acetylcholine
Inorganic signals
Na^+, K^+, Ca^{2+}, (Zn^{2+})
Intermediate speed of signal
Noradrenaline (Norepinephrine)
Dopamine
Serotonin
Nitric oxide
Carbon monoxide
Slowest signals
Neuropeptides (large number >20)
Substance P
Cholecystokinin
Corticotrophin-releasing factor
Melatonin

through to Ca^{2+} metabolism is not thoroughly mapped. The total network of Ca^{2+} action is shown in Fig. 11.

A quite different mineral in vertebrates is the Ca carbonate of otoliths in the ear. They are gravity sensors and in fish are large crystals that grow with age.

Before we turn to the final development of the functional use of the element Ca by humans, we shall describe the coincidental evolution of the types of Ca-binding proteins which are little if at all changed in this last step though variety increases.

8. The evolution of cytoplasmic proteins

Evolution is of systems. Systems' properties do not follow at all directly from knowledge of sequences of DNA because they are dependent on concentrations of components, e.g. of proteins and their interactions in a cooperative dynamic activity. Such cooperative systems of chemicals once set up do not need a code except for repair and reproduction. DNA alone will not secure survival as survival depends on organization utilizing a given required environment. The direction evolution takes therefore is dependent on outside circumstance relative to its necessary interior activity. The involvement of Ca in evolution is then seen as

(1) Rejection by an energized cellular process that allowed the internal cytoplasm to develop including even the code. The proteins used were pumps that became coded. They had a fast response time.

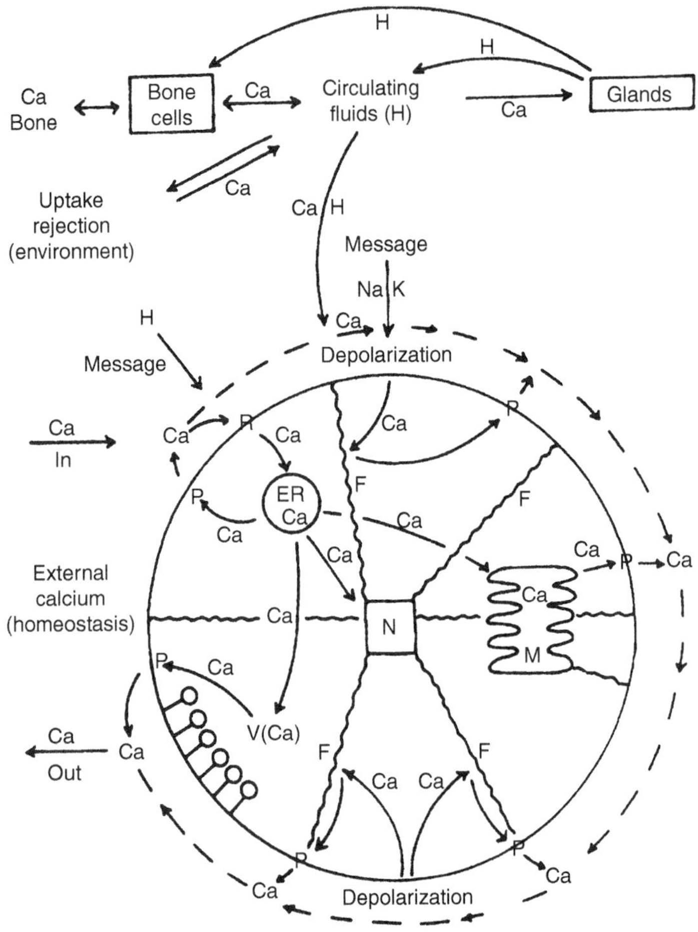

Fig. 11. The total network of Ca^{2+} activities surrounding one cell and leaving aside cell/cell interactions. The complexity is like that of electron flow in a modern computer.

(2) Incidental action of external Ca on outer membranes and proteins outside cells. We describe external proteins later.

(3) The introduction of oxygen into the atmosphere led to single cells with compartments, eucaryotes, of longer life and novel activity in organelles and vesicles. For survival, increased information about their environment and ways in which to respond rapidly were essential. This required new Ca^{2+}-binding proteins (Table 6). In addition, the internal compartments depended on the cytoplasm for proteins of a new kind because the vesicles could contain activities and stores not permitted in the cytoplasm, e.g. Ca^{2+} ions. The need is for an integrated cell network, sensitive to external events that had a fast response. It could not be

Table 8
Evolution of signals related to calcium

Organism	Signal
Procaryote	Simple Ca^{2+} signal to pump genes?
Single-cell eucaryote	Ca^{2+} external to metabolism (energy) Ca external to Ca^{2+} internal IP_3, etc. to Ca^{2+} internal
Multi-cell eucaryotes	Organic messenger release from vesicles Ca^{2+} internal by organic messenger (adrenaline, serotonin and some peptides, NO)
Eucaryotes with nerves	Glutamate, glycine, acetylcholine Neuropeptides

dependent on slow DNA/RNA response; it required a system of fast connections that once made were independent of the central DNA/RNA syntheses. The immediately available connections were those of filamentous proteins and metabolic switches stimulated by fast messages. The original procaryote cells had primitive filaments and had of necessity a strong Ca^{2+} gradient. We observe in eucaryotes event-controlled input of Ca^{2+} to both filaments and to metabolic events, even across organelle membranes so that the Ca^{2+} ion acts as a fast, strong-binding, ion. The growth in these activities is vast in multicellular organisms leading to muscle, organ and nerve communication networks. There are then further additional message carriers, organic transmitters and Na^+/K^+ electrolytic currents up to the brain. To match Ca^{2+} ion speeds required fast protein responses throughout all the families of eucaryotes, and a new group of proteins was required. Outstandingly, we see them in the calmodulins and troponins with variations in the S-100 family [14]. Following this protein evolution is a way of looking at increasing communication and organization so it can be said that the chemistry of Ca and its proteins maps evolution of life, i.e. of diversity of chemotype activity. It is a different reflection of the evolutionary process than that of metabolism changes both dependent on the slow build up of oxygen. Table 8 details the build up of messenger systems in an order and it has continued long after oxygen. Parallels are with the exploitation of Zn and Cu released in quantity only after two billion years of life. Note cellular changes always lag far behind environmental changes – the more so, the more the organism complexity.

A further late development was of extracellular Ca^{2+} proteins, often with oxidized side chains as described in Section 6.2.

8.1. The structures of Ca-binding sites of proteins

It is interesting to look now at the modes of binding of Ca ions by proteins while remembering that both in the cytoplasm and in the extracellular and vesicular fluids the binding sites must select against the much more highly concentrated solutions of Mg^{2+}. This means that no matter where Ca^{2+} binds its protein partners need to give a site some 10^2–10^3 times more powerful than that for Mg^{2+}, i.e. $[Ca^{2+}].\log K_{Ca} >> [Mg^{2+}].\log K_{Mg}$, where K is the binding constant. Now we know how this is brought about in model chelates (see Section 2.1.). A flexible binding site of many donors must suffer from steric or electrostatic repulsion as it produces a binding hole of decreasing radius from 1.0 Å (Ca^{2+}) to 0.6 Å (Mg^{2+}) around a cation (Figs 2 and 3). This is exactly the finding for each protein Ca-binding site. Now each site in isolation may only give rise to a constant of magnitude $K = 10^4 M^{-1}$, and while this is adequate in the vesicles and extracellular fluids, it is not so in the cytoplasm of cells. The ways of increasing binding which appeared in evolution at the time of the eucaryotes appear to be by simple cooperative pairing between sites within the same protein as in calmodulin (gene duplication?) or between single sites and say membrane phospholipids, as in annexins [2,7–10]. The rates of binding and of dissociation of the proteins can be modulated by the mobility of the protein backbone as well as by that of its side chains as is seen by conformational changes [3].

In other chapters of this book, reference will be made to slower processes affected by Ca^{2+} binding. In these cases, it may well be that initial Ca^{2+} binding is strengthened by interaction of the Ca^{2+}-binding protein to other biopolymers, enzymes or even DNA (DREAM). Here direct Ca^{2+}-dependent activity may be sustained for at least minutes so that directly or indirectly (induced phosphorylation of transcription factors) genetic activity can be stimulated, see also phosphatases such as calcineurin. It is probable that short-term as opposed to long-term memory is related to the relative degree to which Ca^{2+} is sustained.

9. Connections to other metal ions

We need to see now the way in which the evolution of Ca^{2+} functions connects to other functions all dependent on the switch from an anaerobic to aerobic environment (Figs 6, 8 and 9). There were in the cytoplasm of anaerobic cells many processes dependent on Mg^{2+} and phosphorylation which became connected to external changes through Ca messages. Note particular the kinases. Prolonged input of Ca^{2+} pulses will adjust the activity of kinases, which in turn link to the genetic machinery, e.g. in fertilization. Again the phosphorylation of ADP in energy capture is switched on by Ca^{2+} entry to the organelles. It is of course Mg^{2+}.ADP and Mg^{2+}.ATP, which form the connection to energy.

The value of Ca associated with oxidation directly is seen in certain messenger systems, e.g. that of NO. The NO-signalling pathway is directly dependent on Fe haem proteins, which have associated calmodulin triggers. Indirectly there are many organic messenger molecules prepared in vesicles by oxidation using Cu enzymes and these organic messengers act cooperatively with Ca. There are also Zn releases from vesicles stimulated by Ca^{2+}. A complex system of regulation switch on by Ca^{2+} which involves Zn^{2+} (Fig. 9) in an enzyme is found in calcineurin, which appears to serve different functions in unicellular yeast and multicellular higher animals [15].

Now the extracellular matrix to which Ca^{2+} binds is also dependent on oxidation by Fe enzymes.

10. Calcium and mankind

The novelty of humans is their use of chemistry outside their body in the making of all kinds of materials and constructions not attached in any way to them. Early in history, humans used Ca-containing rocks, especially marble, as building materials. The Romans are credited with the first use of mortar, cements and possibly concretes. The huge buildings around us owe their stability to the way in which Ca minerals set after various high-temperature treatments, not open to biological systems themselves but in a way this entire endeavour is an extension of the previous uses of Ca linked to structure. The use of Ca ions in message systems is now superseded as humans have discovered various electronic and radiation communication networks. The most striking is in the use of computers but sophisticated as they are they do not respond the way the brain does to external events. Never forget the dependence of this remarkable organ, the brain, on the complex patterns of use to which the Ca ion can be put. The complexity of the use of Ca is now related to a vast variety of organic signals (Tables 7 and 8).

11. Summary

Calcium activity in organisms shows a continuous development in the ecosystem evolution as the oxygen and oxidized chemicals arose in the environment (Table 9). As we tabulate Ca^{2+}-binding proteins from the knowledge of DNA [14] (and subsequent oxidation of protein side chains), what we are seeing is the unavoidable development of organization to counteract chemical change of the environment. Organization itself, both structural and dynamic parts, is demanded by the need to use different chemistries in different connected places given the way the environment changed and provided opportunities. There is therefore only one way for evolution to go utilizing the unavoidable direction of environmental change while capturing more energy in an approach to a final steady state. I believe the major role of Ca^{2+} is to secure cooperation between compartmental activities while assisting response to the environment. In some ways, the ions behave as electrons in an electronic device. This requires more and more sophistication of the free ion fixed local and fluctuational concentrations in different localities and times. The basic control is seen in uptake and carrier proteins linked in places to hormones giving extracellular homeostasis. There is also assistance from the nature of internal bone minerals.

Table 9
Calcium in evolution

Cellular organization	Calcium function
A. Procaryote single cells	Externally – wall crosslinks – random mineralization Internally – very slow signal: swimming No message to genetic code except for sporulation under starvation
B. Early eucaryote single cells (no organelles)	As for procaryotes (A) but external events relayed to internal structures (a) Shape response: contractile devices (b) Metabolic and energy response (c) Controlled external mineralization (e) Control over cell death
C. Late eucaryote single cells + organelles, e.g. see *Acetabularia*	As for (B) but (a) Extra response and energy release from mitochondria and chloroplasts (b) Necessary connection to genetic code through phosphate compounds to give growth, see C(a) (c) Growth pattern linked to constant circulating calcium current around cell edges: differentiation
D. Early multicellular animals (nematodes) and plants are similar but without nerve structures	As for (C) except (a) Link to nerve response as well as chemical response to the environment (b) Control of circulating fluids inside organism by organs, releasing hormones (c) Connective tissue allows build-up of internal mineralization (d) Protective and immune system organizations
E. Late multicellular animals with brains	As for (D) except (a) the brain is a new plastic organ (unlike other organs) and calcium currents around nerve cell tips of the brain are now linked to local growth to give memory Apoptosis

The question arises as to how complexity in the form of novel organization could arise, which is a different question to just that of the evolution of proteins. Protein evolution is by chance mutation, gene duplication and transposition of fragments. All its switches can be described by consideration of chance – random variation that leads to species variety but the change in organization is common to all eucaryote species, plants, animals and fungi. They all use Ca^{2+} communication modes and they differ only in the sophistication of the message system as it develops in animals. Because we consider that it is a matter of necessity, the real drive behind evolution is then not in chance production of related species but ever more effective modes of material turnover and energy use (Fig. 12). How did this use of environmental change impose itself on the simultaneously required development of genetic sequences? In my mind, it is clear that in dynamic systems organization has to

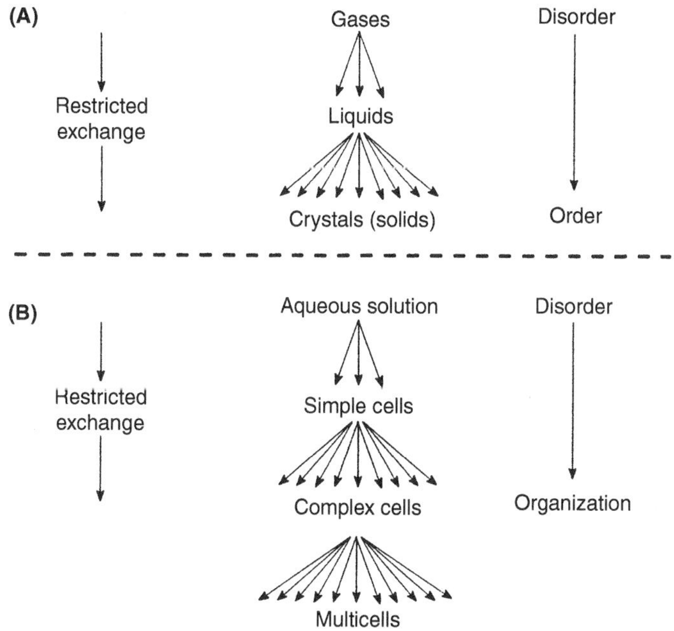

Fig. 12. In the universe, there is a general deviation to disorder but the cooling creates local materials in condensed states (planets) as well as strong sources of radiation (the sun). The effect of radiation is to return material to uncooled disordered states, but if there are interactions in modestly excited states, the mineral can enter into a flow system that can become organized, not ordered.

increase with increase in chemical diversity and will seek ways of doing so towards an end-point. The sequence of introduction of message systems is then also fixed (Tables 8 and 9), as is the way Ca^{2+} functions arise [1].

This chapter does not really require references for it is based on my reading of the work on Ca by the whole scientific community.

References

1. Fraústo da Silva, J.J.R. and Williams, R.J.P. (2001) *The Biological Chemistry of the Elements,* 2nd ed., Oxford University Press, Oxford.
2. Berthon, G. (Ed.) (1995) *Handbook of Metal Ligand Interactions in Biological Fluids,* Marcel Dekker, New York.
3. Fraústo da Silva, J.J.R. and Williams, R.J.P. (2001) *The Biological Chemistry of the Elements,* 2nd ed., Oxford University Press, Oxford, pp. 117–118.
4. Williams, R.J.P. and Fraústo da Silva, J.J.R. (2006) *The Chemistry of Evolution,* Elsevier, Amsterdam.
5. Moller, J.V., Juul, B. and le Maire, M. (1996) *Biochim. Biophys. Acta,* 1286, 1–51.
6. Rutherford, A.W. and Boussac, A. (2004) *Science,* 393, 1782–1784.
7. Carafoli, E. and Klee, C. (Eds) (1999) *Calcium as a Cellular Regulator,* Oxford University Press, Oxford.
8. Pochet, R. (Ed.) (2000) *The Molecular Basis of Calcium Action in Biology and Medicine,* Kluwer Academic Press, Dordrecht, The Netherlands.

9. Vogel, H.J. (Ed.) (2002) *Calcium-Binding Protocols,* Vols 1 and 2, Humana Press, Totawa, N.J., U.S.A.
10. Dominguez, D.C. (2004) *Mol. Microbiol.*, 54, 291–297.
11. Erickson, J. (1992) *An Introduction to Fossils and Minerals, Facts on File*, New York.
12. Chang, V.T-C., Williams, R.J.P., Makishima, A., Belshaw, N.S. and O'nions, R.J. (2004) *Biochim. Biophys. Res. Pub.*, 323, 79–85.
13. Thomson, D'Arcy (1942) *On Growth and Forms,* Vols 1 and 2, Cambridge University Press, Cambridge.
14. Morgan, R.P., Martin-Almedina, S., Gonzalez-Florezm, M.I. and Fernandez, M.P. (2004) *Biochim. Biophys. Acta,* 1742, 133–140.
15. Klee, C. (Ed.) (2003) *(Special Publication on Calcineurin), Biochim. Biophys. Acta,* 311, 1087–1208.

Structural and Functional Aspects of Calcium-Binding Proteins

J. Krebs and M. Michalak (Editors)
Calcium: A Matter of Life or Death

ISSN: 0167-7306/DOI: 10.1016/S0167-7306(06)41003-6

CHAPTER 3

Calcium-binding proteins and the EF-hand principle

Joachim Krebs[1,2] and Claus W. Heizmann[3]

[1]*NMR Based Structural Biology, MPI for Biophysical Chemistry, Am Fassberg 11, D-37077 Göttingen, Germany;*
[2]*Institute of Biochemistry, HPM1, Swiss Federal Institute of Technology, ETH-Hönggerberg, Schafmattstrasse 18, CH-8093 Zürich, Switzerland; Tel.: +49 551 201 2200; Fax: +49 551 201 2202; E-mail: jkrebs@nmr.mpibpc.mpg.de*
[3]*Division of Clinical Chemistry and Biochemistry, Department of Pediatrics, University of Zürich, Steinwiesstrasse 75, CH-8032 Zürich, Switzerland*

Abstract

Calcium is of pivotal importance for many biological processes. It may have a rather static, structure-stabilizing role, or it may participate as one of the second messengers of the cell in signal transduction pathways, fulfilling a more dynamic function. This is made possible by some specific properties of the Ca^{2+} ion (e.g. high dehydration rate, great flexibility in coordinating ligands and largely irregular geometry of the coordination sphere). The control of calcium homeostasis is of central importance for the organism. It is a highly integrated process consisting of a number of hormonally controlled feedback loops and an elaborate system of channels, exchangers, pumps and numerous Ca^{2+}-binding proteins to control Ca^{2+} fluxes into and out of cells or within cells. This chapter describes the different roles of calcium in the regulation of biological functions and the proteins involved in these processes.

Keywords: annexins, C2-domain proteins, calcium-binding proteins, calcium buffers, calcium sensors, calmodulin, centrin, EF-hand principle, non-EF-hand calcium-binding proteins, S100 proteins, second messenger, STIM, troponin C

1. Introduction

Calcium is one of the most common elements on earth and the fifth most abundant in vertebrates. It is not only a major constituent of the skeleton but fulfils a central role in cellular functions as one of the second messengers [1–3]. In minerals as well as in solution, calcium occurs predominantly in a complex form, mostly as calciumphosphate (e.g. hydroxyapatite $[Ca_{10}(PO_4)_6(OH)_2]$), which makes up 60% of the weight of the human skeleton, comprising 99% of the total calcium of a human body. Compared with this amount, the calcium found in the extracellular fluid (ECF) and intracellularly in the cytosol or in other intracellular compartments is almost negligible. In the ECF or in the lumen of intracellular organelles, the calcium concentration is in millimolar (2–5 mM, of which about 50% is unbound, i.e. corresponds to free Ca^{2+} concentration), whereas the cytosolic free Ca^{2+} concentration of a resting cell is about 100–300 nM. This results in a steep concentration gradient of ionized Ca^{2+} across cellular membranes, which is regulated by a variety of channels, pumps, exchangers and other transporting systems controlling the fluxes of Ca^{2+} into and out of the cell and between the various intracellular compartments as described in detail in other parts of this book. On the contrary, calcium homeostasis of the ECF is maintained through a highly integrated and complex endocrine system. This involves the interplay between a receptor sensing the level of Ca^{2+} in ECF (see Section 4.2) and two antagonistic polypeptide hormones, parathyroid hormone (PTH) and calcitonin (CT), and a vitamin D metabolite, 1,25 $(OH)_2D_3$. They regulate the calcium flow into and out of the ECF through acting on target cells of intestine, kidney and bone. In mammals, birds, reptiles and amphibians, PTH prevents calcium of the ECF from falling below a threshold level whereas CT prevents abnormal increases of serum calcium. PTH is also responsible for the formation of 1,25 $(OH)_2D_3$, which acts on specific receptors in the intestine to promote absorption of Ca^{2+}.

When Sidney Ringer [4] at the end of the nineteenth century discovered that calcium plays an essential role regulating the contraction of the heart, the general implication of this landmark observation came ahead of its time. It needed the formulation of the 'second messenger' concept by Earl Sutherland in the 1960s [5,6] to recognize that calcium could play a more general regulatory role. This view was supported by the work of Ebashi and his co-workers [7], who identified troponin C as the regulatory protein of muscle contraction due to its ability to bind Ca^{2+} with high affinity. Thus, troponin C was the first protein recognized as a 'calcium sensor'. However, the discovery of calmodulin (CaM) by Cheung [8] and Kakiuchi [9] was instrumental in realizing the pivotal role of Ca^{2+} as a general regulator of cellular functions. This is made possible due to the tightly controlled steep concentration gradient of ionized Ca^{2+} across cellular membranes. If a cell becomes activated because of an external signal, this often results in up to a 100-fold rise in the intracellular free Ca^{2+} concentration due to the uptake of extracellular Ca^{2+} and/or the release of Ca^{2+} from intracellular stores. These changes of the free Ca^{2+} concentration can cause significant oscillations of Ca^{2+} in the cytosol providing the possibility of signal transduction for a number of different cellular activities such as metabolism, protein phosphorylation and dephosphorylation, fertilization, cell proliferation, division, gene expression and apoptosis, muscle excitation–contraction and stimulus-secretion coupling, to name a few [1–3]. Many of these functions are accomplished through the interaction of Ca^{2+} with specific proteins resulting in modulations of protein–protein interactions due to conformational changes of the Ca^{2+} receptors.

2. *Ca^{2+} ligation*

Ca^{2+} ligation usually occurs through carboxylates (monodentate or bidentate) or neutral oxygen donors because the calcium ion overwhelmingly prefers oxygen as donor groups of ligands. Owing to the great flexibility of calcium in coordination (coordination numbers usually 6–8, but up to 12 are possible) and a largely irregular geometry, both in bond length and in bond angles, calcium has a superior binding property to proteins as compared with Mg^{2+}, which requires a fixed geometry of an octahedron with six coordinating oxygen atoms and a fixed ionic bond distance due to the smaller ionic radius (0.64 Å) as compared with 0.97 Å of calcium. This difference in flexibility of complexation geometry permits calcium a greater versatility in coordinating ligands leading to a higher exchange rate (see Chapter 2, this volume, for more details).

In proteins, calcium is generally bound by seven oxygen atoms in a conformation that is best represented by a pentagonal bipyramid. By contrast, Mg^{2+} ion with its much smaller atomic radius is coordinated by six oxygen atoms in a fixed octahydron type coordination. The superior affinity of calcium for binding proteins over magnesium is reflected in a three orders of magnitude difference in dehydration rate between the two ions, which is the advantage of the loose pentagonal bipyramidal over the tight octahedral packing of the oxygen ligands. This makes Ca^{2+} much more suitable as a signal-transducing factor, especially in the presence of a high excess of Mg^{2+}, the concentration of which is in the mM range on both sides of the cellular membrane.

3. The EF-hand principle

As pointed out in the introduction, the concentration of the intracellular Ca^{2+} in eucaryotic cells is closely regulated to remain below 3×10^{-7} M in a resting cell, whereas outside cells the Ca^{2+} level is 10^{-3} M resulting in a steep concentration gradient across the plasma membrane. The intracellular level of Ca^{2+} can transiently increase because of a response to extracellular signals. In order to function as a second messenger, the Ca^{2+} signal is transduced by a variety of Ca^{2+}-binding proteins. In contrast to the extracellular, low-affinity Ca^{2+}-binding proteins (see Section 4), the intracellular proteins bind Ca^{2+} with high affinity due to a sequential arrangement of about 30 amino acids ligating calcium in a loop flanked by two helical segments. This common helix-loop-helix Ca^{2+}-binding motif, also known as the EF-hand as introduced by Kretsinger [10] is derived from parvalbumin (PV), the first 'EF-hand'-containing calcium-binding protein to have its structure solved [11]. This structure revealed six α-helices, designated A through F; thus, the spatial orientation of the fifth (E) and sixth (F) α-helix of PV enclosing the calcium-binding loop resembles the 'EF-hand' as shown in Fig. 1. PV consists of three homologous EF-hand calcium-binding domains, but only two are able to bind calcium with high affinity. The 'EF-hand' domain later turned out to be an important and highly conserved entity of the intracellular Ca^{2+} receptor proteins to trigger cellular responses. This family of proteins which may contain several copies of the EF-hand motif in one protein is steadily increasing; to date more than 600 have been described (the human genome alone may contain up to 100 EF-hand-containing proteins), but for most of them, a precise function is not known yet [12]. Thus, PV became the prototype of numerous Ca^{2+}-binding proteins with similar structural properties, such as CaM, troponin C, recoverin, S100 proteins and others. To date, more than 50 structures of EF-hand proteins are known (see Chapter 4, this volume).

This family of homologous proteins has probably evolved from a common precursor domain and developed into a tree of numerous different proteins by gene duplications and deletions. Most EF-hand proteins contain an even number of EF-hands, usually tandem copies between 2 and 12 which occur in pairs and are related by a twofold symmetry axis. On the contrary, recently members of a new family have been described

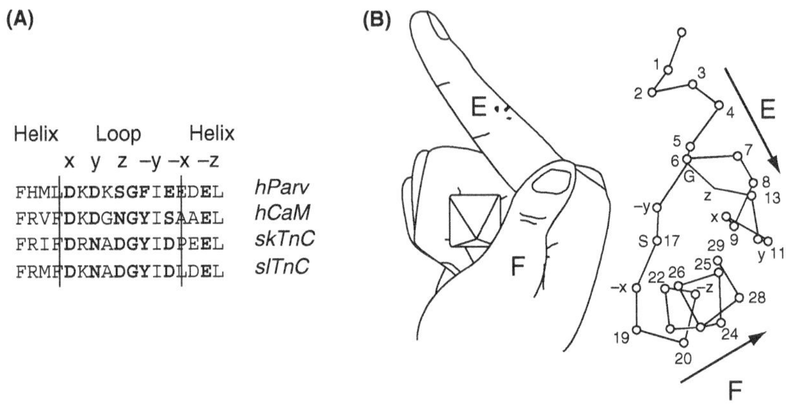

Fig. 1. The EF-hand. **(A)** Alignment of amino acid sequences of Ca^{2+}-binding loop regions of EF-hands from human parvalbumin (PV), calmodulin, skeletal and smooth muscle troponin C showing the conserved residues in single letter code in bold letters. **(B)** The EF-hand as proposed by Bob Kretsinger based on the crystal structure of PV can be represented by the forefinger (helix E) and the thumb (helix F), both enclosing the Ca^{2+}-binding loop, represented by the bent middle finger [adopted from Kretsinger, R.H. (1975) in: *Calcium Transport in Contraction and Secretion* (E. Carafoli, F. Clementi, W. Drabikowski, A. Margreth, eds.), Elsevier, Amsterdam, pp. 469–478].

which contain an uneven number of EF-hands, usually five; therefore this subfamily is called the penta-EF or PEF-family [13], as described in detail in Section 5.1.2.6. However, recently solved structures of some members of this family indicated that the functional entities of these proteins occur in pairs, either by homodimerization or by heterodimerization using the non-functional fifth EF-hand as dimerization domain.

As pointed out in Section 2, all EF-hand domains show a pentagonal–bipyrimidal coordination of the Ca^{2+} ion in the loop flanked by two helices, approximately perpendicular to each other. The residues forming the ligands to Ca^{2+} are highly conserved within a contiguous sequence of 12 residues spanning the loop and the beginning of the second helix. Loop residues in positions 1, 3, 5 and 12 contribute monodentate (1, 3, 5) or bidentate (12) Ca^{2+} ligands through side-chain oxygens, residue 7 through its backbone carbonyl oxygen. Therefore, an invariant glycine residue is in position 6 to permit the sharp bend necessary to ligate Ca^{2+} through the oxygen of a side chain (5) and a backbone carbonyl (7). Residue 9 provides an additional ligand, either directly through an oxygen of its side chain or indirectly through a water molecule. In position 1 of the loop, mostly an aspartate is located (but a glutamate can also be found), whereas in position 12 glutamic acid is invariant. The latter has two reasons:

(1) Both oxygens of the side chain carboxylate ligate to calcium.
(2) Glutamic acid has high α-helix-inducing propensity.

The Ca^{2+}-binding domains usually occur in pairs stabilized by hydrogen-bond bridges between the central residues of adjacent loops which form a minimal antiparallel β-sheet. Owing to these pair-forming Ca^{2+}-binding domains, the Ca^{2+} affinity to these sites and the cooperativity of binding are enhanced. CaM and troponin C contain four

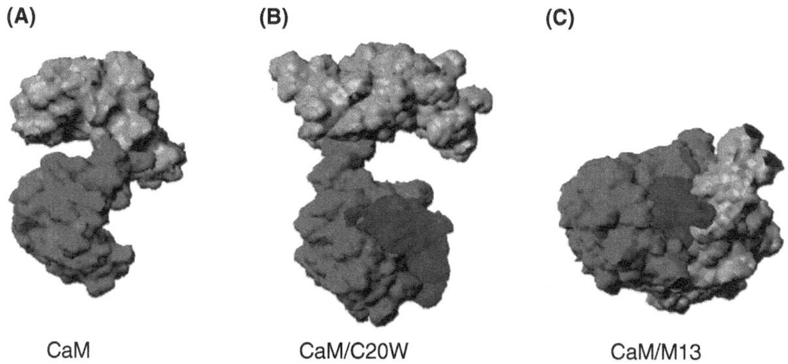

Fig. 2. Surface representation of calcium-bound calmodulin (CaM) and its different binding modes to peptides. The N-terminal half of CaM is in orange, the C-terminal half is in red, and the peptides are in blue. The orientation of the C-terminal half of CaM is the same in all cases. **(A)** Crystal structure of calcium-bound CaM. **(B)** Solution structure of the CaM/C20W complex showing that the peptide C20W, corresponding to the N-terminal portion of the CaM-binding domain of the plasma membrane Ca^{2+}-pump, but lacking the C-terminal hydrophobic anchor residue, binds only to the C-terminal half of CaM. **(C)** The solution structure of CaM/M13 shows a compact globular complex involving both C- and N-terminal halves of CaM in binding the peptide M13 corresponding to the CaM-binding domain of MLCK. (Reprinted with permission from Elshorst et al. [20]) (See Color Plate 4, p. 505).

Ca^{2+}-binding sites. Both proteins display Ca^{2+}-binding characteristics compatible with a 'pair of pairs' model of EF-hands, i.e. the two globular domains at the N- and C-terminal sites each contain a pair of EF-hands, and the domains are connected by a long central helix providing a dumbbell-shaped appearance as revealed by their crystal structures [14–17]. The N- and C-terminal domains bind calcium cooperatively, the C-terminal domain with higher affinity. Recent multinuclear magnetic resonance experiments of the Ca^{2+}-bound form of CaM corroborated the general validity of the dumbbell-shaped structure in solution but indicated the high degree of flexibility of the central part of the helix [18], which is also supported by a high B-factor (temperature factor) in that region in the crystal structure [19]. This flexibility in the centre of the molecule is an important property for the interaction with targets (see Chapter 4, this volume; see also Section 5.1.2.). Upon binding of calcium and subsequent conformational change of the protein, the extended, dumbbell-shaped form of CaM in the absence of a target is transformed into a more compact form upon interacting with a target. The bending of the central helix brings the two domains closely together which enables CaM to engulfe the peptide representing the CaM-binding domain. This property of CaM is documented by numerous examples, some of which will be discussed below in Section 5.1.2 (Fig. 2).

4. *Calcium in the extracellular space*

It was stated in the introduction that calcium in the ECF is tightly controlled to maintain its concentration in a range of 2–5 mM. One of the important functions of Ca^{2+} in the ECF is to stabilize the structure of proteins and to mediate cell–cell or cell–matrix interactions. As a consequence, there are a number of important extracellular processes that require calcium, such as blood clotting, complement activation, cell adhesion or interaction of cell surface receptors such as the Notch receptors or the LDL receptors with their ligands. The Notch receptor (or its homologues) is involved in cell fate decision-making in every metazoan organism so far studied [21], targeted disruption of the Notch gene results in impaired somite formation [22], and a number of human diseases are linked to malfunction of the Notch-signalling pathway [23]. The LDL receptor is responsible for the removal of cholesterol esters from circulating in the blood [24]. Both receptors belong to a family containing calcium-binding epidermal growth factor (EGF)-like domains (cbEGF), which will be discussed in Section 4.1.

An important difference in comparing the Ca^{2+}-binding sites of extracellular and intracellular proteins (e.g. EF-hand proteins) is the spatial arrangement of the ligating groups. As discussed in Section 2, EF-hand-containing Ca^{2+}-binding proteins show a sequential arrangement of their ligating residues in contrast to the extracellular proteins in which these ligands usually are located at distant positions in the amino acid sequence (e.g. see structure of phospholipase A_2 [25]). Therefore, those proteins have a preformed cavity of Ca^{2+}-binding sites with a relative high degree of rigidity with the consequence that the on-rate for Ca^{2+} binding is fairly slow. Because, on the contrary, the off-rate is relatively fast, the affinity of Ca^{2+} for the extracellular proteins is usually low. But owing to the relative high concentration of Ca^{2+} in the ECF, these proteins occur in their Ca^{2+}-bound form and are therefore protected against proteolytic cleavage [26]. The importance of the structure-stabilizing function of calcium is documented for some of these proteins. Mutations in cbEGF domains render them susceptible to proteolysis, which may lead to diseases as shown for fibrillin-1 causing Marfan syndrome [26,27] due to disrupting the pairwise interaction of cbEGF domains. Such pairwise interaction between cbEGF domains which is stabilized by Ca^{2+} may also be important for ligand binding, which is essential for signal transduction of these receptors.

Bone-forming cells, i.e. osteoblasts, synthesize and secrete a number of non-collagenous Ca^{2+}-binding proteins such as osteocalcin, osteopontin and osteonectin which bind to bone minerals such as hydroxyapatite in a calcium-dependent manner (for a review, see Ref. 28). Osteocalcin belongs to the class of Ca^{2+}-binding proteins rich in the unusual amino acid γ-carboxyl glutamic acid [29] mediating the Ca^{2+}-binding property, whereas osteopontin, a glycoprotein, is rich in sequences of aspartic or glutamic acid residues responsible for Ca^{2+} binding. In addition, both proteins contain arginine-glycine-aspartic (R-G-D) acid sequence domains known to be important to mediate binding to cell surface receptors such as integrins. Osteonectin, on the contrary, is an extracellular Ca^{2+}-binding protein that can contain at least one EF-hand type of high-affinity Ca^{2+}-binding sites [30], which are normally typical for intracellular Ca^{2+}-binding proteins. One interesting abnormality of this EF-hand is the fact that it is stabilized by an S-S-bridge which normally does not occur in EF-hand type proteins [31].

Similar to the intracellular Ca^{2+}-binding proteins that are composed of EF-hand domains, most extracellular proteins are assembled from a limited number of domain structures or modules. This is well documented for the calcium-binding domains of a number of extracellular proteins like the γ-carboxyl glutamic acid rich Gla-domain in the vitamin K-dependent clotting factors [32], the cadherin module [33] or the EGF-like domain [34]. The latter is one of the best studied examples and should be described in more detail.

4.1. The calcium-binding EGF domain

The EGF or EGF-like domain is one of the most common modules identified in extracellular proteins [35]. It plays a general role in cell–cell adhesion, blood coagulation and receptor–ligand interaction. EGF modules contain about 40–45 amino acids including six cysteine residues which normally build S-S disulphide bond bridges. A distinct subset of EGF domains containing a calcium-binding domain (cbEGF) has been identified and their structure studied in detail [36,37]. A consensus sequence associated with calcium-binding has been identified as D/N-x-D/N-E/Q-y-D/N-y-Y/F, where x symbolizes a variable amino acid and y a variable sequence of amino acids, and D/N, E/Q and Y/F represent the usual one letter codes for aspartic acid/asparagine, glutamic acid/glutamine or tyrosine/phenylalanine, respectively. One of the striking features of the composition of some cell surface receptors such as the Notch superfamily is that their cbEGF domains are organized as multiple tandem repeats [38] similar to the EF-hand proteins in which the EF-hand domains usually occur in pairs.

The cbEGF domain structures from a variety of receptors have been solved by X-ray crystallography and nuclear magnetic resonance (NMR) spectroscopy [37,39–41]. A superposition of these structures provide evidence for a conserved 'cbEGF core' region [34]. This 'core' structure is stabilized by the three conserved disulphide bonds and by calcium-binding to the N-terminus. The structural analysis of a Ca^{2+}-dependent complex between the cbEGF domain of Notch-1 with a peptide representing the ligand-binding region suggests a rigid, extended conformation exposing the O-fucosylation site Thr466 which may be involved in the regulation of Notch–ligand interaction [37]. The importance of the calcium-binding ability of cbEGF domains is further underlined by a point mutation in the consensus Ca^{2+}-binding region which leads to a loss-of-function phenotype [42]. Such loss-of-functions within cbEGF domains have also been linked to genetic diseases. These can lead to blood clotting due to factor deficiency [43], familial hypercholesterolaemia due to mutations in the LDLR [44], inherited forms of cerebrovascular disorder such as CADASIL due to mutations or deletions in the Notch-signalling receptor [45], or the Marfan syndrome [26,27].

4.2. The extracellular calcium sensor

As stated in the introduction, calcium in the ECF is tightly controlled because of a strict regulation by different hormones keeping the level of calcium concentration within a narrow window. Receptors of the different hormones, i.e. PTH, CT and vitamin D_3, involved in this regulation have been identified, but the molecular basis for sensing the extracellular calcium level was poorly understood. On the basis of several observations, including the demonstration that high levels of extracellular Ca^{2+} and Mg^{2+} stimulated the intracellular accumulation of inositol-polyphosphates [24], it was postulated that a cell surface Ca^{2+}-sensing receptor (CaR) might exist with functional similarities to the Ca^{2+}-mobilizing hormone receptors. In 1993, Brown and co-workers reported the successful cloning and characterization of a 500-kDa protein from bovine parathyroid glands. The protein contained 1085 amino acids and could be identified as a member of the well-characterized family of G-protein-coupled receptors with seven transmembrane spanning domains containing a large, extracellular N-terminal region (comprising about 600 amino acids) and a smaller cytosolic, C-terminal domain [46]. Importantly, the authors could provide evidence that the expressed protein could be activated by the application of extracellular mM Ca^{2+} in the appropriate concentration range or other polycationic agonists. Later it was reported that the CaR could be identified on the surface of many different cell types, especially in cells of those organs that mainly control the level of calcium in the ECF, i.e. in bone-forming and bone-resorbing cells, in parathyroid glands and in cells of kidney and intestine [47]. More recently, CaR has also been found in haematopoetic stem cells for which the CaR may be important for the correct localization of the cells within the bone marrow [48]. The *CaR* gene has probably developed early in evolution because ancestral homologues could be identified in birds [49], amphibians [50] and fishes [51]; in the latter, it may function not only as a sensor for divalent cations, but for salinity in general. The functional unit of CaR is a dimer that is probably stabilized by intermolecular disulphide bonds, because mutations of several cysteins prevent dimer formation [52]. CaR can be glycosylated, and it shares some homologies with metabotropic glutamate receptors, especially in the extracellular domain [47]. The extracellular domain contains low-affinity Ca^{2+}-binding sites characterized by clusters of negatively charged amino acid residues that operate cooperatively to convey to the receptor the necessary sensitivity to distinguish even small fluctuations of calcium concentrations in the ECF [53]. The intracellular C-terminal tail is more than 200 amino acids in length, even if only about 20 amino acids are needed for its biological activity as evidenced by truncation studies [54]. CaR is coupled to intracellular signal transduction pathways through G-proteins, especially to phospholipase C, which leads to the formation of IP_3 and IP_3-dependent pathways. Clinical observations of inheritable hypocalcemic and hypercalcemic syndromes such as familial hypocalciuric hypercalcemia (FHH) have been described long before the discovery of CaR [55–57], which now could be traced back to specific mutations of CaR [58].

4.3. The luminal calcium sensor stromal interaction molecule

The stromal interaction molecule (STIM) is a transmembrane protein of the endoplasmic reticulum (ER) with an EF-hand domain on the luminal site. This protein received much attention recently because a number of results from different laboratories strongly suggest that STIM is the long sought regulator that links the refilling of intracellular calcium stores to plasma membrane capacitative calcium entry channels (see Ref. 59 for a recent review). If intracellular Ca^{2+} stores are depleted of Ca^{2+}, this signals the activation of capacitative calcium entry, which occurs through store-operated or SOC channels [59]. The nature of the SOC channels or the mechanism of their activation in response to the Ca^{2+} depletion of the ER has long been elusive. This mystery seems to be solved since Roos et al. [60] provided strong evidence in Drosophila S2 cells and Liou et al. [61] in HeLa cells that STIM is a key Ca^{2+} sensor to link Ca^{2+} levels in intracellular stores to Ca^{2+} entry through SOC channels to stimulate the Ca^{2+} release-activated Ca^{2+} current (I_{CRAC}) (see Chapter 14, this volume, for more details). On the contrary, CRACM1 [62] or Orai1 [63] has recently been identified as essential for activating CRAC channels. Because it is located in the plasma membrane and according to its amino acid sequence could span the membrane four times, it may represent the channel itself or one of its subunits. Overexpression of either STIM1 or CRACM1 does not amplify significantly I_{CRAC}. However, overexpression of both proteins greatly enhances I_{CRAC} indicating that both proteins are necessary [64–66] (see Chapter 14, this volume, for details).

In mammals, two homologues of STIM exist, STIM1 and STIM2. Although both proteins were originally identified as potential tumour growth suppressors [67] and not suspected to be involved in Ca^{2+} signalling, they have been shown to have a luminal EF-hand close to the N-terminus oriented towards the lumen of the ER. The two proteins can homodimerize or heterodimerize, and they contain several protein–protein interaction domains like a coiled-coil domain in the cytoplasm and a SAM motif (sterile α-motif) in the ER. Interestingly, when key residues in the EF-hand essential for Ca^{2+} binding were mutated to alanine, e.g. D76A [61] or D76AE87A [68], the mutants constitutively activated Ca^{2+} influx even if the Ca^{2+} stores were filled [61] indicating that STIM indeed functions as the Ca^{2+} sensor of Ca^{2+} store depletion. Furthermore, it was found that by depleting the Ca^{2+} stores and thus loosing Ca^{2+} binding by the EF-hand of STIM, the protein translocates to local 'punctal'-like sites close to the plasma membrane suggesting the possibility of direct contact with the Ca^{2+} entry- or SOC-channels thereby triggering – directly or indirectly – Ca^{2+} influx [61,68].

5. *Intracellular calcium-binding proteins*

5.1. *EF-hand-containing proteins*

5.1.1. *Calcium-buffer proteins*

5.1.1.1. Parvalbumin

Parvalbumin is a cytosolic protein mainly found in fast twitch skeletal muscles [69] but is also widely expressed throughout the nervous system [70]. It contains three EF-hand domains, but only the two C-terminal sites are calcium binding. In fact, the ion-binding sites of PV are so-called mixed binding sites because they can bind Mg^{2+} as well [71]. Therefore, under resting conditions, PV is mainly in the Mg^{2+}-bound form inside the cell, even if PV has one of the highest affinities for Ca^{2+} among the EF-hand proteins. If the intracellular Ca^{2+} concentration rises above a certain threshold, Ca^{2+} replaces Mg^{2+} at these sites. Because this is a rather slow process [72] due to the slow dehydration rate of Mg^{2+}, it was long time a controversial issue whether PV should be considered as a Ca^{2+} buffer protein, but detailed localization studies and findings obtained with 'knock-out' mice provided further insights.

Highest concentration of PV is found in fast twitch skeletal muscles in contrast to slow twitch muscles in which the concentration of PV is very low. Interestingly, Heizmann et al. [73] could directly correlate the content of PV in different muscles with their relaxation speed. In fact, if calcium is released from the sarcoplasmic reticulum in a contracting muscle, it binds first to troponin C, even if the affinity for Ca^{2+} is higher for PV. This is because PV is in a Mg^{2+}-bound form if the muscle is at rest and Mg^{2+} has a slow off rate, but during the relaxation phase of the muscle, PV can act like a sponge to take up Ca^{2+} by influencing the decay kinetics of Ca^{2+} transients. This view is underlined by the fact that in PV-deficient mice the rate of Ca^{2+} decay is significantly slower after a fast stimulation of the muscle [74]. In this context, it may also be of interest that in neurons of a number of brain regions, the content of PV seems to correlate with the firing rate of those neurons [75].

5.1.1.2. Calbindin

Two types of calbindin have been described, the small calbindinD9K and the larger calbindinD28K. These proteins are high affinity Ca^{2+}-binding proteins that are widely distributed in different species and organs. These proteins are hormonally controlled by vitamin D (therefore the affix 'D' in the name) in tissues like kidney or intestine, in others like the brain probably not [76]. The highest concentrations of these proteins are found in tissues of calcium transport such as kidney or intestine, but calbindinD28K is especially rich also in the brain, where it fulfils essential neuronal functions by influencing synaptic interactions [77].

CalbindinD9K – or 'intestinal calcium-binding protein' as it was originally called – has two functional EF-hand calcium-binding sites. Ca^{2+}-bound calbindinD9K was the second structure of an EF-hand-containing protein to be solved [78], later also the structure of the apo-form has been solved by NMR [79], and both structures turned out to be very similar, i.e. Ca^{2+} binding to calbindin9K does not induce a significant conformational change and therefore does not expose hydrophobic surfaces upon Ca^{2+} binding like the calcium sensor proteins as described in Section 5.1.2.

CalbindinD9K was also the first calcium-binding protein in which the so-called pseudo EF-hand was characterized [78], which later turned out to be characteristic for the S100 proteins (see Section 5.1.2.5.). Characteristic for the pseudo EF-hand (located at the N-terminus of the protein) is that due to insertion of amino acids into the Ca^{2+}-binding loop Ca^{2+} is ligated to the protein through oxygen atoms of backbone carbonyls which could be detected only by solving the crystal structure [78]. Nevertheless, despite the unusual nature of the pseudo EF-hand site, the Ca^{2+}-binding affinity is not significantly different from the canonical site at the C-terminus [80].

Very recently, the structure of the Ca^{2+}-loaded calbindinD28K has been solved by NMR [81]. Since long time, it was assumed that also calbindinD28K functions solely as a Ca^{2+} buffer, e.g. in kidney distal tubules or in intestinal absorption cells, but recent findings demonstrated that calbindinD28K not only undergoes significant conformational changes upon binding of Ca^{2+} [82] in contrast to calbindinD9K but also interacted with other proteins [83–85] indicating to perform the role of a sensor protein.

CalbindinD28K contains six EF-hand domains but only four are of high affinity for Ca^{2+} [86], these are the loops in domains EF2 and EF6 [81], which do not contain classical consensus sequences of EF-hand loops (for a review, see Ref. 87). According to the solved structure [81], calbindinD28K forms a single, globular domain that is held together by contacts between the EF-hand pairs EF1–EF2, EF3–EF4 and EF5–EF6. By and large, the structures of the EF-hand subdomains resemble those found in other EF-hand-containing proteins. It is remarkable that the overall structure of calbindin28K is significantly different from the structures of Ca^{2+}-bound CaM or troponin C, which have a more extended, dumbbell-like structure (see Section 5.1.2.2). Kojetin et al. [81] also reported on structures of calbindinD28K interacting with peptides corresponding to binding domains of identified targets [83–85]. The authors identified the binding interface composed of hydrophobic residues important for the interaction. However, if one compares the structure of calbindinD28K interacting with binding peptides of targets with those of CaM interacting with peptides of its targets (see Chapter 4, this volume), it is interesting to observe that the interacting peptides do not induce a significant change of conformation of calbindinD28K in contrast to CaM, which changes from an extended structure in the absence of a target to a more globular structure to wrap around the binding domains of its targets. This may support the notion that calbindinD28K does not operate as a traditional Ca^{2+} sensor [88] because the protein can interact with its targets in the absence and in the presence of Ca^{2+} [84,89].

5.1.1.3. Calretinin

Calretinin is a 29-kDa EF-hand Ca^{2+}-binding protein, i.e. of similar size as calbindinD28K and of homologous amino acid sequence. Calretinin was originally identified by cloning of a cDNA from chicken retina [90]. It is abundantly expressed in the central and peripheral nervous system, but also in other organs [91]. The protein contains four EF-hand domains homologous to the first four of calbindinD28K. Like calbindin, calretinin also belongs to the so-called ‘calcium-buffer proteins’, but its function is poorly understood. An interesting observation was recently reported that calretinin – in exchange with calbindinD28K – can be used as a differentiation marker during neuronal development [92]. At the early postmitotic stage of maturing granule cells of the hippocampus, the transient expression of calretinin is characteristic, which later in mature granule cells is exchanged for calbindinD28K [93]. Similarly, calretinin expression is also found in new neurons of the olfactory system, but these cells are still in the process of migrating to their target zones [94]. These findings suggest that calretinin and calbindinD28K are somewhat involved in the maturation and migration process of newly developing neurons, both known to be Ca^{2+}-dependent events.

5.1.2. Calcium sensor proteins

5.1.2.1. Calmodulin

Calmodulin is the most conserved protein of the EF-hand protein family [95,96], it is ubiquitously expressed in all eucaryotic organisms. CaMs of all vertebrates sequenced to date are 100% identical on the protein level. In fact, CaM is ranked among the most conserved proteins known to date sharing it with proteins like histones, actin or ubiquitin [97]. On the contrary, mammalian CaM is encoded by three non-allelic genes *CALM1*, *CALM2* and *CALM3* in humans; the three different genes are dispersed throughout the genome and located on chromosomes 14q24–q31, 2p21.1–p21.3 and 19q13.2–q13.3, respectively [98]. Even if the three genes encode an identical protein, the coding sequences differ maximally [99]. This is in contrast to its related homologue, troponin C, which can differ in sequence in the same organism, depending whether it is of skeletal or cardiac muscle origin [100].

CaM is the modulator that mediates the Ca^{2+} signal to most targets in the cell (for review, see Ref. 96). It regulates numerous fundamental cellular activities such as glycogen metabolism, intracellular motility, calcium transport, cyclic nucleotide metabolism and protein phosphorylation and dephosphorylation, cell cycle progression and gene expression, to name a few. CaM controls these cellular events by activating a number of key enzymes through direct interaction (see Chapter 4, this volume). In some cases, CaM not only regulates the enzyme as an interacting modulator, but by being a direct subunit of the enzyme [101]. CaM exists as a monomer of approximate molecular weight 17 000, it is mainly located in the cytosol, but can be translocated to the nucleus [102], where it is essential for transcription and regulating gene expression [103].

The structure of CaM has been solved for the Ca^{2+}-saturated form [15,17,19] and more recently also for the apoform [104,105]. By binding of Ca^{2+}, CaM alters its conformation so that hydrophobic patches are exposed on its surface [106,107] conferring to the protein the ability to interact with targets. The structures of Ca^{2+}-loaded CaM complexed to binding peptides derived from various targets have been solved by NMR and by X-ray crystallography, respectively (see Chapter 4, this volume). Two main conclusions have emerged from these studies:

(1) The binding of target peptides induces a large conformational change of Ca^{2+}-saturated CaM, which bends from the extended dumbbell shape revealed by X-ray crystallography [15] (Fig. 2A) into a more globular structure in which the two halves of the protein wrap around the target peptide (Fig. 2C). In each binding peptide, two hydrophobic residues, spaced by 8, 12 or even more positions (for a review, see Ref. 108), are key to the interaction with CaM. This view derived from complexes of CaM with peptides corresponding to the various binding domains is corroborated by recent experiments from Trewhella's Lab, which reported results from small-angle X-ray scattering (SAXS) and neutron scattering providing evidence that also when CaM is complexed to an intact enzyme, i.e. MLCK, CaM undergoes a conformational collapse identical to that observed with the peptide corresponding to its binding domain [109].

(2) The CaM-binding peptides, which are mainly random coil in solution, adopt α-helical structures in the complex with CaM, interacting with the latter mostly in an antiparallel, but sometimes also in a parallel manner [110].

Next to the CaM-binding motif just described in which hydrophobic residues spaced by a certain number of amino acids bind to hydrophobic pockets in CaM exposed due to the binding of Ca^{2+}, in recent years a different motif for CaM binding was characterized, the so-called IQ-motif [111], to which CaM can bind even in the absence of Ca^{2+}. In this motif, one finds a conserved isoleucine next to a conserved glutamine and C-terminal to it at least one basic residue, usually arginine with a number of residues in between (usually three or four). A prominent target that is regulated by CaM due to the binding to an IQ-motif is the voltage-dependent calcium channel. It has been known for some time that the voltage-dependent or L-type channel can be regulated by CaM, both by supporting inactivation and facilitation [112]. Entering of Ca^{2+} through the channel can produce Ca^{2+}-dependent inactivation of the channel (CDI) [113], whereas a rise in basal intracellular Ca^{2+} levels can facilitate channel opening (CDF; 112). Both processes require Ca^{2+}–CaM binding to the C-terminal part of the channel [112]. In the C-terminal part of the α1-subunit of the CaV1.2 channel, a putative IQ-motif was identified which turned out to be critical for both processes, CDI and CDF. Recently, the group of Quiocho reported the first crystal structure of CaM bound to a peptide corresponding to the CaM-binding domain of a mammalian CaV1.2 including the IQ-motif [114]. The reported structure shows that CaM binds to the peptide in a parallel fashion [114] goverened by hydrophobic interaction, similar to the structure reported by Osawa et al. [110] for the complex of CaM with the peptide corresponding to its binding domain of the CaM-dependent kinase kinase. Mutating the critical IQ residues to alanine abolished CDI, but not CDF [112]. When Fallon et al. [114] solved the structure of the complex of CaM bound to a peptide in which both, Ile and Gln, had been mutated to alanine, the authors could not detect any significant difference in the structure as compared with the one containing the wildtype peptide, suggesting that the latter structure represents the way how CaM is bound during CDF. However, several studies indicated that the interaction of CaM with CaV1.2 could be more complex [114] and that other sequences of the channel could contribute to the CaM-binding interface. Because Peterson et al. [115] reported that CDI required Ca^{2+} binding only to the C-terminal lobe of CaM, the report by van Petegem et al. [116] on the structure of CaM bound to the peptide corresponding to the CaM-binding domain of the human CaV2.1 revealed some interesting features that may unravel the difference how CaM regulates CDI and CDF. The overall structure was very similar to the one reported by the Quiocho group. However, van Petegem et al. [116] noted some differences in the interaction of the peptide with the N-terminal lobe of CaM, which they interpreted in that way that the N-terminal lobe of CaM can adopt two different conformations which may be the reason for the difference in CaM-dependent regulation between CDI and CDF, especially since Peterson [115] noted the Ca^{2+}-independent interaction of the N-terminal lobe of CaM with CaV2.1 to regulate CDI.

The plasma membrane Ca^{2+}-pump (PMCA) is a CaM target (see Chapter 7, this volume). The enzyme is an essential component of most plasma membranes. It is the product of at least four genes in human [117] with additional isoforms due to alternative splicing. Its CaM-binding domain [118] has the expected propensity to form an amphiphilic helix of basic and apolar amino acids. It functions as an autoinhibitory domain of the pump, binding to two 'receptor' sites near its active centre [119,120]. Relatively unique among CaM targets, the plasma membrane Ca^{2+} pump can be activated by the C-terminal, but not by the N-terminal half of CaM [121] because in other cases of CaM targets isolated CaM fragments either just bind the target but do not activate it [122,123] or the targets are activated by both, the N- and the C-terminal halves of CaM [124,125]. This raises the question of the mode of CaM interaction with the binding domain of the pump, especially which part of the CaM-binding domain of the Ca^{2+} pump is sufficient to interact solely with the C-terminal half of CaM. SAXS studies [126] using complexes of CaM with peptides C20W and C24W, which corresponded to different parts of the CaM-binding domain of PMCA, indicated that the complex of CaM with peptide C24W, which contains both anchoring hydrophobic residues, had a globular shape as observed with other CaM-binding peptides (e.g. see Fig. 2C), whereas the complex of CaM with peptide C20W, which lacks the second hydrophobic anchor, preserved the shape of a flexible tether of CaM [127,20], typical for CaM in solution (compare Fig. 2A and Fig. 2B).

The structure of the CaM/C20W complex (Fig. 2B) may be of special significance to the operation of the Ca^{2+} pump. Differences in binding of CaM with respect to the two halves of the CaM-binding domain resulting in different activation profiles of the pump (including different isoforms due to alternative splicing) agree with the concept of the fine modulation of the pump by CaM as described in the literature [121,128,129]. Thus, the interaction between the C-terminal half of CaM and the N-terminal half of the CaM-binding domain of the pump would be necessary (and sufficient) to release the autoinhibited state of the enzyme, but not for the full cooperativity typical of the system, thus providing a snapshot of the complex trapped in a step along its way to form the collapsed globular structure.

5.1.2.2. Troponin C

As stated in Section 5.1.2.1, CaM is one of the most conserved proteins with 100% identity in all vertebrates, which is ubiquitously expressed in all eucaryotic organisms and interacts with a great variety of targets. On the contrary, the homologous Ca^{2+} sensor of the troponin/tropomyosin system in muscle cells, troponin C, is specific in its function, limited in its distribution and displays a great degree of variation in its amino acid sequences, including tissue specific differences, i.e. troponin C from skeletal muscles and troponin C from cardiac muscles, respectively, are significantly different in their amino acid sequences [130]. In fact, troponin C from skeletal muscles has – like CaM – four functional Ca^{2+}-binding sites [131] in contrast to cardiac troponin C with only three functional EF-hand domains, the most N-terminal EF-hand is not able to bind Ca^{2+} [132].

Muscle contraction is mediated by the troponin/tropomyosin system, which is bound to actin constituting the thin filaments along which the thick filament, built of myosin molecules, is sliding. This process is regulated by binding of Ca^{2+} to troponin C (for reviews, see Refs. 133–135). Troponin is a complex of three proteins, i.e. the tropomyosin-binding component TnT which anchors the troponin complex to the thin filament, the inhibitory protein TnI which in the presence of actin inhibits myosin ATPase at low Ca^{2+} concentrations and the Ca^{2+}-sensitive regulator TnC which removes the TnI inhibition at rising Ca^{2+} concentrations [134]. The C-terminal EF-hand domains are the high-affinity, structural sites capable of binding Ca^{2+} as well as Mg^{2+}, whereas the N-terminal sites are the regulatory sites of lower affinity, they are specific for binding Ca^{2+} [135]. The structural sites are occupied both during contraction and relaxation, whereas the regulatory sites bind Ca^{2+} when the intracellular Ca^{2+} concentration rises during contraction, but releases Ca^{2+} in the relaxed state. By changing the occupancy of the different Ca^{2+}-binding sites of TnC, which subsequently changes its conformation, this influences the interaction between TnC and TnI. In fact, the structural C-terminal domain of TnC is constitutively bound to the N-terminal domain of TnI whereas the regulatory N-terminal domain of TnC interacts with the C-terminal domain of TnI in a Ca^{2+}-dependent manner. By this way of interaction, conformational changes are transmitted to the thin filament and regulate muscle contraction [134,135].

Understanding the details of the mechanism of muscle contraction has long been hampered because of insufficient structural information for troponin. Just recently, a high-resolution crystal structure of the cardiac troponin core domain in its Ca^{2+}-saturated state has been reported by Takeda et al. [136]. Later, Vinogradova et al. [137] reported on the corresponding skeletal troponin complex in its Ca^{2+}-bound state which provided evidence for significant structural differences between the two complexes. The latter complex looked like an elongated, non-compact structure holding a dumbbell-shaped molecule [137]. The N-terminal helix of TnI together with a coiled-coil domain hold the C-terminal part of TnC, of which the central linker forms an elongated α-helix responsible for the dumbbell-shaped nature of TnC within the complex, similar to the structure of isolated Ca^{2+}-saturated TnC as reported by Herzberg et al. [14]. The overall structural organization of the two complexes, the cardiac [136] and the skeletal [137], is similar, but some important details differed: (i) location of the regulatory Ca^{2+}-binding domain; (ii) the conformation of the central linker of TnC; (iii) the structure of the TnI inhibitory segment and (iv) the position and conformation of the N-terminus of TnT [137]. The difference in conformation of the central helix of TnC which is rigid in the skeletal complex, but 'melted' in the cardiac complex, has influence on the local orientation of the N-terminal regulatory Ca^{2+}-binding domain, which is 28 Å apart by comparing the two complexes [137]. From these structural differences, the following conclusion can be made. The main difference concerns the positioning of the N-terminal Ca^{2+}-regulatory domain, which has only one functional Ca^{2+}-binding site in cardiac TnC. This difference is dependent on the orientation of the central helix of TnC, which is much more rigid in skeletal TnC than in cardiac TnC [137] and therefore has a different influence on the TnI inhibitory region which may lead to mechanical differences in the contraction–relaxation cycle between skeletal and cardiac muscles.

5.1.2.3. Centrins

Centrins are CaM-like proteins that have an important function in the organization and duplication of microtubules. The microtubule organizing centre which is a multi-protein complex is essential for directional intracellular transport including the formation of the spindle apparatus during cell division called spindle pole body in yeast or centrosome in animal cells (for review, see Ref. 138). The centrosome consists of two centriolar cylinders with which centrin is associated [139]. The essential function of centrin is underlined by the finding that depletion of centrin leads to loss of microtubular organization [140,141].

Centrin is similar to CaM in size and structure. Like CaM, centrin is also comprised of two structurally independent globular domains connected by a flexible tether, and each domain is composed of two EF-hand type calcium-binding domains. However, in contrast to CaM, most centrins contain one or more non-functional calcium-binding sites, e.g. in cdc31p, the yeast centrin, sites II and III are defective in binding calcium [142].

One of the few identified centrin-binding proteins is Kar1p [143,144]. In yeast, Kar1p is like cdc31p essential for the spindle pole body [143]. It binds to cdc31p at the C-terminal domain in a Ca^{2+}-dependent manner [144]. The structure of the homologous C-terminal domain of centrin from *Chlamydomonas reinhardtii* complexed to a peptide corresponding to the proposed binding domain of Kar1p was recently solved by NMR [145]. The unique calcium-binding properties of centrin in the absence and presence of its target Kar1p [146] suggest that the two proteins are constituitively bound to each other, even at the low Ca^{2+} level of a resting cell with the consequence that the complex of centrin and Kar1p is a permanent component of the spindle pole body as observed for yeast [143]. Because the interaction of centrin with Kar1p is mediated through the C-terminal domain of centrin, the N-terminal domain could function as a Ca^{2+} sensor interacting with other targets in a Ca^{2+}-dependent manner. Such a model as proposed by Hu et al. [146] is reminiscent of the mode of interaction of troponin C with troponin I (see Section 5.1.2.2.). Similar modes of interaction with targets such as channels or receptors, e.g. like the PMCA Ca^{2+} pump, have also been observed for CaM as discussed before [147,148] (see Section 5.1.2.1.).

5.1.2.4. Neuronal calcium sensors

A group of EF-hand proteins that has recently received attention is that of the neuronal calcium sensors (NCS). They are divided into five subfamilies known to date (for review, see Ref. 149, 150). Two of them are expressed in retinal photoreceptors, called recoverins and guanylate cyclase-activating proteins (GCAPs), the three others are frequenins (NCS-1, neurocalcin, hippocalcin), visinin-like proteins (VILIPs) and the Kv channel-interacting proteins (KChIPs) [149].

All NCS proteins possess four EF-hand domains, but not all are able to bind Ca^{2+} [149]. Most NCS proteins are N-terminal myristoylated. In the case of recoverin of which the structure has been extensively studied, it was shown that in the absence of Ca^{2+} the myristoyl group is sequestered in a hydrophobic pocket of the protein which flips out when Ca^{2+} is bound, thereby permitting a reversible association of the protein with membranes [151], a mechanism which is shared with most other NCS proteins [149].

Recoverins and GCAPs have antagonistic roles in phototransduction: recoverin inhibits rhodopsin kinase [152,153], thereby preventing the downregulation of rhodopsin, which in turn prolongs the light response. GCAPs activate the photoreceptor guanylate cyclase at low Ca^{2+} concentrations in the light but inhibit the cyclase in the dark when Ca^{2+} level is rising [154]. Two other NCS families, i.e. frequenins and KChIPs, are supposed to regulate the release of neurotransmitters (frequenins, Ref. 155), the biosynthesis of polyphosphoinositides (frequenin in yeast, Ref. 156) or the activity of Kv channels (KChIPs, Ref. 157), respectively. The KChIP consists of at least four different proteins and additional splice variants expressed in neurons, but also in cardiac myocytes. All four KChIP proteins and their spliced isoforms interact directly with voltage-dependent potassium channels of the Kv4 family [149]. These channels regulate excitability in neurons and in cardiac myocytes; their gating properties are regulated by the KChIP proteins. An interesting subtype is KChIP3 which is identical to downstream regulatory element antagonistic modulator (DREAM) [158], a Ca^{2+}-dependent transcription repressor, as well as to calsenilin [159], which has been described as interacting with presenilin, which – if mutated – can cause certain forms of familial Alzheimer diseases.

The function of the VILIPs is poorly understood, but all isoforms are neuronal specific [149].

5.1.2.5. S100 proteins

S100 proteins (named because they are soluble at 100% ammonium sulphate solution) constitute the largest family within the EF-hand protein superfamily displaying unique properties [160,161]. S100 proteins are small, acidic proteins containing a classical Ca^{2+}-binding EF-hand at the C- and a S100-specific EF-hand at the N-terminus [162–166]. Most S100 proteins show a cell- and tissue-specific expression indicating specialized biological functions. S100 proteins form homodimers and heterodimers and even oligomers, which contribute to their functional diversification. S100 proteins act intracellularly as Ca^{2+}-signalling molecules, but some members are secreted from cells and act in a cytokine-like manner through the receptor for advanced glycation end products (RAGE) or other receptors [167].

Besides Ca^{2+} some S100 proteins display high affinities towards Zn^{2+} and Cu^{2+} ions, which could influence their biochemical activities. S100 proteins are involved in a variety of cellular processes like cell cycle regulation, cell growth, cell differentiation and motility. Interestingly of the 20 human genes, 16 are tightly clustered in region 1q21 of human chromosome. Based on these findings, a nomenclature was introduced (Table 1) and approved by the HUGO nomenclature committee [160,161]. The *S100* gene structures are highly conserved, in most cases, three exons and two introns, of which the first exon is non-coding. Analogous to the human genome, a cluster of *S100* genes exists in region 3F of mouse chromosome 3.

Within the same gene cluster, a related gene family encodes proteins that contain a S100-like domain fused to a larger peptide. These are trichohyalin, filaggrin and

Table 1
Nomenclature and chromosomal location of the S100 genes [161]

Approved gene symbol	Approved gene name	Previous symbols and aliases	Chromosomal location	Sequence accession ID
S100A1	S100 calcium-binding protein A1	S100A, S100-alpha	1q21	NM_006271
S100A2	S100 calcium-binding protein A2	S100L, CaN19	1q21	NM_005978
S100A3	S100 calcium-binding protein A3	S100E	1q21	NM_002960
S100A4	S100 calcium-binding protein A4	Calvasculin, metastasin, murine placental homologue, calcium placental protein (CAPL), MTS1, p9Ka, 18A2, pEL98, 42A	1q21	NM_002961
S100A5	S100 calcium-binding protein A5	S100D	1q21	NM_002962
S100A6	S100 calcium-binding protein A6	Calcyclin (CACY), 2A9, PRA, CABP	1q21	NM_014624
S100A7	S100 calcium-binding protein A7	Psoriasin 1 (PSOR1), S100A7c	1q21	NM_002963
S100A7A	S100 calcium-binding protein A7A	S100A15, S100A7L1	1q21	NM_ 176823
S100A7L2	S100 calcium-binding protein A7-like 2	S100A7b	1q21	–
S100A7P1	S100 calcium-binding protein A7 pseudogene 1	S100A7L3, S100A7d	1q21	–
S100A7P2	S100 calcium-binding protein A7 pseudogene 2	S100A7L4, S100A7e	1q21	–
S100A8	S100 calcium-binding protein A8	Calgranulin A (CAGA), CGLA, P8, MRP8, CFAG, LIAg, 60B8AG, CPIO	1q21	NM_ 002964
S100A9	S100 calcium-binding protein A9	Calgranulin B (CAGB), CGLB, P14, MRP14, CFAG, LIAg, 60B8AG	1q21	NM_ 002965

Table 1
(Continued)

Approved gene symbol	Approved gene name	Previous symbols and aliases	Chromosomal location	Sequence accession ID
S100A10	S100 calcium-binding protein A10	Annexin II ligand (ANX2LG), calpactin I, light polypeptide (CAL1L), p11, CLP11, 42C	1q21	NM_ 002966
S100A11	S100 calcium-binding protein A11	Calgizzarin, S100C	1q21	NM_ 005620
S100A11P	S100 calcium-binding protein A11 pseudogene	S100A14	7q22–q31	–
S100A12	S100 calcium-binding protein A12	Calgranulin C (CAGC), CAAF1, CGRP, p6, ENRAGE	1q21	NM_ 005621
S100A13	S100 calcium-binding protein A13		1q21	NM_005979
S100A14	S100 calcium-binding protein A14	BCMP84, S100A15	1q21	NM_020672
S100A16	S100 calcium-binding protein A16	S100F, DT1P1A7, MGC17528	1q21	NM_ 080388
S100B	S100 calcium-binding protein B	S100-beta	21q22	NM_ 006272
S100G	S100 calcium-binding protein G	Calbindin 3 (CALB3), CaBP9K, CABP1	Xp22	NM_ 004057
S100P	S100 calcium-binding protein P		4p16	NM_ 005980
S100Z	S100 calcium-binding protein Z	S100-zeta	5q13	NM_130772

repetin, which are multidomain proteins involved in epidermal differentiation [168,169].

Recently, S100 proteins have received increasing attention due to their close association with several human diseases including cardiomyopathy, neurodegenerative disorders and cancer, which make them valuable for diagnostic purposes. Therefore, they become increasingly useful as markers to improve clinical management and are considered having a potential as drug targets to improve therapies [170].

5.1.2.5.1. Protein structures, metal binding, and interactions with target proteins All structures determined to date showed an architecture typical for S100 proteins. Each S100 protein is composed of two EF-hand Ca^{2+}-binding domains connected by a central hinge region [162–165]. The C-terminal EF-hand contains the classical Ca^{2+}-binding motif, common to all EF-hand proteins (see Section 3). This Ca^{2+}-binding motif has a typical sequence signature of 12 amino acids as discussed in detail before and is flanked by helices H_{III} and H_{IV}. The N-terminal EF-hand is different from the classical EF-hand motif and is characteristic for the S100 proteins. Therefore, this EF-hand with a 14 amino acid consensus sequence motif, which is flanked by helices H_I and H_{II}, is called the 'S100-specific' or 'pseudo EF-hand' (Fig. 3A and B).

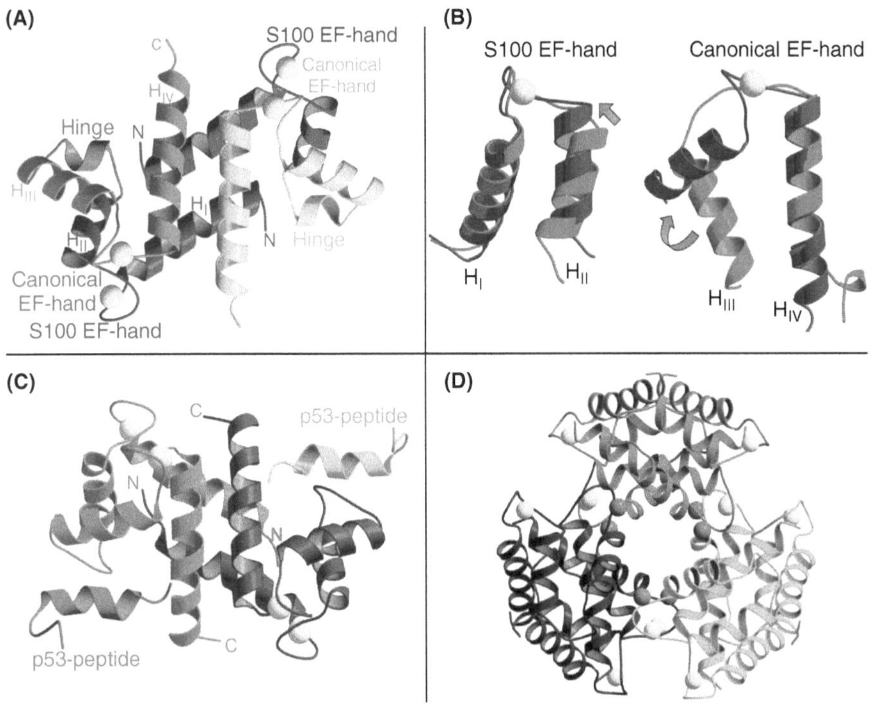

Fig. 3. Three-dimensional structures of S100 proteins. **(A)** Schematic representation of human Ca^{2+}-loaded S100A6 showing the homodimer and bound Ca^{2+} ions (yellow). (PDB code 1K96 [171]). **(B)** Conformational change in EF-hands of S100 proteins. The S100-specific EF-hand is depicted in (left), the canonical EF-hand in (right). The Ca^{2+}-free protein is shown in blue, the Ca^{2+}-loaded form in red. Ca^{2+} ions are shown as yellow spheres. The coordinates were taken from the crystal structures of human S100A6 (PDB codes 1K8U, 1K96; [171]). **(C)** Target binding to S100 proteins. Complex of Ca^{2+}-loaded human S100B dimer with two peptides derived from p53 (PDB code 1DT7; [172]). **(D)** Hexameric structure of Ca^{2+}-loaded S100A12. Three S100A12 dimers (shown in green, blue and magenta) form a hexamer. The Ca^{2+} ions bound to the EF-hands are shown as bright yellow spheres. At the hexamer-forming interface, six additional Ca^{2+} ions are located, (orange spheres) [173]. (Reproduced from the work of Fritz and Heizmann [162], by permission of John Wiley & Sons) (See Color Plate 5, p. 506).

Upon Ca^{2+} binding, S100 proteins undergo a conformational change that is mainly due to a large reorientation of helix H_{III} (Fig. 3B) of the classical EF-hand whereas helix H_{IV} that is engaged in the dimer interface does not move. The N-terminal EF-hand flanked by helices H_I and H_{II} exhibit only minor structural changes upon Ca^{2+} binding. The Ca^{2+}-induced conformational change opens the structure and exposes a wide hydrophobic cleft formed by residues of the hinge region, helix H_{III}, and the C-terminal loop region. This hydrophobic surface represents the interaction site of S100 proteins with their target proteins (Fig. 3C). However, the mode of interaction with different targets is diverse as revealed by the structures of six S100 peptide complexes. In four of these structures, rat or human, S100B was complexed to peptides derived from p53, CapZ and Ndr-kinase. Comparison of the structures revealed differences in the orientation of the peptides and type of interaction with S100B. The structures of the S100A10/annexin II [174] and S100A11/annexin [175] complexes are similar to each other, however, significantly different from the S100 peptide structures.

Some S100 proteins like S100A9 have long and flexible C-terminal extensions that might be required for a target interaction independent of the EF-hand. In general, the C-terminal end exhibits the highest sequence variation and might contribute, therefore, to the specificity of S100 proteins [160].

Multimeric forms of S100 proteins appear to be associated with their extracellular activity. Larger assemblies than dimers were reported for S100A12 (Fig. 3D) [173], S100A4 [176,177] and S100B [178,179] (Ostendorp et al. 2007, submitted). It is proposed that such polymeric forms of S100 proteins trigger aggregation of the receptor RAGE, thereby activating intracellular signal cascades.

Generally, the dimeric S100 proteins bind four Ca^{2+} per dimer ($K_d = \mu M$) with strong positive cooperativity. Besides Ca^{2+}, a number of S100 proteins bind Zn^{2+} with a wide range of affinity ($K_d = 4\,nM$ to $2\,mM$) [180]. Among the S100 proteins, S100A3 displays by far the highest affinity for Zn^{2+} ($K_d = 4\,nM$) and, interestingly, the lowest affinity for Ca^{2+} ($K_d = \sim 20\,mM$) [181], implying that S100A3 functions as a Zn^{2+}-signalling protein rather than a Ca^{2+}-signalling protein. Spectroscopic studies and the crystal structure of metal-free S100A3 [182] allowed the identification of one preformed Zn^{2+}-binding site (distinct from the EF-hand) in the C-terminus of each subunit in which the Zn^{2+} ion is coordinated by one histidine and three cysteine residues. Cu^{2+} was identified as another metal ion that binds to S100 proteins. Binding of four copper ions per homodimer was reported for S100B ($K_d = 0.46\,\mu M$) [183] and S100A5 ($K_d = 4\,\mu M$) [184].

Recently, the structure of the Zn^{2+}-bound S100A2 was characterized in more detail. Site 1 resides at the interface between the subunits of the S100A2 dimer with Zn^{2+} bound by His 17 and Cys21. The binding of Zn^{2+} in site 2 stimulates formation of a novel tetrameric unit not previously recognized in S100 proteins, with coordination of the ion by each of the four Cys2 residues in the tetramer. Binding of Zn^{2+} to site 2 is found to weaken the affinity of S100A2 for Ca^{2+}, lowering the K_d (Ca^{2+}) approximately 40-fold to 2–4 mM. This finding implies that under physiological conditions stoichiometric amounts of Zn^{2+} will render S100A2 into a state that cannot be activated by Ca^{2+}. Thus, our results suggest Zn^{2+}may act to control the cellular activitiy of S100A2 [185].

S100A16 protein is a unique member of the S100 protein family with only one functional Ca^{2+}-binding site located in the C-terminal canonical EF-hand [186]. Surprisingly, the C-terminal EF-hand for Ca^{2+} in the mouse S100A16 is very low; the affinity in the human protein is twofold higher. The weak binding of Ca^{2+} to the C-terminus of the mouse S100A16 may be due to the unusual C-tail (additional Q QEC/S repeats). S100A16 also binds Zn^{2+} and binding studies indicate that Ca^{2+} and Zn^{2+} do not bind to the same sites and that the hydrophobic patches are different.

5.1.2.5.2. Biological functions S100 proteins are involved in a large number of cellular activities such as signal transduction, cell differentiation, regulation of cell motility, transcription and cell cycle progression [160,161,187] through modulation of the activity of target proteins in a Ca^{2+}- (and possibly also in a Zn^{2+}- and Cu^{2+}-) dependent manner.

Understanding the biological function of S100 proteins will crucially depend on the identification of these target proteins. During the last decade, a large number of such possible interactions have been described involving enzymes, cytoskeletal elements as well as transcription factors.

Many of the target protein interactions have been characterized on the biochemical level using in vitro assay systems and are summarized in recent reviews [187–190]. Despite this large amount of biochemical data, very little is known about the actual physiological function of S100 proteins. This can be ascribed to the fact that experiments using cell culture systems and especially whole organisms are still scarce.

New data have been reported about the functions of S100A10 [191–195] and S100A11 [196–199] and of the recently discovered S100A16 [186].

The annexin II in S100A10 complex was found to reorganize the actin cytoskeleton, causing the association of the E-cadherin and nectin systems to form adherent junctions in kidney cells [192]. Furthermore, it has been shown [193] that the annexin A2-S100A10 complex stimulates the conversion of plasminogen to plasmin and that loss of S100A10 from the cell surface of cells results in a dramatic loss of plasmin generation. This suggested that this protein complex is a major regulator of plasmin production.

Recently, the yeast two-hybrid system was applied to identify sensory neuron proteins that interact with the acid-sensing ion channel (ASIC1a) [194]. ASICs are voltage-independent H^+-gated ion channels implicated in mechanosensation, learning and memory, and anxiety-like behaviours.

S100A10 was found to interact physically with ASIC1a. It was suggested that its primary role to regulate ASIC1a activity is to enhance cell surface expression of ASIC1a [194].

Besides this specialized role in enhancing ASIC1a functional expression, S100A10 is also known to traffic the transients receptor potential channels (TRPV5 and TRPV6) to the plasma membrane presenting the facilitation of Ca^{2+} inward currents [191].

Most interestingly, the serotonin 1B receptor was found to co-localize/interact with S100A10, thereby increasing the recruitment of this receptor to the cell surface [195].

Abnormality in serotonin signalling has been implicated in the pathophysiology of depression. S100A10 is decreased in animal models of depression and in brain tissues from depressed patients. S100A10 knock-out mice exhibit a depression-like phenotype and have reduced responsiveness to 5-HT_{1B}. These and other results indicate an important modulation of the 5-HT_{1B} receptor function by S100A10, which can associate with depression-like states.

Progress has also been made in the elucidation of the biological functions of S100A11. It was found to interact with annexin I, thereby facilitating the aggregation of membrane surfaces required for cell vesiculation [196]. It has been demonstrated previously [197–199] that S100A11 is also a key mediator for growth inhibition of normal human epidermal keratinocytes (NHK) triggered by high Ca^{2+} or TGFβ.

The converging pathways for high Ca^{2+}- and TGFβ-induced growth inhibition of NHK cells are illustrated in Fig. 4. Phosphorylation of S100A11 by protein kinase C (PKC)α links these two pathways, thus enabling a more differentiated regulation of

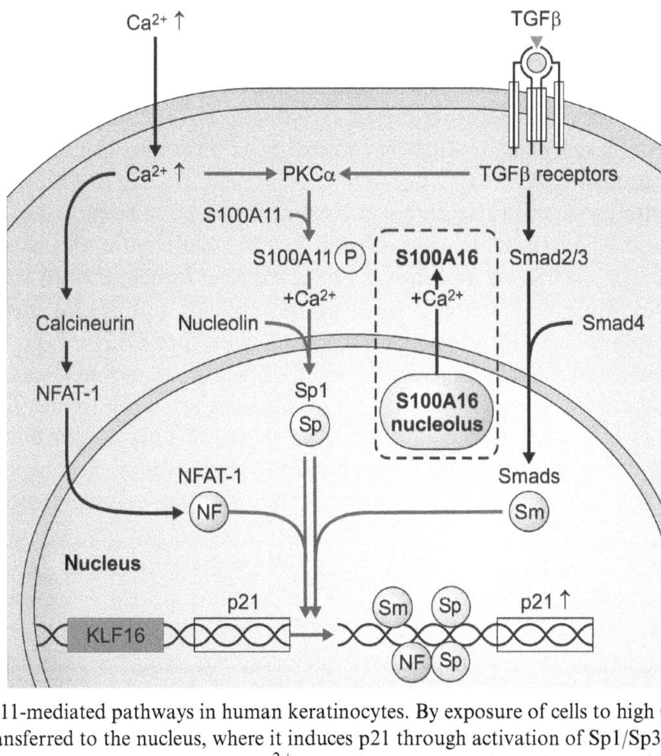

Fig. 4. S100A11-mediated pathways in human keratinocytes. By exposure of cells to high Ca^{2+} or TGFβ, S100A11 is transferred to the nucleus, where it induces p21 through activation of Sp1/Sp3 (for details, see Ref. 199). In other cells, we detected a Ca^{2+}-dependent translocation of S100A16 from the nucleus (nucleolus) to the cytosol, appositely to S100A11 [186] (See Color Plate 6, p. 507).

growth inhibition [199]. Exposure of cells to high Ca^{2+} or TGFβ results in the phosphorylation of S100A11 (^{10}Thr), binds to nucleolin and transfers it to the nucleus. This leads to liberation of Sp1 and the induction of p21. TGFβ activates smad proteins and this leads to an increased affinity of Sp1 to the proximal p21 promoter [198,199].

Recently, we have shown that S100A16 in neuroblastoma cells is translocated to the nucleus (nucleoli) where a co-localization with nucleolin was observed [186]. This suggests that S100A16, together with S100A11, may be involved in the same signalling pathways, as illustrated in Fig. 4.

An increasing number of mouse models, carrying genetically manipulated S100 loci, were generated in recent years to unravel S100 protein functions in vivo. S100B, which is the major S100 protein in the brain, was the first *S100* gene that was genetically engineered. Increased S100B expression in the brain had a positive effect on astrocytosis and neurite proliferation [200]. Moreover, S100B transgenic mice showed enhanced explorative activity, reduced anxiety, impaired learning and reduced memory, which was accompanied by altered synaptic plasticity in the hippocampus [201–204]. In line with these results, an S100B knock-out mouse exhibited strengthened neuronal plasticity in glial cells associated with enhanced spatial and fear memory, suggesting a role for S100B in information processing [205]. In addition, lack of S100B resulted in decreased Ca^{2+}-handling capacity in astrocytes [206], which might be the reason for enhanced epileptogenesis, as demonstrated by these animals [207].

S100A1, mainly expressed in the myocardium, is associated with the ryanodine receptor and with the SERCA/phospholamban complex, which regulates Ca^{2+} homeostasis in sarcoplasmic reticulum [208]. A significant downregulation of S100A1 in end-stage heart failure is indicating a defective excitation/contraction coupling mechanism [209–211].

Upon β-adrenergic stimulation (isoproterenol), we have found significantly prolonged QT_c and ST intervals in 6-month-old S100A1-deficient males when compared with wildtypes ($p < 0.05$). In addition, S100A1-deficient mice of both sexes and ages displayed arrhythmic events after isoproterenol application, but also show behaviour abnormalities, an observation we never made in wildtypes [212].

On a cellular level, the analysis of Ca^{2+} cycling in isolated cardiomyocytes revealed a lesser increase in Ca^{2+} transient amplitudes upon β-adrenergic stimulation (isoproterenol) in mutant than wildtypes cells ($p < 0.05$). This suggests a contribution of aberrant Ca^{2+} handling to the development of arrhythmias in mutant mice.

In order to understand the molecular mechanisms underlying these phenotypic characteristics, we are presently performing transcriptomic as well as proteomic approaches and have identified several candidates involved in cell growth, neuromuscular function and oxidative stress.

The dissection of the molecular pathways leading to the phenotypic characteristics in S100A1-deficient mice will uncover the molecular triggers responsible for the development of long QT-syndromes, arrhythmias and possibly sudden cardiac death.

Recently, Pleger et al. have shown in an animal model that *S100A1* gene therapy is a potential novel treatment strategy to improve contractile function of a diseased heart [213].

It has been reported that S100 proteins can also activate the RAGE [160,167,190,214–216]. RAGE is a member of the immunoglobulin-like cell surface receptor superfamily, expressed at low levels in many adult cell types (Fig. 5). The upregulation of RAGE has been observed in a large number of pathophysiological processes, e.g. immune/inflammatory bone disorders and diabetes [167,217]. High levels of RAGE expression are observed in the central nervous system of embryos suggesting an important role in brain development. RAGE can be activated by several classes of ligands. The first class ligands are advanced glycation endproducts (AGE). AGEs are non-enzymatically glycosylated proteins found under pathological conditions such as diabetes, renal failure or heart dysfunction. The second class of ligands includes amphoterin and members of the S100 protein family.

Engagement of RAGE by its ligands triggers the activation of various cellular pathways involving mitogen-activated protein (MAP) kinases and the transcription factor NF-κB. Activation of RAGE enhances the expression of the receptor itself and initiates a positive feedback loop resulting in sustained RAGE upregulation, which has been suggested to be the starting point of chronic cellular activation and tissue damage. Several studies have succeeded in blocking the RAGE/ligand interactions by specific antibodies (against the receptor RAGE or its ligands), resulting in a suppression of atherosclerosis and improved wound healing in diabetic animal models. A major goal in our Institute is to exactly map the RAGE-ligand binding sites as a basis for developing specific antibodies or drugs to block RAGE–ligand interactions in human diseases.

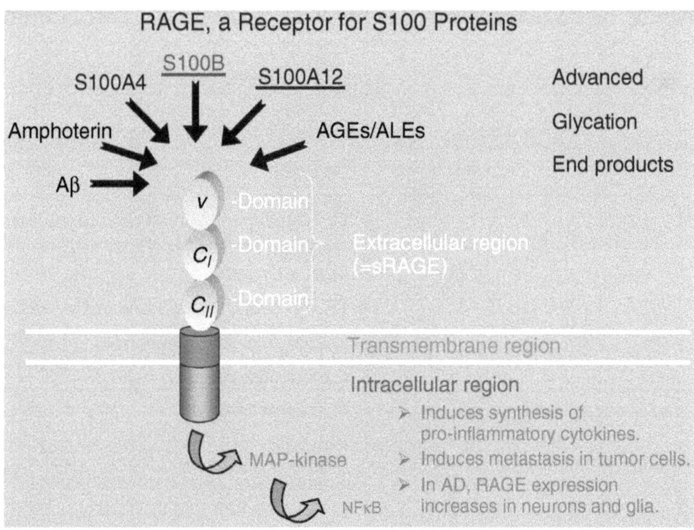

Fig. 5. S100–RAGE signalling. RAGE is a multifunctional receptor of the immunoglobulin family binding a variety of structurally and functionally unrelated ligands. RAGE contains an extracellular domain consisting of 3Ig domains (one V- and two C-types), a transmembrane spanning region and a cytosolic tail. RAGE triggers the activation of key signalling pathways and is involved in several human diseases [167,214] (See Color Plate 7, p. 507).

Compensatory mechanism, such as the counter receptor vβ2 integrin MAC-1 [216], may also be operative in those models.

5.1.2.5.3. S100 proteins and human diseases S100 proteins have attracted great interest because of their association with a great number of human diseases (Table 2) and their use in clinical diagnosis [170].

Cancer: A potential role for S100 proteins in neoplasia stems from the observation that several family members have altered levels of expression in different stages and types of cancer. S100A4 has been implicated in invasion and metastasis, and the prognostic significance of its selective expression in various cancers has been etablished [218–221]. In gastric cancer, the inverse expression of S100A4 in relation to E-cadherin (a tumour suppressor) was found to be a powerful aid in histologic typing and in evaluating the metastasis potential/prognosis of patients with this type of cancer.

It was also demonstrated that S100A4 could act as an angiogenic factor and might induce tumour progression through an extracellular route stimulating angiogenesis. Inhibiting the process of tumour angiogenesis might be possible by either blocking S100A4 secretion or its extracellular function [222].

A prognostic significance of S100A2 in laryngeal squamous-cell carcinoma has also been found allowing discrimination of high- and low-risk patients in the lymph node-negative subgroup and a better adjusted therapy. S100A2 expression together with the methyl-*p*-hydroxyphenylacetate-esterase status allows discrimination of high- and low-risk patients in the lymph node-negative subgroup. These results are of direct clinical relevance in that an aggressive initial treatment of the patients with S100A2-negative tumours would avoid undertreatment, and a much less aggressive treatment would be beneficial for patients with S100A2-positive tumours. S100A2

Table 2

Association of S100 proteins with human diseases [170]

Proteins	**Diseases**
S100B	Developmental brain dysfunction, learning and memory deficits, Alzheimer's disease, blood brain barrier dysfunction
S100A1	Cardiomyopathy, reduced anxiety-related responses
S100A2	Cancer, tumour suppression
S100A4	Cancer, metastasis
S100A6	Cancer, ALS
S100A7	Psoriasis
S100A8/A9	Inflammation, cystic fibrosis Wound healing, juvenile rheumatoid arthritis
S100A10/A11	Cancer
S100A12	Inflammation, Kawasaki disease, Mooren's ulcer
S100P	Cancer

and some other S100 proteins not only could be useful as biomarkers for various types of cancer but also could have a potential as drug targets for more subtle chemotherapies.

Inflammation: The S100A8, A9, and A12 proteins are characterized by a unique expression pattern with strong prevalence in cells of myeloid origin [223–226]. S100A8 and S100A9 tend to form homodimers and heterodimers. The quantity of S100A8/A9 heterocomplexes is higher in neutrophils than in monocytes. S100A8/A9 expression is low in healthy people, whereas in inflammation specific cell populations release homo- or hetero-complexes depending on the phase and type of inflammation. Tests have been developed to detect S100A8 and S100A9 in body fluids of patients with rheumatoid arthritis for the discrimination of active and nonactive osteoarthritis from rheumatoid arthritis. S100A8 and S100A9 are also associated with chronic inflammatory diseases including bowel disease and chronic periodontitis, and both proteins are involved in wound repair by reorganization of the keratin cytoskeleton in the inquired epidermins.

As a consequence of the various proinflammatory properties of S100 proteins, strategies targeting these molecules are a novel option for antiinflammatory therapies, by in vivo administration of S100 antibodies. Another approach is the inhibition of the release of these cytokine-like molecules at sites of inflammation.

Autoimmune diseases: Some S100 proteins have been implicated in autoimmune diseases, including S100B in autoimmune disease of the inner ear and S100A12 in Mooren's ulcer [227] and host parasite defense [228]. S100A12 is abundant in neutrophils and, in its secreted form, is believed to play a role in innate immunity by binding to and targeting foreign antigens for removal by the phagocytic system. Mooren's disease is an autoimmune eye disorder that often presents as a progressive ulceration and destruction of the keratocytes (corneal fibroblasts) in the peripheral cornea of children. The disease has also been observed in patients from India and Nigeria suffering from helminthic infections. In some cases, there is evidence to support an autoimmune process, in which an immune response to human neutrophil calcium-binding proteins, which are associated with parasite proteins, also targets similar proteins present in the cornea of the eye [227].

The S100 protein family constitutes the largest subgroup of the EF-hand family of Ca^{2+}-binding proteins with 20 members discovered to date. S100 proteins have been implicated in pleiotropic Ca^{2+}-dependent cellular events, with specific functions for each of the family members. However, some S100 proteins have also physiologically relevant Zn^{2+} affinities. S100A2 and S100A3, for example, have very low affinities for Ca^{2+} but high affinities for Zn^{2+}, suggesting that Zn^{2+} rather than Ca^{2+} controls their biological activities. In order to understand how the biological functions of S100 proteins are regulated by Zn^{2+} and Ca^{2+}, it will be necessary to persue the determination of the three-dimensional structures of the Zn^{2+}-loaded S100 proteins and to characterize both the distinct mode of Zn^{2+}-binding in the presence of Ca^{2+} and the Zn^{2+}-dependent interaction with target proteins.

In resting cells, S100 proteins are localized in specific cellular compartments from which some of them relocate upon cellular stimulation and even are secreted exerting extracellular, cytokine-like activities. This suggests that translocation might be a temporal and spatial determinant of their interactions with different partner proteins.

Interestingly, different S100 proteins utilize distinct translocation pathways that might lead them to certain subcellular compartments in order to perform their physiological tasks in the same cellular enviroment. Further studies are needed to unravel the mechanisms involved in the translocation and secretion of S100 proteins. It has been found that some S100 proteins (after secretion) can have paracrine effects on neighbouring cells and that the extracellular concentrations play a crucial role in the physiological response. The discovery of the interaction of several S100 proteins with the surface receptor RAGE shed more light on the extracellular functions of these two proteins. Their mode of interaction with RAGE and the question if other S100 protein members act through RAGE or other surface receptors still remains to be investigated. This work is in progress by animal models with inactivated single S100 proteins or deletion mutants of RAGE. Future research activities will also focus on the deregulated expression of *S100* genes, which is a hallmark of a wide range of human diseases, and the applicaton of S100 proteins and antibodies in clinical diagnostics, to further evaluate their prognostic significance to improve clinical management. These proteins are also considered in some cases as drug targets for the inhibition of their intra- and extracellular pathological activities.

5.1.2.6. The penta EF-hand family

The EF-hand motifs in Ca^{2+}-binding proteins are usually tandemly repeated. Owing to the recent determination of the X-ray structure of the small subunits of the calcium-dependent protease calpain [229,230], it was recognized that also uneven numbers of EF-hand domains may exist in calcium-binding proteins. By comparing the primary structures of a number of different calcium-binding proteins, Maki and his co-workers identified several proteins which like calpain contained five EF-hand domains and therefore introduced the name ‘penta EF-hand family’ or PEF family for this subgroup of EF-hand-containing proteins [13]. Members of this family include calpain, sorcin, grancalcin, peflin and ALG-2; the latter was originally identified as ‘apoptosis-linked gene’. Characteristic to all the members of this family is a two amino acid insertion into the loop of the fifth EF-hand domain and therefore – at least under physiological conditions – may not bind calcium anymore (Fig. 6). Instead, the fifth domain may serve as a dimerization domain as originally observed in the calpain structure [231,232] and later confirmed by solving the structures of grancalcin [233] and of ALG-2 [234].

Proteins of the PEF family have been quite conserved during evolution; homologues have been found in invertebrates, plants and fungi. Based on comparison of the gene structures, PEF proteins can be classified into two groups: group I contains ALG-2, peflin and their homologues, and group II is composed of calpain, sorcin and grancalcin [13].

As indicated before calpain is a calcium-dependent, intracellular protease. The mammalian calpain gene superfamily contains 16 genes to date, most of them are ubiquitously expressed [235]. Calpains exist as heterodimers consisting of a large subunit (80 kDa), containing the catalytic site, and of a small regulatory subunit (30 kDa). Recently, Suzuki, Bode and their co-workers determined the structure of the full-length, heterodimeric human calpain II in its Ca^{2+}-free form [232]: the two

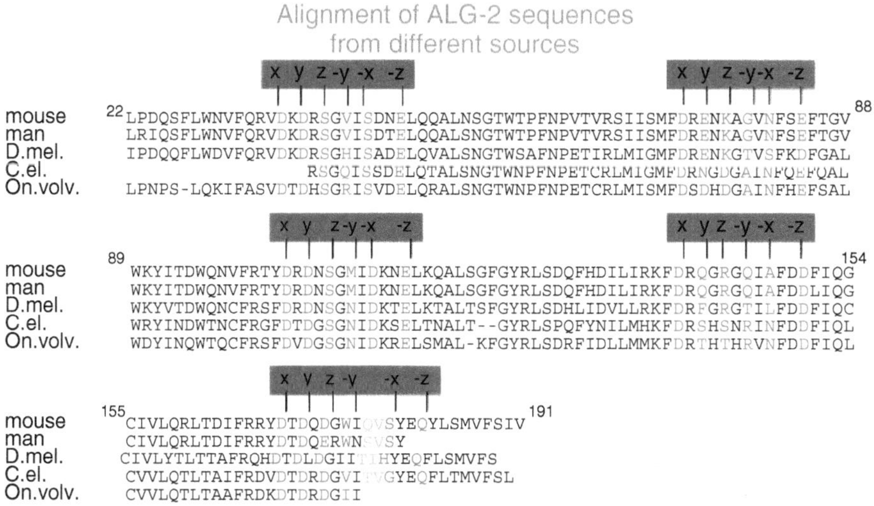

Fig. 6. Alignment of the amino acid sequences of the penta EF-hand Ca^{2+}-binding protein ALG-2 from different organisms. The conserved amino acid residues of the five EF-hand Ca^{2+}-binding loops are indicated in red. In the fifth EF-hand domain, two amino acid residues are inserted (shown in green) between the positions –y and –x. (Reprinted from Krebs 1998, *BioMetals* 11, 375–382 with kind permission of Springer Science and Business Media) (See Color Plate 8, p. 508).

subdomains IIa and IIb of the large subunit comprising the catalytic site are rotated against each other, disrupting the active and the substrate binding sites that would explain the inactivity of calpain in the absence of calcium. The group of Davies determined the structure of calpain I in the Ca^{2+}-bound form [231]. These authors suggested that the cooperative binding of Ca^{2+} to two non-EF hand-type Ca^{2+}-binding sites aligned the active site cleft and converts the protein into an active enzyme. On the contrary, Ca^{2+} binding to the EF-hand type binding sites would contribute to the Ca^{2+} sensitivity of the enzymes as documented by mutational analysis [236]. In addition, both isoforms of calpain are maintained in an inactive form due to the interaction with an endogenous inhibitor, calpastatin, until external signals activate the protease [237].

Increasing evidence indicates that calpain is involved in the regulation of basic cellular processes such as cell proliferation, differentiation and apoptosis due to cytoskeletal remodelling which under pathological conditions could contribute to tissue damage in heart and brain ischaemias as well as neurodegeneration in Alzheimer's disease [238]. In muscles that contain a muscle-specific form of calpain, called p94 or calpain 3, and which is essential for muscle function, a mutated form leads to a specific muscular dystrophy, an autosomal recessive inherited disease [239].

ALG-2 was originally identified as a gene linked to apoptosis [240]. It is a 22-kDa highly conserved protein (Fig. 6), which is ubiquitously expressed. It forms dimers through the fifth EF-hand as documented by the crystal structure [234]. This could provide a new interface for the interaction with possible targets such as AIP or Alix [241,242], which in Xenopus is involved in the maturation process of oocytes [243].

ALG-2 may also interact with peflin [244] or different annexins [245]. All these targets interact with ALG-2 in a calcium-dependent manner indicating that ALG-2 – like CaM – may have calcium-sensing properties. Therefore, the involvement in apoptosis which gave the protein its name would be one, but not the exclusive signal transduction pathway. In addition, it has recently been shown that ALG-2 is not essential for apoptotic responses because by using T-cells from ALG-2-deficient mice [246] apoptotic stimuli could be induced by either stimulating T-cell receptors (TCR) and Fas/CD95 or by glucocorticoids.

Recent experiments carried out in our laboratory indicated that ALG-2 may not only be involved in apoptotic processes but may also play a role during cell proliferation [247]. A significant nuclear concentration of ALG-2 was found in cells prior to cell division in addition to a remarkable increase of ALG-2 expression in highly proliferative cells obtained from cancerous tissues [247,248]. Interestingly, by using a yeast two-hybrid screening to characterize possible interacting partners of ALG-2, we identified the conserved RNA-binding protein RBM22 of unknown function [247,249]. By preparing fluorescent constructs of ALG-2 and RBM22 to characterize their cellular localization, we could demonstrate that in NIH 3T3 cells, RBM22 was almost exclusively localized in the nucleus whereas ALG-2 was localized in the cytosol. However, if we co-injected the two fluorescent constructs into NIH 3T3 cells, it was of interest to note that the two proteins co-localized within the nucleus suggesting a functional connection between the two proteins. By using similar constructs in zebrafish, we observed similar results during development as can be seen from Fig. 7, corroborating the results obtained in tissue cultures [249]. By quantifying the co-injection data, we could demonstrate that on average more than 95% of the two proteins co-localized within the nucleus during development. Different laboratories screened for proteins involved in the regulation of alternative splicing, cell

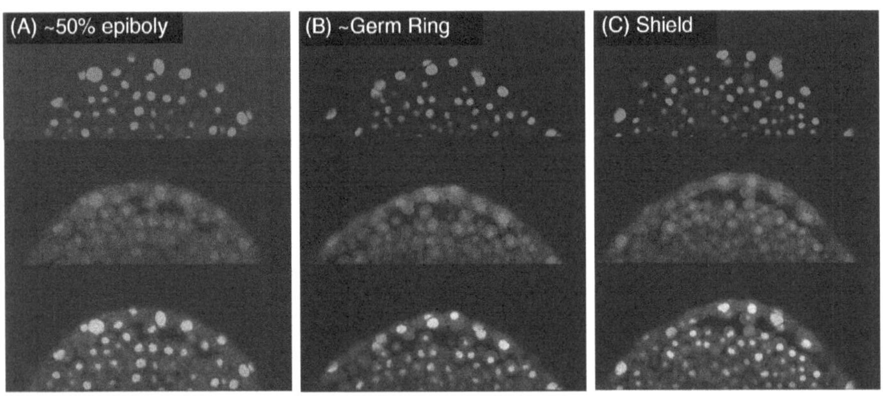

Fig. 7. Co-localization of ALG-2 and RBM22 in the nucleus. Zebrafish embryos have been co-injected with RBM22-EGFP RNA (green fluorescent) and ALG2-mRFP RNA (red fluorescent) at the 1-cell stage. The figure shows a single confocal section at **(A)** 50% epiboly, **(B)** Germ Ring and **(C)** Shield stage of embryonal development. Merging the green and red fluorescent images provides evidence for the co-localization of the two proteins within the nucleus. For details, see Section 5.1.2.6 and Ref. 249 (See Color Plate 9, p. 508).

division and cardiac development; they all identified homologues of RBM22 as an essential protein [250–252]. ALG-2 may thus be an important modulator to link calcium signalling to essential cellular processes such as alternative splicing, cell division and proliferation, and cell death.

5.2. *Non-EF hand Ca^{2+}-binding proteins*

5.2.1. *Annexins*

The family of intracellular Ca^{2+}-binding proteins called annexins are soluble, amphipathic proteins that bind to membranes containing negatively charged phospholipids in a Ca^{2+}-dependent manner (for recent reviews, see Refs 253, 254). They are called annexins because they 'bring' or 'hold together' different cellular structures, in particular membranes. The close to 200 different annexin proteins known to date are widespread in the animal and plant kingdom and have been claimed to be involved in a variety of cellular functions such as interaction with the cytoskeleton, membrane fusion, anticoagulation, signal transduction or phospholipase inhibition. The primary structure of the annexins contains four or eight (annexin A6) conserved repeat units of about 75 amino acids in length, the protein core, which are separated by intervening sequences of variable length. These repeat units probably originated from gene duplications, a view that is corroborated by the evolutionary conservation of the intron–exon boundary positions. Most annexins containing four-repeat units are comprised of 12–15 exons resulting in a quite variable aminoterminal part of the different annexins. Most alternative splicing sites have been located within exons encoding the variable aminoterminal parts of annexins. Because these parts are also the locations containing motifs for binding partners or for posttranslational modifications; thus alternative splicing may contribute to the regulation of annexin function.

It is of interest that a number of targets interacting with the N-terminal parts of annexins belong to the EF-hand protein family. It has been reported that annexin A1 interacts with S100A11 [175] and annexin A2 with S100A10 (also known as p11 [255]), annexin A7 binds to sorcin [256] and annexin A11 interacts with S100A6 (also known as calcyclin [257]). The structural nature of these interactions has been studied in detail, especially for the pair annexin A2/S100A10 [174]. It was shown that the N-terminal part of annexin A2 adopts the conformation of an amphiphatic α-helix binding to the pocket formed by the dimer of S100A10 (p11) [174,258].

After the first determination of an annexin crystal structure (annexin A5) by Huber and his co-workers [259], a number of additional annexin structures from different sources (including those from lower eucaryotes and plants) have been reported (see Ref. 260 for a review). From these structures it can be concluded that the conserved repeat units are packed into an α-helical disk as first described for annexin A5, which is almost entirely α-helical. It consists of five α-helices bundled into a right-handed super-helix. On the basis of this structure, it was proposed that annexin A5 functions as a calcium channel [261], and some experiments using an in vitro reconstituted system seem to support a voltage-gated mechanism. This property is connected with the structure of the annexin core, which is shared by most annexins. Meanwhile for most annexins, a Ca^{2+} channel activity has been demonstrated in artificial bilayer systems, but never in vivo; therefore the physiological relevance of these observations is questionable.

In contrast to the EF-hand-containing Ca^{2+}-binding proteins, the ligands coordinating calcium in the annexins are not adjacent in sequence. Several calcium-binding sites seem to exist in annexins, two invariably in repeats II and IV, one in either repeat I or III, but annexins may contain as many as 10–12 Ca^{2+}-binding sites along the membrane-binding surface of the protein [260]. The sites with the highest Ca^{2+} affinity are structurally related to the Ca^{2+}- and phospholipid-binding site of phospholipase A_2. The calcium ion is heptacoordinated with ligands organized in a pentagonal bipyramidal arrangement with ligating oxygens provided mainly from peptide carbonyls and water molecules, together with a single side-chain oxygen from a distant part of the sequence. Replacement of this latter 'capping' residue, which is usually acidic, with alanine precludes Ca^{2+} binding at the site. Once bound to membranes, many annexins oligomerize to form highly ordered two-dimensional arrays that have been shown to strongly influence in vitro membrane properties, e.g. increased rigidity (e.g. see Ref. 262).

An interesting feature is the role of Trp185 in the Ca^{2+}-binding site of repeat III of annexin A5 [263]. As suggested by the authors, a Ca^{2+}-dependent exposure of Trp185, which is buried within the protein core in the absence of calcium, may facilitate the interaction between annexin A5 and the phopholipid bilayer. This view has been supported by fluorescence data and by mutagenesis experiments in which replacement of Trp185 by alanine decreased the annexin A5 membrane-binding affinity [264].

Annexins have long been known as targets for posttranslational modifications. In fact, annexin A2 was originally isolated as a major substrate of the *src*-encoded protein kinase [265], and annexin A1 was long known to be phosphorylated by the tyrosine kinase activity of the EGF receptor [266], suggesting that annexins A1 and A2 could be used as coupling factors between growth factor receptors and their cellular targets, because these phosphorylations alter the Ca^{2+}-dependent binding of annexin A1 and A2 to membranes. Recent experiments also indicated that annexin A1 influences trafficking of EGF receptor through multivesicular endosomes, a property that is dependent on its phosphorylation by the EGF-receptor tyrosine kinase [267]. Similar observations have also been made for annexins A2 and A6, which can influence endosome morphology and endocytic transport [254] due to unique endosomal targeting sequences within the N-terminal interaction domains. By downregulating annexin A2, the lysosomal transport and internalization of EGF receptors was inhibited [268]. In this context, a recent report may be of interest demonstrating that annexin A2 could bind certain species of RNA which indicates a possible role in RNA transport or export from the nucleus [269], but the physiological relevance of this finding remains to be elucidated.

In spite of extended studies in many laboratories over the years, the precise function of individual annexins is not yet understood. The accumulated data suggest that due to the many ways how annexins interact with different membranes in a Ca^{2+}-dependent manner, annexins may participate in Ca^{2+} signalling as effectors, as mediators or even as regulators, but to clarify their biological activities unambiguously, this has to await further experiments. However, in vivo experiments from knock-out and transgenic mice are beginning to shed some light on annexin function.

5.2.2. C_2-domain proteins

Interaction of proteins, intracellularly or extracellularly, often occurs through different binding modules or domains such as SH2, SH3, WW, PDZ or C_2-domains. These modules are formed by folding domains consisting of 100–150 residues with different binding properties, i.e. SH2-domains interact with phosphotyrosine-containing sequences, SH3- and WW-domains with proline-rich sequences, PDZ-domains with C-terminal sequences to link multiprotein complexes to the cytoskeleton and C_2-domains with phospholipids, some in a Ca^{2+}-dependent, some in a Ca^{2+}-independent manner. C_2-domains consist of about 120–130 amino acid residues. They were first identified in PKC referring to the Ca^{2+}-sensitive site of PKC [270]. Meanwhile more than 200 C_2-domain containing proteins can be identified in current data banks [271].

Most proteins containing C_2-domains are linked either to signal transduction pathways or are involved in membrane traffic. The former proteins generate lipid second messengers (e.g. phospholipase A2, phospholipase C or phosphatidylinositol-3-kinase), phosphorylate proteins (e.g. PKC), or ligate ubiquitin, e.g. Nedd4, a highly conserved family of E3 ligases which next to C2-domains also contains multiple WW-domains for target recognition [272]. Those proteins involved in membrane traffic include for instance the Rab-binding proteins rabphilin and RIM, which are involved in regulating the exocytosis of secretory vesicles. Recently, the crystal structure of the C_2A-domain of RIM2 has been solved [273]. RIMs are localized at presynaptic zones involved in the release of neurotransmitters. They are important for the regulation of presynaptic plasticity by interacting with SNAP-25, a member of the SNARE protein family, and synaptotagmin1, the Ca^{2+} sensor of the membrane fusion machinery [273]. The latter is the best characterized protein of this category and will be discussed in more detail.

Synaptotagmin I is a transmembrane protein belonging to a family of more than 10 members that contain two C_2-domains, C_2A and C_2B, comprising most of the cytoplasmic region of synaptotagmins (for review, see Ref. 274). The C_2-domains contain Ca^{2+}-binding domains of non-EF-hand character. Synaptotagmin I is found in synaptic vesicles and is believed to act as the major Ca^{2+} sensor of exocytosis and neurotransmitter release. Important in this respect is the Ca^{2+}-dependent binding of synaptotagmin to syntaxin, a member of the SNARE family of membrane proteins, involved in vesicle transport mechanisms.

C_2-domains have been suggested to be responsible for binding to membranes in response to Ca^{2+}. Sudhof and his co-workers determined the first structure of a C_2-domain, the C_2A-domain of synaptotagmin I [275]. The C_2-domain consists of a β-sandwich of two four-stranded β-sheets. The eight β-strands are connected by three loops at the top, which bind Ca^{2+} in forms of a cluster, primarily through oxygen of aspartate side chains serving as bidentate ligands, and by four loops at the bottom lacking Ca^{2+}-binding sites. In contrast to Ca^{2+}-binding to EF-hand type of sites which causes substantial conformational changes, Ca^{2+}-binding to sites of C_2-domains leads to structural stabilization rather than backbone rearrangements [276].

Structural determination of other C_2-domains revealed similar designs but significant differences in the topology of the arrangement of the β-strands [277]. It could be shown that the structure of the C_2-domain of PKCβ is very similar to synaptotagmin I (topology I), whereas the topological arrangement of the structure of the C_2-domain of phospholipase Cδ is significantly different (topology II) [276]. Phospholipase Cδ contains several protein modules including EF-hand Ca^{2+}-binding domains, which provides evidence for the existence of proteins containing EF-hand and non-EF-hand Ca^{2+}-binding sites.

In another C_2-domain-containing protein, Nedd4, it was demonstrated that binding of Ca^{2+} to the C_2-domain not only was responsible for the localization of the protein but also for part of its function. Nedd4, the neuronal precursor cell-expressed developmentally downregulated 4 protein, is a multimodular ubiquitin protein ligase [278]. This enzyme is involved in controlling the turnover of membrane proteins. It was shown that localization of Nedd4 to the apical region of polarized epithelial cells was dependent on the Ca^{2+} binding to its C_2-domain [279]. Here an important target of Nedd4 is the Na^+-channel (ENaC), which plays a critical role in Na^+ homeostasis of epithelial cells [280]. However, the C_2-domain of Nedd4 was not required to inhibit EnaC, but multiple WW domains [281].

6. *Conclusions*

In summarizing, several general points can be made:

(1) Calcium as one of the oldest components of organisms plays a central role in biological systems, controling a myriad of key cell processes. It can fulfil a static function in stabilizing structures, or a dynamic function, participating in signal transduction pathways as a second messenger.
(2) Calcium homeostasis in an organism is carefully controlled, involving a variety of systems in the skeleton, in the ECF, and inside cells. Depending on its function, calcium can be complexed in different forms: by hydroxyapatite in the skeleton, by acidic, low-affinity proteins in the ECF and by the high-affinity EF-hand proteins inside cells.
(3) Extracellular and intracellular concentrations of calcium differ by several orders of magnitude. Therefore, cells are exposed to a steep Ca^{2+} gradient across the membranes, which makes it possible for even small changes of membrane permeability to lead to substantial changes of intracellular free Ca^{2+} concentration. Signals can be converted from an extracellular analogue to an intracellular digital form. The control of cellular calcium is maintained by an elaborate system of channels, exchangers and pumps located in the plasma membrane and in intracellular membranes.

(4) The EF-hand proteins play a pivotal role in permitting Ca^{2+} to function as a second messenger. These proteins bind Ca^{2+} with high affinity, selectivity and cooperativity, thereby permitting interaction with targets. More than 600 EF-hand proteins have been identified to date which fulfil the different tasks of calcium-dependent mechanisms (e.g. glycogen metabolism, muscle contraction, excitation-secretion coupling, cell cycle control, gene expression and programmed cell death).
(5) Solving the structure of several key components of calcium controlled pathways helped to understand their mechanism of action.

References

1. Carafoli, E., & Klee, C., eds. *Calcium as a Cellular Regulator,* Oxford University Press, New York, 1999.
2. Carafoli, E., & Krebs, J., eds. *Calcium Homeostasis, Topics in Biological Inorganic Chemistry,* Vol. 3, Springer, Berlin, 2000.
3. Pochet, R., ed. *Calcium. The Molecular Basis of Calcium Action in Biology and Medicine,* Kluwer Academic Publishers, Dordrecht, 2000.
4. Ringer, S. (1883) *J. Physiol.* 4, 29–43.
5. Sutherland, E.W., Oeye, I., & Butcher, R.W. (1965) *Recent Prog. Horm. Res.* 21, 623–646.
6. Sutherland, E.W., & Robison, G.A. (1966) *Pharmacol. Rev.* 18, 145–161.
7. Ebashi, S., & Kodama, A. (1965) *J. Biochem. (Tokyo)* 58, 107–108.
8. Cheung, W.Y. (1970) *Biochem. Biophys. Res. Commun.* 38, 533–538.
9. Kakiuchi, S., Yamazaki, R., & Nakajima, H. (1970) *Proc. Jpn. Acad.* 46, 587–592.
10. Kretsinger, R.H. (1972) *Nat. New Biol.* 240, 85–88.
11. Kretsinger, R.H., & Nockolds, C.E. (1973) *J. Biol. Chem.* 248, 3313–3326.
12. Nakayama, S., Kawasaki, H., & Kretsinger, R.H. (2000) in: *Topics in Biological Inorganic Chemistry 3* (Carafoli, E., & Krebs, J., eds.), pp. 29–58, Springer, Berlin.
13. Maki, M., Narayana, S.V.L., & Hitomi, K. (1997) *Biochem. J.* 328, 718–720.
14. Herzberg, O., & James, M.N. (1985) *Nature* 313, 653–659.
15. Babu, Y.S., Sack, J.S., Greenhough, T.J., Bugg, C.E., Means, A.R., & Cook, W.J. (1985) *Nature* 315, 37–40.
16. Taylor, D.A., Sack, J.S., Maune, J.F., Beckingham, K., & Quiocho, F.A. (1991) *J. Biol. Chem.* 266, 21375–21380.
17. Chattopadhyaya, R., Meador, W.E., Means, A.R., & Quiocho, F.A. (1992) *J. Mol. Biol.* 228, 1177–1192.
18. Barbato, G., Ikura, M., Kay, L.E., Pastor, R.W., & Bax, A. (1992) *Biochemistry* 31, 5269–5278.
19. Babu, Y.S., Bugg, C.E., & Cook, W.J. (1988) *J. Mol. Biol.* 204, 191–204.
20. Elshorst, B., Hennig, M., Försterling, H., Diener, A., Maurer, M., Schulte, P., Schwalbe, H., C. Griesinger and Krebs, J., Schmid, H., Vorherr, T., & Carafoli, E. (1999) *Biochemistry* 38, 12320–12332.
21. Lai, E.C. (2004) *Development* 131, 965–973.
22. Conlon, R.A., Reaume, A.G., & Rossant, J. (1995) *Development* 121, 1533–1545.
23. Harper, J.A., Yuan, J.S., Tan, J.B., Visan, I., & Guidos, C.J. (2003) *Clin. Genet.* 64, 461–472.
24. Brown, E., Enyedi, P., LeBoff, M., Rotberg, J., Preston, J., & Chen, C. (1987) *FEBS Lett.* 218, 113–118.
25. Dijkstra, B.W., Kalk, K.H., Hol, G.W.J., & Dreuth, J. (1981) *J. Mol. Biol.* 147, 93–123.
26. Reinhardt, D.P., Ono, R.N., Notbohm, H., Muller, P.K., Bachinger, H.P., & Sakai, L.Y. (2000) *J. Biol. Chem.* 275, 12339–12345.
27. McGettrick, A.J., Knott, V., Willis, A., & Handford, P.A. (2000) *Hum. Mol. Genet.* 9, 1987–1994.

28. Heinegard, D., & Oldberg, A. (1989) *FASEB J.* 3, 2042–2051.
29. Price, P.A. (1987) in: *Calcium Regulation and Bone Metabolism: Basic and Clinical Aspects* (Cohn, D.V. et al., eds.) Vol.9, pp.419–425, Elsevier Science, Amsterdam.
30. Bolander, M.E., Young, M.F., Fisher, L.W., Yamada, Y., & Termine, J.D. (1988) *Proc. Natl. Acad. Sci. USA* 85, 2919–2923.
31. Engel, J., Taylor, W., Paulsson, M., Sage, H., & Hogan, B. (1987) *Biochemistry* 26, 6958–6965.
32. Drakenberg, T., Forsén, S., & Stenflo, J. (1999) in: *Calcium as a Cellular Regulator* (Carafoli, E., & Klee, C.) pp. 123–151, Oxford University Press, New York.
33. Koch, A.W., Pokutta, S., Lustig, A., & Engel, J. (1997) *Biochemistry* 36, 7697–7705.
34. Downing, A.K., Handford, P.A., & Campbell, I.D. (2000) in: *Calcium Homeostasis, Topics in Biological Inorganic Chemistry,* (E. Carafoli, & J. Krebs, eds.), Vol. 3, Springer, Berlin, pp. 83–99.
35. Campbell, I.D., & Bork, P. (1993) *Curr. Opin. Struct. Biol.* 3, 385–392.
36. Rees, D.J.G., Jones, I.M., Handford, P.A., Walter, S.J., Esnouf, M.P., Smith, K.J., & Brownlee, G.G. (1988) *EMBO J.* 7, 2053–2061.
37. Hambleton, S., Valeyev, N.V., Muranyi, A., Knott, V., Werner, J.M., McMichael, A.J., Handford, P.A., & Downing, A.K. (2004) *Structure* 12, 2173–2183.
38. Rebay, I., Fleming, R.J., Fehon, R.G., Cherbas, L., Cherbas, P., & Artavanis-Tsakonas, S. (1991) *Cell* 67, 687–699.
39. Banner, D.W., D'Arcy, A., Chene, C., Winkler, F.K., Guha, A., Konigsberg, W.H., Nemerson, Y., & Kirchhofer, D. (1996) *Nature* 380, 41.
40. Muranyi, A., Finn, B.E., Gippert, G.P., Forsén, S., Stenflo, J., & Drakenberg, T. (1998) *Biochemistry,* 37, 10605–10615.
41. Kao, Y.-H., Lee, G.F., Wang, Y., Starovasnik, M.A., Kelley, R.F., Spellmann, M.W., & Lerner, L. (1999) *Biochemistry* 38, 7097–7110.
42. De Celis, J.F., Barrio, R., Delarco, R., & Garciabellido, A. (1993) *Proc. Natl. Acad. Sci. USA* 90, 4037–4041.
43. O'Connell, M. (2004) *Semin. Hematol.* 41, 76–81.
44. Kong, W.J., Liu, J., & Jiang, J.D. (2006) *J. Mol. Med.* 84, 29–36.
45. Louvi, A., Arboleda-Velasquez, J.F., & Artavanis-Tsakonas, S. (2006) *Dev. Neurosci.* 28, 5–12.
46. Brown, E.M., Gamba, G., Riccardi, D., Lombardi, M., Butters, R., Kifor, O., Sun, A., Hediger, M.A., Lytton, J., & Hebert, S.C., (1993) *Nature* 366, 575–580.
47. Brown, E.M., & MacLeod, R.J. (2001) *Phys. Rev.* 81, 239–297.
48. Adams, G.B., Chabner, K.T., Alley, I.R., Olson, D.P., Szczepiorkowski, Z.M., Poznansky, M.C., Kos, C.H., Pollak, M.R., Brown, E.M., & Scadden, D.T. (2006) *Nature* 439, 599–603.
49. Diaz, R., Hurwitz, S., Chatopadhyay, N., Pines, M., Yang, Y.H., Kifor, O., Friedman-Einat, M., Butters, R., Hebert, S.C., & Brown E.M. (1997) *Am. J. Physiol.* 273, R1008–R1016.
50. Cima, R.R., Cheng, I., Klingensmith, M.E., Chattopadhyay, N., Kifor, O., Hebert, S.C., Brown, E.M., & Soybel, D.I. (1997) *Am. J. Physiol.* 273, G1051–G1060.
51. Nearing, J., Betka, M., Quinn, S., Hentschel, H., Elger, M., Baum, M., Bai, M., Chattopachyhay, S. C., & Harris, H. W. (2002) *Proc. Natl. Acad. Sci. USA* 99, 9231–9236.
52. Ray, K., Hauschild, B.C., Steinbach, P.J., Goldsmith, P.K., Hauache, O., & Spiegel, A.M. (1999) *J. Biol. Chem.* 274, 27642–27650.
53. Hofer, A.M., & Brown, E.M. (2003) *Nat. Rev. Mol. Cell. Biol.* 4, 530–538.
54. Bai, M., Trivedi, S., & Brown, E.M. (1998) *J. Biol. Chem.* 273, 23605–23610.
55. Attie, M., Gill, J.J., Stock, J., Spiegel, A., Downs, R.J., Levine, M., & Marx, S. (1983) *J. Clin. Invest.* 72, 667–676.
56. Davies, M., Adams, P., & Lumb, G. (1984) *Acta Endocrinol.* 106, 499–504.
57. Law, W.J., & Heath, H. III. (1985) *Ann. Int. Med.* 105, 511–519.
58. Aida, K., Koishi, S., Inoue, M., Nakazato, M., Tawata, M., & Onaya, T. (1995) *J. Clin. Endocrinol. Metab.* 80, 2594–2598.
59. Putney, J.W. (2005) *J. Cell Biol.* 169, 381–382.

60. Roos, J., Digregorio, P.J., Yeromin, A.V., Ohlsen, K., Lioudyno, M., Zhang, S., Safrina, O., Kozak, J.A., Wagner, S.L., Cahalan, M.D., Velicelebi, G., & Staudermann, K.A. (2005) *J. Cell Biol.* 169, 435–445.
61. Liou, J., Kim, M.L., Heo, W.D., Jones, J.T., Myers, J.W., Ferrell, Jr., J.E., & Meyer, T. (2005) *Curr. Biol.* 15, 1235–1241.
62. Vig, M., Peinelt, C., Beck, A., Koomoa, D.L., Rabah, D., Koblan-Huberson, M., Kraft, S., Turner, H., Fleig, A., Penner, R., & Kinet, J.-P. (2006) *Science* 312, 1220–1223.
63. Feske, S., Gwack, Y., Prakriya, M., Srikanth, S., Puppel, S.-H., Tanasa, B., Hogan, P.G., Lewis, R.S., Daly, M., & Rao, A. (2006) *Nature* 441, 179–185.
64. Mercer, J.C., DeHaven, W.I., Smyth, J.T., Wedel, B., Boyles, R.R., Bird, G.S., & Putney, Jr. J.W. (2006) *J. Biol. Chem.* DOI 10.1074/jbc.M604589200.
65. Peinelt, C., Vig, M., Koomoa, D.L., Beck, A., Nadler, M.J.S., Koblan-Huberson, M., Lis, A., Fleig, A., Penner, R., & Kinet, J.-P (2006) *Nat. Cell Biol.* 8, 771–773.
66. Soboloff, J., Spassova, M.A., Tang, X.D., Hewavitharana, T., Xu, W., & Gill, D.L. (2006) *J. Biol. Chem.* 281, 20661–20665.
67. Williams, R.T., Manji, S.S., Parker, N.J., Hancock, M.S., Van, S.L., Eid, J.P., Senior, P.V., Kazenwadel, J.S., Shandala, T., & Saint, R. et al. (2001) *Biochem. J.* 357, 673–685.
68. Spassova, M.A., Soboloff, J., He, L.-P., Xu, W., Dziadek, M.A., & Gill, D.L. (2006) *Proc. Natl. Acad. Sci. USA* 103, 4040–4045.
69. Rome, L.C. (2006) *Annu. Rev. Physiol.* 68, 193–221.
70. Schwaller, B., Meyer, M., & Schiffmann, S. (2002) *Cerebellum* 1, 241–258.
71. Haiech, J., Derancourt, J., Pechere, J.-F., & Demaille, J.G. (1979) *Biochemistry* 18, 2752–2758.
72. Lee, S.H., Schwaller, B., & Neher, E. (2000) *J. Physiol. (London)* 525, 419–432.
73. Heizmann, C.W., Berchtold, M.W., & Rowlerson, A.M. (1982) *Proc. Natl. Acad. Sci. USA* 79, 7243–7247.
74. Schwaller, B., Dick, J., Dhoot, G., Carroll, S., Vrbova, G., Nicotera, P., Pette, D., Wyss, A., Bluethmann, H., Hunziker, W., & Celio, M.R. (1999) *Am. J. Physiol.* 276, C395–C403.
75. Baimbridge, K.G., Celio, M.R., & Rogers, R.H. (1992) *Trends Neurosci.* 15, 303–308.
76. Gross, M., & Kumar, R. (1990) *Am. J. Physiol.* F195–F209.
77. Christakos, S., Barletta, F., Huening, M., Dhawan, P., Liu, Y., Porta, A., & Peng, X. (2003) *J. Cell. Biochem.* 88, 238–244.
78. Szebenyi, D.M.E., Obendorf, S.K., & Moffat, K. (1981) *Nature* 294, 327–332.
79. Skelton, M.J., Kördel, J., & Chazin, W.J. (1995) *J. Mol. Biol.* 249, 441–462.
80. Linse, S., Brodin, P., Drakenberg, T., Thulin, E., Sellers, P., Elmden, K., Grundstrom, T., & Forsén, S. (1987) *Biochemistry* 26, 6723–6735.
81. Kojetin, D.J., Venters, R.A., Kordys, D.R., Thompson, R.J., Kumar, R., & Cavanagh, J. (2006) *Nat. Struct. Mol. Biol.* 13, 641–647.
82. Venters, R.A. Benson, L.M., Craig, T.A., Bagu, J., Paul, K.H., Kordys, D.R., Thompson, R., Naylor, S., Kumar, R., & Cavanagh, J. (2003) *Anal. Biochem.* 317, 59–66.
83. Lutz, W., Frank, E. M., Craig, T. A., Thompson, R., Venters, R. A., Kojetin, D., Cavanagh, J., & Kumar, R. (2003) *Biochem. Biophys. Res. Commun.* 303, 1186–1192.
84. Berggard, T., Szczepankiewicz, O., Thulin, E., & Linse, S. (2002) *J. Biol. Chem.* 277, 41954–41959.
85. Schmidt, H., Schwaller, B., & Eilers, J. (2005) *Proc. Natl. Acad. Sci. USA* 102, 5850–5855.
86. Heizmann, C.W., & Hunziker, W. (1990) in: *Intracellular Calcium Regulation* (Bronner, F., ed.), Alan R. Liss, New York, pp. 211–248.
87. Rigden, D.J., & Galperin, M.Y. (2004) *J. Mol. Biol.* 343, 971–984.
88. Berggard, T., Silow, M., Thulin, E., & Linse, S. (2000) *Biochemistry* 39, 6864–6873.
89. Bellido, T., Huenig, M., Raval-Pandya, M., Manolagas, S.C., & Christakos, S. (2000) *J. Biol. Chem.* 275, 26328–26332.
90. Rogers, J.H., (1987) *J. Cell Biol.* 105, 1343–1353.
91. Heizmann, C.W., & Hunziker, W. (1991) *Trends Biochem. Sci.* 16, 98–103.
92. Kempermann, G., Jessberger, S., Steiner, B., & Kronenberg, G. (2004) *Trends Neurosci.* 27, 447–452.

93. Brandt, M.D. Jessberger, S., Steiner, B., Kronenberg, G., Reuter, K., Bick-Sander, A., von der Behrens, W., & Kempermann, G. (2003) *Mol. Cell. Neurosci.* 24, 603–613.
94. Li, Z., Kato, T., Kawagishi, K., Fukushima, N., Yokouchi, K., & Morizumi, T. (2002) *Neurosci. Res.* 42, 123–132.
95. Krebs, J. (1981) *Cell Calcium*, 2, 295–311.
96. Cohen, P., & Klee, C.B., eds. Calamodulin, in: *Molecular Aspects of Cell Regulation,* Vol. 5, Elsevier, Amsterdam, 1988.
97. Copley, R.R., Schultz, J., Ponting, C.P., & Bork, P. (1999) *Curr. Opin. Struct. Biol.* 9, 408–415.
98. Berchtold, M.W., Egli, R., Rhyner, J.A., Hameister, H., & Strehler, E.E. (1993) *Genomics* 16, 461–465.
99. Fischer, R., Koller, M., Flura, M., Strehler-Page, M.-A., Matthews, S., Krebs, J., Penniston, J.T., Carafoli, E., & Strehler, E.E. (1988) *J. Biol. Chem.* 263, 17055–17062.
100. Sia, S.K., Li, M.X., Spyracopoulos, L., Gagne, S.M., Liu, W., Putkey, J.A., & Sykes, B.D. (1997) *J. Biol. Chem.* 272, 18216–18221.
101. Shenolikar, S., Cohen, P.T., Cohen, P., Nairn, A.C., & Perry, S.V. (1979) *Eur. J. Biochem.* 100, 329–337.
102. Pruschy, M., Ju, Y, Spitz, L., Carafoli, E., & Goldfarb, D.S. (1994) *J. Cell Biol.* 127, 1527–1536.
103. Bers, D.M., & Guo, T. (2005) *Ann. N.Y. Acad. Sci.* 1047, 86–98.
104. Kuboniva, H., Tjandra, N., Grzesiek, Z., Ren, H., Klee, C.B., & Bax, A. (1995) *Nat. Struct. Biol.* 2, 768–776.
105. Zhang, M., Tanaka, T., & Ikura, M. (1995) *Nat. Struct. Biol.* 2, 758–767.
106. La Porte, D.C., Wierman, B.M., & Storm, D.R. (1980) *Biochemistry* 19, 3814–3819.
107. Krebs, J., Buerkler, J., Guerini, D., Brunner, J., & Carafoli, E. (1984) *Biochemistry* 23, 400–403.
108. Rhoads, A.R., & Friedberg, F. (1997) *FASEB J.* 11, 331–340.
109. Heller, W.T., Krueger, J.K., & Trewhella, J. (2003) *Biochemistry* 42, 10279–10588.
110. Osawa, M., Tokumitsu, H., Swindells, M.B., Kurihara, H., Orita, M., Shibanuma, T., Furuya, T., & Ikura, M. (1999) *Nat. Struct. Biol.* 6, 819–824.
111. Yap, K.L., Kim, J., Truong, K., Sherman, M., Yuan, T., & Ikura, M. (2000) *J. Struct. Funct. Genomics* 1, 8–14.
112. Zühlke, R.D., Pitt, G.S., Deisseroth, K., Tsien, R.W., & Reuter, H., (1999) *Nature* 399, 159–162.
113. Dascal, N., Snutch, T.P., Lubbert, H., Davidson, N., & Lester, H.A. (1986) *Science* 231, 1147–1150.
114. Fallon, J., Halling, D.B., Hamilton, S.L., & Quiocho, F.A. (2005) *Structure* 13, 1881–1886.
115. Peterson, B.Z., DeMaria, C.D., Adelman, J.P., & Yue, D.T. (1999) *Neuron* 22, 549–558.
116. van Petegem, F., Chatelain, F.C., & Minor, Jr., D.L., (2005) *Nat. Struct. Biol.* 12, 1108–1115.
117. Stauffer, T.P., Guerini, D., & Carafoli, E. (1995) *J. Biol. Chem.* 270, 12184–12190.
118. James, P., Maeda, M., Fischer, R., Verma, A.K., Krebs, J., Penniston, J.T., & Carafoli, E. (1988) *J. Biol. Chem.* 263, 2905–2910.
119. Falchetto, R., Vorherr, T., Brunner, J., & Carafoli, E. (1991) *J. Biol. Chem.* 266, 2930–2936.
120. Falchetto, R., Vorherr, T., & Carafoli, E. (1992) *Protein Sci.* 1, 1613–1621.
121. Guerini, D., Krebs, J., & Carafoli, E. (1984) *J. Biol. Chem.* 259, 15172–15277.
122. Newton, D.L., Oldewurtel, M.D., Krinks, M.H., Shiloach, J., & Klee, C.B. (1984) *J. Biol. Chem.* 259, 4419–4426.
123. Barth, A., Martin, S.R., & Bayley, P.M. (1998) *J. Biol. Chem.* 273, 2174–2183.
124. Klumpp, S., Guerini, D., Krebs, J., & Schultz, J.E. (1987) *Biochem. Biophys. Res. Commun.* 142, 857–864.
125. Kuznicki, J., Grabarek, Z., Brzeska, H., Drabikowski, W., & Cohen, P. (1981) *FEBS Lett.* 130, 141–145.
126. Kataoka, M., Head, J.F., Vorherr, T., Krebs, J., & Carafoli, E. (1991) *Biochemistry* 30, 6247–6251.
127. Persechini, A., & Kretsinger, R.H. (1988) *J. Biol. Chem.* 263, 12175–12178.
128. Vorherr, T., Quadroni, M., Krebs, J., & Carafoli, E. (1992) *Biochemistry* 31, 8245–8251.
129. Verma, A.K., Enyedi, A., Filoteo, A.G., & Penniston, J.T. (1994) *J. Biol. Chem.* 269, 1687–1691.
130. van Eerd, J.P., & Takahashi, K. (1975) *Biochem. Biophys. Res. Commun.* 64, 122–127.

131. Potter, J.D., & Gergely, J. (1975) *J. Biol. Chem.* 250, 4628–4633.
132. van Eerd, J.P., & Takahashi, K. (1976) *Biochemistry* 15, 1171–1180.
133. Ebashi, S., & Endo, M. (1968) *Prog. Biophys. Mol. Biol.* 18, 123–183.
134. Perry, S.V. (1999) *Mol. Cell. Biochem.* 190, 9–32.
135. Sykes, B.D. (2003) *Nat. Struct. Biol.* 10, 588–589.
136. Takeda, S., Yamashita, A., Maeda, K., & Maeda, Y. (2003) *Nature* 424, 35–41.
137. Vinogradova, M.V., Stone, D.B., Malanina, G.G., Karatzaferi, C., Cooke, R., Mendelson, R.A., & Fletterick, R.J. (2005) *Proc. Natl. Acad. Sci. USA* 102, 5038–5043.
138. Salisbury, J.L. (1995) *Curr. Opin. Cell Biol.* 7, 39–45.
139. Paoletti, A., Moudjou, M., Paintrand, M., Salisbury, J.L., & Bornens, M. (1996) *J. Cell Sci.* 109, 3089–3102.
140. Baum, P., Furlong, C., & Byers, B. (1986) *Proc. Natl. Acad. Sci. USA* 83, 5512–5516.
141. Taillon, B.E., Adler, S.A., Suhan, J.P., & Jarvik, J.W. (1992) *J. Cell Biol.* 119, 1613–1624.
142. Geier, B.M., Wiech, H., & Schiebel, E. (1996) *J. Biol. Chem.* 271, 28366–28374.
143. Biggins, S., & Rose, M.D. (1994) *J. Cell Biol.* 125, 843–852.
144. Spang, A., Courtney, I., Grein, K., Matzner, M., & Schiebel, E. (1995) *J. Cell Biol.* 128, 863–877.
145. Hu, H., & Chazin, W.J. (2003) *J. Mol. Biol.* 330, 473–484.
146. Hu, H., Sheehan, J.H., & Chazin, W.J. (2004) *J. Biol. Chem.* 279, 50895–50903.
147. Ohya, Y., & Botstein, D. (1994) *Genetics* 138, 1041–1054.
148. Schumacher, M.A., Rivard, A.F., Bachinger, H.P., & Adelman, J.P. (2001) *Nature* 410, 1120–1124.
149. Burgoyne, R.D. (2004) *Biochim. Biophys. Acta* 1742, 59–68.
150. Braunewell, K.-H. (2005) *Trends Pharm. Sci.* 26, 345–351.
151. Ames, J.B., Ishima, R., Tanaka, T., Gordon, J.I., Stryer, L., & Ikura, M. (1997) *Nature* 389, 198–202.
152. Senin, I.I., Dean, K.R., Zargarov, A.A., Akhtar, M., & Philippov, P.P. (1997) *Biochem. J.* 321, 551–555.
153. Kawamura, S. (1993) *Nature* 362, 855–857.
154. Gorczyca, W.A., Polans, A.S., Surgucheva, I.G., Subbaraya, I., Baehr, W., & Palczewski, K. (1995) *J. Biol. Chem.* 270, 22029–22036.
155. Pongs, O., Lindemeier, J., Zhu, X.R., Theil, T., Endelkamp, D., Krah-Jentgens, I., Lambrecht, H.-G., Koch, K.W., Schwemer, J., Rivosecchi, R., Mallart, A., Galceran, J., Canal, I., Barbas, J.A., & Ferrus, A. (1993) *Neuron* 11, 15–28.
156. Hendricks, K.B., Wang, B.Q., Schnieders, E.A., & Thorner, J. (1999) *Nat. Cell Biol.* 1, 234–241.
157. An, W.F., Bowlby, M.R., Bett, M., Cao, J., Ling, H.P., Mendoza, G., Hinson, J.W., Mattson, K.I., Strassle, B.W., Trimmer, J.S., & Rhodes, K.J. (2000) *Nature* 403, 553–556.
158. Carrion, A.M., Link, W.A., Ledo, F., Mellstrom, B., Naranjo, & J.R. (1999) *Nature* 398, 80–84.
159. Buxbaum, J.D., Choi, E.K., Luo, Y.X., Lilliehook, C., Crowley, A.C., Merriam, D.E., & Wasco, W. (1998) *Nat. Med.* 4, 1177–1181.
160. Marenholz, I., Heizmann, C.W., & Fritz, G. (2004) *Biochem. Biophys. Res. Commun.* 322, 1111–1122.
161. Marenholz, I., Lovering, R. C., & Heizmann, C. W. (2006) *Biochim. Biophys. Acta Mol. Cell Res.*, 1763, 1282–1283.
162. Fritz, G., & Heizmann, C.W. (2004) *Handbook of Metalloproteins* (A. Messerschmidt, W. Bode, & M. Cygler, eds.), Vol. 3, John Wiley & Sons, pp. 529–540.
163. Bhattacharya, S., Bunick, C.G., & Chazin, W.J. (2004) *Biochim. Biophys. Acta* 1742, 69–79.
164. Vallely, K.M., Rustandi, R.R., Ellis, K.C., Varlamova, O., Bresnick, A.R., & Weber, D.J. (2002) *Biochemistry* 41, 12670–12680.
165. Ikura, M., & Ames, J.B. (2006) *Proc. Natl. Acad. Sci. USA* 103, 1159–1164.
166. Zimmer, D.B., Chaplin, J., Baldwin, A., & Rast, M. (2005) *Cell Mol. Biol. (Noisy-le-grand)* 51, 201–214.
167. Ramasamy, R., Vannucci, S.J., Yan, S.S., Herold, K., Yan, S.F., & Schmidt, A.M. (2005) *Glycobiology* 15, 16R–28R.

168. Marenholz, I., Zirra, M., Fischer, D.F., Backendorf, C., Ziegler, A., & Mischke, D. (2001) *Genome Res.* 11, 341–355.
169. Huber, M., Siegenthaler, G., Mirancea, N., Marenholz, I., Nizetic, D., Breitkreutz, D., Mischke, D., & Hohl, D. (2005) *J. Invest. Dermatol.* 124, 998–1007.
170. Heizmann, C.W. (2005) *J. Pediatr.* 147, 731–738.
171. Otterbein, L.R., Kordowska, J., Witte-Hoffmann, C., Wang, C.L., & Dominguez, R. (2002) *Structure* 10, 557–567.
172. Rustandi, R.R., Baldisseri, D.M., & Weber, D.J. (2000) *Nat. Struct. Biol.* 7, 570–574.
173. Moroz, O.V., Antson, A.A., Dodson, E.J., Burrell, H.J., Grist, S.J., Lloyd, R.M., Maitland, N.J., Dodson, G.G., Wilson, K.S., Lukanidin, E., & Bronstein, I.B. (2002) *Acta Crystallogr. D. Biol. Crystallogr.* 58, 407–413.
174. Rety, S., Sopkova, J., Renouard, M., Osterloh, D., Gerke, V., Tabaries, S., Russo-Marie, F., & Lewit-Bentley, A. (1999) *Nat. Struct. Biol.* 6, 89–95.
175. Rety, S., Osterloh, D., Arie, J.P., Tabaries, S., Seeman, J., Russo-Marie, F., Gerke, V., & Lewit-Bentley, A. (2000) *Structure* 8, 175–184.
176. Novitskaya, V., Grigorian, M., Kriajevska, M., Tarabykina, S., Bronstein, I., Berezin, V., Bock, E., & Lukanidin, E. (2000) *J. Biol. Chem.* 275, 41278–41286.
177. Novitskaya, V., Grigorian, M., Kriajevska, M., Tarabykina, S., Bronstein, I., Berezin, V., Bock, E., & Lukanidin, E. (2000) *J. Biol. Chem.* 41278–41286.
178. Barger, S.W., Wolchok, S.R., & Van Eldik, L.J. (1992) *Biochim. Biophys. Acta* 1160, 105–112.
179. Ostendorp, T., Heizmann, C.W., Kroneck, P.M., & Fritz, G. (2005) *Acta Crystallograph. Sect. F Struct. Biol. Cryst. Commun.* 61, 673–675.
180. Heizmann, C.W., & Cox, J.A. (1998) *Biometals* 11, 383–397.
181. Fritz, G., Heizmann, C.W., & Kroneck, P.M. (1998) *Biochim. Biophys. Acta* 1448, 264–276.
182. Fritz, G., Mittl, P.R., Vasak, M., Grutter, M.G., & Heizmann, C.W. (2002) *J. Biol. Chem.* 277, 33092–33098.
183. Nishikawa, T., Lee, I.S., Shiraishi, N., Ishikawa, T., Ohta, Y., & Nishikimi, M. (1997) *J. Biol. Chem.* 272, 23037–23041.
184. Schafer, B.W., Fritschy, J.M., Murmann, P., Troxler, H., Durussel, I., Heizmann, C.W., & Cox, J.A. (2000) *J. Biol. Chem.* 275, 30623–30630.
185. Koch, M., Bhattacharya, S., Kehl, T., Gimona, M., Vasak, M., Chazin, W., Heizmann, C.W., Kroneck, P.M.H., & Fritz, G. (2007) *Biochim. Biophys. Acta* 1773, 457–470.
186. Sturchler, E., Cox, J.A., Durussel, I., Weibel, M., & Heizmann, C.W. (2006) *J. Biol. Chem.,* 281, 38905–38917.
187. Heizmann, C.W., Fritz, G., & Schafer, B.W. (2002) *Front. Biosci.* 7, d1356–1368.
188. Deloulme, J.C., Gentil, B.J., & Baudier, J. (2003) *Microsc. Res. Tech.* 60, 560–568.
189. Zimmer, D.B., Wright Sadosky, P., & Weber, D.J. (2003) *Microsc. Res. Tech.* 60, 552–559.
190. Donato, R. (2003) *Microsc. Res. Tech.* 60, 540–551.
191. van de Graaf, S.F., Hoenderop, J.G., Gkika, D., Lamers, D., Prenen, J., Rescher, U., Gerke, V., Staub, O., Nilius, B., & Bindels, R.J. (2003) *EMBO J.* 22, 1478–1487.
192. Yamada, A., Irie, K., Hirota, T., Ooshio, T., Fukuhara, A., & Takai, Y. (2005) *J. Biol. Chem.* 280, 6016–6027.
193. Kwon, M., Yoon, C.S., Jeong, W., Rhee, S.G., & Waisman, D.M. (2005) *J. Biol. Chem.* 280, 23584–23592.
194. Donier, E., Rugiero, F., Okuse, K., & Wood, J.N. (2005) *J. Biol. Chem.* 280, 38666–38672.
195. Svenningsson, P., Chergui, K., Rachleff, I., Flajolet, M., Zhang, X., El Yacoubi, M., Vaugeois, J.M., Nomikos, G.G., & Greengard, P. (2006) *Science* 311, 77–80.
196. Dempsey, A.C., Walsh, M.P., & Shaw, G.S. (2003) *Structure* 11, 887–897.
197. Sakaguchi, M., Miyazaki, M., Takaishi, M., Sakaguchi, Y., Makino, E., Kataoka, N., Yamada, H., Namba, M., & Huh, N.H. (2003) *J. Cell Biol.* 163, 825–835.
198. Sakaguchi, M., Miyazaki, M., Sonegawa, H., Kashiwagi, M., Ohba, M., Kuroki, T., Namba, M., & Huh, N.H. (2004) *J. Cell Biol.* 164, 979–984.

199. Sakaguchi, M., Sonegawa, H., Nukui, T., Sakaguchi, Y., Miyazaki, M., Namba, M., & Huh, N.H. (2005) *Proc. Natl. Acad. Sci USA* 102, 13921–13926.

200. Reeves, R.H., Yao, J., Crowley, M.R., Buck, S., Zhang, X., Yarowsky, P., Gearhart, J.D., & Hilt, D.C. (1994) *Proc. Natl. Acad. Sci. USA* 91, 5359–5363.

201. Gerlai, R., Wojtowicz, J.M., Marks, A., & Roder, J. (1995) *Learn. Mem.* 2, 26–39.

202. Gerlai, R., and Roder, J. (1995) *J. Psychiatr. Neurosci.* 20, 105–12.

203. Winocur, G., Roder, J., and Lobaugh, N. (2001) *Neurobiol. Learn. Mem.* 75, 230–43.

204. Bell, K., Shokrian, D., Potenzieri, C., & Whitaker-Azmitia, P.M. (2003) *Neuropsychopharmacology* 28, 1810–1816.

205. Nishiyama, H., Knopfel, T., Endo, S., & Itohara, S. (2002) *Proc. Natl. Acad. Sci. USA* 99, 4037–4042.

206. Xiong, Z., O'Hanlon, D., Becker, L.E., Roder, J., MacDonald, J.F., & Marks, A. (2000) *Exp. Cell Res.* 257, 281–289.

207. Dyck, R.H., Bogoch, I.I., Marks, A., Melvin, N.R., & Teskey, G.C. (2002) *Brain Res. Mol. Brain Res.* 106, 22–29.

208. Kiewitz, R., Acklin, C., Schafer, B.W., Maco, B., Uhrik, B., Wuytack, F., Erne, P., & Heizmann, C.W. (2003) *Biochem. Biophys. Res. Commun.* 306, 550–557.

209. Remppis, A., Greten, T., Schafer, B.W., Hunziker, P., Erne, P., Katus, H.A., & Heizmann, C.W. (1996) *Biochim. Biophys. Acta* 1313, 253–257.

210. Most, P., Remppis, A., Pleger, S.T., Loffler, E., Ehlermann, P., Bernotat, J., Kleuss, C., Heierhorst, J., Ruiz, P., Witt, H., Karczewski, P., Mao, L., Rockman, H.A., Duncan, S.J., Katus, H.A., & Koch, W.J. (2003) *J. Biol. Chem.* 278, 33809–33817.

211. Most, P., Pleger, S.T., Volkers, M., Heidt, B., Boerries, M., Weichenhan, D., Loffler, E., Janssen, P.M., Eckhart, A.D., Martini, J., Williams, M.L., Katus, H.A., Remppis, A., & Koch, W.J. (2004) *J. Clin. Invest.* 114, 1550–1563.

212. Ackermann, G.E., Marenholz, I., Wolfer, D.P., Chan, W.Y., Schäfer, B., Erne, P., & Heizmann, C.W. (2006) *Biochim. Biophys. Acta* 1763, 1307–1319.

213. Pleger, S.T., Remppis, A., Heidt, B., Volkers, M., Chuprun, J.K., Kuhn, M., Zhou, R.H., Gao, E., Szabo, G., Weichenhan, D., Muller, O.J., Eckhart, A.D., Katus, H.A., Koch, W.J., & Most, P. (2005) *Mol. Ther.* 12, 1120–1129.

214. Hofmann, M.A., Drury, S., Fu, C., Qu, W., Taguchi, A., Lu, Y., Avila, C., Kambham, N., Bierhaus, A., Nawroth, P., Neurath, M.F., Slattery, T., Beach, D., McClary, J., Nagashima, M., Morser, J., Stern, D., & Schmidt, A.M. (1999) *Cell* 97, 889–901.

215. Arumugam, T., Simeone, D.M., Van Golen, K., & Logsdon, C.D. (2005) *Clin. Cancer Res.* 11, 5356–5364.

216. Ehlermann, P., Eggers, K., Bierhaus, A., Most, P., Weichenhan, D., Greten, J., Nawroth, P.P., Katus, H.A., & Remppis, A. (2006) *Cardiovasc. Diabetol.* 5, 6.

217. Zhou, Z., Immel, D., Xi, C.X., Bierhaus, A., Feng, X., Mei, L., Nawroth, P., Stern, D.M., & Xiong, W.C. (2006) *J. Exp. Med.* 203, 1067–1080.

218. Jenkinson, S.R., Barraclough, R., West, C.R., & Rudland, P.S. (2004) *Br. J. Cancer* 90, 253–262.

219. Pedersen, M.V., Kohler, L.B., Grigorian, M., Novitskaya, V., Bock, E., Lukanidin, E., & Berezin, V. (2004) *J. Neurosci. Res.* 77, 777–786.

220. Garrett, S.C., Varney, K.M., Weber, D.J., & Bresnick, A.R. (2006) *J. Biol. Chem.* 281, 677–680.

221. Ikoma, N., Yamazaki, H., Abe, Y., Oida, Y., Ohnishi, Y., Suemizu, H., Matsumoto, H., Matsuyama, T., Ohta, Y., Ozawa, A., Ueyama, Y., & Nakamura, M. (2005) *Oncol. Rep.* 14, 633–637.

222. Ambartsumian, N., Klingelhofer, J., Grigorian, M., Christensen, C., Kriajevska, M., Tulchinsky, E., Georgiev, G., Berezin, V., Bock, E., Rygaard, J., Cao, R., Cao, Y., & Lukanidin, E. (2001) *Oncogene* 20, 4685–4695.

223. Nacken, W., Roth, J., Sorg, C., & Kerkhoff, C. (2003) *Microsc. Res. Tech.* 60, 569–580.

224. Roth, J., Vogl, T., Sorg, C., & Sunderkotter, C. (2003) *Trends Immunol.* 24, 155–158.

225. Bozinovski, S., Cross, M., Vlahos, R., Jones, J.E., Hsuu, K., Tessier, P.A., Reynolds, E.C., Hume, D.A., Hamilton, J.A., Geczy, C.L., & Anderson, G.P. (2005) *J. Proteome. Res.* 4, 136–145.

226. Hermani, A., De Servi, B., Medunjanin, S., Tessier, P.A., & Mayer, D. (2006) *Exp. Cell Res.* 312, 184–197.
227. Akpek, E.K., Liu, S.H., Thompson, R., & Gottsch, J.D. (2002) *Invest. Ophthalmol. Vis. Sci.* 43, 2677–2684.
228. Moroz, O.V., Dodson, G.G., Wilson, K.S., Lukanidin, E., & Bronstein, I.B. (2003) *Microsc. Res. Tech.* 60, 581–592.
229. Blanchard, H., Li, Y., Cygler, M., Kay, C.M., Simon, J., Arthur, C., Davies, P.L., & Elce, J.S. (1997) *Nat. Struct. Biol.* 4, 532–538.
230. Lin, G.D., Chattopadhyay, D., Maki, M., Wang, K.K., Carson, M., Jin, L., Yuen, P.W., Takano, E., Hatanaka, M., DeLucas, L.J., & Narayana, S.V. (1997) *Nat. Struct. Biol.* 4, 539–547.
231. Hosfield, C.M., Elce, J.S., Davies, P.L., & Jia, Z. (1999) *EMBO J.* 18, 6880–6889.
232. Strobl, S., Fernandez-Catalan, C., Braun, M., Huber, R., Masumoto, H., Nakagawa, K., Irie, A., Sorimachi, H., Bourenkow, G., Bartunik, H., Suzuki, K., & Bode, W. (2000) *Proc. Natl. Acad. Sci. USA* 97, 588–592.
233. Jia, J., Han, Q., Borregaard, N., Lollike, K., & Cygler, M. (2000) *J. Mol. Biol.* 300, 1271–1281.
234. Jia, J., Tarabykina, S., Hansen, C., Berchtold, M., & Cygler, M. (2001) *Structure* 9, 267–274.
235. Suzuki, K., Hata, S., Kawabata, Y., & Sorimachi, H. (2004) *Diabetes* 53, S12–S18.
236. Dutt, P., Arthur, J.S., Grochulski, P., Cygler, M., & Elce, J.S. (2000) *Biochem. J.* 348, 37–43.
237. Franco, S.J., & Huttenlocher, A. (2005) *J. Cell Sci.* 118, 3829–3838.
238. Zatz, M., & Starling, A. (2005) *N. Engl. J. Med.* 352, 2413–2423.
239. Sorimachi, H., Ono, Y., & Suzuki, K. (2000) *Adv. Exp. Med. Biol.* 481, 383–395.
240. Vito, P., Lacana, E., & D'Adamio, L. (1996) *Science* 271, 521–525.
241. Vito, P., Pellegrini, L., Guiet, C., & D'Adamio, L. (1999) *J. Biol. Chem.* 274, 1533–1540.
242. Missotten, M., Nichols, A., Rieger, K., & Sadoul, R. (1999) *Cell Death Differ.* 6, 124–129.
243. Che, S., El-Hodiri, H.M., Wu, C.-F., Nelman-Gonzalez, M., Weil, M.M., Etkin, L.D., Clark, R.B., & Kuang, J. (1999) *J. Biol. Chem.* 274, 5522–5531.
244. Kitaura, Y., Matsumoto, S., Satoh, H., Hitomi, K., & Maki, M. (2001) *J. Biol. Chem.* 276, 14053–14058.
245. Satoh, H., Shibata, H., Nakano, Y., Kitaura, Y., & Maki, M. (2002) *Biochem. Biophys. Res. Commun.* 291, 1166–1172.
246. Jang, I.K., Hu, R., Lacana, E., D'Adamio, L., & Gu, H. (2002) *Mol. Cell Biol.* 22, 4094–4100.
247. Krebs, J., Saremaslani, P., & Caduff, R. (2002) *Biochim. Biophys. Acta* 1600, 68–73.
248. laCour, J.M., Mollerup, J., Winding, P., Tarabykina, S., Sehested, M., & Berchtold, M.W. (2003) *Am. J. Pathol.* 163, 81–89.
249. Montaville, P., Dai, Y., Cheung, C., Giller, K., Becker, S., Michalak, M., Webb, S.E., Miller, A.L., & Krebs, J. (2006) *Biochim. Biophys. Acta* 1763, 1335–1343.
250. Park, J.W., Parisky, K., Celotto, A.M., Reenan, R.A., & Graveley, B.R. (2004) *Proc. Natl. Acad. Sci. USA* 101, 12792–12797.
251. Kittler, R., Putz, G., Pelletier, L., Poser, I., Heninger, A.-K., Drechsel, D., Fischer, S., Konstantinova, I., Habermann, B., Grabner, H., Yaspo, M.-L., Himmelbauer, H., Korn, B., Neugebauer, K., Pisabarro, M.T., & Buchholz, F. (2004) *Nature* 432, 1036–1040.
252. Kim, Y.-O., Park, S.-J., Balaban, R.S., Nirenberg, M., & Kim, Y.A. (2004) *Proc. Natl. Acad. Sci. USA* 101, 159–164.
253. Seaton, B.A., & Dedman, J.R. (1998) *BioMetals* 11, 399–404.
254. Gerke, V., Creutz, C.E., & Moss, S.E. (2005) *Nat. Mol. Cell Biol.* 6, 449–461.
255. Kube, E., Becker, T., Weber, K., & Gerke, V. (1992) *J. Biol. Chem.* 267, 14175–14182.
256. Brownawell, A.M., & Creutz, C.E. (1997) *J. Biol. Chem.* 272, 12182–12190.
257. Minami, H., Tokumitsu, H., Mizutani, A., Watanabe, Y., Watanabe, M., & Hidaka, H. (1992) *FEBS Lett.* 305, 217–219.
258. Lewit-Bentley, A., Rety, S., Sopkova-de Oliveira-Santos, J., & Gerke, V. (2000) *Cell Biol. Int.* 24, 799–802.

259. Huber, R., Roemisch, J., & Pacques, E.P. (1990) *EMBO J.* 9, 3867–3874.
260. Swairjo, M.A., & Seaton, B.A. (1994) *Annu. Rev. Biophys. Biomol. Struct.* 23, 193–213.
261. Berendes, R., Voges, D., Demange, P., Huber, R., & Burger, A. (1993) *Science* 262, 427–430.
262. Ayala-Sanmartin, J., Gouache, P., & Henry, J.P. (2000) *Biochemistry* 39, 15190–15198.
263. Concha, N.O., Head, J.F., Kaetzel, M.A., Dedman, J.R., & Seaton, B.A. (1993) *Science* 261, 1321–1324.
264. Campos, B., Mo, Y.D., Mealy, T.R., Li, C.W., Swairjo, M.A., Balch, C., Head, J.F., Retzinger, G., Dedman, J.R., & Seaton, B.A. (1998) *Biochemistry* 37, 8004–8010.
265. Kube, E., Weber, K., & Gerke, V. (1991) *Gene* 102, 255–259.
266. Futter, C.E., Felder, S., Schlessinger, J., Ullrich, A., & Hopkins, C.R. (1993) *J. Cell Biol.* 120, 77–83.
267. White, I.J., Bailey, L.M., Aghakhani, M.R., Moss, S.E., & Futter, C.E. (2006) *EMBO J.* 25, 1–12.
268. Mayran, N., Parton, R.G., & Gruenberg, J. (2003) *EMBO J.* 22, 3242–3253.
269. Vedeler, A., & Hollas, H. (2000) *Biochem. J.* 348, 565–572.
270. Nishizuka, Y. (1992) *Science* 258, 607–614.
271. DiNitto, J.P., Cronin, T.C., & Lambright, D.G. (2003) *Sci. STKE* 213, re 16.
272. Shearwin-Whyatt, L., Dalton, H.E., Foot, N., & Kumar, S. (2006) *Bioessays* 28, 617–628.
273. Dai, H., Tomchick, D.R., Garcia, J., Sudhof, T.C., Machius, M., & Rizo, J. (2005) *Biochemistry* 44, 13533–13542.
274. Sudhof, T.C. (2002) *J. Biol. Chem.* 277, 7629–7632.
275. Sutton, R.B., Davletov, B.A., Berghuis, A.M., Sudhof, T.C., & Sprang, S.R. (1995) *Cell* 80, 929–938.
276. Rizo, J., & Sudhof, T.C. (1998) *J. Biol. Chem.* 273, 15879–15882.
277. Nalefski, E.A., & Falke, J.J. (1996) *Protein Sci.* 5, 2375–2390.
278. Ingham, R.J., Gish, G., & Pawson, T. (2004) *Oncogene* 23, 1972–1984.
279. Plant, P.J., Yeger, H., Staub, O., Howard, P., & Rotin, D. (1997) *J. Biol. Chem.* 272, 32329–32336.
280. Malik, B., Yue, O., Yue, G., Chen, X.J., Price, S.R., Mitch, W.E., & Eaton, D.C. (2005) *Am. J. Physiol.* F107–F116.
281. Snyder, P.M., Olson, D.R., McDonald, F.J., & Bucher, D.B. (2001) *J. Biol. Chem.* 276, 28321–28326.

J. Krebs and M. Michalak (Editors)
Calcium: A Matter of Life or Death

ISSN: 0167-7306/DOI: 10.1016/S0167-7306(06)41004-8

CHAPTER 4

Structural aspects of calcium-binding proteins and their interactions with targets

Peter B. Stathopulos[1], James B. Ames[2], and Mitsuhiko Ikura[1,3]

[1]*Division of Signaling Biology, Ontario Cancer Institute and Department of Medical Biophysics, University of Toronto, Toronto, Ontario, Canada M5G 1L7;*
[2]*Chemistry Department, University of California, Davis, CA 95616, USA;*
[3]*Toronto Medical Discovery Tower, MaRS Centre, 4th Floor, Room 4-804, 101 College Street, Toronto, Ontario, Canada M5G 1L7, Tel.: +1 416 581 7550; Fax: +1 416 581 7564;*
E-mail: mikura@uhnres.utoronto.ca

Abstract

Calcium (Ca^{2+}) sensor proteins are critical and ubiquitous mediators of Ca^{2+} messages in eukaryotic cells. These proteins "sense" local changes in intracellular Ca^{2+} concentrations by undergoing a conformational response to binding or release of Ca^{2+} ions. The structural changes intercede interactions with a specific target protein, ultimately regulating its function. Despite the sequence homology between Ca^{2+}-binding motifs (EF-hands), Ca^{2+} sensors show a remarkable diversity in regulatory mechanism and biological function. This chapter reviews the structural aspects of Ca^{2+} sensor proteins and the various mechanisms of target-binding specificity and regulation. A structural survey of calmodulin, neuronal Ca^{2+} sensor proteins, S100 proteins, penta-EF-hand, and several "other" Ca^{2+} sensors [stromal interaction molecule (STIM), Eps15 homology (EH)-domain-containing, nucleobindin, BM40, and calcium-binding protein 40 (CBP40)] reveal some common themes and important disparities in the molecular mechanisms of target recognition. Diversity of target recognition for each Ca^{2+} sensor is considered in terms of two distinct evolutionary mechanisms.

Keywords: BM40, calcium sensor, calmodulin, CBP40, EF-hand, EH domain, NCS, nucleobindin, penta-EF-hand, S100, STIM, structure, target recognition

1. Introduction

Calcium (Ca^{2+}) is a fundamental signaling messenger in a myriad of cellular phenomena including gene transcription, protein folding and degradation, apoptosis, necrosis, and exocytosis, to name a few [1–3]. Changes in local environmental Ca^{2+} concentrations mediate these phenomena through protein–ligand interactions. Proteins play two universal Ca^{2+}-dependent roles in vivo: (i) they act as regulatory sensors and (ii) they function as environmental buffers of Ca^{2+}, modulating local Ca^{2+} concentrations and acting as transporters of this metal [4,5]. In polypeptide terms, Ca^{2+} sensors can be defined as proteins that undergo a conformational response to binding or dissociation of Ca^{2+}; this Ca^{2+}-induced structural transformation results in the regulation of some specific function through protein–protein interactions. The initial Ca^{2+}-induced conformational change to the Ca^{2+} sensor is intermediate to a subsequent structural change occurring upon target association [6]. The archetypal Ca^{2+} sensor is calmodulin (CaM), which modulates the specific function of a host of catalytic and non-catalytic proteins in vivo through Ca^{2+}-induced CaM–target protein interactions [7]. Ca^{2+} buffers are proteins that bind Ca^{2+}, maintaining or shifting local Ca^{2+} concentrations, but not regulating a function through specific protein–protein interactions. A typical Ca^{2+} buffer is parvalbumin, of which the first high-resolution Ca^{2+}-binding motif structure was solved by X-ray crystallography [8].

The principal Ca^{2+}-binding motif of both sensors and buffers is the EF-hand, first elucidated from the parvalbumin atomic structure [8]. It consists of two short α-helical segments linked through a 12 amino acid loop region; this basic motif, described in detail in Chapter 3 (this volume), confers a single Ca^{2+}-binding site liganded through residues in the consensus loop region [5,9,10]. The basic functional and globular domain in the overwhelming majority of Ca^{2+}-binding proteins (CaBPs) is the EF-hand pair [5,9,10]. Pairing of EF-hands results in cooperativity of Ca^{2+} binding, inducing greater overall affinity and Ca^{2+} sensitivity as well as increased protein stability [11]. EF-hand peptides and even isolated binding loops homodimerize in solution adopting native-like tertiary elements when incubated with Ca^{2+}-containing buffers, suggesting these consensus motifs have an innate pairing ability [12–14].

The scope of this chapter includes the structural aspects of CaBPs and the associated target interactions induced by the liganding. Consequently, the work focuses on Ca^{2+} sensor proteins and the conformational changes modulating their protein–protein interactions. Emphasis is placed on the two mechanisms mediating diverse target recognition: (i) conformational plasticity and (ii) genetic polymorphism [6]. These mechanisms are reviewed with respect to four well-studied protein families [CaM, S100, neuronal Ca^{2+} sensor (NCS), and penta-EF-hand]; additionally, some unique Ca^{2+} sensor proteins, which do not fit into any of the above families, are surveyed.

2. *Calmodulin*

The CaM superfamily, also known as the EF-hand superfamily, is the archetypal and major class of Ca^{2+} sensor proteins, defined by the presence of one or more pairs of EF-hands and made up of more than 600 proteins. CaM is different than most proteins of this superfamily in that there is a very high level of primary sequence conservation from humans to lower eukaryotes. Only four known proteins show greater sequence conservation (histone H4, histone H3, actin B, and ubiquitin) [15]. All mammals encode the same CaM protein, and even CaM of a lower order organism such as *Caenorhabditis elegans* shows 96% sequence identity with mammalian [16]. Other proteins of the CaM superfamily typically exist in multiple isoforms with relatively high sequence diversity. For example, troponin C, which contains two pairs of EF-hands, appears in two different isoforms in skeletal and cardiac muscles and numerous different isoforms in lower eukaryotes [17].

2.1. CaM structural topology

CaM is a 148-residue protein (17 kDa) that folds into two EF-hand domains (N- and C-terminal) linked through a central flexible linker (Lys78-Ser81) [18]. The two EF-hand domains are each made up of a pair of EF-hands where all four EF-hands bind Ca^{2+} with high affinity and cooperativity (K_d ~0.1–1 μM) [19]. The N-terminal domain has 46% sequence identity with the C-terminal domain and each pair of EF-hands forms a mini antiparallel β-sheet from the two EF-hand binding loops [5]. The Ca^{2+}-depleted form of CaM adopts an extended dumbbell-like shape as the central helix is comparatively inflexible; moreover, all four EF-hands adopt a "closed" conformation where each E and F helix is antiparallel to one another (Fig. 1A) [20–23]. Upon Ca^{2+} loading, the interhelical angle in each EF-hand is increased thereby forming a hydrophobic cleft in both domains [23–25]. Even in the presence of Ca^{2+}, the central helix linker region remains flexible and non-helical near its midpoint; this flexibility is thought to play a role in CaM plasticity for target recognition (Fig. 1B) (see Section 2.3.) [23,26,27].

2.2. CaM target proteins

The first sensory role of CaM was found in 1970 with the discovery that CaM regulates nucleotide phosphodiesterase [28,29]. Soon thereafter, CaM was found to regulate numerous other proteins including myosin light chain kinase [30], phosphorylase kinase and other protein kinases [31], calcineurin [32], and nitric oxide synthase [33]. CaM also regulates the activity of membrane transporters such as inositol 1,4,5-triphosphate receptor [34], ryanodine receptor, the plasma membrane Ca^{2+} pump, and L-type Ca^{2+} channels [35]. Additionally, CaM plays a role in the regulation of bacterial enzymes such as adenylyl cyclase in *Bacillus anthracis* and *Bordetella pertussis* [36]. The current list of documented CaM targets is greater than 300, and it is certain that this number will continue to climb. A useful summary of many targets of CaM can be found at the CaM target protein database (http://www.calcium.uhnres.utoronto.ca) [6,37].

2.3. Conformational plasticity for target recognition by CaM

When the structure of Ca^{2+}-loaded CaM complexed with a peptide from myosin light chain kinase was solved, the capability of CaM to undergo a large conformational change for target recognition was appreciated [38,39]. This structure revealed that a portion of the central helix region of CaM is highly flexible (residues 78–81); furthermore, this region can be dramatically bent upon binding to a target protein (Fig. 1B) [23,27,40,41]. The flexibility of the central helix domain not only permits a more fitted interaction within domain linker region of CaM, but also allows the two domains of CaM to change independently in order to accommodate the three-dimensional characteristics of the peptide [6]. The CaM–peptide complex also demonstrates that two residues bind simultaneously to the hydrophobic clefts of

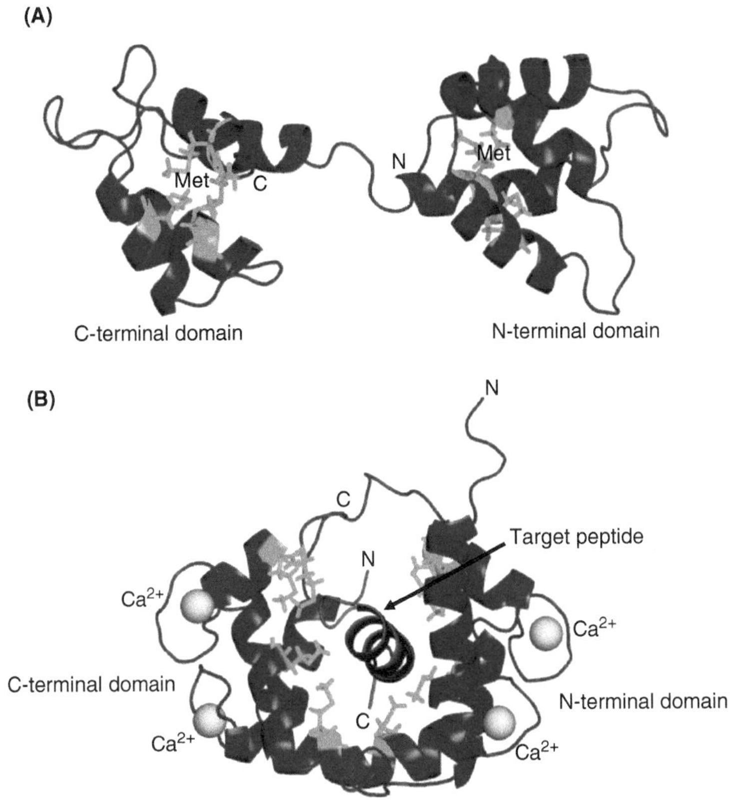

Fig. 1. The effects of Ca^{2+} and target binding on calmodulin (CaM) structure. **(A)** Solution structure of Ca^{2+}-depleted CaM (1DMO.pdb). The blue ribbons show the backbone secondary structural elements of CaM. The green stick representations show the key Met residues (Met36, 51, 71, 72, 109, 124, 144, and 145) involved in "fine-tuning" specific target recognition for CaM. The helices making up each respective EF-hand in the N- and C-terminal domains are in a "closed" antiparallel conformation. The overall shape of apo CaM is dumbbell-like. **(B)** Solution structure of Ca^{2+}-loaded CaM complexed with a peptide from myosin light chain kinase (2BBN.pdb). Upon liganding of Ca^{2+} (yellow spheres), the inter-helical angles in each EF-hand "open" and the central helix region become much more flexible. The target peptide (red ribbon) is accommodated by a drastic shape change in CaM. N and C represent NH_2 and COOH termini, respectively (See Color Plate 10, p. 509).

the N- and C-terminal CaM domains; these clefts are rich in methionine residues (on average, proteins contain 1% methionine enrichment, whereas CaM contains 6%) [6]. The methionine residues confer a microscopic flexibility to the N- and C-terminal hydrophobic clefts of Ca^{2+}-loaded CaM; moreover, the flexibility of clefts allows the accommodation of many different amino acid residues from target proteins (i.e., Trp, Phe, Ile, Leu, Val, and Lys) [6].

Mutations to Leu at different Met sites in CaM significantly alter CaM's ability to stimulate cAMP phosphodiesterase [42]. Additionally, other CaM–target structures demonstrate the vital role of CaM Met residues in the recognition of CaMKII [26],

CaM kinase kinase [43], NO synthase [44], glutamate decarboxylase [45], MARCKS [46], Ca^{2+}-activated K^{+} channel [47], and anthrax adenylyl cyclase [48]. Overall, it is both the macroscopic conformational plasticity afforded by the central helix region of CaM and the important microscopic flexibility ("fine-tuning") of CaM conferred by the methionine residues in the N- and C-terminal hydrophobic clefts (Fig. 1) that allow CaM to recognize and bind the multitude of proteins with high specificity [6].

2.4. Transcriptional regulation of CaM

In mammals, there are three CaM genes that are transcribed into eight different mRNAs; ultimately, all the mRNAs encode identical 148 amino acid proteins [6]. The different CaM mRNAs are targeted into different cellular compartments, allowing for specific enrichment of the protein at different specific sites within the cell [6,49]. Post-translational modifications of CaM include acetylation, trimethylation, carboxymethylation, proteolysis, and phosphorylation; the precise role of these modifications is not known [50–52]. The post-translational modifications and differential expression of CaM and its targets constitute another mechanism by which CaM–target interactions are modulated.

3. Neuronal Ca^{2+} sensor proteins

The NCS proteins are expressed primarily in the central nervous system. Proteins of this family are approximately 200 amino acids in length and encode four EF-hand motifs. Most NCS proteins posses an N-terminal myristoylation consensus sequence [53,54].

3.1. NCS common structural topology

Unlike CaM, which folds into a dumbbell conformation, NCS proteins pack into a single globular fold [6]. Of the four EF-hands, only two or three of the binding loops coordinate Ca^{2+} with μM or sub-μM affinity, depending on the protein [53]. Recoverin is the best structurally characterized NCS protein. In the Ca^{2+}-free state, the recoverin myristoyl group is protected in a hydrophobic pocket of the protein (Fig. 2A); Ca^{2+} loading of EF-hands 2 and 3 results in a conformational switch that mediates the extraction of the buried lipid moiety [53,55,56]. The myristoyl group is free in the Ca^{2+}-loaded state to interact with the membrane (Fig. 2B).

3.2. NCS target proteins

Recoverin is found in retinal rod cells; this protein controls the desensitization of rhodopsin through the regulation of rhodopsin kinase activity [56–59]. Retinal cells also express guanylate cyclase-activating proteins (GCAPs); GCAPs are NCS

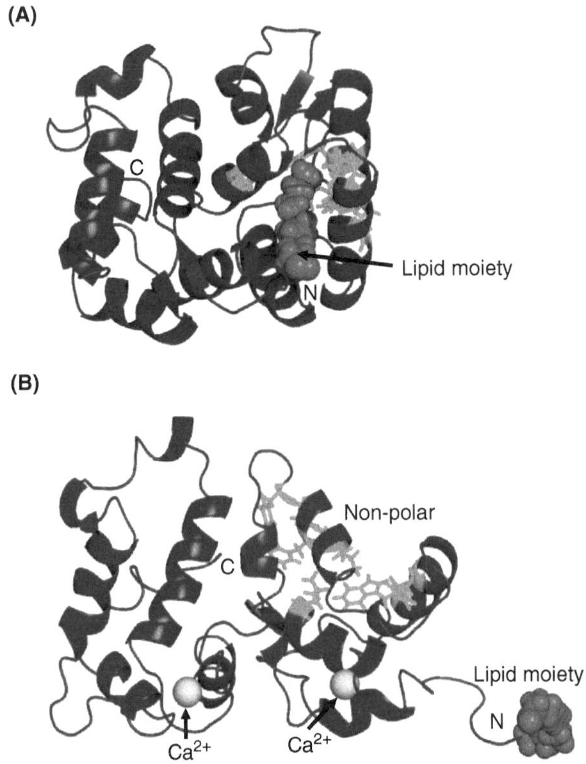

Fig. 2. The effects of Ca^{2+} on the structure of myristoylated recoverin. **(A)** Solution structure of Ca^{2+}-depleted recoverin (1REC.pdb). The blue ribbons show the backbone secondary structural elements of recoverin. The green stick representation show conserved hydrophobic residues (3 × Phe, Tyr, Leu, Trp) providing the main scaffolding for the myristoyl moiety (magenta spheres). **(B)** Solution structure of Ca^{2+}-loaded recoverin (1JSA.pdb). Specific residues on the C-terminal portion of recoverin provide the "fine-tuning" for target specificity in the neuronal Ca^{2+} sensor (NCS) proteins. In the case of recoverin, Ca^{2+} induces a large conformational change which results in the exposure of the hydrophobics and release of the myristoyl group. Interaction of the myristoyl moiety with lipid membranes directs recoverin for the regulation of membrane-bound proteins. N and C represent NH_2 and COOH termini, respectively (See Color Plate 11, p. 510).

proteins that activate guanylate cyclase at low Ca^{2+} levels [60]. Frequenin is expressed in nearly all neuronal tissue types where it regulates neurotransmission and PI(4) kinase activation; frequenin homologs are also expressed in invertebrates [61–64]. K^+ channel-interacting proteins (KChIPs) are expressed in brain and spinal cord tissues serving to regulate the gating of shaker K^+ channels [65]. Neurocalcin is expressed in both brain and retinal tissues and is involved in endocytosis [53,66]. DREAM/calsenelin (KChIP3) is expressed in brain and spinal cord tissues where it regulates K^+ channels, binds presenilin, represses transcription in association with pain, and acts as a proapoptotic factor [67–70]. Hippocalcin is expressed in the brain serving to activate phospholipase D and regulate MAP kinase signaling; hippocalcin also possesses anti-apoptotic activity [6,53,71]. Visinin-like proteins (VILIPs) are also expressed in the brain and retinal tissues activating guanylyl cyclase and trafficking nicotinic receptors [6,53].

3.3. Target recognition of NCS proteins

The N-terminal half of NCS proteins contain conserved hydrophobic residues; in the case of recoverin, these residues are responsible for docking of the myristoyl moiety in the Ca^{2+}-free form [55,56,72]. Exposed hydrophobic residues in Ca^{2+}-loaded recoverin (Phe, Trp, Tyr, and Leu) are directly involved in target recognition, forming intermolecular contacts with targets [73–76]. For example, the crystal structure of KChIP1 bound to a target protein helix shows very similar hydrophobic interactions as observed in the N-terminal CaM interaction with myosin light chain kinase peptide (Fig. 1B) [6,77]. However, unlike CaM, NCS proteins have evolved for recognition of different targets through a mechanism of genetic polymorphism. Specific surface properties are believed to play a role in target recognition specificity and are quite unique for each NCS protein. For example, frequenin displays hydrophobic residues on its C-terminal region that adjoin a hydrophobic crevice in the N-terminal region, forming one elongated hydrophobic channel that is recognized by specific targets [6]. At the same time, recoverin has specific charged residues on the C-terminal half of the protein, while neurocalcin and KChIP1 contain mostly neutral residues [6]. Hence, the C-terminal portion of each individual NCS protein mediates specific target recognition in NCS proteins through differences in charge distribution as well as hydrophobicity, while the N-terminal portion of NCS protein is intimately involved in target scaffolding chiefly through hydrophobic contacts [6].

3.4. Transcriptional regulation of NCS

The expression of NCS proteins occurs in many different cell types and in specific locations within each cell. Recoverin and GCAPS are found in rod and cone retinal cells [78,79]; hippocalcin is localized in hippocampal pyramidal neurons, and VILIP-3 is found in cerebellar Prkinje and granule cells [6,53]. The myristoyl group itself targets many NCS proteins to membranes and promotes the interactions with membrane-bound targets [56,57]. DREAM is exclusively in cerebellar granular cortex [80]. Splice variants of KChIPs result in targeted expression in various sensory and cardiac cells [81]. Tissue-specific expression and N-terminal myristoylation help confer the distinctive regulatory activities of the various NCS proteins [6].

3.5. NCS-like proteins

CIB1 (Ca^{2+} and integrin binding protein; CIB, calmyrin, KIP) is a ubiquitously expressed 22-kDa protein that contains four EF-hand motifs, coordinates two Ca^{2+} ions with μM or sub-μM affinities, and is N-terminally myristoylated [82–86]. CIB1 binding targets include integrin αIIb [87], Rac3 [83], Pax3 [88], presenilin2 [86], and

IP_3R Ca^{2+} release channel [85]. CaBPs are predominantly expressed in brain and retinal tissue and encode four EF-hands; CaBP1 and CaBP2 contain N-myristoylation consensus sequences [89]. CaBP1 exists as a globular homodimer of two 19.4-kDa subunits, regulating neuronal Ca^{2+} channels including IP_3R [90], P/Q-type voltage-gated [91], L-type [92,93], and TRPC5 [94]. Ca^{2+} binding occurs on the C-terminal EF-hands with μM affinity [85,95]. A protozoan flagellum CaBP (FCaBP) that is N-myristoyl and palmitoylated is endogenous to *Trypanosoma cruzi*, binding to the inner flagellar membrane through the acyl moieties [96,97]. The approximately 24-kDa protein encodes four EF-hands; EF-hand 3 binds Ca^{2+} with high affinity (~9 μM) while EF-hand 4 binds Ca^{2+} with a low affinity (~120 μM) [98]. Calcineurin is a heterodimeric Ser/Thr protein phosphatase made up of a catalytic domain (calcineurin A) and a regulatory domain (calcineurin B) [99]. Calcineurin A is approximately 58–64 kDa and is also a CaM sensory target. Calcineurin B is approximately 19 kDa, encodes four EF-hands, and binds two Ca^{2+} with very high affinity (nM) and two additional Ca^{2+} ions with μM affinity [32,100,101]. Calcineurin B is N-terminally myristoylated; however, the lipid moiety is not involved in membrane targeting but is an important stabilizing element in the calcineurin A–calcineurin B heterodimer [102,103]. Every NCS-like protein is N-terminally acylated, encodes four EF-hand motifs (binding between two and four Ca^{2+} ions), and is approximately 20 kDa in the monomeric form; structurally, CIB1 is similar to members of the NCS family (i.e., KChIP1) and to calcineurin B [82]. Despite the similarities mentioned, the functions of NCS-like proteins are all very different and non-overlapping.

4. *S100 protein family*

The proteins that make up the S100 protein family are known to regulate various aspects of the cell cycle, transcription, secretion, cell growth, motility, and differentiation [6,104]. S100 proteins are linked to various pathologies including Alzheimer's, Down syndrome, multiple sclerosis, rheumatoid arthritis, and cancer [6]. Thus far, at least 25 members of the S100 protein family have been identified, sharing 25–65% sequence identity [104]. The large number of S100 proteins in this family implicate a mechanism of genetic polymorphism in the evolution of S100 proteins for multi-target interactions and regulatory function.

4.1. *S100 common structural topology*

S100 proteins are relatively small with a molecular weight of approximately 10–12 kDa per monomer. S100 monomers have an overall acidic character and are made up of two EF-hands; one EF-hand is canonical exhibiting a relatively high Ca^{2+} affinity (K_d ~10–50 μM) [105]. A second EF-hand ("pseudo-EF-hand") in S100 proteins is non-canonical made up with a 14-residue-binding loop that coordinates Ca^{2+} with relatively weak affinity (K_d ~200–500 μM) [104,105]. S100 proteins form homodimers and heterodimers, depending on the relative local concentrations of S100 subunits [104]. The C-terminal canonical EF-hand is separated from the N-terminal "pseudo-EF-hand" by a relatively flexible linker region [106]. Each subunit contributes helix 1 of the pseudo-EF-hand (N-terminal) and helix 2 from the canonical C-terminal EF-hand to the formation of a symmetric dimer interface [104]. The canonical EF-hand on the C-terminus adopts a "closed" conformation in the Ca^{2+}-depleted state (Fig. 3A) [107]. Upon Ca^{2+} loading, the canonical EF-hand undergoes a significant conformational change to an "open" state; the magnitude of opening is similar to the changes observed for CaM (Fig. 3B) [104,108]. The Ca^{2+}-induced structural change results in the exposure of hydrophobic surface [6,104].

4.2. *S100 target proteins*

More than 90 target interactions have been identified for the S100 proteins. S100A1 forms complexes with aldolase C, annexins, SIP1, caldesmon, capZ, GFAP, RAGE, F-actin, p53, SERCA2a, titin, twitchin kinase, and synapsinI [104]. S100A2 binds p53, tropomyosin, and ΔNp63 and is involved in regulation of transcription and tumor suppression [109]. S100A4 and S100A6 are tumor promoting [110,111]. S100B was the first protein identified to interact with p53 [112]. Many S100 proteins interact with cytoskeleton components including actin, myosin, tropomyosin, tubulin, and intermediate filaments, suggesting an important role in cellular architecture [104].

4.3. *Target recognition of S100 proteins*

Helix 2 of the canonical EF-hand in S100 proteins plays a key role in docking of target molecules. Residues near the proximal end of the C-terminus are non-helical in the Ca^{2+}-depleted state; however, upon Ca^{2+} loading, helix 2 adopts an extended structure, thereby accommodating the target (Fig. 3C) [104]. Induction of a helical conformation in the target protein is also important for the hydrophobic interaction between the S100 protein and target (i.e., annexin, p53, and NDR kinase) [6,104,112]. Target helices from p53 and annexin contain an amphipathic hydrophobic face that interacts with a complementary hydrophobic crevice on helix 2 of the canonical EF-hand [6]. S100 proteins differ in their target specificity by encoding different charged residues. For example, S100B contains numerous anionic residues on the protein surface that are balanced by a similar number of cationic residues [6]. S100A10 has fewer charged residues and is more abundant in hydrophobics [6]. Specific orientations of target molecules are dictated by the different distribution of charged and hydrophobic residues. The highly charged environment of S100B may promote a specific target orientation through electrostatic and/or hydrogen bonding, whereas the lack of surface charges in S100A10 result in an altered orientation. The mechanism of target binding and specificity of the S100 family and the overall structure of the S100 family are similar to the NCS protein family. Conserved amino acids within S100 proteins serve as a scaffold for target binding; the hydrophobic scaffold is adapted by specific residue type on the surface of each individual protein for unique target recognition [6].

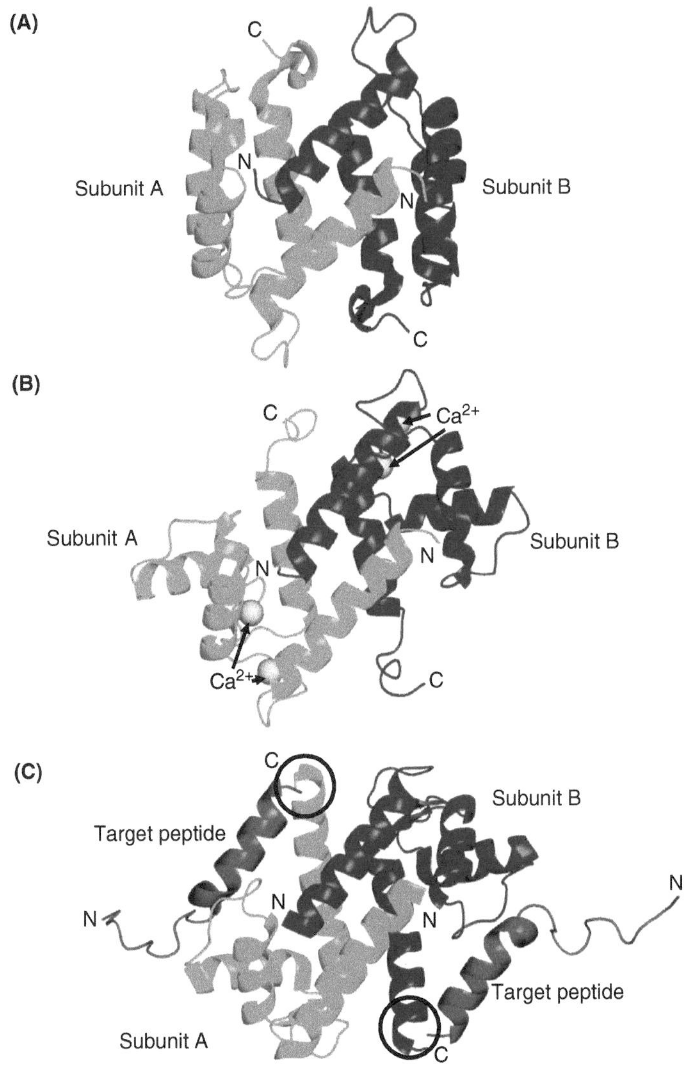

Fig. 3. The effects of Ca^{2+} on the structure and target recognition of S100B. **(A)** Solution structure of homodimeric and Ca^{2+}-depleted S100B (1B4C.pdb). The blue ribbons represent the secondary structural elements defined by the backbone conformation of one subunit (subunit B), while the green ribbon follows the conformation of the dimeric subunits (subunit A). In total, the dimer is made up of four EF-hands where each apo motif is in the "closed" conformation. **(B)** Solution structure of Ca^{2+}-loaded S100B (1QLK.pdb). Each S100B dimer is capable of coordinating four Ca^{2+} ions (yellow spheres). Upon Ca^{2+}-loading, S100B undergoes a large conformational change with EF-hands moving to a relatively "open" conformation. **(C)** Solution structure of Ca^{2+}-loaded S100B complexed with a peptide of the N-terminal domain of NDR kinase (1PSB.pdb). Residues near the C-terminus of each subunit form an extended helical conformation upon Ca^{2+} loading (open circle), thereby accommodating and promoting target interactions. The target protein also undergoes a conformational change upon forming an S100–target protein complex. N and C represent NH_2 and COOH termini, respectively (See Color Plate 12, p. 511).

4.4. *Transcriptional regulation of S100 proteins*

In some cases, different S100 proteins regulate the activity of the same target molecule. This observation taken together with the fact that there is differential expression of the S100 proteins depending on tissue type suggests that transcriptional control of the S100 proteins plays a regulatory role in the diversity of S100 protein sensory function. For example, S100A1 is expressed exclusively in cardiac cells controlling contractility [113]; S100A2, S100A6, and S100B are expressed in the nucleus, involved in the regulation of transcription [109]; S100A7 is extracellularly targeted and has been linked to epidermal inflammatory diseases [114]; S100A8 and S100A9 form heterodimers and act extracellularly as chemotactic molecules in inflammation [115]; S100A10 and S100A11 regulate annexin trafficking to membranes [108,116]; S100B is expressed in brain tissue [5].

5. *The penta-EF-hand family*

EF-hand motifs are generally found in pairs [5,9,10]. As the name of this family suggests, the penta-EF-hand family of proteins encode five EF-hand motifs; however, pairing of the fifth EF-hand occurs through heterodimerization or homodimerization of these proteins. All proteins of this family are made up of eight α-helices, contain variable-length hydrophobic Gly/Pro-rich regions N-terminally, and undergo translocation to membranes in a Ca^{2+}-dependent manner [117]. Proteins making up the penta-EF-hand family include the sorcin, grancalcin, Ca^{2+}-dependent calpain proteases, apoptosis-linked gene-2 (ALG-2), and peflin. The multiple members of this protein family which bind to and regulate diverse targets implicate a mechanism of genetic polymorphism in their evolution. This is best exemplified by grancalcin and sorcin, which exhibit >60% sequence identity but interact with different targets.

5.1. *Sorcin*

Sorcin is a soluble resistance-related CaBP of 22 kDa [117,118]. Upon Ca^{2+} loading, sorcin interacts reversibly to ryanodine receptors and inhibits further Ca^{2+} release of these receptors, which are activated on the sarcoplasmic reticulum in response to low level Ca^{2+} release from L-type Ca^{2+} channels [119–121]. Sorcin serves to rapidly attenuate Ca^{2+} release from ryanodine receptors so that excessive depletion of Ca^{2+} stores does not occur [117]. Other targets of this protein include the α_1-subunit of L-type Ca^{2+} channels in muscle cells, neuronal presenilin-2, and annexin-VII in the brain, differentiating myocytes and erythrocytes [117].

Within each subunit, EF-hand 1 forms a pair with EF-hand 2, and EF-hand 3 pairs with EF-hand 4. Dimerization results in pairing of the fifth EF-hand (Fig. 4A) [118,122]. At physiologically relevant Ca^{2+} concentrations, sorcin binds two Ca^{2+} ions with relatively high affinity (K_d ~1 μM), although there has been some debate over which EF-hands are the physiological Ca^{2+}-binding pair (i.e., EF-hands 1 and 2

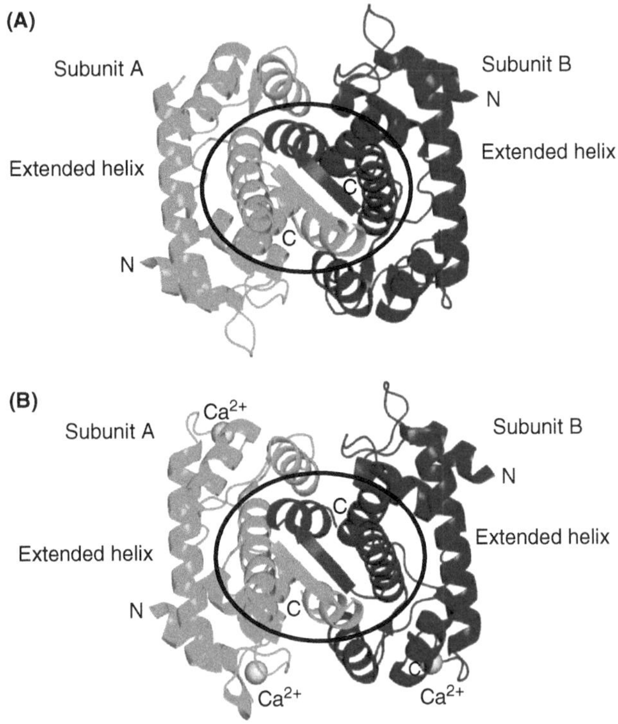

Fig. 4. The effects of Ca^{2+} on the structure of grancalcin. **(A)** Crystal structure of N-terminally truncated (52 residues) grancalcin in the Ca^{2+}-depleted state (1F4Q.pdf). The penta-EF-hand proteins pair all five EF-hands through dimerization. The green ribbons delineate the secondary structure of one grancalcin subunit, while the blue ribbon shows the structure of the homodimerization partner. A short antiparallel β-sheet is formed from the loop region of each EF-hand motif contributed from each subunit. The open circle shows that the interface of grancalcin is made up primarily from the fifth EF-hand of each subunit. A long, "extended helix" makes up the F and E helices of two consecutive EF-hand motifs within each subunit. **(B)** Crystal structure of N-terminally truncated (52 residues) grancalcin in the Ca^{2+}-loaded state (1K94.pdb). Upon Ca^{2+} loading (yellow spheres), only minor structural changes are observed in most penta-EF-hand proteins. Nonetheless, the conformational change propagates through the entire protein to the common Gly/Pro-rich N-terminus; target interactions have also been identified at other specific locations. N and C represent NH_2 and COOH termini, respectively (See Color Plate 13, p. 512).

or EF-hands 2 and 3) [118,122,123]. The binding loop of EF-hand 1 is non-canonical as it only contains 11 residues; EF-hands 2 and 3 contain the consensus 12 residues, while the non-canonical EF-hand 4 has a two-residue deletion and EF-hand 5 has a two-residue insertion [117]. Ca^{2+} loading of EF-hands 1 and 2 results in a relatively small structural change associated with increased hydrophobicity [122,123]. In the case of the ryanodine receptor interactions, the C-terminal (EF-hand-containing) domain of sorcin is important for specific target binding; however, in the case of annexin-IV, the N-terminal Gly/Pro-rich region is critical [118]. Thus, sorcin has evolved multiple domain interaction sites for different target sensory associations that are buttressed by a Ca^{2+}-induced increase in hydrophobicity.

5.2. Grancalcin

After post-translational cleavage of 14 amino acids, grancalcin is a 203 amino acid functional protein of 22 kDa [124]. Ca^{2+} induces grancalcin binding to vesicles and membranes [117,125,126]. L-plastin, another EF-hand-containing protein, is a target of grancalcin. Plastins have actin-bundling activities; hence, the grancalcin/L-plastin interactions, which occur in the absence of Ca^{2+}, may serve to regulate the activity of plastin [124]. Accordingly, grancalcin/L-plastin interactions are inhibited in the presence of Ca^{2+} [124]. An interaction between grancalcin and sorcin has been demonstrated, the significance of which is not known [127].

Each grancalcin monomer has two Ca^{2+}-binding sites on EF-hand 1 and EF-hand 3 [126]. Ca^{2+} affinity has been estimated in the μM range [124,128]. There are only minor structural changes associated with Ca^{2+} loading of grancalcin (Fig. 4B); a change occurs in the orientation of EF-hand 5 which ultimately affects the relative positions of EF-hand 1 of each subunit [125,126]. Binding of Ca^{2+} to EF-hands 1 and 3 reorganizes the loops of the protein. N-terminally truncated grancalcin is much more soluble in the presence of Ca^{2+} than full-length suggesting a role for the N-terminal region in Ca^{2+}-dependent regulatory interactions [126].

5.3. Calpain

Several Ca^{2+}-dependent cysteine proteases possess penta-EF-hand domains; the approximately 80-kDa calpain heavy chains (CAPN1, 2, 3, 8, 9, 11, and 12) and the approximately 28-kDa light chain (CAPN4) each contain penta-EF-hands [117,129]. Calpain heavy and light chains comprise several tissue-specific calpains and two ubiquitously expressed isoforms (calpain 1 and 2) [129]. The ubiquitous isoforms have been labeled μ- and m-calpains because they require μM and mM concentrations of Ca^{2+} for activation, respectively [130]. μ-Calpains and m-calpains are heterodimers of heavy and light chain calpain molecules. The heavy chains are made up of four domains (dI, dII, dIII, and dIV), while the light chains are made up of two additional domains (dV and dVI) [130,131]. dIV of the heavy chain and dVI of the light chain molecules make up most penta-EF-hand dimers. However, recent evidence suggests that the skeletal muscle-specific calpain (calpain 3) homodimerizes through the penta-EF-hand domain of heavy chains (i.e., there is no small subunit in the quaternary structure) [131].

Of the five EF-hands, crystal structures reveal Ca^{2+} binding only on EF-hands 1, 2, and 3. As with the other penta-EF-hand proteins, Ca^{2+} binding seems to cause a relatively small conformational change, localized to the EF-hand-containing domains of these proteases [129]. The largest structural change occurs within EF-hand 1. Ca^{2+} binding may occur in other non-EF-hand-containing domains in order to initiate protease activity of the calpains, as the catalytic residues are located in dII and must undergo a more dramatic realignment for activation than construed by the penta-EF-hand domain upon Ca^{2+} loading [132,133].

Both μ- and m-calpains interact with calpastatin in the presence of Ca^{2+}. Calpastatin inhibits calpain proteolytic activity by competition with other target proteins [134]. Specifically, two different calpastatin domains (A and C) are capable of binding to the penta-EF-hand domains (dIV and dVI); additionally, the central calpastatin domain B binds to dII and dIII of calpain. Gas2 has also been found to inhibit calpain activity [135–138]. The penta-EF-hand dIV of calpain interacts with IκBα in order to facilitate degradation, while dVI targets cytokine receptor γ as a substrate [117,139,140]. Penta-EF-hand domains play important roles in calpain proteolytic initiation, regulation as well as target recognition and binding.

5.4. *Apoptosis-linked gene-2*

ALG-2 is a penta-EF-hand protein that promotes apoptosis independently of caspase-3 [141]. It is made up of 191 amino acids and folds into the typical α-helical penta-EF-hand topology of 22 kDa [141]. Consistent with all members of the penta-EF-hand family, EF-hand 1 is formed from α-helix 1 and 2 and the loop that joins them; α-helix 3 and the half of α-helix 4 make up EF-hand 2; the remaining half of α-helix 4 and α-helix 5 compose EF-hand 3; α-helix 6 and the N-terminal half of α-helix 7 form EF-hand 4; finally, the C-terminal half of α-helix 7 and α-helix 8 compose EF-hand 5 (Fig. 4) [142,143]. Also typical of all proteins of this family, two β-sheets (composed of two short strands each) stabilize each pair of EF-hands within each subunit and EF-hand 5 of two separate macromolecules form an EF-hand pair (and short β-sheet) through dimerization (Fig. 4) [142,143].

ALG-2 binds two Ca^{2+} ions with high affinity, and a third Ca^{2+} ion may be bound less tightly. EF-hand 1 and EF-hand 3 are the two high-affinity sites, while EF-hand 5 is the low-affinity site [142]. Unlike other penta-EF-hand proteins, the binding loop of EF-hand 1 contains the canonical 12 residues (instead of 11 residues) where Ca^{2+} is coordinated in the archetypal pentagonal bypyrimidal geometry [142,143]. However, EF-hand 5 is similar with other penta-EF-hand proteins in that there is a two-residue insertion into the loop connecting adjacent helices.

The best characterized target protein of ALG-2 is ALG-2-interacting protein (AIP1) [117]. Binding with AIP1 occurs in a Ca^{2+}-dependent manner [143]. AIP1 is a 98-kDa protein that plays a regulatory role in cell growth, apoptosis, and signal transduction through protein–protein interactions with ALG-2 as well as other proteins [143]. In the crystal structure of ALG-2, a decapeptide rich in Gly/Pro was found associated to a hydrophobic cleft between the N- and C-termini of ALG-2. The peptide was the likely product of N-terminal proteolysis of full-length ALG-2 [142]. Ca^{2+} loading of ALG-2 results in exposure of the hydrophobic cleft, binding of the N-terminal region to the cleft and a large conformational change (in contrast to the other penta-EF-hand members), ultimately mediating target recognition. Other proteins that interact with ALG-2 include annexin VII and XI, as well as Fas [117,144]. The annexins interact with ALG-2 through their N-terminal extension sequences that are rich in Gly/Pro just as AIP1 [117,145,146].

5.5. *Peflin*

Peflin is distinct from the other members of the penta-EF-hands in that this 30-kDa protein contains the longest N-terminal hydrophobic region [147]. Accordingly, this protein was named peflin due to the fact that it is a penta-EF-hand protein with a long N-terminal hydrophobic domain. The importance of the N-terminal region in penta-EF-hand proteins in membrane targeting and/or target protein interactions has been well documented [117]. Although the precise function of peflin in eukaryotes has yet to be determined, peflin readily associates with ALG-2, suggesting a regulatory role for peflin in ALG-2 function [148,149]. The large hydrophobic N-terminal region of peflin is not involved in this association as N-terminal deletion mutants are capable of interacting with ALG-2 [148,149]. Because other penta-EF-hand proteins heterodimerize through the fifth EF-hand (i.e., calpains), it is likely that ALG-2–peflin associations also occur in this manner [148].

6. *Other*

Currently, there are over 1500 MEDLINE, 4500 NCBI nucleotide, and 180 PDB structure hits associated with "EF-hand." While many of the hits are related to proteins of the four major sub-families discussed above, some of the proteins are somewhat "miscellaneous" in terms of biology, structure, and function. This "miscellaneous" section reviews some of the unique properties associated with these proteins along with the respective mechanism of target interactions.

6.1. *Eps15 homology domain*

The Eps15 homology (EH) domain was first identified as a segment repeated three times in the sequence of the Eps15 protein [150,151]. EH domains are encoded in many proteins associated with endocytosis and vesicle transport [152–157], cytoskeletal function [152,153], and tyrosine kinase-signaling pathways [158]. EH-domain-encoding proteins also exist in lower eukaryotes [159]. Proteins of this family include Eps15, Eps15R, Reps1, POB1, GAMMA-SYN, EHS-1, EDE1, PAN-1, SAGA, and intersectin [159]. The EH domain is a protein–protein interaction module and interacts with target proteins containing the Asn-Pro-Phe (NPF), Asp-Pro-Phe (DPF), Trp/Phe-Trp (W/FW), or Ser-Trp-Gly (SWG) sequences [157,159,160]. Proteins targeted by EH domains include epsine, SCAMP1, Hrb, Hrb1, AP-2 binding protein, Ent1p, Sla1p, RAB, synaptojanin1, NUMB, and yAP180A [159]. The EH domain is made up of approximately 100 amino acids residues; furthermore, within that stretch there are two EF-hand motifs, allowing for EF-hand pairing. However, only a single Ca^{2+} ion is liganded in the two EF-hands [157,161].

Because the Ca^{2+}-binding affinities of these proteins are relatively low compared with archetypal Ca^{2+} sensors (i.e., CaM), the physiological relevance of the Ca^{2+}

binding is in question (intracellular Ca^{2+} concentrations have been estimated at much lower concentrations than the apparent binding affinities of EH domains). Nonetheless, it is clear that Ca^{2+} effects the structure of the EH domain significantly. For example, Ca^{2+} is a necessary structural determinant in the EH domain-containing protein POB1 as the EH domain of POB1 is unstructured and aggregated in the apo state [157]. The solution structure of apo mouse EH1 domain of Eps15 is also considerably different than the Ca^{2+}-loaded human EH2 domain of Eps15 and the Ca^{2+}-loaded mouse EH1 domain suggesting that Ca^{2+} dictates distinctive structure in this class of proteins that ultimately may be related to function [157,159–161].

The general structure of EH domains is characterized by two helix-loop-helix motifs dictated by the two EF-hands. Seven residues involved in packing of the hydrophobic core of EH domains are highly conserved; five of the seven are Phe and form a structural core [160]. The interactions maintained by these core hydrophobic residues allow the different EH domains to adopt similar overall three-dimensional structures. The main structural differences in the EH domain family are the positions of the helices and the position of the Ca^{2+}-free loop. Only three of the four helices superimpose moderately well in known EH domain structures, and the position of the C-terminal Pro-rich regions shows large discrepancies [160].

The general mechanism of the EH-domain-mediated protein–protein interaction seems to be related to conserved hydrophobic residues in a binding pocket formed by two of the four helices in EH domains (helices B and C) [162,163]. These non-polar residues interact with the target consensus sequences (NPF, DPF, W/FW, and SWG). A gate formed by two charged residues over the hydrophobic binding pocket (one anionic and one cationic) plays a role in binding specificity [160,162,163]. The relative distance of these charged residues are different for two EH domain isoforms (Reps1 EH and Eps15 EH2), and conformation is dictated by the neighboring residue-type [160]. Proteins containing EH domains mediate specific protein–protein interactions through different binding pocket conformations that are controlled by different specific residues. This class of EF-hand proteins, therefore, has evolved different target specificity through a mechanism of genetic polymorphism that is analogous to the NCS and S100 proteins.

6.2. Calcium-binding protein 40

Calcium-binding protein 40 (CBP40) was first purified from the acellular slime mould, *Physarum polycephalum* [164–166]. In the presence of μM Ca^{2+}, CBP40 self-associates into large oligomers. These aggregates localize beneath the plasma membrane and form a diffusion barrier between the cytoplasm and the extracellular environment upon plasma membrane damage [165,167]. Normal intracellular Ca^{2+} concentrations for *P. polycephalum* are approximately 0.1 μM; however, upon plasma membrane damage, intracellular Ca^{2+} rises to μM levels. CBP40 senses this increase in Ca^{2+} and undergoes a small conformational change, nonetheless inducing homotypic aggregation and oligomerization [165]. CBP40 may also mediate inhibition of myosin light chain kinase in this mould [164]. CBP40 is a multi-domain protein made up of 355 amino acids that contain four dissimilar EF-hand motifs (approximately 20% sequence identity) following the 218 amino acid N-terminus. The first 32 amino acids on the N-terminus are critical for aggregation as truncation inhibits homotypic interactions [165]. *Physarum polycephalum* encodes an endogenous proteinase that facilitates this cleavage [165]. CBP40 is comprised of three structural domains made up solely of helices. The N-terminal domain folds into two long helices making up a coiled-coil region. An intervening domain immediately adjacent to the N-terminal domain contains six additional helices [167]. The EF-hand domain is located C-terminally [166,167].

The EF-hand domain topology resembles that of the penta-EF-hand family as the two EF-hand pairs are linked by only a single residue. Two of the four EF-hands are high affinity Ca^{2+} sites (EF-hands 1 and 2) and under normal physiological Ca^{2+} concentrations are occupied [165]. Upon Ca^{2+} liganding in the lower affinity sites, the protein undergoes a conformational change that propagates from the EF-hand domain all the way to the coiled-coil region [165,167,168]. In general, the E and F helices of EF-hand-containing proteins adopt one of three orientations relative to one another, and the two EF-hands of a pair assume similar orientations (Figs 1–3). The angle between E and F helices in the "open" conformation is approximately 90°, while helices that are antiparallel to one another are in the "closed" conformation and EF helices that adopt angles in between the "open" and "closed" conformations are "semi-open." One unique structural feature of CBP40 is the fact that EF-hands making up each pair adopt different helix conformations (i.e., EF-hand 1 is "open," 2 is "closed," 3 is "semi-open," and 4 is "closed") [167,168]. Although the precise mechanism of aggregation has yet to be elucidated, residues 1–32 may confer an extended coiled-coil to the protein for helical bundle-induced oligomerization [167]. The Ca^{2+}-induced helix extension is reminiscent of changes observed in S100B (Fig. 3C).

Despite the overall similarity in topology of the EF-hand domain in CPB40 and penta-EF-hand proteins, there is very little primary sequence homology, suggesting distinct evolutionary lineages [167]. However, similar polymorphic events may have resulted in comparable gene duplications as the linker regions are equally short (i.e., 1 residue).

6.3. Nucleobindin (calnuc)

Nucleobindin is capable of playing both universal Ca^{2+}-dependent roles: sensory and buffering. Nucleobindin acts as a Ca^{2+} storage protein in the Golgi; however, it also interacts with several proteins, regulating their function [169–172]. The domain structure of nucleobindin is made up of a signal peptide, DNA-binding region, two EF-hand motifs, and a leucine zipper [173]. On the luminal surface of Golgi membranes, nucleobindin is attached to Gαi3 [174,175]. Nucleobindin is also found attached to Gαi2 and has been identified in the extracellular space [176,177]. As the name suggests, nucleobindin is capable of binding DNA acting as a factor promoting antibody formation against both single and double stranded DNA

[178,179]. Additionally, nucleobindin interacts with both cyclooxygenases 1 and 2 [171]. Nucleobindin also interacts with necdin; necdin is a growth suppressor in neurons and has been found associated with viral oncoproteins and cellular transcription factors [172].

A large Ca^{2+}-induced structural change is associated with Ca^{2+} binding to nucleobindin not limited to the Ca^{2+}-binding region [173]. Accordingly, Ca^{2+} binding results in significant extrinsic stabilization of the protein [173]. The Ca^{2+}-free form of the protein is largely unstructured, while the Ca^{2+}-loaded form retains relatively high solvent accessibility but is significantly more folded [173]. Both EF-hands are thought to bind Ca^{2+} with similar affinity. Despite the binding similarities, only one of the two EF-hands is canonical in terms primary sequence [173].

The dual role of nucleobindin as a Ca^{2+} sensor and buffer points to a functional plasticity. However, the plasticity of nucleobindin is different than the broad conformational plasticity associated with CaM function. The difference is in the requirement of nucleobindin to be extrinsically stabilized by Ca^{2+} in order to acquire a functional fold. The dual role of nucleobindin as a Ca^{2+} buffer and sensor may be facilitated by the requirement for Ca^{2+}-induced stabilization in the cell and the relatively high degree of solvent exposure and dynamic flexibility in the Ca^{2+}-bound form, respectively [173]. The flexibility may be related to the requirement of this protein to interact with many of its targets just as CaM shows a conformational plasticity in target regulation; however, more specific nucleobindin–target structural studies are required to clarify this notion [173].

6.4. BM-40 (osteonectin, SPARC)

BM-40 is a secreted extracellular glycoprotein that plays numerous cellular roles including bone mineralization, wound repair, angiogenesis, modulation of cell–matrix interactions, and embryonic development [180–182]. This protein, which is also known as secreted protein acidic and rich in cysteine (SPARC) and osteonectin, contains three major domains made up of 286 amino acid residues: an N-terminal domain that is rich in glutamate and weakly binds several Ca^{2+} ions, a follastatin homology domain, and a C-terminal Ca^{2+}-binding domain that mediates interactions with collagen [183,184]. Although preliminary studies showed a single EF-hand in the C-terminal Ca^{2+}-binding domain of BM-40, subsequent structural studies elucidated the presence of two EF-hands, each with a coordinated Ca^{2+} ion [182,183,185,186].

Neither of the two EF-hands is canonical as an unusual *cis*-peptide bond is present in the first EF-hand-binding loop, while the second EF-hand-binding loop is circularized by a disulfide bond from two cysteines located on the E and F helices, respectively [182,183,186]. Despite the non-canonical nature of the EF-hands, the pair functions in conjunction with one another, binding Ca^{2+} with high affinity (K_d is sub-μM) and cooperativity. Ca^{2+} is necessary for correct folding of the extracellular domain as short C-terminal EF-hand mutants lacking the ability to coordinate

Ca^{2+} are readily degraded and never secreted in mammalian cells [182]. However, longer mutant constructs containing the follastatin domain are secreted, implying a role for the follastatin homology domain in Ca^{2+} binding and folding of the C-terminal EF-hand domain.

An N-terminal hairpin segment on the follastatin homology domain inhibits endothelial cell proliferation and promotes dissaggregation of focally adhered endothelial cells [183,187–189]. A peptide consisting of the second disulfide-bonded EF-hand from the C-terminal domain binds to endothelial cells and also inhibits proliferation [183,190]. Despite the large distance in sequence space between N-terminal hairpin and the second EF-hand, these fragments elicit similar cellular responses; hence, the BM-40-binding epitope recognized by endothelial cell receptors likely consists of both EF-hands and extends to the β-hairpin of the follastatin homology domain [183]. Structurally, both these regions map to the same side of the protein, opposite to the collagen-binding site of BM-40. The binding site for collagen (collagens I, III, IV, and V) has been mapped to a non-EF-hand helix in the extracellular domain (helix A) [183,184]. The follastatin homology domain also contains a highly basic region suggesting a binding site for glycosaminoglycans; furthermore, heparin is capable of binding to BM-40 [183,191]. Additionally, a high-affinity interaction between BM-40 and platelet-derived growth factor AB and BB has been demonstrated, although the precise regions involved in this interaction remains to be elucidated [192].

BM-40 cannot be considered an archetypal Ca^{2+} sensor because this extracellular protein would presumably be constitutively saturated with Ca^{2+} in its normal environment. Nonetheless, the requirement of Ca^{2+} for folding, stability, and ultimately function, which includes numerous target interactions, suggests that this EF-hand protein should be considered a sensor. The presumed different binding sites on this large protein precludes a mechanism of conformational plasticity at a single region of the protein for target interactions but rather suggests that this protein has evolved for recognition of different targets at different sites of the protein.

6.5. Stromal interaction molecule

Stromal interaction molecule-1 (STIM1) was first identified as a cell–cell adhesion molecule and was subsequently shown to be involved in tumor growth suppression [193–195]. However, recently STIM1 has been implicated as a key molecular player in store-operated Ca^{2+} entry (SOCE) [196,197]. STIM1, a single-pass, type-I transmembrane protein of 685 amino acids, is localized on both the endoplasmic reticulum and plasma membranes [193,198–200]. The putative N-terminal region of STIM1 includes a signal peptide, an EF-hand motif, and a sterile α-motif (SAM) domain; furthermore, the cytosolic C-terminal region consists of two coiled-coil domains, a Pro/Ser-rich region, and a Lys-rich region. The putative EF-hand of STIM1 strongly aligns with the consensus sequence of canonical EF-hands [193,198]. STIM1 binds a single Ca^{2+} ion with low affinity (sub-mM); moreover, binding induces a large conformational change, propagating at least through the majority of the N-terminal region [201].

Mammals encode another STIM molecule (STIM2), which also plays a regulatory role in SOCE [198,202]. The mammalian STIM molecules are highly homologous; the main difference in the domain topology of STIM1 and STIM2 occurs in the C-terminal domain following the coiled-coil region, where the two sequences deviate significantly [198,202]. Whereas STIM1 undergoes a clear redistribution into aggregates near the plasma membrane upon Ca^{2+} store depletion thereby playing a key role in store-operated channel activation, STIM2 only redistributes in conjunction with STIM1 and actually prevents store-operated channel activation when in high concentrations within the aggregates [202]. Accordingly, STIM1 EF-hand mutations abrogating Ca^{2+} binding result in constitutively active store-operated channels, while similar mutations in STIM2 do not [202]. Hence, a role for Ca^{2+} sensing only by STIM1 in the ER (the first ER Ca^{2+} sensor recognized) has been established.

Besides STIM1, the other necessary component for SOCE is Orai1, a plasma membrane-resident protein that likely constitutes a major portion of the store-operated channel pore [203–208]. Protein–protein interactions between Orai1 and STIM1 occur through the cytoplasmic C-terminal coiled-coil region of STIM1 and cytoplasmic region of Orai1 [207,209]. However, in order for the activation interaction to occur, the N-terminal region of STIM1 which is located in the ER lumen must sense a drop in luminal Ca^{2+} levels and undergo redistribution from ER membrane homogeneity to localized aggregation sites near the plasma membrane [196,200,207]. This redistribution facilitates the cytoplasmic interactions between Orai1 and STIM1 activating store-operated channels. The EF-hand and SAM domains likely play a role in the necessary redistribution.

The high sequence homology between STIM1 and STIM2 implicates a mechanism of genetic polymorphism in the function of these proteins. It is interesting that the Ca^{2+} sensing of STIM1 is analogous to that of CBP40 in that one of the targets of STIM1 upon Ca^{2+} depletion is itself in the homotypic aggregation associated with SOCE. However, other targets of STIM1 include STIM2 and Orai1 in the mediation and regulation of SOCE. Therefore, the SAM domain of STIM molecules likely play a role in both homotypic and heterotypic interactions related to SOCE.

7. *Analogies and disparities*

A comparison of the physical and structural aspects of target recognition in the EF-hand sensor proteins reviewed herein is summarized in Table 1. All the proteins discussed in this chapter are linked by the commonality of being made up of one or more EF-hands; however, several other points of interest have been illuminated. First, exposure of hydrophobic area is a key determinant in protein–protein interactions coupled to Ca^{2+} sensor proteins. Hydrophobic interactions provide the main scaffolding for target associations. Second, target specificity is "fine-tuned" by specific residue type on the surface of the protein. In the case of CaM, Met residues provide a microscopic flexibility for accommodation of many different target residues within the same scaffold on each domain. On the contrary, residues of different charge and size play an important role in unique EH-domain-containing, NCS and

Table 1

Comparison of the biophysical and structural aspects Ca^{2+} sensor target recognition

Ca^{2+} sensor protein family	**Molecular weight per monomer (kDa)**	**Molecularity**	**EF-hands per monomer**	**Ca^{2+} affinity (K_d in M)**	**Ca^{2+} atoms bound per monomer**	**Degree of Ca^{2+}-induced structural change**	**Major target scaffolding force**	**Mode of 'fine-tuning' for specificity**	**Mechanism of target diversity**
CaM	17	Monomer	4	10^{-6}–10^{-7}	4	Large: EF-hand domain	Hydrophobic	Main and side-chain plasticity (Met res)	Conformational plasticity
NCS	~22	Monomer-dimer	4	10^{-6}–10^{-7}	2–3	Large: EF-hand domain	Hydrophobic	Surface residue charge and size	Genetic polymorphism
S100	~10–12	Homo-heterodimer	2	10^{-5}–10^{-6}	2	Large: EF-hand domain; helical extension	Hydrophobic	Surface residue charge and size	Genetic polymorphism
Penta-EF-hand	~22–80	Homo-heterodimer	5	10^{-3}–10^{-6}	2–3	Small: EF-hand domain; propagation to other domains	Hydrophobic	Surface residue charge and size; multiple interaction sites	Genetic polymorphism
Others:									
EH-domain	~10–12	Monomer; 3 or more repeats per monomer	2	10^{-3}–10^{-6}	1	Large: required for folding	Hydrophobic	Surface residue charge; multiple interaction sites	Genetic polymorphism

(*Continued*)

Table 1
(Continued)

Ca^{2+} sensor protein family	**Molecular weight per monomer (kDa)**	**Molecularity**	**EF-hands per monomer**	**Ca^{2+} affinity (K_d in M)**	**Ca^{2+} atoms bound per monomer**	**Degree of Ca^{2+}-induced structural change**	**Major target scaffolding force**	**Mode of 'fine-tuning' for specificity**	**Mechanism of target diversity**
BM-40	~32	Monomer	2	10^{-7}	2	Large: required for folding	Hydrophobic	Multiple interaction sites	Genetic polymorphism
CBP40	~40	Monomer-oligomer	4	10^{-7}–10^{-6}	4	Small: EF-hand domain; propagation to other domains; helical extension	Hydrophobic	NA	Genetic polymorphism
Nucleobindin	~63	Monomer	2	10^{-5}	1	Large: required for folding	Hydrophobic	Main chain plasticity; multiple interaction sites	Polymorphism and plasticity
STIM1	~77	NA	1 putative	10^{-4}	1	Large	Hydrophobic	NA	Genetic polymorphism

NA, not available; CaM, calmodulin; NCS, neuronal Ca^{2+} sensor; CBP, calcium-binding protein; STIM, stromal interaction molecule.

S100 protein–target interactions within the same protein region, while BM-40, penta-EF-hand proteins, EH-domain-containing proteins, and STIMs are large multi-domain proteins that contain separate binding regions unique to different targets in distinct domains. Third, the magnitude or utility of the Ca^{2+}-induced conformational change is not fundamentally similar for all Ca^{2+} sensors. For example, penta-EF-hand proteins and CBP40 undergo a relatively small conformational change, CaM and nucleobindin acquire increased dynamic flexibility, and S100, NCS, BM-40, and nucleobindin proteins undergo a relatively large conformational change upon Ca^{2+} loading. Fourth, differential expression of Ca^{2+} sensor proteins is key transcriptional mechanism for regulatory function. For example, CaM has eight different splice variants in eukaryotes targeted to different cell compartments, STIM1 is the only identified ER luminal Ca^{2+} sensor, nucleobindin is an extracellular sensor, and some S100 proteins have evolved to regulate the same target because they are differentially expressed in tissue.

Finally, nature has used both genetic polymorphism and conformational plasticity as diversification mechanisms for target recognition in Ca^{2+} sensor proteins. CaM demonstrates plasticity by shape-adapting as necessitated by specific targets. Alternatively, the NCS, penta-EF-hand, S100, STIM, and EH-domain-containing protein families exhibit a genetically polymorphic nature expressing multiple forms, each "fine-tuned" at the amino-acid level to regulate specific targets. Future structural studies on newly discovered CaBPs will help clarify the significance of the evolutionary differences in Ca^{2+} sensors.

Acknowledgements

This work was made possible through a Canadian Institute of Health Research (CIHR) operating grant (to M.I.). P.B.S. is the recipient of Natural Sciences and Engineering Research Council of Canada. M.I. holds a Canada Research Chair in Cancer Structural Biology.

References

1. Carafoli, E. (2005). *FEBS J* 272, 1073–89.
2. Berridge, M.J., Bootman, M.D. and Roderick, H.L. (2003). *Nat Rev Mol Cell Biol* 4, 517–29.
3. Berridge, M.J., Lipp, P. and Bootman, M.D. (2000). *Nat Rev Mol Cell Biol* 1, 11–21.
4. da Silva, A.C. and Reinach, F.C. (1991). *Trends Biochem Sci* 16, 53–7.
5. Ikura, M. (1996). *Trends Biochem Sci* 21, 14–7.
6. Ikura, M. and Ames, J.B. (2006). *Proc Natl Acad Sci USA* 103, 1159–64. Epub Jan 23, 2006.
7. Hoeflich, K.P. and Ikura, M. (2002). *Cell* 108, 739–42.
8. Kretsinger, R.H. and Nockolds, C.E. (1973). *J Biol Chem* 248, 3313–26.
9. Lewit-Bentley, A. and Rety, S. (2000). *Curr Opin Struct Biol* 10, 637–43.
10. Nelson, M.R. and Chazin, W.J. (1998). *Biometals* 11, 297–318.
11. Linse, S. and Forsen, S. (1995). *Adv Second Messenger Phosphoprotein Res* 30, 89–151.
12. Shaw, G.S., Hodges, R.S. and Sykes, B.D. (1990). *Science* 249, 280–3.

13. Ye, Y., Lee, H.W., Yang, W., Shealy, S.J., Wilkins, A.L., Liu, Z.R., Torshin, I., Harrison, R., Wohlhueter, R. and Yang, J.J. (2001). *Protein Eng* 14, 1001–13.
14. Julenius, K., Robblee, J., Thulin, E., Finn, B.E., Fairman, R. and Linse, S. (2002). *Proteins* 47, 323–33.
15. Copley, R.R., Schultz, J., Ponting, C.P. and Bork, P. (1999). *Curr Opin Struct Biol* 9, 408–15.
16. Kortvely, E. and Gulya, K. (2004). *Life Sci* 74, 1065–70.
17. Kawasaki, H., Nakayama, S. and Kretsinger, R.H. (1998). *Biometals* 11, 277–95.
18. Babu, Y.S., Bugg, C.E. and Cook, W.J. (1988). *J Mol Biol* 204, 191–204.
19. Linse, S., Helmersson, A. and Forsen, S. (1991). *J Biol Chem* 266, 8050–4.
20. Tjandra, N., Kuboniwa, H., Ren, H. and Bax, A. (1995). *Eur J Biochem* 230, 1014–24.
21. Yap, K.L., Ames, J.B., Swindells, M.B. and Ikura, M. (1999). *Proteins* 37, 499–507.
22. Yap, K.L., Ames, J.B., Swindells, M.B. and Ikura, M. (2002). *Methods Mol Biol* 173, 317–24.
23. Chou, J.J., Li, S., Klee, C.B. and Bax, A. (2001). *Nat Struct Biol* 8, 990–7.
24. Osawa, M., Swindells, M.B., Tanikawa, J., Tanaka, T., Mase, T., Furuya, T. and Ikura, M. (1998). *J Mol Biol* 276, 165–76.
25. Lee, A.L., Kinnear, S.A. and Wand, A.J. (2000). *Nat Struct Biol* 7, 72–7.
26. Meador, W.E., Means, A.R. and Quiocho, F.A. (1993). *Science* 262, 1718–21.
27. Barbato, G., Ikura, M., Kay, L.E., Pastor, R.W. and Bax, A. (1992). *Biochemistry* 31, 5269–78.
28. Cheung, W.Y. (1970). *Biochem Biophys Res Commun* 38, 533–8.
29. Kakiuchi, S. and Yamazaki, R. (1970). *Biochem Biophys Res Commun* 41, 1104–10.
30. Yazawa, M. and Yagi, K. (1977). *J Biochem (Tokyo)* 82, 287–9.
31. Hook, S.S. and Means, A.R. (2001). *Annu Rev Pharmacol Toxicol* 41, 471–505.
32. Klee, C.B., Ren, H. and Wang, X. (1998). *J Biol Chem* 273, 13367–70.
33. Stuehr, D.J., Cho, H.J., Kwon, N.S., Weise, M.F. and Nathan, C.F. (1991). *Proc Natl Acad Sci USA* 88, 7773–7.
34. Michikawa, T., Hirota, J., Kawano, S., Hiraoka, M., Yamada, M., Furuichi, T. and Mikoshiba, K. (1999). *Neuron* 23, 799–808.
35. Tang, W., Sencer, S. and Hamilton, S.L. (2002). *Front Biosci* 7, d1583–9.
36. Collier, R.J. and Young, J.A. (2003). *Annu Rev Cell Dev Biol* 19, 45–70.
37. Yap, K.L., Kim, J., Truong, K., Sherman, M., Yuan, T. and Ikura, M. (2000). *J Struct Funct Genomics* 1, 8–14.
38. Meador, W.E., Means, A.R. and Quiocho, F.A. (1992). *Science* 257, 1251–5.
39. Ikura, M., Clore, G.M., Gronenborn, A.M., Zhu, G., Klee, C.B. and Bax, A. (1992). *Science* 256, 632–8.
40. Trewhella, J. (1992). *Cell Calcium* 13, 377–90.
41. Persechini, A. and Kretsinger, R.H. (1988). *J Cardiovasc Pharmacol* 12, S1–12.
42. Zhang, M., Li, M., Wang, J.H. and Vogel, H.J. (1994). *J Biol Chem* 269, 15546–52.
43. Osawa, M., Tokumitsu, H., Swindells, M.B., Kurihara, H., Orita, M., Shibanuma, T., Furuya, T. and Ikura, M. (1999). *Nat Struct Biol* 6, 819–24.
44. Aoyagi, M., Arvai, A.S., Tainer, J.A. and Getzoff, E.D. (2003). *EMBO J* 22, 766–75.
45. Yap, K.L., Yuan, T., Mal, T.K., Vogel, H.J. and Ikura, M. (2003). *J Mol Biol* 328, 193–204.
46. Yamauchi, E., Nakatsu, T., Matsubara, M., Kato, H. and Taniguchi, H. (2003). *Nat Struct Biol* 10, 226–31.
47. Schumacher, M.A., Rivard, A.F., Bachinger, H.P. and Adelman, J.P. (2001). *Nature* 410, 1120–4.
48. Drum, C.L., Yan, S.Z., Bard, J., Shen, Y.Q., Lu, D., Soelaiman, S., Grabarek, Z., Bohm, A. and Tang, W.J. (2002). *Nature* 415, 396–402.
49. Palfi, A., Kortvely, E., Fekete, E., Kovacs, B., Varszegi, S. and Gulya, K. (2002). *Life Sci* 70, 2829–55.
50. Benaim, G. and Villalobo, A. (2002). *Eur J Biochem* 269, 3619–31.
51. Murtaugh, T.J., Rowe, P.M., Vincent, P.L., Wright, L.S. and Siegel, F.L. (1983). *Methods Enzymol* 102, 158–70.

52. Murtaugh, T.J., Wright, L.S. and Siegel, F.L. (1986). *J Neurochem* 47, 164–72.
53. Burgoyne, R. (2004). *Biochim Biophys Acta* 1742, 59–68.
54. Ames, J.B., Tanaka, T., Stryer, L. and Ikura, M. (1996). *Curr Opin Struct Biol* 6, 432–8.
55. Tanaka, T., Ames, J.B., Harvey, T.S., Stryer, L. and Ikura, M. (1995). *Nature* 376, 444–7.
56. Ames, J.B., Ishima, R., Tanaka, T., Gordon, J.I., Stryer, L. and Ikura, M. (1997). *Nature* 389, 198–202.
57. Zozulya, S. and Stryer, L. (1992). *Proc Natl Acad Sci USA* 89, 11569–73.
58. Makino, C.L., Dodd, R.L., Chen, J., Burns, M.E., Roca, A., Simon, M.I. and Baylor, D.A. (2004). *J Gen Physiol* 123, 729–41.
59. Chen, C.K., Inglese, J., Lefkowitz, R.J. and Hurley, J.B. (1995). *J Biol Chem* 270, 18060–6.
60. Palczewski, K., Sokal, I. and Baehr, W. (2004). *Biochem Biophys Res Commun* 322, 1123–30.
61. Pongs, O., Lindemeier, J., Zhu, X.R., Theil, T., Engelkamp, D., Krah-Jentgens, I., Lambrecht, H.G., Koch, K.W., Schwemer, J., Rivosecchi, R. *et al.* (1993). *Neuron* 11, 15–28.
62. Gomez, M., De Castro, E., Guarin, E., Sasakura, H., Kuhara, A., Mori, I., Bartfai, T., Bargmann, C.I. and Nef, P. (2001). *Neuron* 30, 241–8.
63. Hendricks, K.B., Wang, B.Q., Schnieders, E.A. and Thorner, J. (1999). *Nat Cell Biol* 1, 234–41.
64. Hamasaki-Katagiri, N., Molchanova, T., Takeda, K. and Ames, J.B. (2004). *J Biol Chem* 279, 12744–54. Epub Jan 13, 2004.
65. An, W.F., Bowlby, M.R., Betty, M., Cao, J., Ling, H.P., Mendoza, G., Hinson, J.W., Mattsson, K.I., Strassle, B.W., Trimmer, J.S. *et al.* (2000). *Nature* 403, 553–6.
66. Hidaka, H. and Okazaki, K. (1993). *Neurosci Res* 16, 73–7.
67. Lilliehook, C., Bozdagi, O., Yao, J., Gomez-Ramirez, M., Zaidi, N.F., Wasco, W., Gandy, S., Santucci, A.C., Haroutunian, V., Huntley, G.W. *et al.* (2003). *J Neurosci* 23, 9097–106.
68. Cheng, H.Y., Pitcher, G.M., Laviolette, S.R., Whishaw, I.Q., Tong, K.I., Kockeritz, L.K., Wada, T., Joza, N.A., Crackower, M., Goncalves, J. *et al.* (2002). *Cell* 108, 31–43.
69. Buxbaum, J.D., Choi, E.K., Luo, Y., Lilliehook, C., Crowley, A.C., Merriam, D.E. and Wasco, W. (1998). *Nat Med* 4, 1177–81.
70. Carrion, A.M., Link, W.A., Ledo, F., Mellstrom, B. and Naranjo, J.R. (1999). *Nature* 398, 80–4.
71. Kobayashi, M., Takamatsu, K., Saitoh, S., Miura, M. and Noguchi, T. (1992). *Biochem Biophys Res Commun* 189, 511–7.
72. Ames, J.B., Hamasaki, N. and Molchanova, T. (2002). *Biochemistry* 41, 5776–87.
73. Ermilov, A.N., Olshevskaya, E.V. and Dizhoor, A.M. (2001). *J Biol Chem* 276, 48143–8. Epub Oct 2, 2001.
74. Krylov, D.M., Niemi, G.A., Dizhoor, A.M. and Hurley, J.B. (1999). *J Biol Chem* 274, 10833–9.
75. Tachibanaki, S., Nanda, K., Sasaki, K., Ozaki, K. and Kawamura, S. (2000). *J Biol Chem* 275, 3313–9.
76. Olshevskaya, E.V., Boikov, S., Ermilov, A., Krylov, D., Hurley, J.B. and Dizhoor, A.M. (1999). *J Biol Chem* 274, 10823–32.
77. Zhou, W., Qian, Y., Kunjilwar, K., Pfaffinger, P.J. and Choe, S. (2004). *Neuron* 41, 573–86.
78. Dizhoor, A.M., Ray, S., Kumar, S., Niemi, G., Spencer, M., Brolley, D., Walsh, K.A., Philipov, P.P., Hurley, J.B. and Stryer, L. (1991). *Science* 251, 915–8.
79. Palczewski, K., Subbaraya, I., Gorczyca, W.A., Helekar, B.S., Ruiz, C.C., Ohguro, H., Huang, J., Zhao, X., Crabb, J.W., Johnson, R.S. *et al.* (1994). *Neuron* 13, 395–404.
80. Hammond, P.I., Craig, T.A., Kumar, R. and Brimijoin, S. (2003). *Brain Res Mol Brain Res* 111, 104–10.
81. Kuo, H.C., Cheng, C.F., Clark, R.B., Lin, J.J., Lin, J.L., Hoshijima, M., Nguyen-Tran, V.T., Gu, Y., Ikeda, Y., Chu, P.H. *et al.* (2001). *Cell* 107, 801–13.
82. Gentry, H.R., Singer, A.U., Betts, L., Yang, C., Ferrara, J.D., Sondek, J. and Parise, L.V. (2005). *J Biol Chem* 280, 8407–15. Epub Dec 1, 2004.
83. Haataja, L., Kaartinen, V., Groffen, J. and Heisterkamp, N. (2002). *J Biol Chem* 277, 8321–8. Epub Dec 27, 2001.

84. Yamniuk, A.P., Nguyen, L.T., Hoang, T.T. and Vogel, H.J. (2004). *Biochemistry* 43, 2558–68.
85. White, C., Yang, J., Monteiro, M.J. and Foskett, J.K. (2006). *J Biol Chem* 281, 20825–33. Epub May 24, 2006.
86. Stabler, S.M., Ostrowski, L.L., Janicki, S.M. and Monteiro, M.J. (1999). *J Cell Biol* 145, 1277–92.
87. Naik, U.P., Patel, P.M. and Parise, L.V. (1997). *J Biol Chem* 272, 4651–4.
88. Hollenbach, A.D., McPherson, C.J., Lagutina, I. and Grosveld, G. (2002). *Biochim Biophys Acta* 1574, 321–8.
89. Haeseleer, F., Sokal, I., Verlinde, C.L., Erdjument-Bromage, H., Tempst, P., Pronin, A.N., Benovic, J.L., Fariss, R.N. and Palczewski, K. (2000). *J Biol Chem* 275, 1247–60.
90. Yang, J., McBride, S., Mak, D.O., Vardi, N., Palczewski, K., Haeseleer, F. and Foskett, J.K. (2002). *Proc Natl Acad Sci USA* 99, 7711–6.
91. Lee, A., Westenbroek, R.E., Haeseleer, F., Palczewski, K., Scheuer, T. and Catterall, W.A. (2002). *Nat Neurosci* 5, 210–7.
92. Zhou, H., Kim, S.A., Kirk, E.A., Tippens, A.L., Sun, H., Haeseleer, F. and Lee, A. (2004). *J Neurosci* 24, 4698–708.
93. Zhou, H., Yu, K., McCoy, K.L. and Lee, A. (2005). *J Biol Chem* 280, 29612–9. Epub Jun 26, 2005.
94. Kinoshita-Kawada, M., Tang, J., Xiao, R., Kaneko, S., Foskett, J.K. and Zhu, M.X. (2005). *Pflugers Arch* 450, 345–54. Epub May 14, 2005.
95. Wingard, J.N., Chan, J., Bosanac, I., Haeseleer, F., Palczewski, K., Ikura, M. and Ames, J.B. (2005). *J Biol Chem* 280, 37461–70. Epub Sep 7, 2005.
96. Engman, D.M., Krause, K.H., Blumin, J.H., Kim, K.S., Kirchhoff, L.V. and Donelson, J.E. (1989). *J Biol Chem* 264, 18627–31.
97. Godsel, L.M. and Engman, D.M. (1999). *EMBO J* 18, 2057–65.
98. Buchanan, K.T., Ames, J.B., Asfaw, S.H., Wingard, J.N., Olson, C.L., Campana, P.T., Araujo, A.P. and Engman, D.M. (2005). *J Biol Chem* 280, 40104–11. Epub Sep 7, 2005.
99. Klee, C.B., Draetta, G.F. and Hubbard, M.J. (1988). *Adv Enzymol Relat Areas Mol Biol* 61, 149–200.
100. Kakalis, L.T., Kennedy, M., Sikkink, R., Rusnak, F. and Armitage, I.M. (1995). *FEBS Lett* 362, 55–8.
101. Feng, B. and Stemmer, P.M. (1999). *Biochemistry* 38, 12481–9.
102. Aitken, A., Cohen, P., Santikarn, S., Williams, D.H., Calder, A.G., Smith, A. and Klee, C.B. (1982). *FEBS Lett* 150, 314–8.
103. Kennedy, M.T., Brockman, H. and Rusnak, F. (1996). *J Biol Chem* 271, 26517–21.
104. Santamaria-Kisiel, L., Rintala-Dempsey, A.C. and Shaw, G.S. (2006). *Biochem J* 396, 201–14.
105. Donato, R. (1986). *Cell Calcium* 7, 123–45.
106. Schafer, B.W. and Heizmann, C.W. (1996). *Trends Biochem Sci* 21, 134–40.
107. Dempsey, A.C., Walsh, M.P. and Shaw, G.S. (2003). *Structure* 11, 887–97.
108. Rety, S., Osterloh, D., Arie, J.P., Tabaries, S., Seeman, J., Russo-Marie, F., Gerke, V. and Lewit-Bentley, A. (2000). *Structure* 8, 175–84.
109. Wicki, R., Franz, C., Scholl, F.A., Heizmann, C.W. and Schafer, B.W. (1997). *Cell Calcium* 22, 243–54.
110. Schmidt-Hansen, B., Klingelhofer, J., Grum-Schwensen, B., Christensen, A., Andresen, S., Kruse, C., Hansen, T., Ambartsumian, N., Lukanidin, E. and Grigorian, M. (2004). *J Biol Chem* 279, 24498–504. Epub Mar 26, 2004.
111. Nowotny, M., Spiechowicz, M., Jastrzebska, B., Filipek, A., Kitagawa, K. and Kuznicki, J. (2003). *J Biol Chem* 278, 26923–8. Epub May 13, 2003.
112. Rustandi, R.R., Baldisseri, D.M., Inman, K.G., Nizner, P., Hamilton, S.M., Landar, A., Zimmer, D.B. and Weber, D.J. (2002). *Biochemistry* 41, 788–96.
113. Du, X.J., Cole, T.J., Tenis, N., Gao, X.M., Kontgen, F., Kemp, B.E. and Heierhorst, J. (2002). *Mol Cell Biol* 22, 2821–9.

114. Broome, A.M., Ryan, D. and Eckert, R.L. (2003). *J Histochem Cytochem* 51, 675–85.
115. Newton, R.A. and Hogg, N. (1998). *J Immunol* 160, 1427–35.
116. Rety, S., Sopkova, J., Renouard, M., Osterloh, D., Gerke, V., Tabaries, S., Russo-Marie, F. and Lewit-Bentley, A. (1999). *Nat Struct Biol* 6, 89–95.
117. Maki, M., Kitaura, Y., Satoh, H., Ohkouchi, S. and Shibata, H. (2002). *Biochim Biophys Acta* 1600, 51–60.
118. Xie, X., Dwyer, M.D., Swenson, L., Parker, M.H. and Botfield, M.C. (2001). *Protein Sci* 10, 2419–25.
119. Valdivia, H.H. (1998). *Trends Pharmacol Sci* 19, 479–82.
120. Meyers, M.B., Pickel, V.M., Sheu, S.S., Sharma, V.K., Scotto, K.W. and Fishman, G.I. (1995). *J Biol Chem* 270, 26411–8.
121. Meyers, M.B., Zamparelli, C., Verzili, D., Dicker, A.P., Blanck, T.J. and Chiancone, E. (1995). *FEBS Lett* 357, 230–4.
122. Zamparelli, C., Ilari, A., Verzili, D., Vecchini, P. and Chiancone, E. (1997). *FEBS Lett* 409, 1–6.
123. Mella, M., Colotti, G., Zamparelli, C., Verzili, D., Ilari, A. and Chiancone, E. (2003). *J Biol Chem* 278, 24921–8. Epub Apr 23, 2003.
124. Lollike, K., Johnsen, A.H., Durussel, I., Borregaard, N. and Cox, J.A. (2001). *J Biol Chem* 276, 17762–9. Epub Feb 13, 2001.
125. Jia, J., Han, Q., Borregaard, N., Lollike, K. and Cygler, M. (2000). *J Mol Biol* 300, 1271–81.
126. Jia, J., Borregaard, N., Lollike, K. and Cygler, M. (2001). *Acta Crystallogr D Biol Crystallogr* 57, 1843–9. Epub Nov 21, 2001.
127. Hansen, C., Tarabykina, S., la Cour, J.M., Lollike, K. and Berchtold, M.W. (2003). *FEBS Lett* 545, 151–4.
128. Lollike, K., Sorensen, O., Bundgaard, J.R., Segal, A.W., Boyhan, A. and Borregaard, N. (1995). *J Immunol Methods* 185, 1–8.
129. Goll, D.E., Thompson, V.F., Li, H., Wei, W. and Cong, J. (2003). *Physiol Rev* 83, 731–801.
130. Strobl, S., Fernandez-Catalan, C., Braun, M., Huber, R., Masumoto, H., Nakagawa, K., Irie, A., Sorimachi, H., Bourenkow, G., Bartunik, H., Suzuki, K. and Bode, W. (2000). *Proc Natl Acad Sci USA* 97, 588–92.
131. Ravulapalli, R., Diaz, B.G., Campbell, R.L. and Davies, P.L. (2005). *Biochem J* 388, 585–91.
132. Moldoveanu, T., Hosfield, C.M., Lim, D., Elce, J.S., Jia, Z. and Davies, P.L. (2002). *Cell* 108, 649–60.
133. Moldoveanu, T., Jia, Z. and Davies, P.L. (2004). *J Biol Chem* 279, 6106–14. Epub Oct 27, 2003.
134. Maki, M., Takano, E., Osawa, T., Ooi, T., Murachi, T. and Hatanaka, M. (1988). *J Biol Chem* 263, 10254–61.
135. Ma, H., Yang, H.Q., Takano, E., Hatanaka, M. and Maki, M. (1994). *J Biol Chem* 269, 24430–6.
136. Maki, M., Bagci, H., Hamaguchi, K., Ueda, M., Murachi, T. and Hatanaka, M. (1989). *J Biol Chem* 264, 18866–9.
137. Yang, H.Q., Ma, H., Takano, E., Hatanaka, M. and Maki, M. (1994). *J Biol Chem* 269, 18977–84.
138. Croall, D.E. and McGrody, K.S. (1994). *Biochemistry* 33, 13223–30.
139. Shumway, S.D., Maki, M. and Miyamoto, S. (1999). *J Biol Chem* 274, 30874–81.
140. Benetti, R., Del Sal, G., Monte, M., Paroni, G., Brancolini, C. and Schneider, C. (2001). *EMBO J* 20, 2702–14.
141. Lo, K.W., Zhang, Q., Li, M. and Zhang, M. (1999). *Biochemistry* 38, 7498–508.
142. Jia, J., Tarabykina, S., Hansen, C., Berchtold, M. and Cygler, M. (2001). *Structure* 9, 267–75.
143. Subramanian, L., Crabb, J.W., Cox, J., Durussel, I., Walker, T.M., van Ginkel, P.R., Bhattacharya, S., Dellaria, J.M., Palczewski, K. and Polans, A.S. (2004). *Biochemistry* 43, 11175–86.
144. Jung, Y.S., Kim, K.S., Kim, K.D., Lim, J.S., Kim, J.W. and Kim, E. (2001). *Biochem Biophys Res Commun* 288, 420–6.
145. Satoh, H., Shibata, H., Nakano, Y., Kitaura, Y. and Maki, M. (2002). *Biochem Biophys Res Commun* 291, 1166–72.

146. Satoh, H., Nakano, Y., Shibata, H. and Maki, M. (2002). *Biochim Biophys Acta* 1600, 61–7.
147. Kitaura, Y., Watanabe, M., Satoh, H., Kawai, T., Hitomi, K. and Maki, M. (1999). *Biochem Biophys Res Commun* 263, 68–75.
148. Kitaura, Y., Matsumoto, S., Satoh, H., Hitomi, K. and Maki, M. (2001). *J Biol Chem* 276, 14053–8. Epub Feb 1, 2001.
149. Kitaura, Y., Satoh, H., Takahashi, H., Shibata, H. and Maki, M. (2002). *Arch Biochem Biophys* 399, 12–8.
150. Wong, W.T., Schumacher, C., Salcini, A.E., Romano, A., Castagnino, P., Pelicci, P.G. and Di Fiore, P.P. (1995). *Proc Natl Acad Sci USA* 92, 9530–4.
151. Fazioli, F., Minichiello, L., Matoskova, B., Wong, W.T. and Di Fiore, P.P. (1993). *Mol Cell Biol* 13, 5814–28.
152. Benedetti, H., Raths, S., Crausaz, F. and Riezman, H. (1994). *Mol Biol Cell* 5, 1023–37.
153. Wendland, B., McCaffery, J.M., Xiao, Q. and Emr, S.D. (1996). *J Cell Biol* 135, 1485–500.
154. Carbone, R., Fre, S., Iannolo, G., Belleudi, F., Mancini, P., Pelicci, P.G., Torrisi, M.R. and Di Fiore, P.P. (1997). *Cancer Res* 57, 5498–504.
155. Benmerah, A., Lamaze, C., Begue, B., Schmid, S.L., Dautry-Varsat, A. and Cerf-Bensussan, N. (1998). *J Cell Biol* 140, 1055–62.
156. Roos, J. and Kelly, R.B. (1998). *J Biol Chem* 273, 19108–19.
157. Koshiba, S., Kigawa, T., Iwahara, J., Kikuchi, A. and Yokoyama, S. (1999). *FEBS Lett* 442, 138–42.
158. Yamaguchi, A., Urano, T., Goi, T. and Feig, L.A. (1997). *J Biol Chem* 272, 31230–4.
159. Confalonieri, S. and Di Fiore, P.P. (2002). *FEBS Lett* 513, 24–9.
160. Kim, S., Cullis, D.N., Feig, L.A. and Baleja, J.D. (2001). *Biochemistry* 40, 6776–85.
161. Whitehead, B., Tessari, M., Carotenuto, A., van Bergen en Henegouwen, P.M. and Vuister, G.W. (1999). *Biochemistry* 38, 11271–7.
162. de Beer, T., Carter, R.E., Lobel-Rice, K.E., Sorkin, A. and Overduin, M. (1998). *Science* 281, 1357–60.
163. de Beer, T., Hoofnagle, A.N., Enmon, J.L., Bowers, R.C., Yamabhai, M., Kay, B.K. and Overduin, M. (2000). *Nat Struct Biol* 7, 1018–22.
164. Okagaki, T., Ishikawa, R. and Kohama, K. (1991). *Biochem Biophys Res Commun* 176, 564–70.
165. Nakamura, A., Okagaki, T., Takagi, T., Nakashima, K., Yazawa, M. and Kohama, K. (2000). *Biochemistry* 39, 3827–34.
166. Nakayama, S. and Kretsinger, R.H. (1994). *Annu Rev Biophys Biomol Struct* 23, 473–507.
167. Iwasaki, W., Sasaki, H., Nakamura, A., Kohama, K. and Tanokura, M. (2003). *Structure* 11, 75–85.
168. Iwasaki, W., Sasaki, H., Nakamura, A., Kohama, K. and Tanokura, M. (1999). *J Biochem (Tokyo)* 126, 7–9.
169. Lin, P., Yao, Y., Hofmeister, R., Tsien, R.Y. and Farquhar, M.G. (1999). *J Cell Biol* 145, 279–89.
170. Lin, P., Le-Niculescu, H., Hofmeister, R., McCaffery, J.M., Jin, M., Hennemann, H., McQuistan, T., De Vries, L. and Farquhar, M.G. (1998). *J Cell Biol* 141, 1515–27.
171. Ballif, B.A., Mincek, N.V., Barratt, J.T., Wilson, M.L. and Simmons, D.L. (1996). *Proc Natl Acad Sci USA* 93, 5544–9.
172. Taniguchi, N., Taniura, H., Niinobe, M., Takayama, C., Tominaga-Yoshino, K., Ogura, A. and Yoshikawa, K. (2000). *J Biol Chem* 275, 31674–81.
173. de Alba, E. and Tjandra, N. (2004). *Biochemistry* 43, 10039–49.
174. Lin, P., Fischer, T., Weiss, T. and Farquhar, M.G. (2000). *Proc Natl Acad Sci USA* 97, 674–9.
175. Weiss, T.S., Chamberlain, C.E., Takeda, T., Lin, P., Hahn, K.M. and Farquhar, M.G. (2001). *Proc Natl Acad Sci USA* 98, 14961–6.
176. Mochizuki, N., Hibi, M., Kanai, Y. and Insel, P.A. (1995). *FEBS Lett* 373, 155–8.
177. Lavoie, C., Meerloo, T., Lin, P. and Farquhar, M.G. (2002). *Mol Endocrinol* 16, 2462–74.
178. Kanai, Y., Takeda, O., Miura, K., Amagai, M., Kaneko, T., Kubota, T., Tanuma, S. and Kurosawa, Y. (1995). *Immunol Lett* 45, 35–42.
179. Kanai, Y., Kyuwa, S., Miura, K. and Kurosawa, Y. (1995). *Immunol Lett* 46, 207–14.

180. Sage, E.H. (1997). *Nat Med* 3, 144–6.
181. Lane, T.F. and Sage, E.H. (1994). *FASEB J* 8, 163–73.
182. Busch, E., Hohenester, E., Timpl, R., Paulsson, M., Maurer, P. (2000). *J Biol Chem* 275, 25508–15.
183. Hohenester, E., Maurer, P. and Timpl, R. (1997). *EMBO J* 16, 3778–86.
184. Sasaki, T., Hohenester, E., Gohring, W. and Timpl, R. (1998). *EMBO J* 17, 1625–34.
185. Engel, J., Taylor, W., Paulsson, M., Sage, H. and Hogan, B. (1987) *Biochemistry* 26, 6958–65.
186. Hohenester, E., Maurer, P., Hohenadl, C., Timpl, R., Jansonius, J.N. and Engel, J. (1996). *Nat Struct Biol* 3, 67–73.
187. Funk, S.E. and Sage, E.H. (1991). *Proc Natl Acad Sci USA* 88, 2648–52.
188. Funk, S.E. and Sage, E.H. (1993). *J Cell Physiol* 154, 53–63.
189. Murphy-Ullrich, J.E., Lane, T.F., Pallero, M.A. and Sage, E.H. (1995). *J Cell Biochem* 57, 341–50.
190. Yost, J.C. and Sage, E.H. (1993). *J Biol Chem* 268, 25790–6.
191. Nakamura, T., Sugino, K., Titani, K. and Sugino, H. (1991). *J Biol Chem* 266, 19432–7.
192. Raines, E.W., Lane, T.F., Iruela-Arispe, M.L., Ross, R. and Sage, E.H. (1992). *Proc Natl Acad Sci USA* 89, 1281–5.
193. Williams, R.T., Senior, P.V., Van Stekelenburg, L., Layton, J.E., Smith, P.J. and Dziadek, M.A. (2002). *Biochim Biophys Acta* 1596, 131–7.
194. Oritani, K. and Kincade, P.W. (1996). *J Cell Biol* 134, 771–82.
195. Manji, S.S., Parker, N.J., Williams, R.T., van Stekelenburg, L., Pearson, R.B., Dziadek, M. and Smith, P.J. (2000). *Biochim Biophys Acta* 1481, 147–55.
196. Liou, J., Kim, M.L., Heo, W.D., Jones, J.T., Myers, J.W., Ferrell, J.E., Jr. and Meyer, T. (2005). *Curr Biol* 15, 1235–41.
197. Roos, J., DiGregorio, P.J., Yeromin, A.V., Ohlsen, K., Lioudyno, M., Zhang, S., Safrina, O., Kozak, J.A., Wagner, S.L., Cahalan, M.D., Velicelebi, G. and Stauderman, K.A. (2005). *J Cell Biol* 169, 435–45. Epub May 2, 2005.
198. Williams, R.T., Manji, S.S., Parker, N.J., Hancock, M.S., Van Stekelenburg, L., Eid, J.P., Senior, P.V., Kazenwadel, J.S., Shandala, T., Saint, R., Smith, P.J. and Dziadek, M.A. (2001). *Biochem J* 357, 673–85.
199. Zhang, S.L., Yu, Y., Roos, J., Kozak, J.A., Deerinck, T.J., Ellisman, M.H., Stauderman, K.A. and Cahalan, M.D. (2005). *Nature* 437, 902–5.
200. Spassova, M.A., Soboloff, J., He, L.P., Xu, W., Dziadek, M.A. and Gill, D.L. (2006). *Proc Natl Acad Sci USA* 103, 4040–5. Epub Mar 6, 2006.
201. Stathopulos, P.B., Li, G.-Y., Plevin, M.J., Ames, J.B. and Ikura, M. (2006). *J Biol Chem* 281, 35855–62. Epub Oct 3, 2006.
202. Soboloff, J., Spassova, M.A., Hewavitharana, T., He, L.P., Xu, W., Johnstone, L.S., Dziadek, M.A. and Gill, D.L. (2006). *Curr Biol* 16, 1465–70.
203. Peinelt, C., Vig, M., Koomoa, D.L., Beck, A., Nadler, M.J., Koblan-Huberson, M., Lis, A., Fleig, A., Penner, R. and Kinet, J.P. (2006). *Nat Cell Biol* 8, 771–3. Epub May 30, 2006.
204. Zhang, S.L., Yeromin, A.V., Zhang, X.H., Yu, Y., Safrina, O., Penna, A., Roos, J., Stauderman, K.A. and Cahalan, M.D. (2006). *Proc Natl Acad Sci USA* 103, 9357–62. Epub Jun 2, 2006.
205. Soboloff, J., Spassova, M.A., Tang, X.D., Hewavitharana, T., Xu, W. and Gill, D.L. (2006). *J Biol Chem* 281, 20661–5. Epub June 9, 2006.
206. Prakriya, M., Feske, S., Gwack, Y., Srikanth, S., Rao, A. and Hogan, P.G. (2006). *Nature Epub Aug 20, 2006*.
207. Mercer, J.C., Dehaven, W.I., Smyth, J.T., Wedel, B., Boyles, R.R., Bird, G.S. and Putney, J.W., Jr. (2006). *J Biol Chem* 281, 24979–90. Epub June 28, 2006.
208. Feske, S., Gwack, Y., Prakriya, M., Srikanth, S., Puppel, S.H., Tanasa, B., Hogan, P.G., Lewis, R.S., Daly, M. and Rao, A. (2006). *Nature* 441, 179–85. Epub Apr 2, 2006.
209. Huang, G.N., Zeng, W., Kim, J.Y., Yuan, J.P., Han, L., Muallem, S. and Worley, P.F. (2006). *Nat Cell Biol* 8, 1003–10. Epub Aug 13, 2006.

Calcium Homeostasis of Cells and Organelles

J. Krebs and M. Michalak (Editors)
Calcium: A Matter of Life or Death

ISSN: 0167-7306/DOI: 10.1016/S0167-7306(06)41005-X

CHAPTER 5

Voltage-gated calcium channels, calcium signaling, and channelopathies

Erika S. Piedras-Rentería[1], Curtis F. Barrett[2], Yu-Qing Cao[3], and Richard W. Tsien[2]

[1]*Department of Physiology, Loyola University Medical Center, Stritch School of Medicine, Maywood, IL 60156-5500, USA;*
[2]*Department of Molecular and Cellular Physiology, Stanford University Medical Center, Stanford, CA 94305-5345, USA, Tel.: +1 650 725 7564; Fax: +1 650 725 8021, E-mail: rwtsien@leland.stanford.edu;*
[3]*Washington University Pain Center, Basic Research Division, Department of Anesthesiology, Washington University School of Medicine, St. Louis, MO 63110, USA*

Abstract

Voltage-gated calcium channels mediate intracellular Ca^{2+} influx and are absolutely critical for the physiological function of excitable cells in brain, skeletal, cardiac and smooth muscle, endocrine glands, and other tissues. These membrane proteins translate membrane depolarization signals into rapid, local increases in cytoplasmic calcium concentration that can trigger key events such as excitability, exocytosis, contraction, migration, and gene transcription. This chapter reviews basic structural features and functional properties of Ca^{2+} channels that confer their remarkable selectivity, modulation properties, and responses to therapeutic drugs. The diversity of Ca^{2+} channels is considered in the context of their varied functional roles, their molecular subunits, and the dysfunction of these subunits in disease states. The latter part of this chapter reviews key physiological processes using particular channel types to exemplify their function and discusses recent information obtained from genetic models and diseases that arise from genetic and acquired abnormalities in Ca^{2+} channel function.

Keywords: voltage-gated Ca^{2+} channels, VGCC, Ca_v1 channel, Ca_v2 channel, Ca_v3 channel, excitation–contraction coupling, excitation–transcription coupling, channelopathy

1. Basic structural features of voltage-gated Ca^{2+} channels

Voltage-gated calcium channels (sometimes called VGCCs, but here abbreviated simply as Ca^{2+} channels) are members of a superfamily of voltage-gated ion channels, which also includes sodium and potassium channels (Na^+ and K^+ channels). Expressed in a wide variety of cell types, Ca^{2+} channels transduce changes in membrane potential (depolarization) to increased intracellular [Ca^{2+}], which in turn triggers diverse cell biological processes including contraction, secretion, migration, proliferation, and transcription (Fig. 1) (for review, see Refs 1,2–4). By transferring cations, Ca^{2+} channels also

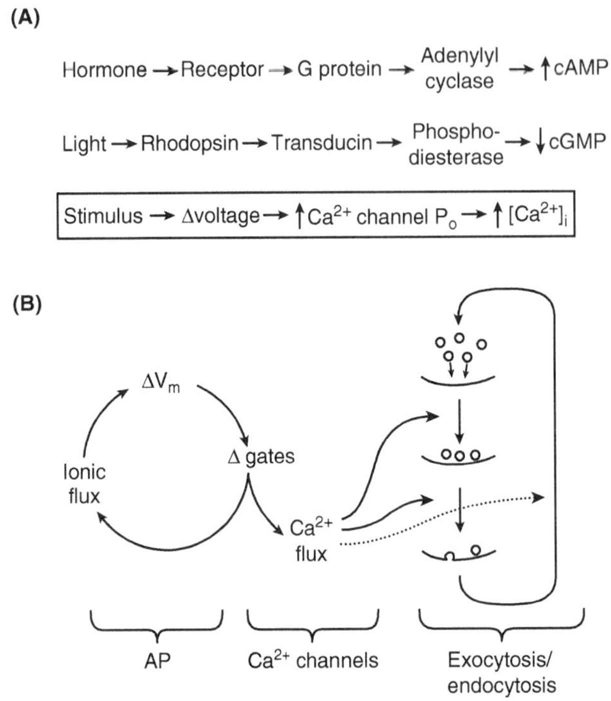

Fig. 1. (**A**) Ca^{2+} channels can be modeled as key signal transducers, analogous to G-protein-coupled receptors and other famous membrane-associated signaling systems. In this pathway, the stimulus (e.g., a sensory input) leads to depolarization of the plasma membrane, which in turn activates voltage-gated Ca^{2+} channels. Ca^{2+} influx through these channels serves as a diffusible second messenger to trigger downstream cellular events (e.g., contraction, secretion, and transcription). (**B**) In synaptic transmission, Ca^{2+} forms the link between the classical Hodgkin cycle (as depicted in Ref. 2) and synaptic vesicle recycling. AP, action potential at the presynaptic terminal; V_m, membrane potential.

provide a depolarizing ionic current that can support action potentials entirely on its own, as in cardiac pacemaker or smooth muscle cells [5,6]. More generally, in a wide variety of excitable cells, Ca^{2+} channels work alongside Na^+ channels to generate specific aspects of the overall electrical activity, including rhythm-generating depolarizations or action potential-prolonging plateau phases [7,8].

Ca^{2+} channels are composed of several proteins, which assemble to form heteromultimers [9–12]. At the heart of the channel is the α_1 subunit (Fig. 2). Composed of four repeating pseudosubunit domains (I–IV) within a single protein ranging from approximately 170 to 250 kDa, the α_1 subunit contains the minimum necessary components to form a functional channel. Within each repeating domain are six putative transmembrane-spanning helices, S1 through S6. In each domain, the fourth helix (IS4, IIS4, etc.) contains the voltage sensor, in the form of positively charged amino acids, and the S5–S6 region forms the P-loop, making the pore. Found in association with the α_1 subunit are several auxiliary subunits (the α_2 and β,γ,δ subunits), which have diverse effects on the channel's properties, including expression, trafficking, and altered kinetics of activation and/or inactivation [14].

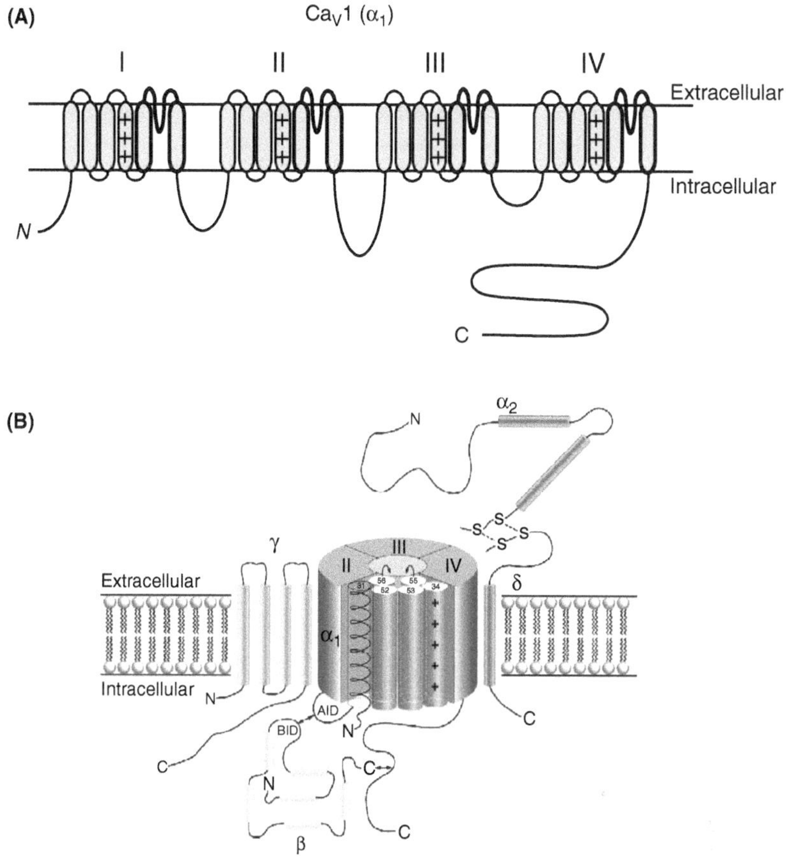

Fig. 2. (**A**) Predicted topology of the α_1 subunit. Each one of four repeats contains six putative membrane-spanning helices (S1 through S6). Helices S4 contain positively charged amino acids critical for channel activation–deactivation gating. The P-loop (composed of S5–S6) involved in Ca^{2+} selectivity and permeation is indicated in bold. (**B**) Schematic representation of multi-heteromeric voltage-gated Ca^{2+} channels. Disulfide bonds between the α_2 and the δ subunits are indicated. AID, α_1-interacting domain, which interacts with the auxiliary subunit; BID, β-interacting domain (Reprinted with permission from Randall and Benham [13]) (See Color Plate 14, p. 513).

Given the wide range of downstream effects mediated by Ca^{2+} entry, it is perhaps not surprising that a rich diversity of Ca^{2+} channels have evolved, with individual channel types specialized for, but not exclusively limited to, specific biological roles. These channels can be roughly classified according to their voltage threshold for activation; high-voltage–activated (HVA) channels require strong depolarization, whereas low-voltage–activated (LVA) channels activate with relatively little change in membrane potential. In addition, channels can be classified by their pharmacological sensitivity.

The rich diversity of Ca^{2+} channel function is rooted in their molecular architecture. Ten different genes encoding α_1 subunits have been identified. These can be divided

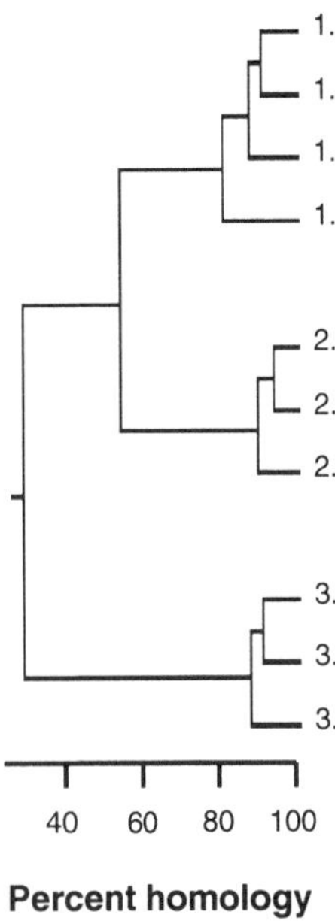

Ca_V	α_1	Type	Distribution	Chromosome	Channelopathy
1.1	S	L	Skeletal muscle	1q31–q32	HypoPP[1], MH[2]
1.2	C	L	Cardiac & smooth muscle, neuronal	12p13.3	TS[3], severe arrhythmia disorder
1.3	D	L	Endocrine, neuronal	3p14.3	
1.4	F	L	Retinal	Xp11.23	CSNB2[4], "RDNZA"[5]
2.1	A	P/Q	Neuronal , endocrine	19p13.1–13.2	EA2[6], SCA6[7], FHM1[8], epilepsy
2.2	B	N	Neuronal	9q34	
2.3	E	R	Neuronal, endocrine	1q25–q31	
3.1	G	T	Cardiac, neuronal	17q22	CAE[9], idiopathic generalized epilepsy, ASD[10]
3.2	H	T	Cardiac, neuronal	16p13.3	
3.3	I	T	Neuronal	22q12.3–13.2	

[1]Hypokalemic periodic paralysis
[2]Malignant hypothermia
[3]Timothy syndrome
[4]Incomplete X-linked congenital stationary night blindness
[5]Retinal disorder with associated autism in a New Zealand family
[6]Episodic ataxia type 2
[7]Spinocerebellar ataxia type 6
[8]Familial hemiplegic migraine type 1
[9]Childhood absence epilepsy (note: non-Mendelian inheritance)
[10]Autism spectrum disorder

Fig. 3. Homology table for the Ca_V family of voltage-gated calcium channels. Shown are the currently accepted nomenclature [15] and the corresponding conventional nomenclature [1]. Channelopathies associated with the respective channels are indicated Singh et al., 2007 reported the detection of channelopathies linked to the CACNA1G gene (T-type channel a1G subunit, $Ca_V3.1$), using mutational analysis in a cohort of Japanese and Hispanic patients with idiopathic generalized epilepsies (IGE) [366].

into three subfamilies based on molecular homology (Fig. 3): HVA channels include $Ca_V1.1$–1.4 (α_{1S}, α_{1C}, α_{1D}, and α_{1F}, respectively) and $Ca_V2.1$–2.3 (α_{1A}, α_{1B}, and α_{1E}, respectively); LVA channels are composed of $Ca_V3.1$–3.3 ($\alpha_{1G–I}$, respectively). Extensive alternative splicing of several α_1 genes provides additional diversity (in some cases, most notably $Ca_v1.3$, alternative splicing can alter voltage dependence strongly enough to blur the traditional HVA–LVA distinction [16]). Finally, the accessory subunits are encoded by multiple genes: four β subunits, three α_2 subunits, and eight γ subunits have been identified. The accessory subunits are not the main focus of this chapter but are gaining increasing attention as modifiers of channel function [12,17–19] and as targets of the therapeutic agents gabapentin and pregabalin [20].

2. *Brief overview of fundamental functional properties of Ca^{2+} channels*

Ca^{2+} channels respond to changes in membrane potential to allow Ca^{2+} to enter the cell. Membrane potential variations induce movement of the voltage sensor in the α_1 subunit, which results in a conformation change that leads to gating of the channel. Ion channel gating includes opening (or activation) and closing (deactivation) of the permeation pathway. Typically, Ca^{2+} channels activate and deactivate very rapidly in response to depolarization and repolarization, respectively. During maintained depolarization (e.g., during a cardiac action potential), Ca^{2+} channels can enter a refractory non-conducting state by a mechanism known as inactivation, which serves as a negative feedback mechanism to prevent cytotoxicity and maintain Ca^{2+} homeostasis [21–23]. While inactivation is a general property of Ca^{2+} channels, the kinetics of development and removal of inactivation vary widely [24].

One of the signature properties of Ca^{2+} channels is their exquisite selectivity for Ca^{2+} ions, a necessary feature for mediating Ca^{2+} influx under physiological conditions, where Ca^{2+} is greatly outnumbered by monovalent cations. Yet despite this high selectivity, Ca^{2+} channels permeate Ca^{2+} at a very high rate, sufficient to cause a large increase in $[Ca^{2+}]_i$ (>1 μM) in a very localized domain (~1 μm) near the mouth of the open channel. The flux through Ca^{2+} channels can be manipulated by the application of divalent ion blockers such as cadmium (Cd^{2+}), cobalt (Co^{2+}), or nickel (Ni^{2+}), and a wide array of organic compounds that have been developed as experimental inhibitors or therapeutic agents. Naturally occurring peptide toxins have also been invaluable tools for the dissection of the contributions of specific types of Ca^{2+} channels. Finally, Ca^{2+} channels are often found in tight association with Ca^{2+}-dependent signaling molecules [e.g., transmitter release machinery, calmodulin (CaM) and Ca^{2+}/CaM-dependent kinases, ryanodine receptors (RyR), etc.]. These interactions can support high temporal and spatial fidelity of coupling between excitation and downstream effector mechanisms while allowing modulation to fine-tune Ca^{2+} signaling.

Current progress in understanding the structural basis of these basic functional properties is reviewed in the next section.

3. *Classifications of Ca^{2+} channels*

3.1. *Ca_V1 (L-type) Ca^{2+} channels*

L-type Ca^{2+} channels are generally considered to be HVA. However, this distinction is not absolute, as it is important to note that, under some circumstances, L-type Ca^{2+} channels may exhibit LVA properties [25]. L-type Ca^{2+} channels are generally identified by a strong sensitivity to the 1,4-dihydropyridine (DHP) class of drugs, which includes both antagonists (e.g., nifedipine and nimodipine) and agonists (e.g., Bay K 8644) [26–28].

There are four members of the Ca_V1 subfamily, and pathophysiological states have been linked to mutations (channelopathies) in three channels: $Ca_V1.1$, 1.2, and 1.4. The nature of these mutations is discussed in detail in Section 5.

3.2. *Ca_V2 (N-, P/Q- and R-type) Ca^{2+} channels*

A second subfamily of Ca^{2+} channels shares over 50% homology with the L-type Ca^{2+} channels and is also classified as HVA. Ca_V2 channels include P/Q-, N-, and R-type Ca^{2+} channels, each with unique pharmacological and biophysical properties [13,22,29].

Whereas Ca_V1 channels are often uniquely privileged in mediating excitation–contraction (e–c) and excitation–transcription coupling, Ca_V2 channels generally mediate excitation–secretion (e–s) coupling at the presynaptic nerve terminal. There are some notable exceptions [30].

Despite the >90% homology in this subfamily, channelopathies have only been identified in $Ca_V2.1$ (P/Q-type) Ca^{2+} channels; these mutations range from missense and nonsense mutations to frame-shifts to polyglutamine expansions, as discussed in Section 5. It is unclear why $Ca_V2.2$ and $Ca_V2.3$ channels appear to be spared; however, one possibility is that mutations in these channels are more tolerated.

3.3. Ca_V3 (T-type) Ca^{2+} channels

The T-type Ca^{2+} channels share the least homology with Ca_V1 and Ca_V2 channels, and are classified as low-voltage activated (LVA), activating at voltages as negative as −70 mV [31]. Rather than coupling excitation directly to Ca^{2+}-dependent signaling mechanisms, Ca_V3 channels help shape excitability, playing important roles in neuronal development and activity, as well as cardiac function [32–36].

To date, three Ca_V3 genes in humans have been identified: $Ca_V3.1$ (α_{1G}), $Ca_V3.2$ (α_{1H}), and $Ca_V3.3$ (α_{1I}) [31,37,38]. Unlike Ca_V1 and Ca_V2 channels, which require auxiliary subunits for proper expression, trafficking, and biophysical properties, native T-type Ca^{2+} channels do not appear to require any accessory subunits (39,40, but see Refs 41,42).

Because of their critical role in neuronal excitability, it is not surprising that an increasing number of mutations in $Ca_V3.2$ have been linked to childhood absence epilepsy (CAE) [43]. Less obvious is the recent finding that mutations in this same protein correlate with a prevalence for autism spectrum disorder (ASD) [44]. These channelopathies are discussed in greater detail in Section 5.

4. Structural basis of key functions of Ca^{2+} channels

There is considerable interest in understanding the molecular determinants of the essential properties of voltage-gated Ca^{2+} channels, not only the generic features common to the entire family of channels but also those specific features that endow individual types of Ca^{2+} channels with their own distinct character. Structure–function studies have advanced considerably through the analysis of recombinant channels. The following sections are adapted from a book chapter co-authored with David Wheeler [5].

4.1. Activation

Changes in membrane potential, typically in the depolarizing direction, cause many kinds of ion channels to open, a process termed activation. As in the case of most other voltage-gated channels, activation of Ca^{2+} channels occurs more quickly and is more complete with larger depolarizations. A positively charged transmembrane segment (S4) has been found in each of the four homologous repeats of Ca^{2+}

channels [45] and is very similar to S4 segments in Na^+ and K^+ channels [46] where S4 has been firmly established as part of the voltage-sensing mechanism [47–50]. Analysis of Ca^{2+} channel gating is less extensive than that for the other ion channels, but Ca^{2+} channel chimeras have provided insights into specific contributions of individual repeats and motifs within them. The determinants of the rate of activation – slow, skeletal muscle-like activation as opposed to rapid, cardiac-like activation – were first traced to repeat I [51], then further localized to the membrane spanning segment IS3 and the external linker between IS3 and IS4 [52]. Mutant DHP receptors with α_{1S} sequence in this region activated relatively slowly, whereas mutants with the α_{1C} sequence in the same region activated relatively rapidly. More recent analysis has focused on how charge movements associated with Ca^{2+} channel gating are affected by mutations in α_1 subunits. There are many unanswered questions about the nature of activation gating in Ca^{2+} channels: the nature of S4 motion, its relationship to opening of the activation gate, and the location of that gate relative to the permeation pathway (for a stimulating review, see Ref. 22).

4.2. *Voltage-dependent inactivation*

Inactivation refers to the closing of channels during maintained depolarization, another fairly general property of Ca^{2+} channels. The speed of inactivation varies widely, ranging from very slow (hardly visible during depolarizations lasting seconds), as in the inner segments of photoreceptors, to relatively rapid (complete within tens of milliseconds), as in the case of certain Ca^{2+} channels found on nerve terminals [53]. In most cases, the underlying mechanism of inactivation is dependent on depolarization per se. The rate of this voltage-dependent inactivation (VDI) is strongly dependent on a region in the first repeat, including IS6 and extracellular and cytoplasmic residues on either side of it [54]. Residues in the S6 transmembrane segments of other repeats may also be influential [21,55]. The mechanism of channel closing may resemble "C-type" inactivation in K^+ channels, a concerted pinching of the permeation pathway by the S6 segments [56,57]. A "hinged lid" mechanism involving the I–II loop has also been put forward [58,59].

4.3. *Ca^{2+}-dependent inactivation*

Some Ca^{2+} channels, such as L-type channels, are subject to Ca^{2+}-dependent inactivation (CDI) along with VDI [60]. First discovered in *Paramecium* [61], CDI is an important negative feedback property that helps limit voltage-gated Ca^{2+} entry as intracellular $[Ca^{2+}]_i$ increases beyond a critical level [62]. Ca^{2+} influx through one channel can promote the inactivation of another adjacent channel, without a generalized elevation of bulk intracellular Ca^{2+} concentration, an instructive example of localized Ca^{2+} signaling [63]. Intracellular application of the Ca^{2+} chelator BAPTA (1,2-bis(o-aminophenoxy)ethane-N,N,N′,N′–tetraacetic acid) greatly diminishes such negative interactions within Ca^{2+} channel pairs. CDI has been proposed to transpire by a Ca^{2+}-induced shift of channel gating to a low open-probability mode, distinguished by a >100-fold slowing of entry into the open state [64].

As a powerful auto-feedback phenomenon, CDI has been the focus of many years of mechanistic investigation. CDI of L-type channels was observed in planar bilayers and could be reversed in the complete absence of protein kinases and ATP, excluding channel dephosphorylation as the sole controlling event in CDI [65]. Kinases and phosphatases are clearly important for other aspects of Ca^{2+} channel modulation [66]. The rapid kinetics of CDI initiation provide additional weight to the idea that CDI is initiated by the direct binding of Ca^{2+} to a Ca^{2+} sensor closely associated with the channel.

Initially, attention was focused on a consensus EF-hand (putative Ca^{2+}-binding motif) in the cytoplasmic carboxy terminal region downstream of repeat IV [67,68]. However, mutations that disrupted the EF-hand failed to inhibit CDI [69], thus directing attention to other domains of the L-type channel. Meanwhile, Reuter and colleagues [70] had found that portions of the C-terminal tail, beyond the EF hand region, also participated in CDI. Replacement of an ~80 amino acid segment of one particular splice variant speeded inactivation by >8-fold and also eliminated the Ca^{2+} dependence. The amino acid sequence that enabled CDI included a consensus "IQ motif," first identified as a CaM-binding motif in neuromodulin by Daniel Storm and colleagues [71] and characterized in myosins by Cheney and Mooseker [72]. The IQ motif in L-type channels was first spotted by Karl Deisseroth in our group (see Ref. 73). Composed of the isoleucine-glutamine combination and downstream basic residues, IQ motifs are typically able to bind CaM. Indeed, expression of dominant-negative CaM (CaM with key EF-hand Ca^{2+}-binding sites disabled) eliminated CDI [74–75]. Mutation of IQ to AQ or EQ abolished CDI [74]. Furthermore, CaM showed strong, one-to-one, Ca^{2+}-dependent interactions with a peptide containing the IQ sequence from the L-type channel [74].

The ability of dominant-negative CaM to prevent CDI implies that CaM must be tethered to the L-type channel, consistent with the rapid (millisecond) time scale of CDI development. Several groups have contributed studies on structural determinants of such tethering [e.g., 21,77–79]. The consensus view is that, even in the channel's resting state, the Ca^{2+} sensor CaM is poised to start signaling once Ca^{2+} entry begins.

4.4. *Ca^{2+}-dependent facilitation*

A fascinating feature of many Ca^{2+} channels is that they display two distinct gating properties that are regulated by the very ion that permeates the channel: not only CDI but also calcium-dependent facilitation (CDF). In the case of cardiac L-type channels, for example, CDI helps determine the length of the cardiac action potential plateau, and CDF contributes to the positive force–frequency relationship of the cardiac contraction. Both negative and positive feedback mechanisms are important for regulation of the electromechanical activity of the heart. The coexistence of CDI and CDF also holds for members of the Ca_V2 family, most notably $Ca_V2.1$ [80,81] but is less extensively studied and will not be discussed further in this review.

Interestingly, as with CDI, CDF also depends on CaM binding to the IQ motif [74]. This further raises the stakes for understanding the molecular details of this interaction. The structure of the Ca^{2+}/CaM–IQ peptide complex (Fig. 4) has been solved by two groups [82,83], who reached mutually consistent and partially overlapping conclusions. Key points of their elegant analysis may be summarized as follows. First, the interaction between CaM and the IQ motif differs significantly from CaM interactions with well-known partners such as Ca^{2+}/CaM-dependent protein kinase II (CaMKII) or myosin light chain kinase. Second, the side chain of the critical isoleucine is buried by the C-terminal lobe (C-lobe) of CaM, and replacement of IQ with AA has little effect on the structure of the peptide–Ca^{2+}/CaM complex. Third, the structures do not explain how CDI and CDF coexist, but provide a framework for interpreting previous structure activity analysis.

How can the same Ca^{2+}/CaM–IQ complex be involved in both forms of Ca^{2+}-dependent modulation? Presumably, this is achieved by controlling very different effector mechanisms (Fig. 5). While CDF depends on CaM binding to the IQ domain, it is slower to develop than CDI and probably requires more steps, including

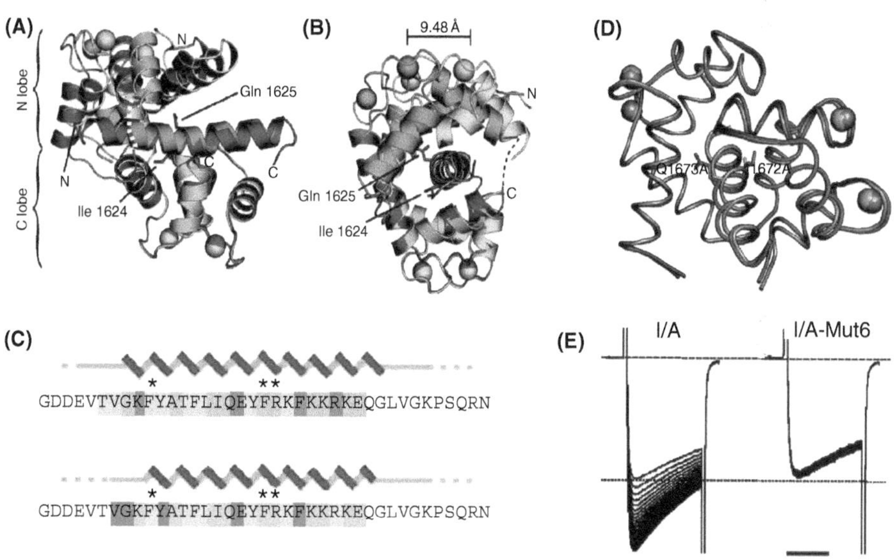

Fig. 4. High-resolution structures of Ca^{2+}/calmodulin (CaM) binding to IQ domain. (**A**) Ribbon diagram of the complex. Green, CaM Ca^{2+}/N lobe; blue, Ca^{2+}/C lobe; red, IQ domain, with the IQ residues in stick representation. Darker shades, complex A; lighter shades, complex C. (**B**) Right angle rotation of (A). (**C**) Amino acid sequence put into the context of the three-dimensional structure. Zigzag represents α-helical regions; straight line, non-α-helical residues. IQ peptide residues contacting CaM (4 Å) are subdivided into those interacting with the Ca^{2+}/N lobe (green), Ca^{2+}/C lobe (cyan), and both lobes (purple) [82]. (**D**) A very similar structure [83] demonstrates that replacement of IQ with AA, which abolishes Ca^{2+}-dependent inactivation and enhances calcium-dependent facilitation (CDF), does not significantly alter the structure of the Ca^{2+}/CaM-IQ complex. (**E**) CDF, measured with Ca2+ ions as charge carrier in *Xenopus* oocytes, shows prominent facilitation when the isoleucine of the IQ motif is replaced by an alanine (I/A, left), but not with the additional alanine replacement of six residues upstream of the IQ motif, TVGKFY (mut6), which were implicated in interaction of Ca^{2+}/CaM-dependent protein kinase II with the $Ca_V1.2$ C-terminal domain [84]. Similar abolition of CDF was independently obtained by alanine replacement of three amino acids indicated by the asterisks [82] (See Color Plate 15, p. 514).

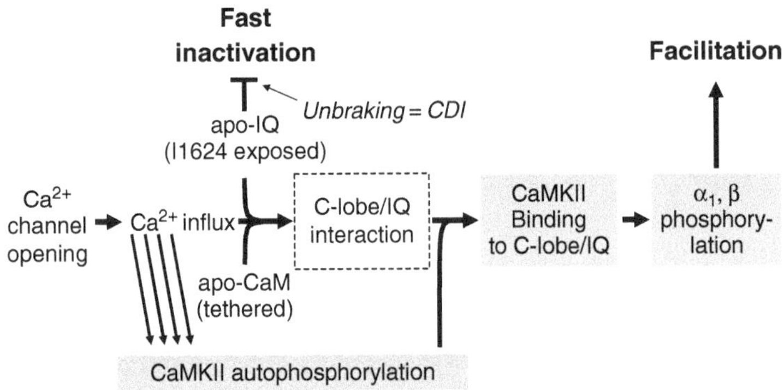

Fig. 5. "Push-Pull" hypothesis for Ca^{2+}-dependent modulation of L-type Ca^{2+} channels (C.F. Barrett and R.W. Tsien, unpublished). A tentative scheme to reconcile Ca^{2+}-dependent inactivation (CDI), calcium-dependent facilitation (CDF), and their joint dependence on calmodulin (CaM) binding to the IQ motif [74]. CaM-less IQ motif (apo-IQ) acts to retard fast inactivation, relying in part on the bulky isobutyl side chain of the isoleucine of the IQ motif [85,86]. Ca^{2+} influx allows the C-lobe of Ca^{2+}/CaM to bind to apo-IQ (outlined box), burying the isobutyl side chain, thereby acting to pull away the brake on fast inactivation. Thus, CDI is caused by disinhibition of an apo-IQ brake rather than promotion of a Ca^{2+}/CaM-IQ accelerator. CDF involves the biochemical steps indicated by shaded boxes, which are pushed forward by Ca^{2+}/CaM binding. During ongoing activity, Ca^{2+}/CaM-dependent protein kinase II (CaMKII) is tethered to $Ca_V1.2$. Ca2+/CaM binding to the C-interface, in combination with CaMKII interaction with the residues on the N-lobe interacting interface of the IQ motif [82,84], and promotes kinase phosphorylation of sites on the $Ca_V1.2$ (α_{1C}) [66,87] and β subunits [88], thus leading to facilitation of channel activity.

the activation of CaMKII [89,90], persistent tethering of CaMKII to the $Ca_V1.2$ C-terminus [84], and CaMKII phosphorylation of target sites. Recent studies have implicated residues important for the binding of CaMKII, some immediately N-terminal to the IQ motif, and identified possible phosphorylation sites on both $Ca_V1.2$ itself [66,87] and auxiliary subunit β_2 [88].

It has been proposed that the effector mechanism for CDI depends critically on the nature of the amino acid side chain at position 1624, normally occupied by the isoleucine of the IQ motif. Only amino acids with bulky hydrophobic side chains were able to preserve CDI [85]. This is consistent with the notion that the isoleucine side chain acts like a ligand to modify the gating state [86]. Here, we propose that the side group acts strictly as a braking mechanism that opposes fast inactivation rather than favoring it. In this scenario, binding of Ca^{2+}/CaM to the IQ motif masks the isoleucine side group and thereby removes the brake. Thus, as the active, uncomplexed IQ motif is sequestered by Ca^{2+}/CaM, CDI takes place through disinhibition.

In contrast to CDI, CDF was more tolerant of amino acid substitutions for the isoleucine, so that even an alanine was as good as phenylalanine, leucine, or valine [85]. Indeed, the double-mutant I1624A/Q1625A not only abolished CDI but also revealed or promoted a very prominent component of CDF. We propose that the same Ca^{2+}/CaM binding that removes the brake on fast inactivation also sets in motion CDF. However, the structural requirement is merely that C-lobe binds to its cognate interface, leaving the other surface of the IQ motif (N interface) free to interact with the effector machinery. The obvious candidate is CaMKII itself, because its binding depends on residues within the N interface.

4.5. Selectivity, permeation, and block by divalent cations and protons

As mentioned in the introduction, Ca^{2+} channels operate with great efficiency in allowing Ca^{2+} permeation with rapid turnover rates while also showing exquisite selectivity for Ca^{2+} over other more abundant extracellular ions such as Na^+ [2,91,92]. Permeation through open Ca^{2+} channels has been studied with patch-clamp techniques, often with ~100 mM external barium (Ba^{2+}) to increase the unitary current size. Under these conditions, the Ca^{2+} channel type with the largest Ba^{2+} conductance (L-type, ~25 pS) shows a unitary current amplitude of approximately –1.6 pA at zero membrane potential, corresponding to a transfer rate of 5 million Ba^{2+} ions per second [93]. Recordings from L-type channels at physiological levels of external Ca^{2+} (2 mM) yield a much small unitary conductance, 2.4 pS [94]. Selectivity of voltage-gated Ca^{2+} channels for Ca^{2+} ions over monovalent cations is >1000-fold [93], so that Ca^{2+} is the main charge carrier even when it is greatly outnumbered by others ions, as under normal physiological conditions. As the Ca^{2+} channel pore is relatively large (~6 A diameter) [95], the selectivity cannot be explained by molecular sieving [91].

Mutagenesis studies have revealed that the Ca^{2+} channel pore contains a single locus of high affinity binding within the pore, which can either bind a single Ca^{2+} ion with high affinity ($K_d \sim 1\ \mu M$) or multiple divalent cations with lower affinity [96]. This locus comprises four highly conserved glutamate residues, one from each of the pore-lining S5–S6 linkers of the α_1 subunit repeats [96–99]. The localization of the Ca^{2+} interaction to a specific cluster of carboxylates fits in with earlier hypotheses about the mechanism of selectivity and permeation [100–103]. Smaller cations such as Na^+ and K^+ are thought to bind weakly and permeate rapidly in the absence of Ca^{2+} but are rejected when Ca^{2+} occupies the high-affinity site. Ca^{2+} fluxes become appreciable at millimolar levels of $[Ca^{2+}]_o$, when mass action drives more than one Ca^{2+} ion into the pore. The flux rate for Ca^{2+} is then dependent on negative interactions between individual divalent cations within the pore (either through electrostatic repulsion or competition for a limited number of negatively charged oxygen groups). Recent reviews bring into sharp focus the remaining questions about how the negative interactions are brought about [92,104,105]. In the absence of Ca^{2+}, larger divalent cations such as Ba^{2+} and strontium (Sr^{2+}) permeate the channel better than Ca^{2+} because they do not bind as tightly to the high-affinity sites. Ca^{2+} channels are generally impermeant to magnesium (Mg^{2+}), probably because of its slow rate of dehydration. At supra-physiological concentrations, however, Mg^{2+} is capable of blocking Ca^{2+} flow through the channel.

Blockade of Ca^{2+} channels can also occur with larger divalent cations such as Cd^{2+} and Co^{2+}, which potently inhibit Ca^{2+} influx by binding more tightly than Ca^{2+} to the high-affinity site [106]. The order of potency of block depends somewhat on the subtype of Ca^{2+} channel. As with the binding of Ca^{2+}, the block of L-type channels by other divalent cations is also strongly reduced in channels with changes in the conserved glutamates [96,97]. Open Ca^{2+} channels are also blocked by H^+ ions, which titrate the glutamates that support Ca^{2+} selectivity. Mutagenesis experiments have demonstrated that the carboxylate side chains from repeats I, II, and III act together to form the proton-binding site [107,108]. The coordinated action of the multiple oxygen groups creates a site with much higher affinity for protons than a single carboxylate alone. Acidification also decreases the degree of Ca^{2+} channel opening by shifting the voltage dependence of gating toward more depolarized potentials [109].

4.6. Responsiveness to drugs and toxins

Another interesting characteristic of voltage-gated Ca^{2+} channels is their ability to respond to drugs and toxins [8]. Sensitivity to neurotoxins, many derived from the venoms of spiders and marine snails, is an important earmark of several individual types of voltage-gated Ca^{2+} channels [110,111]. In the case of ω-conotoxin-GVIA (ω-CTx-GVIA), the potent and selective blocker of N-type channels, high-affinity binding seems to involve extracellular channel domains located in all four repeats of the $Ca_V2.2$ (α_{1B}) subunit, but particularly in repeats I and III [112]. Specific differences between $Ca_V2.2$ and $Ca_V2.1$ have been localized to individual amino acids in loops near the S5–S6 region. Thus, the available evidence is consistent with a scenario in which the toxin straddles the mouth of the pore. The precise mechanism of block is not known, but observations of competition with permeant cations [113,114] suggest the possibility of direct occlusion of the permeation pathway. Further analysis may allow the rigid toxin to be used as a kind of molecular caliper to gain information about the outer aspects of the channel. In this context, it is noteworthy that blockade of N-type channels by ω-CTx-GVIA, ω-Conotoxin-MVIIA (ω-CTx-MVIIA), and related toxins is strongly voltage dependent as a result of enhancement of block by channel inactivation [115]. The dependence of toxin block on channel gating is a familiar phenomenon in Ca^{2+} channels; for example, ω-Agatoxin-IVA (ω-Aga-IVA) block of P/Q-type channels is strongly antagonized by channel activation [116].

The pore region has also been implicated in the block of L-type Ca^{2+} channels by DHPs and other small organic agents. This was suggested by early photoaffinity labeling experiments in which photo-reactive DHPs labeled a putative extracellular loop between segments IIIS5 and IIIS6 and transmembrane segment IVS6 [117]. General agreement has now been reached that the IVS6 segment is critical for the action of DHPs and phenylalkylamines [118]. Some of the most compelling evidence comes from experiments where DHP sensitivity is transferred to chimeras based on the $Ca_V2.1$ (α_{1A}) subunit [119]. These reinforce the idea that the minimum sequence for DHP sensitivity includes segments IIIS5 and IIIS6 and the connecting linker, as well as the IVS5–IVS6 linker plus segment IVS6. Interestingly, the DHP-responsive α_{1A} chimera still retains sensitivity to ω-Aga-IVA and ω-CTx-MVIIC.

The idea that organic agents interact at or near the pore has been further buttressed by consideration of the Ca^{2+} dependence of drug binding [120,121]. The high-affinity Ca^{2+} site that regulates DHP binding can be eliminated by replacing the conserved pore glutamate in repeat III with a lysine residue. This finding and other data indicate that high-affinity DHP binding is dependent on Ca^{2+} coordination by the same glutamate residues, which form the selectivity locus [99]. A three-state model has been proposed whereby affinity for DHPs is promoted by the presence of one bound Ca^{2+} ion within the pore as opposed to zero or two [121].

5. *Vital biological functions of specific Ca^{2+} channels in relation to genetic and acquired diseases*

The remainder of this chapter will be directed toward the multiple functions of voltage-gated Ca^{2+} channels. Since, the mid- 1990s, there has been a steep increase in knowledge about diseases that arise from genetic and acquired abnormalities in Ca^{2+} channel function. This provides us an opportunity to combine perspectives on how Ca^{2+} channels support vital biological functions and how alterations in Ca^{2+} channel properties give rise to pathophysiology. We will use particular channel types to illustrate how key physiological processes are carried out without implying that the channel type plays an exclusive role. With little exception, individual Ca^{2+} channel types serve multiple functions, and particular functions can be carried out by multiple kinds of Ca^{2+} channel.

5.1. *Excitation–contraction coupling exemplified by function of $Ca_V1.1$ (α_{1S}) L-type channels*

L-type Ca^{2+} channels are essential participants in e–c coupling in skeletal, cardiac, and smooth muscle, although other channel types may play a supporting role in some of these cells [122]. A striking example of how L-type channels contribute to e–c coupling is provided by $Ca_V1.1$ (α_{1S}) channels in skeletal muscle. Composed of α_{1S}, β_{1a}, γ_1, and $\alpha_{2\delta-1}$ subunits, these channels are largely localized to the transverse tubule system where they support e–c coupling by mediating depolarization-induced Ca^{2+} release from the sarcoplasmic reticulum [123]. Ca^{2+} entry through the L-type channel is not immediately required for skeletal muscle contraction (reviewed in Refs 124,125), in contrast to cardiac muscle, where Ca^{2+} entry is essential [126,127]. Interestingly, blockade of L-type channels in skeletal muscle by organic Ca^{2+} antagonists completely inhibits contraction [128]. The explanation of these findings centers on gating charge movement in the T-tubule membrane, which was known to be essential for intracellular Ca^{2+} release [129]. DHPs eliminate charge movement, thereby blocking skeletal muscle contraction [123]. These findings helped establish that DHP-sensitive L-type Ca^{2+} channels act as voltage sensors to link T-tubule depolarization to intracellular Ca^{2+} release through RyR channels in the sarcoplasmic reticulum.

The structural basis of the L-type Ca^{2+} channel-RyR signaling was clarified in elegant experiments by Tanabe, Numa, Beam, and their colleagues. The cloning of the DHP receptor protein from skeletal muscle led immediately to its identification as a voltage-gated channel [45], and expression of the cloned DHP receptor in dysgenic skeletal muscle myotubes could restore electrically evoked contractility in these formerly non-responsive cells [130], along with L-type Ca^{2+} current [130,131] and gating charge movement [132]. While the skeletal DHP receptor ($Ca_V1.1$) allowed contraction even in the absence of extracellular Ca^{2+}, the cardiac L-type Ca^{2+} channel ($Ca_V1.2$) restored contractility only with Ca^{2+} entry [133]. The structural basis of the skeletal-type e–c coupling was investigated with molecular chimeras. By inserting pieces of the $Ca_V1.1$ gene into a $Ca_V1.2$ background, Tanabe et al. [133] showed that the key domain was the intracellular loop joining repeats II and III of $Ca_V1.1$. More recently, other groups have shown that purified II–III loop fragments can directly activate the RyR [134,135] and that this region may contain phosphorylation sites for the regulation of e–c coupling [136]. Kurt Beam and colleagues [137] have further shown that the L-type channels and RyRs communicate in the retrograde as well as orthograde direction. A stretch of ~50 amino acids in the II–III loop of the $Ca_V1.1$ subunit is critical for both kinds of signaling [138]. Similar retrograde signaling from RyRs for regulation of L-type channels (presumably $Ca_V1.2$ rather than $Ca_V1.1$) has been reported in neurons [139]. The potent interactions between Ca^{2+} channels in closely apposed membranes provide a powerful but intricate system for Ca^{2+} signal processing.

5.1.1. $Ca_V1.1$ mutations linked to hypokalemic periodic paralysis

Three missense mutations in CaV1.1 have been linked to hypokalemic periodic paralysis (HypoPP), a dominantly inherited disorder with episodic weakness accompanied by reduced serum potassium levels. In these mutations, an arginine is replaced by either a histidine or glycine, causing the loss of a positive charge within the highly conserved voltage-sensing segments of the protein (IIS4 and IVS4). (It is interesting to note that a closely related disease, HypoPP type 2 (HypoPP2), also involves charge neutralization in analogous S4 segments but in a voltage-gated sodium channel rather than a Ca^{2+} channel [140,141]).

The α_{1S} mutations have been carefully examined in both cultured myotubes and exogenous expression systems [142–145], but no clear pattern has emerged to provide a satisfactory account for the pathology of HypoPP. In some cases, a gain of function was reported [142,143,146], whereas in others, the effect was described as a clear loss of function [142,144,145]. Thus, the connection between altered channel function and HypoPP remains unclear. Indeed, it is conceivable that the disease might arise from defects not directly associated with channel function per se, for example, altered expression, subcellular distribution, or signaling to the sarcoplasmic reticulum.

5.1.2. $Ca_V1.1$ mutations linked to malignant hyperthermia

Malignant hypothermia (MH), a skeletal muscle disorder, is one of the main causes of anesthesia-related death. Susceptibility to MH has been generally associated with mutations in the RyR [147–149]. However, a single mutation in the $Ca_V1.1$ channel has been linked to MH [150]. Interestingly, the mutation is an arginine-to-histidine substitution, similar to the mutations identified in HypoPP, but within the intracellular loop connecting domains III and IV rather than in an S4 segment. How this mutation gives rise to MH is currently unclear.

5.2. *Excitation–transcription coupling and changes in gene expression exemplified by $Ca_V1.2$ (α_{1C}) and $Ca_V1.3$ (α_{1D}) L-type Ca^{2+} channels*

$Ca_V1.2$ (α_{1C}) channels play a critical role in linking membrane depolarization to cellular contractility in cardiac myocytes, in partial analogy with the operation of $Ca_V1.1$ channels in skeletal muscle. As described above, $Ca_V1.1$ channels link depolarization to Ca^{2+} efflux from sarcoplasmic reticulum, with plasmalemmal Ca^{2+} entry not required for contraction. In contrast, contraction of cardiac muscle requires Ca^{2+} influx through $Ca_V1.2$ channels at the plasma membrane and subsequent Ca^{2+}-induced Ca^{2+} release [126,127].

$Ca_V1.2$ (α_{1C}) channels also perform the vital function of coupling surface membrane excitation to nuclear gene transcription. In neurons, and perhaps heart cells as well, $Ca_V1.2$ (α_{1C}) channels perform this function in parallel with $Ca_V1.3$ (α_{1D}) channels [151,152]. The properties of $Ca_V1.2$ and $Ca_V1.3$ are relatively similar with regard to channel regulation by CDI and pharmacological blockade with DHPs. Indeed, until recently, these two channels were generally grouped together under the generic term of "neuronal L-type channels." However, recent studies emphasize significant biophysical differences: $Ca_V1.3$ channels are activated at more negative potentials and are less sensitive to DHPs than $Ca_V1.2$ channels [16]. Furthermore, $Ca_V1.3$ channels are reversibly inhibited by ω-CTx-GVIA [153], a peptide blocker of N-type channels. While investigators wait for the discovery of agents that selectively inhibit $Ca_V1.2$ or $Ca_V1.3$, knowledge of their partially independent, partially overlapping functional roles may be gleaned from recent gene deletion experiments [154].

As a group, L-type channels work downstream of extracellular factors that influence cell growth and activity by depolarizing the membranes of their target cells [155]. The membrane depolarization opens voltage-gated Ca^{2+} channels, and the resulting influx of Ca^{2+} can trigger gene transcription [156]. L-type Ca^{2+} channels were originally found to be important in this cascade because agonists of these channels such as Bay K 8644 could induce the expression of several protooncogenes in the absence of other stimuli [157]. Indeed, the mode and location of Ca^{2+} entry may be important to how the Ca^{2+} signal is interpreted by the cell [65,158]. Some recent studies have shed light on the cascade of events, which follow influx of Ca^{2+} through L-type channels. Much of the work has focused on how Ca^{2+} entry is involved in activation of gene expression through the cyclic–adenosine monophosphate (cAMP) and Ca^{2+} response element (CRE) and its nuclear binding protein (CREB) [159,160]. The interaction of CREB with the CRE is facilitated by CREB phosphorylation at serine-133 [161], catalyzed by several kinases including Ca^{2+}/CaM kinases II and IV, cAMP-dependent protein kinase [162], and others. Thus, rises in $[Ca^{2+}]_i$ can act either directly, through Ca^{2+}/CaM and its dependent kinases, or indirectly, by stimulating Ca^{2+}/CaM-sensitive adenylyl cyclase leading to increased cAMP levels. Later studies showed that Ca^{2+} entry through L-type channels can trigger CREB phosphorylation [163–165] and that Ca^{2+} probably binds to a target molecule within 1 μm of the point of entry [166]. The diversity of signaling mechanisms can be illustrated by a few examples (Fig. 6). Signaling to the nucleus may involve translocation of a transcription factor, NF-AT [167], a Ca^{2+}-binding protein, CaM itself [164], or diffusion of Ca^{2+}, to interact with the nuclear Ca^{2+}-binding protein DREAM [168], which itself regulates transcription. In the first two instances, L-type channels appear to be privileged relative to other mechanisms of Ca^{2+} entry or release. The nature of the local signal transduction apparatus is still not completely understood, although there is compelling evidence that the binding of Ca^{2+}/CaM to the α_1 C-terminal IQ motif is one necessary condition [169]. To promote CREB phosphorylation, an additional requirement is that the L-type channel be associated with key cytoplasmic proteins. The long C-terminal $Ca_V1.3$ splice variant appears to use its C-terminal amino acids to recognize and bind to the postsynaptic adaptor protein Shank [170]. In parallel findings on $Ca_V1.2$, a PDZ interaction sequence (VSNL) at the C-terminal end of the α_1 subunit is critical for L-type channels to couple to CREB phosphorylation [171]. Taken together, these studies suggest that appropriate localization of L-type channels is critical in coupling to downstream

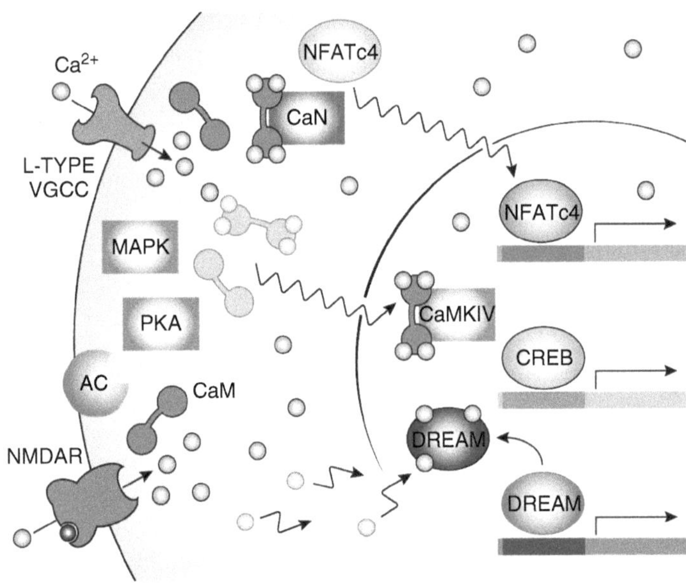

Fig. 6. A schematic representation of the early stages (0–5 min) of activity-dependent gene expression. Following rises in intracellular Ca^{2+}, DREAM dissociates from DNA resulting in the lifting of transcriptional repression. Within seconds of Ca^{2+} entry through L-type Ca^{2+} channels and NMDA receptors, calmodulin (CaM) translocates to the nucleus, supporting CREB phosphorylation through activation of CaMKIV. Nearly as rapid, NFATc4 also undergoes translocation to the nucleus following its dephosphorylation by calcineurin (CaN). Other pathways not shown here include the MAPK/PKA pathways, which exert their influence over a longer time scale. Here, VGCC denotes voltage-gated Ca^{2+} channels, principally L-type. (Reprinted with permission from Deisseroth et al. [151]) (See Color Plate 16, p. 515).

signaling, but leave unanswered the detailed nature of the signaling machinery that puts L-type channels in a privileged position for signaling to the nucleus, relative to other channel types.

5.2.1. $Ca_V1.2$ mutations linked to Timothy syndrome

Timothy syndrome (TS) is a rare childhood multisystem disorder [172–174]. TS patients exhibit several developmental defects, including syndactyly (webbed extremities) and facial dysmorphism, as well as cardiac arrhythmia, including prolonged QT intervals. Moreover, TS patients often display cognitive deficits and test positive for autism. Recently, TS was shown to arise from a mutation in exon 8 of $Ca_V1.2$, leading to a glycine-to-arginine substitution at the cytoplasmic border of IS6 [175]. It is interesting to note that exon 8 is expressed in most but not all splice variants of $Ca_V1.2$ [176]; in some splice variants, exon 8a is expressed, where mutations analogous to the TS mutation have been shown to give rise to severe arrhythmia disorder [177]. Northern and dot blot analyses have shown that both exons 8 and 8a are widely expressed in various tissues, including heart and brain [175,177].

The mutations in exons 8 and 8a appear in regions previously identified as "hotspots" for VDI; mutations in this and homologous regions have been shown to slow inactivation, leading to prolonged channel activity [178]. It is therefore not surprising that the mutations linked to TS and severe arrhythmia disorder dramatically slow inactivation [175,177], manifesting as an apparent gain-of-function phenotype. Specifically, the mutations appear to affect the hinge of an inactivation lid formed by the I–II loop, the intracellular linker between domains I and II [58,59]. Thus, the TS mutation has been interpreted as a "hinge-opathy." Although it is clear to see the cause–effect relationship between prolonging cardiac Ca^{2+} channel activity and subsequent prolonged QT interval, the link between the channelopathy and the effects in the central nervous system (CNS) are less clear.

In addition to its effects on inactivation, the TS mutation has also been described as a "phosphorylopathy", as the arginine substitution as position 406 creates an aberrant site for phosphorylation by the Ca^{2+}/CaM kinase CaMKII [66]. Analysis of single-channel activity by cell-attached patch showed that TS channels reside in a gating mode characterized by long channel openings, reminiscent of the pattern of gating induced by CaMKII phosphorylation of wild-type (WT) $Ca_V1.2$ channels [90], known as "mode 2" [101], leading to the hypothesis that the TS mutation causes a gain-of-function phenotype by increasing open probability. It remains to be tested whether the mutations in exon 8a which cause severe arrhythmia disorder (but not TS and autism) also create a phosphorylopathy.

It should be noted that both the hingeopathy and the phosphorylopathy hypotheses are not necessarily mutually exclusive. Although the defect in inactivation is likely to occur in all cells where the channel is expressed, the aberrant phosphorylation might be cell or tissue specific. Moreover, the identity of the accessory subunit associated with the channel might determine the degree to which the mutation affects inactivation. For example, in cardiac myocytes α_{1C} preferentially associates with the β_2 subunit [179], which confers slow inactivation compared with other subunits [180].

Although no channelopathy has yet been discovered for $Ca_V1.3$, defects in these L-type channels would be expected to result in hearing deficiency and cardiac rhythm disturbances, based on recent studies in $Ca_V1.3$ knockout (KO) mice [181–183].

5.3. *Excitation–secretion coupling: generic properties exemplified by $Ca_V1.4$ (α_{1F}) L-type Ca^{2+} channels*

The most commonly studied role of Ca^{2+} is its ability to trigger neurotransmitter release. The importance of Ca^{2+} ions in the release of neurotransmitter has been appreciated for more than 60 years [184]. Seminal work from Katz [185] and colleagues in neurons (and parallel contributions from W.W. Douglas in endocrine cells [186]) demonstrated that Ca^{2+} ions exert a strong influence on the amount of neurotransmitter that is released from nerve terminals and other secretory systems. The action of Ca^{2+} ions in the regulation of neurotransmission was shown to be cooperative, in a manner consistent with the binding of four or more Ca^{2+} ions to a receptor to trigger release [187]. The importance of Ca^{2+} action in the nerve terminal was further supported by the observation that the injection of Ca^{2+} into the terminal triggered the release of transmitter at the squid giant synapse [188]. Subsequently, the Ca^{2+}-sensitive protein, aequorin, was used to show that presynaptic $[Ca^{2+}]_i$ increases during neurotransmission [189].

Studies using simultaneous voltage clamp of the presynaptic terminal and postsynaptic axon of the squid giant synapse provided direct measurements of the Ca^{2+} currents in the presynaptic membrane, which trigger the release of neurotransmitter [190,191]. Ongoing issues include the identification of presynaptic Ca^{2+} channels and clarification of the functional consequences of their diversity (for review, see Refs 110,192–196). All of the Ca_V2 family members have been found to participate in transmitter release.

Although the majority of studies of neurotransmitter release have failed to identify a role for L-type Ca^{2+} channels [192], this subtype has been implicated in a few specialized forms of exocytosis. For example, activation of L-type channels is required for zona pellucida-induced exocytosis from the acrosome of mammalian sperm [197]. L-type channels also seem to play a role in mediating hormone release from endocrine cells. Inhibition of L-type Ca^{2+} channels reduces insulin secretion from pancreatic β cells [198,199], oxytocin, and vasopressin release from the neurohypophysis [200], luteinizing hormone-releasing hormone release from the bovine infundibulum [201], and catecholamine release from adrenal chromaffin cells [202]. L-type channels also seem to play an important role in supporting release of GABA (γ-aminobutyric acid) from retinal bipolar cells [203,204], as well as dynorphin release from dendritic domains of hippocampal neurons [205]. In some cases, L-type channels may function to release excitatory amino acid transmitters, in response to particular patterns of activity [206], in cells that exhibit graded potentials [207], during extended depolarizations with high K^+, or under the experimental influence of the DHP agonist Bay K 8644 (see Ref. 208).

Over the last decade, it has become increasingly clear that L-type (Ca_V1) family members play a critical role in supporting secretion in specific neurons in both the auditory and the visual pathways [209]. For example, sensory nerve terminals of hair cells, the mechanosensory cells that support hearing and balance, use L-type channels of the $Ca_V1.3$ (α_{1D}) class to control glutamate release [210]. Likewise, in the visual pathway, $Ca_V1.3$ channels are also employed by presynaptic terminals of cone photoreceptors (not to be confused with rods, which largely rely on another Ca_V1 channel family member).

It is not completely clear why certain nerve terminals in these sensory pathways employ L-type channels and not a member of the Ca_V2 class as seen at the bulk of synapses in the CNS. Immunocytochemistry in the cochlea shows the presence of several types of Ca^{2+} channels [211]. Similarly, all varieties of α_1 subunit can be found within various cells of the retina [209]. This further begs the question of whether there exist general rules for determining which kinds of Ca^{2+} channels are used for particular cellular functions. One perspective emphasizes that neurons in sensory pathways are subject to unusual functional demands and, in some cases, rarely stop releasing neurotransmitter. This need can be met by employing channels that become activated at relatively negative potentials (matched to the range of voltages easily achieved by non-selective sensory transduction channels) and that resist inactivation during maintained depolarizations. How Ca_V1 family members do away with inactivation is an area of ongoing research, but one possible scenario involves alternative splicing of the $Ca_V1.3$ (α_{1D}) gene (see Ref. 212, see also Ref. 213).

5.3.1. Channelopathies in $Ca_V1.4$ L-type Ca^{2+} channels in retina

A fourth gene encoding a Ca_V1 Ca^{2+} channel was identified in 1998. Called α_{1F} [having been identified after the gene encoding α_{1E} ($Ca_V2.3$) R-type channels], this gene showed high homology to other members of the L class of Ca^{2+} channels [214]. Unlike the other nine Ca^{2+} channel Ca_V genes, $Ca_V1.4$ was discovered by positional cloning (in an attempt to map the chromosomal region mutated in patients exhibiting a form of night blindness). Thus, this is an interesting example of how a channelopathy has led to the discovery of the affected gene. Although $Ca_V1.4$ has been shown definitively to express primarily in the retina [215], its function in this tissue and others is not completely understood. However, targeted disruption leads to profound defects in Ca^{2+} signaling, synaptic transmission, and cellular organization in the mouse retina [216]. Moreover, recent experiments have verified that $Ca_V1.4$ is a bona fide L-type Ca^{2+} channel, with hallmark sensitivity to DHPs [215].

X-linked congenital stationary night blindness (CSNB) is a recessive non-progressive retinal disorder, which includes night blindness, impaired visual acuity, and myopia or hypermetropia. In complete CSNB, patients have undetectable rod function; patients with incomplete CSNB (termed CSNB2) exhibit impaired but measurable rod function [217]. The locus for CSNB2 has been mapped to the *CACNA1F* gene, which encodes the retinal $Ca_V1.4$ L-type Ca^{2+} channel [214]. Whereas the mutations in $Ca_V1.1$ and $Ca_V1.2$ described above are missense mutations, the mutations linked to CSNB2 include missense and nonsense mutations, as well as frame-shifting insertions or deletions, leading to premature stop condons [214,218].

Recently, an X-linked retinal disorder similar to CSNB2 but with notably different clinical features was characterized in a New Zealand family [219]. Genetic screening implicated a single missense mutation, leading to an isoleucine-to-threonine substitution in the highly conserved cytoplasmic region of IIS6 [220]. Unlike the putative loss-of-function mutations associated with CSNB2, the I745T mutation was shown to confer a gain of function. When expressed in recombinant cells, the mutant channels displayed significantly impaired inactivation. Indeed, the phenotype is highly reminiscent of the TS mutation in $Ca_V1.2$ discussed above. Beyond its effects on inactivation, the I745T mutation also causes a dramatic shift in voltage-dependent activation to more negative potentials, rendering the channel much more sensitive to voltage. In this respect, the functional consequences for $Ca_V1.4$ were different than those for $Ca_V1.2$ channels affected by TS, which displayed no shift in voltage dependence of activation [175].

It is particularly interesting to note that several individuals bearing the I745T mutation in $Ca_V1.4$ also showed autistic features [220]. Although the mutations in $Ca_V1.2$ and $Ca_V1.4$ appear in homologous regions of the protein (IS6 and IIS6, respectively), $Ca_V1.4$ does not appear to be expressed in the brain, at least not at detectable levels [215]. Thus, the link between a mutation in a retinal Ca^{2+} channel and impaired intellectual function and autism is unclear.

5.4. Excitation–secretion coupling at CNS nerve terminals exemplified by function of $Ca_V2.1$ P/Q-type Ca^{2+} channels

Ca^{2+} channels from the Ca_V2 subfamily are the primary types responsible for e–s coupling at fast excitatory and inhibitory synapses in the CNS. Interestingly, just as the II–III loop of the Ca_V1 channel interacts with the Ca^{2+} channel's effector for contraction, the II–III loop of the Ca_V2 channel engages in interactions with protein components of the secretory apparatus [221]. Initially, the binding of $Ca_V2.2$ to SNARE proteins was hypothesized as important for precise channel localization at the active zone [221], but more recent evidence suggests that the regulation of channel trafficking to the nerve terminal may be more critical [222]. Anchoring of Ca_V2 channels to elements in the active zone has been elegantly described in morphological views of motor nerve terminals [223] and analysis of protein–protein interactions with molecular components of the presynaptic meshwork, including the proteins Mint1 and CASK [224].

Which of the Ca_V2 channels triggers neurotransmission? Neurotransmitter release at most CNS synapses is generally mediated through the joint action of multiple types of Ca_V2 family of channels, most prominently, P/Q- and N-type [192,225]. This seems to be the case for large presynaptic structures [226,227] as well as typical $\sim 1\,\mu m$ CNS nerve terminals [195,228,229]. R-type channels also participate in triggering exocytosis in those instances where its potential contribution has been carefully explored [230]. Thus, all three members of the Ca_V2 subfamily play a significant role in fast central neurotransmission. The biophysical properties of these channels (e.g., their rapid activation upon strong depolarization and brisk deactivation following repolarization) are well suited to respond to typical presynaptic action potentials, which reach strongly positive overshoots but have brief duration, on the order of a millisecond [231].

The detailed composition of Ca^{2+} channel subtypes that mediate transmission can vary widely among individual excitatory nerve terminals [232,233]. Heterogeneity in the employment of Ca^{2+} channels has also been reported at inhibitory synapses. At some hippocampal inhibitory synapses, GABA release is mediated entirely by N-type or by P/Q-type Ca^{2+} channels [234]. Differences in the reliance on various Ca_V2 family members take on additional significance when synaptic circuits are the target of pharmacological and possibly therapeutic intervention. For example, in the dorsal horn of the spinal cord, center stage in the circuitry of pain transmission/modulation, inhibitory synaptic transmission is dominated by P/Q-type Ca^{2+} channels; in contrast, N-type channels are more closely associated with glutamate release from the peptidergic primary sensory neurons that convey nociceptive information to the dorsal horn [228,235].

5.4.1. Mutations in $Ca_V2.1$ (P/Q-type) Ca^{2+} channels can give rise to migraine

Familial hemiplegic migraine type 1 (FHM1) is a rare hereditary form of migraine with aura, likely caused by abnormal neurotransmission as well as ionic homeostasis in the CNS [236]. Since the identification of the FHM1 gene in 1996 [237], more than 20 missense mutations in $Ca_V2.1$ have been associated with the disease [238]. However, the mechanism by which these mutations affect P/Q-channel properties and ultimately contribute to the migraine phenotype is still not well understood. In heterologous expression systems, some FHM1 mutant channels displayed increased Ca^{2+} influx at the single-channel level [239–241]. Introducing the R192Q mutation into the mouse $Ca_V2.1$ gene through homologous recombination resulted in increased P/Q-type current density as well as a decreased threshold for cortical spreading depression (CSD), a progressive slow wave of neuronal activity followed by silencing, believed to be the substrate for migraine aura. Consistent with this finding, mice with spontaneous loss-of-function $Ca_V2.1$ mutations exhibit increased CSD threshold [242]. Thus, it seems clear that increased Ca^{2+} influx through P/Q-type channels leads to an increased likelihood of CSD and perhaps increased susceptibility to migraine. However, it is still a matter of debate whether aura is a causative precursor to migraine pain or merely parallels its development [243,244].

In contrast to the above-mentioned data, neurons from $Ca_V2.1$ knockout (KO) mice expressing human α_{1A} subunits bearing FHM1 mutations consistently displayed reduced P/Q-type current density relative to those expressing wild-type (WT) human α_{1A} [239,245]. One mutation (T666M) was studied at the level of gating currents and was found to express and traffic to the membrane normally but exhibited impaired gating [246]. Furthermore, mutant P/Q-type channels are impaired in mediating both excitatory as well as inhibitory neurotransmission in accordance with their deficiency in supporting whole-cell Ca^{2+} currents [245,247]. Moreover, synapses expressing mutant channels exhibit altered sensitivity to modulation by G-protein-coupled receptors [247,248]. Finally, expression of FHM1 mutant channels in WT neurons reduces the synaptic contribution of P/Q-type channels, consistent with the dominant inheritance of FHM1. These results are consistent with previous functional studies, which demonstrated that attenuation of P/Q-type channel activity leads to hypersensitivity of dorsal horn neurons and consequently enhanced pain transmission [228,249–251].

It is noteworthy that there is a remarkable clinical heterogeneity among families with the same FHM1 mutation (T666M) [252]. Therefore, one must exercise caution when attempting to characterize FHM1 mutations as simply loss- or gain-of-function. Indeed, both could give rise to a similar phenotype; precedence for this comes from $Na_V1.1$, a voltage-gated sodium channel, in which both loss- and gain-of-function mutations result in an epilepsy phenotype [253].

5.4.2. Mutations in $Ca_V2.1$ Ca^{2+} channels may also cause episodic ataxia 2 and epilepsy

Mutations in *CACNA1A*, the gene encoding $Ca_V2.1$ channels, have also been linked to episodic ataxia type 2 (EA2), a syndrome characterized by episodes of impaired coordination and balance [254]. In addition to nonsense mutations that result in truncated channel proteins, there have been reports of missense mutations [254], a multiple base-pair insertion [255], and intronic mutations that caused abberant splicing [256]. Unlike the conflicting results concerning FHM1 mutations, essentially all EA2 mutations studied lead to severely reduced or abolished P/Q-type currents. A dominant negative effect of some EA2 mutations on protein trafficking has also been reported [257–259]. Also of possible relevance to the effect of EA2 missense mutants, dominant negative effects on neurotransmission have been observed for FHM1 mutants coexpressed with WT channels [245,247]. The likely interpretation is that $Ca_V2.1$ channels compete for a limited number of presynaptic "slots."

Loss-of-function mutations of $Ca_V2.1$ have also been identified in epileptic patients [260,261]. Interestingly, there is substantial overlap in clinical symptoms for individuals diagnosed with episodic ataxia, epilepsy, or hemiplegic migraine.

5.4.3. Spinocerebellar ataxia type 6: a channelopathy or "glutaminopathy"?

Spinocerebellar ataxia type 6 (SCA6) is characterized by late-onset cerebellar ataxia and, in particular, loss of cerebellar Purkinje neurons [262,263]. SCA6 is a polyglutamine disease that results from an expansion of CAG trinucleotide repeats [264] within exon 47 of the "long" carboxyl-terminal isoform of $Ca_V2.1$ (α_{1A}) [265–269]. However, unlike other polyglutamine diseases, the number of extra CAG repeats in SCA6 is relatively small. Affected individuals can have as few as 19 repeats [270,271]; in contrast, Huntington's disease, another polyglutamine disease, requires at least 42 repeats [272].

Despite many studies, the molecular basis of SCA6 remains incompletely understood. Biophysical analyses of heterologous $Ca_V2.1$ channels bearing increased polyglutamine repeats showed conflicting results. Two studies reported alterations in current magnitude consistent with a gain of function [273,274], whereas two others reported a loss of function [275,276]. The discrepancies might have arisen from the use of different experimental systems, auxiliary subunits, and/or α_{1A} splice variants, and species. Whether SCA6 can be considered a *bona fide* channelopathy remains uncertain.

In contrast, features of the disease as a "glutaminopathy" are becoming increasingly clear. Mutant proteins form cytoplasmic and nuclear aggregates in cerebellar Purkinje neurons from SCA6 patients [263,277]. Both WT and SCA6 $Ca_V2.1$ proteins are cleaved, yielding a 75-kDa C-terminal fragment that appears to confer toxicity [278]; the SCA6 C-terminal fragment produced no higher intrinsic toxicity than the WT fragment but was more resistant to proteolysis [278]. Furthermore, a recent study demonstrated that α_{1A} is ubiquitinylated, a process dependent on the α_{1A} C-terminus, and that coexpression of ubiquitin in heterologous cells decreases current density of WT α_{1A} but not SCA6 mutant subunits [279]. Finally, the Gomez laboratory has shown that the ~75-kDa C-terminal cleavage products contain several nuclear localization signals, which facilitate their translocation to the nucleus, suggesting a possible role as a transcription factor; once in the nucleus, the SCA6 C-terminus is cytotoxic [280]. Taken together, these studies demonstrate possible pathophysiological effects of the polyglutamine-expanded C-terminal domain, independent of any alteration in P/Q-type channel function.

5.5. *Ca^{2+} channels in nociception and as targets for pain-alleviating agents, exemplified by $Ca_V2.2$ (N-type) Ca^{2+} channels*

The diversity of Ca^{2+} channels provides an array of potential targets for the design of new analgesic drugs. Both N- and T-type channels have been implicated in the initiation and/or maintenance of the chronic pain state based on their anatomical localization and functional studies with pharmacological tools as well as gene/mRNA knockdown approaches [281,282]. Among the interventions, the analgesic effect of N-type channel blockers is the most extensively studied.

N-type channels have been shown to play an important role in mediating painful neurotransmission from first-order primary afferent neurons (PANs) to neurons in the dorsal horn of the spinal cord. N-type channels are densely expressed in the superficial laminae (I and IIo) of the spinal cord dorsal horn, to which the central terminals of peptidergic PANs project [283–285]. Furthermore, electrophysiological studies have demonstrated prominent N-type current in small-diameter nociceptors in dorsal root ganglia (DRG) [286]. N-type blockers have been found to reduce evoked release from PAN terminals, for glutamate as well as the neuropeptides CGRP and substance P [287–289].

Importantly, $Ca_V2.2$ expression is increased in DRG neurons following nerve injury or tissue inflammation, implicating a role for N-type channels in pain transmission under pathological conditions [290,291]. This makes N-type Ca^{2+} channels an appealing therapeutic target in the treatment of neuropathic pain, a large percentage of which is insensitive to opioid analgesics [292]. Indeed, spinal administration of the N-type channel peptide blocker ω-CTx-GVIA reduces noxious stimulus-evoked activity in dorsal horn neurons of normal and nerve-injured rats [293].

Moreover, the sensitivity to ω-CTx-GVIA blockade was significantly enhanced following nerve injury [293]. Furthermore, in various behavioral models of neuropathic pain, spinal delivery of N-type blockers significantly attenuated hyperalgesia (increased sensitivity to noxious stimuli) and allodynia (painful response to non-noxious stimuli) in response to mechanical, chemical, or thermal insult [294]. Consistent with the outcome of spinal administration, the application of ω-CTx-GVIA to the DRG reduced the ectopic activity of injured PANs, believed to be the cause of spontaneous pain [295]. Perineural administration of N-type blockers also showed antinociceptive effect [296]. In the case of inflammatory pain, spinal delivery of N-type blockers inhibited the hyper-excitability of dorsal horn neurons as well as hyperalgesia responses in animal models [297,298].

Available genetic evidence agrees well with pharmacological evidence for the importance of N-type channels in pain behavior. Indeed, all three strains of α_{1B} KO mice exhibit attenuated nociceptive responses following either nerve or tissue injury [299–301], emphasizing once again the contribution of N-type channels to the development of the chronic pain state.

In 2004, drug regulatory agencies in the USA and Europe granted approval to ziconotide (Prialt®), a reversible peptide blocker of N-type Ca^{2+} channels, for use in the treatment of intractable pain. Prialt® represents one of only a few successful attempts to develop an analgesic through a highly mechanism-selective drug development effort [302–304]. In humans, spinal infusion of Prialt leads to significant pain relief in patients suffering from various chronic intractable pain associated with cancer, AIDS, and neuropathies. Importantly, Prialt does not lead to tolerance and works in patients who no longer respond to opioid drugs [303]. Prialt is chemically equivalent to the conopeptide ω-CTx-MVIIA. Other conotoxins as well as small molecule N-type blockers are currently under development, with the goal of achieving higher efficacy, a better safety profile, and easier delivery routes [282,302]. The recent discovery of DRG-specific $Ca_V2.2$ splice variants opened the possibility of developing drugs, which preferentially target N-type channels in PANs [305]. Furthermore, the internalization of N-type channels in DRG neurons in response to capsaicin as well as the activation of G-protein-coupled receptors suggests a novel mechanism of presynaptic inhibition [306–308].

5.6. Multifunctional effects of Ca^{2+} channels exemplified by $Ca_V2.3$ (R-type) Ca^{2+} channels

R-type or $Ca_V2.3$ Ca^{2+} channels play important roles in various systems, including the CNS and peripheral nervous system, as well as the endocrine, cardiovascular, reproductive, and gastrointestinal systems [309]. In the CNS, the properties and localization of R-type channels allow them to contribute to Ca^{2+} entry at presynaptic sites and to help support various forms of synaptic plasticity. Studies in mossy fiber synapses from $Ca_V2.3$ KO mice showed significant impairment of post-tetanic potentiation and long-term potentiation [194]. Importantly, these channels also mediate action potential generation at dendritic arbors and Ca^{2+} influx in dendritic spines [310]. R-type channels can also contribute to synaptic transmission in select synapses in the cerebellum and in the calyx of held [195,230].

As with $Ca_V2.2$, no human channelopathies have as yet been associated with $Ca_V2.3$. However, studies from KO models have yielded insights into the function of these channels. Similarly, the discovery that the spider toxin SNX-482 blocks the R-type current mediated by $Ca_V2.3$ has helped define the role of this channel [311–315]. $Ca_V2.3$ KO mice display altered fear behavior, suggesting that these channels are functionally important in the central amygdala and the limbic system in general [313]. These mice also display altered pain responses, enhanced morphine analgesia, and reduced morphine tolerance, suggesting that blockade of R-type channels may be a suitable target in opioid pain therapy [311,316]. R-type channels also appear to confer protection during ischemic neuronal injury [317]. Further information on the possible involvement of R-type channels in disease is discussed in the Sections 5.6.1–5.6.3.

5.6.1. $Ca_V2.3$ (R-type) Ca^{2+} channels in systemic glucose tolerance

Glucose tolerance is mediated by the secretion of glucagon and insulin from pancreatic cells. The role of R-type channels in this process was demonstrated by ablation of $Ca_V2.3$ channels and pharmacological analysis. Loss of $Ca_V2.3$ resulted in complete suppression of the second phase of insulin secretion, which accounts for ~20% of total glucose-evoked insulin secretion. The glucose-induced glucagon response was concomitantly impaired [312,318].

5.6.2. $Ca_V2.3$ Ca^{2+} channels and cardiac arrhythmias

A critical role for R-type channels in normal impulse generation and conduction in the heart has been demonstrated by studies of cardiovascular function in isolated prenatal hearts lacking $Ca_V2.3$ and by in vivo telemetry in adult KO mice. An increased variability within heartbeat episodes in isolated $Ca_V2.3$ null hearts suggested arrhythmia, which could also be elicited in WT mice treated with the R-type blocker SNX-482. Moreover, ECG recordings from these mice showed subsidiary escape rhythm, atrioventricular (AV) conduction disturbances, alterations in the QRS wave morphology, and altered atrial activation [319,320]. In addition, time-domain analysis showed increased heart rate and coefficient of variance (CV) [319,320]. Autonomic block experiments demonstrated that the increased heart rate in $Ca_V2.3$ KO mice could be ascribed to increased sympathetic tonus in these animals (probably because of the enhanced anxiety levels in these mouse strains [313]), which suggest a role for R-type channel activity in heart rate, despite the channels' relatively low contribution to Ca^{2+} current influx compared with other Ca^{2+} channels in neurons from sympathetic and parasympathetic ganglia [321,322]. However, autonomic block did not completely eliminate the increased heart rate CV and did not affect ectopic atrial/AV escape rhythms and QRS dysmorphology, suggesting that $Ca_V2.3$ do contribute to murine intrinsic heart function.

5.6.3. $Ca_V2.3$ Ca^{2+} channels and seizure susceptibility

Multiple lines of evidence suggest that R-type Ca^{2+} channels play a major role in generating plateau potentials and after depolarizations in hippocampal neurons

[309]. Recent studies have demonstrated that these channels contribute to epileptiform bursts of activity in the hippocampus [323,324]. Additional support for these notions has come recently from studies of $Ca_V2.3$ KO mice: electrocorticographic and deep intrahippocampal recordings showed dramatically reduced susceptibility to pentylenetetrazol-induced seizures [309].

$Ca_V2.3$ channels are indirectly involved in human juvenile myoclonic epilepsy (JMC), the most frequent cause of hereditary grand mal seizures. The genetically modified protein is EFHC1, or EF-hand domain (C-terminal) containing 1, a recently identified protein that contains an EF-hand motif [325]. Missense mutations in EFHC1 prevent the binding of EFHC1 to $Ca_V2.3$ and thereby disrupt its ability to facilitate Ca^{2+} influx through R-type channels, which is believed to promote normal neuronal apoptosis in the hippocampus. Thus, one interesting hypothesis is that hindering apoptosis leads to increased neuronal density and hyperexcitable circuits in patients with JMC. This needs to be reconciled with the evidence that $Ca_V2.3$ activity promotes epileptiform activity and suggests that the involvement of $Ca_V2.3$ in hippocampal apoptosis may be more complex than initially thought. Alternatively, it is also possible that binding of EFHC1 to partners other than $Ca_V2.3$ subunits mediates the seizure phenotype, thus leaving the role of $Ca_V2.3$ undetermined.

5.7. *Ca^{2+} entry in support of electrogenesis and Ca^{2+} regulation exemplified by Ca_V3 (T-type) Ca^{2+} channels*

There has been a rapid growth of knowledge about the biophysical properties, functional roles, and regulation of T-type Ca^{2+} channels, summarized in a book [6]. There is consensus that T-type Ca^{2+} channels play important roles in neuronal development and activity, cellular proliferation, nociception, and cardiac function [32–36,326–328]. For example, the role of T-type channels in pacemaker activity of the sinoatrial node has been clarified recently, thanks to the use of mouse models. Delays in AV conduction and bradycardia have been revealed by intracardiac recordings and telemetric electrocardiogram analysis of $Ca_V3.1$ KO mice [329].

Various abnormalities occurring in humans have been recently linked to T-type channels and are briefly discussed in Sections 5.7.1–5.7.4.

5.7.1. *Ca_V3 Ca^{2+} channels and pain*

Along with N-type channels, T-type channels have been implicated in pain transmission and modulation of pain processing at various anatomic levels [327]. All three subtypes of T-type channels ($Ca_V3.1$–3.3; $\alpha_{1G–I}$, respectively) are expressed in neurons of the pain pathway. Specifically, antisense oligonucleotide knockdown of $Ca_V3.2$ leads to attenuated nociceptive response after nerve injury [326]. Local injection of the T-type blocker mibefradil also produced an anti-nociceptive effect [330,331]. On the contrary, $Ca_V3.1$ KO mice displayed hypersensitivity upon application of noxious visceral stimuli, suggesting an antinociceptive function of $Ca_V3.1$ at the supraspinal level [332].

5.7.2. Ca_V3 Ca^{2+} channels and idiopathic generalized epilepsy

While the relevance of Ca^{2+} channels to epilepsy is well known, T-type channels are particularly important, given their abundance in brain structures associated with seizures, such as the thalamus and neocortex [333,334]. $Ca_V3.1$ expression is robust in thalamocortical neurons, whereas $Ca_V3.2$ and $Ca_V3.3$ are abundant in the reticular nucleus of the thalamus. All three channels are expressed in the hippocampus. Although $Ca_V3.1$ and $Ca_V3.3$ have been eliminated as candidates for childhood epilepsy in genetic studies in the Chinese Han population [335,336], a direct link between mutations in $Ca_V3.2$ (α_{1H}) and generalized spike-wave epilepsy was recently reported. At least 12 different single-nucleotide polymorphisms (SNPs) were found in the α_{1H} gene from patients with childhood absence epilepsy (CAE) [337–340]. These SNPs are predicted to result in abnormal T-type channel activity, and further studies in heterologous systems revealed that the majority of SNPs lead to altered channel gating. Some were predicted to yield a gain of function, for example, increased neuronal firing in thalamocortical circuits. However, three of the newly discovered SNPs were predicted to result in decreased neuronal firing. Thus, as seen with some other channelopathies, there is no clear pattern of cause and effect to relate the molecular defect to the pathophysiology. Accordingly, alternative hypotheses to explain the epileptic phenotype have been proposed, and these include altered gene expression or interactions with other channel modulatory proteins. It should also be noted that the SNPs do not segregate with a single type of idiopathic generalized epilepsy and could be detected in normal individuals [43,341]. Interestingly, the combination of a CAE-specific SNP and an SNP also found in normal individuals generated channel behavior different than that generated by either SNP alone. This points to the importance of sequence background, further suggesting that common polymorphisms can act to increase the likelihood of disease. All in all, it seems clear that $Ca_V3.2$ is a susceptibility gene in CAE and idiopathic generalized epilepsies.

5.7.3. $Ca_V3.2$ Ca^{2+} channels and ASD

A series of four distinct missense mutations in $Ca_V3.2$ (α_{1H}) T-type channels were recently associated with autism spectrum disorder (ASD) [44]. Although there is no clear causal link between a molecular dysfunction of T-type channels and ASD, it is worth noting that $Ca_V3.2$ channels are abundant in the limbic system, which often displays abnormalities in autistic patients. When examined in a heterologous expression system, each of the missense mutations resulted in a net loss of channel function, although they also gave rise to a slight slowing of channel inactivation [44]. Paradoxically, these effects are qualitatively opposite to those associated with L-type mutations in ASD, in which the mutations lead to increased Ca^{2+} influx. This suggests a more intricate scenario, including an alteration of the relative strength of excitatory and inhibitory inputs, which causes neuronal circuits to malfunction [342,343].

5.7.4. Ca_V3 Ca^{2+} channels and cancer

The prevalence of T-type channels in proliferating cells and transformed cell lines led to the notion of a link between these channels and cancer. Recent studies have described an association between T-type channels and proliferation in mammary, brain, leukemic, colorectal, gastric, and prostate tumors [328,344]. Moreover, hypermethylation of CpG islands has been consistently detected in the human *CACNA1G* gene ($Ca_V3.1$) in several tumors, including pancreatic, gastric, colon, and colorectal cancers, as well as leukemia [345–347]. The silencing of these GC-rich sequences by DNA methylation in the T-type channel promoter region has led to the hypothesis that this gene could act as a tumor suppressor in various cancers.

5.8. Acquired Ca^{2+} channelopathies

The majority of Ca^{2+} channelopathies arise from mutations directly affecting a Ca^{2+} channel gene. However, channelopathies have been identified in which the channel genes and proteins themselves are unaltered, but rather, their function is affected by aberrantly produced antibodies. These acquired channelopathies can have profound effects, leading to pathophysiology.

Lambert–Eaton myasthenic syndrome (LEMS) is an acquired channelopathy in which antibodies reactive to Ca^{2+} channels are produced secondary to small-cell lung carcinoma [348–351]. These antibodies can enter the neuromuscular junction, where they affect synaptic transmission [352,353]. Moreover, the phenotype could be reproduced in mice by passive transfer of human LEMS antibodies [354]. An examination of these antibodies has shown that they inhibit $Ca_V2.1$ currents in small-cell lung cancer cells [355], in acutely dissociated neurons [356,357]; moreover, LEMS antibodies have been shown to block recombinant $Ca_V2.1$ channels [357–359]. Finally, studies have shown that LEMS antibodies affect channel function by targeting the highly conserved pore-forming P-loop of the α_1 subunit [356]. Thus, LEMS can be considered a bona fide channelopathy.

Antibodies produced secondary to small-cell lung cancers have also been implicated in paraneoplastic cerebellar ataxia (PCA), in which the antibodies enter the CNS and affect cerebellar function. PCA patients have high titers of anti-Ca^{2+} channel antibodies [360–363], as well a selective loss of cerebellar neurons [364,365]. PCA may therefore also be classified as an acquired channelopathy.

6. Conclusions

The field of voltage-gated Ca^{2+} channels has witnessed remarkable advances since their initial discovery. Beginning with biophysical characterization of channel activity, and progressing through biochemical purification and molecular cloning, research in this area has led to a remarkable consensus about various Ca^{2+} channel types and their main distinguishing features. The evolved diversity of Ca^{2+} channels seems well matched to the large range of vital cellular processes that they support. Recognition of Ca^{2+} channel diversity has also contributed to clarifying the relationship between channel structure and functional properties such as gating, modulation, permeation, and pharmacological blockade. The uncovering of Ca^{2+} channel mutants has tested our concepts about these functions and, in some instances, led to clarification of disease pathophysiology. It is satisfying to highlight these developments with specific examples of Ca^{2+} channel function and dysfunction. However, many new challenges must now be faced to bridge the gap between the isolation of mutations in Ca^{2+} channel genes and understanding how they give rise to malfunctioning cells and organs in complex diseases. We can expect the study of Ca^{2+} channelopathies to generate a feedback effect on the basic science of Ca^{2+} channels and, one hopes, insights into possible therapeutic strategies for the broader disease entities that they represent.

Acknowledgments

Erika Piedras-Rentería is supported by AHASDG 0335142N. Richard Tsien receives support from NIH grants NS24067 and GM058234. We thank David Wheeler and Jeremy Bergsman for their input and Yaping Joyce Liao for help with the literature on acquired channelopathies.

References

1. Tsien, R.W., Tsien, R.Y. (1990). *Annu Rev Cell Biol* 6:715–60.
2. Hille, B., *Ionic Channels of Excitable Membranes,* Second edn. 1992, Sinauer Associates: Sunderland, MA.
3. Berridge, M.J. (1998). *Neuron* 21:13–26.
4. Carafoli, E., Klee, C., eds., *Calcium as a Cellular Regulator.* 1999, Oxford University Press: New York.
5. Tsien, R.W., Wheeler, D.B., in *Calcium as a Cellular Regulator,* E. Carafoli and C.B. Klee, eds. 1999, Oxford University Press: New York. pp. 171–99.
6. Tsien, R.W., Clozel, J.-P., Nargeot, J., eds., *Low-Voltage-Activated T-type Calcium Channels.* 1998, Tattenhall: Chester, UK.
7. Zamponi, G.W., ed., *Voltage-Gated Calcium Channels.* 2003, Plenum: New York, NY.
8. McDonough, S.I., ed., *Calcium Channel Pharmacology.* 2004, Plenum/Kluwer: Boston.
9. Catterall, W.A. (2000). *Annu Rev Cell Dev Biol* 16:521–55.
10. Campbell, K.P., Leung, A.T., Sharp, A.H. (1988). *Trends Neurosci* 11:425–30.
11. Glossmann, H., Striessnig, J. (1990). *Rev Physiol Biochem Pharmacol* 114:1–105.
12. Dolphin, A.C. (2006). *Br J Pharmacol* 147 (Suppl 1):S56–62.
13. Randall, A., Benham, C.D. (1999). *Mol Cell Neurosci* 14:255–72.
14. Walker, D., De Waard, M. (1998). *Trends Neurosci* 21:148–54.
15. Ertel, E.A., Campbell, K.P., Harpold, M.M., Hofmann, F., Mori, Y., Perez-Reyes, E., Schwartz, A., Snutch, T.P., Tanabe, T., Birnbaumer, L., Tsien, R.W., Catterall, W.A. (2000). *Neuron* 25:533–5.
16. Lipscombe, D., Helton, T.D., Xu, W. (2004). *J Neurophysiol* 92:2633–41.
17. Black, J.L., 3rd. (2003). *J Bioenerg Biomembr* 35:649–60.
18. Dolphin, A.C. (2003). *J Bioenerg Biomembr* 35:599–620.
19. Felix, R. (1999). *Receptors Channels* 6:351–62.
20. Bian, F., Li, Z., Offord, J., Davis, M.D., McCormick, J., Taylor, C.P., Walker, L.C. (2006). *Brain Res* 1075:68–80.
21. Morad, M., Soldatov, N. (2005). *Cell Calcium* 38:223–31.

22. Jones, S.W. (2003). *J Bioenerg Biomembr* 35:461–75.
23. Varadi, G., Mori, Y., Mikala, G., Schwartz, A. (1995). *Trends Pharmacol Sci* 16:43–9.
24. Cens, T., Rousset, M., Leyris, J.P., Fesquet, P., Charnet, P. (2006). *Prog Biophys Mol Biol* 90:104–17.
25. Avery, R.A., Johnston, D. (1996). *J Neurosci* 16:5567–82.
26. Nowycky, M.C., Fox, A.P., Tsien, R.W. (1985). *Nature* 316:440–3.
27. Nilius, B., Hess, P., Lansman, J.B., Tsien, R.W. (1985). *Nature* 316:443–6.
28. Bean, B.P. (1985). *J Gen Physiol* 86:1–30.
29. Randall, A., Tsien, R.W. (1995). *J Neurosci* 15:2995–3012.
30. Sutton, K.G., McRory, J.E., Guthrie, H., Murphy, T.H., Snutch, T.P. (1999). *Nature* 401:800–4.
31. Huguenard, J.R. (1996). *Annu Rev Physiol* 58:329–48.
32. Chemin, J., Nargeot, J., Lory, P. (2002). *J Neurosci* 22:6856–62.
33. Cribbs, L.L., Lee, J.H., Yang, J., Satin, J., Zhang, Y., Daud, A., Barclay, J., Williamson, M.P., Fox, M., Rees, M., Perez-Reyes, E. (1998). *Circ Res* 83:103–9.
34. Perez-Reyes, E. (2004). *Mol Interv* 4:16–8.
35. Son, W.Y., Han, C.T., Lee, J.H., Jung, K.Y., Lee, H.M., Choo, Y.K. (2002). *Dev Growth Differ* 44:181–90.
36. Yunker, A.M., McEnery, M.W. (2003). *J Bioenerg Biomembr* 35:533–75.
37. Lee, J.H., Daud, A.N., Cribbs, L.L., Lacerda, A.E., Pereverzev, A., Klöckner, U., Schneider, T., Perez-Reyes, E. (1999). *J Neurosci* 19:1912–21.
38. Perez-Reyes, E., Cribbs, L.L., Daud, A., Lacerda, A.E., Barclay, J., Williamson, M.P., Fox, M., Rees, M., Lee, J.H. (1998). *Nature* 391:896–900.
39. Lambert, R.C., Maulet, Y., Mouton, J., Beattie, R., Volsen, S., De Waard, M., Feltz, A. (1997). *J Neurosci* 17:6621–8.
40. Leuranguer, V., Bourinet, E., Lory, P., Nargeot, J. (1998). *Neuropharmacology*. 37:701–8.
41. Lacinova, L., Klugbauer, N. (2004). *Arch Biochem Biophys* 425:207–13.
42. Hansen, J.P., Chen, R.S., Larsen, J.K., Chu, P.J., Janes, D.M., Weis, K.E., Best, P.M. (2004). *J Mol Cell Cardiol* 37:1147–58.
43. Chen, Y., Lu, J., Pan, H., Zhang, Y., Wu, H., Xu, K., Liu, X., Jiang, Y., Bao, X., Yao, Z., Ding, K., Lo, W.H., Qiang, B., Chan, P., Shen, Y., Wu, X. (2003). *Ann Neurol* 54:239–43.
44. Splawski, I., Yoo, D.S., Stotz, S.C., Cherry, A., Clapham, D.E., Keating, M.T. (2006). *J Biol Chem* 281:22085–91.
45. Tanabe, T., Takeshima, H., Mikami, A., Flockerzi, V., Takahashi, H., Kangawa, K., Kojima, M., Matsuo, H., Hirose, T., Numa, S. (1987). *Nature* 328:313–8.
46. Jan, L.Y., Jan, Y.N. (1989). *Cell* 56:13–25.
47. Stuhmer, W., Conti, F., Suzuki, H., Wang, X.D., Noda, M., Yahagi, N., Kubo, H., Numa, S. (1989). *Nature* 339:597–603.
48. Papazian, D.M., Timpe, L.C., Jan, Y.N., Jan, L.Y. (1991). *Nature* 349:305–10.
49. Yang, N., Horn, R. (1995). *Neuron* 15:213–8.
50. Larsson, H.P., Baker, O.S., Dhillon, D.S., Isacoff, E.Y. (1996). *Neuron* 16:387–397.
51. Tanabe, T., Adams, B.A., Numa, S., Beam, K.G. (1991). *Nature* 352:800–3.
52. Nakai, J., Adams, B.A., Imoto, K., Beam, K.G. (1994). *Proc Natl Acad Sci USA*. 91:1014–8.
53. Lemos, J.R., Nowycky, M.C. (1989). *Neuron* 2:1419–26.
54. Zhang, J.-F., Ellinor, P.T., Aldrich, R.W., Tsien, R.W. (1994). *Nature* 372:97–100.
55. Hering, S., Berjukow, S., Sokolov, S., Marksteiner, R., Weiss, R.G., Kraus, R., Timin, E.N. (2000). *J Physiol* 528 (Pt 2):237–49.
56. Hoshi, T., Zagotta, W.N., Aldrich, R.W. (1991). *Neuron* 7:547–56.
57. Liu, Y., Jurman, M.E., Yellen, G. (1996). *Neuron* 16:859–867
58. Cens, T., Restituito, S., Galas, S., Charnet, P. (1999). *J Biol Chem* 274:5483–90.
59. Stotz, S.C., Hamid, J., Spaetgens, R.L., Jarvis, S.E., Zamponi, G.W. (2000). *J Biol Chem* 275:24575–82.

60. Lee, K.S., Marban, E., Tsien, R.W. (1985). *J Physiol* 364:395–411.
61. Brehm, P., Eckert, R. (1978). *Science* 202:1203–6.
62. Chad, J.E., Eckert, R. (1986). *J Physiol (Lond)* 378:31–51.
63. Imredy, J.P., Yue, D.T. (1992). *Neuron* 9:197–207.
64. Imredy, J.P., Yue, D.T. (1994). *Neuron* 12:1301–18.
65. Haack, J.A., Rosenberg, R.L. (1994). *Biophys J* 66:1051–60.
66. Erxleben, C., Liao, Y., Gentile, S., Chin, D., Gomez-Alegria, C., Mori, Y., Birnbaumer, L., Armstrong, D.L. (2006). *Proc Natl Acad Sci USA* 103:3932–7.
67. Babitch, J. (1990). *Nature* 346:321–2.
68. De Leon, M., Wang, Y., Jones, L., Perez-Reyes, E., Wei, X., Soong, T.W., Snutch, T.P., Yue, D.T. (1995). *Science* 270:1502–6.
69. Zhou, J., Olcese, R., Qin, N., Noceti, F., Birnbaumer, L., Stefani, E. (1997). *Proc Natl Acad Sci USA* 94:2301–5.
70. Soldatov, N.M., Zuhlke, R.D., Bouron, A., Reuter, H. (1997). *J Biol Chem* 272:3560–6.
71. Alexander, K.A., Wakim, B.T., Doyle, G.S., Walsh, K.A., Storm, D.R. (1988). *J Biol Chem* 263:7544–9.
72. Cheney, R.E., Mooseker, M.S. (1992). *Curr Opin Cell Biol* 4:27–35.
73. Zuhlke, R.D., Reuter, H. (1998). *Proc Natl Acad Sci USA* 95:3287–94.
74. Zühlke, R.D., Pitt, G.S., Deisseroth, K., Tsien, R.W., Reuter, H. (1999). *Nature* 399:159–62.
75. Peterson, B.Z., DeMaria, C.D., Adelman, J.P., Yue, D.T. (1999). *Neuron* 22:549–58.
76. Qin, N., Olcese, R., Bransby, M., Lin, T., Birnbaumer, L. (1999). *Proc Natl Acad Sci USA* 96:2435–8.
77. Pitt, G.S., Zuhlke, R.D., Hudmon, A., Schulman, H., Reuter, H., Tsien, R.W. (2001). *J Biol Chem* 276:30794–802.
78. Erickson, M.G., Liang, H., Mori, M.X., Yue, D.T. (2003). *Neuron* 39:97–107.
79. Xiong, L., Kleerekoper, Q.K., He, R., Putkey, J.A., Hamilton, S.L. (2005). *J Biol Chem* 280:7070–9.
80. Halling, D.B., Aracena-Parks, P., Hamilton, S.L. (2005). *Sci STKE2005:re15.*
81. DeMaria, C.D., Soong, T.W., Alseikhan, B.A., Alvania, R.S., Yue, D.T. (2001). *Nature* 411:484–9.
82. Van Petegem, F., Chatelain, F.C., Minor, D.L., Jr. (2005). *Nat Struct Mol Biol* 12:1108–15.
83. Fallon, J.L., Halling, D.B., Hamilton, S.L., Quiocho, F.A. (2005). *Structure* 13:1881–6.
84. Hudmon, A., Schulman, H., Kim, J., Maltez, J.M., Tsien, R.W., Pitt, G.S. (2005). *J Cell Biol* 171:537–47.
85. Zuhlke, R.D., Pitt, G.S., Tsien, R.W., Reuter, H. (2000). *J Biol Chem* 275:21121–9.
86. Kim, J., Ghosh, S., Nunziato, D.A., Pitt, G.S. (2004). *Neuron* 41:745–54.
87. Lee, T.S., Karl, R., Moosmang, S., Lenhardt, P., Klugbauer, N., Hofmann, F., Kleppisch, T., Welling, A. (2006). *J Biol Chem* 281:25560–7.
88. Grueter, C.E., Abiria, S.A., Dzhura, I., Wu, Y., Ham, A.J., Mohler, P.J., Anderson, M.E., Colbran, R.J. (2006). *Mol Cell* 23:641–50.
89. Yuan, W., Bers, D.M. (1994). *Am J Physiol* 267:H982–93.
90. Dzhura, I., Wu, Y., Colbran, R.J., Balser, J.R., Anderson, M.E. (2000). *Nat Cell Biol* 2:173–7.
91. Tsien, R.W., Hess, P., McCleskey, E.W., Rosenberg, R.L. (1987). *Annu Rev Biophys Biophys Chem* 16:265–90.
92. Sather, W.A., in *Voltage-Gated Calcium Channels*, G.W. Zamponi, ed. 2003, Plenum: New York. pp. 205–18.
93. Hess, P., Lansman, J.B., Tsien, R.W. (1986). *J Gen Physiol* 88:293–319.
94. Church, P.J., Stanley, E.F. (1996). *J Physiol* 496 (Pt 1):59–68.
95. McCleskey, E.W., Almers, W. (1985). *Proc Natl Acad Sci USA* 82:7149–53.
96. Ellinor, P.T., Yang, J., Sather, W.A., Zhang, J.-F., Tsien, R.W. (1995). *Neuron* 15:1121–32.
97. Kim, M.S., Morii, T., Sun, L.X., Imoto, K., Mori, Y. (1993). *FEBS Lett* 318:145–148.
98. Tang, S., Mikala, G., Bahinski, A., Yatani, A., Varadi, G., Schwartz, A. (1993). *J Biol Chem* 268:13026–9.

99. Yang, J., Ellinor, P.T., Sather, W.A., Zhang, J.F., Tsien, R.W. (1993). *Nature* 366:158–61.
100. Almers, W., McCleskey, E.W. (1984). *J Physiol* 353:565–83.
101. Hess, P., Lansman, J.B., Tsien, R.W. (1984). *Nature* 311:538–44.
102. Armstrong, C.M., Neyton, J. (1992). *Ann NY Acad Sci* 635:18–25.
103. Kuo, C.C., Hess, P. (1993). *J Physiol* 466:629–55.
104. McCleskey, E.W. (1999). *J Gen Physiol* 113:765–72.
105. Sather, W.A., McCleskey, E.W. (2003). *Annu Rev Physiol* 65:133–59.
106. Lansman, J.B., Hess, P., Tsien, R.W. (1986). *J Gen Physiol* 88:321–47.
107. Chen, X.H., Tsien, R.W. (1997). *J Biol Chem* 272:30002–8.
108. Chen, X.-H., Bezprozvanny, I., Tsien, R.W. (1997). *J Gen Physiol* 108: 363–74.
109. Klockner, U., Isenberg, G. (1994). *J Gen Physiol* 103:665–78.
110. Olivera, B.M., Miljanich, G.P., Ramachandran, J., Adams, M.E. (1994). *Annu Rev Biochem* 63:823–67.
111. McDonough, S.I., in *Calcium Channel Pharmacology*, S.I. McDonough, ed. 2003, Kluwer Academic Press: Boston.
112. Ellinor, P.T., Zhang, J.-F., Horne, W.A., Tsien, R.W. (1994). *Nature* 372:272–5.
113. Boland, L.M., Morrill, J.A., Bean, B.P. (1994). *J Neurosci* 14:5011–27.
114. McDonough, S.I., Swartz, K.J., Mintz, I.M., Boland, L.M., Bean, B.P. (1996). *J Neurosci* 16:2612–23.
115. Stocker, J.W., Nadasdi, L., Aldrich, R.W., Tsien, R.W. (1997). *J Neurosci* 17:3002–13.
116. Mintz, I.M., Venema, V.J., Swiderek, K.M., Lee, T.D., Bean, B.P., Adams, M.E. (1992). *Nature* 355:827–9.
117. Nakayama, H., Taki, M., Striessnig, J., Glossmann, H., Catterall, W.A., Kanaoka, Y. (1991). *Proc Natl Acad Sci USA* 88:9203–7.
118. Schuster, A., Lacinova, L., Klugbauer, N., Ito, H., Birnbaumer, L., Hofmann, F. (1996). *EMBO J* 15:2365–2370.
119. Grabner, M., Wang, Z., Hering, S., Striessnig, J., Glossmann, H. (1996). *Neuron* 16:207–18.
120. Mitterdorfer, J., Sinnegger, M.J., Grabner, M., Striessnig, J., Glossmann, H. (1995). *Biochemistry* 34:9350–5.
121. Peterson, B.Z., Catterall, W.A. (1995). *J Biol Chem* 270:18201–4.
122. Zhou, Z., January, C.T. (1998). *Biophys J* 74:1830–9.
123. Rios, E., Brum, G. (1987). *Nature* 325:717–20.
124. Miller, R.J., Freedman, S.B. (1984). *Life Sci* 34:1205–21.
125. Dirksen, R.T., Beam, K.G. (1999). *J Gen Physiol* 114:393–403.
126. Näbauer, M., Callewaert, G., Cleemann, L., Morad, M. (1989). *Science* 244:800–3.
127. Bers, D.M. (2002). *Nature* 415:198–205.
128. Eisenberg, R.S., McCarthy, R.T., Milton, R.L. (1983). *J Physiol (Lond)* 341:495–505.
129. Schneider, M.F., Chandler, W.K. (1973). *Nature* 242:244–6.
130. Tanabe, T., Beam, K.G., Powell, J.A., Numa, S. (1988). *Nature* 336:134–9.
131. Garcia, J., Tanabe, T., Beam, K.G. (1994). *J Gen Physiol* 103:125–47.
132. Adams, B.A., Tanabe, T., Mikami, A., Numa, S., Beam, K.G. (1990). *Nature* 346:569–72.
133. Tanabe, T., Beam, K.G., Adams, B.A., Niidome, T., Numa, S. (1990). *Nature* 346:567–9.
134. Lu, X., Xu, L., Meissner, G. (1994). *J Biol Chem* 269:6511–6.
135. el-Hayek, R., Antoniu, B., Wang, J., Hamilton, S.L., Ikemoto, N. (1995). *J Biol Chem* 270:22116–8.
136. Lu, X., Xu, L., Meissner, G. (1995). *J Biol Chem* 270:18459–64.
137. Nakai, J., Dirksen, R.T., Nguyen, H.T., Pessah, I.N., Beam, K.G., Allen, P.D. (1996). *Nature* 380:72–5.
138. Grabner, M., Dirksen, R.T., Suda, N., Beam, K.G. (1999). *J Biol Chem* 274:21913–9.
139. Chavis, P., Fagni, L., Lansman, J.B., Bockaert, J. (1996). *Nature* 382:719–22.

140. Jurkat-Rott, K., Mitrovic, N., Hang, C., Kouzmekine, A., Iaizzo, P., Herzog, J., Lerche, H., Nicole, S., Vale-Santos, J., Chauveau, D., Fontaine, B., Lehmann-Horn, F. (2000). *Proc Natl Acad Sci USA* 97:9549–54.
141. Carle, T., Lhuillier, L., Luce, S., Sternberg, D., Devuyst, O., Fontaine, B., Tabti, N. (2006). *Biochem Biophys Res Commun* 348:653–61.
142. Morrill, J.A., Cannon, S.C. (1999). *J Physiol* 520 (Pt 2):321–36.
143. Lerche, H., Klugbauer, N., Lehmann-Horn, F., Hofmann, F., Melzer, W. (1996). *Pflugers Arch* 431:461–3.
144. Lapie, P., Goudet, C., Nargeot, J., Fontaine, B., Lory, P. (1996). *FEBS Lett* 382:244–8.
145. Morrill, J.A., Brown, R.H., Jr., Cannon, S.C. (1998). *J Neurosci* 18:10320–34.
146. Jurkat-Rott, K., Uetz, U., Pika-Hartlaub, U., Powell, J., Fontaine, B., Melzer, W., Lehmann-Horn, F. (1998). *FEBS Lett* 423:198–204.
147. Keating, K.E., Quane, K.A., Manning, B.M., Lehane, M., Hartung, E., Censier, K., Urwyler, A., Klausnitzer, M., Muller, C.R., Heffron, J.J., et al. (1994). *Hum Mol Genet* 3:1855–8.
148. Gillard, E.F., Otsu, K., Fujii, J., Duff, C., de Leon, S., Khanna, V.K., Britt, B.A., Worton, R.G., MacLennan, D.H. (1992). *Genomics* 13:1247–54.
149. Galli, L., Orrico, A., Lorenzini, S., Censini, S., Falciani, M., Covacci, A., Tegazzin, V., Sorrentino, V. (2006). *Hum Mutat* 27:830.
150. Monnier, N., Procaccio, V., Stieglitz, P., Lunardi, J. (1997). *Am J Hum Genet* 60:1316–25.
151. Deisseroth, K., Mermelstein, P.G., Xia, H., Tsien, R.W. (2003). *Curr Opin Neurobiol* 13:354–65.
152. Dolmetsch, R. (2003). *Sci STKE* 2003:PE4
153. Williams, M.E., Feldman, D.H., McCue, A.F., Brenner, R., Velicelebi, G., Ellis, S.B., Harpold, M.M. (1992). *Neuron* 8:71–84.
154. Striessnig, J., Koschak, A., in *Voltage-Gated Calcium Channels*, G.W. Zamponi, ed. 2003, Plenum: New York. pp. 346–72.
155. Hill, C.S., Treisman, R. (1995). *Cell* 80:199–211.
156. Morgan, J.I., Curran, T. (1989). *Trends Neurosci* 12:459–62.
157. Morgan, J.I., Curran, T. (1988). *Cell Calcium* 9:303–11.
158. Ghosh, A., Ginty, D.D., Bading, H., Greenberg, M.E. (1994). *J Neurobiol* 25:294–303.
159. Montminy, M.R., Bilezikjian, L.M. (1987). *Nature* 328:175–8.
160. Hoeffler, J.P., Meyer, T.E., Yun, Y., Jameson, J.L., Habener, J.F. (1988). *Science* 242:1430–3.
161. Gonzalez, G.A., Montminy, M.R. (1989). *Cell* 59:675–80.
162. Greenberg, M.E., Thompson, M.A., Sheng, M. (1992). *J Physiol (Paris)* 86:99–108.
163. Yoshida, K., Imaki, J., Matsuda, H., Hagiwara, M. (1995). *J Neurochem* 65:1499–504.
164. Deisseroth, K., Heist, E.K., Tsien, R.W. (1998). *Nature* 392:198–202.
165. Rajadhyaksha, A., Barczak, A., Macías, W., Leveque, J.C., Lewis, S.E., Konradi, C. (1999). *J Neurosci* 19:6348–59.
166. Deisseroth, K., Bito, H., Tsien, R.W. (1996). *Neuron* 16:89–101.
167. Graef, I.A., Mermelstein, P.G., Stankunas, K., Neilson, J.R., Deisseroth, K., Tsien, R.W., Crabtree, G.R. (1999). *Nature* 401:703–8.
168. Carrion, A.M., Link, W.A., Ledo, F., Mellstrom, B., Naranjo, J.R. (1999). *Nature* 398:80–4.
169. Dolmetsch, R.E., Pajvani, U., Fife, K., Spotts, J.M., Greenberg, M.E. (2001). *Science* 294:333–9.
170. Zhang, H., Maximov, A., Fu, Y., Xu, F., Tang, T.S., Tkatch, T., Surmeier, D.J., Bezprozvanny, I. (2005). *J Neurosci* 25:1037–49.
171. Weick, J.P., Groth, R.D., Isaksen, A.L., Mermelstein, P.G. (2003). *J Neurosci* 23:3446–56.
172. Marks, M.L., Trippel, D.L., Keating, M.T. (1995). *Am J Cardiol* 76:744–5.
173. Marks, M.L., Whisler, S.L., Clericuzio, C., Keating, M. (1995). *J Am Coll Cardiol* 25:59–64.
174. Reichenbach, H., Meister, E.M., Theile, H. (1992). *Kinderarztl Prax* 60:54–6.

175. Splawski, I., Timothy, K.W., Sharpe, L.M., Decher, N., Kumar, P., Bloise, R., Napolitano, C., Schwartz, P.J., Joseph, R.M., Condouris, K., Tager-Flusberg, H., Priori, S.G., Sanguinetti, M.C., Keating, M.T. (2004). *Cell* 119:19–31.
176. Zuhlke, R.D., Bouron, A., Soldatov, N.M., Reuter, H. (1998). *FEBS Lett* 427:220–4.
177. Splawski, I., Timothy, K.W., Decher, N., Kumar, P., Sachse, F.B., Beggs, A.H., Sanguinetti, M.C., Keating, M.T. (2005). *Proc Natl Acad Sci USA* 102:8089–96, discussion 8086–8
178. Shi, C., Soldatov, N.M. (2002). *J Biol Chem* 277:6813–21.
179. Murakami, M., Ohba, T., Takahashi, Y., Watanabe, H., Miyoshi, I., Nakayama, S., Ono, K., Ito, H., Iijima, T. (2006). *J Mol Cell Cardiol* 41:115–25.
180. Sather, W.A., Tanabe, T., Zhang, J.F., Mori, Y., Adams, M.E., Tsien, R.W. (1993). *Neuron* 11:291–303.
181. Sidi, S., Busch-Nentwich, E., Friedrich, R., Schoenberger, U., Nicolson, T. (2004). *J Neurosci* 24:4213–23.
182. Dou, H., Vazquez, A.E., Namkung, Y., Chu, H., Cardell, E.L., Nie, L., Parson, S., Shin, H.S., Yamoah, E.N. (2004). *J Assoc Res Otolaryngol* 5:215–26.
183. Platzer, J., Engel, J., Schrott-Fischer, A., Stephan, K., Bova, S., Chen, H., Zheng, H., Striessnig, J. (2000). *Cell* 102:89–97.
184. Feng, T.P. (1936). *Chin J Physiol* 10:513–28.
185. Katz, B., *The Release of Neural Transmitter Substances*. 1969, Liverpool University Press: Liverpool. pp. 60.
186. Douglas, W.W., Rubin, R.P. (1963). *J Physiol* 167:288–310.
187. Dodge, F., Jr., Rahamimoff, R. (1967). *J Physiol* 193:419–32.
188. Miledi, R. (1973). *Proc R Soc Lond (Biol)* 183:421–5.
189. Llinás, R., Nicholson, C. (1975). *Proc Natl Acad Sci USA* 72:187–90.
190. Llinás, R., Steinberg, I.Z., Walton, K. (1981). *Biophys J* 33:323–51.
191. Augustine, G.J., Charlton, M.P., Smith, S.J. (1985). *J Physiol (Lond)* 367:163–81.
192. Dunlap, K., Luebke, J.I., Turner, T.J. (1995). *Trends Neurosci* 18:89–98.
193. Reuter, H. (1996). *Curr Opin Neurobiol* 6:331–7.
194. Dietrich, D., Kirschstein, T., Kukley, M., Pereverzev, A., von der Brelie, C., Schneider, T., Beck, H. (2003). *Neuron* 39:483–96.
195. Mintz, I.M., Sabatini, B.L., Regehr, W.G. (1995). *Neuron* 15:675–88.
196. Wu, L.G., Westenbroek, R.E., Borst, J.G., Catterall, W.A., Sakmann, B. (1999). *J Neurosci* 19:726–36.
197. Florman, H.M., Corron, M.E., Kim, T.D., Babcock, D.F. (1992). *Dev Biol* 152:304–14.
198. Ashcroft, F.M., Proks, P., Smith, P.A., Ammala, C., Bokvist, K., Rorsman, P. (1994). *J Cell Biochem* 55:54–65.
199. Bokvist, K., Eliasson, L., Ammala, C., Renstrom, E., Rorsman, P. (1995). *EMBO J* 14:50–7.
200. Lemos, J.R., Nowycky, M.C. (1989). *Neuron* 2:1419–26.
201. Dippel, W.W., Chen, P.L., McArthur, N.H., Harms, P.G. (1995). *Domest Anim Endocrinol* 12:349–54.
202. Lopez, M.G., Albillos, A., de la Fuente, M.T., Borges, R., Gandia, L., Carbone, E., Garcia, A.G., Artalejo, A.R. (1994). *Pflugers Arch* 427:348–54.
203. Maguire, G., Maple, B., Lukasiewicz, P., Werblin, F. (1989). *Proc Natl Acad Sci USA* 86:10144–7.
204. Duarte, C.B., Ferreira, I.L., Santos, P.F., Oliveira, C.R., Carvalho, A.P. (1992). *Brain Res* 591:27–32.
205. Simmons, M.L., Terman, G.W., Gibbs, S.M., Chavkin, C. (1995). *Neuron* 14:1265–72.
206. Bonci, A., Grillner, P., Mercuri, N.B., Bernardi, G. (1998). *J Neurosci* 18:6693–703.
207. Schmitz, Y., Witkovsky, P. (1997). *Neuroscience* 78:1209–16.
208. Sabria, J., Pastor, C., Clos, M.V., Garcia, A., Badia, A. (1995). *J Neurochem* 64:2567–71.
209. Kelly, M., In *Voltage-Gated Calcium Channels*, Zamponi, GW, ed. 2003, Plenum: New York, NY.

210. Kollmar, R., Montgomery, L.G., Fak, J., Henry, L.J., Hudspeth, A.J. (1997). *Proc Natl Acad Sci USA* 94:14883–8.
211. Lopez, I., Ishiyama, G., Acuna, D., Ishiyama, A., Baloh, R.W. (2003). *Cell Tissue Res* 313:177–86.
212. Kollmar, R., Fak, J., Montgomery, L.G., Hudspeth, A.J. (1997). *Proc Natl Acad Sci USA* 94:14889–93.
213. Singh, A., Hamedinger, D., Hoda, J.C., Gebhart, M., Koschak, A., Romanin, C., Striessnig, J. (2006). *Nat Neurosci* 9:1108–16.
214. Bech-Hansen, N.T., Naylor, M.J., Maybaum, T.A., Pearce, W.G., Koop, B., Fishman, G.A., Mets, M., Musarella, M.A., Boycott, K.M. (1998). *Nat Genet* 19:264–7.
215. McRory, J.E., Hamid, J., Doering, C.J., Garcia, E., Parker, R., Hamming, K., Chen, L., Hildebrand, M., Beedle, A.M., Feldcamp, L., Zamponi, G.W., Snutch, T.P. (2004). *J Neurosci* 24:1707–18.
216. Mansergh, F., Orton, N.C., Vessey, J.P., Lalonde, M.R., Stell, W.K., Tremblay, F., Barnes, S., Rancourt, D.E., Bech-Hansen, N.T. (2005). *Hum Mol Genet* 14:3035–46.
217. Carr, R.E. (1974). *Trans Am Ophthalmol Soc* 72:448–87.
218. Strom, T.M., Nyakatura, G., Apfelstedt-Sylla, E., Hellebrand, H., Lorenz, B., Weber, B.H., Wutz, K., Gutwillinger, N., Ruther, K., Drescher, B., Sauer, C., Zrenner, E., Meitinger, T., Rosenthal, A., Meindl, A. (1998). *Nat Genet* 19:260–3.
219. Hope, C.I., Sharp, D.M., Hemara-Wahanui, A., Sissingh, J.I., Lundon, P., Mitchell, E.A., Maw, M.A., Clover, G.M. (2005). *Clin Experiment Ophthalmol.* 33:129–36.
220. Hemara-Wahanui, A., Berjukow, S., Hope, C.I., Dearden, P.K., Wu, S.B., Wilson-Wheeler, J., Sharp, D.M., Lundon-Treweek, P., Clover, G.M., Hoda, J.C., Striessnig, J., Marksteiner, R., Hering, S., Maw, M.A. (2005). *Proc Natl Acad Sci USA* 102:7553–8.
221. Sheng, Z.H., Rettig, J., Takahashi, M., Catterall, W.A. (1994). *Neuron* 13:1303–13.
222. Mochida, S., Westenbroek, R.E., Yokoyama, C.T., Zhong, H., Myers, S.J., Scheuer, T., Itoh, K., Catterall, W.A. (2003). *Proc Natl Acad Sci USA* 100:2819–24.
223. Harlow, M.L., Ress, D., Stoschek, A., Marshall, R.M., McMahan, U.J. (2001). *Nature.* 409:479–84.
224. Maximov, A., Bezprozvanny, I. (2002). *J Neurosci* 22:6939–52.
225. Wheeler, D.B., Randall, A., Sather, W.A., Tsien, R.W. (1995). *Prog Brain Res* 105:65–78.
226. Yawo, H. (1993). *Ann NY Acad Sci* 707:379–81.
227. Regehr, W.G., Mintz, I.M. (1994). *Neuron* 12:605–13.
228. Takahashi, T., Momiyama, A. (1993). *Nature* 366:156–8.
229. Wu, L.G., Saggau, P. (1994). *J Neurosci* 14:5613–22.
230. Wu, L.G., Borst, J.G., Sakmann, B. (1998). *Proc Natl Acad Sci USA* 95:4720–5.
231. Borst, J.G., Sakmann, B. (1998). *J Physiol* 506 (Pt 1):143–57.
232. Reid, C.A., Clements, J.D., Bekkers, J.M. (1997). *J Neurosci* 17:2738–45.
233. Reuter, H. (1995). *Neuron* 14:773–9.
234. Poncer, J.C., McKinney, R.A., Gahwiler, B.H., Thompson, S.M. (1997). *Neuron* 18:463–72.
235. Heinke, B., Balzer, E., Sandkuhler, J. (2004). *Eur J Neurosci* 19:103–11.
236. Sanchez-Del-Rio, M., Reuter, U., Moskowitz, M.A. (2006). *Curr Opin Neurol* 19:294–8.
237. Ophoff, R.A., Terwindt, G.M., Vergouwe, M.N., van Eijk, R., Oefner, P.J., Hoffman, S.M.G., Lamerdin, J.E., Mohrenweiser, H.W., Bulman, D.E., Ferrari, M., Haan, J., Lindhout, D., van Ommen, G.-J.B., Hofker, M.H., Ferrari, M.D., Frants, R.R. (1996). *Cell* 87:543–52.
238. De Vries, B., Haan, J., Frants, R.R., Van den Maagdenberg, A.M., Ferrari, M.D. (2006). *Headache* 46:1059–68.
239. Tottene, A., Fellin, T., Pagnutti, S., Luvisetto, S., Striessnig, J., Fletcher, C., Pietrobon, D. (2002). *Proc Natl Acad Sci USA* 99:13284–9.
240. Pietrobon, D., Striessnig, J. (2003). *Nat Rev Neurosci* 4:386–98.
241. Van Den Maagdenberg, A.M., Pietrobon, D., Pizzorusso, T., Kaja, S., Broos, L.A., Cesetti, T., Van De Ven, R.C., Tottene, A., Van Der Kaa, J., Plomp, J.J., Frants, R.R., Ferrari, M.D. (2004). *Neuron* 41:701–10.

242. Ayata, C., Shimizu-Sasamata, M., Lo, E.H., Noebels, J.L., Moskowitz, M.A. (2000). *Neuroscience* 95:639–45.

243. Goadsby, P.J. (2001). *Ann Neurol* 49:4–6.

244. Goadsby, P.J. (2004). *Neuron* 41:679–80.

245. Cao, Y.Q., Piedras-Renteria, E.S., Smith, G.B., Chen, G., Harata, N.C., Tsien, R.W. (2004). *Neuron* 43:387–400.

246. Barrett, C.F., Cao, Y.Q., Tsien, R.W. (2005). *J Biol Chem* 280:24064–71.

247. Cao, Y.Q., Tsien, R.W. (2005). *Proc Natl Acad Sci USA* 102:2590–5.

248. Zhou, Y.D., Turner, T.J., Dunlap, K. (2003). *J Physiol* 547:497–507.

249. Knight, Y.E., Bartsch, T., Kaube, H., Goadsby, P.J. (2002). *J Neurosci* 22(RC213):1–6.

250. Ogasawara, M., Kurihara, T., Hu, Q., Tanabe, T. (2001). *FEBS Lett* 508:181–6.

251. Ebersberger, A., Portz, S., Meissner, W., Schaible, H.G., Richter, F. (2004). *Cephalagia* 24:250–61.

252. Kors, E.E., Haan, J., Giffin, N.J., Pazdera, L., Schnittger, C., Lennox, G.G., Terwindt, G.M., Vermeulen, F.L., Van den Maagdenberg, A.M., Frants, R.R., Ferrari, M.D. (2003). *Arch Neurol* 60:684–8.

253. Goadsby, P.J., Kullmann, D.M. (2005). *Lancet* 366:345–6.

254. Jen, J., Kim, G.W., Baloh, R.W. (2004). *Neurology* 62:17–22.

255. Imbrici, P., Eunson, L.H., Graves, T.D., Bhatia, K.P., Wadia, N.H., Kullmann, D.M., Hanna, M.G. (2005). *Neurology* 65:944–6.

256. Wan, J., Carr, J.R., Baloh, R.W., Jen, J.C. (2005). *Ann Neurol* 57:131–5.

257. Jeng, C.J., Chen, Y.T., Chen, Y.W., Tang, C.Y. (2006). *Am J Physiol Cell Physiol* 290:C1209–20.

258. Wan, J., Khanna, R., Sandusky, M., Papazian, D.M., Jen, J.C., Baloh, R.W. (2005). *Neurology* 64:2090–7.

259. Page, K.M., Heblich, F., Davies, A., Butcher, A.J., Leroy, J., Bertaso, F., Pratt, W.S., Dolphin, A.C. (2004). *J Neurosci* 24:5400–9.

260. Imbrici, P., Jaffe, S.L., Eunson, L.H., Davies, N.P., Herd, C., Robertson, R., Kullmann, D.M., Hanna, M.G. (2004). *Brain* 127:2682–92.

261. Jouvenceau, A., Eunson, L.H., Spauschus, A., Ramesh, V., Zuberi, S.M., Kullmann, D.M., Hanna, M.G. (2001). *Lancet* 358:801–7.

262. Gomez, C.M., Thompson, R.M., Gammack, J.T., Perlman, S.L., Dobyns, W.B., Truwit, C.L., Zee, D.S., Clark, H.B., Anderson, J.H. (1997). *Ann Neurol* 42:933–50.

263. Ishikawa, K., Fujigasaki, H., Saegusa, H., Ohkoshi, N., Shoji, S., Tanabe, T., Misuzawa, H.Abstract for Second International Conference on Unstable Microsatellites and Human Disease, University of North Carolina at Chapel Hill, April 17–19.

264. Zoghbi, H.Y. (1997). *Neurology* 49:1196–9.

265. Krovetz, H.S., Helton, T.D., Crews, A.L., Horne, W.A. (2000). *J Neurosci* 20:7564–70.

266. Ligon, B., Boyd, A.E., 3rd, Dunlap, K. (1998). *J Biol Chem* 273:13905–11.

267. Mori, Y., Friedrich, T., Kim, M.S., Mikami, A., Nakai, J., Ruth, P., Bosse, E., Hofmann, F., Flockerzi, V., Furuichi, T., Mikoshiba, K., Imoto, K., Tanabe, T., Numa, S. (1991). *Nature* 350:398–402.

268. Soong, T.W., DeMaria, C.D., Alvania, R.S., Zweifel, L.S., Liang, M.C., Mittman, S., Agnew, W.S., Yue, D.T. (2002). *J Neurosci* 22:10142–52.

269. Zhuchenko, O., Bailey, J., Bonnen, P., Ashizawa, T., Stockton, D.W., Amos, C., Dobyns, W.B., Subramony, S.H., Zoghbi, H.Y., Lee, C.C. (1997). *Nat Genet* 15:62–9.

270. Frontali, M. (2001). *Brain Res Bull* 56:227–31.

271. Zoghbi, H.Y., Orr, H.T. (2000). *Annu Rev Neurosci* 23:217–47.

272. Group, T.H.s.D.C.R. (1993). *Cell* 72:971–83.

273. Restituito, S., Thompson, R.M., Eliet, J., Raike, R.S., Riedl, M., Charnet, P., Gomez, C.M. (2000). *J Neurosci* 20:6394–403.

274. Piedras-Renteria, E.S., Watase, K., Harata, N., Zhuchenko, O., Zoghbi, H.Y., Lee, C.C., Tsien, R.W. (2001). *J Neurosci* 21:9185–93.
275. Matsuyama, Z., Wakamori, M., Mori, Y., Kawakami, H., Nakamura, S., Imoto, K. (1999). *J Neurosci* 19:RC14.
276. Toru, S., Murakoshi, T., Ishikawa, K., Saegusa, H., Fujigasaki, H., Uchihara, T., Nagayama, S., Osanai, M., Mizusawa, H., Tanabe, T. (2000). *J Biol Chem* 275:10893–8.
277. Ishikawa, K., Owada, K., Ishida, K., Fujigasaki, H., Shun Li, M., Tsunemi, T., Ohkoshi, N., Toru, S., Mizutani, T., Hayashi, M., Arai, N., Hasegawa, K., Kawanami, T., Kato, T., Makifuchi, T., Shoji, S., Tanabe, T., Mizusawa, H. (2001). *Neurology* 56:1753–6.
278. Kubodera, T., Yokota, T., Ohwada, K., Ishikawa, K., Miura, H., Matsuoka, T., Mizusawa, H. (2003). *Neurosci Lett* 341:74–8.
279. Chen, H., Piedras-Renteria, E.S. In *Society for Neuroscience Annual Meeting. 2006. Atlanta, GA.*
280. Kordasiewicz, H.B., Thompson, R.M., Clark, H.B., Gomez, C.M. (2006). *Hum Mol Genet* 15:1587–99.
281. Bourinet, E., Zamponi, G.W. (2005). *Curr Top Med Chem* 5:539–46.
282. McGivern, J.G. (2006). *Drug Discov Today* 11:245–53.
283. Westenbroek, R.E., Hoskins, L., Catterall, W.A. (1998). *J Neurosci* 18:6319–30.
284. Kerr, L.M., Filloux, F., Olivera, B.M., Jackson, H., Wamsley, J.K. (1988). *Eur J Pharmacol* 146:181–3.
285. Gohil, K., Bell, J.R., Ramachandran, J., Miljanich, G.P. (1994). *Brain Res* 653:258–66.
286. Scroggs, R.S., Fox, A.P. (1992). *J Neurosci* 12:1789–801.
287. Gruner, W., Silva, L.R. (1994). *J Neurosci* 14:2800–8.
288. Holz, G.G.t., Dunlap, K., Kream, R.M. (1988). *J Neurosci* 8:463–71.
289. Santicioli, P., Del Bianco, E., Tramontana, M., Geppetti, P., Maggi, C.A. (1992). *Neurosci Lett* 136:161–4.
290. Cizkova, D., Marsala, J., Lukacova, N., Marsala, M., Jergova, S., Orendacova, J., Yaksh, T.L. (2002). *Exp Brain Res* 147:456–63.
291. Yokoyama, K., Kurihara, T., Makita, K., Tanabe, T. (2003). *Anesthesiology* 99:1364–70.
292. Rice, A.S., Hill, R.G. (2006). *Annu Rev Med* 57:535–51.
293. Matthews, E.A., Dickenson, A.H. (2001). *Pain* 92:235–46.
294. Yaksh, T.L. (2006). *J Pain* 7:S13–30.
295. Liu, X., Zhou, J.L., Chung, K., Chung, J.M. (2001). *Brain Res* 900:119–27.
296. Xiao, W.H., Bennett, G.J. (1995). *J Pharmacol Exp Ther* 274:666–72.
297. Vanegas, H., Schaible, H. (2000). *Pain* 85:9–18.
298. McGivern, J.G., McDonough, S.I. (2004). *Curr Drug Targets CNS Neurol Disord* 3:457–78.
299. Saegusa, H., Kurihara, T., Zong, S., Kazuno, A., Matsuda, Y., Nonaka, T., Han, W., Toriyama, H., Tanabe, T. (2001). *EMBO J* 20:2349–56.
300. Hatakeyama, S., Wakamori, M., Ino, M., Miyamoto, N., Takahashi, E., Yoshinaga, T., Sawada, K., Imoto, K., Tanaka, I., Yoshizawa, T., Nishizawa, Y., Mori, Y., Niidome, T., Shoji, S. (2001). *Neuroreport* 12:2423–7.
301. Kim, C., Jun, K., Lee, T., Kim, S.S., McEnery, M.W., Chin, H., Kim, H.L., Park, J.M., Kim, D.K., Jung, S.J., Kim, J., Shin, H.S. (2001). *Mol Cell Neurosci* 18:235–45.
302. Snutch, T.P. (2005). *NeuroRx* 2:662–70.
303. Miljanich, G.P. (2004). *Curr Med Chem* 11:3029–40.
304. Terlau, H., Olivera, B.M. (2004). *Physiol Rev* 84:41–68.
305. Bell, T.J., Thaler, C., Castiglioni, A.J., Helton, T.D., Lipscombe, D. (2004). *Neuron* 41:127–38.
306. Altier, C., Khosravani, H., Evans, R.M., Hameed, S., Peloquin, J.B., Vartian, B.A., Chen, L., Beedle, A.M., Ferguson, S.S., Mezghrani, A., Dubel, S.J., Bourinet, E., McRory, J.E., Zamponi, G.W. (2006). *Nat Neurosci* 9:31–40.
307. Wu, Z.Z., Chen, S.R., Pan, H.L. (2005). *J Biol Chem* 280:18142–51.

308. Tombler, E., Cabanilla, N.J., Carman, P., Permaul, N., Hall, J.J., Richman, R.W., Lee, J., Rodriguez, J., Felsenfeld, D.P., Hennigan, R.F., Diverse-Pierluissi, M.A. (2006). *J Biol Chem* 281:1827–39.

309. Weiergraber, M., Henry, M., Krieger, A., Kamp, M., Radhakrishnan, K., Hescheler, J., Schneider, T. (2006). *Epilepsia* 47:839–50.

310. Yasuda, R., Sabatini, B.L., Svoboda, K. (2003). *Nat Neurosci* 6:948–55.

311. Saegusa, H., Kurihara, T., Zong, S., Minowa, O., Kazuno, A., Han, W., Matsuda, Y., Yamanaka, H., Osanai, M., Noda, T., Tanabe, T. (2000). *Proc Natl Acad Sci USA* 97:6132–7.

312. Pereverzev, A., Mikhna, M., Vajna, R., Gissel, C., Henry, M., Weiergraber, M., Hescheler, J., Smyth, N., Schneider, T. (2002). *Mol Endocrinol* 16:884–95.

313. Lee, S.C., Choi, S., Lee, T., Kim, H.L., Chin, H., Shin, H.S. (2002). *Proc Natl Acad Sci USA* 99:3276–81.

314. Newcomb, R., Szoke, B., Palma, A., Wang, G., Chen, X., Hopkins, W., Cong, R., Miller, J., Urge, L., Tarczy-Hornoch, K., Loo, J.A., Dooley, D.J., Nadasdi, L., Tsien, R.W., Lemos, J., Miljanich, G. (1998). *Biochemistry* 37:15353–62.

315. Wilson, S.M., Toth, P.T., Oh, S.B., Gillard, S.E., Volsen, S., Ren, D., Philipson, L.H., Lee, E.C., Fletcher, C.F., Tessarollo, L., Copeland, N.G., Jenkins, N.A., Miller, R.J. (2000). *J Neurosci* 20:8566–71.

316. Yokoyama, K., Kurihara, T., Saegusa, H., Zong, S., Makita, K., Tanabe, T. (2004). *Eur J Neurosci* 20:3516–9.

317. Toriyama, H., Wang, L., Saegusa, H., Zong, S., Osanai, M., Murakoshi, T., Noda, T., Ohno, K., Tanabe, T. (2002). *Neuroreport* 13:261–5.

318. Jing, X., Li, D.Q., Olofsson, C.S., Salehi, A., Surve, V.V., Caballero, J., Ivarsson, R., Lundquist, I., Pereverzev, A., Schneider, T., Rorsman, P., Renstrom, E. (2005). *J Clin Invest* 115:146–54.

319. Weiergraber, M., Henry, M., Sudkamp, M., de Vivie, E.R., Hescheler, J., Schneider, T. (2005). *Basic Res Cardiol* 100:1–13.

320. Lu, Z.J., Pereverzev, A., Liu, H.L., Weiergraber, M., Henry, M., Krieger, A., Smyth, N., Hescheler, J., Schneider, T. (2004). *Cell Physiol Biochem* 14:11–22.

321. Hong, S.J., Chang, C.C. (1995). *Br J Pharmacol* 116:1577–82.

322. Jeong, S.W., Wurster, R.D. (1997). *J Neurophysiol* 77:1769–78.

323. Kuzmiski, J.B., MacVicar, B.A. (2001). *J Neurosci* 21:8707–14.

324. Metz, A.E., Jarsky, T., Martina, M., Spruston, N. (2005). *J Neurosci* 25:5763–73.

325. Suzuki, T., Delgado-Escueta, A.V., Aguan, K., Alonso, M.E., Shi, J., Hara, Y., Nishida, M., Numata, T., Medina, M.T., Takeuchi, T., Morita, R., Bai, D., Ganesh, S., Sugimoto, Y., Inazawa, J., Bailey, J.N., Ochoa, A., Jara-Prado, A., Rasmussen, A., Ramos-Peek, J., Cordova, S., Rubio-Donnadieu, F., Inoue, Y., Osawa, M., Kaneko, S., Oguni, H., Mori, Y., Yamakawa, K. (2004). *Nat Genet* 36:842–9.

326. Bourinet, E., Alloui, A., Monteil, A., Barrere, C., Couette, B., Poirot, O., Pages, A., McRory, J., Snutch, T.P., Eschalier, A., Nargeot, J. (2005). *EMBO J* 24:315–24.

327. Jevtovic-Todorovic, V., Todorovic, S.M. (2006). *Cell Calcium* 40:197–203.

328. Lory, P., Bidaud, I., Chemin, J. (2006). *Cell Calcium* 40:135–46.

329. Mangoni, M.E., Traboulsie, A., Leoni, A.L., Couette, B., Marger, L., Le Quang, K., Kupfer, E., Cohen-Solal, A., Vilar, J., Shin, H.S., Escande, D., Charpentier, F., Nargeot, J., Lory, P. (2006). *Circ Res* 98:1422–30.

330. Dogrul, A., Gardell, L.R., Ossipov, M.H., Tulunay, F.C., Lai, J., Porreca, F. (2003). *Pain.* 105:159–68.

331. Todorovic, S.M., Meyenburg, A., Jevtovic-Todorovic, V. (2002). *Brain Res* 951:336–40.

332. Kim, D., Park, D., Choi, S., Lee, S., Sun, M., Kim, C., Shin, H.S. (2003). *Science* 302:117–9.

333. Talley, E.M., Cribbs, L.L., Lee, J.H., Daud, A., Perez-Reyes, E., Bayliss, D.A. (1999). *J Neurosci* 19:1895–911.

334. Talley, E.M., Solorzano, G., Depaulis, A., Perez-Reyes, E., Bayliss, D.A. (2000). *Brain Res Mol Brain Res* 75:159–65.
335. Wang, J., Zhang, Y., Liang, J., Pan, H., Wu, H., Xu, K., Liu, X., Jiang, Y., Shen, Y., Wu, X. (2006). *Pediatr Neurol* 35:187–190.
336. Chen, Y., Lu, J., Zhang, Y., Pan, H., Wu, H., Xu, K., Liu, X., Jiang, Y., Bao, X., Zhou, J., Liu, W., Shi, G., Shen, Y., Wu, X. (2003). *Neurosci Lett* 341:29–32.
337. Shin, H.S. (2006). *Cell Calcium* 40:191–6.
338. Vitko, I., Chen, Y., Arias, J.M., Shen, Y., Wu, X.R., Perez-Reyes, E. (2005). *J Neurosci* 25:4844–55.
339. Khosravani, H., Zamponi, G.W. (2006). *Physiol Rev* 86:941–66.
340. Peloquin, J.B., Khosravani, H., Barr, W., Bladen, C., Evans, R., Mezeyova, J., Parker, D., Snutch, T.P., McRory, J.E., Zamponi, G.W. (2006). *Epilepsia* 47:655–8.
341. Heron, S.E., Phillips, H.A., Mulley, J.C., Mazarib, A., Neufeld, M.Y., Berkovic, S.F., Scheffer, I.E. (2004). *Ann Neurol* 55:595–6.
342. Rubenstein, J.L., Merzenich, M.M. (2003). *Genes Brain Behav* 2:255–67.
343. Rippon, G., Brock, J., Brown, C., Boucher, J. (2006). *Int J Psychophysiol*
344. Panner, A., Wurster, R.D. (2006). *Cell Calcium* 40:253–9.
345. Shen, L., Ahuja, N., Shen, Y., Habib, N.A., Toyota, M., Rashid, A., Issa, J.P. (2002). *J Natl Cancer Inst* 94:755–61.
346. Toyota, M., Ho, C., Ohe-Toyota, M., Baylin, S.B., Issa, J.P. (1999). *Cancer Res* 59:4535–41.
347. Ueki, T., Toyota, M., Sohn, T., Yeo, C.J., Issa, J.P., Hruban, R.H., Goggins, M. (2000). *Cancer Res* 60:1835–9.
348. Lang, B., Pinto, A., Giovannini, F., Newsom-Davis, J., Vincent, A. (2003). *Ann NY Acad Sci.* 998:187–95.
349. Lang, B., Waterman, S., Pinto, A., Jones, D., Moss, F., Boot, J., Brust, P., Williams, M., Stauderman, K., Harpold, M., Motomura, M., Moll, J.W., Vincent, A., Newsom-Davis, J. (1998). *Ann NY Acad Sci* 841:596–605.
350. Lennon, V.A., Kryzer, T.J., Griesmann, G.E., O'Suilleabhain, P.E., Windebank, A.J., Woppmann, A., Miljanich, G.P., Lambert, E.H. (1995). *N Engl J Med* 332:1467–74.
351. Pelucchi, A., Ciceri, E., Clementi, F., Marazzini, L., Foresi, A., Sher, E. (1993). *Am Rev Respir Dis* 147:1229–32.
352. Dalmau, J., Gultekin, H.S., Posner, J.B. (1999). *Brain Pathol* 9:275–84.
353. Vincent, A., Lang, B., Newsom-Davis, J. (1989). *Trends Neurosci* 12:496–502.
354. Fukunaga, H., Engel, A.G., Lang, B., Newsom-Davis, J., Vincent, A. (1983). *Proc Natl Acad Sci USA* 80:7636–40.
355. Barry, E.L., Viglione, M.P., Kim, Y.I., Froehner, S.C. (1995). *J Neurosci* 15:274–83.
356. Garcia, K.D., Mynlieff, M., Sanders, D.B., Beam, K.G., Walrond, J.P. (1996). *Proc Natl Acad Sci USA* 93:9264–9.
357. Pinto, A., Gillard, S., Moss, F., Whyte, K., Brust, P., Williams, M., Stauderman, K., Harpold, M., Lang, B., Newsom-Davis, J., Bleakman, D., Lodge, D., Boot, J. (1998). *Proc Natl Acad Sci USA* 95:8328–33.
358. Pinto, A., Iwasa, K., Newland, C., Newsom-Davis, J., Lang, B. (2002). *Muscle Nerve* 25:715–24.
359. Pinto, A., Moss, F., Lang, B., Boot, J., Brust, P., Williams, M., Stauderman, K., Harpold, M., Newsom-Davis, J. (1998). *Ann NY Acad Sci* 841:687–90.
360. Mason, W.P., Graus, F., Lang, B., Honnorat, J., Delattre, J.Y., Valldeoriola, F., Antoine, J.C., Rosenblum, M.K., Rosenfeld, M.R., Newsom-Davis, J., Posner, J.B., Dalmau, J. (1997). *Brain* 120 (Pt 8):1279–300.
361. Wirtz, P.W., Lang, B., Graus, F., van den Maagdenberg, A.M., Saiz, A., de Koning Gans, P.A., Twijnstra, A., Verschuuren, J.J. (2005). *J Neuroimmunol* 164:161–5.
362. Graus, F., Lang, B., Pozo-Rosich, P., Saiz, A., Casamitjana, R., Vincent, A. (2002). *Neurology* 59:764–6.

363. Clouston, P.D., Saper, C.B., Arbizu, T., Johnston, I., Lang, B., Newsom-Davis, J., Posner, J.B. (1992). *Neurology* 42:1944–50.

364. Satoyoshi, E., Kowa, H., Fukunaga, N. (1973). *Neurology* 23:764–8.

365. Fukuda, T., Motomura, M., Nakao, Y., Shiraishi, H., Yoshimura, T., Iwanaga, K., Tsujihata, M., Eguchi, K. (2003). *Ann Neurol* 53:21–8.

366. Singh, B., Monteil, A., Bidaud, I., Sugimoto, Y., Suzuki, T., Hamano, S.I., Oguni, H., Osawa, M., Alonso, M.E., Delgado-Escueta A.V., Inoue, Y., Yasui-Furukori, N., Kaneko, S., Lory, P., Yamakawa, K. (2007). *Hum Mutat.* 28(5): 524–5 [Epub ahead of print].

J. Krebs and M. Michalak (Editors)
Calcium: A Matter of Life or Death

ISSN: 0167-7306/DOI: 10.1016/S0167-7306(06)41006-1

CHAPTER 6

Exchangers and Ca^{2+} signaling

Joachim Krebs

NMR Based Structural Biology, MPI for Biophysical Chemistry, Am Fassberg 11, D-37077 Göttingen, Germany; Institute of Biochemistry, HPM1, Swiss Federal Institute of Technology, ETH-Hönggerberg, Schafmattstrasse 18, CH-8093 Zürich, Switzerland; Tel.: +49 551 201 2200; Fax: +49 551 201 2202; E-mail: jkrebs@nmr.mpibpc.mpg.de

Abstract

The Na^+/Ca^{2+} exchanger (NCX) is a transmembrane protein of the plasma membrane involved in calcium homeostasis. It is a low-affinity, high-capacity electrogenic antiporter that extrudes Ca^{2+} against its concentration gradient at the expense of the inward-directed sodium gradient. Two different families of NCX exist, one independent of K^+ (NCX), the other one, Na^+/Ca^{2+}(K) exchanger (NCKX), also transports K^+ together with Ca^{2+} in exchange for Na^+. The exchanger is especially important in excitable cells, where it is necessary to eject large amounts of Ca^{2+} on a regular basis. In mammals, three NCX and six NCKX genes have been described, including several alternative splicing isoforms. NCX and NCKX are also present in non-mammalian vertebrates and in invertebrates. Structural and functional properties, regulation of the proteins, and tissue-specific expression of the different isoforms will be discussed in this chapter.

Keywords: α-repeats, alternative splicing, CBD1, CBD2, loop structure, NCKX, NCX, NMR, XIP, X-ray

1. Introduction

Ca^{2+} is essential for all cells. It is important for the origin of new cell life at fertilization (see Chapter 16, Chun and Santella, this volume) as well as for cells going into apoptosis (see Chapter 18, Pozzan and Rizzuto, this volume). Calcium homeostasis of eukaryotic cells is maintained and tightly controlled by a number of channels, pumps, and exchangers to keep the intracellular free calcium concentration $[Ca^{2+}]_I$ in resting cells in the nanomolar range. The steep concentration gradient of calcium across the plasma membrane or intracellular membranes of organelles, with millimolar Ca^{2+} outside cells or within organelles, helps Ca^{2+} to fulfill its manifold functions as a second messenger.

The Na^+/Ca^{2+} exchanger (NCX) was discovered and characterized in different laboratories during the 1960s when it was realized that a countertransport mechanism must be present in plasma membranes of excitable and non-excitable cells, which exchanges Na^+ and Ca^{2+} ions across the membrane [1–4]. However, detailed studies on the properties of NCX have only been possible after the Laboratory of Ken Philipson reported the first successful molecular cloning, purification, and expression of an NCX isoform [5,6]. This chapter will concentrate on the most recent findings on structural and functional properties of the exchanger. For more detailed reviews, the reader is referred to several recent publications [7–10].

NCX is found in most cells of vertebrates and invertebrates [11]. The NCX is a member of a large superfamily of membrane proteins [12], which is defined by the presence of sequence motifs known as α-repeats (Fig. 1). The α-repeats are homology regions within the transmembrane segments of these proteins [13]. The importance of the α-repeats for ion translocation by NCX1 and other membrane transporters has been indicated by mutational analysis [14]. Based on the presence of the highly conserved α-repeats, Cai and Lytton [11] have included NCX, NCKX, the H^+/Ca^{2+} exchanger CAX (cation exchanger) and the cation/Ca^{2+} exchanger CCX in the cation/calcium exchanger superfamily.

The NCX plays a pivotal role in maintaining calcium homeostasis in cardiac myocytes, neurons, kidney cells, and a variety of other cells. These proteins are fast Ca^{2+} transporting systems of high capacity but have lower Ca^{2+} affinity than the Ca^{2+} pumps of plasma membranes. Basically, two different families of NCXs have been characterized: the prototype, K^+-independent exchanger-denominated NCX-denominated NCX, is found in many cells and catalyzes the electrogenic exchange of three Na^+ for one Ca^{2+} [15]. As mentioned, in mammals three distinct gene members of the NCX family have been described: NCX1 [6], NCX2 [16], and NCX3 [17]. The three different families share about 70% overall amino acid identity [17]. The amino acid sequences define single-chain proteins with molecular masses in the range of 120 kDa.

The second family comprises K^+-dependent NCXs denominated as NCKX. They were originally described in the rods and cones of the retina but were later identified also in other cells. This exchanger catalyzes the electrogenic exchange of four Na^+ for one

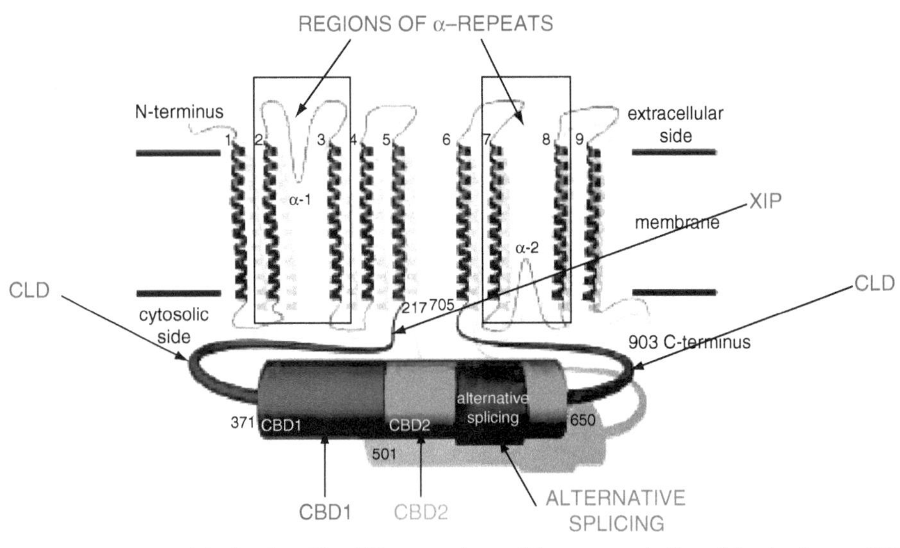

Fig. 1. Topology model of NCX. The different regions of interest are indicated in the figure. CBD, calcium-binding domain; CLD, catenin-like domain; and XIP, inhibitory peptide of the exchanger. The figure has been adopted from Fig. 1 of Ref. [47] with permission from Elsevier (See Color Plate 17, p. 516).

Ca^{2+} and one K^+ [18–20] and differs from NCX in its strict K^+ dependence, its lower Ca^{2+}-transport rate, and its difference in amino acid sequences. The molecular masses are much smaller than those of the NCX family (with the exception of NCKX1) [10,15]. To date, six different genes encoding the NCKX family have been described in mammals: NCKX1 [21,22], NCKX2 [23], NCKX3 [24], NCKX4 [25], NCKX5 [26], and NCKX6 [27]. In contrast to the other members of the family, which have been characterized in some detail, NCKX5 has only been identified in the GenBank (Acc. No. AB085629).

2. *The NCX family of NCXs*

The best studied example of the NCX family is NCX1, originally found in heart sarcolemmal membranes but later also in many other tissues [7,28,29]. The electrogenic transport of one Ca^{2+} in exchange for three Na^+ has been established long ago [30,31], see also Ref. [15]. This was made experimentally possible by a special preparation of Na^+-loaded sarcolemmal vesicles, which permitted Reeves and Sutko the quantitative measurement of radioactively labeled $^{45}Ca^{2+}$ uptake [32]. Ca^{2+} is extruded across the membrane against its concentration gradient at the expense of the energy stored in the transmembrane Na^+ gradient, which is in turn regulated by the Na^+/K^+-ATPase. As the exchange is electrogenic, it is influenced by the membrane potential. Therefore, the exchanger can basically operate in both directions, that is, depending on the electrochemical gradients of Ca^{2+} and Na^+ and on the membrane potential, Ca^{2+} is either extruded from the cell in exchange for Na^+ or Ca^{2+} is taken up in exchange for extruded Na^+.

In 1989, Hilgemann [33] published an improvement of the conventional patch-clamping technique by introducing the so-called giant patch technique, in which giant cardiac sarcolemmal patches are excised to measure quantitatively the current of the exchanger by exploiting its electrogenicity [34]. Thus, it was possible to show that in heart, the exchanger was able to extrude Ca^{2+} with a 10-fold higher rate as compared to the rates of the plasma membrane calcium (PMCA) pump [35], indicating that the exchanger has a significant higher capacity to extrude calcium than the pump. PMCA, due to its higher affinity for Ca^{2+}, is instead responsible for the fine-tuning of the cellular Ca^{2+} concentration [10]. Furthermore, using the giant patch technique, Hilgemann in collaboration with Philipson [36] has collected convincing evidence for a sequential mechanism of Na^+/Ca^{2+} exchange, that is, charges are moved in consecutive steps across the membrane with the result that only after Na^+ has moved across the membrane, Ca^{2+} can move in the opposite direction.

2.1. *Structural aspects*

The cDNA encoding NCX1 was the first clone of a NCX to be isolated in 1990 [6]. Analysis of the nucleotide sequence of this cDNA revealed that NCX is a protein of a molecular mass of 120 kDa with originally proposed 12 transmembrane domains. A large cytosolic loop (>500 amino acids) was also predicted for NCX1 connecting transmembrane domains 5 and 6 [6]. Subsequently, the first putative transmembrane domain of NCX was identified as a signal peptide containing a recognition site for a signal peptidase in all three NCX isoforms [6,16,17]. Therefore, the modified model predicted that the

N-terminus of NCX would be located extracellular, whereas the C-terminus would be cytosolic (Fig. 1). Later on, the model was further revised based on the reactivity of accessible cysteines introduced into the protein by site-specific mutations [37]. These results suggested only nine transmembrane segments in NCX, a now widely accepted topology model [37,38] (Fig. 1). In a more three-dimensional (3D) view of the arrangement and proximity of the different transmembrane segments, Ren et al. [39] have provided evidence for the close neighborhood of transmembrane segments 1, 2, and 6.

As documented for other transporters, the portion of the molecule essential for the counter-transport of the ions across the membrane is associated with the transmembrane segments of NCX. Similarly, it was shown that the large cytosolic loop is not essential for the transport, because more than 80% could be removed without significant loss of transport activity [40].

A region of 20 amino acids downstream of amino acid 250 of the cardiac exchanger (NCX1) resembled a calmodulin-binding domain [41]. The exchanger was never shown to be regulated by calmodulin directly, but this sequence could play an autoinhibitory role as that of calmodulin-binding domains of different calmodulin-dependent enzymes. Indeed, a synthetic peptide corresponding to the 20 amino acid sequence of this region inhibited NCX exchange activity [40]. This region was thus called exchange inhibitory peptide (XIP; Fig. 1) [41]. By using different deletion mutants, the target sequence for XIP was located within Ca^{2+}-binding domain (CBD2) (amino acid sequence 562–685 of NCX1 according to Ref. [40]; Fig. 1), whereas the amino acid sequence responsible for Na^{+}-dependent inactivation of the exchanger was located upstream of CBD1, within the cytosolic loop [40]. Interestingly, when a chimera of the cardiac and the kidney exchanger was used [40], no difference in the inhibitory effect of XIP was found, indicating that sequence differences between the two exchangers due to alternative splicing had no influence on the inhibitory activity [40]. This suggests that the sequence responsible for the inhibition by XIP (which is part of the CBD2 domain [40]) should be common to the different spliced isoforms. These data also provide evidence that the intracellular loop is largely responsible for the regulation of the exchanger's activity including a region that binds Ca^{2+} with high affinity [42,43]. This Ca^{2+}-binding segment, that is, amino acids 371–508, designated as Calx-β by Schwarz and Benzer [44] during their identification and characterization of the NCX gene in *Drosophila* comprises the first of two tandem domains. Several aspartate residues are important for Ca^{2+} binding within this segment, which induces substantial conformational changes [43,45] as demonstrated by mutational analysis [42,43].

Recently, high-resolution structures of either the primary CBD1 [46] or both binding domains, CBD1 and 2 [47], were determined by either X-ray crystallography [46] or nuclear magnetic resonance (NMR) [47]. Both domains are located within the large intracellular loop of NCX and comprise the two tandem Calx-β domains [46], located N-terminally to a region of alternative splicing (see Section 2.2) (Fig. 1). Both structural studies have provided evidence that in the presence of calcium, CBD1 and CBD2 show an immunoglobulin-like fold [47] (Fig. 2A and B). This fold, which is a common core structure of many other proteins including cadherins or integrins [48], forms a β-sandwich motif by two antiparallel β-sheets consisting of strands A–G (Fig. 2A and B). Both—X-ray and NMR—structures superimpose well [46]. NMR experiments [47] show that CBD1 binds Ca^{2+} with high affinity inducing a substantial conformational change, whereas CBD2 binds Ca^{2+} with significant lower affinity preserving the fold structure as observed

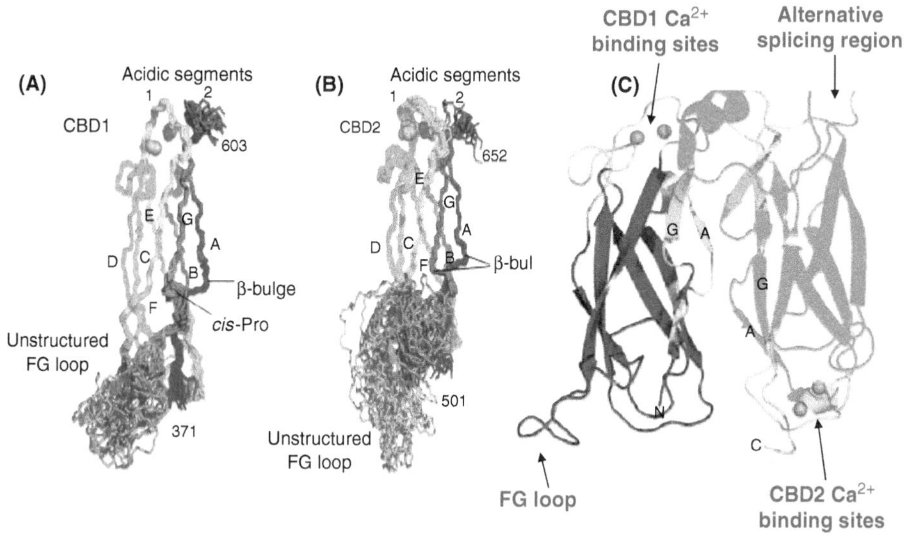

Fig. 2. Structures of the Ca^{2+}-binding domains CBD1 (**A**) and CBD2 (**B**) obtained by an overlay of 20 nuclear magnetic resonance (NMR)-derived structures. (**C**) A possible interaction between CBD1 and CBD2. Regions of interest as discussed in the text are indicated in the figure. CBD, calcium-binding domain. The figure was adopted from Figs 2 and 5 of Ref. [47] with permission from Elsevier (See Color Plate 18, p. 516).

in the absence of Ca^{2+} [47]. The same studies further indicated that the high-affinity Ca^{2+} sensor CBD1 and the low affinity sensor CBD2 may build a functional heterodimer along the interface formed by the strands A and G in an antiparallel arrangement of the CBDs [47] (Fig. 2C). In order to convey the regulatory properties of the cytosolic loop to the ion translocation of the transmembrane domains of the exchanger, Hilge et al. [47] made another interesting proposal. Owing to homology of the N-terminal (residues 217–370 including the Na^+-inactivation domain) and C-terminal amino acid sequences (residues 651–705) with α-catenin, these authors suggested that these regions form a third domain designated as catenin-like domain (CLD; Fig. 1). As in the model proposed by Hilge et al. [47], CBD1 and CBD2 are arranged in an antiparallel fashion (Fig. 2C), the N- and C-terminal amino acid sequences of the loop would be in close proximity and could indeed form such a domain to transmit the regulatory signals to the ion transport sites.

Dunn et al. [49] have shown that high intracellular Ca^{2+} can alleviate intracellular Na^+-dependent inactivation of the exchanger. As there are several NCX isoforms due to alternative splicing [50] (see Section 2.2), it was interesting to observe that Ca^{2+} alleviation of the Na^+-dependent inactivation of the exchanger could be detected only for isoforms containing exon A, not B [51]. The important amino acid sequence difference between the two exons, which are mutually exclusive [50], is found in a short acidic segment that is important for Ca^{2+} binding to CBD2. Substitution of D^{578} in exon A (which participates in Ca^{2+} ligation, see Ref. 47) by an arginine in exon B is apparently sufficient to prevent alleviation of Na^+-dependent inactivation because of binding of Ca^{2+} to CBD2. These results support the view proposed by Hilge et al. [47], who suggested that CBD2 is close to the part of CLD upstream of CBD1, which is responsible for the Na^+-dependent inactivation (Fig. 2C). In exon B variants, rearrangement of the Ca^{2+}-binding site could alter both its Ca^{2+}-binding property and its interaction with CLD. It would be of interest to show whether the basic autoinhibitory peptide XIP [40] (Fig. 1) could bind to CBD2, especially to the short acidic segment mentioned above, as it was reported [40] that the deletion mutant Δ562–685, which includes most of the sequence of CBD2, is not inhibited by XIP. It would also be interesting to see whether phosphorylation of Y^{581} by a tyrosine kinase could influence the Ca^{2+}-binding property of CBD2. Based on the solved NMR structure of CBD2 Hilge et al. [47] pointed out that Y^{581} – which is conserved in both exons A and B – is easily accessible.

Another interesting observation based on the NMR structure of the loop [47] is the presence of the unstructured region connecting strands F and G, the FG loop (Fig. 2), which corresponds to amino acid sequences 468–482 in CBD1 and to 599–627 in CBD2, respectively. Most of the FG loop amino acid sequence of the CBD1 is not present in CalX1.1, the Drosophila exchanger, which is inhibited rather than activated by μM Ca^{2+} [52]. This suggests that the FG loop in CBD1 is involved in Ca^{2+}-dependent exchanger activation. This view is supported by recent findings made by Bano et al. [53], who studied calpain-dependent cleavage of NCX under the conditions of brain ischemia. These authors reported that calpain cleaved the exchanger NCX3 in regions corresponding to the FG loop leading to cell death due to Ca^{2+} overload [53]. On the other hand, in CBD2, the FG loop amino acid sequence also includes the region of alternative splicing. This arrangement may permit tissue-specific modulation of the exchanger activity.

Surprisingly, the X-ray structure of CBD1 [46] revealed the presence of four Ca^{2+} bound in a cluster that may have substantial functional consequences. By contrast, Hilge et al. [47] reported that only two Ca^{2+} were bound to CBD1, even if they determined the same amino acids binding Ca^{2+} like those described by Nicoll et al. [46]. The latter authors argue that the indirect way in which Hilge et al. [47] predicted the Ca^{2+}-binding sites in the NMR structure may have led to a positional average as compared to those sites observed in the crystal structure [46]. One may add another caveat. As Hilge et al. [47] assigned the two Ca^{2+}-binding sites by including pseudo-contact shift data using spectra obtained in the presence of Yb^{3+}, this may hamper the determination of the exact Ca^{2+}-binding sites, as it has been often observed that lanthanides and calcium do not occupy the same sites in calcium-binding proteins. On the other hand, if the novel Ca^{2+}-binding site that consists of four Ca^{2+} ions arranged in a planar cluster as reported by Nicoll et al. [46] is correct, this would be an example for the high flexibility of calcium coordination, because in the X-ray structure, Ca1 and Ca4 are penta-coordinated, whereas Ca2 and Ca3 are hexa- and hepta-coordinated, respectively [46]. Interestingly, a similar cluster arrangement for Ca^{2+} has been predicted for other CBDs, that is, for the C_2 domains of synaptotagmin I and phospholipase C [54]. In the C_2 domains of synaptotagmin [55] and phospholipase C [56], the β-sandwich fold permits variable loops widely separated within the amino acid sequence of the protein to facilitate the binding of Ca^{2+} ions in a cluster.

2.2. Alternative splicing and regulation of tissue-specific expression

The structures of CBD2 or CBD12 reported by Hilge et al. [47] included the region of alternative splicing [50], which is part of CBD2 (Fig. 1). This region combines six small exons (labeled A–F) in a variable manner. Exons A and B are mutually exclusive to maintain an open reading frame [57], C through F are "cassette" exons [50], that is, any combination could potentially result in a functional transcript. Exon A is normally found in excitable tissues in contrast to exon B, which is almost exclusively found in non-excitable tissues [58]. In this context, it is interesting to note that transcripts containing exon A function more efficiently over a wider voltage range than those containing exon B [59]. The shortest splice variant AD was mainly detected in brain, the splice variant BD in kidney, whereas ACDEF, which constitutes the longest sequence, was found in heart and skeletal muscles [50,57]. At least 17 different splice variants have been described, which are expressed in a tissue-specific manner. Most isoforms have been reported for NCX1 (of which maximally 16 different splice variants can be produced, because all alternatively spliced transcripts that are functional must contain exon D) [59], a few have been described for NCX3, but none so far for NCX2 [50]. The physiological importance of the different splice variants is obscure.

An issue that is still controversial is the regulation of the exchanger by phosphorylation. Stimulation of the exchanger by ATP is an old observation [60] and has been attributed to kinase-mediated phosphorylation [61], but evidence for a cAMP-dependent phosphorylation could not be documented even if a highly conserved protein kinase A consensus site was reported [6]. On the other hand, it has been shown that in *Xenopus laevis*, the cardiac NCX is regulated through a β-adrenergic cAMP-dependent pathway [62]. Iwata et al. [63] reported that the amino acid sequence of the frog cardiac exchanger contains an additional exon comprising nine amino acids that has the feature of a P-loop (or Walker A) motif (A/GxxxxGKS/T). Later, Shuba et al. [64] reported that deletion of this motif abolished the cAMP-dependent regulation of the frog exchanger. Recently, He et al. [65] have demonstrated that mutation of serine in the conserved protein kinase A consensus site to glycine (S374G) significantly reduced the cAMP-mediated regulation of the exchanger.

Details of transcription have been best described for the NCX1 gene. NCX1 is expressed in all mammalian tissues studied so far. The highest level of transcription and amount of protein have been found in heart ventricles in contrast to lung tissues in which both message and protein are very low. This difference in tissue expression might be explained by the fact that NCX1 is controlled by three independent promoters that differ in their potency in various tissues [66–68]. NCX2 and NCX3 are only expressed in brain and skeletal muscles [16,17]. In the neonatal brain, 70% of the total NCX transcripts come from the NCX1 gene indicating its importance during embryonic development [69]. This is further supported by the findings reported by Reuter et al. [70] that NCX1 deficiency in mice is embryonally lethal. After birth, the developing rat brain shows a constant increase in NCX2 transcript levels [69]. In rat cerebellar granule cells containing seven NCX1, two NCX3 splice variants, and one NCX2 transcript (for NCX2 no splice variants have been described to date) [50], the expression of the different isoforms changes substantially in a Ca^{2+}-dependent manner during maturation [71]. Most striking is the behavior of the NCX2 transcript that is rapidly down-regulated after membrane depolarization in a calcineurin-dependent pathway [71,72].

Recently, Philipson's group [73,74] reported a cardiac-specific knock out (KO) of the NCX in mice. Surprisingly, the mice were viable and survived till adulthood. Even more surprisingly, these KO mice showed an almost normal cardiac performance that raises the question whether NCX is an essential component of the cardiac excitation–contraction coupling. The main observation [74] was that in the absence of NCX, the ventricular action potential (AP) was shortened, which, together with a decrease of the current provided by the L-type Ca^{2+} channel (probably due to a Ca^{2+}-dependent inactivation mechanism) [75], was critical to maintain Ca^{2+} homeostasis and contractility in the absence of NCX [73]. In an attempt to elucidate the compensating mechanism responsible for the accelerated AP repolarization, it was found that the transient outward current was increased by almost 80%, accompanied by an increased expression of the K^+ channel interacting protein and of Kv4.2, the subunit of the K^+ channel generating the transient outward current. Pott et al. [74] used a computer model developed by Bondarenko et al. [76] to determine the relative importance of the different components contributing to the shape and kinetics of AP. According to this model, the influence of the inward Ca^{2+} current and the absence of the NCX exchange current on the duration of the AP was relatively small in contrast to the transient outward current that substantially accelerated the repolarization of AP. The authors therefore discuss the possibility that the up-regulation of the transient outward current may be an important regulatory mechanism to prevent Ca^{2+} overload if myocytes have a reduced Ca^{2+} extrusion capacity [74].

3. *The K^+-dependent NCX: NCKX*

A second family of NCXs, the NCKX family, is strictly dependent on potassium. NCKX exchangers extrude Ca^{2+} from the cell by using both the inward sodium and the outward potassium gradient. Like the NCX family, NCKX also functions electrogenically by extruding one Ca^{2+} and one K^+ for four Na^+.

The NCKX protein was first purified from bovine retinal rod outer segments by Cook and Kaupp [77] and originally thought to resemble an isoform of the NCX specific for the rod photoreceptors. The protein is a single polypeptide of molecular mass 220 kDa, which was later recognized as the first member of a new family of cation exchangers [78,79]. Cloning of the rod NCKX1 from bovine retina revealed an amino acid sequence of 1216 residues [22]. To date, five NCKX genes have been discovered in mammals [23–27], indicating a broader expression pattern in different tissues (for a recent review see Ref. 26), but related isoforms of NCKX genes have also been described for lower organisms like *Drosophila* [80], *Caenorhabditis elegans* [81], or sea urchin [82].

The best studied example of the NCKX genes is NCKX1 from rod photoreceptors of the retina [26] (in cone photoreceptors, NCKX2 has been found) [83]. After Ca^{2+} has entered the cells in the dark through a light-sensitive channel gated by cyclic GMP, it is extruded by the exchanger. This reduces the cytosolic Ca^{2+} concentrations to the nanomolar range and indicates a higher Ca^{2+} sensitivity than NCX, which transports Ca^{2+} with high capacity but low affinity [35]. Even if there is little amino acid sequence homology between members of the NCX and the NCKX families (with the exception of the α-repeat regions), it is believed that the two exchangers share a similar membrane topology. The NCKX protein is proposed to consist of either 10 or 11 transmembrane regions with a large cytosolic loop connecting transmembrane segments 5 and 6. The

N- and C-termini are either extracellular (even numbers of transmembrane domains) or the N-terminus is extracellular and the C-terminus cytosolic (odd numbers of transmembrane domains) [84–86] (for a review see [26]; Fig. 1). The extracellular localization of the C-terminus is supported by antibody binding to an epitope inserted at the C-terminus [86]. In contrast to NCX, NCKX proteins contain a large, N-terminal hydrophilic loop in the extracellular space following signal peptide cleavage [86]. Deletion of the signal sequence has not affected NCKX protein synthesis but disrupted plasma membrane targeting [87]. The two large hydrophilic loops contain no sequence elements conserved among the different NCKX isoforms [86]. In analogy to the NCX protein, the hydrophilic loops of the NCKX1 protein can be deleted without affecting the Ca^{2+}-transport activity of the transmembrane part or its K^+ dependence [81,86]. An interesting observation was recently made [88] on the amino acids that are important for the K^+ dependence of the NCKX exchangers. After testing more than 100 different amino acids of the sequence of NCKX2, Kang et al. [88] have identified a single amino acid (D^{575} of human NCKX2) for which substitution by asparagine removed the K^+ dependence of NCKX. According to the current topology model of NCKX, D^{575} should be located within the transmembrane segment 9 [88]. This aspartic acid is conserved among all mammalian NCKX genes studied so far. Interestingly, the authors have also reported that an asparagine is found in this position in all three NCX genes that transport Ca^{2+} independently of K^+ [88]. If D^{575} was substituted by glutamic acid, Ca^{2+} transport was eliminated indicating considerable steric constraints at this position. In another topological study from Schnetkamp's group [86], the spatial proximity of the two α-repeat regions was investigated by site-directed disulfide mapping of residues contributing to the Ca^{2+}- and K^+-binding pocket. In this study, Kinjo et al. [86] identified two critical acidic residues, E^{188} located in the α1-repeat (transmembrane segment 2) and D^{548} in α2-repeat (transmembrane segment 8), which could form a disulfide cross-link by substituting with cysteine. Several other hydrophilic residues located within these transmembrane segments could be shown to be in close proximity indicating that transmembrane segments 2, 8, and 9 are close in space and bring together E^{188}, D^{548}, and D^{575} to form a single Ca^{2+} and K^+ pocket in the middle of the phospholipid bilayer [86,88].

As discussed in Section 2 for the NCX protein, the cytosolic loop contains two Ca^{2+}-binding sites to regulate Ca^{2+} transport. Even if there is no amino acid sequence homology between the cytosolic loops of NCX and NCKX, it was suggested by Schnetkamp early on that the NCKX protein should contain a Ca^{2+} sensory region on its cytoplasmic domain to regulate Ca^{2+} extrusion [89]. Recently, Seiler et al. [90] have indeed provided evidence for a high-affinity Ca^{2+}-binding site on the cytoplasmic loop of bovine NCKX1. In gel overlay experiments using purified portions of the cytosolic loop expressed in bacteria, the authors identified a sequence in the N-terminal portion of the loop that bound radioactive Ca^{2+} with high affinity [90]. In addition, the same sequence conferred K^+ sensitivity to the Ca^{2+} transport by introducing this domain into the appropriate site of cardiac NCX [90]. Unfortunately, experiments carried out to determine K^+ transport have been inconclusive [90]. Those results could have been very informative in view of the findings reported recently by Kang et al. [88] that D^{575} located within transmembrane domain 9 is critical for the K^+ sensitivity of NCKX, it is replaced by asparagine in all NCX exchanger genes to render them K^+-independent [88].

4. Conclusions

This chapter described structural and functional aspects of the plasma membrane NCXs, NCX and NCKX. These proteins are important for Ca^{2+} homeostasis of the cell. Both work electrogenically using the downhill Na^{+} gradient (and in addition the K^{+} outward gradient in NCKX) to export Ca^{2+} out of the cell against its concentration gradient. Both exchangers are fully reversible, the direction of transport being determined by the membrane potential. The exchanger is a protein of high transport capacity but has lower Ca^{2+} affinity than the plasma membrane Ca^{2+} pump. However, NCKX1 can sense Ca^{2+} in the nanomolar range [26]. Several gene products exist in both families, some of them in various isoforms because of alternative splicing. Regulation of the proteins is still poorly understood, even if some interesting properties have been described. A very recent observation has shed light on the importance of NCKX2, which is abundantly expressed in cortical neurons: mice deficient in NCKX2 showed a profound loss of long-term potentiation and an increase in long-term depression at hippocampal synapses with the consequence of clear deficits in spatial working memory [91]. A more detailed insight into the structure/function relationship will have to await for the full elucidation of the 3D structure of the two NCX types. The genetic alterations of those proteins may provide a rationale for the existence of tissue-specific alternative splicing isoforms.

References

1. Baker, P.F. & Blaustein, M.P. (1968) *Biochim. Biophys. Acta* 150, 167–170.
2. Reuter, H. & Seitz, N. (1968) *J. Physiol.* 195, 451–470.
3. Baker, P.F., Blaustein, M.P., Hodgkin, A.L., & Steinhardt, R.A. (1969) *J. Physiol.* 200, 431–458.
4. Martin, D.L. & De Luca, H.F. (1969) *Am. J. Physiol.* 216, 1351–1359.
5. Philipson, K.D., Longoni, S., & Ward, R. (1988) *Biochim. Biophys. Acta* 945, 298–306.
6. Nicoll, D.A., Longoni, S., & Philipson, K.D. (1990) *Science* 250, 562–565.
7. Philipson, K.D. & Nicoll, D.A. (2000) *Annu. Rev. Physiol.* 62, 111–132.
8. Lytton, Schnetkamp, Hryshko, Blaustein (eds) (2002) *Ann. N. Y. Acad. Sci. 976, 1–549.*
9. Hilgemann, D.W. (2004) *Am. J. Physiol. Cell. Physiol.* 287, C1167–C1172.
10. Guerini, D., Coletto, L., & Carafoli, E. (2005) *Cell Calcium* 38, 281–289.
11. Cai, X. & Lytton, J. (2004) *Mol. Biol. Evol.* 21, 1692–1703.
12. Paulsen, I.T., Nguyen, L., Sliwinski, M.K., Rabus, R., & Saier, M.H. (2000) *J. Mol. Biol.* 301, 75–100.
13. Schwarz, E.M. & Benzer, S. (1997) *Proc. Natl. Acad. Sci. USA* 94, 10249–10254.
14. Nicoll, D.A., Hryshko, L.V., Matsuoka, S., Frank, J.S., & Philipson, K.D. (1996) *J. Biol. Chem.* 271, 13385–13391.
15. Blaustein, M.P. & Lederer, W.J. (1999) *Physiol. Rev.* 79, 763–854.
16. Li, Z., Matsuoka, S., Hryshko, L.V., Nicoll, D.A., Bersohn, M.M., Burke, E.P., Lifton, R.P., & Philipson, K.D. (1994) *J. Biol. Chem.* 269, 17434–17439.
17. Nicoll, D.A., Quednau, B.D., Qui, Z., Xia, Y.R., Lusis, A.J., & Philipson, K.D. (1996) *J. Biol. Chem.* 271, 24914–24921.
18. Cervetto, L., Lagnado, L., Perry, R.J., Robinson, D.W., McNaughton, P.A. (1989) *Nature* 337, 740–743.
19. Schnetkamp, P.P., Basu, D.K., & Szerenesai, R.T. (1989) *Am. J. Physiol.* 257, C153–C157.
20. Dong, H., Dunn, J., & Lytton, J. (2001) *Biophys. J.* 82, 1943–1952.
21. Reilander, H., Achilles, A., Friedel, U., Maul, G., Lottspeich, F., & Cook, N.J. (1992) *EMBO J.* 11, 1689–1695.

22. Tucker, J.E., Winkfein, R.J., Cooper, C.B., & Schnetkamp, P.P.M. (1998) *IOVS* 39, 435–440.
23. Tsoi, M., Rhee, K.H., Bungard, D., Li, X.F., Lee, S.L., Auer, R.N., & Lytton, J. (1998) *J. Biol. Chem.* 273, 4155–4162.
24. Kraev, A., Quednau, B.D., Leach, S., Li, X.F., Dong, H., Winkfein, R., Perizzolo, M., Cai, X., Yang, R., Philipson, K.D., & Lytton, J. (2001) *J. Biol. Chem.* 276, 23161–23172.
25. Li, X.-F., Kraev, A.S., & Lytton, J. (2002) *J. Biol. Chem.* 277, 48410–48417.
26. Schnetkamp, P.P. (2004) *Pfluegers Arch.* 447, 683–688.
27. Cai, X. & Lytton, J. (2004) *J. Biol. Chem.* 279, 5867–5876.
28. Philipson, K.D. & Nicoll, D.A. (1992) *Curr. Opin. Cell Biol.* 4, 678–683.
29. Guerini, D. (1998) *Biometals* 11, 319–330.
30. Reeves, J.P. & Sutko, J.L. (1980) *Science* 208, 1461–1464.
31. Reeves, J.P. & Hale, C.C. (1984) *J. Biol. Chem.* 259, 7733–7739.
32. Reeves, J.P. & Sutko, J.L. (1979) *Proc. Natl. Acad. Sci. USA* 76, 590–594.
33. Hilgemann, D.W. (1989) *Pfluegers Arch.* 415, 247–249.
34. Hilgemann, D.W. (1990) *Nature* 344, 242–245.
35. Cannell, M.B. (1991) *Ann. N. Y. Acad. Sci.* 639, 428–443.
36. Hilgemann, D.W., Nicoll, D.A., & Philipson, K.D. (1991) *Nature,* 352, 715–718.
37. Nicoll, D.A., Ottolia, M., Lu, L., Lu, Y., & Philipson, K.D. (1999) *J. Biol. Chem.* 274, 910–917.
38. Iwamoto, T., Nakamura, T.Y., Pan, Y., Uehara, A., Imanaga, I., & Shikegawa, M. (1999) *FEBS Lett.* 446, 264–268.
39. Ren, X., Nicoll, D.A., & Philipson, K.D. (2006) *J. Biol. Chem.* 281, 22808–22814.
40. Matsuoka, S., Nicoll, D.A., Reilly, R.F., Hilgemann, D.W., & Philipson, K.D. (1993) *Proc. Natl. Acad. Sci. USA* 90, 3870–3874.
41. Li, Z., Nicoll, D.A., Collins, A., Hilgemann, D.W., Filoteo, A.G., Penniston, J.T., Weiss, J.N., Tomich, J.M., & Philipson, K.D. (1991) *J. Biol. Chem.* 266, 1014–1020.
42. Matsuoka, S., Nicoll, D.A., Hryshko, L.V., Levitsky, D.O., Weiss, J.N., & Philipson, K.D. (1995) *J. Gen. Physiol.* 105, 403–420.
43. Levitsky, D.O., Nicoll, D.A., & Philipson, K.D. (1994) *J. Biol. Chem.* 269, 22847–22852.
44. Schwarz, E.M. & Benzer, S. (1997) *Proc. Natl. Acad. Sci. USA* 94, 10249–10254.
45. Levitsky, D.O., Fraysse, B., Leoty, C., Nicoll, D.A., & Philipson, K.D. (1996) *Mol. Cell. Biochem.* 160–161, 27–32.
46. Nicoll, D.A., Sawaya, M.R., Kwon, S., Cascio, D., Philipson, K.D., & Abramson, J. (2006) *J. Biol. Chem.* 281, 21577–21581.
47. Hilge, M., Aelen, J., & Vuister, G.W. (2006) *Mol. Cell* 22, 15–25.
48. Halaby, D.M., Poupon, A., & Mornon, J. (1999) *Protein Eng.* 12, 563–571.
49. Dunn, J., Elias, C.L., Le, H.D., Omelchenko, A., Hryshko, L.V., & Philipson, K.D. (2002) *J. Biol. Chem.* 277, 33957–33962.
50. Kofuji, P., Lederer, W.J., & Schulze, D.H. (1994) *J. Biol. Chem.* 269, 5145–5149.
51. Dyck, C., Omelchenko, A., Elias, C.L., Quednau, B.D., Philipson, K.D., Hnatowich, M., & Hryshko, L.V. (1999) *J. Gen. Physiol.* 114, 701–711.
52. Hryshko, L.V., Matsuoka, S., Nicoll, D.A., Weiss, J.N., Schwarz, E.M., Benzer, S., & Philipson, K.D. (1996) *J. Gen. Physiol.* 108, 67–74.
53. Bano, D., Young, K.W., Guerin, C.J., Lefeuvre, R., Rothwell, N.J., Naldini, L., Rizzuto, R., Carafoli, E., & Nicotera, P. (2005) *Cell* 120, 275–285.
54. Ubach, J., Zhang, X., Shao, X., Sudhof, T.C., & Rizo, J. (1998) *EMBO J.* 17, 3921–3930.
55. Dai, H., Shin, O.H., Machius, M., Tomchik, D.R., Sudhof, T.C., & Rizo, J. (2004) *Nat. Struct. Mol. Biol.* 11, 844–849.
56. Essen, L.O., Perisic, O., Cheung, R., Katan, M., & Williams, R.L. (1996) *Nature* 380, 595–602.
57. Quednau, B.D., Nicoll, D.A., & Philipson, K.D. (1997) *Am. J. Physiol.* 272, C1250–C1261.
58. Philipson, K.D., Nicoll, D.A., Ottolia, M., Quednau, B.D., Reuter, H., John, S., & Qiu, Z. (2002) *Ann. N. Y. Acad. Sci.* 976, 1–10.

59. Schulze, D.H., Polumuri, S.K., Gille, T., & Ruknudin, A. (2002) *Ann. N. Y. Acad. Sci.* 976, 187–196.
60. Blaustein, M.P. & Santiago, E.M. (1977) *Biophys. J.* 20, 79–111.
61. Iwamoto, T., Pan, Y., Wakabayashi, S., Imagawa, T., Yamanaka, H.I., & Shigekawa, M. (1996) *J. Biol. Chem.* 271, 13609–13615.
62. Fan, J., Shuba, Y.M., & Morad, M. (1996) *Proc. Natl. Acad. Sci. USA* 93, 5527–5532.
63. Iwata, T., Kraev, A., Guerini, D., & Carafoli, E. (1996) *Ann. N. Y. Acad. Sci.* 779, 37–45.
64. Shuba, Y.M., Iwata, T., Naidenov, V.G., Oz, M., Sandberg, K., Kraev, A., Carafoli, E., & Morad, M. (1998) *J. Biol. Chem.* 273, 18819–18825.
65. He, L.-P., Cleemann, L., Soldatov, N.M., & Morad, M. (2003) *J. Physiol.* 548.3, 677–689.
66. Lee, S.L., Yu, A.S., & Lytton, J. (1994) *J. Biol. Chem.* 269, 14849–14852.
67. Nicholas, S.B., Yang, W., Lee, S.L., Zhu, H., Philipson, K.D., & Lytton, J. (1998) *Am. J. Physiol.* 274, 217–232.
68. Scheller, T., Kraev, A., Skinner, S., & Carafoli, E. (1998) *J. Biol. Chem.* 273, 7643–7649.
69. Polumuri, S.K., Ruknudin, A., McCarthy, M.M., Perrot-Sinal, T.S., & Schulze, D.H. (2002) *Ann. N. Y. Acad. Sci.* 976, 67–72.
70. Reuter, H., Henderson, S.A., Han, T., Mottino, G.A., Frank, J.S., Ross, R.S., Goldhaber, J.I., & Philipson, K.D. (2003) *Cell Calcium* 34, 19–26.
71. Carafoli, E., Genazzani, A., & Guerini, D. (1999) *Biochem. Biophys. Res. Commun.* 266, 624–632.
72. Li, L., Guerini, D., & Carafoli, E. (2000) *J. Biol. Chem.* 275, 20903–20910.
73. Henderson, S.A., Goldhaber, J.I., So, J.M., Han, T., Motter, C., Ngo, A., Chantawansri, C., Ritter, M.R., Friedlander, M., Nicoll, D.A., Frank, J.S., Jordan, M.C., Roos, K.P., Ross, R.S., & Philipson, K.D. (2004) *Circ. Res.* 95, 604–611.
74. Pott, C., Ren, X., Tran, D.X., Yang, M.-J., Henderson, S., Jordan, M.C., Roos, K.P., Garfinkel, A., Philipson, K.D., & Goldhaber, J.I. (2007) *Am. J. Physiol. Cell Physiol.* 292, C968–973.
75. Pott, C., Yip, M., Goldhaber, J.I., & Philipson, K.D. (2007) *Biophys. J.* 92, 1431–1437.
76. Bondarenko, V.E., Szigeti, G.P., Bett, G.C., Kim, S.J., & Rasmusson, R.L. (2004) *Am. J. Physiol. Heart Circ. Physiol.* 287, 1378–1403.
77. Cook, N.J. & Kaupp, U.B. (1988) *J. Biol. Chem.* 263, 11382–11388.
78. Cervetto, L., Lagnado, L., Perry, R.J., Robinson, D.W., & McNaughton, P.A. (1989) *Nature* 337, 740–743.
79. Schnetkamp, P.P.M., Basu, D.K., & Szerencsei, R.T. (1989) *Am. J. Physiol.* 257, C153–C157.
80. Haug-Kollet, K., Pearson, B., Webel, R., Szerencsei, R.T., Winkfein, R.J., Schnetkamp, P.P., & Colley, N.J. (1999) *J. Cell Biol.* 147, 659–670.
81. Szerencsei, R.T., Tucker, J.E., Cooper, C.B., Winkfein, R.J., Farrell, P.J., Iatrou, K., & Schnetkamp, P.P. (2000) *J. Biol. Chem.* 275, 669–676.
82. Su, Y.H. & Vacquier, V.D. (2002) *Proc. Natl. Acad. Sci. USA* 99, 6743–6748.
83. Prrinsen, C.F.M., Szerencsei, R.T., & Schnetkamp, P.P.M. (2000) *J. Neurosci.* 20, 1424–1434.
84. Cai, X., Zhang, K., & Lytton, J. (2002) *J. Biol. Chem.* 277, 48923–48930.
85. Kinjo, T.G., Szerencsei, R.T., Winkfein, R., Kang, K.J., & Schnetkamp, P.P.M. (2003) *Biochemistry* 42, 2485–2491.
86. Kinjo, T.G., Kang, K.J., Szerencsei, R.T., Winkfein, R.J., & Schnetkamp, P.P.M. (2005) *Biochemistry* 44, 7787–7795.
87. Kang, K.-J. & Schnetkamp, P.P.M. (2003) *Biochemistry* 42, 9438–9445.
88. Kang, K.J., Shibukawa, Y., Szerencsei, R.T., & Schnetkamp, P.P. (2005) *J. Biol. Chem.* 280, 6834–6839.
89. Schnetkamp, P.P.M. (1995) *J. Biol. Chem.* 270, 13231–13239.
90. Seiler, E.P., Guerini, D., Guidi, F., & Carafoli, E. (2000) *Eur. J. Biochem.* 267, 2461–2472.
91. Li, X.-F., Kiedrowski, L., Tremblay, F., Fernandez, F.R., Perizzolo, M., Winkfein, R.J., Turner, R.W., Bains, J.S., Rancourt, D.E., & Lytton, J. (2006) *J. Biol. Chem.* 281, 6273–6282.

J. Krebs and M. Michalak (Editors)
Calcium: A Matter of Life or Death

ISSN: 0167-7306/DOI: 10.1016/S0167-7306(06)41007-3

CHAPTER 7

The plasma membrane calcium pump

Claudia Ortega*, Saida Ortolano*, and Ernesto Carafoli

Venetian Institute of Molecular Medicine and Department of Biochemistry, University of Padova, Viale Colombo 3, 35121 Padova, Italy, Tel.: +39 49 792 3241; Fax: +39 049 792 3250; E-mail: ernesto.carafoli@unipd.it

Abstract

The plasma membrane Ca^{2+}-ATPase (PMCA) is the high Ca^{2+}-affinity system that maintains the intracellular free Ca^{2+} at the appropriately low level in all eukaryotic cells. It is a member of the P-type ATPases family, which is regulated by a number of factors/processes, chiefly, by the interaction with calmodulin. In mammals, it is the product of four independent genes: two of the gene products are ubiquitous (PMCA1 and 4) and two are tissue-restricted (essentially neuronal). A number of alternative splicing processes increase the multiplicity of isoforms. The rationale for the existence of so many PMCA pump variants is not understood but may reflect the interaction with isoform-specific regulatory protein partners. Several such partners have been identified, and recent work has shown that one of them (protein 14.3.3ε) only interacts with some of the pump isoforms. Genetic defects of the PMCAs are beginning to be identified. The most interesting affects PMCA2, which is abundantly expressed in the outer hair cells of the inner ear. A number of mutations in the PMCA2 pump gene have been shown to produce hereditary hearing loss in mice and in humans.

Keywords: Ca^{2+} pumps, hereditary deafness, pump interactors, splicing isoforms

1. Introduction

A family of membrane-intrinsic Ca^{2+}-transporting proteins that operate with various transporting modes maintain cell Ca^{2+} at the low free concentration demanded by its signalling function. The ATPase mode is uniquely capable of interacting with Ca^{2+} with high affinity, thus, the fine-tuning of cell Ca^{2+} is performed by calcium-ATPases pumps that are located in the sarco(endo)plasmic reticulum Ca^{2+}-ATPase (SERCA), in the Golgi membranes [secretory pathway Ca^{2+}-transport ATPase (SPCA)] and in the plasma membrane Ca^{2+}-transport ATPase (PMCA). The three Ca^{2+} pumps belong to the P-type pump family and thus share most mechanistic properties. However, they also exhibit significant functional and structural differences. The present article focuses on the PMCA pump, so it will touch only briefly on traditional aspects, e.g. history or the details of the enzyme reaction. For these

*These authors have contributed equally to the work

aspects a number of comprehensive recent reviews could be consulted [1,2] The focus of the contribution will be aspects of the pump that past review articles have considered in passing and/or which have come to the forefront only recently.

2. *General aspects*

The pump was discovered by H.J. Schatzmann 40 years ago [3] as a system responsible for the active extrusion of Ca^{2+} from red cells. Later on, Schatzmann and Vincenzi [4] demonstrated that Ca^{2+}, when transported, activated the hydrolysis of ATP at the internal surface of the membrane. Following the initial observations on the red cell membrane, a large number of studies on numerous cell types characterized the ATPase as a bona fide member of the family of transport ATPases that were later classified as P-type pumps [5,6]. These pumps are characterized by the formation of an energized intermediate in which the γ-phosphate of ATP is transferred to an invariant Asp in the active site of the enzyme, resulting in the formation of a high-energy acyl-phosphate. As in all other P-type pumps, the reaction cycle of the calcium pumps oscillates between two conformational states: the E1 state, in which the enzymes have high affinity for Ca^{2+}, and the E2 state, in which the enzymes have much lower Ca^{2+} affinity. Ca^{2+} is bound to the cytosolic domain of the enzyme in the high-affinity conformation (E1) prior to the phosphorylation of the catalytic aspartic acid by ATP. A conformational change of the phosphorylated enzyme (E1P $\rightarrow$ E2P) then exposes bound Ca^{2+} to the extracellular/lumenal side of the membrane and promotes its release prior to the cleavage of the phosphorylated intermediate, after that, the pump returns to the E1 conformation.

As mentioned above, three P-type Ca^{2+} pumps are now recognized: The Golgi-membrane SPCA pump is the most recent entry in the field and has not yet been studied as extensively as the other pumps – thus, most of the discussion on general properties in the next paragraphs will concentrate on the pumps of the plasmamembrane (PMCA) and of the SERCA, for which considerable information is available. The two pumps share basic properties but also have significant differences. For instance, the SERCA pump transports two Ca^{2+} ions per molecule of hydrolyzed ATP, whereas the Ca^{2+}/ATP stoichiometry of the PMCA pump is 1:1. Both pumps exchange Ca^{2+} for H^+: early reports on the reconstituted PMCA pump had indicated an electroneutral Ca^{2+}/H^+ exchanger [7], but more recently, work has instead convincingly shown partial electrogenicity, i.e. one H^+ is exchanged for one Ca^{2+} [8]. The single Ca^{2+}-binding site of the PMCA pump, which is thought to correspond to site II in the SERCA pump, is likely to be located on the cytosolic site of the membrane prior to the translocation step and to face the exterior of the cell at its end. After releasing Ca^{2+} to the exterior of the cell, the pump becomes dephosphorylated and eventually returns to the E1 conformation. The process is fully reversible; the phosphorylated intermediate can be formed, under appropriate conditions, from inorganic phosphate as well.

As with all other P-type pumps, the PMCA pump is rapidly and irreversibly inhibited by micromolar concentrations of the phosphate analogue orthovanadate, which binds to the E2 conformer of the pump, thereby preventing the E2/E1 transition [9].

The other general inhibitors of the P-type pumps are lanthanide ions that have been shown to displace Ca^{2+} by combining with its high-affinity binding sites on the SERCA pump [10,11]. As they allow phosphorylation to occur, they could replace Ca^{2+} as the metal ion that activates the phosphorylation as suggested for the SERCA pump [11,12]. As expected, lanthanides also inhibit the PMCA pump; however, at variance with the SERCA pump (and other P-type pumps), they markedly increase the steady-state level of the phosphoenzyme [13]. The effect is attributed to the inhibition of the hydrolysis of the phosphorylated intermediate by La^{3+}: it is very advantageous, as it permits the recognition of the PMCA pump in autoradiographic gels of membrane preparations containing much higher amounts of the SERCA or other P-type pumps.

The peptide caloxin 2A1 has been recently reported to inhibit the PMCA pump [14] by binding to putative extracellular domain 2 of isoform 1 (see Section 5 for information on the membrane topology of the pump), which is one of the two PMCA isoforms of erythrocytes. Another component of the caloxin's peptide family, caloxin 1A1, apparently blocks the pump by perturbing its first extracellular domain. The caloxin findings suggest that transmembrane domains 1 and 2 play a role in the reaction cycle of the PMCA pump [15].

The PMCA pump is essential for the control of the cytosolic Ca^{2+} concentration in all eukaryotic nonmuscle cells but in muscle cells, particularly heart cells, is likely to be of minor importance with respect to the much more abundant SERCA pump and to the high-transport capacity sodium calcium exchanger of the plasma membrane. In eukaryotic cells, the amount of PMCA pump is thought not to exceed the level of 0.1% of the total membrane protein. This value may perhaps be underestimated in neurons, where the level of expression of the pump is apparently higher than in the other cells types [16].

3. *Isolation and purification of the calcium pump*

The purification of the PMCA pump was first reported in 1979 by Niggli et al. [17] using calmodulin-affinity chromatography. The decision to use calmodulin column was prompted by previous work on erythrocyte membranes that had shown that calmodulin activated the pump [18,19]. Red blood cell membranes were solubilized with Triton X-100 and stabilized with phosphatidylserine, and the preparation was applied to the column. Most of the membrane proteins, including the Mg^{2+}-dependent ATPase activity, were eliminated with Ca^{2+}-containing buffers. Ethylenediaminetetraacetic acid then eluted from the column the Ca^{2+}-dependent ATPase as a major band of about 125 KDa accompanied by a minor band of about 205 KDa. Both bands were phosphorylated by ATP in the presence of Ca^{2+} and Mg^{2+}: the heavier band was later shown to be a dimer of ATPase. The PMCA was then successfully reconstituted as an active pump in liposomes. A crucial advance that resulted from the work on liposomes was the detailed analysis of the role of acidic phospholipids as activators. In the first purification experiments, the pump had already been maximally active in the absence of calmodulin. This was so because the calmodulin column had been saturated with the acidic phospholipid phosphatidylserine, which was later found to activate the pump alternatively to calmodulin (see Section 4.). Replacement of phosphatidylserine with the zwitterionic phospholipid phosphatidylcholine soon produced purified preparations that were fully responsive to calmodulin. Another important result of the early work on the purified PMCA was the precise definition of the pattern of proteolysis, especially that by calpain [20]. The first calpain cleavage was found to remove about half of the calmodulin-binding domain and the portion of the pump downstream of it. A second cut then truncated the pump at the beginning of the calmodulin-binding domain. The fully active 124-kDa peptide resulting from the two calpain cuts could be easily purified, because it was not retained by calmodulin columns and was later used to demonstrate that the calmodulin-binding sequence acted as an autoinhibitory domain. It bound to two sites in the cytosolic domain of the pump close to the active centre (see Section 4.), maintaining the pump inactive [21,22]. The irreversible activation of the pump by calpain was thus evidently due to the removal of the autoinhibitory region, i.e., the calmodulin-binding domain. It may have significance in pathological states of the cell, when the resting level of free Ca^{2+} remains elevated for long periods, thus activating calpain [23].

4. *Cloning of the pump and recognition of isoforms*

The primary structure of the pump was deduced in 1988 using rat brain [24] and human teratoma [25] cDNA libraries. Two rat sequences contained 1176 (129.5 KDa) and 1198 (132.6 KDa) amino acids, and the human sequence contained 1220 residues (137.7 KDa) corresponding to that of the latter rat isoform. All regions of functional importance present in other ion-transporting ATPases were soon identified. The translated sequence had the calmodulin-binding in the carboxyl terminal cytosolic tail. It also had a domain near the amino terminus very rich in basic aminoacids that was soon found to have a role in the interaction with activatory acidic phospholipids (see Section 5).

It soon emerged that several PMCA isoforms were encoded by a multigene family and that additional isoform variability was produced by alternative RNA splicing of each gene transcript.

All regions important to the catalytic function of P-type pumps were found to be highly conserved in the PMCA sequences, and, as expected, no isoform diversity was detected in these regions that are essential to the catalytic function of the pump. The structural motifs responsible for the membrane-folding pattern of the P-type pumps also tend to be conserved, i.e., isoform diversity was not found to involve the general arrangement of the transmembrane domains of the pump.

5. *The plasma membrane Ca^{2+} pumps: structural and regulatory characteristics*

The cloning work predicted the PMCAs to contain 10 membrane-spanning segments (TM), with the NH_2 and COOH termini located on the cytosolic side of the membrane. The bulk of the protein mass faces the cytosol and consists of three principal domains: the intracellular loop between TM 2 and 3, a large unit between TM domains 4 and 5 and the extended tail following the 10th transmembrane domain. The first intracellular loop between membrane-spanning domains 2 and 3 corresponds to the transduction or actuator domain of the SERCA pump [26] and is thought to play a role in the long-range transmission of conformational changes occurring during the transport cycle. The large cytosolic region (about 400 residues) between transmembrane segments 4 and 5 contains the catalytic domain (the ATP-binding site and the invariant aspartate that forms the acyl phosphate intermediate). Finally, the extended COOH-terminal tail is the major regulatory domain of the PMCAs. On the basis of sequence comparison and computer modelling (unpublished), the overall 3D structure of the PMCAs resembles that of the SERCA pump. However, differences between the two types of calcium pumps are also evident: the COOH-terminal tail is much shorter in the SERCA than in the PMCA pumps, and the PMCAs contain a basic sequence (see Section 4) in the transduction domain that binds regulatory acidic phospholipids [27].

Long before the first PMCA cDNA was cloned, it was realized that the pump was tightly regulated by Ca^{2+} bound to the modulator protein calmodulin (see Section 4), which decreased the K_d of the pump for Ca^{2+} from 10–20 μM in the resting state to <1 μM: calmodulin interacts with the C-terminal tail of the pump [28]. In the absence of calmodulin, and as long as the Ca^{2+} concentration in the vicinity of the pumps is low (<50–100 nM), the PMCAs remain in an autoinhibited state. This is so because the calmodulin-binding domain in the C-terminal tail makes intramolecular contacts with the first and second cytosolic loop of the pump, thus presumably preventing substrate access to the active site [21,22]. Ca^{2+}-loaded calmodulin deinhibits the pump by removing the C-terminal tail of the pump from the intramolecular binding sites, increasing Ca^{2+} affinity and raising V_{max}. [29]. This regulatory mechanism is similar to that of other Ca^{2+}–calmodulin-dependent enzymes. In the case of one of these enzymes, calmodulin-dependent protein kinase I, the crystal structure has convincingly shown that in the absence of Ca^{2+}–calmodulin the COOH-terminal regulatory domain (which overlaps with the calmodulin-binding domain) interacts with multiple sites in the catalytic core, preventing substrate access and obstructing ATP binding [30] exactly as predicted for the case of the PMCA pump. One additional comment is in order on the interplay between the PMCA pump and calmodulin: unique among calmodulin targets, the pump can be activated by the C-terminal half of calmodulin. As the latter has peculiarly high Ca^{2+} affinity, it is probably Ca^{2+} saturated even at the low level of Ca^{2+} in the resting cell. It could thus interact with the N-terminal half of the calmodulin-binding domain, maintaining the pump in a low state of activity even under conditions of resting cytosolic Ca^{2+} [31].

In addition to calmodulin, kinase-mediated phosphorylations also activate the pump. The effect was first identified in the pump of heart sarcolemma membranes and attributed to the cAMP protein kinase (PKA). The phosphorylation step increased both the V_{max} and the Ca^{2+} affinity of the pump, decreasing the K_m (Ca^{2+}) to about 1 μM in the absence of calmodulin. The consensus site for PKA Lys-Arg-Asn-Ser-Ser was identified downstream of the calmodulin-binding domain in the C-terminal region of one of the isoforms of the pump (PMCA1) [32].

The problem of activation by protein kinase C (PKC) is more complex. Stimulation of the pump by this kinase has been described by several authors [33–39], also in the solubilized, purified erythrocyte Ca^{2+} pump [40]. The phosphorylation occurs in the C-terminal region, involving both serine and threonine residues, one of the phosphorylation sites being a threonine residue within the calmodulin-binding domain. Based on peptide data [41], phosphorylation would be expected to prevent binding of calmodulin to the pump while causing a calmodulin-like activation [42].

As mentioned above, PMCA pump stimulation is also induced by acidic phospholipids and polyunsaturated fatty acids. The observation that acidic phospholipids are more effective than calmodulin, i.e. they decrease the K_m (Ca^{2+}) to values lower than those induced by calmodulin [43], has led to the suggestion that acidic phospholipids and calmodulin activate the Ca^{2+}-ATPase by separate mechanisms involving different binding sites. Work with distinct proteolytic fragments of the PMCA pump and with synthetic peptides has indeed revealed two separate phospholipid-binding regions in the pump: one corresponding to the calmodulin-binding domain, which has basic character [44], and the other, as mentioned, situated in the first cytosolic loop [27]. The suggestion that a substantial portion of the phospholipid stimulation of the pump was apparently mediated by the calmodulin-binding domain is in line with the observation that also other calmodulin-regulated proteins, i.e., neuromodulin and myosin light-chain kinase, are activated by acidic phospholipids.

The phospholipid effect could be important in vivo, as the ATPase in the membrane is presumably surrounded by an amount of phosphatidylserine adequate for about 50% of maximal activation [45]. However, the concentration of phosphatidylserine, and of the other abundant acidic phospholipids in the plasma membrane, is presumably invariant. By contrast, the concentration of phosphatidylinositol (PI) and of its two phosphorylated products (PIP1 and PIP2) in the membrane is modulated in response to first messenger challenges to the cell, making them more attractive as potential modulators. Thus, an activation/deactivation cycle of the pump, linked to the metabolism of PI and designed to modulate the export of Ca^{2+} from the cell as demanded by stimuli that promote intracellular Ca^{2+} release, appears possible [46].

In addition to mediating the regulation by Ca^{2+}–calmodulin, the C-terminal region of the PMCAs has also been shown to be affected by the irreversible activator calpain (see Section 3) [20,47,48] and to mediate the interaction with PDZ (Postsynaptic density protein; Drosophila disc large tumor suppressor; Zo-1 protein)-domain proteins. The interaction may be important for the local organization of Ca^{2+}-signalling domains at the plasma membrane and/or for anchoring Ca^{2+}-regulatory complexes to the cytoskeleton. The PMCAs are also activated by self-association [49]. Even if self-association still permits to bind Ca^{2+}–calmodulin [50] in the oligomerized state there is no further activation by calmodulin. The functional significance of the dimerization (oligomerization) of the PMCAs, which is mediated by the calmodulin-binding domain [51], is a controversial point. An increase of Ca^{2+}-ATPase activity at high enzyme concentrations has been attributed to pump oligomerization [50]. However, more recent findings [52] suggest that the dimerization does not correlate with the function of the pump or has minimal effects on it [53].

6. *Isoforms of the PMCA pump*

Four basic PMCA isoforms are encoded by four independent genes named ATP2B1–ATP2B4 that map, respectively, to the following human chromosomal loci: 12q21–q23, 3p25–p26, Xq28 and 1q25–q32 [54–57].

The size of the four human genes ranges from about 60 kb (PMCA3) to over 350 kb (PMCA2) [58–61]; however, the number and size of protein-coding exons are highly conserved among the four genes, suggesting a common origin by duplication of an ancestor gene. In all genes, the first exon encodes a 5′-untranslated sequence that precedes the ATG-starting codon in the second exon. The transcribed proteins have a homology of about 70–80% and maintain the general folding features. Residues differ among isoforms especially in the N-terminal protrusion and the C-terminal tail.

PMCA1 and PMCA4 are ubiquitously distributed in animal cells and are presumed to carry out housekeeping functions to maintain Ca^{2+} homeostasis. PMCA2 and 3 are tissue-restricted and are found in particular in the nervous system, suggesting functional specificity. PMCA1 appears to be ubiquitously expressed since the earliest states of development, as highlighted in studies of mice-pump phenotypes. PMCA4 is widely expressed in human tissues and supports PMCA1 in maintaining Ca^{2+} homeostasis. In human cells PMCA4 is particularly abundant in kidney, erythrocytes and testis, where it is the principal isoform.

In the nervous system, PMCA2 and PMCA3 are present with regional and developmental distribution differences. PMCA2 is abundant in the Purkinje cells of cerebellum and in the outer hair cells of the inner ear [62,63], whereas PMCA3 is abundant in the habenula and in the choroid plexuses [64,65]. However, traces of PMCA3 are also present in skeletal muscle [66]. Hippocampus cells, instead, are particularly rich in PMCA1, suggesting an additional tissue-specific function of this ubiquitous isoform. PMCA2 is also abundantly expressed in lactating mammary glands, and significant levels of this isoform have also been found in the kidney. During development, isoforms 2 and 3 appear at about day 12.5 of gestation and vary in levels and time distribution: PMCA2 starts to be detected in high levels in the dorsal root ganglia and in the retinoblasts of the developing eye. PMCA3 could be important for the correct development of the organs, as it is widely expressed in tissues from day 12.5 up to day 16.5 of gestation, when it starts to be selectively expressed in cell types [67].

Information on the differential functional properties of the four basic PMCA isoforms is scarce. The calmodulin affinity of the two tissue-specific variants, PMCA2 and PMCA3, is much higher than that of the two ubiquitously expressed isoforms PMCA1 and PMCA4 [16], and the sensitivity to calpain is particularly elevated in PMCA1 [68]. When tested in the native intracellular environment, the two recombinant ubiquitous isoforms are far less effective in controlling temporary Ca^{2+} increases induced by the challenging of model cells overexpressing the pumps with agonists that produce Ca^{2+} transients than the two tissue-specific isoforms PMCA2 and PMCA3 [69].

7. *Splicing variants*

RNAs encoding for plasma membrane calcium pumps are subjected to alternative splicing [70] at two different sites, considerably increasing the number of pumps variants. About 20 of the many putative variants have been identified, some, however, only at the RNA level. The physiological rationale for the existence of such a great number of tissue-specific variants is one of the open problems in the PMCA pump area.

Supplementary exons are alternatively transcribed in the proximity of the two main regulatory domains of PMCAs. Splice site A is located upstream of the phospholipids-binding domain in the first intracellular loop of the pump, and splice site C is placed inside the calmodulin-binding domain in the C terminal cytosolic tail of the protein. A third splice site, splicing site B, is now considered an aberrant splice variant. Alternative splices at sites A and C have been extensively characterized for human and rat isoforms and reviewed in numerous papers [71–73]. A possible additional variant has so far only been identified in amphibians [74]. The scheme of Figure 1 summarizes the variants so far identified for each isoform.

Two different nomenclature systems have been developed to identify the splicing variants [75,76] (Table 1). According to the most commonly used one, insertions in splice A are addressed with letters from *z* to *w*, according to the increased number of inserted exons: for instance, in variant *w* of PMCA2 three exons are included, in variant *x* only one exon of 42 bp is transcribed, in variant *y* two exons, one of 33 and the other of 60 bp are inserted and in variant *z* all supplementary exons are excluded. At splice site C, the alternatively transcribed exons are identified by letters from *a* to *e*. At variance with site A, at splice site C alternative exons are fully or partially transcribed determining changes of the reading frame and the premature truncation of the protein [1].

PMCA1 is characterized by the presence in all identified mature transcripts of the sequence of an exon of 40 bp, which is usually alternatively included in the other isoforms; therefore all known PMCA1s correspond to the *x* variant. PMCA2 presents a more complex pattern of possible splice variants for site A. Up to three different exons (33, 60 and 42 bp) can be transcribed. So far, only four of the eight possible combinations have been identified. Variant *y* has been so far identified in rat but not in humans. Variant *v*, in which a fourth transcribed exon is inserted at splice site A, has been found only in bullfrog. PMCA3 and 4 exist only as variants *x* or *z*, including or excluding an exon of 42 or 36 nucleotides, respectively.

Splicing at site C occurs with different degrees of complexity for each isoform; however, common options can be identified. A multitude of variants are generated by exon-internal splice sites, read-through into adjoining introns and inclusion or omission of entire exons. As mentioned, such a combination of events usually leads to reading frame changes with the consequent insertion of premature-stop codons and modification of the C terminal structure of the pump. The most common modification determined by total or partial insertion of exons is the truncation of the calmodulin-binding domain, with changes in the regulatory mechanism of the pumps. Unmodified (no inserts) C-terminal proteins are addressed as *b* variants. In general, for PMCA1 and PMCA4, a single exon is involved in alternative splices at site C, whereas PMCA2 and PMCA3 include or exclude partially or completely two exons.

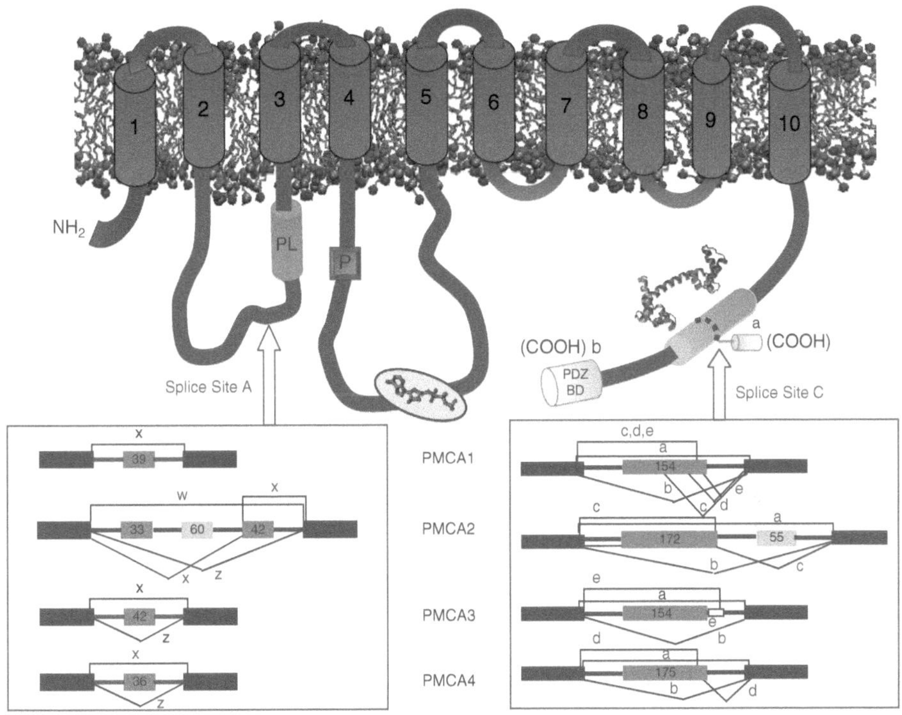

Fig. 1. Membrane topology domains and splicing variants of the pump. The 10 putative transmembrane regions are numbered and indicated by red boxes. The acidic phospholipid (PL)-sensitive region preceding transmembrane domain 3 and the C-terminal calmodulin-binding domain (next to the 3D structure of calmodulin) is shown as orange cylinders. The P in the red square indicates the site of aspartyl-phosphate formation during pump function. The ATP-binding domain is indicated by the gray ellipse with the ATP structure inside. The putative C-terminal PDZ-binding domain (PDZ-BD) is shown as a shaded yellow box at the extreme C-terminus. The sites of alternative RNA splicing are indicated under the topology structure and are labelled as sites A and C. The exon structure of the region affected by alternative splicing is shown for each of the four plasma membrane Ca^{2+}-ATPase (PMCA) genes. Constitutively spliced exons are shown as dark blue boxes and alternatively inserted exons in light blue. The sizes of alternatively spliced exons are given as nucleotide numbers; the splicing options are indicated by connecting lines (See Color Plate 19, p. 517).

The tissue distribution and the targeting to plasma membrane domains of the splice isoforms have not been determined in detail, but it is known that the ubiquitous variants PMCA1xb and PMCA4b predominantly localize in the basolateral part of the plasma membranes of polarized cells. PMCA2wb is expressed in mammary gland cells and localizes instead in the apical section of the cell membrane. Splicing *w* probably has a general role in the membrane sorting of PMCA variants: in polarized cell lines, the splicing *w* at site A is a determinant for the targeting of the pumps to the apical plasma membrane [77]. Interestingly, PMCA2wa is selectively expressed in cochlear and vestibular stereocilia, as shown by immunoprecipitation from purified hair bundles of rat utricles [78]. Recently,

Table 1
Nomenclature table

	Original nomenclature [75]	Alternative nomenclature [76]
Splice site A	z	AI
	x	AII
	y	AIV
	w	AIII
	v	AV
Splice site C	a	CII
	b	CI
	c	CIII
	d	CIV
	e	CV
	f	CIV

it has also been demonstrated that the combination of splice *w* at site A and *a* at site C is essential for the correct targeting of the protein to the stereocilia [79].

8. *Protein interactors of PMCA pumps*

Naturally, the established function of PMCAs is to expel Ca^{2+} from cytosol. But an additional role for the pump has been recently suggested, on the basis of newly discovered interactions of the pumps with molecules involved in signalling pathways [80,81]: the pump would function as a modulator of signal transduction. For example, in cardiomyocites, where the Na/Ca^{2+} exchanger is far more important than the PMCA pump in the extrusion of Ca^{2+}, PMCA4 has been shown to form a ternary complex with α1-syntropin and nitric oxide synthase-1 (NOS-1). The physical interaction with NOS-1 induces a negative regulation of the NO production, and the overexpression of α1-syntropin synergistically increases the inhibitory effect. NOS-1 binds to the PDZ-domain in the C-terminal tail of PMCA, whereas the α1-syntropin interacts with the distal region of the large intracellular loop of the protein (residues 652–840) [82].

The same region of the main intracellular loop of PMCA4 binds the tumour suppressor Ras-associated factor 1 (RASSF1), a Ras effector protein involved in H-Ras-mediated apoptosis. Co-expression of RASSF1 and PMCA4b inhibits the epidermal growth factor-dependent activation of Ras [83].

Another interactor of the pump that binds to its second intracellular loop (residues 501–575) is calcineurin, whose modulation of the transcriptional activity of nuclear factor of activated T-cells (NFAT) is inhibited by the interaction. PMCA would therefore be an upstream effector of calcineurin/NFAT signalling, acting as a regulator of differentiation and adaptation of different tissues [84]. It would seem logical to think that the interaction of the pump with partners should be isoform specific, and the search of isoform-specific interactors has recently been identified. The 14-3-3ε protein has been found to interact with PMCA4 but not with PMCA2 when co-expressed in model cells. The interaction that occurs at the N-terminal portion of the pump is inhibitory [85].

Interactors of the PMCA pumps were also identified using the C-terminal tail of the pump as bait in yeast double screening experiments. These include members of the membrane-associated guanylate-kinase family (MAGUK), cytoskeletal proteins, the Na^+/H^+ exchanger regulatory factor-2 (NHERF2), the Ca^{2+}-/calmodulin-dependent serine protein kinase (CASK) and the novel protein PMCA-interacting single-PDZ protein (PISP). CASK co-precipitates with PMCA4 from brain and kidney tissues lysates. The CASK protein is a co-activator of promoters containing transcription of T-elements. Overexpression of PMCA down-regulates T-element-dependent reporter activity [80].

As for PISP, it transiently interacts with PMCAb variants and plays a role in the sorting of the pump to or from the plasma membrane [86].

PMCA4b interacts with the PDZ domains of several MAGUKs, a result that has led to the suggestion that C-terminal splicing would generate pump isoforms that may be differentially recruited to multifunctional protein complexes involved in Ca^{2+} regulation [87].

PMCA2b and PMCA4b have been used to explore the association with NHERF1 and NHERF2. The interaction with the latter was found to be specific for PMCA2b, i.e., it did not occur with PMCA4b. PMCA2b apparently preferred NHERF2 over NHERF1 as an interactor. The interaction may provide a link between the Ca^{2+} pump and the actin cytoskeletal network, potentially stabilizing the pump in a particular membrane microdomain, possibly allowing its regulation by co-assembled protein kinase A. Alternatively, NHERF2-mediated co-clustering of multiple PMCA2b molecules may facilitate their oligomerization, which has been shown in a previous section to lead to calmodulin-independent pump activation [88].

One more example of an attractive partner protein for *b* variants of all PMCAs through their PDZ domain is the Ania 3/Homer protein. The Homer family of proteins has gained attention as a component of the post-synaptic density and is involved in the coupling of NMDA glutamate receptors to the emptying of Ca^{2+} stores. Co-expressed Ania 3 co-localized with PMCA at the plasma membrane of polarized madin-darby canine kidney (MDCK) epithelial cells and endogenous Ania3/Homer and PMCA2 co-expressed in the soma and dendrites of primary hippocampal neurons [89]. Figure 2 offers a comprehensive view of partners of PMCA pumps identified so far and of their approximate sites of interaction.

9. *PMCA pump and disease*

9.1. *PMCA pump knockout mice*

Knockout mice have been developed and the phenotypes extensively studied for all isoforms of the PMCA pumps with the exception of PMCA3. As this isoform is widely expressed in tissues of developing embryos, its role may be critical for normal development of gestation. Hopefully, PMCA3 gene ablation studies will be soon forthcoming.

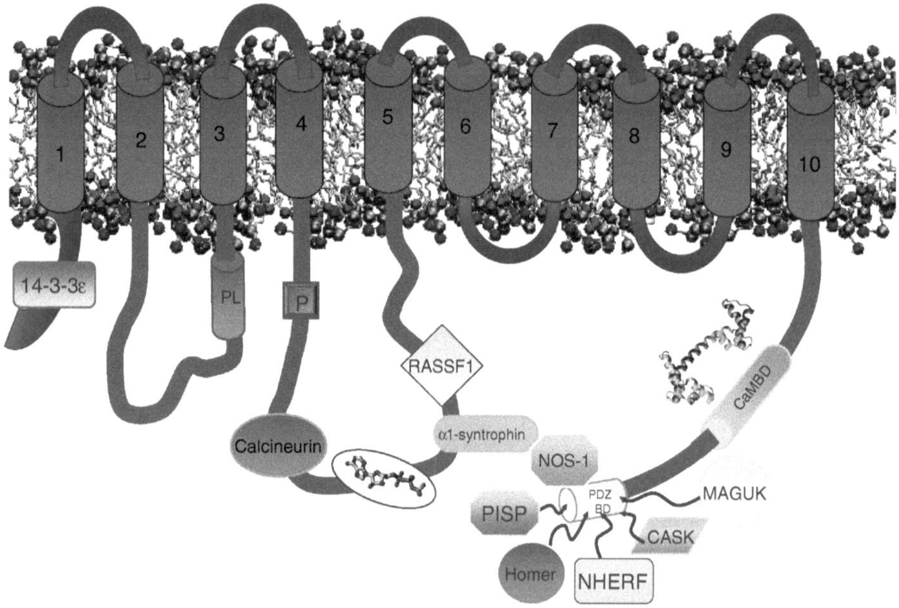

Fig. 2. Interactors of plasma membrane Ca^{2+}-ATPase (PMCA) pumps. The cartoon shows the approximate sites of interaction of the PMCA isoforms with ligands (drawn in coloured boxes). The essential domains of the pump are designated as in the legend for Fig. 1. The 14-3-3ε protein binds to the N-terminal domain of PMCA (first 90 residues) [84]. Calcineurin binds to the amino acid region 501–575 In the main intracellular loop, the region is close to the ATP-binding domain [83], Ras-associated factor 1 (RASSF1) and α1-syntrophin interact with residues 652–840 [81,82]. As shown, most ligands interact with the C-terminal region of the pumps, interacting with the PDZ-binding domain (PDZ-BD) at the end of the C-terminal tail [nitric oxide synthase-1 (NOS-1), PMCA-interacting single-PDZ protein (PISP), Na^+/H^+ exchanger regulatory factor (NHERF), Homer, Ca^{2+}-/calmodulin-dependent serine protein kinase (CASK) and membrane-associated guanylate-kinase family (MAGUK) family] [79,81,85–88]. Note how syntrophin binds to NOS-1 to form a ternary complex with the pump [81] (See Color Plate 20, p. 518).

PMCA1 was disrupted by targeting the catalytic phosphorylation site, but homozygous knockouts resulted in embryonic lethality. Null-mutant embryos were identified up to day 3 of gestation but not during the period of organogenesis. The inability to generate adult live animals underlines the important role of this housekeeping isoform from the earliest ages of development. On the other hand, heterozygous mutants did not present a pathological phenotype, even if the smooth muscle of blood vessels (the portal vein was examined) appeared apoptotic. As this smooth muscle does not express the other ubiquitous isoform PMCA4, the absence of PMCA1 gene on one allele was evidently inefficiently compensated [90].

Mice subjected to the targeted ablation of the PMCA4 survived and appeared outwardly healthy. Histological analysis of organs provided no evidence of major tissue alterations or in vivo cell death [91]. Thus, despite its ubiquitous expression, PMCA4 appears to be less critical than PMCA1 in the maintenance Ca^{2+} homeostasis. A major phenotypic alteration was nonetheless detected, and this was male infertility. Sperm was unable to achieve efficient hyperactivated motility and was unable to reach and fertilize the egg. This was evidently due to the fact that isoform 4 represents 90% of all PMCA pumps expressed in testis cells [92].

PMCA 4 was also found to be crucial in modulating calcium signals in B lymphocytes. Triggering of the B-cell receptor in null-mutant mice caused an enhanced transient increase in cytosolic Ca^{2+}, which was attenuated by CD22 (a B-cell-restricted transmembrane glycoprotein) through stimulation of Ca^{2+}efflux. The observation suggests a direct interaction among PMCA4, CD22 and SHP-1 tyrosine phosphatase [93].

The analysis of the phenotypes of PMCA2 knockout mice has revealed interesting characteristics. Though the animals appeared fairly normal at birth, they started to present balance impairment around day 10 [94]. Recording of the auditory brain response revealed that they were deaf, and the analysis of the vestibular inner ear showed the absence of otoconia. It was also observed that sensory hair cells started to degenerate after day 10. The most severely affected animals also presented partial loss of nerve cells. Unexpectedly, Purkinje neurons, even if they abundantly express PMCA2, were unaltered, and the cerebella maintained normal organization and cell distribution. Thus, the balance defect in knockout mice was not due to a general (cerebellar) neuronal impairment, but it was rather determined by the absence of otoconia that sense gravity and equilibrium changes in the vestibular system.

Similar phenotypes were observed in mice presenting naturally occurring mutations of the PMCA2 gene. The first historically identified animals with mutated sequence of the pump were the deafwaddler mice [63]. The deafwaddler mutation will be described in more detail in the next section but is mentioned here because one of the deafwaddler mutations generates a null mutant. Indeed, deafwaddlers carry either a point mutation (G283S) in the first intracellular loop of the PMCA2 protein (dfwj) or a frame shift mutation at codon 471 generating a null mutant (dfw2j and dfw3j). Subsequently, another point mutation of the PMCA2 gene (K412E) was identified in the deaf Wriggle Sagami mouse. The aminoacidic substitution in the fourth transmembrane domain determined an altered membrane targeting of the pumps that was therefore no longer expressed in the stereocilia [95].

The PMCA2 knockout mouse thus shows that this isoform has a crucial role in the auditory system. Interestingly, in the outer hair cells, PMCA2 is specifically and selectively expressed in the stereocilia. PMCA1 and PMCA4 are also present in these sensory cells but are confined to the basolateral section of the plasma membrane.

9.2. Hereditary deafness and other disease conditions

Studies aimed at characterizing disease features caused by genetic defects of the PMCA proteins, as well as by variations of gene transcription levels have now begun to appear [96], but for a long time, no human disease was directly linked to a defect in any of the genes of the four different isoforms of the PMCA pump. Only very recently, two mutations in the gene encoding for PMCA2 have been identified in patients with hearing impairment. Schultz et al. have described a hypofunctional variant of *ATP2B2* that introduced a V586M substitution in the vicinity of the active site of the protein and enhanced the auditory defect in patients with an accompanying homozygous mutation in the gene of Cadherin 23 [97]. Another mutation in the PMCA2 gene, leading to a G293S substitution, has been observed by our group in a child with hearing impairment in an Italian family (unpublished observations). It is of interest that the G293S mutation involves the same two aminoacids only 10 residues downstream the site of the deafwaddler mutation. These findings, together with the results of the studies on knockout mice for the *ATP2b2* gene described in the preceding section, highlight the essential role of PMCA2 in mammalian hearing. Apparently, no other PMCA isoform can adequately compensate for the loss of PMCA2, or of PMCA2 activity, in sensory hair cells. The PMCA2 gene mutations identified so far as causative of hereditary deafness are summarized in Figure 3. A particularly interesting PMCA2 mutation has recently been found in a mouse with hereditary hearing loss. The phenotype of the mouse and the biochemistry of the mutated pump are presently being analyzed (unpublished work in collaboration with Karen Steel and her co-workers). The special interest of the mutation lies in the fact that it affects a transmembrane domain of the pump, yet, the pump is correctly delivered to the plasma membrane.

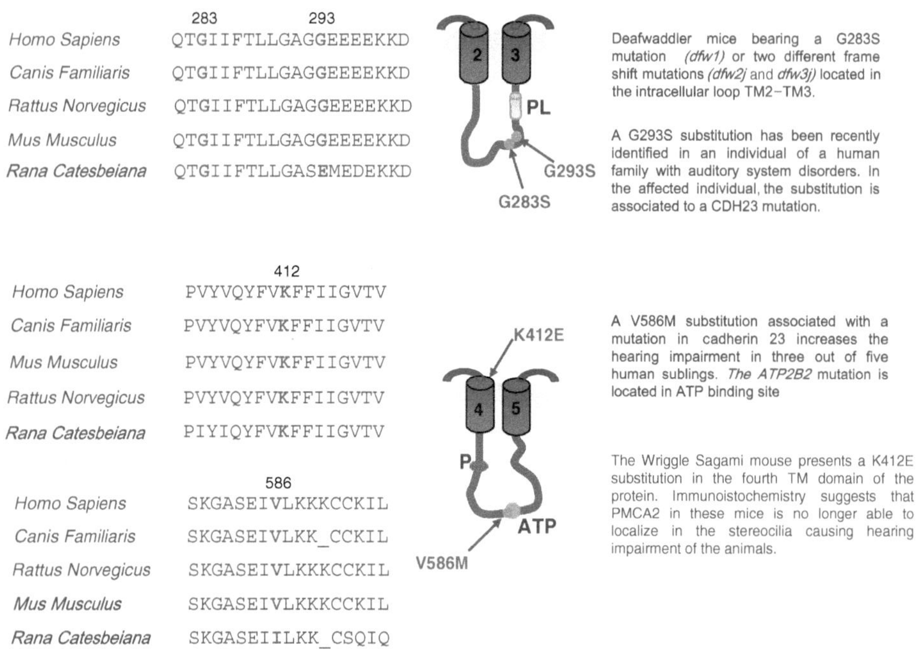

Fig. 3. Deafness-associated mutations of plasma membrane Ca^{2+}-ATPase 2 (PMCA2). Sequence alignment showing the mutation sites in the PMCA2 genes known so far. They are given on the left in the local sequences for different species. Amino acid substitutions are highlighted in red, and the number of the residues given. The localization of the mutations is marked by arrows in the topography structures, and a summarized description of the related phenotypes is given at the side (See Color Plate 21, p. 519).

In most cell types, non-genetic alterations of PMCA activity contribute to pathological states. Modification of expression levels or abnormal proteolysis has been observed in different organs and systems. In neuronal differentiation, for instance, up-regulation of calcium channels to trigger calcium-dependent events is associated with PMCA up-regulation to respond to the increased demand for calcium extrusion [98]. In particular, PMCA2 and three isoforms, specifically expressed in the central nervous system, are up-regulated during the development of granule neurons of cerebellum [99]. Isoforms PMCA2 and PMCA3 are normally up-regulated in neuronal development, as shown in the neuroblastoma cell line IMR-32 [100]. PMCA4 is less involved in neuronal differentiation and may even be down-regulated [101], whereas PMCA1 is a determinant for neurite extension as shown in PC-6 neuronal cells [102], where knocking down of the PMCA1 gene by siRNA leads to the loss of nerve growth factor-mediated neurite elongation [103].

Alterations of calcium homeostasis play a particularly important role in brain ageing, brain damage by ischemia/anoxia, and several neurodegenerative diseases. Free radicals affect membrane Ca^{2+} transport by acting directly on the PMCA or by deranging the lipidic bilayer that hosts it. In Alzheimer patients' neurons, calcium homeostasis is altered. The amyloid beta-peptide (Abeta), which accumulates in AD brain, is neurotoxic: its addition to cultured hippocampal neurons caused altered calcium homeostasis preceded by the impairment of Na–K-ATPase and Ca-ATPase activities [104].

Cardiac hypertrophy and myopathy may also be associated with abnormalities in calcium homeostasis linked to modifications of the PMCA pump. Sarcolemmal membranes from rats with pressure and hypertension-related cardiac hypertrophy presented increased activity of the PMCA. This is likely to be an adaptive reaction to facilitate the removal of Ca^{2+} from myocardial cells during the development of hyperthropy [105]. In atherosclerosis and restenosis the regulation of apotosis is controlled by the transcription factor C-myb, whose binding to the PMCA1 gene promoter represses pump transcription in vascular smooth cells muscle at G1/S cell cycle interface [106].

Pancreatic islet and beta cells of pancreas contain a unique combination of PMCA isoforms, predominantly PMCA1b, 2b, 3a, 3c and 4a. Evidently, functionally distinct PMCA isoforms participate in Ca^{2+} homeostasis in insulin-secreting cells [107,108]. The PMCA activity of pancreatic islets is suppressed by high glucose and is therefore altered in diabetes [109].

In kidney, the expression of PMCA isoforms can be selectively modulated by pathophysiological stimuli. At variance with the constant expression level of PMCA1 and PMCA4 isoforms, the expression of PMCA2 and PMCA3 in tubules has been shown to be depressed in hypercalciuric rats (induced by feeding with a low-phosphate diet) in comparison with hypocalciuric rats. In brain and liver, however, PMCA2 and PMCA3 were unchanged, suggesting a specific role for these isoforms in tubular Ca^{2+} reabsortion [110,111]. Familial hypercalcemia has been associated with global defects of PMCA function [112].

Finally, PMCA2 mRNA has been detected in higher level in some breast cancer cell lines, where it may be overexpressed by a factor of up to 100-fold [113].

10. Conclusions

This review of the PMCA field has brought to the forefront topics that have traditionally been difficult to rationalize, e.g., the existence of so many gene products and spliced isoforms, or that have only gained prominence recently, e.g., the matter of the protein partners of the pump, or finally, that are new and can be expected to attract increasing attention in the future: e.g., the genetic pathology of the pump. One attractive aspect of the pump that has recently begun to emerge, and that distinguishes it from the other two Ca^{2+} pumps, is the possibility that in addition to the obvious role in the process of Ca^{2+} extrusion, it may have the additional function of regulating cell signalling. The tissue-specific expression and function of the different PMCA isoforms and splicing variants have also been given special attention, especially in the light of new findings that have shown that PMCA pumps are not only critical to the maintenance of cytosolic Ca^{2+} concentration but also in shaping the Ca^{2+} signal in a spatially and temporally defined manner. PMCA pumps have been with us for 40 years, and for a long time, knowledge on them has progressed with difficulty: clearly, they are hard enzymes to work with. Now, at long last, the wind appears to have shifted: one may expect with some confidence a period of rapid progress and of exciting findings in a number of new directions.

References

1. Strehler, E., Zacharias, D. (2001) *Physiol. Rev.* 8, 121–150.
2. Guerini, D., Carafoli, E. (1999) in: E. Carafoli and C. Klee (Eds), *Calcium as a Cellular Regulator*, Oxford University Press, New York, 249–278.
3. Schatzmann, H.J. (1966) *Experientia* 22(6), 364–365.
4. Vincenzi, F.F., Schatzmann, H.J. (1967) *Helv. Physiol. Pharmacol. Acta* 25(2), CR233–CR234.
5. Pedersen, P.L., Carafoli, E. (1987) *Trends Biochem. Sci.* 12, 146–150.
6. Pedersen, P.L., Carafoli, E. (1987) *Trends Biochem. Sci.* 12, 186–189.
7. Niggli, V., Sigel, E., Carafoli, E. (1982) *J. Biol. Chem.* 10, 2350–2356.
8. Hao, L., Rigaud, J.L., Inesi, G. (1994) *J. Biol. Chem.* 269, 14268–14275.
9. Barrabin, H., Garrahan, P.J., Rega, A.F. (1980) *Biochim. Biophys. Acta* 600, 796–804.
10. Scott, T.L. (1984) *J. Biol .Chem.* 259, 4035–4037.
11. Squier, T.C., Bigelow, D.J., Fernandez-Belda, F.J., deMeis, L., Inesi, G. (1990) *J. Biol. Chem.* 265, 13713–13720.
12. Asturias, F.J., Fischetti, R.F., Blasie, J.K. (1994) *Biophys. J.* 66, 1665–1677.
13. Szasz, I., Sarkadi, B., Schubert, A., Gardos, G. (1978) *Biochim. Biophys. Acta* 512(2), 331–340.
14. Chaudhary, J., Walia, M., Matharu, J., Escher, E., Grover, A.K. (2001) *Am. J. Physiol. Cell Physiol.* 280, C1027–C1030.
15. Pande, J., Mallhi, K.K., Grover, A.K. (2005) *Eur. J. Pharmacol.* 508, 1–6.
16. Hilfiker, H., Guerini, D., Carafoli, E. (1994) *J. Biol. Chem.* 269, 26178–26183.
17. Niggli, V., Penniston, J.T., Carafoli, E. (1979) *J. Biol. Chem.* 254, 9955–9958.
18. Gopinath, R.M., Vincenzi, F.F. (1977) *Biochem. Biophys. Res. Commun.* 77, 1203–1209.
19. Jarrett, H.W., Penniston, J.T. (1977) *Biochem. Biophys. Res. Commun.* 77, 1210–1216.
20. James, P., Vorherr, T., Krebs, J., Morelli, A., Castello, G., McCormick, D.J., Penniston, J.T., De Flora, A., Carafoli, E. (1989) *J. Biol. Chem.* 264, 8289–8296.

21. Falchetto, R., Vorherr, T., Brunner, J., Carafoli, E. (1991) *J. Biol. Chem.* 266, 2930–2936.
22. Falchetto, R., Vorherr, T., Carafoli, E. (1992) *Protein Sci.* 12, 1613–1621.
23. Salamino, F., Sparatore, B., Melloni, E., Michetti, M., Viotti, P.L., Pontremoli, S., Carafoli, E. (1994) *Cell Calcium* 15, 28–35.
24. Shull, G.E., Greeb, J. (1988) *J. Biol. Chem.* 263, 8646–8657.
25. Verma, A.K., Filoteo, A.G., Stanford, D.R., Wieben, E.D., Penniston, J.T., Strehler, E.E., Fischer, R., Heim, R., Vogel, G., Mathews, S., et al. (1988) *J. Biol. Chem.* 263, 14152–14159.
26. Toyoshima, C., Nakasako, M., Nomura, H., Ogawa, H. (2000) *Nature* 405, 647–655.
27. Zvaritch, E., James, P., Vorherr, T., Falchetto, R., Modyanov, N., Carafoli, E. (1990) *Biochemistry* 29, 8070–8076.
28. James, P., Maeda, M., Fischer, R., Verma, A.K., Krebs, J., Penniston, J.T., Carafoli, E. (1988) *J. Biol. Chem.* 263, 2905–2910.
29. Enyedi, A., Vorherr, T., James, P., McCormick, D.J., Filoteo, A.G., Carafoli, E., Penniston, J.T. (1989) *J. Biol. Chem.* 264, 12313–12321.
30. Goldberg, J., Nairn, A.C., Kuriyan, J. (1996) *Cell* 84, 875–887.
31. Elshorst, B., Hennig, M., Forsterling, H., Diener, A., Maurer, M., Sculte, P., Schwalbe, H., Griesinger, C., Krebs, J., Schmid, H., Vorherr, T., Carafoli, E. (1999) *Biochemistry* 38, 12320–12332.
32. James, P.H., Pruschy, M., Vorherr, T.E., Penniston, J.T., Carafoli, E. (1989) *Biochemistry* 28, 4253–4258.
33. Smallwood, J.I., Gugi, B., Rasmussen, H. (1988) *J. Biol. Chem.* 263, 2195–2202.
34. Furukawa, K., Tawada, Y., Shigekawa, M. (1989) *J. Biol. Chem.* 264, 4844–4849.
35. Ogurusu, T., Wakabayashi, S., Furukawa, K., Tawada-Iwata, Y., Imagawa, T., Shigekawa, M. (1990) *Biochemistry (Tokyo)* 108, 222–229.
36. Fukuda, T., Ogurusu, T., Furukawa, K., Shigekawa, M. (1990) *Biochemistry (Tokyo)* 108, 629–634.
37. Kuo, T.H., Wang, K.K., Carlock, L., Diglio, C., Tsang, W. (1991) *Biol. Chem.* 266, 2520–2525.
38. Qu, Y., Torchia, J., Sen, A.K. (1992) *Can. J. Physiol. Pharmacol.* 70, 1230–1235.
39. Wright, L.C., Chen, S., Roufogalis, B.D. (1993) *Arch. Biochem. Biophys.* 306, 277–284.
40. Wang, K.K., Wright, L.C., Machan, C.L., Allen, B.G., Conigrave, A.D., Roufogalis, B.D. (1991) *J. Biol. Chem.* 266, 9078–9085.
41. Hofmann, F., Anagli, J., Carafoli, E., Vorherr, T. (1994) *J. Biol. Chem.* 269, 24298–24303.
42. Farrar, Y.J., Vanaman, T.C., Slevin, J.T. (1994) *FASEB J.* 8, A1390.
43. Enyedi, A., Flura, M., Sarkadi, B., Gardos, G., Carafoli, E. (1987) *J. Biol. Chem.* 262, 6425–6430.
44. Brodin, P., Falchetto, R., Vorherr, T., Carafoli, E. (1992) *Eur. J. Biochem.* 204, 939–946.
45. Niggli, V., Adunyah, E.S., Carafoli, E. (1981) *J. Biol. Chem.* 256, 8588–8592.
46. Penniston, J.T. (1982) *Ann. N. Y. Acad. Sci.* 402, 296–303.
47. Wang, K.K., Villalobo, A., Roufogalis, B.D. (1992) *Trends Cell Biol.* 2, 46–52.
48. Hofmann, F., James, P., Vorherr, T., Carafoli, E. (1993) *J. Biol. Chem.* 268, 10252–10259.
49. Kosk-Kosicka, D., Bzdega, T. (1988) *J. Biol. Chem.* 263, 18184–18189.
50. Kosk-Kosicka, D., Bzdega, T., Johnson, J.D. (1990) *Biochemistry* 29, 1875–1879.
51. Vorherr, T., Kessler, T., Hofmann, F., Carafoli, E. (1991) *J. Biol. Chem.* 266, 22–27.
52. Bredeston, L.M., Rega, A.F. (1999) *Biochim. Biophys. Acta* 1420, 57–62.
53. Levi, V., Rossi, J.P., Castello, P.R., Gonzalez Flecha, F.L. (2000) *FEBS Lett.* 483, 99–103.
54. Olson, S., Wang, M.G., Carafoli, E., Strehler, E.E., McBride, O.W. (1991) *Genomics* 9, 629–641.
55. Brandt, P., Ibrahim, E., Bruns, G.A.P., Neve, R.L. (1992) *Genomics* 14, 484–487.
56. Latif, F., Duh, F.M., Gnarra, J., Tory, K., Kuzmin, I., Yao, M., Stackhouse, T., Modi, W., Geil, L., Schmidt, L., Li, H., Orcutt, M.L., Maher, E., Richards, F., Phipps, M., Ferguson-Smith, M., Le Paslier, D., Linehan, W.M., Zbar, B., Lerman, M.I. (1993) *Cancer Res.* 53, 861–867.
57. Wang, M.G., Yi, H., Hilfiker, H., Carafoli, E., Strehler, E.E., McBride, O.W. (1994) *Cytogenet. Cell Genet.* 67, 41–45.

58. Brandt, P., Neve, R.L., Kammesheidt, A., Rhoads, R.E., Vanaman, T.C. (1992) *J. Biol. Chem.* 267, 4367–4385.
59. Heim, R., Hug, M., Iwata, T., Strehler, E.E., Carafoli, E. (1992) *Eur. J. Biochem.* 205, 333–340.
60. Kuzmin, I., Stackhouse, T., Latif, F., Duh, F.M., Geil, L., Gnarra, J., Yao, M., Li, H., Tory, K., Le Paslier, D., Chumakov, I., Cohen, D., Chinault, A.C., Linehan, W.M., Lerman, M.I., Zbar, B. (1994) *Cancer Res.* 54, 2486–2491.
61. Brown, B.J., Hilfiker, H., DeMarco, S.J., Zacharias, D.A., Greenwood, T.M., Guerini, D., Strehler, E.E. (1996) *Biochim. Biophys. Acta* 1283, 10–13.
62. Furuta, H., Luo, L., Hepler, K., Ryan, A.F. (1998) *Hear. Res.* 123, 10–26.
63. Street, V.A., McKee-Johnson, J.W., Fonseca, R.C., Tempel, B.L., Noben-Trauth, K. (1998) *Nature Genet.* 19, 390–394.
64. Stahl, W.L., Eakin, T.J., Owens, J.W.M., Breininger, J.F., Filuk, P.E., Anderson, W.R. (1992) *Mol. Brain Res.* 16, 223–231.
65. Stauffer, T.P., Guerini, D., Celio, M.R., Carafoli, E. (1997) *Brain Res.* 748, 21–29.
66. Greeb, J., Shull, G.E. (1989) *J. Biol. Chem.* 264, 18569–18576.
67. Zacharias, D.A., Kappen, C. (1999) *Biochim. Biophys. Acta* 1428, 397–405.
68. Guerini, D., Pan, B., Carafoli, E. (2003) *J. Biol. Chem.* 278, 38141–38148.
69. Brini, M., Coletto, L., Pierobon, N., Kraev, N., Guerini, D., Carafoli, E. (2003) *J. Biol. Chem.* 278, 24500–24508.
70. Strehler, E.E., Strehler-Page, M.A., Vogel, G., Carafoli, E. (1989) *Proc. Natl. Acad. Sci. U.S.A.* 86, 6908–6912.
71. Burk, S.E., Shull, G.E. (1992) *J. Biol. Chem.* 267, 19683–19690.
72. Stauffer, T.P., Hilfiker, H., Carafoli, E., Strehler, E.E. (1993) *J. Biol. Chem.* 268, 25993–26003.
73. Zacharias, D.A., Dalrymple, S.J, Strehler, E.E. (1995) *Mol. Brain Res.* 28, 263–272.
74. Dumont, R.A., Lins, U., Filoteo, A.G., Penniston, J.T., Kachar, B., Gillespie, P.G. (2001) *J. Neurosci.* 21, 5066–5078.
75. Keeton, T.P, Burk, S.E., Shull, G.E. (1993) *J. Biol. Chem.* 268, 2740–2748.
76. Carafoli, E. (1994) *FASEB J.* 8, 993–1002.
77. Chicka, M., Strehler, E.E. (2003) *J. Biol. Chem.* 278, 18464–18470.
78. Hill, J.K., Williams, D.E., LeMasurier, M., Dumont, R.A., Strehler, E.E., Gillespie, P.G. (2006) *J. Neurosci.* 26, 6172–6180.
79. Grati, M., Aggarwal, N., Strehler, E.E., Wenthold, R.J. (2006) *J. Cell. Sci.* 119, 2995–3007.
80. Schuh, K., Uldrijan, S., Gambaryan, S., Roethlein, N., Neyses, L (2003) *J. Biol. Chem.* 278, 9778–9783.
81. Kim, E., DeMarco, S.J., Marfatia, S.M., Chishti, A.H., Sheng, M., Strehler, E.E. (1998) *J. Biol. Chem.* 273, 1591–1595.
82. Williams, J.C., Armesilla, A.L., Mohamed, T.M.A., Hagarty, C.L., McIntyre, F.H., Schomburg, S., Zaki, A.O., Oceandy, D., Cartwright, E.J., Buch, M.H., Emerson, M., Neyes, L (2006) *J. Biol. Chem.* 281, 23341–23348.
83. Armesilla, A.L., Williams, J.C., Buch, M.H., Pickard, A., Emerson, M., Cartwright, E.J., Oceandy, D., Vos Michele, D., Gillies, S., Clark, G.J., Neyses, L. (2004) *J. Biol. Chem.* 279, 31318–31328.
84. Buch, M.H., Pickard, A., Rodriguez, A., Gillies, S., Maass, A.H., Emeron, M., Cartwright, E.J., Williams, J.C., Oceandy, D., Redondo, J.M., Neyses, L., Armesilla, A.L., (2005) *J. Biol. Chem.* 280, 29479–29487.
85. Rimessi, A., Coletto, L., Pinton, P., Rizzuto, R., Brini, M., Carafoli, E. (2005) *J. Biol. Chem.,* 280, 37195–37203.
86. Goellner, G.M., DeMarco, S.J., Strehler, E.E. (2003) *Ann. N. Y. Acad. Sci.* 986, 461–471.
87. DeMarco, S.J., Strehler, E. (2001) *J. Biol. Chem.* 276, 21594–21600.
88. DeMarco, S.J., Chicka, M.C., Strehler, E. (2002) *J. Biol. Chem.* 277, 10506–10511.

89. Sgambato-Faure, V., Xiong, Y., Berke, J.D., Hyman, S.E., Strehler, E.E. (2006) *Biochem. Biophys. Res. Commun.* 343, 630–637.
90. Okunade, G.W., Miller, M.L., Pyne, G.J., Sutliff, R.L., O'Connor, K.T., Neumann, J.C., Andringa, A., Miller, D.A., Prasad, V., Doetschman, T., Paul, R.J., Shull, G.E. (2004) *J. Biol. Chem.* 279, 33742–33750.
91. Schuh, K., Cartwright, E.J., Jankevics, E., Bundschu, K., Liebermann, J., Williams, J.C., Armesilla, A.L., Emerson, M., Oceandy, D., Knobeloch, K.P., Neyses, L. (2004) *J. Biol. Chem.* 279, 28220–28229.
92. Prasad, V., Okunade, G.W., Miller, M.L., Shull, G.E. (2004) *Biochem. Biophys. Res. Commun.* 322, 1192–1203.
93. Chen, J., McLean, P.A., Neel, B.G., Okunade, G., Shull, G.E., Wortis, H.H. (2004) *Nat. Immunol.* 5, 651–657.
94. Kozel, P.J., Friedman, R.A., Erway, L.C., Yamoah, E.N., Liu, L.H., Riddle, T., Duffy, J.J., Doetschman, T., Miller, M.L., Cardell, E.L., Shull, G.E. (1998) *J. Biol. Chem.* 273, 18693–18696.
95. Takahashi, K., Kitamura, K. (1999) *Biochem. Biophys. Res. Commun.* 261, 773–778.
96. Lethotsky, J., Kaplan, P., Murin, R., Raeymaekers, L. (2002) *Front. Biosci.* 7, d53–d84.
97. Schultz, J.M., Yang, Y., Caride, A.J., Filoteo, A., Penheither, A.R., Lagziel, A., Morell, R.J., Mohiddin, S.A., Fananapazir, L., Madeo, A.C., Penniston, J.T., Griffith, A. (2005) *N. Engl. J. Med.* 352, 1557–1564.
98. Berridge, M.J., Bootman, M.D., Lipp, P. (1998) *Nature* 395, 645–648.
99. Guerini, D., Garcia-Martin, E., Gerber, A., Volbracht, C., Leist, M., Merino, C.G., Carafoli, E. (1999) *J. Biol. Chem.* 274, 1667–1676.
100. Usachev, Y.M., Toutenhoofd, S.L., Goellner, G.M., Strehler, E.E., Thayer, S.A. (2001) *J. Neurochem.* 76, 1756–1765.
101. Guerini, D., Wang, X., Li, L., Genazzani, A., Carafoli, E. (2000) *J. Biol. Chem.* 275, 3706–3712.
102. Brandt, P.C., Vanaman, T.C. (2000) *J. Biol. Chem.* 275, 24534–24539.
103. Brandt, P.C., Sisken, J.E., Neve, R.L., Vanaman, T.C. (1996) *Proc. Natl. Acad. Sci. U.S.A.* 93, 13843–13848.
104. Mark, R.J., Hensley, K., Butterfield, D.A., Mattson, M.P. (1995) *J. Neurosci.* 15, 6239–6249.
105. Nakanishi, H., Makino, N., Hata, T., Matsui, H., Yano, K., Yanaga, T. (1989) *Am. J. Physiol.* 257, H349–H356.
106. Afroze, T., Husain, M. (2000) *J. Biol. Chem.* 275, 9062–9069.
107. Garcia, M.E., Del Zotto, H., Caride, A.J., Filoteo, A.G., Penniston, J.T., Rossi, J.P., Gagliardino, J.J. (2002) *J. Membr. Biol.* 185, 17–23.
108. Kamagate, A., Herchuelz, A., Bollen, A., Van Eylen, F. (2000) *Cell Calcium* 4, 231–246.
109. Ximenes, H.M., Kamagate, A., Van Eylen, F., Carpinelli, A., Herchuelz, A. (2003) *J. Biol. Chem.* 278, 22956–22963.
110. Glendenning, P.T., Ratajczak, T., Dick, I.M., Prince, R.L. (2000) *Arch. Biochem. Biophys.* 380, 126–132.
111. Caride, A.J., Chini, E.N., Penniston, J.T., Dousa, T.P. (1999) *Kidney* 56, 1818–1825.
112. Donahue, H.J., Penniston, J.T., Heath, H.J. (1989) *J. Clin. Endocrinol. Metab.* 68, 893–898.
113. Lee, W.J., Roberts-Thomson, S.J., Montheith, G.R. (2005) *Biochem. Biophys. Res. Commun.* 337, 779–783.

J. Krebs and M. Michalak (Editors)
Calcium: A Matter of Life or Death

ISSN: 0167-7306/DOI: 10.1016/S0167-7306(06)41008-5

CHAPTER 8

Endoplasmic reticulum dynamics and calcium signaling

Allison Kraus and Marek Michalak

Department of Biochemistry, University of Alberta, Edmonton, Alberta, Canada T6G 2H7, Tel.: +1 780 492 2256; Fax: +1 780 492 0886; E-mailmarek.michalak@ualberta.ca

Abstract

Ca^{2+} is an important signaling molecule in the cytosol of the cell. Emerging evidence suggests that Ca^{2+} might also play a signaling role in other regions of the cell. For example, agonist-induced fluctuations in free Ca^{2+} concentration in the endoplasmic reticulum (ER) can affect many functions of the ER, including protein and lipid synthesis, chaperone interactions, induction of ER stress, and apoptosis. Changes in ER luminal Ca^{2+} concentration also impact on the ER to nucleus, to plasma membrane, and to mitochondria communication. It appears, therefore, that Ca^{2+} may play an important signaling role within the ER.

Keywords: calcium, calcium binding proteins, calreticulin, chaperones, ER, protein folding

1. Introduction

The endoplasmic reticulum (ER) is a dynamic organelle of the cell integral to many cellular functions including protein and lipid synthesis, protein folding and post-translational modification, and Ca^{2+} homeostasis. The ER is responsible for the storage of the majority of intracellular Ca^{2+}, and not surprisingly, changes in ER Ca^{2+} homeostasis influence many cellular pathways. ER functions are also highly dependent on Ca^{2+} dynamics in the ER lumen. For example, the high Ca^{2+} concentration in the ER is required for protein folding and glycoprotein processing [1,2], whereas reduced ER luminal Ca^{2+} triggers store-operated Ca^{2+} influx [3]. The ER Ca^{2+} stores affect virtually every cellular function including protein folding and synthesis, muscle contraction and relaxation, embryogenesis and subsequent development, cell differentiation and proliferation, transcription factor activation, secretion, gene expression, learning and memory, membrane excitability, energy metabolism, cell cycle progression, and apoptosis [4].

The ER Ca^{2+} is tightly regulated by Ca^{2+} release from the ER by the inositol 1,4,5-trisphosphate receptor ($InsP_3R$) and ryanodine receptor (RyR) [5,6] and replenishment of the stores by sarcoplasmic/endoplasmic reticulum Ca^{2+}-ATPase (SERCA) [5]. There is also a leak pathway for Ca^{2+}, some of which appears to take place through the translocon responsible for threading newly synthesized proteins into the ER [7].

This pore can be blocked at the cytoplasmic surface by ribosome and by BiP in the lumen of the ER [7]. SERCA pumps Ca^{2+} back into the ER lumen. SERCA is regulated by both calreticulin and calnexin, indicating the inextricable nature of Ca^{2+} signaling and protein folding and subsequently the involvement of ER-resident chaperones in Ca^{2+} signaling and homeostasis [8,9].

In this review, we focus on the signaling role of ER luminal Ca^{2+} with major emphasis on its impact on many cellular functions. We focus on the regulation of ER luminal Ca^{2+}, Ca^{2+} buffers, and a role of ER Ca^{2+} as a signaling molecule. It is likely that agonist-induced changes in ER luminal Ca^{2+} concentration play a signaling role. This would be similar to that described for cytosolic Ca^{2+}. Understanding of ER luminal dynamics is critical, because impaired function of the ER is associated with many severe disease including vascular diseases, neurodegenerative diseases, prion diseases, and cancer.

2. Ca^{2+} *buffering in the ER lumen*

Ca^{2+} is stored in the ER bound to Ca^{2+} binding/buffering resident chaperones [10–12]. As the total Ca^{2+} concentration in the ER is in excess of 2 mM while the free Ca^{2+} concentration ranges from 50 to 500 μM [11–13], maintenance of this unique Ca^{2+} balance with Ca^{2+} buffering proteins is critical. Almost 20 years ago, the link between ER Ca^{2+} and protein sorting, the secretory pathway, and ER morphology was made [14]. ER-resident proteins normally retained in the ER, including BiP, calreticulin, and protein disulphide isomerase (PDI), are secreted when ER stores are depleted of Ca^{2+} in 3T3 and primary murine fibroblastoid LB1 and CJ8 cell lines [14]. Concurrently, with the increased secretion of ER-resident proteins, Ca^{2+} depletion of the stores affects ER morphology. The ER becomes more vesicular or tubulovesicular as compared with the perinuclear cisternae arrangement in control cells [14]. As ER Ca^{2+} stores are of significant consequence in numerous cellular pathways, it is, therefore, fitting that Ca^{2+} has effects on ER morphology and secretory properties.

2.1. Ca^{2+} *buffering and ER-resident chaperones*

There are a number of ER-resident chaperones that bind large quantities of Ca^{2+} and therefore influence Ca^{2+} buffering and Ca^{2+} signaling in the ER [10,11]. Importantly, almost all of these Ca^{2+} buffering proteins are involved in the protein folding and quality control of the secretory pathway.

2.1.1. Calreticulin, Grp94, and BiP

Calreticulin, glucose-regulated protein 94 (Grp94), and BiP (immunoglobulin binding protein, Grp78) are the most abundant Ca^{2+} buffering proteins found in the lumen of the ER. These proteins bind Ca^{2+} with high capacity at their highly acidic C-terminal domains [10]. Ca^{2+} binding to calreticulin, Grp94, BiP, and other ER-resident proteins appears to play a dual role: buffering of the ER Ca^{2+} and modulation of

Ca^{2+}-dependent functions of these chaperones [15]. Therefore, Ca^{2+} levels in the ER are of central importance in dictating the function of resident proteins.

Additional evidence for a critical role of ER luminal Ca^{2+} comes from studies on calreticulin-deficient mice and embryonic stem cells [16–18]. Calreticulin deficiency is embryonically lethal because of impaired cardiac development [17]. Cells derived from calreticulin-deficient embryos have impaired ability to handle Ca^{2+} as well as compromised protein folding and quality control [17,19–21]. ER Ca^{2+} capacity and Ca^{2+}-dependent signaling pathways are significantly affected in the absence of calreticulin. Furthermore, Ca^{2+}-dependent transcriptional processes are impaired during cardiac development in calreticulin-deficient mice [22–24]. In the absence of calreticulin, ER Ca^{2+} stores are partially depleted, and agonist-induced Ca^{2+} release from ER is inhibited [17]. Reduced Ca^{2+} release from the ER of calreticulin-deficient cells is not sufficient to activate many Ca^{2+}-dependent processes including the activation of calcineurin. Activation of calcineurin is essential for nuclear translocation and function of many transcriptional factors including the cardiac-specific factor MEF2C [22]. Over-expression of constitutively active calcineurin in heart of calreticulin-deficient mouse rescues these animals from embryonic lethality [25], indicating that Ca^{2+} release from the ER is required for the activation of calcineurin and documenting a physiologically relevant relationship between calreticulin and calcineurin-dependent pathways. These results demonstrate the importance of both calreticulin and calcineurin in the Ca^{2+}- and ER-dependent signaling pathways that are essential for normal cardiac development [23,24].

Next to calreticulin, Grp94 (94 kDa) is one of the most abundant ER-resident proteins, constituting 5–10% of total ER luminal protein [26,27]. The C-terminal domain of Grp94 contains approximately 19 Ca^{2+} binding sites, 4 of which have high affinity ($K_d = 2\,\mu M$) and low capacity (1 mol of Ca^{2+} per mol of protein) and 15 of which have low affinity ($K_d = 600\,\mu M$) but high capacity (10 mol of Ca^{2+} per mol of protein) [28]. BiP/Grp78 is a 78-kDa ER luminal, Ca^{2+} binding chaperone [29,30]. It is a monomeric protein with two distinct functional domains: an ATP-binding domain and a peptide-binding domain. Although BiP/Grp78 has a relatively low capacity for binding Ca^{2+} (1–2 mol of Ca^{2+} per mol of protein), it may contribute as much as 25% of the total Ca^{2+} storage capacity of the ER, and its elevated expression causes an appreciable increase in ER Ca^{2+} storage capacity [30].

2.1.2. PDI-like proteins

A large family of PDI-like ER-resident proteins contributes to Ca^{2+} buffering in the ER [31]. For example, the C-terminal of PDI has a high capacity (approximately 20 mol of Ca^{2+} per mol of protein) but weak affinity ($K_d = 2\text{–}5\,mM$) Ca^{2+} binding site [32]. ER calcistorin/PDI is a calsequestrin-/PDI-like ER-resident protein with high capacity, low-affinity Ca^{2+} binding sites [33]. It binds 23 mol of Ca^{2+} per mol of protein with low affinity ($K_d = 1\,mM$). ERp72, a 72-kDa member of the PDI family [34], contains C-terminal stretches of negatively charged amino acids involved in binding of 12 mol of Ca^{2+} per mol of protein [35]. Another important ER-resident chaperone containing a PDI-like thioredoxin domain is ERp44. The protein is involved in the control of oxidative protein folding [36]. In addition, ERp44 senses changes in the ER luminal environment and modulates activity of the $InsP_3$ receptor in a Ca^{2+}-dependent manner [37]. In contrast, ERp57, a PDI-like protein involved in quality control of the secretory pathway, does not contain an obvious Ca^{2+} binding site, and its contribution to ER Ca^{2+} buffering, if any, needs further investigation [38,39].

2.1.3. *Calsequestrin*

One of the best examples of a highly specialized luminal Ca^{2+} buffering protein with high Ca^{2+} capacity is calsequestrin. The protein is similar to calreticulin, and it is the most abundant Ca^{2+} buffering protein in the sarcoplasmic reticulum (SR) although being only a minor component of ER [40]. Calsequestrin binds approximately 50 mol of Ca^{2+} per mol of protein with low affinity ($K_d = 1$ mM). Calsequestrin associates with triadin and junctin in the lumen of the SR, regulating the RyR [5]. A major function of calsequestrin is buffering free Ca^{2+} inside the SR. The protein is responsible for lowering Ca^{2+} concentration and thus decreasing the Ca^{2+} concentration gradient against which SERCA must work [5]. Ca^{2+} buffering by calsequestrin may also impact on muscle function in a different way [41,42]. For example, Ca^{2+} binding to calsequestrin affects Ca^{2+} wave generation in cardiac muscle [43] and the function of RyR receptor [44]. Clearly, Ca^{2+} buffering in the SR by calsequestrin plays a physiologically important role in muscle function and pathology.

2.1.4. *Reticulocalbins*

Considering that free ER luminal Ca^{2+} fluctuates from <50 (empty stores) to 500 μM (full stores), it is surprising that several ER-resident proteins have relatively high-affinity Ca^{2+} binding sites. Furthermore, many of them contain EF-hand-like Ca^{2+} binding sites that bind Ca^{2+} with a very high affinity ($K_d = 1$ μM). These proteins are a new subset of the EF-hand superfamily of proteins. The function of these high-affinity binding sites is not obvious, given the high concentrations of Ca^{2+} in the lumen of the ER. It is thought that they might play a structural role or that they might be responsible for the regulation/maintenance of specific protein–protein interactions. This class of ER luminal Ca^{2+} binding proteins includes reticulocalbin, Cab45, calumenin, ERC-55/TCBP-49/E6BP, crocalbin/CBP-50, and calreticulin [45–49]. These Ca^{2+} binding proteins have also been found in the Golgi and may, therefore, play an important role in the secretory pathway [48,49]. With the exception of calreticulin, they all contain six or seven EF-hand type, high-affinity Ca^{2+} binding sites and are considered to be members of the reticulocalbin family. The function of the high-affinity Ca^{2+} binding sites in these proteins is not clear, but they may be involved in Ca^{2+}-dependent processes.

2.2. *Ca^{2+} sensing by the ER*

The ER luminal environment is heterogeneous and well equipped to handle continuous agonist-dependent fluctuations of intraluminal Ca^{2+}. Ca^{2+} binding chaperones bind Ca^{2+} with different affinities, thus affecting ER/SR Ca^{2+} capacity and inferring the ability to react to rapid changes in the ER luminal Ca^{2+} concentration. These proteins modulate many functions of the ER including Ca^{2+} transport across the membrane as well as protein and lipid synthesis. Ca^{2+} in the ER lumen also plays a critical role in signaling to other cellular organelles including the plasma membrane, mitochondria, and nucleus. Store-operated Ca^{2+} influx is one of the most studied Ca^{2+} signaling processes involving communication between the ER lumen and the plasma membrane, that is, Ca^{2+} depletion of the ER triggers opening of a store-operated Ca^{2+} channel (SOC) in the plasma membrane [50]. In many cell types, SOC carries a highly Ca^{2+}-selective, non-voltage-gated, inwardly rectifying current termed I_{CRAC} [50].

2.2.1. Stromal-interacting molecule

How the ER senses changes in the ER luminal Ca^{2+} concentration and how this information is communicated to the plasma membrane is not entirely clear. Recently, a high throughput RNAi screen identified stromal-interacting molecule (STIM1) to be required for SOC [51,52]. STIM1 is a type I integral membrane protein of the ER and the plasma membrane [53]. The protein contains an N-terminal domain oriented toward the lumen of the ER and a C-terminus in the cytoplasm. The luminal domain of STIM1 contains an EF-hand domain, and it was postulated to function as a Ca^{2+} sensor in the ER lumen [52]. The cytoplasmic domain of the protein contains coiled–coiled protein–protein interaction region and a sterile α-motif. Point mutation of the N-terminal EF-hand transforms the Ca^{2+} release activated Ca^{2+} channel (CRAC) current (I_{CRAC}) into a constitutively active Ca^{2+} store-independent mode. Mutations in the EF-hand and cytoplasmic C-terminus of STIM1 alter CRAC channels pharmacological profile and inactivation properties [54]. These are exciting observations because STIM1 may function as the missing link between Ca^{2+} store depletion and SOC influx. Most importantly, the protein may serve as a Ca^{2+} sensor that translocates upon store depletion to the plasma membrane to activate I_{CRAC} channels [51–53,55,56]. It remains to be established whether STIM1 can report changes in the ER luminal Ca^{2+} concentration to other cellular compartments, that is, mitochondria, and nucleus.

2.2.2. SOC

The molecular nature of store-operated Ca^{2+}-selective channels in the plasma membrane has remained a mystery for many years. Recent findings have implicated Orai1 protein as having essential roles in store-operated Ca^{2+} entry across the plasma membrane [55,56]. The severe combined immune deficiency (SCID) patients are homozygous for a single missense mutation in Orai1, and they have no measurable SOC [55]. Expression of wild-type Orai1 in SCID T cells fully restores store-operated Ca^{2+} influx and the I_{CRAC}. Most importantly, over-expression of STIM1 and Orai1 together results in an approximate 20-fold increase in store-operated Ca^{2+} entry and Ca^{2+}-selective current demonstrating that these two proteins are involved in both the signaling and permeation mechanisms of Ca^{2+}-selective store-operated Ca^{2+} entry [55,57,58]. As pointed out earlier, in addition to SOC, there are a number of ER functions that are influenced by changes in ER Ca^{2+} homeostasis. It is conceivable that STIM1 might be a universal Ca^{2+} sensor in the lumen of ER affecting many functions of this organelle including protein and lipid synthesis and transcriptional regulation of metabolic and developmental pathways to name a few.

3. *ER, a multifunctional signaling organelle*

The ER plays an important role in protein folding and quality control, the unfolded protein response (UPR), apoptosis, and signal transduction to many intracellular destinations, including to the nucleus, the plasma membrane, the Golgi, and the mitochondria. It is also an important place for lipid and protein steroid synthesis. Its role as a Ca^{2+} storage tank cannot be alienated from its role in these diverse processes. Although the ER lumen appears continuous as indicated by fluorescence recovery after photobleaching (FRAP) studies of a luminal elastase-green fluorescent protein fusion protein [59], the distribution of ER chaperones and Ca^{2+} channels supports that it is spatially and functionally heterogeneous [60]. Electron energy loss imaging studies indicate that Ca^{2+} in the ER has a high heterogeneous distribution, with Ca^{2+} positive lumen being short distances from areas of undetectable Ca^{2+} levels [61]. The heterogeneous distribution of ER Ca^{2+} handling proteins has an impact on the generation of Ca^{2+} signals [60]. Interestingly, high-capacity Ca^{2+} binding proteins, calreticulin and Grp94, colocalize in the ER lumen, whereas calreticulin and PDI show overlapping yet distinct localizations [60]. In human oocytes, calreticulin localizes predominantly in the cell cortex, and the molecular chaperone calnexin localizes to the cell cortex but in a trilaminar arrangement, whereas calsequestrin is distributed throughout the cell [62]. This demonstrates that in oocytes, there are two distinct Ca^{2+} storage compartments: one with high levels of calreticulin (and the $InsP_3$ receptor) and one with calsequestrin (and the RyR) [62]. Considering the functional heterogeneity of Ca^{2+} pools in the luminally connected ER, it is interesting to note that upon cytosolic Ca^{2+} increase by treatment with thapsigargin or ionomycin, there is a reversible fragmentation of the ER but not the connected nuclear envelope [59]. This further demonstrates the functional heterogeneity of the ER.

3.1. *ER Ca^{2+} and lipid synthesis*

Animals coordinate lipid homeostasis by several mechanisms involving ER-dependent regulation of transcription, lipid synthesis and secretion of lipid-processing enzymes, lipid uptake, and extracellular hydrolysis [63–65]. For example, synthesis and assembly of lipoprotein particles takes place in the ER [66]. ER is also involved in sensing and synthesis of cholesterol as well as in transcriptional regulation of lipid metabolic pathways [64]. Changes in the ER luminal Ca^{2+} affect synthesis and targeting of lipid-processing molecules involved in intracellular cholesterol homeostasis [lipid lipoprotein particle (LDL) receptor, HMG-CoA reductase]. 3-hydroxy-3-methyl-glutaryl-CoA (HMG-CoA) reductase is a rate-limiting enzyme in cholesterol synthesis and an ER integral membrane protein. The LDL receptor is involved in the cellular internalization of LDL. The protein is synthesized in the ER and targeted to the plasma membrane. Maturation, folding, and secretion of certain lipases are also affected by ER Ca^{2+} homeostasis [67]. For example, lipoprotein lipase (produced mainly by adipocytes), a lipase responsible for hydrolysis of very lipid lipoprotein particle (VLDL) triacylglycride, is synthesized in the ER lumen, secreted and functional only if properly folded [63,67–69]. This process requires ER luminal Ca^{2+} [63]. Similarly, a hepatic lipase (produced by hepatocytes) responsible for hydrolysis of heavy density lipoprotein particle (HDL) triacylglyceride (TAG) is synthesized in the ER and secreted [65]. These are excellent examples of a functional coupling between protein folding and ER Ca^{2+} homeostasis.

Sterol regulatory element-binding proteins (SREBPs) are ER integral membrane proteins that, upon specific cleavage, release a transcription factor that activates lipid metabolism genes [64,70]. Therefore, SREBPs play the role of "master regulator" in pathways involved in lipid synthesis, secretion, and uptake [64]. The N-terminal cytoplasmic domain of SREBP is cleaved by a sterol-regulated protease [64,71]. The cleaved portion of the proteins then dimerizes, and the dimer translocates to the nucleus to bind to the sterol regulatory elements of genes that encode proteins necessary for lipid biosynthesis and lipid uptake [64,71]. Proteolytic cleavage of SREBP by site-1 protease requires high ER luminal and Golgi Ca^{2+} concentration [72,73] linking ER luminal Ca^{2+} to sophisticated machinery involved in regulation of lipid and cholesterol metabolism. Interestingly, SREBP target genes encode Ca^{2+}-dependent enzymes, including the Ca^{2+}/calmodulin-dependent protein kinase 1D [74].

3.2. ER and mitochondria

ER membrane is proximal to other organelles of the cell. For example, electron micrographs indicate the close spatial distribution of mitochondria and ER [75]. This proximity allows localized pockets of Ca^{2+} to be released from the ER and to be taken up by the mitochondria. High-resolution 3D images of the mitochondria and ER demonstrate that mitochondria are a largely interconnected dynamic "tubular" network with areas in very close apposition to the ER [76]. These sites of close contact likely harbor microdomains of high Ca^{2+} concentration from $InsP_3$-mediated release, allowing rapid uptake of large amounts of Ca^{2+} into the mitochondria [76]. Accumulation of Ca^{2+} in the mitochondria is a rapid process despite low-affinity transport mechanisms [77]. The ER as a heterozygous organelle allows localized events to take place, allowing a complex level of spatial control. The importance of Ca^{2+} release in localized areas to act on target molecules and organelles, such as uptake by the mitochondria, is likely due to the promiscuous nature of Ca^{2+} and Ca^{2+}-dependent signaling events. Communication between ER and mitochondria is of vital importance for cell survival as it has a major impact on energy metabolism.

The relationship between mitochondria and ER becomes yet more intricate when we consider SOC. Mitochondria have been shown to buffer Ca^{2+} in the cytosol, an event necessary for SOC and ER store replenishment [78,79]. Its proximity to ER membrane and its Ca^{2+} channels, as well as to the plasma membrane, makes it an ideal candidate for such a role. It has been demonstrated that under physiological conditions, mitochondrial Ca^{2+} uptake is required for $InsP_3$ to activate I_{CRAC} [80]. Mitochondria and SERCA pumps are thought to compete for the removal of Ca^{2+}, a form of antagonism that determines the extent of store depletion and therefore I_{CRAC} [80]. Furthermore, mitochondrial Ca^{2+} buffering increases the dynamic range of $InsP_3$ needed to activate SOC [81]. During agonist-induced Ca^{2+} release from the ER, Ca^{2+} entry through SOC for ER store replenishment transits the mitochondria [79]. The mitochondria are needed to maintain microdomains of low Ca^{2+} necessary to sustain SOC [78]. Trans-mitochondrial flux of Ca^{2+} might direct Ca^{2+} toward SERCA and thus promote ER store refilling [79].

The importance of spatial proximity between the ER and the mitochondria comes to light when we consider apoptotic induction in cells. A key player in apoptosis is Bcl-2, a 26-kDa homolog to the *Caenorhabditis elegans* anti-apoptotic protein cell death-9 (CED-9). Considering the role of Ca^{2+} in apoptosis, it is not surprising that Bcl-2 has an effect on Ca^{2+} concentration within the ER. High expression of Bcl-2 in mouse lymphoma A20 cells and rat embryo fibroblasts (R6) cells leads to a reduction in thapsigargin-induced Ca^{2+} release [82–84]. Bcl-2 expression also reduces the free Ca^{2+} concentration within the ER [82–84], increases the Ca^{2+} leak, and reduces the steady-state Ca^{2+} levels [83]. The rate of Ca^{2+} influx activated by ER store depletion is also reduced [83]. Considering that Bcl-2 is found in the mitochondria, the ER, and the nuclear envelope, its role and effect on ER luminal Ca^{2+} must also be coordinated with its Ca^{2+}-dependent role in other organelles throughout the cell.

4. *Protein folding and ER chaperones*

Protein folding in the ER is a highly orchestrated process involving a variety of different players. Quality control describes the differential sorting of newly synthesized proteins of the ER according to their conformation [85]. Properly folded and mature proteins can be secreted from the ER, a process that separates them from misfolded/unfolded conformational variants [85]. Quality control in the ER is dependent on such chaperones as calnexin, calreticulin, BiP, PDI-like proteins Grp94, and oxidoreductases ERp57, ERp72, and likely many others.

Protein retention in the ER in an unfolded state results from the exposure of hydrophobic patches that preferentially associate with molecular chaperones or retention signals, such as RxR (Arginine-any amino acid-Arginine) that are exposed prior to assembly of a functional protein oligomer [86]. Other signals such as the cytoplasmic DxE (Aspartic acid-any amino acid-Glutamic acid) signal result in the accelerated transport of molecules from the ER. Mutations of the acidic residues of DxE to alanine in glycoprotein of the vesicular stomatitis virus (VSV-G) had no effect on the maturation, folding, or oligomerization of the protein, indicating a further level of complexity in the determination of protein retention or export apart from molecular chaperones [87]. The retention of a unfolded viral glycoprotein is not restricted to the ER lumen but involves a circuit between the ER, the intermediate compartment, and the Golgi [88].

4.1. Calnexin and calreticulin, ER lectin-like chaperones

Calnexin and calreticulin are two similar lectin-like chaperones involved in the folding of glycosylated proteins. Calnexin is a type I integral membrane protein, whereas calreticulin is a soluble, ER luminal protein. The 3D structure of a soluble, luminal domain of calnexin indicates that it consists of a globular, Ca^{2+} binding domain responsible for interaction of mono-glycosylated substrate (N-domain), as well as an extended arm containing conserved proline residues and four tandem repeats (P-domain) [89]. Calreticulin resides in the lumen of the ER and shares 30% amino acid sequence identity and 41% amino acid sequence similarity with calnexin [90]. Structural modeling and the nuclear magnetic resonance structure of the P-domain of calreticulin indicate that protein also consists of a globular N-domain, a P-domain arm with three tandem repeats, and a highly acidic C-domain [91,92]. The N-domain is known to bind Zn^{2+}, whereas the C-domain binds Ca^{2+} with low affinity and high capacity, consistent with calreticulin's role in Ca^{2+} homeostasis [19]. Both calreticulin and calnexin interact with ERp57 on the tip of the extended P-domain in a Zn^{2+}-dependent manner [93–95].

Calnexin and calreticulin bind the glucose α-1,3 mannose glycosidic bond on high-mannose containing asparagine-linked oligosaccharides in a Ca^{2+}-dependent manner [85]. Two *N*-acetylglucosamine and nine mannoses with three terminal glucoses are assembled and transferred as a core oligosaccharide onto asparagine residues of the consensus site (Asn-X-Ser/Thr, where X is any amino acid except Pro) of a nascent polypeptide. The three terminal glucoses are trimmed by glucosidase I and II, and reglycosylation takes place by uridine 5′-diphosphate (UDP)-glucose : glycoprotein transferase (UGGT). UGGT can discriminate between folded and unfolded proteins [96], ensuring that only unfolded proteins are reglycosylated. A terminal mannose can be trimmed by ER mannosidases. As monoglycosylated oligosaccharides interact with calreticulin and calnexin, the deglycosylation–glycosylation cycle ensures that unfolded proteins remain bound to molecular chaperones until they are properly folded [85]. The calnexin–calreticulin, degylcosylation–glycosylation cycle ensures the folding and quality control of nascent glycoproteins; however, when proteins remain unfolded or are damaged, they undergo ER-associated degradation (ERAD). ERAD ensures that misfolded and damaged proteins are degraded, as well as normal proteins maintain turnover rates.

Recent studies of *Drosophila* calnexin suggest that the protein may play a role in regulation of Ca^{2+} homeostasis. In *Drosophila* calnexin mutants with a premature stop codon at glutamine 182 or tryptophan 119, there is an elevated and a sustained cytosolic Ca^{2+} concentration in the photoreceptor cells following light stimulation [97]. This might be due to a Ca^{2+} buffering property of the C-terminal cytosolic domain of calnexin. Studies on *Xenopus* oocytes revealed that there might be functional complexes between calnexin and SERCA2b, resulting in the inhibition of intracellular Ca^{2+} oscillations [8]. Calnexin interacts with the C-terminus of SERCA2b in a phosphorylation-dependent manner modulating Ca^{2+} signaling and controlling Ca^{2+}-sensitive chaperone functions in the ER [8]. It is tempting to speculate that calnexin's proximity to Ca^{2+} transporting molecules such as SERCA would make it an ideal candidate for the transmission of molecular signals from the cytosol to the ER lumen, in particular the transmission of molecular events directly related to Ca^{2+}. For instance, saturation of Ca^{2+} binding sites on the cytosolic tail of calnexin (i.e., rise in cytosolic Ca^{2+} levels) may infer subtle structural changes or lead to recruitment of different binding partners to induce SERCA-driven uptake of Ca^{2+}. Interestingly, mathematical models have demonstrated that cytosolic Ca^{2+} oscillations can be controlled by the effect of ER luminal Ca^{2+} on Ca^{2+} uptake [98].

5. *ER stress and UPR*

Accumulation of unfolded proteins in the ER lumen results in a specific ER stress response [99], termed UPR. UPR results in ER-nucleus and ER-plasma membrane signaling, attenuation of protein transcription and translation, mRNA degradation, and up-regulation of chaperone expression [100–103]. The function of UPR is to limit the damage to the cell by adapting to the situation causing ER stress. However, increased UPR caused by prolonged stress may trigger additional damage by inducing apoptosis and leading to metabolic modifications [103–106].

The UPR mechanism involves transcriptional activation of chaperone genes by ER membrane-associated activating transcription factor-6 (ATF-6) in conjunction with the ER membrane kinase, endoribonuclease inositol-requiring kinase 1 (IRE1), and dsRNA-activated protein kinase-like ER kinase (PERK) [100–103]. As a key element of UPR, ER luminal chaperone BiP associates with all three currently identified sensors of UPR (ATF-6, IRE1, and PERK) in their inactive states [100–103]. How are these molecules activated in the ER? BiP associates with IRE1, PERK, and ATF-6 molecules in the absence of ER stress. However, upon accumulation of unfolded proteins, BiP dissociates from IRE1, PERK, and ATF-6, and it is recruited by unfolded proteins [102,103,107–109]. This results in activation of ATF-6, IRE1, and PERK. Shuttling between unfolded proteins and UPR sensors serves as a back-regulation mechanism as the amount of sites occupied by BiP on PERK, IRE1, or ATF-6 is determined by the amount of unfolded proteins in the ER [110]. A role of ER luminal Ca^{2+} in activation of PERK, IRE1, or ATF-6 has not been investigated in detail. However, prolonged Ca^{2+} depletion of ER leads to full induction of UPR and to activation of PERK, IRE1, and ATF-6.

ATF-6 is a type II integral membrane protein that is subject to proteolysis upon the accumulation of unfolded proteins in the ER. ER stress-induced cleavage of ATF-6 is carried out by site-1 protease and site-2 protease in the Golgi [111]. In coordination with ATF-4 and X-box-binding protein 1 (XBP1), the action of proteolytically processed ATF-6 increases the expression of ER-resident molecules, such as BiP, Grp94, calreticulin, and calnexin [102].

Activated PERK phosphorylates eIF2α on Ser51 [112]. Phosphorylation of eIF2α results in the attenuation of general translation, providing an immediate response to unfolded protein accumulation on the translational level [112]. Concomitant with PERK activation and translational attenuation is the activation of IRE1. Mammalian systems have two homologs of the *Saccharomyces cerevisiae* IRE1, Ire1α, and Ire1β. IRE1 is a type I integral membrane ER protein with an ER luminal N-terminal domain that senses the ER stress signal and a C-terminal cytoplasmic tail with a Ser/Thr kinase domain and ribonuclease activity [113]. Accumulation of unfolded proteins in the ER lumen and subsequent dissociation of BiP from IRE1 promote autophosphorylation and dimerization of IRE1, which in turn activates its cytoplasmic RNase activity. Activated IRE1 splices a 26-base intron from XBP1 mRNA, resulting in translation of a potent UPR transcription factor in the form of XBP1 with an altered C-terminus [113]. Another consequence of induction of UPR may be degradation of mRNA encoding secreted and membrane proteins [114]. Recognition of mRNA encoding secreted protein is based on the presence and translation of a short signal sequence. This indicates that there is signaling across the ER membrane where the translocating nascent polypeptide, in a process possibly mediated through IRE1, signals to destroy its encoding mRNA [115].

Although molecules involved in UPR have not yet been linked to Ca^{2+} signaling, the fact that UPR is induced upon Ca^{2+} depletion underlines the relationship between the two pathways. An interesting observation is that, similar to ATF-6, SREBP is also cleaved by site-1 protease and site-2 protease [64]. This suggests a potential cross-talk between pathways of UPR and lipid metabolism involving ER luminal Ca^{2+} signaling.

6. *Ca^{2+} signaling in the ER*

As mentioned earlier, the release and reuptake of Ca^{2+} by the ER results in a continuous fluctuation of the ER luminal Ca^{2+} concentration. Under resting conditions (Fig. 1A), there is a relatively low concentration of free Ca^{2+} in the lumen of the ER (50–400 μM). These conditions support normal protein and lipid synthesis and posttranslational modification of newly synthesized proteins. BiP is associated with IRE1, PERK, and ATF-6, silencing the UPR (Fig. 1A).

Agonist-induced Ca^{2+} release from the ER leads to activation of STIM1 and subsequent communication to plasma membrane to open SOC (Fig. 1B). This provides an influx of fresh Ca^{2+} to replenish the stores as well as a supply of Ca^{2+} for sustained activation of many cytoplasmic pathways involving kinases and phosphatases. Prolonged and uncontrolled Ca^{2+} depletion of the ER and reduction of free ER luminal Ca^{2+} below 20 μM leads to severe consequences (Fig. 1B). One of the best characterized consequences of sustained ER Ca^{2+} depletion is activation of the UPR. This is accomplished by activation of IRE1, PERK, and ATF-6, accumulation of unfolded proteins, up-regulation of ER-resident chaperones, attenuation of translation, and degradation of mRNA (Fig. 1B). Protein synthesis and quality control are also compromised, and SREBP-dependent regulation of lipid metabolism might be modified. Prolonged ER stress conditions also lead to activation of apoptosis and cell death.

Recent findings indicate that prolonged ER Ca^{2+} overload may also have a negative impact on ER function and cellular signaling (Fig. 1C). Increased buffering of Ca^{2+} in the ER lumen may lead to congenital problems. For example, up-regulation

(A)

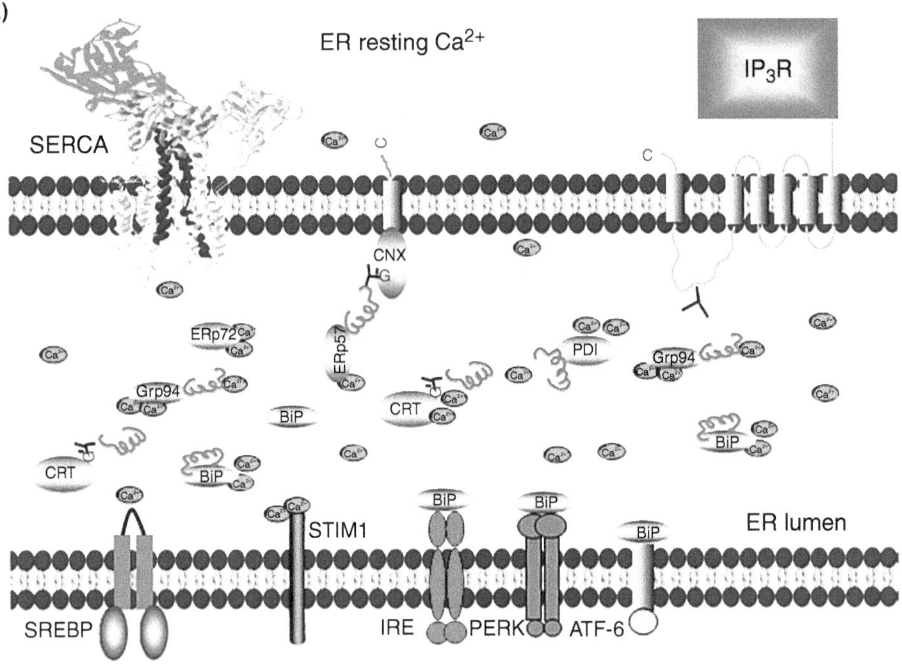

Fig. 1. **(A)** Ca^{2+} dynamics in the lumen of the endoplasmic reticulum (ER). This diagram shows a model of events occurring in response to agonist-activated Ca^{2+} fluctuations in the lumen of the ER. These events affect numerous cellular functions. **A** represents ER Ca^{2+} resting conditions. Under these conditions, there are no major changes in the ER luminal Ca^{2+} concentration (See Color Plate 22, p. 520).

of calreticulin and sustained increase in the ER Ca^{2+} content result in impaired development of pacemaker cells and severe arrhythmias. Increased ER luminal Ca^{2+} also affects $InsP_3$ receptor function and may modulate lipid metabolism. In conclusion, fluctuation of the ER luminal Ca^{2+} concentration impacts on many cellular functions and controls communication between ER and other cellular compartments.

7. *Impact of ER signaling on disease*

What might be physiologically relevant consequences of the ER Ca^{2+} signaling? Changes in the ER luminal environment have been implicated in many severe pathologies (Table 1). For example, calreticulin deficiency in mice is embryonic lethal at E14.5 because of defective cardiac development [17]. Ca^{2+} release by the $InsP_3$-dependent pathway was impaired in calreticulin-deficient cells, as was the translocation of the nuclear factor of activated T-cells (NF-AT) transcription factor [17]. The phenotype of calreticulin-deficient cells is directly relevant to its role as a Ca^{2+} buffering protein and less so as a molecular chaperone. Calreticulin over-expression in embryonic cardiac pacemaker cells results in a dramatic reduction in the beating activity of differentiating

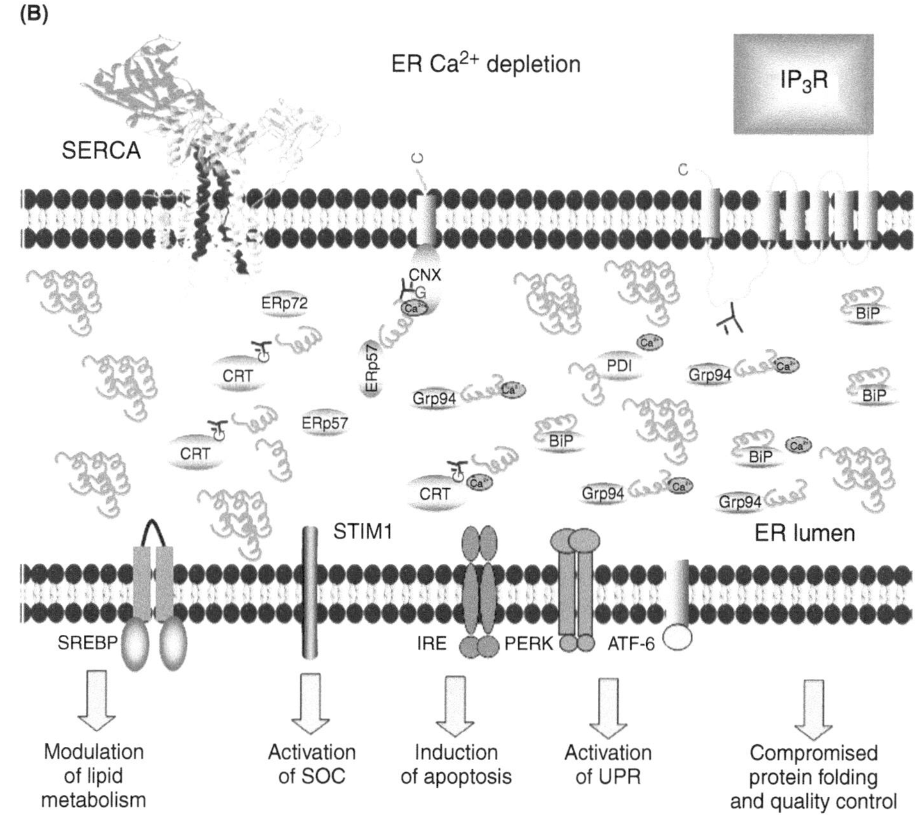

Fig. 1. **(B)** represents ER Ca^{2+} depletion conditions, where ER luminal Ca^{2+} is below 20 μM. Under these conditions, there is activation of store-operated Ca^{2+} channel (SOC) through Ca^{2+} sensing by stromal-interacting molecule (STIM). Prolonged ER Ca^{2+} depletion leads to activation of unfolded protein response (UPR), accumulation of unfolded proteins, activation of dsRNA-activated protein kinase-like ER kinase (PERK), inositol-requiring kinase (IRE1), and activating transcription factor-6 (ATF-6), apoptosis, and cell death. Function of sterol regulatory element-binding protein (SREBP) may also be affected under ER Ca^{2+} depletion. Sustained increase in ER luminal Ca^{2+} (See Color Plate 23, p. 521).

cardiomyocytes [127]. This indicates that the proper regulation of ER luminal Ca^{2+} is crucial for the generation of early pacemaker activity in developing embryoid bodies [128]. At the animal level, severe cardiac pathology develops when calreticulin is over-expressed in the heart of postnatal mice. This phenotype includes sinus bradycardia, atrio-ventricular (AV) node dysfunction, and progressive prolongation of the P-R interval, succeeded by complete heart block and sudden death [18].

The underlying cause of SCID syndrome was recently identified as a missense mutation in Orai1 [56]. SCID pathology in the form of mutant Orai1 manifests as a substantial defect in store-operated Ca^{2+} entry [56]. Considering that depletion of ER luminal Ca^{2+} leads to SOC activation, one must consider the relationship between STIM1, the ER-localized putative Ca^{2+} sensor, and Orai1. Impaired ER function has recently been identified as the common denominator in neuronal cell injury of various

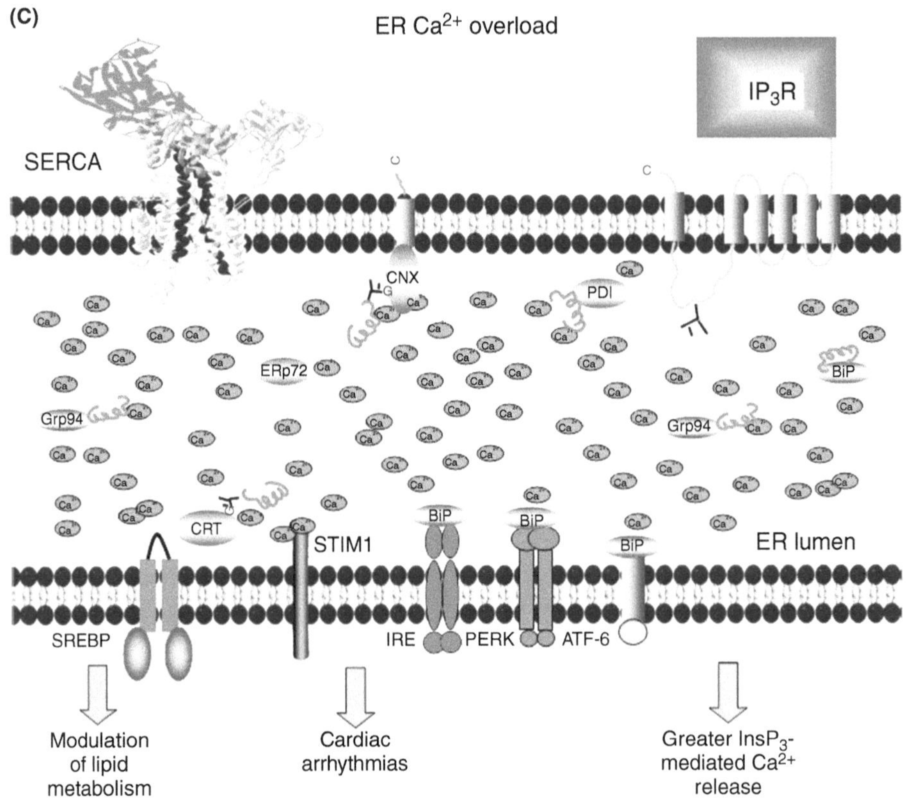

Fig. 1. (C) leads to congenital problems including cardiac arrhythmias. It also affects inositol 1,4,5-trisphosphate ($InsP_3$) receptor function and may impact on lipid metabolism. CNX, calnexin; CRT, calreticulin; PDI, protein disulphide isomerase; and SERCA, sarcoplasmic/ER Ca^{2+}-ATPase (See Color Plate 24, p. 522).

pathologies [116]. As proper ER function is analogous to Ca^{2+} homeostasis in the ER lumen, Ca^{2+} becomes the common factor. It has been hypothesized that cerebral ischemia causes ER dysfunction leading to ER Ca^{2+} depletion [116].

Alzheimer's disease is a progressive neurodegenerative disease resulting in a decrease of cognitive function, including memory loss and reasoning ability, as well as learning and language skills. Disease progression of Alzheimer's is accompanied by the formation of twisted neuronal cytoskeleton of hyperphosphorylated aggregates of the microtubule-associated protein TAU known as neurofibrillary tangles, as well as the extracellular aggregation of the amyloid β (Aβ) peptide in the formation of amyloid plaques [117,129]. Aβ peptides are generated from a precursor known as the Aβ peptide precursor (APP). Two endoproteolytic-processing events result in the liberation of Aβ. An amino-terminal cleavage by β-secretase, a membrane-bound aspartyl protease, takes place resulting in the release of a large secreted fragment and a membrane-associated carboxyl-terminal fragment. This C-terminal fragment can be further processed by γ-secretase, generating an APP intracellular domain fragment

Table 1
Selected pathologies associated with impaired Ca^{2+} homeostasis in the endoplasmic reticulum (ER)

Pathology	**Molecular event**	**Effect on ER Ca^{2+}**	**Reference**
Severe combined immune deficiency (SCID) syndrome	Missense mutation of Orai1	Defect in store-operated Ca^{2+} entry	[56]
Complete heart block	Calreticulin over-expression in postnatal heart	Increase in ER Ca^{2+} levels	[18]
Alzheimer's disease	1. Increase in Aβ production, leading to the formation of plaques 2. Mutant presenilin	Both depletion and ER Ca^{2+} overload may play a role	[117–119]
Darier's disease	Impaired up-regulation of p21WAF1, leading to skin lesions	Depletion of ER Ca^{2+} stores	[120,121]
Primary immunodeficiency		Inactive SOC	[122]
Acute pancreatitis	Premature intracellular digestive enzyme activation	Increase in cytosolic Ca^{2+} levels due to ER Ca^{2+} release and SOC activation	[123,124]
Chronic myeloid leukemia	Bcr-Abl, constitutively active tyrosine kinase	Decrease in free releasable ER Ca^{2+} and decrease in SOC	[125]
Cerebral ischemia		Depletion of ER Ca^{2+} stores	[116] and references therein
Prion disease	PrPSc treatment induces fast, dose-dependent Ca^{2+} release from ER	Induces Ca^{2+} release from ER	[126]

SOC, store-operated Ca^{2+} channel.

and the Aβ-containing fragment [117,129]. Ca^{2+} dysregulation and neurodegeneration in Alzheimer's disease had been suggested by Khachaturian [130]. In 1990s, it was demonstrated that when HEK293 cells expressing the Aβ precursor, APP, are subjected to caffeine-induced stimulation of the RyR, the resulting Ca^{2+} release leads to an increase of Aβ production [118]. Caffeine treatment together with thapsigargin administration resulted in an increase in Aβ production larger than Caffeine or thapsigragin individually [118]. Depletion of ER Ca^{2+} stores must be of greater importance in the production of Aβ than increases in cytosolic Ca^{2+} [117]. The exact mechanism by which ER luminal Ca^{2+} releases contributes to the production of Aβ remains unclear. Another player in the Ca^{2+} dysregulation aspect of Alzheimer's pathology are the presenilins, integral membrane proteins that reside largely in the ER [131]. Mutations in PS1 and PS2 that have been studied thus far show disrupted Ca^{2+} signaling [117]. The presence of mutated presenilins may lead to increased Ca^{2+} stores in the lumen of the

ER, which generates a greater driving force for Ca^{2+} efflux through $InsP_3$ channels [117]. Although this Ca^{2+} overload hypothesis is popular, another recently published study indicates that a flavin adenine dinucleotide (FAD)-linked PS2 mutation, T122R, reduced ER luminal Ca^{2+} concentration by 20% [119]. Although the discrepancy may be due to a myriad of factors, the underlying message of the importance of ER Ca^{2+} regulation is the same. The importance of ER Ca^{2+} homeostasis is supported clinically. For example, individuals from AD families who remain symptom-free long after the expected onset of clinical symptoms have Ca^{2+} responses that are within the "normal" range [132]. Whether Ca^{2+} alterations precede other molecular markers of Alzheimer's disease or whether Ca^{2+} dyshomeostasis is a result of upstream altered molecules is a matter of debate. However, the fact that altered ER luminal Ca^{2+} is again at the forefront of mechanisms underlying disease is no coincidence.

Acknowledgments

Work in our laboratory was supported by the CIHR and AHFMR. A.K. is a recipient of CIHR Studentship and M.M. is a CIHR Senior Investigator.

References

1. Lodish, H.F., Kong, N., and Wikstrom, L. (1992) *J. Biol. Chem.* 267, 12753–12760.
2. Kuznetsov, G., Brostrom, M.A., and Brostrom, C.O. (1992) *J. Biol. Chem.* 267, 3932–3939.
3. Putney, J.W., Jr. and Bird, G.S. (1993) *Endocr. Rev.* 14, 610–631.
4. Berridge, M.J., Bootman, M.D., and Roderick, H.L. (2003) *Nat. Rev. Mol. Cell Biol.* 4, 517–529.
5. MacLennan, D.H. (2000) *Eur. J. Biochem.* 267, 5291–5297.
6. Taylor, C.W. and Laude, A.J. (2002) *Cell Calcium* 32, 321–334.
7. Lomax, R.B., Camello, C., Van Coppenolle, F., Petersen, O.H., and Tepikin, A.V. (2002) *J. Biol. Chem.* 277, 26479–26485.
8. Roderick, H.L., Lechleiter, J.D., and Camacho, P. (2000) *J. Cell Biol.* 149, 1235–1248.
9. John, L.M., Lechleiter, J.D., and Camacho, P. (1998) *J. Cell Biol.* 142, 963–973.
10. Fliegel, L., Burns, K., Wlasichuk, K., and Michalak, M. (1989) *Biochem. Cell Biol.* 67, 696–702.
11. Meldolesi, J. and Pozzan, T. (1998) *Trends Biochem. Sci.* 23, 10–14.
12. Corbett, E.F. and Michalak, M. (2000) *Trends Biochem. Sci.* 25, 307–311.
13. Demaurex, N., and Frieden, M. (2003) *Cell Calcium* 34, 109–119.
14. Booth, C. and Koch, G.E.L. (1989) *Cell* 59, 729–737.
15. Corbett, E.F., Oikawa, K., Francois, P., Tessier, D.C., Kay, C., Bergeron, J.J., Thomas, D.Y., Krause, K.H., and Michalak, M. (1999) *J. Biol. Chem.* 274, 6203–6211.
16. Li, J., Puceat, M., Perez-Terzic, C., Mery, A., Nakamura, K., Michalak, M., Krause, K.-H., and Jaconi, M.E. (2002) *J. Cell Biol.* 158, 103–113.
17. Mesaeli, N., Nakamura, K., Zvaritch, E., Dickie, P., Dziak, E., Krause, K.-H., Opas, M., MacLennan, D.H., and Michalak, M. (1999) *J. Cell Biol.* 144, 857–868.
18. Nakamura, K., Robertson, M., Liu, G., Dickie, P., Guo, J.Q., Duff, H.J., Opas, M., Kavanagh, K., and Michalak, M. (2001) *J. Clin. Invest.* 107, 1245–1253.
19. Nakamura, K., Zuppini, A., Arnaudeau, S., Lynch, J., Ahsan, I., Krause, R., Papp, S., De Smedt, H., Parys, J.B., Müller-Esterl, W., Lew, D.P., Krause, K.-H., Demaurex, N., Opas, M., and Michalak, M. (2001) *J. Cell Biol.* 154, 961–972.

20. Knee, R., Ahsan, I., Mesaeli, N., Kaufman, R.J., and Michalak, M. (2003) *Biochem. Biophys. Res. Commun.* 304, 661–666.
21. Molinari, M., Eriksson, K.K., Calanca, V., Galli, C., Cresswell, P., Michalak, M., and Helenius, A. (2004) *Mol. Cell* 13, 125–135.
22. Lynch, J., Guo, L., Gelebart, P., Chilibeck, K., Xu, J., Molkentin, J.D., Agellon, L.B., and Michalak, M. (2005) *J Cell Biol.* 170, 37–47.
23. Groenendyk, J., Lynch, J., and Michalak, M. (2004) *Mol. Cells* 17, 383–389.
24. Lynch, J.M., Chilibeck, K., Qui, Y., and Michalak, M. (2006) *Trends Cardiovasc. Med.* 16, 65–69.
25. Guo, L., Nakamura, K., Lynch, J., Opas, M., Olson, E.N., Agellon, L.B., and Michalak, M. (2002) *J. Biol. Chem.* 277, 50776–50779.
26. Koch, G., Smith, M., Macer, D., Webster, P., and Mortara, R. (1986) *J. Cell Sci.* 86, 217–232.
27. Nicchitta, C.V. (1998) *Curr. Opin. Immunol.* 10, 103–109.
28. Van, P.N., Peter, F., and Soling, H.-D. (1989) *J. Biol. Chem.* 264, 17494–17501.
29. Haas, I.G. and Wabl, M. (1983) *Nature* 306, 387–389.
30. Lievremont, J.P., Rizzuto, R., Hendershot, L., and Meldolesi, J. (1997) *J. Biol. Chem.* 272, 30873–33089.
31. Lucero, H.A., Lebeche, D., and Kaminer, B. (1994) *J. Biol. Chem.* 269, 23112–23119.
32. Lebeche, D., Lucero, H.A., and Kaminer, B. (1994) *Biochem. Biophys. Res. Commun.* 202, 556–561.
33. Lucero, H.A., and Kaminer, B. (1999) *J. Biol. Chem.* 274, 3243–3251.
34. Nigam, S.K., Goldberg, A.L., Ho, S., Rohde, M.F., Bush, K.T., and Sherman, M. (1994) *J. Biol. Chem.* 269, 1744–1749.
35. Lucero, H.A., Lebeche, D., and Kaminer, B. (1998) *J. Biol. Chem.* 273, 9857–9863.
36. Anelli, T., Alessio, M., Mezghrani, A., Simmen, T., Talamo, F., Bachi, A., and Sitia, R. (2002) *EMBO J.* 21, 835–844.
37. Higo, T., Hattori, M., Nakamura, T., Natsume, T., Michikawa, T., and Mikoshiba, K. (2005) *Cell* 120, 85–98.
38. Solda, T., Garbi, N., Hammerling, G.J., and Molinari, M. (2006) *J. Biol. Chem.* 281, 6219–6226.
39. Zhu, X., Peng, J., Chen, D., Liu, X., Ye, L., Iijima, H., Kadavil, K., Lencer, W.I., and Blumberg, R.S. (2005) *J. Immunol.* 175, 967–976.
40. Milner, R.E., Baksh, S., Shemanko, C., Carpenter, M.R., Smillie, L., Vance, J.E., Opas, M., and Michalak, M. (1991) *J. Biol. Chem.* 266, 7155–7165.
41. Terentyev, D., Nori, A., Santoro, M., Viatchenko-Karpinski, S., Kubalova, Z., Gyorke, I., Terentyeva, R., Vedamoorthyrao, S., Blom, N.A., Valle, G., Napolitano, C., Williams, S.C., Volpe, P., Priori, S.G., and Gyorke, S. (2006) *Circ. Res.* 98, 1151–1158.
42. Wang, Y., Xu, L., Duan, H., Pasek, D.A., Eu, J.P., and Meissner, G. (2006) *J. Biol. Chem.* 281, 15572–15581.
43. Kubalova, Z., Gyorke, I., Terentyeva, R., Viatchenko-Karpinski, S., Terentyev, D., Williams, S.C., and Gyorke, S. (2004) *J. Physiol.* 561, 515–524.
44. Beard, N.A., Casarotto, M.G., Wei, L., Varsanyi, M., Laver, D.R., and Dulhunty, A.F. (2005) *Biophys. J.* 88, 3444–3454.
45. Ozawa, M. (1995) *J. Biochem.* 118, 154–160.
46. Yabe, D., Nakamura, T., Kanazawa, N., Tashiro, K., and Honjo, T. (1997) *J. Biol. Chem.* 272, 18232–18239.
47. Vorum, H., Liu, X.D., Madsen, P., Rasmussen, H.H., and Honore, B. (1998) *Bba. Protein. Struct. Mol. Enzyme* 1386, 121–131.
48. Scherer, P.E., Lederkremer, G.Z., Williams, S., Fogliano, M., Baldini, G., and Lodish, H.F. (1996) *J. Cell Biol.* 133, 257–268.
49. Lin, P., Le-Niculescu, H., Hofmeister, R., McCaffery, J.M., Jin, M., Hennemann, H., McQuistan, T., De Vries, L., and Farquhar, M.G. (1998) *J. Cell Biol.* 141, 1515–1527.
50. Parekh, A.B. and Putney, J.W., Jr. (2005) *Physiol. Rev.* 85, 757–810.

51. Liou, J., Kim, M.L., Heo, W.D., Jones, J.T., Myers, J.W., Ferrell, J.E., Jr., and Meyer, T. (2005) *Curr. Biol.* 15, 1235–1241.
52. Roos, J., DiGregorio, P.J., Yeromin, A.V., Ohlsen, K., Lioudyno, M., Zhang, S., Safrina, O., Kozak, J.A., Wagner, S.L., Cahalan, M.D., Velicelebi, G., and Stauderman, K.A. (2005) *J. Cell Biol.* 169, 435–445.
53. Zhang, S.L., Yu, Y., Roos, J., Kozak, J.A., Deerinck, T.J., Ellisman, M.H., Stauderman, K.A., and Cahalan, M.D. (2005) *Nature* 437, 902–905.
54. Spassova, M.A., Soboloff, J., He, L.P., Xu, W., Dziadek, M.A., and Gill, D.L. (2006) *Proc. Natl. Acad. Sci. U.S.A.* 103, 4040–4045.
55. Mercer, J.C., Dehaven, W.I., Smyth, J.T., Wedel, B., Boyles, R.R., Bird, G.S., and Putney, J.W., Jr. (2006) *J. Biol. Chem.* 281, 24979–24990.
56. Feske, S., Gwack, Y., Prakriya, M., Srikanth, S., Puppel, S.H., Tanasa, B., Hogan, P.G., Lewis, R.S., Daly, M., and Rao, A. (2006) *Nature* 441, 179–185.
57. Soboloff, J., Spassova, M.A., Tang, X.D., Hewavitharana, T., Xu, W., and Gill, D.L. (2006) *J. Biol. Chem.* 281, 20661–20665.
58. Peinelt, C., Vig, M., Koomoa, D.L., Beck, A., Nadler, M.J., Koblan-Huberson, M., Lis, A., Fleig, A., Penner, R., and Kinet, J.P. (2006) *Nat. Cell Biol.* 8, 771–773.
59. Subramanian, K. and Meyer, T. (1997) *Cell* 89, 963–971.
60. Papp, S., Dziak, E., Michalak, M., and Opas, M. (2003) *J. Cell Biol.* 160, 475–479.
61. Pezzati, R., Bossi, M., Podini, P., Meldolesi, J., and Grohovaz, F. (1997) *Mol. Biol. Cell* 8, 1501–1512.
62. Balakier, H., Dziak, E., Sojecki, A., Librach, C., Michalak, M., and Opas, M. (2002) *Hum. Reprod.* 17, 2938–2947.
63. Mead, J.R., Irvine, S.A., and Ramji, D.P. (2002) *J. Mol. Med.* 80, 753–769.
64. Rawson, R.B. (2003) *Nat. Rev. Mol. Cell Biol.* 4, 631–640.
65. Perret, B., Mabile, L., Martinez, L., Terce, F., Barbaras, R., and Collet, X. (2002) *J. Lipid Res.* 43, 1163–1169.
66. Stillemark, P., Boren, J., Andersson, M., Larsson, T., Rustaeus, S., Karlsson, K.A., and Olofsson, S.O. (2000) *J. Biol. Chem.* 275, 10506–10513.
67. Ben-Zeev, O., Mao, H.Z., and Doolittle, M.H. (2002) *J. Biol. Chem.* 277, 10727–10738.
68. Zhang, R., Leslie, B., Boudreaux, J.P., Frey, D., and Reisin, E. (2003) *Am. J. Med. Sci.* 325, 202–208.
69. Briquet-Laugier, V., Ben-Zeev, O., White, A., and Doolittle, M.H. (1999) *J. Lipid Res.* 40, 2044–2058.
70. Horton, J.D., Goldstein, J.L., and Brown, M.S. (2002) *J. Clin. Invest.* 109, 1125–1131.
71. Brown, M.S. and Goldstein, J.L. (1997) *Cell* 89, 331–340.
72. Cheng, D., Espenshade, P.J., Slaughter, C.A., Jaen, J.C., Brown, M.S., and Goldstein, J.L. (1999) *J. Biol. Chem.* 274, 22805–22812.
73. Sampath, H. and Ntambi, J.M. (2005) *Annu. Rev. Nutr.* 25, 317–340.
74. Maxwell, K.N., Soccio, R.E., Duncan, E.M., Sehayek, E., and Breslow, J.L. (2003) *J. Lipid Res.* 44, 2109–2119.
75. Ardail, D., Gasnier, F., Lerme, F., Simonot, C., Louisot, P., and Gateau-Roesch, O. (1993) *J. Biol. Chem.* 268, 25985–25992.
76. Rizzuto, R., Pinton, P., Carrington, W., Fay, F.S., Fogarty, K.E., Lifshitz, L.M., Tuft, R.A., and Pozzan, T. (1998) *Science* 280, 1763–1766.
77. Pozzan, T., Rizzuto, R., Volpe, P., and Meldolesi, J. (1994) *Physiol. Rev.* 74, 595–636.
78. Hoth, M., Button, D.C., and Lewis, R.S. (2000) *Proc. Natl. Acad. Sci. U.S.A.* 97, 10607–10612.
79. Malli, R., Frieden, M., Osibow, K., Zoratti, C., Mayer, M., Demaurex, N., and Graier, W.F. (2003) *J. Biol. Chem.* 278, 44769–44779.
80. Parekh, A.B. (2003) *J. Physiol.* 547, 333–348.
81. Gilabert, J.A., Bakowski, D., and Parekh, A.B. (2001) *EMBO J.* 20, 2672–2679.

82. Foyouzi-Youssefi, R., Arnaudeau, S., Borner, C., Kelley, W.L., Tschopp, J., Lew, D.P., Demaurex, N., and Krause, K.H. (2000) *Proc. Natl. Acad. Sci. U.S.A.* 97, 5723–5738.
83. Pinton, P., Ferrari, D., Magalhaes, P., Schulze-Osthoff, K., Di Virgilio, F., Pozzan, T., and Rizzuto, R. (2000) *J. Cell Biol.* 148, 857–862.
84. Pinton, P., Ferrari, D., Rapizzi, E., Virgilio, F.D., Pozzan, T., and Rizzuto, R. (2001) *EMBO J.* 20, 2690–2701.
85. Williams, D.B. (2006) *J. Cell Sci.* 119, 615–623.
86. Michelsen, K., Yuan, H., and Schwappach, B. (2005) *EMBO Rep.* 6, 717–722.
87. Nishimura, N. and Balch, W.E. (1997) *Science* 277, 556–558.
88. Hammond, C. and Helenius, A. (1994) *Science* 266, 456–458.
89. Schrag, J.D., Bergeron, J.J.M., Li, Y., Borisova, S., Hahn, M., Thomas, D.Y., and Cygler, M. (2001) *Mol. Cell* 8, 633–644.
90. Groenendyk, J. and Michalak, M. (2005) *Acta Biochim. Pol.* 52, 381–395.
91. Ellgaard, L., Riek, R., Braun, D., Herrmann, T., Helenius, A., and Wuthrich, K. (2001) *FEBS Lett.* 488, 69–73.
92. Michalak, M., Robert Parker, J.M., and Opas, M. (2002) *Cell Calcium* 32, 269–278.
93. Frickel, E.M., Riek, R., Jelesarov, I., Helenius, A., Wuthrich, K., and Ellgaard, L. (2002) *Proc. Natl. Acad. Sci. U.S.A.* 99, 1954–1959.
94. Leach, M.R., Cohen-Doyle, M.F., Thomas, D.Y., and Williams, D.B. (2002) *J. Biol. Chem.* 277, 29686–29697.
95. Martin, V., Groenendyk, J., Steiner, S.S., Guo, L., Dabrowska, M., Parker, J.M., Muller-Esterl, W., Opas, M., and Michalak, M. (2006) *J. Biol. Chem.* 281, 2338–2346.
96. Trombetta, E.S. and Parodi, A.J. (2003) *Annu. Rev. Cell Dev. Biol., 649–676.*
97. Rosenbaum, E.E., Hardie, R.C., and Colley, N.J. (2006) *Neuron* 49, 229–241.
98. Yano, K., Petersen, O.H., and Tepikin, A.V. (2004) *Biochem. J.* 383, 353–360.
99. Rutkowski, D.T. and Kaufman, R.J. (2004) *Trends Cell Biol.* 14, 20–28.
100. Schroder, M. and Kaufman, R.J. (2005) *Annu. Rev. Biochem.* 74, 739–789.
101. Shen, X., Zhang, K., and Kaufman, R.J. (2004) *J. Chem. Neuroanat.* 28, 79–92.
102. Wu, J. and Kaufman, R.J. (2006) *Cell Death Differ.* 13, 374–384.
103. Bertolotti, A., Zhang, Y., Hendershot, L.M., Harding, H.P., and Ron, D. (2000) *Nat. Cell Biol.* 2, 326–332.
104. Harding, H.P., Zeng, H., Zhang, Y., Jungries, R., Chung, P., Plesken, H., Sabatini, D.D., and Ron, D. (2001) *Mol. Cell* 7, 1153–1163.
105. Schroder, M., and Kaufman, R.J. (2006) *Curr. Mol. Med.* 6, 5–36.
106. Ma, Y. and Hendershot, L.M. (2004) *J. Chem. Neuroanat.* 28, 51–65.
107. Ma, D. and Jan, L.Y. (2002) *Curr. Opin. Neurobiol.* 12, 287–292.
108. Liu, C.Y., Xu, Z., and Kaufman, R.J. (2003) *J. Biol. Chem.* 278, 17680–17687.
109. Shen, J., Chen, X., Hendershot, L., and Prywes, R. (2002) *Dev. Cell* 3, 99–111.
110. Brostrom, M.A. and Brostrom, C.O. (2003) *Cell Calcium* 34, 345–363.
111. Ye, J., Rawson, R.B., Komuro, R., Chen, X., Dave, U.P., Prywes, R., Brown, M.S., and Goldstein, J.L. (2000) *Mol. Cell* 6, 1355–1364.
112. Harding, H.P., Zhang, Y., and Ron, D. (1999) *Nature* 397, 271–274.
113. Lee, K., Tirasophon, W., Shen, X., Michalak, M., Prywes, R., Okada, T., Yoshida, H., Mori, K., and Kaufman, R.J. (2002) *Genes Dev.* 16, 452–466.
114. Hollien, J. and Weissman, J.S. (2006) *Science* 313, 104–107.
115. Ron, D. (2006) *Science* 313, 52–53.
116. Paschen, W. and Mengesdorf, T. (2005) *Cell Calcium* 38, 409–415.
117. LaFerla, F.M. (2002) *Nat. Rev. Neurosci.* 3, 862–872.
118. Querfurth, H.W., Jiang, J., Geiger, J.D., and Selkoe, D.J. (1997) *J. Neurochem.* 69, 1580–1591.

119. Zatti, G., Burgo, A., Giacomello, M., Barbiero, L., Ghidoni, R., Sinigaglia, G., Florean, C., Bagnoli, S., Binetti, G., Sorbi, S., Pizzo, P., and Fasolato, C. (2006) *Cell Calcium* 39, 539–550.
120. Byrne, C.R. (2006) *J. Invest. Dermatol.* 126, 702–703.
121. Muller, E.J., Caldelari, R., Kolly, C., Williamson, L., Baumann, D., Richard, G., Jensen, P., Girling, P., Delprincipe, F., Wyder, M., Balmer, V., and Suter, M.M. (2006) *J. Invest. Dermatol.* 126, 721–731.
122. Partiseti, M., Le Deist, F., Hivroz, C., Fischer, A., Korn, H., and Choquet, D. (1994) *J. Biol. Chem.* 269, 32327–32335.
123. Parekh, A.B. (2000) *Proc. Natl. Acad. Sci. U.S.A.* 97, 12933–12934.
124. Raraty, M., Ward, J., Erdemli, G., Vaillant, C., Neoptolemos, J.P., Sutton, R., and Petersen, O.H. (2000) *Proc. Natl. Acad. Sci. U.S.A.* 97, 13126–13131.
125. Piwocka, K., Vejda, S., Cotter, T.G., O'Sullivan, G.C., and McKenna, S.L. (2006) *Blood* 107, 4003–4010.
126. Hetz, C., Russelakis-Carneiro, M., Maundrell, K., Castilla, J., and Soto, C. (2003) *EMBO J.* 22, 5435–5445.
127. Mery, A., Aimond, F., Menard, C., Mikoshiba, K., Michalak, M., and Puceat, M. (2005) *Mol. Biol. Cell* 16, 2414–2423.
128. Mery, L., Mesaeli, N., Michalak, M., Opas, M., Lew, D.P., and Krause, K.-H. (1996) *J. Biol. Chem.* 271, 9332–9339.
129. Gandy, S. (2005) *J. Clin. Invest.* 115, 1121–1129.
130. Khachaturian, Z.S. (1989) *Ann. N. Y. Acad. Sci.* 568, 1–4.
131. Cook, D.G., Sung, J.C., Golde, T.E., Felsenstein, K.M., Wojczyk, B.S., Tanzi, R.E., Lee, V.M., and Doms, R.W. (1996) *Proc. Natl. Acad. Sci. U.S.A.* 93, 9223–9228.
132. Etcheberrigaray, R., Hirashima, N., Nee, L., Prince, J., Govoni, S., Racchi, M., Tanzi, R.E., and Alkon, D.L. (1998) *Neurobiol. Dis.* 5, 37–45.

J. Krebs and M. Michalak (Editors)
Calcium: A Matter of Life or Death

ISSN: 0167-7306/DOI: 10.1016/S0167-7306(06)41009-7

CHAPTER 9

Structural aspects of ion pumping by Ca^{2+}-ATPase of sarcoplasmic reticulum

Chikashi Toyoshima

Institute of Molecular and Cellular Biosciences, The University of Tokyo, 1-1-1 Yayoi, Bunkyo-ku, Tokyo 113-0032, Japan, Tel.: +81 3 5841 8492; Fax: +81 3 5841 8491; E-mail: ct@iam.u-tokyo.ac.jp

Abstract

Ca^{2+}-ATPase of muscle sarcoplasmic reticulum (SR) is an ATP-powered Ca^{2+} pump that establishes >10 000-fold concentration gradient across the membrane for Ca^{2+} used for muscle contraction. Its crystal structures have been determined for seven different states that cover almost the entire reaction cycle. Presented here is a brief structural account of the ion pumping process, in which a series of very large domain rearrangements takes place.

Keywords: Ca^{2+}-ATPase, Ca^{2+} pump, crystal structure, membrane protein, muscle contraction, sarcoplasmic reticulum

1. Introduction

Ca^{2+}-ATPase of skeletal muscle sarcoplasmic reticulum (SR, SERCA1a), an integral membrane protein of M_r 110K comprising a single polypeptide of 994 amino acid residues, is an ATP-powered Ca^{2+} pump. It was first identified in the 'relaxing factor' of muscle contraction and gave rise to the calcium theory that Ca^{2+} is a fundamental and ubiquitous factor in the regulation of intracellular processes [1]. In muscle contraction, Ca^{2+} is released from SR into muscle cells through Ca^{2+} release channel. Ca^{2+}-ATPase then pumps back the released Ca^{2+} into the SR to cause relaxation. This pump runs as long as ATP and Ca^{2+} are present in the cytoplasm and establishes a $>10^4$-fold concentration gradient across membranes. SERCA1 is both structurally and functionally the best characterized member of the P-type (or E1/E2-type) ion translocating ATPases. According to the classical E1/E2 theory [2,3], transmembrane Ca^{2+}-binding sites have high affinity and face the cytoplasm in E1 and have low affinity and face the lumen of SR (or extracellular side) in E2. Actual transfer of bound Ca^{2+} is thought to take place between two phosphorylated intermediates, E1P and E2P, in exchange of H^+ from the lumenal side (inset in Fig. 1 for a simplified reaction scheme).

Several types of Ca^{2+}-ATPases exist in different tissues [3]. Although all transfer Ca^{2+} from the cytoplasm to the opposite side of the membrane and countertransport H^+, they appear to have different stoichiometry of Ca^{2+}: H^+: ATP. It is well established that SERCA1a can transfer two Ca^{2+} per ATP hydrolyzed in the

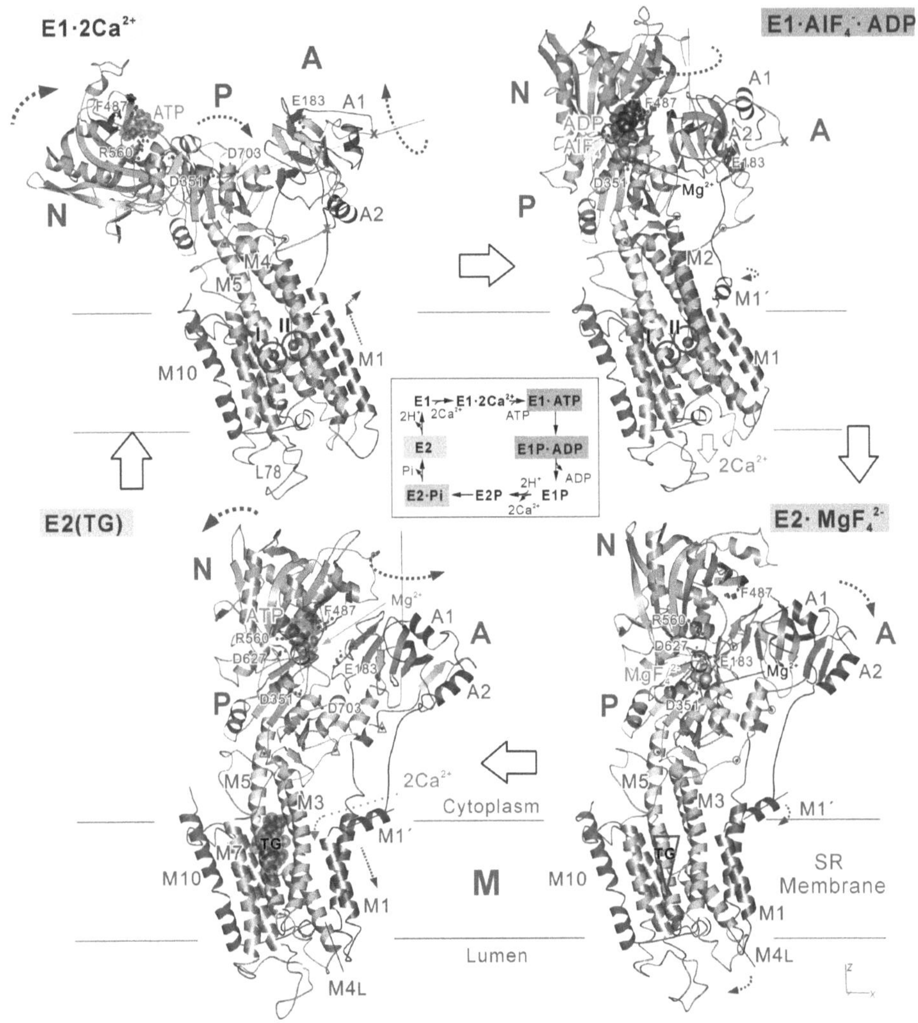

Fig. 1. Front views [parallel to the membrane (x–y) plane] of Ca^{2+}-ATPase in four different states and a simplified reaction scheme (showing only the forward direction), in which different colours correspond to the respective structures presented here. Colours change gradually from the amino terminus (blue) to the carboxy terminus (red). Purple spheres (numbered and circled) represent bound Ca^{2+}. Three cytoplasmic domains (A, N and P), the α-helices in the A-domain (A1 and A2) and those in the transmembrane domain (M1, M4L, M5 and M10) are indicated. M1′ is an amphipathic part of the M1 helix lying on the bilayer surface. Digestion sites are shown in small circles (protected), triangles (partially protected) and crosses (not protected) with proteinase K (orange) and trypsin (T2 site, brown). Docked ATP and thapsigargin (TG) are shown in transparent space fill. Several key residues – E183 (A), F487 and R560 (N, ATP binding), D351 (phosphorylation site), D627 and D703 (P) – are shown in ball-and-stick representation. PDB accession codes are 1SU4 (E1·$2Ca^{2+}$), 1WPE (E1·AlF_x·ADP), 1WPG (E2·$MgF_4{}^{2-}$) and 1IWO (E2(TG)). Atomic co-ordinates of the aligned models are available at the author's web site (http://www.iam.u-tokyo.ac.jp/StrBiol/resource/res.html) (See Color Plate 25, p. 523).

forward direction [4] and two to three H^+ in the opposite direction [5]. Therefore, active transport of Ca^{2+} is an electrogenic process, although no H^+ gradient is built up across the SR membrane, because it is leaky to H^+. This Ca^{2+}/H^+ exchange may cause pathological pH effects with plasma membrane Ca^{2+}-ATPase [6].

Because Ca^{2+} is so fundamental in regulation of biological processes, malfunctioning of Ca^{2+} pumps is likely to be deleterious to living organisms. Therefore, although Darier and Hailey–Hailey diseases are well known [7], human diseases related to Ca^{2+}-ATPases are not many. Conversely, Ca^{2+}-ATPases can be utilized for killing cells. For instance, cancer cells may be killed by delivering a potent inhibitor of this pump, such as thapsigargin (TG) [8]. Artemisinin, a well-known anti-malarial drug, inhibits a malarial Ca^{2+}-ATPase [9]. In cardiac muscles, phospholamban and sarcolipin work as regulators of Ca^{2+}-ATPase and are becoming therapeutic targets [10].

We crystallized Ca^{2+}-ATPase from rabbit white skeletal muscle (SERCA1a) in phospholipid bilayer and solved the first atomic structure (E1·$2Ca^{2+}$ form) in 2000 [11]. Now, seven crystal structures have been determined of this ATPase in six different states that approximately cover the entire reaction cycle (Table 1; Fig. 1) [11–18]. We also carried out all-atom molecular dynamics simulations for wild type and some mutants to understand the functional roles of critical residues [19]. As a result, we can now propose a fairly detailed scenario of ion pumping and answer important questions such as (i) what are the roles of ATP [13] and phosphorylation [14], (ii) why such large domain movements are necessary [14] and (iii) why H^+ countertransport is necessary despite that the SR membrane is leaky to H^+ [15]. An earlier overview of Ca^{2+}-ATPase structures is found in [20] and more general ones on P-type ATPases in [3] and [21]. Atomic co-ordinates of the aligned structures, movies on the structural changes during the reaction cycle and some results of molecular dynamics simulations are available in the author's homepage (http://www.iam.u-tokyo.ac.jp/StrBiol/resource/res.html). A very extensive database for mutations (http://www.fi.au.dk/jpa/smd/) is now made available by J. P. Andersen.

Table 1
Current status of the crystallography of SERCA1a

State	Substrate analogues[a]	Other ligands[b]	Resolution (current)[c] Å	PDB ID
E1·$2Ca^{2+}$	TNP-AMP	Ca^{2+}	2.4 (2.3)	1SU4 [11]
E1·ATP	AMPPCP	Ca^{2+} and Mg^{2+}	2.9 (2.5)	1T5S [16] and 1VFP [13]
E1:P·ADP[d]	ADP and AlF_x	Ca^{2+} and Mg^{2+}	2.7 (2.4)	1WPE [14] and 1T5T [16]
E2:P[d]	AlF_4^-	Mg^{2+} and TG[b]	*3.0* (2.7)	1XP5 [17]
E2·Pi[d]	MgF_4^{2-}	Mg^{2+} and TG	2.3	1WPG [14]
E2		TG	3.1 (2.3)	1IWO [12] and 2C8L [18]
		TG + BHQ[b]	2.4	2AGV [15]
E2·ATP	AMPPCP	Mg^{2+} and TG	2.5	2D88 [18] and 2DQS

[a]AlF_x and MgF_4^{2-} are stable analogues of phosphate.
[b]Thapsigargin (TG) and 2,5-di-*tert*-butyl-1,4-dihydroxybenzene (BHQ) are potent inhibitors of SERCA1, binding to different parts of the transmembrane domain [15].
[c]Numbers in the parentheses represent unpublished results in the author's laboratory.
[d]The colon denotes here a transition state, whereas the dot denotes a product state.

2. *Architecture of Ca^{2+}-ATPase*

SERCA1a consists of three cytoplasmic domains (A, actuator; N, nucleotide binding and P, phosphorylation), 10 transmembrane (M1–M10) helices and small lumenal loops (Fig. 1). As described below, the A-domain, connected to the M1–M3 helices with rather long linkers, works as the actuator of the transmembrane-gating mechanism that regulates Ca^{2+} binding and release. The A-domain contains one of the signature sequences ^{181}TGES (Thr-Gly-Glu-Ser) motif [3], which plays an important role in dephosphorylation [22,23]. The P-domain contains the phosphorylation residue Asp351, magnesium co-ordinating residue Asp703 and many other critical residues that characterize the P-type ATPase as a member of the haloacid dehalogenase superfamily [24]. These critical residues are also shared by bacterial two-component regulators [25] that have a different folding pattern. The N-domain, a long insertion between two parts forming the P-domain, contains the residues for adenosine binding (Phe487 is shown in Fig. 1) and those critical for bridging the N- and P-domains (Arg560 is shown in Fig. 1) [13,16]. These three domains are well separated in the $E1 \cdot 2Ca^{2+}$ crystal structure but gather to form a compact headpiece in the other states (Fig. 1).

There are 10 transmembrane helices, and some of them (M2–M5) have long cytoplasmic extensions (Fig. 1). In particular, M5 is 60 Å long and extends from the lumenal surface to the end of the P-domain, working as the spine of the molecule. Two (M4 and M6) helices have proline residues at the middle and are partly unwound throughout the whole reaction cycle. M6 and M7 are far apart and connected by a cytosolic loop (L67) of approximately 20 residues. L67 runs along the bottom of the P-domain and is hydrogen bonded to M5 (Fig. 1). This loop appears to be very rigid and presumably restricts the movement of the M5 helix. M4–M6 and M8 contain the residues directly co-ordinating two Ca^{2+} (Fig. 2). The amino acid sequence is well conserved for M4–M6 but not for M8 even within the members of closely related P-type ATPases, such as $Na^{+}K^{+}$- and $H^{+}K^{+}$-ATPases [3]. All helices from M1 to M6 move considerably during the reaction cycle, whereas M7–M10 helices do not move (Fig. 1), apparently acting as a membrane anchor. In SERCA1, the middle part of M7 containing a GXXG (Gly-xxx-xxx-Gly) motif [26] makes a tight couple with M5 and allows the bending of M5. A short helical segment between M9 and M10 appears to serve as a socket for the lumenal end of the M2 helix. M7–M10 are thought to have a specialized function in each subfamily of P-type ATPases and are in fact absent in heavy metal pumps [3].

3. *Transmembrane Ca^{2+}-binding sites*

It is well established that SERCA1 has two high-affinity transmembrane Ca^{2+}-binding sites, and the binding is co-operative [4]. There was much confusion as to their locations, partly because these Ca^{2+}-binding sites do not appear to bind lanthanides. Debates still exist as to the presence of low-affinity sites, in particular, in the lumenal region. However, no support for the presence of low-affinity sites has

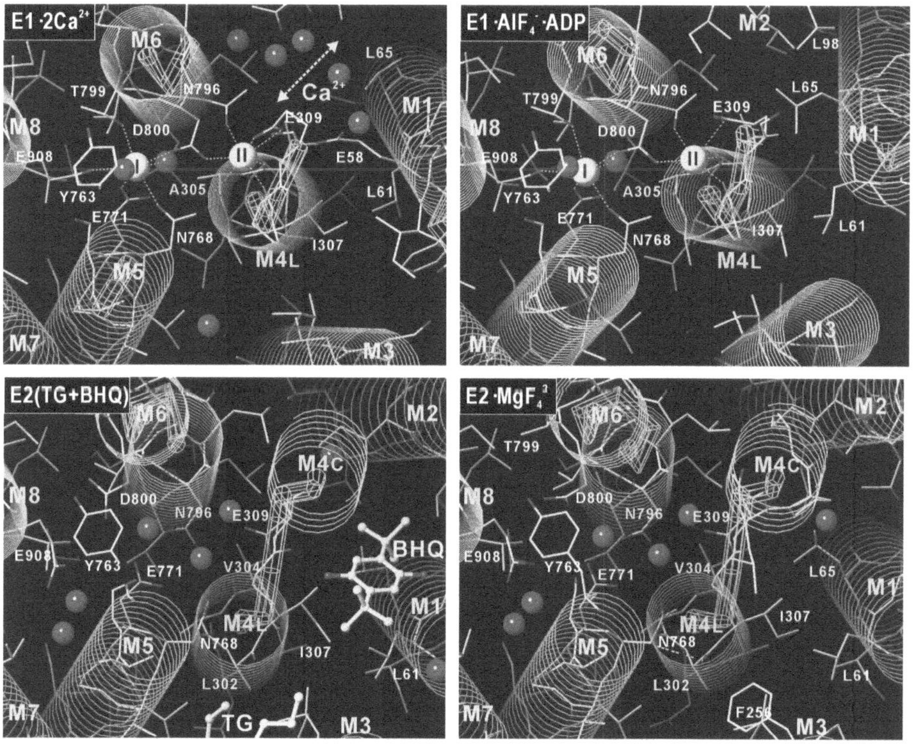

Fig. 2. Details of the transmembrane Ca^{2+}-binding sites in four different states viewed approximately normal to the membrane. Ca^{2+} appears as cyan spheres and water molecules as red spheres. Note that site II Ca^{2+} is exchangeable with those in the cytoplasm in E1·$2Ca^{2+}$ by conformation change of the E309 side chain but not in E1·AlF_4^-·ADP, as the space is occupied by L65 (M1) and other hydrophobic side chains. Two inhibitors, TG and 2,5-di-*tert*-butyl-1,4-dihydroxybenzene (BHQ), are shown in ball-and-stick in E2(TG + BHQ) (See Color Plate 26, p. 524).

been obtained by X-ray crystallography, although the crystals were formed in the presence of 10 mM Ca^{2+} [11].

The two Ca^{2+}-binding sites (I and II) are located side by side near the cytoplasmic surface of the lipid bilayer (Figs 1 and 2). However, the binding of two Ca^{2+} is sequential and co-operative. Site I, the binding site for the first Ca^{2+}, is located in a space surrounded by the M5, M6 and M8 helices and formed entirely by side chain oxygen atoms and two water molecules (Fig. 2). Asn768, Glu771 (M5), Thr799, Asp800 (M6) and Glu908 (M8) contribute to this site. Site II is nearly 'on' the M4 helix and located approximately 3 Å closer to the cytoplasmic surface than site I (Figs 1 and 2). The M4 helix is partly unwound (between Ile307-Gly310) and provides three main chain carbonyls for co-ordination (Fig. 2). Asn796 and Asp800 (M6) provide one side chain oxygen, whereas Glu309 provides two side chain oxygen atoms to cap the bound Ca^{2+} (Fig. 2). No water molecule contributes to site II. This arrangement of oxygen atoms is reminiscent of the EF-hand motif.

Thus, both sites have seven co-ordination but of different characteristics. Asp800, on the unwound part of M6, is the only residue that contributes to both sites (Fig. 2). Any substitutions to site I residues, except for the Glu908 to Gln mutation, totally abolish the binding of Ca^{2+}. In contrast, even double mutations of Glu309 and Asn796 leave 50% Ca^{2+} binding [27]. Taken together, these mutagenesis experiments indicate that site II is the binding site for the second Ca^{2+}. Further mutagenesis studies demonstrated that Glu309 is in fact the gating residue for cytoplasmic Ca^{2+} [28]. It is well known that the second Ca^{2+} is exchangeable with those in the cytoplasm without affecting the first Ca^{2+}. The Asn796Ala mutant kept such property. In contrast, even the first Ca^{2+} exchanged rapidly with the Glu309Gln mutant. Thus, the first Ca^{2+} will meet Glu309 and enter the binding cavity through improperly formed site II (otherwise it will be trapped there). The binding of the first Ca^{2+} to site I will position Asp800 properly. This will rotate M6 and may straighten the M5 helix (compare E2(TG) and E1·2Ca^{2+} in Figs 1 and 2) and alter the arrangement of oxygen atoms in site II to form a higher affinity binding site. Finally, the carboxyl of Glu309 side chain will cap site II Ca^{2+}, and the binding signal will be transmitted to the phosphorylation site some 50 Å away.

A key question is, then, how the Ca^{2+}-binding sites and phosphorylation site communicate with each other. In this regard, it is important to note that the cytoplasmic end of M5 is integrated into the P-domain near the phosphorylation site and hydrogen bonded to the M4 helix forming a short β-strand (Fig. 3) [12]. The P-domain is also connected to M3 with critical hydrogen bonds. Thus, the events that occur at either site can be mechanically transmitted to the other site [12,20].

4. Scenario of ion pumping

Based on seven crystal structures listed in Table 1, we can now propose a fairly detailed scenario of ion pumping. We do not yet have exact analogues of E1P and E2P. However, as shown by limited proteolysis [29,30], domain organizations of E1·AMPPCP and E1·AlF_4^-·ADP crystal structures are virtually the same [13,16] and expected to be very similar to that of E1P. Also, E2·AlF_4^- and E2·MgF_4^{2-} have virtually identical domain organizations [14,17] that will be similar to E2P, except that the lumenal gate is not fully open in either crystal structure to allow access of Ca^{2+} from the lumenal side [31]. Thus, we can describe the entire reaction cycle with four principal structures presented in Fig. 4. They show how the affinity of the transmembrane Ca^{2+}-binding sites is altered, and how the lumenal gate is opened and closed by events that occur around the phosphorylation site (see also movies available at the author's web site: http://www.iam.u-tokyo.ac.jp/StrBiol/resource/res.html).

Ca^{2+} binding to the ATPase straightens the M5 helix and breaks the closed configuration of the three cytoplasmic domains to allow more inclination of the N-domain. ATP can bind to the N-domain in the absence of Ca^{2+} but cannot reach Asp351, the phosphorylation residue (Fig. 1). Two Ca^{2+} enter the high-affinity sites through the gating residue Glu309 on M4. Because there is enough space around Glu309, the cytoplasmic gate can still open, and site II Ca^{2+} is exchangeable with

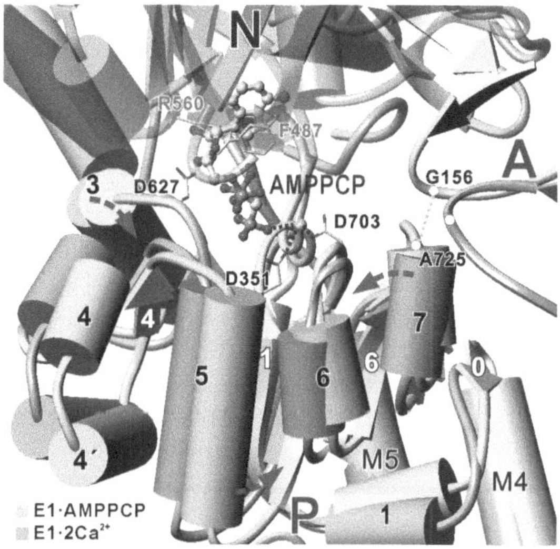

Fig. 3. Structural changes caused by the binding of ATP and Mg^{2+} to the P-domain. AMPPCP, a non-hydrolysable analogue of ATP, is used here; the bound metal in the crystal structure (small green sphere) is most likely Ca^{2+} rather than Mg^{2+}. Two crystal structures are fitted with the 10 residues at the N-terminal end of the M5 helix. The N-domain in E1·AMPPCP is shown transparent. Note that the P-domain is bent (arrows in red broken lines) by co-ordination of the metal by the γ-phosphate, D351 and D703 (shown in stick model) and that the A-domain tilts because of the inclination of the P7 helix (See Color Plate 27, p. 525).

those in the cytoplasm. The M1 helix is deeply embedded in the lipid bilayer, stabilized indirectly by the bound Ca^{2+}.

ATP cross-links the P- and N-domains, so that the γ-phosphate of ATP and Mg^{2+} bind to the P-domain to bend it (Fig. 3). The N-domain is fixed in a highly inclined position and makes contact with the A-domain to form a mechanical couple between them in preparation for the next main event, the rotation of the A-domain in the E1P–E2P transition. The bending of the P-domain tilts the A-domain by approximately 30° (Fig. 3), placing strain on the link between the A-domain and the M3 helix, as demonstrated by protection against proteinase K attack [29] (Fig. 1). This strain appears to be the driving force for the A-domain rotation in the next step, as the proteolysis at Glu243 in a loop connecting M3 and the A-domain prevents the E1P → E2P transition [32]. At the same time, the M1 helix is pulled up and bent so that the top of the transmembrane part (in particular Leu65; Fig. 2) closes the cytoplasmic gate of the Ca^{2+}-binding sites, by occupying the space around Glu309. Thus, the conformation of the Glu309 side chain is fixed, and two Ca^{2+} are occluded in the transmembrane-binding sites. M1 and M2 now form a V-shaped structure that moves as a rigid body until the end of the reaction cycle and transmits the movements of the A-domain to other transmembrane helices, primarily the lumenal half of M4 (M4L, Fig. 1).

Phosphoryl transfer to Asp351 allows the dissociation of ADP, which triggers the opening of the N- and P-domain interface [14]. The A-domain rotates horizontally by 110° so that the TGES loop of the A-domain wedges into the gap formed between the

N- and P-domains. Thus, the TGES loop occupies the space where ADP was to prevent the binding of ADP and to shield the aspartylphosphate from bulk water and interacts with the phosphorylation site with the help of the Mg^{2+} bound to the P-domain. This A-domain rotation causes a drastic rearrangement of the transmembrane helices M1–M6, including a large downward movement of M4, sharp bending of M5 toward M1 (Fig. 1) and rotation of M6 (Fig. 2), which destroy the Ca^{2+}-binding sites [12,20]. The lower sections of M1 and M2 push against M4L, opening the lumenal gate and releasing the bound Ca^{2+} into the lumen. This will allow protons and water molecules to enter and stabilize the empty Ca^{2+}-binding sites (Fig. 2) [15].

The TGES loop of the A-domain now fixes a particular water molecule and catalyzes its attack on the aspartylphosphate. Glu183 plays a particularly important role here [22,23], presumably working as a general acid/base catalyst. The release of the phosphate and Mg^{2+} relaxes the P-domain. This in turn releases the M1 and M2 helices so that M4L closes the lumenal gate completely. The top amphipathic part of M1 (M1′) forms a part of a cytoplasmic access funnel leading to Glu309, the side chain of which becomes stabilized by hydrogen bonding with Asn796 (Fig. 2). By release of protons to the cytoplasm, Glu309 is supposed to change the side chain conformation and catch Ca^{2+}.

Obviously, the scenario is still incomplete. For example, we do not know the structural changes that take place when the first Ca^{2+} binds. We do not have the structure in which the lumenal gate is fully opened and do knot know exactly when the lumenal gate closes. These require further studies.

5. *Conclusions*

As described, the P- and N-domains change interfaces and thereby control the orientation of the A-domain, the actuator of transmembrane gates. ATP, phosphate, Mg^{2+} and Ca^{2+} are the modifiers of the interfaces (Fig. 4). Energy barriers between the principal intermediates appear to be comparable to the thermal energy, as the key events, for example, the rotation of the A-domain, occur when ADP is released. That will be the reason why nearly 100% efficiency of energy utilization is possible. It seems that the overall reaction is made possible by suppressing backward reaction. In fact, in the crystal structures, we have seen such devices integrated into Ca^{2+}-ATPase [14].

We now also understand, to some extent, why the structure of Ca^{2+}-ATPase has to be so. For instance, closed configurations of the cytoplasmic domains are necessary, because Ca^{2+}-ATPase changes domain interfaces to orient the A-domain for regulation of the gating mechanism. In the E2 state, the closed configuration appears to be important in two other aspects: (i) restriction of the delivery of ATP to the phosphorylation site and (ii) restriction of the thermal movements of transmembrane helices. As ATP can bind to the ATPase even in the E2 state and SERCA1 has no mechanism for regulating the reaction cycle other than Ca^{2+} itself, the delivery of ATP γ-phosphate to the phosphorylation site has to be restricted physically. If the A-domain fluctuates too much by thermal energy, Ca^{2+} transferred into the lumen of SR will leak into the cytoplasm. Therefore, the A-domain has to be fixed by other

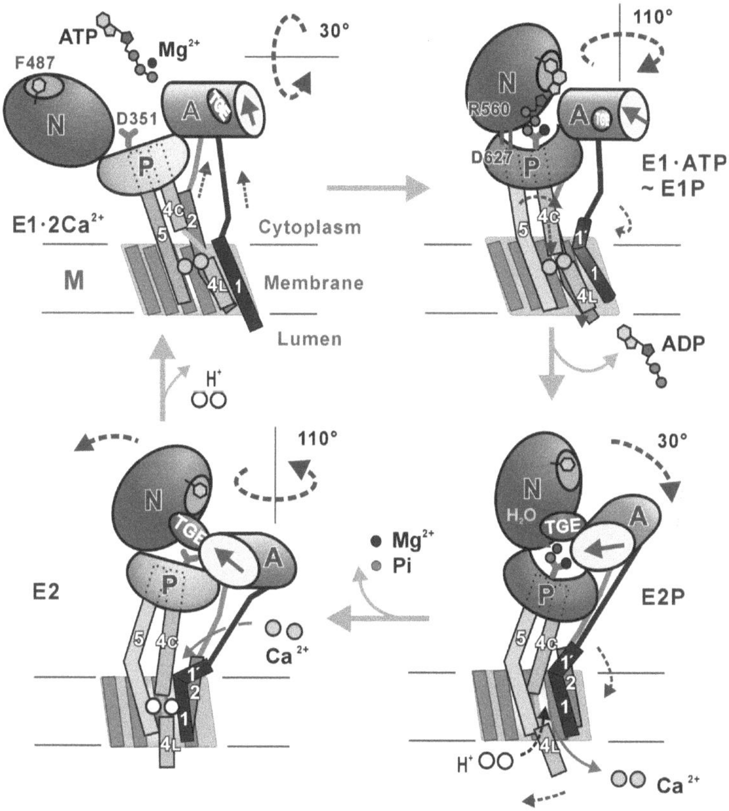

Fig. 4. A cartoon illustrating the structural changes of the Ca^{2+}-ATPase during the reaction cycle based on the crystal structures in the five different states [14] (See Color Plate 28, p. 526).

cytoplasmic domains, unless Ca^{2+} itself fixes the transmembrane helices. Even so, such leakage does occur, though small, in the absence of Ca^{2+}, and TG suppresses it. The link between the A-domain and the M1 helix has to be flexible to provide tolerance to the gating machinery for thermal fluctuation. Yet, its length is important [33], presumably because the size of the A-domain movement is set just enough to surpass the thermal fluctuation. Perhaps we are beginning to understand that ion pumps are working in thermally fluctuating world, yet utilizing thermal energy effectively for ion transport.

Acknowledgements

This work was supported in parts by grant-in-aid for Creative Scientific Research from the Ministry of Education, Culture, Sports, Science and Technology 14GS0308, the Japan New Energy and Industry Technology Development Organization and the Human Frontier Science Program. I thank my colleagues in Tokyo, Baltimore, Sarclay and Cape Town, and in particular, I thank Hiromi Nomura, who prepared nearly all the crystals, and Yuji Sugita, who gave us new insights by molecular dynamics simulation. I am greatly indebted to Giuseppe Inesi in many aspects of this work.

References

1. Ebashi, S. and Lipman, F. (1962) *J. Cell Biol.* 14, 389–400.
2. de Meis, L. and Vianna, A.L. (1979) *Annu. Rev. Biochem.* 48, 275–292.
3. Møller, J.V., Juul, B. and le Maire, M. (1996) *Biochim. Biophys. Acta* 1286, 1–51.
4. Inesi, G., Kurzmack, M., Coan, C. and Lewis, D.E. (1980) *J. Biol. Chem.* 255, 3025–3031.
5. Yu, X., Carroll, S., Rigaud, J.L. and Inesi, G. (1993) *Biophys. J.* 64, 1232–1242.
6. Mata, A.M. and Sepulveda, M.R. (2005) *Brain Res. Brain Res. Rev.* 49, 398–405.
7. Foggia, L. and Hovnanian, A. (2004) *Am. J. Med. Genet. C Semin. Med. Genet.* 131C, 20–31.
8. Denmeade, S.R. and Isaacs, J.T. (2005) *Cancer Biol. Ther.* 4, 14–22.
9. Eckstein-Ludwig, U., Webb, R.J., Van Goethem, I.D., East, J.M., Lee, A.G., Kimura, M., O'Neill, P.M., Bray, P.G., Ward, S.A. and Krishna, S. (2003) *Nature* 424, 957–961.
10. MacLennan, D.H. and Kranias, E.G. (2003) *Nat. Rev. Mol. Cell Biol.* 4, 566–577.
11. Toyoshima, C., Nakasako, M., Nomura, H. and Ogawa, H. (2000) *Nature* 405, 647–655.
12. Toyoshima, C. and Nomura, H. (2002) *Nature* 418, 605–611.
13. Toyoshima, C. and Mizutani, T. (2004) *Nature* 430, 529–535.
14. Toyoshima, C., Nomura, H. and Tsuda, T. (2004) *Nature* 432, 361–368.
15. Obara, K., Miyashita, N., Xu, C., Toyoshima, I., Sugita, Y., Inesi, G. and Toyoshima, C. (2005) *Proc. Natl. Acad. Sci. U.S.A.* 102, 14489–14496.
16. Sørensen, T.L., Møller, J.V. and Nissen, P. (2004) *Science* 304, 1672–1675.
17. Olesen, C., Sørensen, T.L., Nielsen, R.C., Møller, J.V. and Nissen, P. (2004) *Science* 306, 2251–2255.
18. Jensen, A.M., Sørensen, T.L., Olesen, C., Møller, J.V. and Nissen, P. (2006) *EMBO J* 25, 2305–2314.
19. Sugita, Y., Miyashita, N., Ikeguchi, M., Kidera, A. and Toyoshima, C. (2005) *J. Am. Chem. Soc.* 127, 6150–6151.
20. Toyoshima, C. and Inesi, G. (2004) *Annu. Rev. Biochem.* 73, 269–292.
21. Kühlbrandt, W. (2004) *Nat. Rev. Mol. Cell Biol.* 5, 282–295.
22. Clausen, J.D., Vilsen, B., McIntosh, D.B., Einholm, A.P. and Andersen, J.P. (2004) *Proc. Natl. Acad. Sci. U.S.A.* 101, 2776–2781.
23. Ma, H., Lewis, D., Xu, C., Inesi, G. and Toyoshima, C. (2005) *Biochemistry* 44, 8090–8100.
24. Aravind, L., Galperin, M.Y. and Koonin, E.V. (1998) *Trends Biochem. Sci.* 23, 127–129.
25. Johnson, L.N. and Lewis, R.J. (2001) *Chem. Rev.* 101, 2209–2242.
26. Senes, A., Ubarretxena-Belandia, I. and Engelman, D.M. (2001) *Proc. Natl. Acad. Sci. U.S.A.* 98, 9056–9061.
27. Zhang, Z., Lewis, D., Strock, C., Inesi, G., Nakasako, M., Nomura, H. and Toyoshima, C. (2000) *Biochemistry* 39, 8758–8767.
28. Inesi, G., Ma, H., Lewis, D. and Xu, C. (2004) *J. Biol. Chem.* 279, 31629–31637.
29. Danko, S., Daiho, T., Yamasaki, K., Kamidochi, M., Suzuki, H. and Toyoshima, C. (2001) *FEBS Lett.* 489, 277–282.
30. Danko, S., Yamasaki, K., Daiho, T., Suzuki, H. and Toyoshima, C. (2001) *FEBS Lett.* 505, 129–135.
31. Picard, M., Toyoshima, C. and Champeil, P. (2006) *J. Biol. Chem.* 281, 3360–3369.
32. Lenoir, G., Picard, M., Gauron, C., Montigny, C., Le Marechal, P., Falson, P., le Maire, M., Moller, J.V. and Champeil, P. (2004) *J. Biol. Chem.* 279, 9156–9166.
33. Daiho, T., Yamasaki, K., Wang, G., Danko, S., Iizuka, H. and Suzuki, H. (2003) *J. Biol. Chem.* 278, 39197–39204.

J. Krebs and M. Michalak (Editors)
Calcium: A Matter of Life or Death

ISSN: 0167-7306/DOI: 10.1016/S0167-7306(06)41010-3

CHAPTER 10

Ca^{2+}/Mn^{2+} pumps of the Golgi apparatus and Hailey–Hailey disease

Leonard Dode, Jo Vanoevelen, Ludwig Missiaen, Luc Raeymaekers, and Frank Wuytack

Laboratorium voor Fysiologie, K.U. Leuven, Campus Gasthuisberg O/N, Herestraat 49 Bus 802, B3000 Leuven, Belgium, Tel.: +32 16 345720; Fax: +32 16 345991; E-mail: frank.wuytack@med.kuleuven.be

Abstract

All compartments of the secretory pathway require a sufficiently high luminal Ca^{2+} (and Mn^{2+}) concentration ($[Ca^{2+}]_l$) to carry out their respective functions. The Ca^{2+}-uptake mechanism of the endoplasmic reticulum (ER) consists of the well-characterized sarco(endo)plasmic reticulum Ca^{2+}-transport ATPases (SERCAs), whereas the Golgi complex depends on both SERCAs and on the much less characterized secretory pathway Ca^{2+}-transport ATPases (SPCAs). SPCAs (encoded by *ATP2C1* and *ATP2C2* in humans) are Golgi-localized pumps that transport both Ca^{2+} and Mn^{2+} with high affinity. The role of intra-Golgi Ca^{2+}/Mn^{2+} homeostasis in cellular function and its dependence on SPCAs is discussed, with special emphasis on the expression, catalytic properties, and the link between the defective *ATP2C1* gene and Hailey–Hailey disease (HHD) (an autosomal dominant skin disorder).

Keywords: *ATP2C1*, *ATP2C2*, calcium, Golgi, Hailey–Hailey, manganese, SERCA, SPCA

1. Introduction

The various functions executed by the intracellular Ca^{2+} pools of the endoplasmic reticulum (ER) and the secretory pathway can be divided in two broad categories. On the one hand, the intracellular organelles represent a dynamic source of activator Ca^{2+} that can be released into the cytosol by the regulated opening of specific Ca^{2+} channels. On the other hand, a sufficiently high luminal Ca^{2+} concentration ($[Ca^{2+}]_l$) in these organelles is required for the proper synthesis, post-translational processing, and trafficking of newly formed proteins (reviewed in [1,2]). Both types of functions and the role played by Ca^{2+} pumps in their execution, particularly the role of secretory pathway Ca^{2+}-transport ATPases (SPCAs), will be addressed in this chapter.

The loading of ER with Ca^{2+} is accomplished by sarco(endo)plasmic reticulum Ca^{2+}-transport ATPases (SERCA) Ca^{2+} pumps, the best-characterized group of the large family of P-type Ca^{2+}-transport ATPases (see also Section 4.). Human SERCAs are encoded by three genes, named *ATP2A1-3*, encoding respectively SERCA1-3.

SERCA2 is a housekeeping enzyme, whereas SERCA3 and especially SERCA1 have a more restricted expression pattern (reviewed in [2]).

Ca^{2+} pumps typically localizing in the Golgi and belonging to the related SPCA subfamily have only recently been studied. The first recognized member was the yeast *Saccharomyces cerevisiae* homolog [3–5], named less appropriately Pmr1 (for plasma membrane ATPase-related). Two genes, *ATP2C1* and *ATP2C2*, encode, respectively, SPCA1 and SPCA2 in humans [6–10]. SPCA1 is the housekeeping form [10,11], whereas SPCA2, judging from its expression pattern and biochemical properties, appears to execute a more specialized function in specific cell types [9,10,12]. In contrast to SERCAs, SPCAs also efficiently transport Mn^{2+} [13]. As will be described in Section 2, Mn^{2+} ions are essential for many enzymatic activities within the Golgi compartment. In addition, the SPCA pumps also contribute to Mn^{2+} detoxification, especially in yeast.

The recent discovery that haploinsufficiency of human SERCA2 or SPCA1 pumps results in similar skin diseases, respectively, Darier disease [14] and Hailey–Hailey disease (HHD) [6,7], has further highlighted the critical role of these housekeeping Ca^{2+} pumps in cellular Ca^{2+} homeostasis. However, these discoveries have generated more questions than they have answered. Although the similar outcome of defects of these two pumps is strongly suggestive of a dysfunction of Ca^{2+}-dependent processing within the secretory pathway, it is not known precisely which process is affected, nor is it known which molecular components are affected by this defective activity. Therefore, we thought it useful to include in this chapter an overview of the various processes that might in principle be altered by a disturbance in the SPCA function, that is, the transport of either Ca^{2+} or Mn^{2+}.

2. *Cellular functions of Ca^{2+} and Mn^{2+} in the Golgi apparatus*

A high $[Ca^{2+}]_l$ is maintained throughout the secretory pathway, starting already in the ER with the Ca^{2+} requirement for the proper synthesis and processing of secreted proteins. This Ca^{2+}-rich environment is preserved during further transit in the ER-Golgi intermediate compartment, in the various Golgi subcompartments, *trans*-Golgi network, and finally in the secretory vesicles. This Ca^{2+} is indispensable not only for Golgi-specific enzymatic activity but in recent studies has also been implicated in protein trafficking through the Golgi apparatus. In addition to Ca^{2+}, many Golgi enzymes also require Mn^{2+} for optimal activity. In mammalian cells, this luminal Mn^{2+} is provided by the SPCAs, whereas Ca^{2+} appears to be accumulated by both SERCAs and SPCAs. The role of SPCAs in the Golgi functioning has first been described for the yeast homolog, named Pmr1. In yeast, this function could be even more important than in mammalian cells because yeast cells lack SERCA pumps [3,4].

2.1. *Role of Ca^{2+} and Mn^{2+} in Golgi-specific enzymatic activities*

The Golgi apparatus carries out a wide diversity of transfer reactions that assist in the synthesis of complex carbohydrates, glycoproteins, and glycolipids. The most important groups of these enzymes are the glycosyltransferases (GT) and the sulfotransferases. Although not all the numerous different transferases have been studied extensively, it is clear that many of them require bivalent ions for their activity, preferentially Ca^{2+} and/or Mn^{2+} [15,16].

In addition, the pro-hormone convertases (PCs) activity and the casein kinase of the Golgi have also been found to require a sufficiently elevated Ca^{2+} concentration [17].

2.1.1. Glycosyltransferases

The various GT synthesize oligosaccharides, glycolipids, and glycoproteins by adding a range of different sugars to an acceptor substrate. These enzymes have been classified into >65 different families depending on both sequence similarity and substrate/product chemistry [18,19] (indications of families will be based on the numbering used therein). Although there is little sequence conservation between the different families, many share a DxD motif. The members of these families require divalent cations for optimal activity [20]. In most enzymatic tests, species of divalent cations have been added separately. In these assays, Mn^{2+} is often more effective than Ca^{2+}, but maximum activation usually still requires millimolar Mn^{2+} concentrations. As the Mn^{2+} concentration in the Golgi is not known, it is not certain whether optimally activating concentrations really prevail in the luminal milieu. Some reports show, however, that Ca^{2+} and Mn^{2+} may act synergistically when added together.

The DxD motif that confers sensitivity to a divalent cation is found in galactosyltransferases (GalTs), acetylglucosaminyltransferases, acetylgalactosaminyltransferases, and α-1,3-fucosyltransferases. The α-1,2-fucosyltransferases (family GT11) and sialyltransferases (family GT29) lacking a DxD motif do not require divalent metals. The reported crystal structures of Mn^{2+}-dependent transferases show one or two aspartate side chains of the DxD motif contributing to the Mn^{2+} coordination.

Among the Golgi enzymes, the divalent cation dependence has most thoroughly been studied for the GalTs belonging to the families GT6 and GT7. These enzymes catalyze the transfer of galactose to various acceptor substrates through β1–4, β1–3, α1–3, and α1–4 linkages. The best studied member, β1–4 GalT (family GT7), transfers galactose from Uridine diphosphate-α-D-galactose to *N*-acetylglucosamine or, specifically in the lactating mammary gland, to free glucose to produce lactose. Mn^{2+} is a cofactor in the reaction. The metal ion is partially coordinated by the enzyme (site I) and partially by the substrate. The presence of a second ion binding site in the enzyme which is not activated by Mn^{2+} (site II) is suggested by biochemical assays [21] and by the crystal structure [22]. Studies on the isolated enzyme show that Ca^{2+} is a coactivator at this site, requiring millimolar concentrations for a full effect. In the presence of Ca^{2+}, the effect of Mn^{2+} at site I is maximal at about $10\,\mu M$.

Other Golgi GT that have been shown to depend on Mn^{2+} ions for maximum activity include *N*-acetylglucosaminyltransferases of family GT13 [23], *N*-acetylgalactosaminyltransferases of family GT27 responsible for mucin-type O-linked glycosylation of Ser/Thr side chains [24], mannosyltransferase of family GT71 [20], glucuronyltansferase (family GT43, reviewed in [25]), and fucosyltransferase of the GT10 family [26].

Given the fact that SPCA-type pumps play an important role in the accumulation of Ca^{2+} and Mn^{2+} into the Golgi apparatus, it can be expected that SPCA-dependent cation transport is indispensable for the maintenance of essential Golgi functions. In line with this view, it has been found that cultured cells with a deficiency of SPCA1 induced by small interfering RNA (sRNAi) exhibit diminished Ca^{2+} uptake in a subfraction of the Golgi [27] and show defects in glycan processing of secreted thyroglobulin [28]. In addition, the SPCA1 deficiency rendered the cells hypersensitive to ER stress.

2.1.2. Sulfotransferases

Golgi-resident sulfotransferases transfer a sulfuryl group from the substrate 3′-phosphoadenosine 5′-phosphosulfate to a tyrosyl residue in a protein or to the alcohol (or amino) group of complex glycoconjugates (reviewed in [29]). The enhancement of the basal activity by Mn^{2+} has been demonstrated for both tyrosylprotein [30,31] and carbohydrate sulfotransferases [32–34].

Sulfation is an important post-translational modification of mucins in addition to glycosylation [35]. In humans, some sulfotransferase isoforms active on mucins present a restricted tissue distribution, for example, in colon and small intestine [36] or in lung [37], that is, in tissues rich in mucin-secreting cells. As will be made clear in Section 3.2.2., this distribution correlates with the expression levels of SPCAs.

2.1.3. Casein kinase, calcium, and milk production

The kinase that phosphorylates the major milk protein casein also resides in the lumen of the Golgi apparatus of mammary gland cells. Its molecular identity has not yet been fully elucidated [38,39]. It should, however, be distinguished from another group of well-characterized protein kinases, the members of which are found only in the cytosol and not in the Golgi lumen. The latter are currently designated casein kinase 1 and 2, because their enzymatic activity has been initially detected using casein as a substrate.

It is not known whether a mammary gland-specific casein kinase exists or if it is widely expressed. A report that the rat liver Golgi expresses a casein kinase similar to the casein kinase of lactating mammary gland could indicate that the Golgi kinase may not be exclusively dedicated to casein phosphorylation but rather represents a pleiotropic kinase catalyzing the observed phosphorylation of a wide diversity of secreted proteins carrying out various functions [40].

When depleted of Ca^{2+}, isolated mammary gland Golgi vesicles show little casein kinase activity. Addition of Ca^{2+} (in the millimolar range) or Mn^{2+} (at 10-fold lower concentrations) to Ca^{2+}-depleted Golgi causes a substantial increase in the kinase activity [41]. Ionomycin-induced Ca^{2+} depletion of intact acini has no effect on an early stage of casein phosphorylation but results in the partial inhibition of a later phase [42]. However, because removal of Ca^{2+} by ionomycin is suboptimal in acidic compartments, the possibility remains that in these experiments, the downstream Golgi compartments particularly contained some residual Ca^{2+}. Thus, at least the optimal casein phosphorylation requires Ca^{2+} and/or Mn^{2+}.

Another important aspect of the physiology of mammary gland cells is that phosphorylation is important to allow casein to bind Ca^{2+} and to deliver this element in large quantities into the milk (30 mM in cow milk). It remains unknown how much of this milk Ca^{2+} is contributed by the Golgi itself and what fraction is delivered by the more downstream secretory compartments or by other cellular transport processes. Reinhardt et al. [43,44] observed, firstly, that several Ca^{2+}-motive ATPases expressed in the lactating rat mammary gland such as SERCA2, SERCA3, SPCA1, and, particularly, the plasma membrane Ca^{2+}-transport ATPases (PMCAs) are strongly up-regulated and, secondly, that PMCA2$^{-/-}$ mice produced milk containing 60% less Ca^{2+} relative to wild-type mice. SPCAs are unique among these transport ATPases in that up-regulation already occurs 1 week before parturition [45]. Also SPCA2 in mammary gland is up-regulated during lactation in rat [46].

2.1.4. *Proteolytic processing*

Proteolytic processing by PCs is another Golgi-localized Ca^{2+}-dependent process in the regulated secretory pathway. Furin, PC1 and PC2 are the best-characterized members of this large enzyme family [17]. Furin and PC1 require millimolar Ca^{2+} for half-maximal activation of pro-hormone cleavage, whereas PC2 requires only approximately 5–100 μM Ca^{2+} [47–50]. Davidson et al. [47] suggested that in pancreatic β-cells, the activity with high Ca^{2+} affinity residues in the Golgi, whereas that with low Ca^{2+} affinity would be located in the insulin-containing secretory granules.

In addition, other less well-defined Golgi-localized convertases have been found to depend on Ca^{2+}, for example, proteases that process plasma protein precursors in liver cells [51], and a furin-like convertase responsible for propeptide cleavage of beta-site amyloid precursor protein-cleavage enzyme (BACE), the Alzheimer β-secretase [52].

2.2. *Role of luminal Ca^{2+} in trafficking and secretion of proteins*

The high $[Ca^{2+}]_l$ not only is indispensable for Golgi-specific enzymatic activity, as described in Section 2.1., but has in recent studies also been implicated in vesicle-mediated protein trafficking through the Golgi apparatus. The implication of a rise in cytosolic Ca^{2+} concentration ($[Ca^{2+}]_c$) in the final stage of secretion, that is, the fusion of secretory vesicles with the plasma membrane, is well established. Recent evidence suggests that regulation of anterograde and retrograde transport between the Golgi stacks also involves a source of Ca^{2+}. The finding that a fast Ca^{2+} buffer – like BAPTA (1,2-bis(2-aminophenoxy)ethane-N,N,N′,N′-tetraacetic acid) – is more effective than the slower Ca^{2+} buffer EGTA (ethylene glycol-bis(beta-aminoethylether)-N,N,N′,N′-tetraacetic acid) in blocking this traffic suggests that this regulator Ca^{2+} is delivered locally, possibly by a constant release or leakage from the Golgi itself [53–56]. However, the implications of these data should also be considered in the light of an alternative to the vesicle trafficking model, that is, the Golgi maturation model that recently gained more additional support from experiments on yeast cells [57].

Proteins are secreted through one or a combination of two pathways: a constitutive pathway that is not regulated and that does not require storage of the cargo and a regulated pathway that involves intermediate packaging and storage of secretory products. It has been well established that the Ca^{2+} concentration inside secretory granules of the regulated pathway, such as mucin-, hormone-, or neurotransmitter-containing vesicles, is much higher than that in the cytoplasm. The high $[Ca^{2+}]_l$ acts here as an important cofactor in the condensation of the cargo [58,59]. It is less clear whether a similarly steep Ca^{2+} gradient is maintained in vesicles of the constitutive pathway [60]. The high $[Ca^{2+}]_l$ in the secretory vesicles is probably already established when they are budding off from the *trans*-Golgi network. However, further dynamic regulation of the vesicular Ca^{2+} content may occur through specific release channels and Ca^{2+}-uptake mechanisms. Part of the Ca^{2+} uptake depends on the acidic luminal milieu and uses H^+ gradient-dependent Ca^{2+} loading, either directly through Ca^{2+}/H^+ exchangers [61,62] or through a combination of H^+/Na^+ and Na^+/Ca^{2+} exchangers [63]. In addition, ATP-dependent Ca^{2+} pumps also load Ca^{2+}. The latter system would contribute about 20% in insulin-containing vesicles of pancreatic β-cells and would be catalyzed by SPCAs (schematically represented in Fig. 1) [64,65].

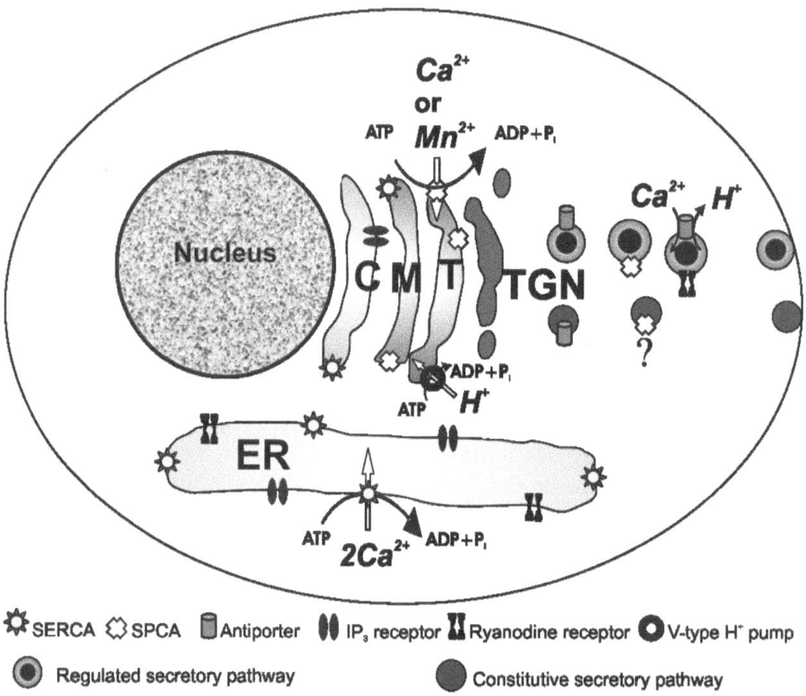

Fig. 1. Schematic overview of the organization of Ca^{2+} and Mn^{2+} uptake into the compartments of the secretory pathway in animal cells. ER, endoplasmic reticulum; C, *cis*-Golgi; M, *medial*-Golgi; T, *trans*-Golgi; and TGN, *trans*-Golgi network. The largest Ca^{2+} store is represented by ER, which is loaded with Ca^{2+} by sarco(endo)plasmic reticulum Ca^{2+}-transport ATPase (SERCA) pumps and depleted of Ca^{2+} through inositol 1,4,5-trisphosphate (IP_3) receptor and ryanodine receptor channels. The Golgi is loaded with Ca^{2+} by both SERCAs and secretory pathway Ca^{2+}-transport ATPases (SPCAs) and partially emptied by IP_3 receptors. Cytosolic Mn^{2+} is accumulated within the Golgi by SPCAs, too. Protons (H^+) are accumulated by the V-type ATPases into the Golgi and secretory pathway (see text for details).

ATP-dependent Ca^{2+} uptake appears to be relatively more important in brain synaptic vesicles [61] and in chromaffin granules [63]. Based on the sensitivity to thapsigargin (TG), Mahapatra et al. [63] ascribed this pathway to the activity of SERCA pumps. However, at the high concentration of 10-μM TG used in this study, additional inhibition of the SPCA-type pumps cannot be excluded [12].

Besides the role of $[Ca^{2+}]_l$ in the secretory pathway as an indispensable cofactor or regulator of enzymes and sorting processes, the secretory pathway could contribute to the removal of Ca^{2+} and Mn^{2+} from the cell. Secretion of Ca^{2+} could be especially important in yeast, which lacks the well-defined plasma membrane Ca^{2+} extrusion pumps.

In animal cells, secretion of Ca^{2+} from the cells through the secretory pathway, either of largely unbound Ca^{2+} or of Ca^{2+} in combination with Ca^{2+} binding proteins, may be important in altering the composition of the cell surroundings. A typical example of a central role of the regulated pathway is the Golgi of lactating mammary gland cells (see Section 2.1.3.). However, the constitutive pathway of all cells studied also secretes Ca^{2+}-binding proteins into the extracellular milieu, specifically proteins of the nucleobindin family. These are encoded by two genes, whose protein products are NUCB1 (also named calnuc) and NUCB2 (also named p54/NEFA). Both are EF-hand-type Ca^{2+}-binding proteins structurally related to the ER protein calreticulin and, like calreticulin, are almost ubiquitously distributed in the cell. NUCB1 is enriched in the *cis*-Golgi lumen but is detected also in the cytosol, the ER, the nucleus, and the extracellular space [66]. In bone, NUCB1 protein secreted by bone cells could play a role as a modulator of matrix maturation in the mineralization process [67]. The intracellular localization of NUCB2 is variable. In fibroblasts, NUCB2 colocalizes with the *medial*-Golgi marker mannosidase II [68]. NUCB2 in brain neurons is found in the ER and in the nuclear envelope [69]. In human vascular endothelial cells (HUVECs), NUCB2 is distributed in intracytoplasmic vesicles, which do not colocalize with Golgi markers or with endosomes [70]. HUVECs and airway epithelial cells secrete NUCB2 in association with exosome-like vesicles. The release of NUCB2, but not NUCB1, has been found to be required for the extracellular release of tumor necrosis factor receptor 1 through the formation of a Ca^{2+}-dependent complex with aminopeptidase regulator of tumor necrosis factor receptor 1 shedding (ARTS-1). Secreted receptors inhibit tumor necrosis factor bioactivity by competing with cell surface tumor necrosis factor receptors.

3. *Role of the Golgi complex in cellular Ca^{2+} and/or Mn^{2+} regulation*

Several factors add up to make the study of Golgi Ca^{2+} homeostasis a difficult enterprise, particularly the lack of a SPCA-specific inhibitor and the compartmentalization of the Golgi in various subregions presenting different properties. Visualization of the different subcompartments by optical microscopy is not straightforward because of the small size of the organelle and because marker proteins often do not show an all-or-none distribution over the various regions. Two complicating factors hamper an accurate estimation of $[Ca^{2+}]_l$ in the Golgi. First, the low luminal pH in the Golgi compartment, that decreases even further in the downstream direction, interferes with fluorescent probes. Second, the reducing luminal environment interferes with the correct folding of some Ca^{2+} reporter proteins that can adequately fold in the cytosol. It is also noteworthy that the higher cholesterol content, particularly in the downstream subcompartments, can be expected to result in their functional disruption by saponin concomitant with the permeabilization of the surface membrane. Membrane permeabilization is often used with the intention to gain direct access to intact intracellular Ca^{2+}-transporting organelles.

3.1. Role of the Golgi Pmr1 in Ca^{2+} and Mn^{2+} homeostasis in yeast and implications for resistance to oxidative stress

The link between SPCA-type transporters and Mn^{2+} metabolism has first been described for the *S. cerevisiae* homolog Pmr1, based on the observation that oxidative damage in cells lacking superoxide dismutase (SOD) is suppressed in Pmr1 mutants [71]. Cells lacking a functional Pmr1 accumulate Mn^{2+}, which can substitute for SOD by scavenging reactive oxygen species. Mutant cells are also extremely sensitive to manganese toxicity, demonstrating that Pmr1 is important for Mn^{2+} detoxification [71]. These mutant cells also show a disturbed Ca^{2+} homeostasis, particularly in a low Ca^{2+} medium [72,73]. Pmr1 mutants of *Schizosaccharomyces pombe* present altered activity of calcineurin through effects on Mn^{2+} homeostasis, demonstrating that in these cells calcineurin senses not only Ca^{2+} but also Mn^{2+} [74]. Because calcineurin is a cytosolic protein, these observations thus show that yeast Pmr1 plays a role not only in maintaining Golgi luminal Mn^{2+} but also in the regulation of the cytoplasmic levels of this ion, like it appears to do so for Ca^{2+}. Cells lacking Pmr1 show an up to 16-fold increased $[Ca^{2+}]_c$ [72]. Yeast cells lack SERCA pumps. Ca^{2+} sequestration is accomplished by Pmr1 together with the Pmc1 Ca^{2+} pump localized in the vacuolar membrane.

Interestingly, the basic consequences of *PMR1* mutation observed in yeast are also observed with SPCA knock-down in the multicellular organism *Caenorhabditis elegans*. Manipulated animals are highly sensitive to EGTA and Mn^{2+} and become resistant to oxidative stress, probably due to increased levels of cytosolic Mn^{2+} [75]. Similar experiments on higher organisms would require strict control of the Mn^{2+} content of the diet and would be more difficult to interpret because of the higher complexity. However, the observation of decreased stress sensitivity associated with SPCA down-regulation cannot be extrapolated to higher vertebrates and humans, because defects in SPCA, as in HHD patients, generally result in higher stress sensitivity (see Section 5.).

3.2. Golgi-resident proteins involved in Ca^{2+} regulation

The question whether the Golgi can function as a dynamic Ca^{2+} store capable of functionally significant effects on cytosolic Ca^{2+} depends on the Ca^{2+} storage capacity of the organelle and on the presence of Ca^{2+} release channels and Ca^{2+} pumps.

3.2.1. *Ca^{2+} storage capacity of the Golgi*

Ion microscopy has since long demonstrated a high concentration of Ca^{2+} in the Golgi apparatus. Saponin-permeabilized cells loaded with Ca^{2+} in the presence of oxalate show the highest Ca^{2+} signal in the region of the Golgi [76]. Also, intact cells cryogenically prepared for the preservation of ion distribution showed the highest Ca^{2+} concentration in the Golgi region [77]. Using a high-resolution technique applied to PC12 cells, Pezzati et al. [78] observed a strong signal that was uniform along the *cis-trans* axis. Remarkably, weaker signals were recorded in the *trans*-Golgi network, whereas secretion granules were strongly positive for Ca^{2+} [78].

The Golgi complex occupies a much smaller cellular volume than the ER. However, like the ER, its Ca^{2+} buffering capacity is increased by the presence of Ca^{2+}-binding proteins, including the many Ca^{2+}-dependent enzymes mentioned above. Nucleobindins (see also Section 2.2.) were reported to be involved in establishing an agonist-mobilizable Ca^{2+} store in the Golgi lumen [79]. Interestingly, an apparently compensating increase in NUCB1 expression is seen in COS-7 cells overexpressing the SPCA1 Ca^{2+}-transport ATPase [80].

In addition, the Golgi houses members of the CREC (Cab45/reticulocalbin/ERC45/calumenin) family of EF-hand proteins (see [81] for review). Some of these proteins, like reticulocalbin and ERC 55 (endoplasmic-reticulum Ca^{2+}-binding protein of 55 KDa), are typical ER proteins. Another member, Cab45, is strictly localized to the Golgi lumen [82], whereas calumenin is distributed throughout the whole secretory pathway and is also secreted into the medium [83].

Quite unexpectedly, substantial amounts of the typical ER protein GRP94 (endoplasmin) have been detected in the Golgi apparatus, where it is phosphorylated by casein kinase [84].

3.2.2. *Ca^{2+} pumps of the secretory pathway*

It is a general observation that the SPCA in invertebrates and the SPCA1 isoform of higher vertebrates are localized in the Golgi (Fig. 2). However, it has been observed in many laboratories that Ca^{2+} uptake in the Golgi apparatus is partially inhibited by low concentrations of TG, thus establishing the fact that also SERCAs contribute to the Ca^{2+} uptake in this compartment [27,79,85–87]. In contrast to SERCAs, SPCA pumps also catalyze the uptake of Mn^{2+}. The relative contribution of SERCA and SPCA pumps to the uptake of Ca^{2+} into the Golgi is cell-type dependent. Because these studies used Golgi-targeted aequorin as a probe, the values refer to free $[Ca^{2+}]_l$, which are not necessarily directly correlated with the total uptake. Among various cultured cells, the highest SPCA-dependent Ca^{2+} uptake (about 67%) was observed in human keratinocytes [88], which fits with the fact that haploinsufficiency of *ATP2C1* mainly affects the epidermis (see Section 5.).

Because SERCAs are typically found in the ER and SPCAs in the downstream compartments of the secretory pathway, it is tempting to speculate that the SERCA pumps responsible for the TG-sensitive Ca^{2+} uptake in the Golgi are preferentially

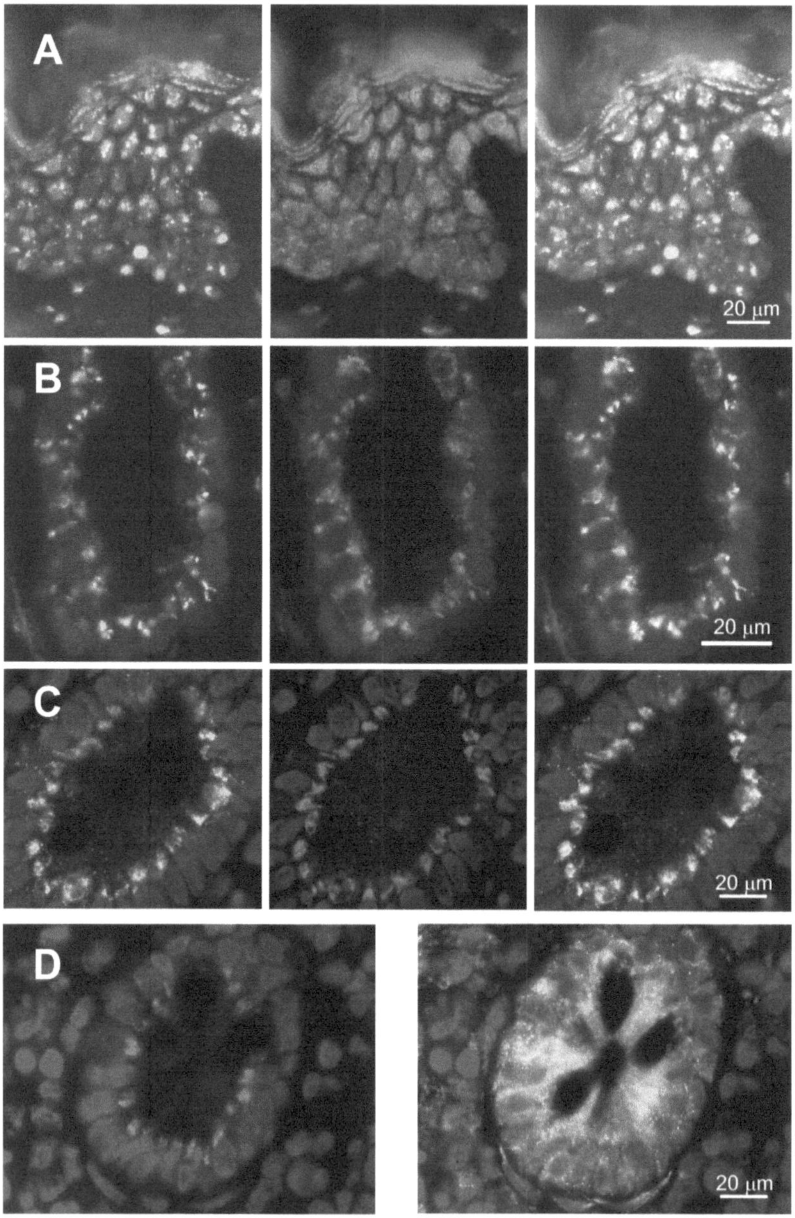

Fig. 2. Immunocytochemical localization of secretory pathway Ca^{2+}-transport ATPase 1 (SPCA1) in human epidermis (**A**), prostate (**B**), colon (**C**), and of SPCA2 in human colon (**D**). In **A–C**, sections were triple labeled for *trans*-Golgi network 46 (TGN46) as a Golgi marker (green, left panels), for SPCA1 (red, middle panels) and with DAPI (4′,6-diamidino-2-phenylindole, dihydrochloride) to stain the nuclei (blue). The overlay is shown in the right panels. The section in **D** was stained for SPCA2 (red, left) and for SERCA2b (green, right), demonstrating the more restricted localization of SPCA2 as compared with SERCA2 (See Color Plate 29, p. 527).

localized in the upstream *cis*- and *medial*-Golgi (Fig. 1). Behne et al. [89] indeed observed an enrichment of SERCA and SPCA1, respectively, in *cis*- and *trans*-Golgi membranes isolated from human keratinocytes. Another approach to investigate this problem is to follow the effect of brefeldin-A on the subcellular distribution of SPCA in living cells. By blocking anterograde vesicle transport, brefeldin-A mainly causes the *cis*- and *medial*-Golgi, and to a lesser extent the *trans*-Golgi, to redistribute into the ER [90]. In agreement with the data of Behne et al. [89], we have observed that as a consequence of brefeldin-A treatment, GFP-tagged SPCA overexpressed in COS-1 cells partially redistributes into the cytosol but still preserves a partial perinuclear distribution pattern (unpublished data; Fig. 3). This reaction of overexpressed SPCA should be confirmed for the endogenous pump. The incomplete redistribution of overexpressed SPCA shows resemblance to the partial resistance to brefeldin-A of the subcellular distribution of the Menkes copper transport protein [91]. The differential localization of the Ca^{2+} pumps to either the proximal or distal regions of the secretory pathway may be determined by the cholesterol gradient present between ER and *trans*-Golgi network. The cholesterol content is low in the ER and increases in the distal direction. Membranes of the *trans*-Golgi compartments are therefore thicker and more rigid. Both parameters are important determinants of Ca^{2+}-ATPase function and possibly of their localization as well [92].

SPCA1 is a housekeeping enzyme and has been detected in most cell types, including neurons and the various supporting cell types in the brain [93]. However,

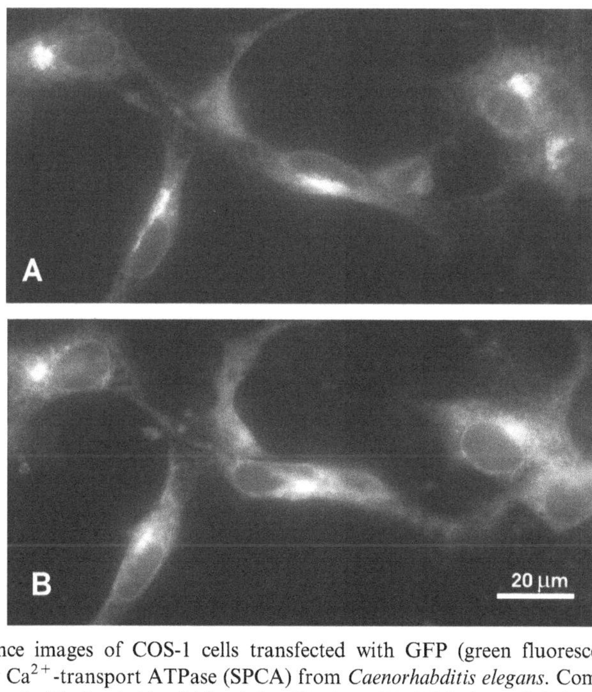

Fig. 3. Fluorescence images of COS-1 cells transfected with GFP (green fluorescent protein)-tagged secretory pathway Ca^{2+}-transport ATPase (SPCA) from *Caenorhabditis elegans*. Compared with control cells (**A**), cells treated with 5 µg/ml brefeldin-A for 60 min at 37 °C (**B**) show slightly more staining of the cytosol but retain a considerable fraction of the signal at a Golgi-like juxtanuclear position.

different laboratories have obtained partially divergent results on the relative expression level in various tissues. Wootton et al. [94] observed much higher expression levels, both at the mRNA and at the protein level, in rat brain and testis than in the other tissues examined, including lung and liver. Such relatively high expression in brain and testis was not observed at the mRNA level in humans [6,10]. Wootton et al. [94] also observed a high TG-resistant Ca^{2+}-uptake activity in rat aortic smooth-muscle cells. The cell-type-dependent expression of SPCA1 in various animal species thus merits further study.

Special mention should be made of the peculiar position of SPCA1 in human spermatozoa. SPCA1 is localized in an area behind the nucleus and extending into the midpiece. It may be the only intracellular Ca^{2+} pump in these cells, because both functional and immunocytochemical tests failed to demonstrate the presence of SERCAs [95].

More detailed information is also needed on the expression pattern of the second related gene, *ATP2C2*, encoding the SPCA2 pump. The available data all indicate that SPCA2 expression in human tissues is more heterogeneous than that of SPCA1. Vanoevelen et al. [10] have shown that SPCA2 mRNA is most abundant in the various segments of the gastrointestinal tract, trachea, salivary gland, thyroid and mammary gland, and prostate. Similarly, the highest SPCA2 mRNA levels have been observed in gastrointestinal and airway tracts of rat, pig, and human [46]. Interestingly, SPCA2 levels in the mammary gland increase during lactation [46]. Together with the observation that in human colon SPCA2 is mainly expressed in the mucin-secreting goblet cells [46] (Fig. 2), these data suggest a correlation between the expression of SPCA2 and the presence of a regulated secretory pathway. Keratinocytes may form an exception, because cells isolated from human foreskin, which were allowed to proliferate in culture and then induced to differentiate, should be added to the list of human cell types expressing SPCA2 mRNA [10]. However, this picture of the SPCA2 mRNA expression pattern should be complemented with a more detailed analysis at the protein level because, with the presently available antibodies, Vanoevelen et al. [10] could demonstrate the expression of SPCA2 protein only in colon. In addition, these observations should be contrasted with those of Xiang et al. [9], who noted instead a prominent SPCA2 protein expression in human brain and testis.

A further, not yet resolved, question concerns the precise subcellular localization of SPCA2. In human goblet cells, the subcellular distribution of SPCA2 looks very similar to that of SPCA1, and both pumps colocalize with Golgi markers in a compact structure near the apical pole of the nucleus [10] (Fig. 2). Again, these observations contrast with those on cultured mouse hippocampal neurons, which show a punctate distribution of SPCA2 staining in the cell body and in the dendrites that partially colocalize with the *trans*-Golgi marker TGN38 [9]. Xiang et al. [9] conclude that SPCA2 in hippocampal neurons is localized in downstream, post-Golgi segments of the secretory pathway.

3.2.3. Ca^{2+} release channels of secretory pathway membranes

Pinton et al. [86] were the first to demonstrate an inositol 1,4,5-trisphosphate (IP_3)-induced Ca^{2+} release in Golgi by measuring $[Ca^{2+}]_l$ through an aequorin construct that contains a Golgi-targeting sequence. The IP_3-induced Ca^{2+} release appears incomplete, even at high doses of the agonist. Interestingly, the agonist-sensitive fraction corresponds to the pool loaded by pumps that are sensitive to SERCA inhibitors, whereas the agonist-insensitive fraction is dependent on pumps other than SERCAs (Fig. 1) [85,87]. The SERCA-independent compartment is probably loaded by SPCAs, because it is absent in HeLa cells treated with SPCA1-specific sRNAi [27].

Up till now, the presence of ryanodine receptors or other Ca^{2+}-release mechanisms in the Golgi apparatus could not be demonstrated. However, insulin-containing vesicles of pancreatic β-cells do express functional ryanodine receptors that are sensitive to NAADP, whereas IP_3 receptors are absent [96]. The presence of typical ER Ca^{2+} channels and pumps in post-ER compartments further complicates the question as to which signals regulate the proper targeting of these membrane proteins [97].

3.3. Does the Golgi contribute to the regulation of cytoplasmic Ca^{2+} in mammalian cells?

The lack of specific SPCA inhibitors has up till now prevented the obtaining of a clear picture on the specific contribution of the secretory pathway pumps to the cellular Ca^{2+} homeostasis. Our limited knowledge is based on overexpression experiments, on the use of RNA interference method to down-regulate endogenous SPCA, or on observations made on keratinocytes obtained from HHD patients.

Overexpression of SPCA1 in COS-1 cells alters cytoplasmic Ca^{2+} signals. Stimulation of purinergic receptors with ATP results in baseline Ca^{2+} spikes in these transfected cells, in the presence as well as in the absence of a functional ER. After reintroduction of Ca^{2+} to Ca^{2+}-depleted cells, the latency period preceding the rise of cytosolic Ca^{2+} is much prolonged in SPCA1-transfected cells, suggesting that the overexpression of SPCA pumps results in faster intracellular Ca^{2+} sequestration [98]. This result is in agreement with measurements of $[Ca^{2+}]_l$ using Golgi-targeted aequorin, which demonstrate that Ca^{2+} uptake into the Golgi is enhanced in cells transfected with DNA encoding SPCA [87]. As mentioned in Section 3.2.1., overexpression of SPCA also results in increased expression of the Ca^{2+}-storage protein NUCB1 [80].

To complement these data, showing that overexpression of SPCA1 results in the appearance of Ca^{2+} oscillations in otherwise quiescent cells, Van Baelen et al. [27] investigated the effect of disrupting the expression of SPCA1 on agonist-elicited Ca^{2+} oscillations in HeLa cells. Cells lacking SPCA1 can still set up baseline Ca^{2+} spiking. However, fewer cells responded to the agonist with Ca^{2+} oscillations and showed a single $[Ca^{2+}]_c$ peak instead, demonstrating that SPCA1 is not absolutely needed but contributes to the shaping of the cytosolic Ca^{2+} signals. Disruption of the Golgi structure with brefeldin-A on the other hand resulted more often in responses involving Ca^{2+} oscillations [99]. Also in pancreatic β-cells depleted of SPCA1, the shape and the duration of Ca^{2+} oscillations, in response to glucose, is modified [65]. The strongest effect of interference with SPCA activity has been observed in spermatozoa, where progesterone-induced Ca^{2+} oscillations are inhibited [95]. This relative important role of SPCA1 in spermatozoa could be related to the absence of SERCAs in these cells, as mentioned above. Taken together, with the notable exception of the more pronounced effects observed in human spermatozoa, these results thus indicate rather subtle effects of the SPCA1-containing Ca^{2+} compartment on cytosolic Ca^{2+} regulation in cells that respond to stimulation with Ca^{2+} oscillations.

Possible effects of SPCA1 on $[Ca^{2+}]_c$ in keratinocytes will be discussed in Section 5.3.

4. *Molecular characterization of the secretory pathway Ca^{2+}-ATPases*

SPCAs, SERCAs, and PMCAs represent the Ca^{2+}-transporting pumps of the Golgi apparatus and downstream secretory pathway, the sarcoplasmic or ER, and the plasma membrane, respectively. The abovementioned single-subunit integral membrane enzymes (*E*) together with the gastric pump (H^+- and K^+-ATPase), the sodium pump (Na^+- and K^+-ATPase), and other transport ATPases are collectively known as the P-type ATPases. The term "P-type" describes the compulsory formation, upon ATP hydrolysis, of a high-energy phosphorylated intermediate ($E \sim P(X)_n$, where X is the transported ion and n is the number of transported ions) during the course of ion transport across the membranes. The high-energy covalent bond is formed between the hydrolyzed γ-phosphate moiety of ATP and an invariable aspartyl residue (e.g., **Asp**$^{\mathbf{351}}$ in human SERCAs) from the sequence **Asp**-Lys-Thr-Gly-Thr-[Leu/Ile/Val/Met]-[Thr/Ile/Ser], which represents the highly conserved phosphorylation motif of ion-motive ATPases.

In view of their structural determinants and Ca^{2+}-transport ability (see Sections 4.1. and 4.2.), SPCA pumps indeed resemble SERCAs. In addition, SPCAs also elicit distinct properties: (i) unmatched specificity and sensitivity toward Mn^{2+} accumulation in the Golgi lumen, (ii) transport of only one ion (Ca^{2+} or Mn^{2+}) per each hydrolyzed ATP molecule, thus resembling the PMCAs in this respect, and (iii) different cellular localization. Furthermore, proton binding and countertransport events also accompany the active transport of Ca^{2+} mediated by SERCAs and PMCAs [100]. Until recently, it was not clear whether SPCAs would countertransport luminal protons from the Golgi compartments to the cytosol (see Section 4.2.).

Except for the number of transported ions, the SPCA reaction cycle (Fig. 4) is similar to that of SERCAs. It consists of several reaction steps associated with (i) ion binding and translocation and (ii) ATP hydrolysis. This is accomplished by the reversible cycling of the ion-motive ATPase between its E_1 and E_2 conformational states in line with the currently held E_1/E_2 catalytic model [101]. Accordingly, the two conformational states display different affinities toward the transported ion irrespective of the phosphorylation status: the E_1 states display binding sites with high affinity for Ca^{2+} or Mn^{2+} at the cytoplasmic side and occlude tightly the respective ion, whereas the E_2 states are characterized by luminally oriented binding sites with low affinity for either ion. Upon binding of the ion in the transmembrane region of SPCA (event described by forward reactions 1 and 2 in Fig. 4), several

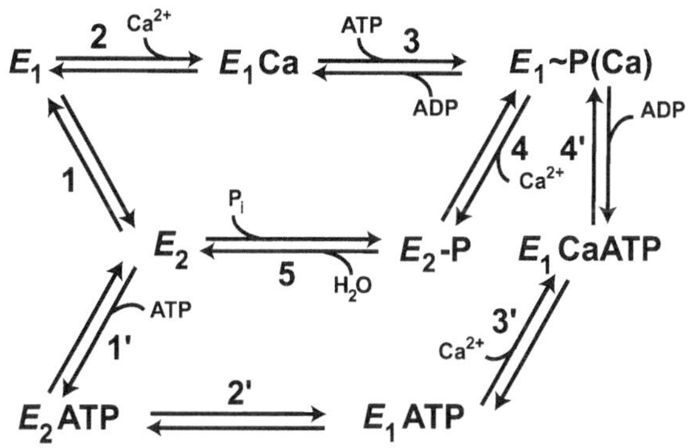

Fig. 4. Secretory pathway Ca^{2+}-transport ATPase (SPCA) reaction cycle. SPCAs transport only one Ca^{2+} or Mn^{2+} per each hydrolyzed ATP through several steps involving a coupling between ion binding and ATP hydrolysis events. The reaction steps can be interpreted on the basis of the model proposed by De Meis and Vianna [101], which involves the reversible trafficking of the enzyme between E_1-type conformations displaying high-affinity ion-binding sites at the cytoplasmic site (e.g., E_1, E_1Ca, E_1ATP, and E_1CaATP) or sequestering the transported ion in an occluded state such as in the $E_1 \sim$P(Ca) intermediate and E_2-type forms exhibiting low-affinity ion-binding sites at the luminal side (E_2, E_2ATP, and E_2-P). The most representative intermediates of the catalytic cycle are shown, and it is noteworthy that, for many of them, crystal structures have been elucidated for the related sarco(endo)plasmic reticulum Ca^{2+}-transport ATPase 1 (SERCA1) pump that transports two Ca^{2+}/ATP [102–106]. Toyoshima and Nomura [127] proposed that in SERCA1, the ion transport and ATP utilization would follow the reversible route indicated by reactions 1–5, which would imply large movements of the cytosolic domains. Recently, Jensen *et al.* [105] proposed an alternate route (reactions 1′–3′) by taking into account the fact that the E_2ATP and E_1Ca_2ATP crystals display slightly different compact structures, thus requiring less ample domain movements. This proposal is justified by the increased cytosolic ATP concentration that normally saturates all of the intermediates shown in the catalytic cycle. It should be noted that reaction steps 3′ and 4′ are equivalent to reaction 3. The transition from E_1Ca to $E_1 \sim$P(Ca) also implies the existence of an E_1CaATP intermediate, which for simplicity is not shown.

conformational changes take place, thus enabling the ATP phosphorylation of an aspartate residue (forward reaction 3). In addition, the resulting high-energy phosphoenzyme, $E_1 \sim P(Ca^{2+})$ or $E_1 \sim P(Mn^{2+})$, containing the occluded ion is ADP-sensitive, meaning that in presence of a high ADP/ATP ratio, it can function in the reverse mode (reverse reaction 3) with the consequent enzyme dephosphorylation and ATP regeneration. The release of Ca^{2+} into the Golgi lumen depends on the rate-limiting step of the catalytic cycle, which is represented by the transformation of either the $E_1 \sim P(Ca^{2+})$ or the $E_1 \sim P(Mn^{2+})$ intermediate to an ADP-insensitive low-energy E_2-P phosphoenzyme (forward reaction 4). Hydrolysis of E_2-P and regeneration of the Ca^{2+}- or Mn^{2+}-free E_2 enzyme (forward reaction 5) complete the reversible cycle. Proton countertransport by SERCAs is associated with this partial reaction of the catalytic cycle. On the other hand, recent results [11,12] have unambiguously demonstrated that in SPCAs, this reaction step is not activated upon binding of protons or even potassium ions (see also Section 4.2.).

It should be noted that several of the reaction intermediates can now be distinguished experimentally by direct structural analyses [102–106]. Various functional biochemical studies, among which some are based on the pump's ability to phosphorylate with

[γ-^{32}P]ATP or inorganic phosphate ([^{32}P]H_3PO_4 or shortly $^{32}P_i$) in the presence and absence of Ca^{2+} (or Mn^{2+}) have been developed and extensively used to characterize the wild-type skeletal muscle SERCA1a and its mutants both in steady-state and transient kinetic conditions [107–109]. Additionally, this methodology has been successfully employed to analyze in vitro the human SERCA2 [110] and SERCA3 [111] isoforms, the SPCA1 [11] and SPCA2 [12] isoenzymes, and finally several SERCA2 [110] and SPCA1 [8] mutants detected in patients affected by Darier or Hailey–Hailey skin diseases. For further details, the reader is directed to recent reviews dealing with the crystallization of the functionally different SERCA1a conformations [106] and the steady-state and transient kinetic analysis of Ca^{2+} pumps [112].

4.1. Gene structure and protein organization

Chronologically, the yeast Pmr1 was the first SPCA family member to be identified as an intracellular Ca^{2+}-ATPase, whose expression and function have been proven of paramount importance for the yeast *medial*-Golgi compartment [3,4,16]. Genetic mapping revealed that *PMR1* is located on the left arm of chromosome VII [3]. Similar to most genes in *S. cerevisiae* [113], *PMR1* is intronless, too. The length and molecular weight of yeast SPCA calculated from the predicted amino acid sequence are, respectively, 950 amino acids (aa) and 104 kDa, respectively, thus closely resembling those of the mammalian SERCA pumps, for example, the skeletal muscle 110-kDa SERCA1a enzyme is 994-aa long. Subsequently, mutants of yeast SPCA were engineered, and their functional properties were investigated by functional complementation of a yeast strain lacking any of the endogenous Ca^{2+}-ATPases. This procedure has provided insights into the structure-function relationship by identifying amino acid residues that are critical for cation sensitivity and transport ([114] and references therein). SPCA homologs have been reported in other invertebrate and vertebrate organisms including *C. elegans* [13], *Drosophila melanogaster* (PID accession number 7296577), rat [115], and human [6,7,9,10].

The *C. elegans* SPCA gene located on chromosome 1 spans a region of >19 kb of genomic DNA and consists of 12 exons (from −3 to 9), and its transcripts are alternatively and *trans*-spliced at their 5′-end [13]. The ATG codon is invariably in exon 1. Although three alternative polyadenylation signals can be used in the 3′-untranslated region of the primary gene transcript, no alternative splicing event could be detected at this site [13]. Therefore, it appears that the SPCA gene transcription in *C. elegans* leads only to one SPCA protein containing 901 aa [13].

Relative to its worm homolog, the exon/intron layout of the *D. melanogaster* SPCA gene (GenBank™ accession number AC014929) is less elaborate, with only three intervening sequences separating the four existing exons. It is noteworthy that the boundary occurring between exons 2 and 3 in *D. melanogaster* is conserved as the exon 5/6 junction in the *C. elegans* gene. Interestingly, despite losing several introns during evolution, the *D. melanogaster* SPCA gene still encodes a 901-aa protein.

Although already discovered in the early 1990s, the rat SPCA gene [115] has received little attention [80], mainly because the emphasis on mammalian SPCAs was soon shifted to the corresponding human gene (*ATP2C1*) immediately after it has been documented by two independent reports [6,7] that *ATP2C1* represents the gene defective in the skin disorder known as the HHD (see Section 5.). Additionally, a second human SPCA gene was recently discovered (*ATP2C2*) coding for SPCA2 [9,10]. Corresponding SPCA2 genes have also been found in amphibians, birds, and mammals, but not in fish. Vertebrates, from fish to mammals, appear to have the most complete set of introns. Interestingly, an inspection of the *ATP2C1* ortholog in the primitive metazoan *Nematostella vectensis* (starlet sea anemone) at www.stellabase.org [116] shows a nearly complete conservation of the intron positions with respect to those of the vertebrate genes.

ATP2C1 is localized at the cytogenetic position 24.1 on the *q* arm of human chromosome 3 and consists of 28 exons [6]. In comparison to the *C. elegans* and *D. melanogaster* genes, the human SPCA1 gene presents the most elaborate exon/intron layout with only five junctions conserved between the human and worm genes and only two between the human and fly genes. The transcription-initiation site (+1) is located 259 nucleotides (nt) upstream of the translation-start (AUG) codon [117]. Functional promoter assays have been conducted upon transient transfection of several chimeric reporter gene constructs in cells belonging to the human keratinocyte HaCaT cell line. Two critically important segments were required to initiate the transcription of *ATP2C1* in HaCaT cells, namely the −347 to −311 and +22 to +76 DNA stretches located in a rather GC-rich region [117]. In silico sequence analysis of the 5′-flanking region revealed the absence of a classical TATA box and the presence of highly conserved *cis*-acting-binding motifs for several transcription factors such as the Sp1, YY1, NF-κB, and those belonging to the Ets family. YY1 and Sp1 factors were both documented to bind to the region from +21 to +57, and the *ATP2C1* promoter is transcriptionally activated upon their overexpression in normal human keratinocytes [117]. The modulatory effect of Sp1 on the expression of *ATP2C1* is not surprising, because both genes are expressed in all cell types, in agreement with their "housekeeping" cellular functions. Furthermore, when stimulated to differentiate in the presence of excess Ca^{2+} (1.2 mM), normal keratinocytes but not keratinocytes derived from HHD patients respond with an increased accumulation of both nuclear Sp1 protein and Sp1-induced *ATP2C1* mRNA levels [117].

Concerning the 3′-end of the gene, it is now documented that the optional exons 27 (containing the two 5′-internal splice donor sites D_1 and D_2 and further downstream the translation-stop codon for *ATP2C1a*) and 28 (containing one translation-stop codon for both *ATP2C1b* and *ATP2C1d* splice variants and a more upstream one for *ATP2C1c*) can be alternatively spliced to generate four transcripts, from *ATP2C1a* to *ATP2C1d* [8]. In turn, each *ATP2C1* transcript gives rise to a distinct SPCA1 isoform [8]. A similar alternative splicing mechanism acting at the 3′-end of the gene's pre-mRNA and involving the use of an internal 5′-end splice donor site was documented for the production of SERCA3 and PMCA1 isoforms that are, respectively, encoded by *ATP2A3* [118–120] and *ATP2B1* (reviewed in [121,122]). For *ATP2C1*, the transcript *ATP2C1a* contains the entire optional exon 27 and its stop codon, whereas transcripts *ATP2C1b* and *ATP2C1d* contain exon 28 and, respectively, the partial portions of exon 27 up to D_1 and D_2 splice donor sites.

Total exclusion of exon 27 gives rise to the splice variant *ATP2C1c*. The resulting SPCA1 isoforms (SPCA1a, 919 aa; SPCA1b, 939 aa; SPCA1c, 888 aa; and SPCA1d, 949 aa) are characterized by the incorporation of distinct C-termini (the C-terminal peptides differ in length and specific amino acid sequence). The significance of having such isoform diversity is not yet clear, although some functional differences may exist between the SPCA1 isoforms (see Section 4.2.).

The second human SPCA gene (*ATP2C2*) is found at the cytogenetic position 22.1 on the *q* arm of chromosome 16. Spanning over a genomic region of 95.5 kb, the SPCA2 gene consists of 27 exons [10]. *ATP2C1* and *ATP2C2* display similar exon/intron layouts with conserved junctions at the interface between the constitutively spliced exons. Furthermore, the intron/exon organizations of *ATP2C1* and *ATP2C2* bear no resemblance to those of SERCAs or PMCAs. In contrast to *ATP2C1*, there is no evidence of alternative processing for the *ATP2C2* primary transcript, and therefore, only one SPCA2 isoform of 946 aa is generated [9,10]. So far, no attempts to functionally characterize the transcriptional activation and regulation of the *ATP2C2* promoter and thus to explain the restricted tissue and cellular expression pattern of *ATP2C2* have been reported.

In pairwise comparisons, human SPCA1 exhibits 49, 59, and 64% overall amino acid sequence identity with *S. cerevisiae* SPCA, *C. elegans* SPCA, and human SPCA2, respectively. Owing to lack of any structural data and/or extensive mutational studies, most of the current knowledge about the putative three-dimensional organization of SPCAs comes as insights from the structure-function investigations of SERCA1a mutants [2,123–125] and from the recently elucidated high-resolution structures generated by X-ray crystallography of SERCA1a crystals in the Ca^{2+}-free inhibitor-bound E_2 and Ca^{2+}-bound E_1 forms [102–105]. In addition, several of the reported atomic structures have been elucidated either in the presence or absence of a bound nucleotide or a non-hydrolyzable ATP analog [102–105]. The different crystals corresponding to specific conformational states have been frozen using specific compounds with stabilizing effects such as inhibitors for pump's activity, nucleotide analogs, or calcium ions. The specific SERCA inhibitor employed to stabilize the E_2 form crystals was TG, either alone (in most of the structural studies) or together with 2,5-di-*tert*-butyl-1,4-dihydroxybenzene (*t*-BHQ) [104]. Interestingly, neither TG nor *t*-BHQ is able to inhibit SPCAs, and to date, there is no specific inhibitor known to totally suppress the activity of SPCAs.

Comparative analysis of the deduced amino acid sequences of SPCAs and SERCAs revealed that the following key architectural elements are conserved between the two groups of ion-motive ATPases: the 10 hydrophobic helices (M1–M10) forming the membrane-spanning domain as well as the nucleotide-binding (N), actuator (A), and phosphorylation (P) domains making up the large cytosolic head of the molecule (Fig. 5). In the Ca^{2+}-bound E_1 form as originally described by Toyoshima et al. [126], the cytosolic N- and A-domains are found largely separated from the P-domain, and therefore, large intramolecular movements of the cytosolic and membrane domains were expected to take place in order to assume the very compact Ca^{2+}-free E_2 form [127] during the transport cycle. Recently, both the well-known modulatory and catalytic effects of ATP on the SERCA1a pump were investigated at

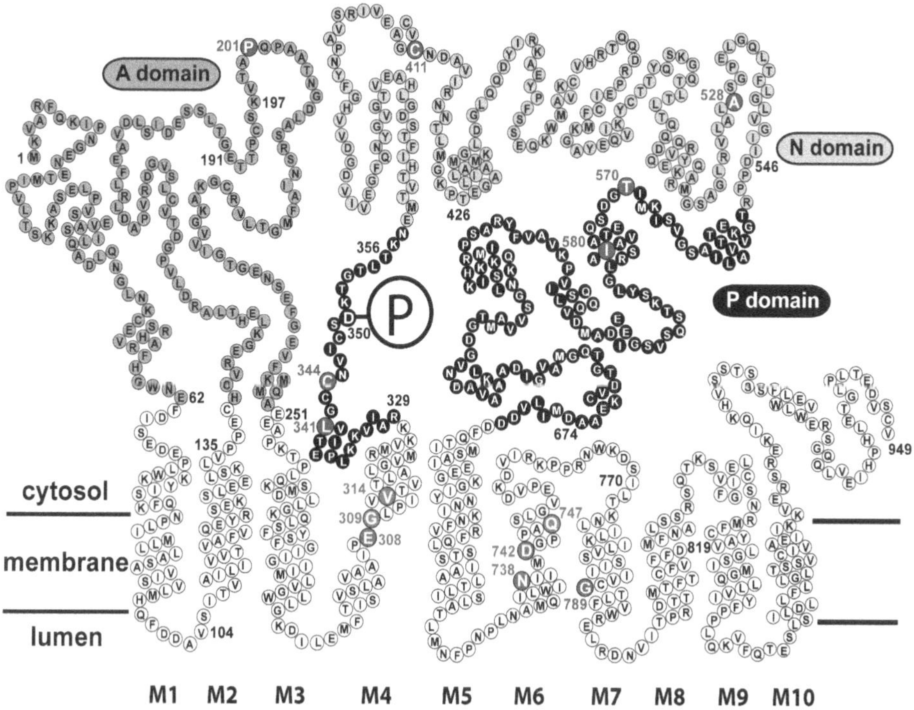

Fig. 5. Schematic presentation of a planar model showing the primary and putative secondary structure of the human secretory pathway Ca^{2+}-transport ATPase 1d (SPCA1d) isoform highlighting various structural determinants and some of the Hailey–Hailey disease (HHD) mutants studied. The model is based on the crystal structure of sarco(endo)plasmic reticulum Ca^{2+}-transport ATPase 1a (SERCA1a) in the Ca^{2+}-bound E_1 form (reviewed in [102]) and adapted from the SERCA2b model described by Dode et al. [110]. Each amino acid residue is indicated by its single-letter code inside its corresponding circle. Stacked diagonal rows of three or four residues indicate α-helices, whereas the β-strands are shown as ladder-type residue arrangements. The loops connecting different secondary structures are shown as linear sections. The membrane portion of SPCA1d contains ten putative membrane-spanning helices, M1–M10. Critical amino acid residues involved in binding either Ca^{2+} or Mn^{2+} are shown in M4 and M6 as blue circles with white letters, whereas those involved in conferring SPCAs their well-known selectivity to Mn^{2+} binding and transport are depicted as green circles containing white letters. The phosphorylation site represented by Asp^{350} (with the phosphate group also indicated) is highlighted by inverted color. Some of the amino acids substituted in HHD patients analyzed in previous studies [8,150,151] are highlighted as red circles with white letters (Pro^{201} → Leu, Leu^{341} → Pro, Cys^{344} → Tyr, Cys^{411} → Arg, Ala^{528} → Pro, Ile^{580} → Val, Thr^{570} → Ile, and Gly^{789} → Arg). Other HHD mutations include Gly^{309} → Cys and Asp^{742} → Tyr [8], where the Gly and Asp residues are highlighted as green and blue circles, respectively, because they are structurally involved in ion recognition and binding processes. A-domain, actuator domain (azure circles); P-domain, phosphorylation domain (inverted color circles); and N-domain, nucleotide-binding domain (orange circles) (See Color Plate 30, p. 528).

the structural level [105]. In the presence of a non-hydrolyzable ATP analog, the structure of the TG-stabilized E_2-ATP state is rather compact, and it represents an initial stage of the E_2 deprotonation modulated by ATP prior to the binding of Ca^{2+} at the cytoplasmic side of SERCA. Deprotonation is required for SERCAs prior to the initiation of a new catalytic cycle. Jensen et al. [105] proposed an alternative mechanism (indicated by the alternative route encompassing the reactions 1′–3′ in Fig. 4) involving small conformational changes. Accordingly, Ca^{2+} activation of the phosphorylation reaction can proceed directly from the compact Ca^{2+}-free E_2-ATP state to another relatively compact Ca^{2+}-bound E_1-ATP form. As outlined below, SPCAs are defective in proton binding and countertransport [11,12]. Therefore, ATP is not required to modulate the deprotonation event in SPCAs. Also, in light of the structural findings characterizing the E_2-ATP state of SERCA1a, the modulatory effect of ATP on SPCAs may be disputed, because the recently identified SERCA1a-specific amino acid residue (Glu^{439}), involved in the modulatory ATP binding, is replaced for instance in human SPCA1 and SPCA2 isozymes by Lys^{324} and Gln^{455}, respectively. However, it cannot be ruled out that ATP may still exert a modulatory effect on SPCAs before cytoplasmic Ca^{2+} is bound into the pump's membrane domain or at other stages of the catalytic cycle shown in Fig. 1. Of course, ATP is a key catalytic substrate for the formation of the phosphorylated intermediate $E_1 \sim P$ with occluded Ca^{2+} in both SPCAs and SERCAs [11,12], in agreement with the large conservation of N-domain in both types of ion-transport ATPases.

SERCAs and SPCAs share the same phosphorylation signature motif in the P-domain, namely the S**D**KTGTLT sequence (Fig. 5), where the phosphorylation residue shown in bold is as follows: (i) Asp^{351} in SERCAs; (ii) Asp^{371}, Asp^{336}, and Asp^{333} in SPCAs from *S. cerevisiae*, *C. elegans*, and *D. melanogaster*, respectively; and (iii) Asp^{350} and Asp^{379} in human SPCA1 and SPCA2, respectively. Site-directed mutagenesis of the phosphorylation site ($Asp^{371} \rightarrow$ Glu and $Asp^{371} \rightarrow$ Asn) has led to yeast mutant enzymes completely lacking the Ca^{2+}-transporting activity despite their good expression levels and proper targeting to the Golgi compartments [5]. Additionally, the loop (D^{601}PPR) connecting the N- and P-domains in SERCA1a is entirely conserved as D^{546}PPR in SPCA1 isoforms and as D^{576}PPR in SPCA2. Other highly conserved residues critical for the pump function are those of a 4-aa long loop in the A-domain, which must move upward to contact the phosphorylation site during the E_2-P dephosphorylation step, that is, the T^{181}GES loop in SERCA1a and the corresponding T^{189}GET and T^{220}GEA loops in human SPCA1 and SPCA2, respectively.

It is obvious from the amino acid sequence alignment of human SPCA1 and rabbit SERCA1a and from the comparison of the superimposed structures of rabbit SERCA1 in the E_1 conformation (PDB identification codes 1SU4 and 2C9M, which replace the old 1EUL code) and that of human SPCA1 isoforms (the latter being generated by SWISS-MODEL [128] based on the structural coordinates of the SERCA1a crystal structure) that several other SERCA-specific sequence elements are distinct in SPCAs. In general, SPCA1 is a shorter and more compact molecule than SERCA1a. For instance, SPCA1 proteins display shorter sequence segments linking the membrane-spanning helices M1 and M2 (loop 1–2), M3 and M4 (loop 3–4), and M7 and M8 (loop 7–8). Several loops present in the N-domain of SERCA1a are absent in SPCA1. On the other hand, the N-terminus (belonging to A-domain) is longer in SPCA1 isoforms relative to SERCA1. This N-terminal part contains an interesting structural element (Q^{39}ADLQNGLNKCE) and an unpaired

EF-hand-like motif that is a helix-loop-helix structure able to bind Ca^{2+} and thus modulate the ion transport. This EF-hand-like motif is different from the actual binding and transporting amino acid residues in the M-domain. It has been suggested recently that this motif appears to modulate indeed the ion transport in the yeast *S. cerevisiae* [129]. Unpublished results from our group indicate that the EF-hand-like motif derived from the human SPCA1d, but not the less-conserved one from the human SPCA2 (C^{70}VDLHTGLSEFS, where the conserved residues are underlined), is also able, at least in vitro, to bind Ca^{2+} as suggested by Ca^{2+}-overlay experiments.

An important architectural component of ion-motive ATPases is the membrane-spanning section of the pump involved in the ion entry and translocation pathway. The transmembrane helices incorporate the residues involved in the selectivity, coordination, and transport of Ca^{2+} and Mn^{2+}. For SERCA1, structural and extensive mutational works have unambiguously established that the Ca^{2+}-liganding side chains in the high-affinity Ca^{2+}-binding/occlusion sites are supplied by the following residues: Glu^{309} in M4; Asn^{768} and Glu^{771} in M5; Asn^{796}, Thr^{799}, and Asn^{800} in M6; and Glu^{908} in M8 [123,124,126]. These residues are organized in two sites: site I (Asn^{768}, Glu^{771}, Thr^{799}, and Glu^{908}) and site II (Glu^{309}, Asn^{796}, and Asp^{800}), each site being responsible for the binding of one Ca^{2+} [123,124,126]. Sequence comparison between SERCA1 and any SPCA revealed that most likely only the site corresponding to site II in SERCA1—Glu^{329}, Asn^{774}, and Asp^{778} in yeast SPCA; Glu^{308}, Asn^{738}, and Asp^{742} in human SPCA1 isoforms; and, Glu^{337}, Asn^{768}, and Asp^{772} in the human SPCA2 pump—is conserved. Functional evidence gathered from the analysis of mutant SPCA pumps confirmed the importance of these residues conserved at site II for the transport of Ca^{2+} and Mn^{2+} in yeast [130]. Similar analyses are yet to be performed for the human SPCAs. Interestingly, the conserved use of site II for ion binding and translocation is a property defining both SPCAs and PMCAs. The existence of one-ion-binding site may also indicate that only one ion (Ca^{2+} or Mn^{2+}) is transported by SPCAs per each hydrolyzed ATP.

It has been demonstrated that the yeast SPCA owes its selectivity for Mn^{2+} binding and transport to the critical conformation-sensitive packing interactions at the cytoplasmic interface between the side chains of Gln^{783} in M6 and Val^{335} in M4 [131,132]. These residues are conserved in the human SPCA1 isoforms as Gln^{747} and Val^{314} and in human SPCA2 as Gln^{777} and Val^{343}. In addition, our recent functional work on several human SPCA1 mutants has also suggested that Gly^{309} in M4 (corresponding to Gly^{330} in yeast SPCA and Gly^{338} in human SPCA2) may play an important part in conferring human SPCA1 its selectivity toward Mn^{2+} binding at the cytoplasmic side [8].

The N-terminus of human SPCA2 is 32 aa longer than that of human SPCA1 pumps [9], and as part of the A-domain, it may be involved in possible interactions and domain rearrangements modulating ion affinity as recently proposed for the copper-ATPases defective in Wilson and Menkes diseases [133]. The C-terminus of SPCA2 also seems more divergent relative to SPCA1a. Xiang et al. [9] pointed out the existence of two motifs in this region: (i) the dileucine (L^{925}L) motif, a putative retrieval signal from the plasma membrane, and (ii) the extreme C-terminal P^{943}EDV stretch, a putative type III PDZ-binding motif [134,135]. It will be of great interest to determine whether or not these two C-terminal motifs can be potentially involved in localization and trafficking of SPCA2 between the *trans*-Golgi network (TGN) stacks, TGN-derived vesicles, and even plasma membrane. For other ion-motive ATPases containing similar motifs, such as the Menkes- and Wilson-type copper-ATPases, the ion-dependent Golgi to vesicular trafficking is an important event [136].

4.2. Transport properties

Initially, the SPCA protein was the subject of several functional studies in yeast, which have all established the importance of SPCA in maintaining the cellular homeostasis of Ca^{2+} and Mn^{2+} in *S. cerevisiae* [5,129–132]. It is noteworthy to mention that yeast cells lack the intracellular Ca^{2+} pumps corresponding to the mammalian SERCAs, thus the removal of cytosolic Ca^{2+} and Mn^{2+} relies heavily on the ion-binding and transport properties of SPCA [137]. $^{45}Ca^{2+}$ transport [129] and Ca^{2+}-activated ATPase [131] assays made it clear that in *S. cerevisiae*, the Golgi-localized SPCA displays an unusually high apparent affinity (decreased $K_{0.5}$) for Ca^{2+} ($K_{0.5} = 0.070\,\mu M$) relative to, for instance, the mammalian SERCA2b isoform ($K_{0.5} = 0.11-0.13\,\mu M$), the Ca^{2+} transporter in the ER with the highest known apparent affinity for Ca^{2+} [110]. Other studies have reported that the human SPCA1a [138], *C. elegans* SPCA [13], human SPCA1d [8], and human SPCA2 [10] are able to selectively transport Ca^{2+} with rather high affinities ($K_{0.5} = 0.20-0.26\,\mu M$) but still several-fold lower than that of yeast SPCA. One report also claimed that human SPCA2 could actually display a much lower apparent affinity for Ca^{2+} ($K_{0.5} = 1.35\,\mu M$) relative to human SPCA1 and yeast SPCA [9]. As discussed on page 251, the apparent affinities for Ca^{2+} of human SPCA pumps have been recently comprehensively reevaluated [11,12], and in view of these new data, it appears that the methods previously employed to assess the Ca^{2+} uptake on pooled Golgi fractions isolated from yeast [5] and Ca^{2+} fluxes on permeabilized COS-1 cells [13] have been most likely underestimating the affinity for Ca^{2+}. Mn^{2+} is also transported with high affinity by yeast [131], worm [13], and human SPCAs [8–10]. Mn^{2+} and Ca^{2+} are competitively inhibiting each other's transport. As mentioned in the previous section, the amino acid residues involved in the binding of Ca^{2+} and Mn^{2+}, in determining the Mn^{2+} selectivity, and in the active phosphorylation by ATP have been pinpointed mainly for the yeast SPCA by functional site-directed mutagenesis [130–132].

Despite these efforts, the characteristics of the overall and partial reactions defining the catalytic cycle (Fig. 1) have not been investigated in any detail for SPCAs until very recently, when Dode et al. [11,12] have documented for the first time the kinetic differences (i) between SERCA1a and the human SPCA1 isoforms; (ii) between the human SPCA1 isoforms; and (iii) between the human SPCA1 and SPCA2 isozymes. In addition, the ability of SPCA1c to function as a pump was addressed as well [11]. The functionality of SPCA1c isoform was questioned because, firstly, the alternative splicing of the *ATP2C1* primary transcript would generate a pump with a truncated transmembrane segment M10 [8], and, secondly, SPCA1c cannot functionally compensate for the absence of SPCA1a, SPCA1b, and SPCA1d in a HHD patient carrying an *ATP2C1* mutation located in the splice acceptor site of intron 26 [139]. The resulting effect of mutation would be an impaired splicing of exon 27, that is, the exon coding for the rest of M10. The hypothesis that indeed the truncation of M10 could lead to an improperly folded protein likely sensitive to enhanced cellular degradation has been confirmed experimentally upon heterologous expression of human SPCA1c in HEK-293 cells: SPCA1c was expressed at very low levels and could not transport Ca^{2+} [11]. The fact that SPCA1c may still be able to bind Ca^{2+} cannot be excluded, but the ion transport across the membrane is definitely compromised because, as recently shown [11], SPCA1c cannot be phosphorylated, and therefore, it cannot couple the energy from the energy-donating ATP molecule to the vectorial ion transport.

SPCAs and SERCAs display distinct kinetic parameters that in turn are responsible for the dramatic changes in (i) the concentration dependences for various substrates (Ca^{2+}, ATP, and inorganic phosphate) and inhibitors (vanadate or TG); (ii) the maximum turnover rate; and (iii) the sensitivity to modulation by protons and potassium ions of the E_2-P dephosphorylation [11,12].

By employing more sensitive methods (e.g., those based on the enzyme's ability to phosphorylate in the presence of radioactive [γ-^{32}P]ATP), it has been shown [11] that the functionally active human SPCA1a, SPCA1b, and SPCA1d isoforms display essentially extremely high apparent affinities for Ca^{2+} activation at the cytosolic site ($K_{0.5} = 0.009-0.010\,\mu M$) relative to SERCA1a ($K_{0.5} = 0.284\,\mu M$) or yeast SPCA ($K_{0.5} = 0.070\,\mu M$). Additionally, an increased apparent affinity for cytosolic Ca^{2+} ($K_{0.5} = 0.025\,\mu M$) in activation of the phosphorylation activity (forward reaction 3 in Fig. 4) has also been determined for the human SPCA2 pump relative to SERCA1a, but the apparent affinity for Ca^{2+} displayed by SPCA2 is still 2.5-fold lower than that of the human SPCA1d isoform [12]. Furthermore, the Hill coefficient values deduced from the Ca^{2+} titration of phosphorylation also support the presence of a single-ion-binding site per SPCA molecule displaying a high affinity toward the transported ion at the cytoplasmic side. Steady-state and transient kinetic analyses have demonstrated that the high apparent affinity for Ca^{2+} displayed by SPCA1 isoforms is not determined by a true increase of affinity relative to SERCA1a, because the rates in the Ca^{2+} binding transition (forward reactions 1 and 2) and Ca^{2+} dissociation (reverse reaction 2) of the SPCA1 variants were not significantly different from those of SERCA1a. It has been established that, in comparison to SERCA1a, SPCA1 isoforms owe their increased apparent affinity for Ca^{2+} to a "kinetic effect" represented by the slower rate of $E_1 \sim P(Ca^{2+})$ to E_2-P transition (forward reaction 4), which for SERCAs constitutes the most critical rate-limiting step of the cycle in physiological conditions (substrate concentrations and pH). On the other hand, relative to SPCA1d, the 2.5-fold lower apparent affinity for Ca^{2+} displayed by SPCA2 can be explained by an enhancement of the rate of $E_1 \sim P(Ca^{2+})$ to E_2-P conformational step and also by a significant (7-fold) decrease in the rate of the Ca^{2+}-binding transition. Nevertheless, the structural elements responsible for the observed changes in the rates of the abovementioned partial reactions are yet to be identified. The

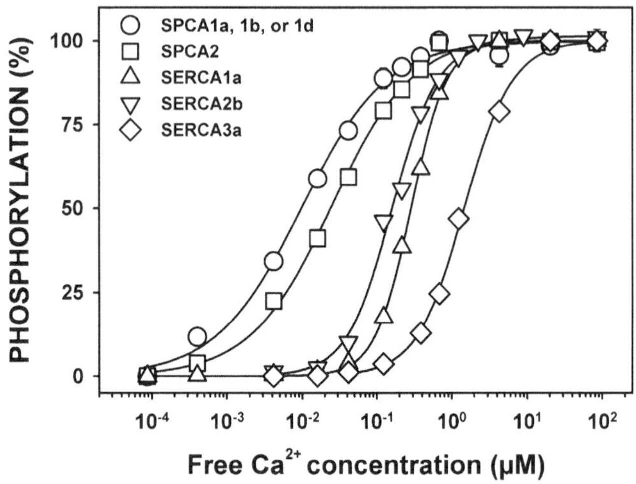

Fig. 6. Ca^{2+}-dependence of phosphorylation by [γ-^{32}P]ATP. The indicated Ca^{2+}-ATPases were heterologously expressed in HEK-293 cells. The phosphorylation activity was assayed on microsomes isolated from the transfected HEK-293 cells. Phosphorylation of the indicated proteins was carried out at 25°C for 5 s in a medium containing 40 mM of MOPS (3-(N-morpholino) propanesulfonic acid)/Tris, pH 7.0, 80 mM of KCl, 5 mM of $MgCl_2$, 1 mM of EGTA, 5 μM of [γ-^{32}P]ATP, and various concentrations of $CaCl_2$ to set the free Ca^{2+} concentration as indicated, followed by acid quenching. Phosphorylation of sarco(endo)plasmic reticulum Ca^{2+}-transport ATPase 1a (SERCA1a), SERCA2b, or SERCA3a was performed in the absence of thapsigargin, whereas for microsomes containing secretory pathway Ca^{2+}-transport ATPases (SPCAs), the phosphorylation medium contained 0.1 μM of thapsigargin needed to block the endogenous SERCA2b activity. The data have been normalized separately by taking as 100% the maximum phosphorylation level reached, and the lines show the best fits of the Hill equation. The respective $K_{0.5}$ values and Hill coefficients are as follows: SPCA1a, SPCA1b, or SPCA1d, 0.010 μM and 0.80; SPCA2, 0.025 μM and 0.80; SERCA1a, 0.280 μM and 1.81; SERCA2b, 0.160 μM and 1.45; and SERCA3a, 1.47 μM and 1.36. This figure was compiled from data previously published (for SPCAs and the other SERCAs) in Refs [11,12,107,110,111] or unpublished (in the case of SERCA2b).

differences in the apparent affinity for Ca^{2+} between the various SERCAs and SPCAs are illustrated in Fig. 6.

The fact that the $E_1 \sim P(Ca^{2+})$ to E_2-P conformational transition represents indeed an important rate-limiting step can be experimentally documented in SPCAs by the use of the calcium ionophore A23187, which relieves the back inhibition on this transition exerted by Ca^{2+} accumulated inside the Golgi lumen [11,12]. But even in these favorable conditions, the maximal turnover rates of the ATPase activity for SPCA1a (20.27/s), SPCA1b (22.50/s), SPCA1d (27.44/s), and SPCA2 (40.7/s) isozymes were, respectively, 4.7-, 5.8-, 6.4- and 3.2-fold lower than that of SERCA1a (130/s). It should be noted that SPCA2 is characterized by an almost 2-fold enhanced catalytic turnover rate relative to the human SPCA1 isoforms [12]. Again, the lower overall rates of ATPase activity for SPCA1 pumps relative to SERCA1a are caused by the slower transition of $E_1 \sim P(Ca^{2+})$ to E_2-P. Interestingly, both the rate of ATP hydrolysis and the rate of $E_1 \sim P(Ca^{2+})$ to E_2-P have been found to increase with the length of the alternatively spliced C terminus, that is in the order SPCA1a, SPCA1b, and SPCA1d [11].

This represents the first evidence for functional differences between the human SPCA1 isoforms, but the physiological significance of this finding and, implicitly, of the SPCA1 isoform diversity remains to be further elucidated. For SPCA2, the faster ATP hydrolysis rate is due to a approximately 3-fold enhancement of the $E_1 \sim P(Ca^{2+})$ to E_2-P transition rate relative to SPCA1d.

In comparison to SERCA1a, SPCA1 isoforms are characterized by slower rates of phosphorylation (forward reaction 3), but the analysis of the ATP concentration dependence of the initial phosphorylation rate has demonstrated that the reduced phosphorylation rate is a V_{max} effect rather than a consequence of reduced affinity for the catalytic substrate [11]. The catalytic properties of the partial reaction step 4 are decreased in both directions, because also the phosphoenzyme processing in the presence of ADP (reverse reaction 3) was slower in the SPCA1 isoforms relative to SERCA1a. For SPCA2, both rates are increased, thus showing improved catalytic abilities in both directions of the reaction 3 [12]. The increased rate of phosphorylation in SPCA2 represents a V_{max} effect rather than an intrinsic increased affinity for ATP [12].

When present in the E_2 conformation, the SPCA1 pumps display other phenotypic properties that are distinct from those of SERCA1a, such as the reduced apparent affinities of E_2 to inorganic phosphate and vanadate (a phosphate analog) and reduced sensitivity (K_d increased) toward TG inhibition (e.g., SPCA1d has a K_d value of 28 μM versus 0.031 nM for SERCA1a) [12]. The reduced apparent affinity for inorganic phosphate can be unambiguously explained by the markedly enhanced rate of the E_2-P dephosphorylation in SPCA1 isoforms as recently described [11]. The structural elements contributing to the destabilization of E_2-P may also be responsible for the destabilization of vanadate-bound E_2 form. In SERCA1a, the amino acid residue Phe^{256} is essential in binding TG, the SERCA-specific inhibitor. In SPCAs, this residue is not conserved, thus explaining most likely the reduced inhibitory effect of TG on SPCA1 isoforms. For SPCA2, the even more reduced apparent affinities for inorganic phosphate and vanadate are explained by the faster rate of E_2-P dephosphorylation relative to SPCA1d [12]. The combination of this kinetic effect with the decreased rate of E_2 to E_1 observed for SPCA2 relative to SPCA1d also explains the surprising increased TG sensitivity of SPCA2 (K_d = 2 μM), because the E_2 form that binds TG will accumulate to a higher extent in SPCA2 than in SPCA1d [12].

The E_2-P dephosphorylation step is constitutively highly activated in SPCAs [11,12]. Moreover, in contrast to SERCA1a, this step in SPCAs does not require either proton- or K^+-induced stimulation [11,12]. Our results promote the concept that SPCAs do not countertransport luminal protons of the acidic Golgi compartments in exchange for Ca^{2+} or Mn^{2+}. The role of luminal protons in stabilizing the empty Ca^{2+} pocket immediately after calcium ions were delivered into the ER/SR lumen by SERCAs has been structurally documented [104]. SPCAs have less potentially charged residues capable for H^+ binding in their Ca^{2+}-binding pocket (e.g., Glu^{308} and Asp^{819} in SPCA1, and Glu^{337}, and Asp^{850} in SPCA2) than SERCA1a (Glu^{58}, Glu^{309}, Glu^{771}, and Glu^{908}). From the cellular point of view, protons are needed in the lumen of Golgi apparatus to support several vital functions, and therefore, they are actively accumulated by the V-type pumps and kept inside by the tight Golgi membranes. Therefore, it appears that SPCAs are particularly suited to accumulate Ca^{2+} and Mn^{2+} into the Golgi lumen and adapted to prevent H^+ from leaving the Golgi compartment through the pump itself.

Both SPCA1 and SPCA2 share the insensitivity to K^+ modulation [11,12]. In SERCA1, the negatively charged residue Glu^{732} is essential in conferring SERCA1 its sensitivity to K^+ stimulation [140]. This residue is replaced by Asp^{674} in SPCA1 and by Asn^{704} in SPCA2. It appears that the lack of a K^+-induced effect on SPCAs is mainly due to the slight shortening of the side chain and not to the absence of a negative charge.

The insensitivity to regulation by K^+ and pH of the constitutively fast E_2-P dephosphorylation of SPCAs represents undoubtedly a novel and unique SPCA-specific feature that is not shared by SERCAs.

It is known from SERCA1 that the conformational changes in A-domain and its intimate associations with P- and N-domains play a crucial role in controlling the transmembrane gating during ion translocation and E_2-P dephosphorylation events. Interestingly, the kinetic properties of the two aforementioned steps appear to be distinct between SPCAs and SERCA1a enzymes. Different mechanistic influences of A-domain on the luminal gating may account for the observed distinct kinetics as proposed below. The increased rate limitation imposed on the Ca^{2+} translocation step ($E_1 \sim P(Ca^{2+}) \rightarrow E_2$-P) and enhancement of the ensuing hydrolysis (E_2-P $\rightarrow E_2$) reported [11,12] for SPCA1 enzymes (and explaining the phenotypic characteristics displayed by SPCA1 enzymes) are consistent with a Ca^{2+}-ATPase characterized by (i) a delayed opening of the luminal gate (essentially affecting Ca^{2+} translocation and release into the Golgi lumen) and (ii) an accelerated closure of the luminal gate (caused by the faster release of inorganic phosphate). Involvement of A-domain in mediating these particular changes in the luminal gating state is supported by numerous sequence differences between, for instance, SPCA1 and SERCA1 enzymes in A-domain and its linkers to transmembrane segments M1, M2, and M3. In SERCA1, the rotation of A-domain and the movements of M1 during the transition from $E_1 \sim P(Ca_2)$ to E_2-P rely on the critical loop E^{40}LPAEEGKS (linking M1 with A-domain) [141], which, in SPCA1 enzymes, corresponds to the less conserved $\underline{E}^{62}$FDIS$\underline{E}$DEP element (conserved residues are underlined here and below). Although in SERCA1a, the M1 and M2 move together as a rigid (V-shaped) body as a consequence of the A-domain rotation, this movement is predicted to be less flexible in SPCA1 proteins because of a shorter luminal loop between M1 and M2. Further hindrance to the rotation of A-domain could be brought about by the shorter luminal loop linking M3 with M4. It was shown in SERCA1 that Tyr^{122} and Glu^{123} found at the top part of M2 (which is connected to A-domain) form hydrogen bonding and salt interactions with Arg^{324} and Arg^{334} of the top part of M4 protruding in the P-domain [126]. These interactions appear to be important for the E_2 to $E_1(Ca^{2+})_2$, $E_1 \sim P(Ca^{2+})_2$ to E_2-P, and E_2-P to E_2 transitions [142]. In addition, Arg^{324} forms hydrogen bonds with Asn^{101} and Gln^{108} on M2. Although the aforementioned M4 residues from SERCA1 are conserved in SPCA1 enzymes as Arg^{323} and Lys^{333}, their interaction with the residues from M2 seems to be compromised, because Leu^{134} and Val^{135} substitute Tyr^{122} and Glu^{123} in SPCA1 enzymes, respectively, and finally, the amino acid residue corresponding to Asn[101] in SERCA1 is not found in the SPCA1 enzymes. The importance of the linker segment connecting A-domain with M3 ("A-M3 linker") for the $E_1 \sim P(Ca^{2+})_2$ to E_2-P transition and for E_2-P hydrolysis was demonstrated by mutagenesis studies of SERCA1 [143,144]. It is noteworthy that the primary structure of SPCA1 isoforms differs considerably from that of SERCA1 in the A-M3 linker region. In SERCA1, the importance of the intactness of the sequence (T[242]EQD) connecting A-domain with M3 for the $E_1 \sim P(Ca^{2+})_2$ to E_2-P transition was recently demonstrated by proteolysis analysis [145], and interestingly, this sequence is replaced in SPCA1 isoforms by the poorly conserved E[251]EAP stretch, which might change the strain exerted on the top part of M3. The outermost loop (K[189]HTEPVPDPRAVNQDKK) of A-domain in SERCA1 (known to be making intimate contact with P-domain and to be involved in the processing and hydrolysis of E_2-P) is replaced by K[197]VTAPQPAATNGD**L**ASRS in SPCA1, where **L** represents an amino acid insertion. In addition, the A-domain is enlarged by an extension of 22 amino acid residues found at the extreme N-terminus of the SPCA1 pumps. In relation to the insensitivity of E_2-P dephosphorylation to K^+ in SPCA1 isoforms, it is noteworthy that in SERCA1a, the modulatory effect of K^+ is likely mediated through a long-range interaction with the catalytic site that involves the A-M3 linker and the residue Q[244], which as mentioned above is replaced by alanine in SPCA1 isoforms [140,146]. The functional effect of all the abovementioned amino acid sequence differences between SPCA1 and SERCA1 on the activity of SPCA1 needs to be further explored by functional site-directed mutagenesis.

The specific catalytic properties of the SPCA1 pumps (low turnover rate and high apparent affinity for Ca^{2+}) could in principle represent necessary adaptations to function in the Golgi compartments, where a higher rate of ion transport is not required, because the depletion of Ca^{2+} occurs only during the slower process of downstream trafficking of Ca^{2+}-rich vesicles. On the other hand, a higher turnover rate may be required for SPCA2 in secretory cells, where the trafficking of Ca^{2+}-rich vesicles must be faster than in non-secretory cells. The relatively high apparent affinity for Ca^{2+} displayed by SPCAs ($K_{0.5}$ values below the resting $[Ca^{2+}]_c$) ensures that the refilling process of the Golgi with cytosolic Ca^{2+} is carried out continuously even in the absence of transient rises in $[Ca^{2+}]_c$.

5. *Role of SPCA in keratinocytes and in HHD*

5.1. *HHD*

HHD (OMIM 169600) is a blistering skin disorder, histologically characterized by the loss of adhesion between suprabasal keratinocytes (acantholysis) and abnormal keratinization of the epidermis. The onset of the disease is situated in the third or fourth decade of life and typically manifests as uncomfortable erosions or lesions mainly at sites of sweating or sites exposed to friction or UV.

HHD is an autosomal dominant disorder, caused by mutations inactivating one allele of *ATP2C1*, encoding SPCA1 [6,7]. HHD and its link to the SPCA1 pump have been the subject of several recent reviews [2,147–149]. At present, close to 100 mutations are known (for review and references see [149]). HHD shows clinical and histological similarities to Darier disease (OMIM 124200), another genodermatosis, which is caused by a mutation in the *ATP2A2* gene encoding the ER Ca^{2+} pump SERCA2. Ultrastructural analysis of acantholytic cells of HHD patients reveals perinuclear aggregates of keratin filaments that have retracted from desmosomes. In Darier disease, acantholysis is less pronounced, whereas dyskeratosis is more pronounced.

Several of the HHD SPCA1 mutations have been introduced into the human SPCA1d isoform and functionally investigated following expression in COS-1 cells [8,150]. Despite normal levels of mRNA and correct targeting to the Golgi compartments, more than half of the investigated mutants displayed low levels of protein expression (e.g., $Leu^{341} \rightarrow Pro$, $Cys^{344} \rightarrow Tyr$, $Cys^{411} \rightarrow Arg$, $Ala^{528} \rightarrow Pro$, $Thr^{570} \rightarrow Ile$, and $Gly^{789} \rightarrow Arg$), which is in good agreement with the markedly reduced SPCA1 expression observed in keratinocytes of HHD patients. Other mutants were characterized by lack of ion transport that could be pinpointed to defects in Ca^{2+} and Mn^{2+} binding ($Asp^{742} \rightarrow Tyr$ and $Gly^{309} \rightarrow Cys$) or to the inability of the phosphoenzyme intermediate to undergo the major $E_1 \sim P(Ca^{2+})$ to E_2-P conformational transition ($Ile^{580} \rightarrow Val$). Surprisingly, the $Pro^{201} \rightarrow Leu$ mutant displayed wild-type properties with respect to its intracellular targeting and catalytic abilities [8]. The latter mutant, when expressed in yeast [151], is able to restore the cellular phenotypes of a yeast strain lacking *PMR1*. Further detailed functional analysis is still needed for this mutant to explain the HHD defect, although it cannot be ruled out that this mutation is indeed a benign polymorphism and that the actual defect might be associated with the non-coding parts of *ATP2C1*.

In an unusual case, HHD patients showed more severely affected skin lesions in a mosaic form, presented as unilateral segmental skin regions along the lines of Blaschko. These patterned skin lesions correspond to clones of keratinocytes descending during embryogenesis from a common precursor that as a result of a somatic mutation had lost the only remaining wild-type *ATP2C1* allele. The exacerbated areas were therefore found superimposed on the ordinary symmetrical germ-line transmitted HHD phenotype. The first symptoms in these patches were noted already at the age of 3 months [152]. This study demonstrates that the complete absence of functional SPCA1, although aggravating the phenotype by interfering with keratinocyte differentiation, still allows keratinocytes to proliferate and to survive at least for some time.

The precise link between haploinsufficiency of functional SPCA1 and the occurrence of lesions that are limited to the skin has not yet been resolved nor is it known for haploinsufficiency of SERCA2 in case of the related Darier disease. Therefore, more information is needed on the cellular and molecular defects in the diseased keratinocytes and how these are affected by Ca^{2+} and by the disturbance of Ca^{2+} pump function. Because SPCA1 also transports Mn^{2+}, defects in Mn^{2+} homeostasis may contribute to the HHD symptoms [153]. However, the similarity between HHD and Darier disease, despite the poor Mn^{2+}-transporting activity of SERCA2 as compared to SPCA1 [154], argues against a major role of Mn^{2+} ions.

The limitation of our current knowledge on HHD is further illustrated by our inability to localize the major defect in ion homeostasis in the lumen of the secretory pathway, in the cytosol, or even in the extracellular environment. Before reviewing the experimental data, we will first outline the rationale for each of these three possibilities.

It is clear from the survey presented above that decreased SPCA activity results in lower $[Ca^{2+}]_l$ in the Golgi and possibly in post-Golgi secretory compartments and that this altered luminal environment may affect a multitude of enzymatic and sorting processes. Abnormal post-translational processing of various components, particularly of key molecules involved in cell-to-cell adhesion, could then predispose the affected cells to develop the clinical symptoms. In order to explain why the symptoms of HHD are almost completely restricted to skin, one needs to invoke additional factors, like the disturbed glycosylation state of components involved in keratinocyte-specific cell-to-cell contacts or modifications of specific keratins among others. Alternatively, non-cutaneous tissue could be protected by compensatory mechanisms that are lacking in keratinocytes.

In view of the relatively important contribution of the SPCA1 pump relative to the SERCA2 for the removal of Ca^{2+} from the cytosol of normal keratinocytes (see Section 3.2.2.), any condition leading to an increased Ca^{2+} influx into the cytosol could potentially overwhelm the Ca^{2+} accumulation system in face of a decreased number of SPCA1 pumps. It is, in this context, also important to consider that the mere loss of one functional *ATP2C1* allele on its own is not sufficient to cause the HHD symptoms. Indeed, the skin of HHD patients is typically symptom-free during the first decades of life, and even thereafter, the symptoms are only seen in certain areas of the skin subject to friction, local warming, or sweating. Furthermore, the symptoms can be elicited in a matter of minutes by locally rubbing the skin. Clearly, other external factors must be taken into account to explain the phenotype. A possible involvement of stretch- or heat-sensitive Ca^{2+} influx channels, possibly of the TRP (transient receptor potential) type, can be considered.

Third, also extracellular Ca^{2+} could be involved because of the well-known fact that extracellular Ca^{2+} concentration ($[Ca^{2+}]_e$) strongly influences keratinocyte differentiation, combined with the concept of the existence of a steep Ca^{2+} gradient in the epidermis. Evidence related to the possible involvement of SPCA1 in the pathways connected to differentiation and to the generation of the epidermal Ca^{2+} gradient will be discussed in Sections 5.2 and 5.3.

5.2. Role of Ca^{2+} in the epidermis and in keratinocyte differentiation

Each epidermal cell layer is defined by a specific state of differentiation of keratinocytes, characterized by the expression of distinct proteins or protein isoforms (for review see [155]). Undifferentiated keratinocytes originate in the basal layer. Daughter cells exiting the basal layer give rise to differentiated cells as they move upward, forming consecutively the spinous, granular, and cornified layers. The mechanical strength and the permeability barrier of the epidermis is mainly attained by extensive remodeling in the granular layer and in the transition zone to the dead epidermis, by the formation of multiprotein assemblies and associated lipids, held together by disulfide bridges and covalent cross-links formed by transglutaminases that are Ca^{2+}-dependent enzymes [156]. In the basal and spinous layers, strength is mainly provided by desmosomes and adherens junctions. Tight junctions are expressed mainly in the granular layer [157].

Primary keratinocytes cultured at a $[Ca^{2+}]_e$ below 0.1 mM proliferate rapidly and fail to develop intercellular contacts. The increase of $[Ca^{2+}]_e$ above 0.1 mM halts proliferation and suffices to trigger differentiation. A receptor sensing $[Ca^{2+}]_e$ controls the differentiation pathway (for review see [158]). Remarkably, this Ca^{2+} receptor would also be expressed in the Golgi. Indeed, according to one report, signaling complexes comprising IP_3 receptor, PLCγ1, and SPCA1 can be immunoprecipitated from *trans*-Golgi fractions of keratinocytes [159].

In normal epidermis, cell proliferation is confined to the basal layer where the local $[Ca^{2+}]_e$ is kept low [160]. Proton-induced X-ray emission (PIXE) [161] and Ca^{2+}-precipitation cytochemical analyses [160] showed that within normal epidermis, a steep gradient of $[Ca^{2+}]$ exists with a low concentration at the actively proliferating basal keratinocyte layer and a high one at the outer granular layer of the epidermis. This gradient not only regulates keratinocyte differentiation [160] but also is important for maintaining the normal epidermal permeability barrier [162]. The barrier can quickly be disrupted upon wounding. The rapid drop in local $[Ca^{2+}]_e$ would then stimulate the keratinocyte proliferation necessary for the cutaneous wound repair, and it also promotes the ensuing upward migration of the newly formed suprabasal keratinocytes. The latter effect relies on decreased binding of Ca^{2+}-dependent cell surface molecules involved in cell–cell and cell–matrix interactions belonging to the E-cadherin or integrin family [163].

It is not known by what mechanisms and structures the epidermal Ca^{2+} gradient is formed and maintained. It would appear that tight junctions are indispensable structures to prevent the collapse of ion gradients if these gradients occur in the extracellular space (see also Section 5.4.). It has been shown that tight junctions are important for epidermal barrier formation in mice [164], but to our knowledge, direct information on a possible link to the epidermal Ca^{2+} gradient is lacking.

5.3. Role of SPCA1 and the Golgi in keratinocyte Ca^{2+} homeostasis

As mentioned in Section 3.2.2., the relative contribution of SPCA1 and SERCA Ca^{2+} pumps to the Ca^{2+} uptake in the Golgi apparatus is cell-type dependent. Interestingly, among various cell types tested, reliance on SPCA1 was highest (roughly two-thirds of the Ca^{2+} uptake into the Golgi is resistant to TG) in human keratinocytes [88]. This observation is in line with the reported high expression level of SPCA1 mRNA in these cells [6]. However, like in all other cell types tested, the compartments that are loaded with Ca^{2+} by SPCAs were insensitive to agonist stimulation and hence are unlikely to take part in cytosolic Ca^{2+} signaling [87].

It should be mentioned that there is some controversy about the distribution of SPCA1 in human epidermis. Yoshida et al. [153] report a specific localization of SPCA1 in the basal layer of normal epidermis, with low expression levels in differentiated keratinocytes in suprabasal layers. This result appears to be at variance with the immunohistochemical data of Porgpermdee et al. [165] and also with our own observations (Fig. 2), showing a clear labeling of both basal and suprabasal cells.

HHD keratinocytes have been used to study the effect of reduced SPCA1 function on Ca^{2+} regulation, however, with conflicting results. Hu et al. [6] observed an elevated basal $[Ca^{2+}]_c$ in diseased keratinocytes, whereas basal levels remained unchanged in the experiments of Leinonen et al. [166]. Interestingly, the latter authors also report a smaller rise in $[Ca^{2+}]_c$ induced by TG in HHD keratinocytes. This observation suggests that the SERCA-dependent Ca^{2+} store is smaller in these cells. It would therefore be worthwhile to find out whether SERCA levels are decreased in HHD keratinocytes, which would help to explain the similarities between Darier and HHD disease. Inversely, also the expression of SPCA1 might be changed in Darier keratinocytes: Foggia et al. [167] reported that SPCA1 is up-regulated and may partially compensate for the lower SERCA activity. On the other hand, Porgpermdee et al. [165] observed a moderate decrease of SPCA1 levels only in the affected skin of Darier patients and not in the unaffected skin.

5.4. Does SPCA1 activity influence keratinocyte differentiation?

That SPCA1 could play a central role in the generation of the epidermal Ca^{2+} gradient is suggested by its absence in both affected and unaffected HHD epidermis [89]. The contribution of active Ca^{2+}-transport ATPases in the generation of this intraepidermal Ca^{2+} gradient has, however, been disputed by others who state that passive mechanisms alone can account for Ca^{2+} gradient formation [162]. Hence, the exact relationship and, a fortiori, the existence of a possible link between the pronounced effect of Ca^{2+} on keratinocyte proliferation, the extent of the intraepidermal Ca^{2+} gradient and the expression of SPCA1 remains as yet unclear. First, PIXE analysis does not allow to assess what fraction of the signal represents intracellular and what part extracellular Ca^{2+}, but Ca^{2+} precipitation studies demonstrated that the increase in Ca^{2+} in the outer epidermis is mainly extracellular [168]. Second, quantifications of the relative expression of SPCA1 from immunohistochemical analyses are difficult and necessarily very arbitrarily. Nevertheless, Porgpermdee et al. [165] reported a moderate reduction (about 50%) of SPCA1 in uninvolved HHD epidermis and a clear (70–80%) reduction in involved HHD epidermis versus normal epidermis. Third, it is unclear whether or not SPCA1 should be considered as an epidermal differentiation marker that is controlled by $[Ca^{2+}]_e$. According to one view, a rise in $[Ca^{2+}]_e$ would, at least in normal keratinocytes, result in a concomitant increase in $[Ca^{2+}]_c$, followed by the translocation of Sp1 from the cytosol to the nucleus where it *trans*-activates *cis*-acting enhancing elements in the *ATP2C1* promoter and thereby markedly increases SPCA1 mRNA levels. In HHD keratinocytes, the rise of $[Ca^{2+}]_c$ is suppressed, the translocation of Sp1 to the nucleus is blunted, and the transcriptional up-regulation of SPCA1 mRNA is impaired [117]. Also, Mayuzumi et al. [169] report an increased SPCA1 expression after a shift of the keratinocytes to high $[Ca^{2+}]_e$. But in other reports, either no change [153] or a decrease of SPCA1 [28] is seen when $[Ca^{2+}]_e$ is increased.

Yoshida et al. [153] correlate their observation of a preferential localization of SPCA1 in the undifferentiated cells of the basal layer (already discussed above) with the induction of differentiation markers upon suppression of SPCA1 expression. This effect could be mimicked by addition of the ionophore Br-A23187, which has a higher affinity for Mn^{2+} than for Ca^{2+}, but not with the classical Ca^{2+} ionophore A23187. As a possible link between Mn^{2+} levels and differentiation, the authors suggest the requirement of integrin-dependent binding processes for Mn^{2+} [170], combined with the significance of integrin-mediated signaling in the regulation of keratinocyte differentiation [171].

6. Conclusions

The recent studies of SPCAs have considerably advanced our understanding of their role in the Ca^{2+} and Mn^{2+} homeostasis in the Golgi and downstream secretory pathway. A selection of the most important contributions would include (i) the finding that SPCA pumps are able to transport Mn^{2+} along with Ca^{2+}; (ii) the discovery of the link between the *ATP2C1* gene encoding SPCA1 and HHD; (iii) the functional characterization of the SPCAs demonstrating their high affinity for both Mn^{2+} and Ca^{2+} relative to SERCAs; (iv) their insensitivity to luminal protons, thus rendering these transporters a better adaptation to the acidic milieu of the Golgi; and (v) the discovery of the second gene *ATP2C2* encoding SPCA2, a pump with higher turnover rate relative to SPCA1.

Inevitably, many new questions have surfaced. A major unsolved problem is why haploinsufficiency of *ATP2C1* in HHD only affects the skin. Another challenging question concerns the specific role of SPCA2 in cellular ion homeostasis in view of their restricted tissue distribution and the aforementioned catalytic properties. Despite these and other shortcomings in our understanding, SPCAs have recently been established as important players in the regulation of cellular ion homeostasis along with SERCAs and PMCAs.

References

1. Berridge, M.J., Lipp, P. and Bootman, M.D. (2000) *Nat. Rev. Mol. Cell Biol.* 1, 11–21.
2. Wuytack, F., Raeymaekers, L. and Missiaen, L. (2002) *Cell Calcium* 32, 279–305.
3. Rudolph, H.K., Antebi, A., Fink, G.R., Buckley, C.M., Dorman, T.E., LeVitre, J., Davidow, L.S., Mao, J.I. and Moir, D.T. (1989) *Cell* 58, 133–145.
4. Antebi, A. and Fink, G.R. (1992) *Mol. Biol. Cell* 3, 633–654.
5. Sorin, A., Rosas, G. and Rao, R. (1997) *J. Biol. Chem.* 272, 9895–9901.
6. Hu, Z., Bonifas, J.M., Beech, J., Bench, G., Shigihara, T., Ogawa, H., Ikeda, S., Mauro, T. and Epstein, E.H., Jr. (2000) *Nat. Genet.* 24, 61–65.

7. Sudbrak, R., Brown, J., Dobson-Stone, C., Carter, S., Ramser, J., White, J., Healy, E., Dissanayake, M., Larregue, M., Perrussel, M., Lehrach, H., Munro, C.S., Strachan, T., Burge, S., Hovnanian, A. and Monaco, A.P. (2000) *Hum. Mol. Genet.* 12, 1131–1140.
8. Fairclough, R.J., Dode, L., Vanoevelen, J., Andersen, J.P., Missiaen, L., Raeymaekers, L., Wuytack, F. and Hovnanian, A. (2003) *J. Biol. Chem.* 278, 24721–24730.
9. Xiang, M., Mohamalawari, D. and Rao, R. (2005) *J. Biol. Chem.* 280, 11608–11614.
10. Vanoevelen, J., Dode, L., Van Baelen, K., Fairclough, R.J., Missiaen, L., Raeymaekers, L. and Wuytack, F. (2005) *J. Biol. Chem.* 280, 22800–22808.
11. Dode, L., Andersen, J.P., Raeymaekers, L., Missiaen, L., Vilsen, B. and Wuytack, F. (2005) *J. Biol. Chem.* 280, 39124–39134.
12. Dode, L., Andersen, J.P., Vanoevelen, J., Raeymaekers, L., Missiaen, L., Vilsen, B. and Wuytack, F. (2006) *J. Biol. Chem.* 281, 3182–3189.
13. Van Baelen, K., Vanoevelen, J., Missiaen, L., Raeymaekers, L. and Wuytack, F. (2001) *J. Biol. Chem.* 276, 10683–10691.
14. Sakuntabhai, A., Ruiz-Perez, V., Carter, S., Jacobsen, N., Burge, S., Monk, S., Smith, M., Munro, C.S., O'Donovan, M., Craddock, N., Kucherlapati, R., Rees, J.L., Owen, M., Lathrop, G.M., Monaco, A.P., Strachan, T. and Hovnanian, A. (1999) *Nat. Genet.* 21, 271–277.
15. Kaufman, R.J., Swaroop, M. and Murtha-Riel, P. (1994) *Biochemistry* 33, 9813–9819.
16. Dürr, G., Strayle, J., Plemper, R., Elbs, S., Klee, S.K., Catty, P., Wolf, D.H. and Rudolph, H.K. (1998) *Mol. Biol. Cell* 9, 1149–1162.
17. Creemers, J.W., Jackson, R.S. and Hutton, J.C. (1998) *Semin. Cell Dev. Biol.* 9, 3–10.
18. Coutinho, P.M. and Henrissat, B. (1999) (http://www.cazy.org). In: H.J. Gilbert, G. Davies, B. Henrissat and B. Svensson (Eds), *Recent Advances in C.arbohydrate Bioengineering*, The Royal Society of Chemistry, Cambridge, pp. 3–12.
19. Coutinho, P.M., Deleury, E., Davies, G.J. and Henrissat, B. (2003) *J. Mol. Biol.* 328, 307–317.
20. Wiggins, C.A. and Munro, S. (1998) *Proc. Natl. Acad. Sci. U.S.A.* 95, 7945–7950.
21. Powell, J.T. and Brew, K. (1976) *J. Biol. Chem.* 251, 3645–3652.
22. Boeggeman, E. and Qasba, P.K. (2002) *Glycobiology* 12, 395–407.
23. Nishikawa, Y., Pegg, W., Paulsen, H. and Schachter, H. (1988) *J. Biol. Chem.* 263, 8270–8281.
24. Fritz, T.A., Hurley, J.H., Trinh, L.B., Shiloach, J. and Tabak, L.A. (2004) *Proc. Natl. Acad. Sci. U.S.A.* 101, 15307–15312.
25. Negishi, M., Dong, J., Darden, T.A., Pedersen, L.G. and Pedersen, L.C. (2003) *Biochem. Biophys. Res. Commun.* 303, 393–398.
26. Palma, A.S., Morais, V.A. and Coelho, A.V. Costa J. (2004) *Biometals* 17, 35–43.
27. Van Baelen, K., Vanoevelen, J., Callewaert, G., Parys, J.B., De Smedt, H., Raeymaekers, L., Rizzuto, R., Missiaen, L. and Wuytack, F. (2003) *Biochem. Biophys. Res. Commun.* 306, 430–436.
28. Ramos-Castaneda, J., Park, Y.N., Liu, M., Hauser, K., Rudolph, H., Shull, G.E., Jonkman, M.F., Mori, K., Ikeda, S., Ogawa, H. and Arvan, P. (2005) *J. Biol. Chem.* 280, 9467–9473.
29. Negishi, M., Pedersen, L.G., Petrotchenko, E., Shevtsov, S., Gorokhov, A., Kakuta, Y. and Pedersen, L.C. (2001) *Arch. Biochem. Biophys.* 390, 149–157.
30. Rens-Domiano, S. and Roth, J.A. (1989) *J. Biol. Chem.* 264, 899–905.
31. Mishiro, E., Liu, M.Y., Sakakibara, Y., Suizo, M. and Liu, M.C. (2004) *Biochem. Cell Biol.* 82, 295–303.
32. Lo-Guidice, J.-M., Périni, J.-M., Lafitte, J.-J., Ducourouble, M.-P., Roussel, P. and Lamblin, G. (1995) *J. Biol. Chem.* 270, 27544–27550.
33. Spiro, R.G., Yasumoto, Y. and Bhoyroo, V. (1996) *Biochem. J.* 319, 209–216.
34. Seko, A., Sumiya, J.-I. and Yamashita, K. (2005) *Biochem. J.* 391, 77–85.
35. Brockhausen, I. (2003) *Biochem. Soc. Trans.* 31, 318–325.
36. Lee, J.K., Bhakta, S., Rosen, S.D. and Hemmerich, S. (1999) *Biochem. Biophys. Res. Commun.* 263, 543–549.

37. Degroote, S., Lo-Guidice, J.M., Strecker, G., Ducourouble, M.P., Roussel, P. and Lamblin, G. (1997) *J. Biol. Chem.* 272, 29493–29501.
38. Duncan, J.S. and Wilkinson, M.C. Burgoyne R.D. (2000) *Biochem J.* 350, 463–468.
39. Tibaldi, E., Arrigoni, G., Brunati, A.M., James, P. and Pinna, L.A. (2006) *Cell Mol. Life Sci.* 63, 378–389.
40. Lasa, M., Marin, O. and Pinna, L.A. (1997) *Eur. J. Biochem.* 243, 719–725.
41. West, D.W. and Clegg, R.A. (1984) *Biochem. J.* 219, 181–187.
42. Burgoyne, R.D., Duncan, J.S. and Sudlow, A.W. (1998) *Biochem. Soc. Symp.* 63, 91–100.
43. Reinhardt, T.A., Filoteo, A.G., Penniston, J.T. and Horst, R.L. (2000) *Am. J. Physiol.* 279, C1595–C1602.
44. Reinhardt, T.A., Lippolis, J.D., Shull, G.E. and Horst, R.L. (2004) *J. Biol. Chem.* 279, 42369–42373.
45. Reinhardt, T.A. and Horst, R.L. (1999) *Am. J. Physiol.* 276, C796–C802.
46. Dmitriev, R.I., Pestov, N.B., Korneenko, T.V., Kostina, M.B. and Shakhparonov, M.I. (2005) *J. Gen. Physiol.* 126, 71a–72a.
47. Davidson, H.W., Rhodes, C.J. and Hutton, J.C. (1988) *Nature* 333, 93–96.
48. Austin, C.D. and Shields, D. (1996) *J. Biol. Chem.* 271, 1194–1199.
49. Molloy, S.S., Bresnahan, P.A., Leppla, S.H. and Klimpel, K.R. Thomas G. (1992) *J. Biol. Chem.* 267, 16396–16402.
50. Zhou, Y. and Lindberg, I. (1993) *J. Biol. Chem.* 268, 5615–5623.
51. Oda, K. (1992) *J. Biol. Chem.* 267, 7465–7471.
52. Bennett, B.D., Denis, P., Haniu, M., Teplow, D.B., Kahn, S., Louis, J.C., Citron, M. and Vassar, R. (2000) *J. Biol. Chem.* 275, 37712–37717.
53. Ivessa, N.E., De Lemos-Chiarandini, C., Gravotta, D., Sabatini, D.D. and Kreibich, G. (1995) *J. Biol. Chem.* 270, 25960–25967.
54. Porat, A. and Elazar, Z. (2000) *J. Biol. Chem.* 275, 29233–29237.
55. Ahluwalia, J.P., Topp, J.D., Weirather, K. and Zimmerman, M. Stamnes M. (2001) *J. Biol. Chem.* 276, 34148–34155.
56. Chen, J.L., Ahluwalia, J.P. and Stamnes, M. (2002) *J. Biol. Chem.* 277, 35682–35687.
57. Matsuura-Tokita, K., Takeuchi, M., Ichihara, A., Mikuriya, K. and Nakano, A. (2006) *Nature* 441, 1007–1010.
58. Verdugo, P. (1990) *Annu. Rev. Physiol.* 52, 157–176.
59. Wightman, R.M., Troyer, K.P., Mundorf, M.L. and Catahan, R. (2002) *Ann. N. Y. Acad. Sci.* 971, 620–626.
60. Meldolesi, J. and Grohovaz, F. (2001) *Cell Calcium* 30, 1–8.
61. Goncalves, P.P., Meireles, S.M., Neves, P. and Vale, M.G. (1999) *Mol. Brain Res.* 71, 178–184.
62. Moreno, A., Lobaton, C.D., Santodomingo, J., Vay, L., Hernandez-SanMiguel, E., Rizzuto, R., Montero, M. and Alvarez, J. (2005) *Cell Calcium* 37, 555–564.
63. Mahapatra, N.R., Mahata, M., Hazra, P.P., McDonough, P.M., O'Connor, D.T. and Mahata, S.K. (2004) *J. Biol. Chem.* 279, 51107–51121.
64. Mitchell, K.J., Pinton, P., Varadi, A., Tacchetti, C., Ainscow, E.K., Pozzan, T., Rizzuto, R. and Rutter, G.A. (2001) *J. Cell Biol.* 155, 41–51.
65. Mitchell, K.J., Tsuboi, T. and Rutter, G. (2004) *Diabetes* 53, 393–400.
66. Lavoie, C., Meerloo, T., Lin, P. and Farquhar, M.G. (2002) *Mol. Endocrinol.* 16, 2462–2474.
67. Petersson, U., Somogyi, E., Reinholt, F.P., Karlsson, T., Sugars, R.V. and Wendel, M. (2004) *Bone* 34, 949–960.
68. Morel-Huaux, V.M., Pypaert, M., Wouters, S., Tartakoff, A.M., Jurgan, U., Gevaert, K. and Courtoy, P.J. (2002) *Eur. J. Cell Biol.* 81, 87–100.
69. Taniguchi, N., Taniura, H., Niinobe, M., Takayama, C., Tominaga-Yoshino, K., Ogura, A. and Yoshikawa, K. (2000) *J. Biol. Chem.* 275, 31674–31681.
70. Islam, A., Adamik, B., Hawari, F.I., Ma, G., Rouhani, F.N., Zhang, J. and Levine, S.J. (2006) *J. Biol. Chem.* 281, 6860–6873.

71. Lapinskas, P.J., Cunningham, K.W., Liu, X.F., Fink, G.R. and Culotta, V.C. (1995) *Mol. Cell. Biol.* 15, 1382–1388.
72. Halachmi, D. and Eilam, Y. (1996) *FEBS Lett.* 392, 194–200.
73. Cortés, J.C., Katoh-Fukui, R., Moto, K., Ribas, J.C. and Ishiguro, J. (2004) *Eukaryot. Cell* 3, 1124–1135.
74. Maeda, T., Sugiura, R., Kita, A., Saito, M., Deng, L., He, Y., Yabin, L., Fujita, Y., Takegawa, K., Shuntoh, H. and Kuno, T. (2004) *Genes Cells* 9, 71–82.
75. Cho, J.H., Ko, K.M., Singaravelu, G. and Ahnn, J. (2005) *FEBS Lett.* 579, 778–782.
76. Kowarski, D., Shuman, H., Somlyo, A.P. and Somlyo, A.V. (1985) *J. Physiol.* 366, 153–175.
77. Chandra, S., Kable, E.P., Morrison, G.H. and Webb, W.W. (1991) *J. Cell Sci.* 100, 747–752.
78. Pezzati, R., Bossi, M., Podini, P., Meldolesi, J. and Grohovaz, F. (1997) *Mol. Biol. Cell* 8, 1501–1512.
79. Lin, P., Yao, Y., Hofmeister, R., Tsien, R.Y. and Farquhar, M.G. (1999) *J. Cell Biol.* 145, 279–289.
80. Reinhardt, T.A., Horst, R.L. and Waters, W.R. (2004) *Am. J. Physiol.* 286, C164–169.
81. Honore, B. and Vorum, H. (2000) *FEBS Lett.* 466, 11–18.
82. Scherer, P.E., Lederkremer, G.Z., Williams, S., Fogliano, M., Baldini, G. and Lodish, H.F. (1996) *J. Cell Biol.* 133, 257–268.
83. Vorum, H., Hager, H., Christensen, B.M., Nielsen, S. and Honore, B. (1999) *Exp. Cell Res.* 248, 473–481.
84. Brunati, A.M., Contri, A., Muenchbach, M., James, P., Marin, O. and Pinna, L.A. (2000) *FEBS Lett.* 471, 151–155.
85. Taylor, R.S., Jones, S.M., Dahl, R.H., Nordeen, M.H. and Howell, K.E. (1997) *Mol. Biol. Cell* 8, 1911–1931.
86. Pinton, P., Pozzan, T. and Rizzuto, R. (1998) *EMBO J.* 17, 5298–5308.
87. Vanoevelen, J., Raeymaekers, L., Parys, J.B., De Smedt, H., Van Baelen, K., Callewaert, G. and Wuytack, F. Missiaen L. (2004) *Cell Calcium* 35, 115–121.
88. Callewaert, G., Parys, J.B., De Smedt, H., Raeymaekers, L., Wuytack, F., Vanoevelen, J., Van Baelen, K., Simoni, A., Rizzuto, R. and Missiaen, L. (2003) *Cell Calcium* 34, 157–162.
89. Behne, M.J., Tu, C.L., Aronchik, I., Epstein, E., Bench, G., Bikle, D.D., Pozzan, T. and Mauro, T.M. (2003) *J. Invest. Dermatol.* 121, 688–694.
90. Chege, N.W. and Pfeffer, S.R. (1990) *J. Cell Biol.* 111, 893–899.
91. Dierick, H.A., Adam, A.N., Escara-Wilke, J.F. and Glover, T.W. (1997) *Hum. Mol. Genet.* 6, 409–416.
92. Li, Y., Ge, M., Ciani, L., Kuriakose, G., Westover, E.J., Dura, M., Covey, D.F., Freed, J.H., Maxfield, F.R., Lytton, J. and Tabas, I. (2004) *J. Biol. Chem.* 279, 37030–37039.
93. Murin, R., Verleysdonk, S., Raeymaekers, L., Kaplan, P. and Lehotsky J. (2006) *Cell Mol. Neurobiol.* 26, 1353–1363.
94. Wootton, L.L., Argent, C.C., Wheatley, M. and Michelangeli, F. (2004) *Biochim. Biophys. Acta* 1664, 189–197.
95. Harper, C., Wootton, L., Michelangeli, F., Lefièvre, L., Barratt, C. and Publicover, S. (2005) *J. Cell Sci.* 118, 1673–1685.
96. Mitchell, K.J., Lai, F.A. and Rutter, G.A. (2003) *J. Biol. Chem.* 278, 11057–11064.
97. Parker, A.K., Gergely, F.V. and Taylor, C.W. (2004) *J. Biol. Chem.* 279, 23797–23805.
98. Missiaen, L., Vanoevelen, J., Van Acker, K., Raeymaekers, L., Parys, J.B., Callewaert, G., Wuytack, F. and De Smedt, H. (2002) *Biochem. Biophys. Res. Commun.* 294, 249–253.
99. Vanoevelen, J., Raeymaekers, L., Dode, L., Parys, J.B., De Smedt, H., Callewaert, G., Wuytack, F. and Missiaen, L. (2005) *Cell Calcium* 38, 489–495.
100. Wakabayashi, S., Ogurusu, T. and Shigekawa, M. (1987) *J. Biol. Chem.* 262, 9121–9129.
101. De Meis, L. and Vianna, A.L. (1979) *Annu. Rev. Biochem.* 48, 275–292.
102. Toyoshima, C. and Inesi, G. (2004) *Annu. Rev. Biochem.* 73, 269–292.

103. Møller, J.V., Nissen, P., Sørensen, T.L.-M. and le Maire, M. (2005) *Curr. Opin. Struct. Biol.* 15, 387–393.
104. Obara, K., Miyashita, N., Xu, C., Toyoshima, I., Sugita, Y., Inesi, G. and Toyoshima, C. (2005) *Proc. Natl. Acad. Sci. U.S.A.* 102, 14489–14496.
105. Jensen, A.-M.L., Sørensen, T.L.-M., Olesen, C., Møller, J.V. and Nissen, P. (2006) *EMBO J.* 25, 2305–2314.
106. Sørensen, T.L.-M., Olesen, C., Jensen, A.-M.L., Møller, J.V. and Nissen, P. (2006) *J. Biotechnol.* 124, 704–716.
107. Sørensen, T., Vilsen, B. and Andersen, J.P. (1997) *J. Biol. Chem.* 272, 30244–30253.
108. Sørensen, T.L.-M., Dupont, Y., Vilsen, B. and Andersen, J.P. (2000) *J. Biol. Chem.* 275, 5400–5408.
109. Andersen, J.P., Sørensen, T.L.-M., Povlsen, K. and Vilsen, B. (2001) *J. Biol. Chem.* 276, 23312–23321.
110. Dode, L., Andersen, J.P., Leslie, N., Dhitavat, J., Vilsen, B. and Hovnanian, A. (2003) *J. Biol. Chem.* 278, 47877–47889.
111. Dode, L., Vilsen, B., Van Baelen, K., Wuytack, F., Clausen, J.D. and Andersen, J.P. (2002) *J. Biol. Chem.* 277, 45579–45591.
112. Dode, L., Raeymaekers, L., Missiaen, L., Vilsen, B., Andersen, J.P. and Wuytack, F. (2006) In: Putney, J.W. Jr. (Ed.), *Calcium Signaling*, Second Edition, Methods in Signal Transduction Series, CRC Press, Boca Raton, FL, chapter 15, pp. 335–385.
113. Lopez, P.J. and Séraphin, B. (2000) *Nucleic Acids Res.* 28, 85–86.
114. Ton, V.-K. and Rao, R. (2004) *Am. J. Physiol.* 287, C580–C589.
115. Gunteski-Hamblin, A.M., Clarke, D.M. and Shull, G.E. (1992) *Biochemistry* 31, 7600–7608.
116. Sullivan, J.C., Ryan, J.F., Watson, J.A., Webb, J., Mullikin, J.C., Rokhsar, D. and Finnerty, J.R. (2006) *Nucleic Acids Res.* 34, D495–D499.
117. Kawada, H., Nishiyama, C., Takagi, A., Tokura, T., Nakano, N., Maeda, K., Mayuzumi, N., Ikeda, S., Okumura, K. and Ogawa, H. (2005) *J. Invest. Dermatol.* 124, 1206–1214.
118. Dode, L., De Greef, C., Mountian, I., Attard, M., Town, M.M., Casteels, R. and Wuytack, F. (1998) *J. Biol. Chem.* 273, 13982–13994.
119. Martin, V., Bredoux, R., Corvazie, E., van Gorp, R., Kovács, T., Gélébart, P. and Enouf, J. (2002) *J. Biol. Chem.* 277, 24442–24452.
120. Bobe, R., Bredoux, R., Corvazier, E., Andersen, J.P., Clausen, J.D., Dode, L., Kovács, T. and Enouf, J. (2004) *J. Biol. Chem.* 279, 24297–24306.
121. Strehler, E.E. and Zacharias, D.A. (2001) *Physiol. Rev.* 81, 21–50.
122. Strehler, E.E. and Treiman, M. (2004) *Curr. Mol. Med.* 4, 323–335.
123. Clarke, D.M., Loo, T.W., Inesi, G. and MacLennan, D.H. (1989) *Nature* 339, 476–478.
124. Andersen, J.P. (1995) *Biosci. Rep.* 15, 243–261.
125. MacLennan, D.H., Rice, W.J. and Green, N.M. (1997) *J. Biol. Chem.* 272, 28815–28818.
126. Toyoshima, C., Nakaso, M., Nomura, H. and Ogawa, H. (2000) *Nature* 405, 647–655.
127. Toyoshima, C. and Nomura, H. (2002) *Nature* 418, 605–611.
128. Schwede, T., Kopp, J., Guex, N. and Peitsch, M.C. (2003) *Nucleic Acids Res.* 31, 3381–3385.
129. Wei, Y., Marchi, V., Wang, R. and Rao, R. (1999) *Biochemistry* 38, 14534–14541.
130. Wei, Y., Chen, J., Rosaa, G., Tompkins, D.A., Holt, A. and Rao, R. (2000) *J. Biol. Chem.* 275, 23927–23932.
131. Mandal, D., Woolf, T.B. and Rao, R. (2000) *J. Biol. Chem.* 275, 23933–23938.
132. Mandal, D., Rulli, S.J. and Rao, R. (2000) *J. Biol. Chem.* 278, 35292–35298.
133. Huster, D. and Lutsenko, S. (2003) *J. Biol. Chem.* 278, 32212–32218.
134. Heilker, R., Manning-Krieg, U., Zuber, J.F. and Spiess, M. (1996) *EMBO J.* 15, 2893–2899.
135. Wang, P., Wang, X. and Pei, D. (2004) *J. Biol. Chem.* 279, 20461–20470.
136. Greenough, M., Pase, L., Voskoboinik, I., Petris, M.J., O'Brien, A.W. and Camakaris, J. (2004) *Am. J. Physiol.* 287, C1463–C1471.
137. Marchi, V., Sorin, A., Wei, Y. and Rao, R. (1999) *FEBS Lett.* 454, 181–186.
138. Ton, V.-K., Mandal, D., Vahadji, C. and Rao, R. (2002) *J. Biol. Chem.* 277, 6422–6427.

139. Dobson-Stone, C., Fairclough, R., Dunne, E., Brown, J., Dissanayake, M., Munro, C.S., Strachan, T., Burge, S., Sudbrak, R., Monaco, A.P. and Hovnanian, A. (2002) *J. Invest. Dermatol.* 118, 338–343.
140. Sørensen, T.L.-M., Clausen, J.D., Jensen, A.-M.L., Vilsen, B., Møller, J.V., Andersen, J.P. and Nissen, P. (2004) *J. Biol. Chem.* 279, 46355–46358.
141. Daiho, T., Yamasaki, K., Wang, G., Danko, S., Iizuka, H. and Suzuki, H. (2003) *J. Biol. Chem.* 278, 39197–39204.
142. Yamasaki, K., Daiho, T., Danko, S. and Suzuki, H. (2004) *J. Biol. Chem.* 279, 2202–2210.
143. Andersen, J.P., Vilsen, B., Leberer, E. and MacLennan, D.H. (1989) *J. Biol. Chem.* 264, 21018–21023.
144. Andersen, J.P. and Vilsen, B. (1993) *Biochemistry* 32, 10015–10020.
145. Møller, J.V., Lenoir, G., Marchand, C., Montigny, C., le Maire, M., Toyoshima, C., Juul, B.S. and Champeil, P. (2002) *J. Biol. Chem.* 277, 38647–38659.
146. Olesen, C., Sørensen, T.L., Nielsen, R.C., Møller, J.V. and Nissen, P. (2004) *Science* 306, 2251–2255.
147. Foggia, L. and Hovnanian, A. (2004) *Am. J. Med. Genet.* 131C, 20–31.
148. Missiaen, L., Van Acker, K., Van Baelen, K., Raeymaekers, L., Wuytack, F., Parys, J.B., De Smedt, H., Vanoevelen, J., Dode, L., Rizzuto, R. and Callewaert, G. (2004) *Cell Calcium* 36, 479–487.
149. Van Baelen, K., Dode, L., Vanoevelen, J., Callewaert, G., De Smedt, H., Missiaen, L., Parys, J.B., Raeymaekers, L. and Wuytack, F. (2004) *Biochim. Biophys. Acta* 1742, 103–112.
150. Fairclough, R.J., Lonie, L., Van Baelen, K., Haftek, M., Munro, C.S., Burge, S.M. and Hovnanian, A. (2004) *J. Invest. Dermatol.* 123, 67–71.
151. Ton, V.-K. and Rao, R. (2004) *J. Invest. Dermatol.* 123, 1192–1194.
152. Poblete-Gutiérrez, P., Wiederholt, T., Konig, A., Jugert, F.K., Marquardt, Y., Rubben, A., Merk, H.F., Happle, R. and Frank, J. (2004) *J. Clin. Invest.* 114, 1467–1474.
153. Yoshida, M., Yamasaki, K., Daiho, T., Iizuka, H. and Suzuki, H. (2006) *J. Dermatol. Sci.* 43, 21–33.
154. Gomes da Costa, A. and Madeira, V.M. (1986) *Arch. Biochem. Biophys.* 249, 199–206.
155. Eckert, R.L., Crish, J.F. and Robinson, N.A. (1997) *Physiol. Rev.* 77, 397–424.
156. Eckert, R.L., Sturniolo, M.T., Broome, A.M., Ruse, M. and Rorke, E.A. (2005) *J. Invest. Dermatol.* 124, 481–492.
157. Morita, K. and Miyachi, Y. (2003) *J. Dermatol. Sci.* 31, 81–89.
158. Tu, C.L. and Bikle, D.D. (2003) *J. Invest. Dermatol.* 121, abstract 0725
159. Tu, C.L., Oda, Y., Komuves, L. and Bikle, D.D. (2004) *Cell Calcium* 35, 265–273.
160. Elias, P.M., Ahn, S.K., Denda, M., Brown, B.E., Crumrine, D., Kimutai, L.K., Komuves, L., Lee, S.H. and Feingold, K.R. (2002) *J. Invest. Dermatol.* 119, 1128–1136.
161. Mauro, T., Bench, G., Sidderas-Haddad, E., Feingold, K., Elias, P. and Cullander, C. (1998) *J. Invest. Dermatol.* 111, 1198–1201.
162. Elias, P.M., Ahn, S.K., Brown, B.E., Crumrine, D. and Feingold, K.R. (2002) *J. Invest. Dermatol.* 119, 1269–1274.
163. Grzesiak, J.J. and Pierschbacher, M.D. (1995) *J. Clin. Invest.* 95, 227–233.
164. Furuse, M., Hata, M., Furuse, K., Yoshida, Y., Haratake, A., Sugitani, Y., Noda, T., Kubo, A. and Tsukita, S. (2002) *J. Cell Biol.* 156, 1099–1111.
165. Porgpermdee, S., Yu, X., Takagi, A., Mayuzumi, N., Ogawa, H. and Ikeda, S. (2005) *J. Dermatol. Sci.* 40, 137–140.
166. Leinonen, P.T., Myllyla, R.M., Hagg, P.M., Tuukkanen, J., Koivunen, J., Peltonen, S., Oikarinen, A., Korkiamaki, T. and Peltone, J. (2005) *Br. J. Dermatol.* 153, 113–117.
167. Foggia, L., Aronchik, I., Aberg, K., Brown, B., Hovnanian, A. and Mauro, T.M. (2006) *J. Cell Sci.* 119, 671–679.
168. Menon, G.K., Grayson, S. and Elias, P.M. (1985) *J. Invest. Dermatol.* 84, 508–512.
169. Mayuzumi, N., Ikeda, S., Kawada, H. and Ogawa, H. (2005) *Br. J. Dermatol.* 152, 920–924.
170. Banères, J.L., Roquet, F. and Martin, A. Parello J. (2000) *J. Biol. Chem.* 275, 5888–5903.
171. Levy, L., Broad, S., Diekmann, D., Evans, R.D. and Watt, F.M. (2000) *Mol. Biol. Cell* 11, 453–466.

J. Krebs and M. Michalak (Editors)
Calcium: A Matter of Life or Death

ISSN: 0167-7306/DOI: 10.1016/S0167-7306(06)41011-5

CHAPTER 11

IP_3 receptors and their role in cell function

Katsuhiko Mikoshiba

The Institute of Medical Science, The University of Tokyo, and RIKEN, Brain Science Institute, Calcium Oscillation Project, ICORP-SORST, JST, 4-6-1 Shirokanedai, Minato-ku, Tokyo 108-8639, Japan, Tel.: +81 3 5449 5316; Fax: +81 3 5449 5420; E-mail: mikosiba@ims.u-tokyo.ac.jp

Abstract

Inositol 1,4,5-trisphosphate receptor (IP_3R) is a Ca^{2+} release channel localized on the endoplasmic reticulum (ER) and plays an important role in various cell functions. IP_3R was discovered as a developmentally regulated glyco-phosphoprotein missing in cerebellar mutant mice. Recent studies using loss of function analysis indicate that IP_3R is involved in fertilization, early development, neuronal plasticity, and other cell functions. IP_3R works like a scaffold protein associating with various molecules that may regulate the function of IP_3R. IP_3 works not only to release Ca^{2+} through the channel pore of IP_3R but also to release IP_3R-binding protein released with IP_3 (IRBIT) from the IP_3-binding core. IRBIT binds to and activates pancreas-type Na^+/HCO_3^- cotransporter 1 that is important for regulating acid–base balance. Electron microscopic (EM) studies show the IP_3R has allosteric property to change reversibly its form from square to windmill in the presence of Ca^{2+}. Cryo-EM analysis of IP_3R shows a balloon-like structure with holes on the surface and a large cavity inside, this structure is convenient for IP_3R to associate various molecules to be regulated. It is also found that ER carrying IP_3R moves along microtubules in addition to reticular ER. All these data suggests that the IP_3R/Ca^{2+} channel works as a "signaling center" inside cells by associating with many molecules like a scaffolding protein presumably forming a "calcio-signalosome."

Keywords: calcium, CARP, endoplasmic reticulum, ERp44, IP_3, IP_3 indicator 4.1N, IP_3R, IRBIT, Na^+/HCO_3^- cotransporter 1, Na,K-ATPase

1. Introduction

Inositol 1,4,5-trisphosphate (IP_3) is a second messenger produced through phosphoinositide turnover in response to many extracellular stimuli (hormones, growth factors, neurotransmitters, neutrophins, odorants, light, etc.). IP_3 induces Ca^{2+} release from intracellular Ca^{2+} stores, such as the endoplasmic reticulum (ER), and controls a variety of Ca^{2+}-dependent cell functions (cell proliferation, differentiation, fertilization, embryonic development, secretion, muscular contraction, immune responses, brain functions, chemical senses, light transduction, etc.). The IP_3 receptor (IP_3R) is an IP_3-gated Ca^{2+} release channel.

In this review, I first present the brief history of discovery of IP_3R discovered as a developmentally regulated P_{400} protein missing in cerebellar mutant mice. Then, I present the important role of IP_3R during development. I also present unique 3D structural and biophysical property of IP_3R. These data help readers to understand that IP_3R works as a "signaling center" for a variety of cell functions by interacting with various Ca^{2+}-related molecules like a scaffolding protein forming a "Calcio-signalsome."

2. *IP_3R identified from the developmental study of the cerebellar mutant mice*

IP_3 was found as a second messenger to increase cytosolic Ca^{2+} [1] (Fig. 1), when the Ca^{2+} store was only reported as a nonmitochondrial store, and it was not known whether the IP_3R is an IP_3-binding protein or whether the IP_3R is a Ca^{2+} release channel. Many researchers were attempting to identify the target molecule of IP_3 [2,3]. We had been working on a protein (P400) whose expression increases during development but which is greatly decreased in cerebellar mutant mice where Purkinje cells are deficient or spines of Purkinje cells are all absent [4–6]. We discovered that this developmentally regulated P400 protein is an IP_3R and determined its complete primary sequence [7,8]. The IP_3R was found to form a channel by incorporating the purified IP_3-binding protein into a lipid bilayer [9]. Liposomes that incorporate IP_3R express Ca^{2+}-releasing activity [10,11]. Overexpression of IP_3R1 enhances both

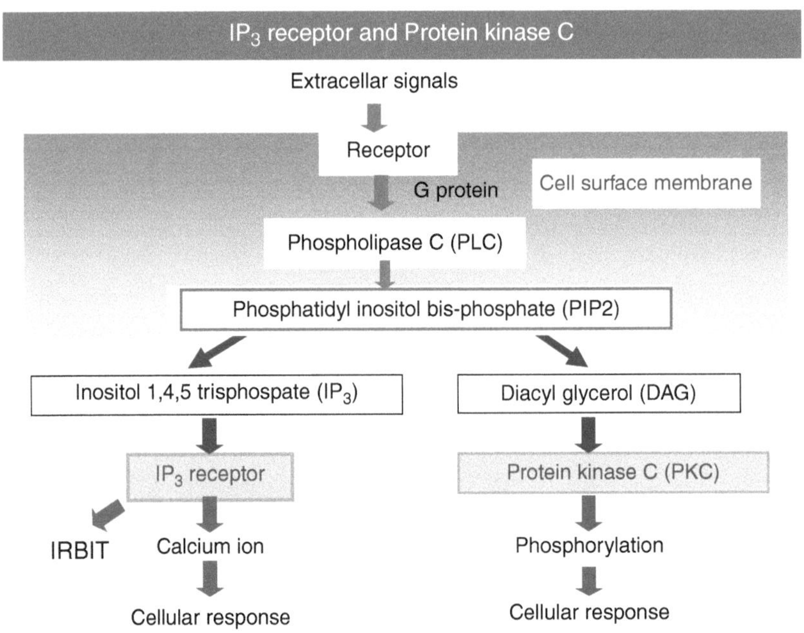

Fig. 1. Signaling pathways of Protein Kinase C and IP_3 receptor/Ca^{2+} channel. Both IP_3 and DAG are produced from the PIP2 by activated PLC. Recently, it is found that the role of IP_3 is not only to release Ca^{2+} but also to release IRBIT which activates pNBC1 (pancreas type Na^+/HCO_3^- cotransporter 1) (see more in detail in Fig. 5 and text in the Section of 5: Two roles of IP_3 on IP_3R function).

IP_3-binding activity and Ca^{2+}-releasing activity [12]. Immunogold method shows that P_{400} protein/IP_3R is located on ER [13]. All these data indicate that the IP_3R is an IP_3-gated Ca^{2+} release channel located on ER.

3. *Study of the role of IP_3R in development by analyzing the loss of function mutation*

3.1. *Loss of function mutations of IP_3R1 in mouse*

There are two types of mutations: one is produced by gene-targeting technique (knock-out mouse), and another is a spontaneous mutant mouse. Both show a similar abnormal phenotype of epileptic seizures. The survival rate of homozygous mice at the 10th postnatal day is about 20% of the control. They usually die by about 3 weeks postnatally [14]. The heterozygous mice behave and survive normally but have an impairment of motor coordination in the rotating rod test [15]. After about 14 postnatal days, repetitive tonic or tonic-clonic seizures prevail, and opisthotonus-like postures are observed. Encephalograms of IP_3R1-deficient mice show paroxysmal polyspike activities [14]. Intraperitoneal injection of anticonvulsants, pentobarbital, or diazepam eliminates the seizures but ataxic movements become prominent [14]. It is clear that the epileptic seizures or cerebellar ataxia has been caused by the absence of IP_3R1.

Electrophysiological studies in IP_3R1-deficient mice show various abnormalities in neuronal plasticity. Long-term depression (LTD) at the parallel fiber (PF)-Purkinje cell synapse in the cerebellum—a cellular basis for cerebellar function, that is, motor learning [16]—is induced by a conjunctive stimulation of PF and climbing fiber synapses. Inhibition of mGluR [17–19] or Protein Kinase C (PKC) [19–21] results in the blockade of LTD. The work with IP_3R1-deficient mice shows absence of LTD in the cerebellum and function-blocking antibody against IP_3R1 also inhibits LTD [22]. These results clearly show that IP_3R1 is necessary for the induction of LTD. Long-term potentiation (LTP) in the hippocampus is a typical example to study neural plasticity. LTP is persistent synaptic enhancement induced by a brief period of high-frequency electrical stimulation of afferents [23,24]. LTD, the second form of long-term synaptic change in efficacy, is induced by low-frequency afferent stimulation (LFS) [24]. LFS reverses a pre-established LTP, a phenomenon that is termed depotentiation [25,26]. LFS applied before tetanus suppresses LTP ("LTP suppression") [27,28]. A short tetanus protocol evokes a much larger amplitude LTP in the IP_3R1-deficient mice than in the control. Depotentiation or LTP suppression is attenuated in the mutant mice, the mean magnitude of the responses after delivery of LFS or tetanus being significantly greater than in wild-type mice [28]. The facilitation of LTP induction, the attenuation of depotentiation, and LTP suppression seen in the IP_3R1-deficient mice indicate that IP_3R1 plays an important role in synaptic plasticity in hippocampal CA1 neurons. In addition, we found in hippocampal CA1 pyramidal cells that IP_3R1 is involved in the determination of polarity and input specificity of activity-induced synaptic modification [29].

IP_3R1 is highly enriched in smooth muscle. Smooth muscle from wild-type mice generates slow waves that in turn initiate spike potentials, whereas those from mutant mice are either quiescent or generate irregular bursts of spike potentials [30]. Therefore, IP_3R1 is involved in the generation of slow waves but not in the generation of action potentials. The autosomal recessive *opisthotonos* (*opt*) mutant that has a mutation in IP_3R1 is found to display epileptic seizure. A genomic deletion that alters the IP_3R1 protein in *opt* mice has been found [31,32]. A genomic deletion in the gene results in the removal of two exons from the IP_3R1 mRNA but does not interrupt the translational reading frame [32]. The altered protein has lost several modulatory sites and is present at markedly reduced levels in *opt* homozygotes.

3.2. Role of IP_3R2 and IP_3R3 in exocrine secretion

Mice with a single-gene disruption of IP_3R2 and IP_3R3 do not show distinct abnormality, whereas double gene knockout mice die of starvation within 1 week following the end of the weaning period [33]. As fluid secretion is mediated by IP_3-induced Ca^{2+} release (IICR) in the salivary gland, intracellular Ca^{2+} signaling is examined using dissociated acinar cells. $[Ca^{2+}]_i$ increases induced by muscarinic acetylcholine receptor (mAChR) stimulation are severely impaired in the acinar cells prepared from the double knockout mice, whereas those in mice with a single deficiency of IP_3R2 or IP_3R3 are only slightly reduced. The histological examination of pancreatic tissue demonstrates an abnormal accumulation of zymogen granules in the cytosol of pancreatic acinar cells in double knockout mice (Fig. 2). By contrast, acinar cells in mice with a single deficiency of IP_3R2 or IP_3R3 exhibit a polarized distribution of zymogen granules in the apical region, similar to the control. Pilocarpine-induced saliva secretion is severely impaired in the double knockout mice, suggesting an indispensable role of IP_3R2 and IP_3R3 in saliva secretion. Acinar cells dissociated from the double mutants fail to show increases in the $[Ca^{2+}]_i$ in response to mAChR or cholecystokinin receptor stimulation. The severe impairment of secretory functions in the salivary gland and the pancreas in the IP_3R2/IP_3R3 double knockout mice is due to lack of Ca^{2+} signaling in these glands. In vitro experiments using dissociated acinar cells demonstrated that secretion of digestive enzymes (such as amylase and lipase) in response to mAChR stimulation is abolished. These results show that IP_3R2 and IP_3R3 play crucial roles in exocrine function of the pancreas. Double mutants had reduced body weights and lower blood glucose levels than their wild-type littermates. Defects in digestion could account for the malnourishment in these mice. Therefore, IP_3R2 and IP_3R3 are responsible for the physiological function in exocrine cells such as the salivary gland and the pancreas [33].

3.3. IP_3R mutants in Drosophila melanogaster and Caenorhabditis elegans

In *D. melanogaster* and *C. elegans*, only one IP_3R gene exists and is important in development [34–36].

3.3.1. Drosophila

In the *Drosophila* IP_3R (*itpr*) mutant, maternal IP_3R mRNA is sufficient for progression through the embryonic stages [34–36], but larval organs show

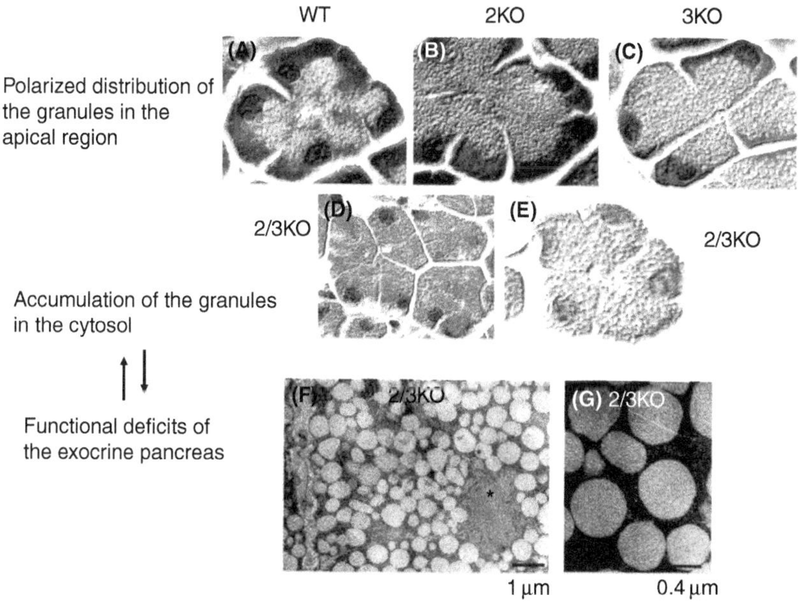

Fig. 2. Role of IP_3 receptor type two (IP_3R2) and type 3 (IP_3R3) in exocrine secretion. Pancreatic acinar cells show polarized structure. In normal acinar cells, granules are accumulated at the apical pole. Single IP_3R deficient mice of type 2 (2KO) and type 3 (3KO) have cytoplasmic area with no secretion granules at the basal region. However, double deficient mice of type 2 and type 3 (2/3K0) show secretion granules fully in all the cytoplasm (See Color Plate 31, p. 529).

asynchronous and defective cell divisions [37]. The imaginal discs arrest early and fail to differentiate [38].

3.3.2. C. elegans

The lethal mutant of *C. elegans* [*dec-4* (*n2559*)] fails to initiate the defecation motor program [39], has no observable defecation cycle, and as a result is severely constipated. Overexpression of *itr-1* speeds up the defecation cycle. IP_3R controls these periodic muscle contractions nonautonomously from the intestine. Ca^{2+} levels oscillate with the same period as the defecation cycle, and peak level precedes muscle contraction, but the mutant shows no oscillation.

3.4. Fertilization and IP_3R

In egg activation, eggs of all the species show Ca^{2+} wave or oscillation (Fig. 3). An intracellular Ca^{2+} transient is critical for the initiation of several events related to egg activation and cell cycle control [40–44]. In sea urchin eggs, Ca^{2+} transients are involved in the exocytosis of cortical granules [45,46] to prevent polyspermy. They are also involved in the cell cycle in early cleavage [43,47,48] and nuclear envelope breakdown [49,50]. Role of IP_3R in fertilization was first demonstrated in the hamster egg using IP_3R-specific function-blocking monoclonal (mAb) antibody

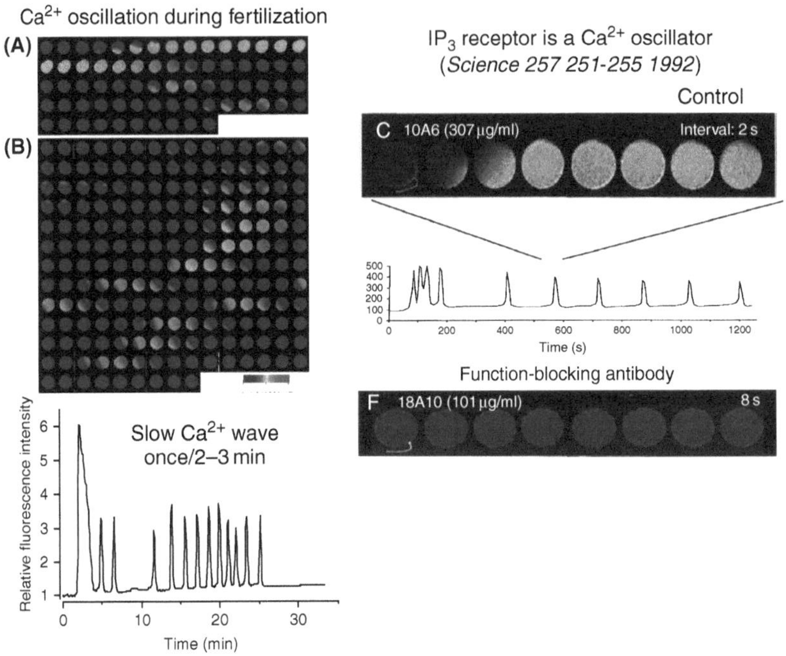

Fig. 3. Role of IP_3 receptor type 1 (IP_3R1) in fertilization and in Ca^{2+} oscillation. Left panel: Ca^{2+} oscillation with very slow frequency (once in 2–3 min) is observed during fertilization in ascidian egg. Right panel: When function-blocking antibody (18A10) is introduced in hamster eggs, Ca^{2+} oscillation and fertilization is blocked even though the sperm is attached to the egg (left side of the bottom panel) which suggests that IP_3 and IP_3R1 is required for fertilization and Ca^{2+} oscillation (See Color Plate 32, p. 530).

18A10 [42]. Role of IP_3R is also demonstrated in starfish oocyte, by injecting IP_3 sponge (a high-affinity IP_3-binding sequence) [51]. Fertilized hamster eggs exhibit repetitive Ca^{2+} transients as well as the Ca^{2+} wave in each response [41]. The inhibition of peak $[Ca^{2+}]_i$ is dependent on the intracellular concentration of mAb 18A10. The inhibition curve shifts to the right, and its slope is reduced with an increasing concentration of mAb 18A10. The regenerative Ca^{2+} release induced by IP_3 is completely blocked by mAb 18A10, but the inhibitory effect of mAb 18A10 is compensated by increasing doses of IP_3.

3.5. Role of IP_3R in dorso-ventral axis formation

Lithium chloride is postulated to be an inhibitor of the Phosphatidylinositol (PI) cycle. Lithium inhibits inositol monophosphate phosphatase and inositol polyphosphate 1-phosphatase (for a review see Ref. [52]). Lithium affects dorso-ventral axis formation in *Xenopus laevis*, resulting in duplication of an axis in a stage-specific manner. At the early cleavage stage, lithium induces dorsalization and causes a reduction of posterior structure [53–56]. Late application (gastrula stage) causes ventralization and reduction of the anterior structures [57]. Lithium completely rescues dorsal structures in

UV-irradiated ventralized embryos [58]. The amount of embryonic IP_3 was decreased after lithium treatment [59]. The results suggests that IP_3 plays an essential role in transducing ventral signal in the process of mesoderm induction, in support of the "inositol depletion hypothesis." Ventral injection of the function-blocking antibody against *Xenopus* IP_3R1 at the four-cell stage induced the formation of a secondary axis, whereas dorsal injection of the antibody or the control IgG showed no effects [60]. The ectopic axis-inducing activity of the antibody corresponded with the ability to block IICR. It is known that glycogen synthase kinase 3β (GSK 3β) blockage also causes dorsalization of the embryo. As GSK 3β is another target for lithium [61,62], the effect of lithium might result from the combined inhibition of both the IP_3/Ca^{2+} signaling and GSK 3β. Calcineurin and nuclear factor of activated T-cells (NF-AT) are presumed to be downstream targets of IP_3R/Ca^{2+} signaling, thus serving as ventralizing signals in body axis formation. Overexpression of a dominant negative form of NF-AT RNA mutant leads to induction of an ectopic dorsal axis [63]. The ectopic dorsal axis formation caused by blockage of IP_3R is rescued by NF-AT.

4. *Three-dimensional structure of IP_3Rs*

4.1. *X-ray crystallographic analysis of three-dimensional structure*

Crystal structures of the IP_3-binding core in complex with IP_3 were first reported in the structural analysis of IP_3R at 2.2 Å resolution [64] (Fig. 4). The IP_3-binding core forms an asymmetric boomerang-like structure consisting of an amino-terminal β-trefoil domain and a carboxyterminal α-helical domain containing an armadillo repeat-like fold. The cleft formed by the two domains exposes a cluster of arginine and lysine residues that coordinate the three phosphoryl groups of IP_3 [64]. Displaying a shape akin to a hammer, the suppressor region contains a "head" subdomain forming the β-trefoil fold and an "arm" subdomain possessing a helix-turn-helix structure that protrudes from the globular head subdomain.

The suppressor domain at the N-terminal region of IP_3-binding core was also crystallized at 1.8 Å resolutions [65]. Site-directed mutagenesis studies provide evidence for the involvement of a large conserved surface area on the head subdomain in the suppression of IP_3 binding to the IP_3-binding core domain. This conserved region is in close proximity to the previously proposed binding sites of Homer, RACK1, calmodulin, and CaBP1 [65].

4.2. *Dynamic allosteric structural change of IP_3R1*

The structure observed by electron microscopy of IP_3R1 was reported by several groups [66–71]. It proved to be a major biochemical effort to purify the receptor in a physiological state capable of allosteric structural changes. The purified IP_3R1 particles revealed two distinct structures with 4-fold symmetry: a windmill-like structure and a square-like structure (Fig. 4). Ca^{2+} reversibly promoted a transition from the

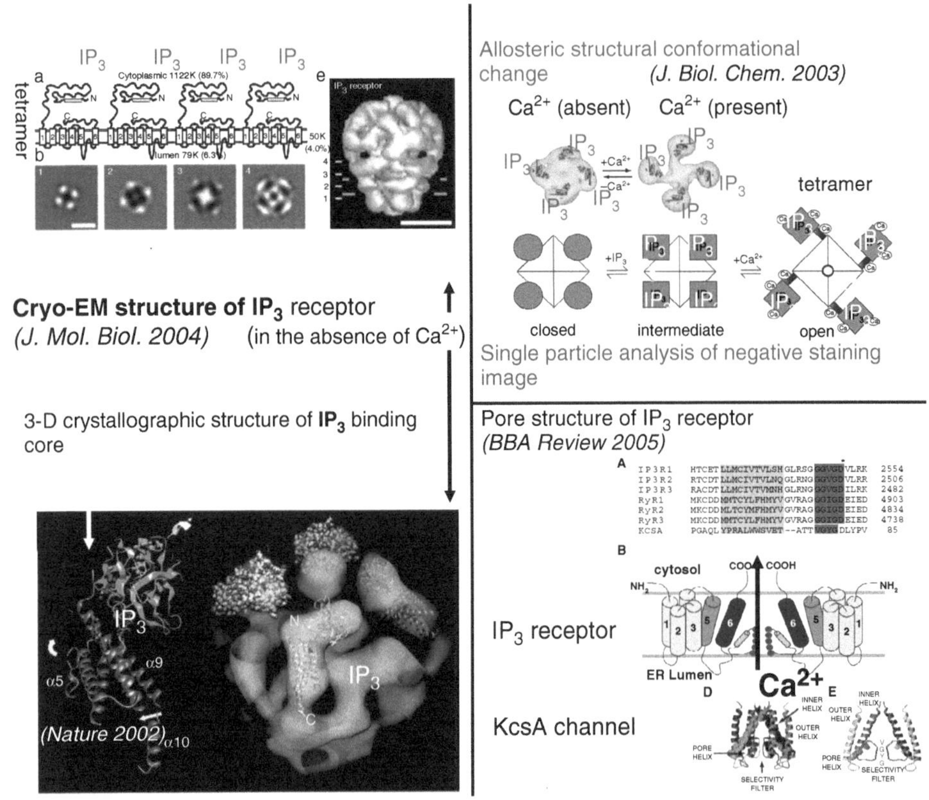

Fig. 4. Three dimensional structure of IP_3 receptor (IP_3R). Left panel: Cryo-EM structure of IP_3R1 in the absence of Ca^{2+}. The three dimensional X-ray crystallographic structure of IP_3 binding core (left side of the lower panel) is aligned on the L-angle part of the IP_3R1 (right side of the lower panel). Upper right panel: The allosteric conformational change in the presence and absence of Ca^{2+}. Upper figure is obtained from the single particle analysis of negative staining of purified IP_3R1. Lower right panel: The similarity of pore structure of IP_3R, Ryanodine receptor and KcsA channel (See Color Plate 33, p. 531).

square-like structure to the windmill-like structure with relocation of four peripheral IP_3-binding domains [67,68]. These data suggest that the Ca^{2+}-specific conformational change allosterically regulates the IP_3-gated channel opening within the IP_3R [67].

4.3. Cryo-EM study of IP_3R

Cryo-EM study has been performed on the ligand-free form of the IP_3R1 purified from mouse cerebella. The three-dimensional structure was based on a single-particle technique using an originally designed electron microscope equipped with a helium-cooled specimen stage and an automatic particle-picking system. The shape of the density map obtained at 15 Å resolution is reminiscent of a hot air balloon, with the spherical cytoplasmic domain (diameter of 175 Å) representing the balloon and the square-shaped luminal domain (side length of 96 Å) representing the basket [71] (Fig. 4).

The structure of the density map consists of two layers. The outer hot air balloon-shaped shell forms many holes and cavities, whereas the inner shell is composed of a continuous square-shaped tubular density. There is a prominent vacant space over the inner tubular density similar to the voltage-gated Na^+ channels [71].

5. *Two roles of IP_3 for IP_3R function*

5.1. *IP_3R-binding protein released with IP_3, a pseudo ligand of IP_3, regulates IP_3-induced Ca^{2+} release*

IP_3R-binding protein released with IP_3 (IRBIT) interacts with IP_3Rs and is released upon IP_3 binding to IP_3R1 (Fig. 5). IRBIT was purified from a high salt extract of crude rat brain microsomes with IP_3 elution using an affinity column to which the large N-terminal cytoplasmic region of IP_3R1 is immobilized (residues 1–2217) [72]. IRBIT (530 amino acids long) has a domain homologous to *S*-adenosylhomocysteine hydrolase in the C-terminal region and, in the N-terminal portion, a 104 amino acid appendage containing multiple potential phosphorylation sites. This N-terminal region of IRBIT is essential for interaction. The IRBIT-binding region of IP_3R1 has been mapped to the IP_3-binding core. IP_3 dissociates IRBIT from IP_3R1. Alkaline phosphatase treatment abolishes the interaction, suggesting that phosphates are essential for binding. As IRBIT binds to the same site as IP_3, role of IRBIT is to regulate IP_3-induced Ca^{2+} release [73].

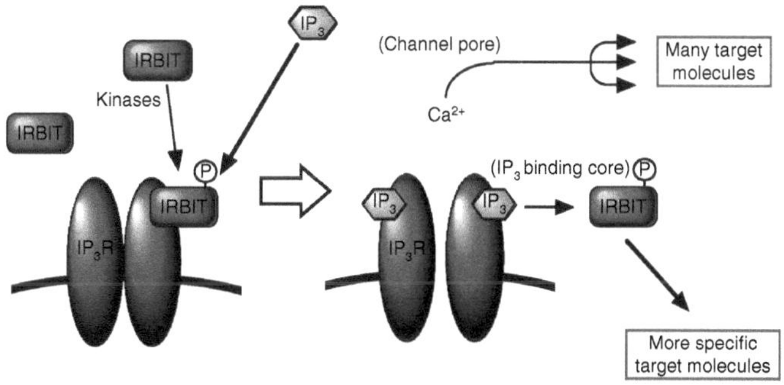

Fig. 5. IRBIT (IP_3R binding protein released with inositol 1,4,5-trisphosphate) interacts with IP_3 binding core of IP_3R and is released by high concentration of IP_3. There are two roles of IRBIT: (1) Modulation of the Ca^{2+} oscillation frequency by competing with IP_3. (2) Third messenger with respect to IP_3 to bind and activate pancreas Na^+/HCO_3^- cotransporter 1 to regulate acid-base balance.

5.2. IRBIT binds to Na^+/HCO_3^- cotransporter 1 to regulate acid–base balance

As IRBIT is released from IP_3R when IP_3 binds to the receptor, this raises the possibility that IRBIT acts as a signaling molecule downstream from IP_3R. Na^+/HCO_3^- cotransporter 1 (NBC1) is identified as an IRBIT-binding protein [74] (Fig. 5). NBC plays an important role in the regulation of intracellular pH. As NBC1 transports more than one (two or three) HCO_3^- per Na^+, the transport mediated by NBC1 is associated with a net movement of negative charge across the membrane. NBC1 has two isoforms, the pancreatic-type pNBC1 and the type mainly expressed in kidney, kNBC1, because of two major splicing sites. pNBC1 and kNBC1 are 93% identical but with N-terminal 85 amino acids of pNBC1 replaced by 41 distinct amino acids in kNBC1. The tissue distributions of these NBC1s are quite different. pNBC1 is predominantly expressed in pancreas, with lower levels of expression in several other tissues. kNBC1 is predominantly expressed in kidney. The difference of tissue distributions indicates the characteristic difference between pNBC1 and kNBC1. IRBIT specially interacts with pNBC1 but not with kNBC1. IRBIT binds to N-terminal pNBC1-specific domain. Interaction between IRBIT and pNBC1 is dependent on phosphorylation of multiple serine residues of IRBIT. IRBIT potentiates the transport activity of pNBC1 but not that of kNBC1 in *Xenopus* oocyte system. These findings strongly suggest that pNBC is a target molecule of IRBIT and raise the possibility that the regulation through IRBIT might explain the characteristic difference between pNBC1 and kNBC.

6. Development of a new IP_3 indicator, IP_3R-based IP_3 sensor

Observation of cytosolic IP_3 dynamics during Ca^{2+} spikes helps us better understand the mechanism responsible for the generation of Ca^{2+} spikes. IP_3 indicator was developed from the detailed analysis of the IP_3-binding core as follows: The first important clue to use the IP_3-binding core is that crystals were obtained only in the presence of IP_3, suggesting that the structure is different in the presence or absence of IP_3. There are a couple of reports to construct an IP_3 indicator [75–77]. But no one has ever used them to study the mechanism underlying Ca^{2+} spike generation in living cells. A newly developed IP_3 sensor named IP_3R-based IP_3 sensor (IRIS), not interfering with Ca^{2+} dynamics using fluorescence resonance energy transfer (FRET), enables to monitor the spatiotemporal dynamics of IP_3 and Ca^{2+} [78] (Fig. 6). IRIS is the name of a goddess in Greek mythology. IRIS is also the name of a flower. IP_3 starts to increase at a relative constant rate before Ca^{2+} rises (see Fig. 6). The IP_3 content does not return to its basal level during the intervals between spikes. IP_3 gradually accumulates in the cytosol with little or no fluctuation during cytosolic Ca^{2+} oscillation, thus Ca^{2+} spikes occur without cytosolic IP_3 spikes. Therefore, IP_3R works as an analog-to-pulse converter (pulse generator).

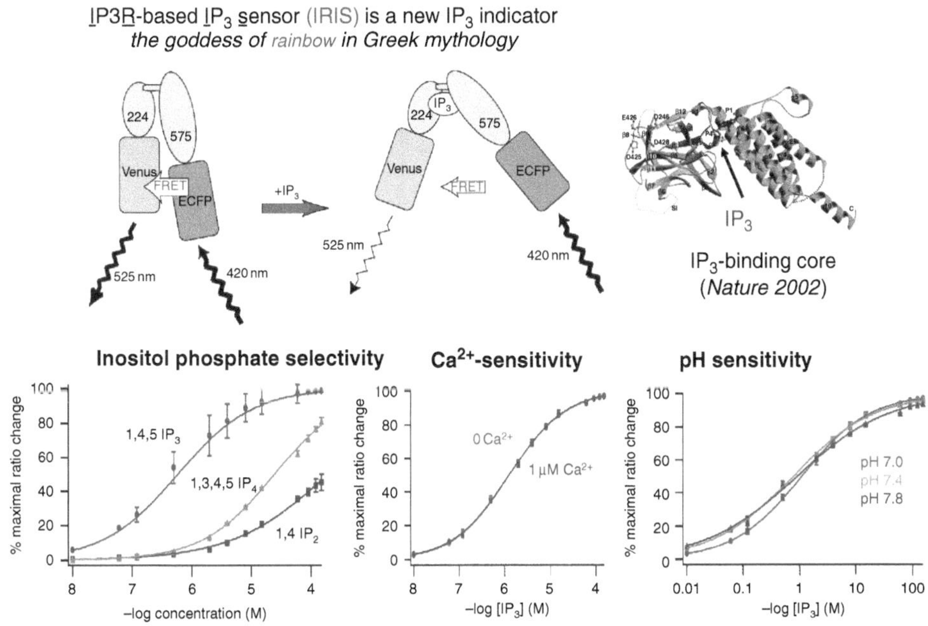

Fig. 6. Development of IP_3 indicator and its inositol specificity, Ca^{2+} sensitivity and pH sensitivity. Upper panel: The new indicator is named IRIS (IP_3R-based IP_3 sensor) using IP_3 binding core (right panel) by FRET technique. (IRIS is the 'Goddess of rainbow' in Greek mythology). Lower panel: IRIS shows inositol specificity (left panel). Ca^{2+} (central panel) and pH (right panel) changes do not influence the detection of IP_3 by IRIS.

7. Identification and characterization of IP_3R-binding proteins

IP_3R works as a signaling center forming a calcio-signalsome like scaffolding proteins by interacting with many molecules. Chromogramin is reported to bind to the channel region to regulate channel activity [79]. Carbonic anhydrase-related protein (CARP) is found to bind to a central part between the IP_3-binding core and channel region and regulates channel activity [80]. Cytochrome-c was recently found to bind to the C-terminal region [81]. Huntington (Htt)-associated protein (HAPIA) was also found to bind to the C-terminus of IP_3R1 [82]. IP_3R1 activation by IP_3 is sensitized by polyQ expansion of Htt caused by Hungtinton disease. Protein phosphatases (PPI and PP2A) [83,84], RACK1 [85], Ankyrin [86], and Homer [87] are all reported to associate with IP_3R. The physiological relevance of these interactions is still unclear.

7.1. ERp44, a redox sensor in the ER lumen

Ca^{2+} signaling and redox regulation are considered to be independent phenomena. However, we found that ERp44, an ER luminal protein of the thioredoxin family,

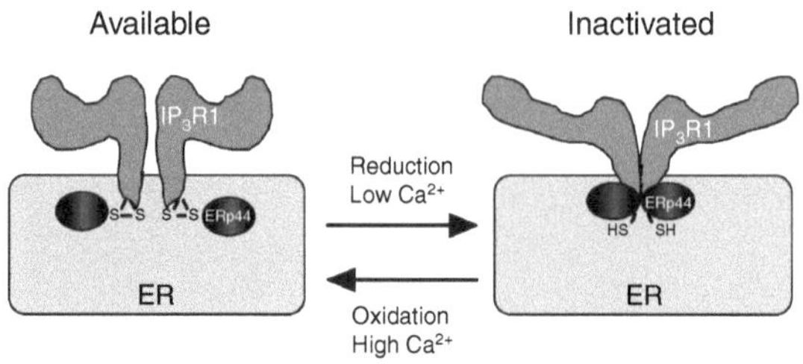

Fig. 7. Discovery of a new redox (oxido-reduction) sensor. ERP44 (thioredoxin family) senses oxido-reduction state in the ER lumen and regulates IP_3R1 activity in a Ca^{2+} dependent manner.

directly interacts with the third luminal loop of IP_3R1 dependent on pH, Ca^{2+} concentration, and redox state (Fig. 7). The presence of free-cysteine residues in the loop is essential for binding. Ca^{2+} imaging and single-channel recording of IP_3R1 in a planar lipid bilayer demonstrated that Ca^{2+} release channel activity of IP_3R1 is directly inhibited by ERp44 [88]. It is, therefore, clear that ERp44 senses the environment in the ER lumen and modulates IP_3R1 activity.

7.2. 4.1N, a protein involved in translocation of the IP_3R to the plasma membrane and in lateral diffusion

4.1N is a homolog to erythrocyte protein 4.IR. 4.1N was identified as a binding molecule for the C-terminal cytoplasmic tail of IP_3R1 using a yeast two-hybrid system [89,90]. 4.1N and IP_3R1 associate in both subconfluent and confluent Madin–Darby canine kidney (MDCK) cells, a polarized epithelial cell line. In subconfluent MDCK cells, 4.1N is distributed in the cytoplasm and the nucleus; IP_3R1 is localized in the cytoplasm. In confluent MDCK cells, both 4.1N and IP_3R1 are predominantly translocated to the basolateral membrane domain, and other ER marker proteins are still present in the cytoplasm. There is a controversy in the binding site in the IP_3R1 structure between the two groups. Maximov et al. [89] narrowed down the binding site to an 87 amino acid stretch in the C-terminal tail (2590–2676aa; CTM1, cytoplasmic tail middle 1), whereas Zhang et al. [90] confined the binding site of IP_3R1 in the C-terminal 14 amino acids (2736–2749aa; CTT14aa). The conclusion is that both CTT14aa and CTM1 fragments have binding affinity for 4.1N binding in the same order as the full length tetramer form of IP_3R1. GFP-IP_3R-ΔCTT14aa did not bind to 4.1N despite including CTM1 in the immunoprecipitation experiment [90], and GFP-IP_3R1-ΔCTT14aa did not show 4.1N-dependent translocation in confluent MDCK cells [90]. D_{eff} of GFP-IP_3R1-ΔCTT14aa was larger than that of GFP-IP_3R1 and as fast as that of GFP-IP_3R3, which lacks CTT14aa sequence. When GFP-IP_3R3 was fused with a CTT14aa fragment, it became able to bind to 4.1N, and lateral diffusion became subjected to actin-dependent regulation [91]. Therefore, the CTT14aa sequence is necessary and sufficient for the IP_3R1–4.1N interaction, and when IP_3R1 is in its native form, 4.1N may not be able to access to CTM1 [92]. A monoclonal antibody 18A10, the epitope of which lies over CTT14aa, binds to native IP_3R1, thereby blocking its channel function [93]. This supports the idea that the CTT14aa region of IP_3R1 is exposed and capable of associating with other proteins.

7.3. *Na,K-ATPase*

The Na,K-ATPase is an integral plasma membrane protein that enables the electrochemical gradient across the plasma membrane in all mammalian cells. Ouabain is a steroid derivative that binds specifically to Na,K-ATPase. Recently, the ouabain/Na,K-ATPase-complex is considered to act as a signal transducer and transcription activator [94–97] modulating cell growth [98,99], apoptosis [100], and cell motility [101]. These pathways are considered to involve release of calcium (Ca^{2+}) from intracellular stores through the IP_3R. Results from recent studies indicate that the Na,K-ATPase tethers the $InsP_3R$ into a Ca^{2+}-regulatory complex [96,97]. The N-terminal tail of the Na,K-ATPase catalytic α-subunit is found to bind to the N-terminus of the $InsP_3R$. Interaction between Na,K-ATPase and $InsP_3R$ modulates the Ca^{2+} oscillatory signal that serves to protect the cell from apoptosis [102].

7.4. *CARP*

CARP is originally identified as one of the Purkinje cell-specific genes [103]. CARP is composed of 291 amino acids. It is composed of a central carbonic anhydrase motif but lacks carbonic anhydrase activity because catalytic zinc coordinating residues are absent [104]. CARP is identified by yeast two-hybrid system to bind to the modulatory domain of IP_3R1 (amino acids 1387–1647) [80]. CARP inhibits IP_3-binding to IP_3R1 by reducing the affinity for IP_3. As IP_3 sensitivity for IICR in Purkinje cells is low compared with other tissues, this low sensitivity would be due to coexpression of CARP with IP_3R in Purkinje cells and its inhibitory effects on IP_3 binding.

8. *Cell biological analysis of intracellular dynamics of IP_3R*

IP_3R1 is highly expressed in neurons. In Purkinje cells in the cerebellum, enormous amount of IP_3R1 is distributed in soma, dendrites, and spines. IP_3R1 is considered to be localized on ER. It is interesting to understand the dynamic property of IP_3R1 in neurons, how IP_3R1 is synthesized, how it is transported to various parts of neurons, and how it works for the neuronal plasticity.

8.1. *Transport of IP_3R1 as vesicular ER on microtubules*

The ER is thought of as a continuous meshwork structure. Fluorescent protein-tagged ER proteins were expressed in cultured mouse hippocampal neurons to monitor their movements using time-lapse microscopy. In addition to the reticular

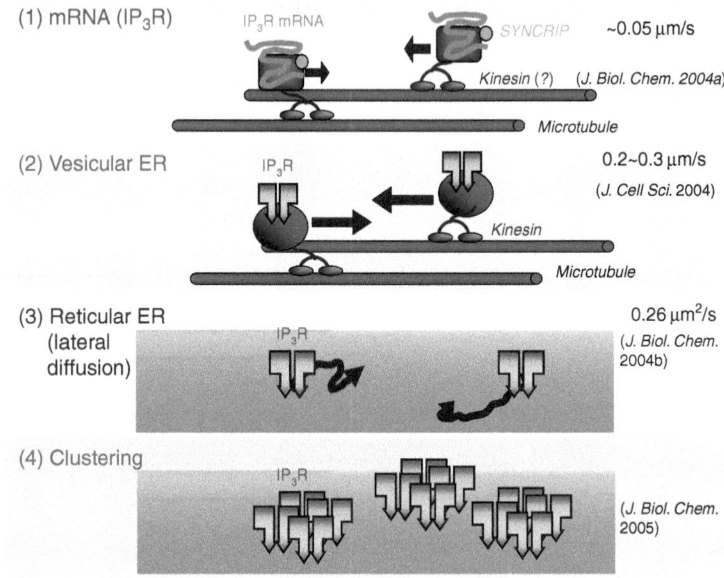

Fig. 8. Four different ways of IP_3R trafficking inside the cell (see Section 8).

ER, vesicular ER is observed and is capable of taking up and releasing Ca^{2+} [105] (Fig. 8). The vesicular ER moves rapidly along dendrites in both anterograde and retrograde directions at a velocity of 0.2–0.3 μm/s (Fig. 8). Depolymerization of microtubules, overexpression of dominant-negative kinesin, and kinesin depletion by antisense technology reduced the number and velocity of the moving vesicles [105]. Kinesin may therefore drive the transport of vesicular ER along microtubules. The rapid transport of the Ca^{2+}-releasable ER vesicles would contribute to the rapid supply of fresh ER proteins to the distal part of the cells, for example, dendrite in neurons, or to regulate the spatio-terminal pattern of intracellular Ca^{2+} signaling.

8.2. Transport of IP_3R mRNA within an mRNA granule

The mRNAs for IP_3R1 [106], β-actin, Arc, alpha subunit of Ca^{2+}/CaM kinase II, glycine receptors, glutamate receptors, and MAP2 have been reported to be located in dendrites of neurons. mRNAs are transported to the dendrites as components of a ribonucleoprotein complex, termed mRNA granules. mRNA granules contain ribosomes and other components of the translational machinery [107–109] as well as fragile X mental retardation protein [110] and Staufen [111]. Recently, the protein SYNCRIP (Synaptotagmin-binding cytoplasmic RNA-interacting protein) was found to be a component of the complex as well [112]. The 3′-untranslated region of IP_3R1 mRNA is cotransported with SYNCRIP and Staufen within the dendrites of hippocampal neurons at about 0.05 μm/s (Fig. 8). Treatment with nocodazole significantly inhibited the movement of GFP-SYNCRIP [112] in neurons. Protein synthesis may occur at any part of neuronal dendrites depending on the stimulus condition for neuronal plasticity.

9. *Conclusions*

9.1. *IP_3R is an intracellular signaling center*

IP_3R is an IP_3-gated Ca^{2+} channel (Fig. 9). Recent biochemical study has revealed that IP_3R works like a scaffolding protein by associating with so many molecules [113,114] such as 4.1N, Homer, Ankyrin, protein phosphatases, PP1 and PP2A, CARP, HAP1A, ERp44, and chromogranin (for a recent review see Ref. [115]). It should be noted that the role of IP_3 is not only to release Ca^{2+} through the channel pore but also to release IRBIT from the IP_3-binding core. The role of IRBIT is to regulate IP_3-induced Ca^{2+} release by competing with IP_3 [73]. Second function of IRBIT may work as a third messenger to second messenger IP_3. One of the target molecules of IRBIT is found to be Na^+-bicarbonate cotransporter 1 that regulates intracellular and extracellular acid–base balance [74]. IP_3Rs have various

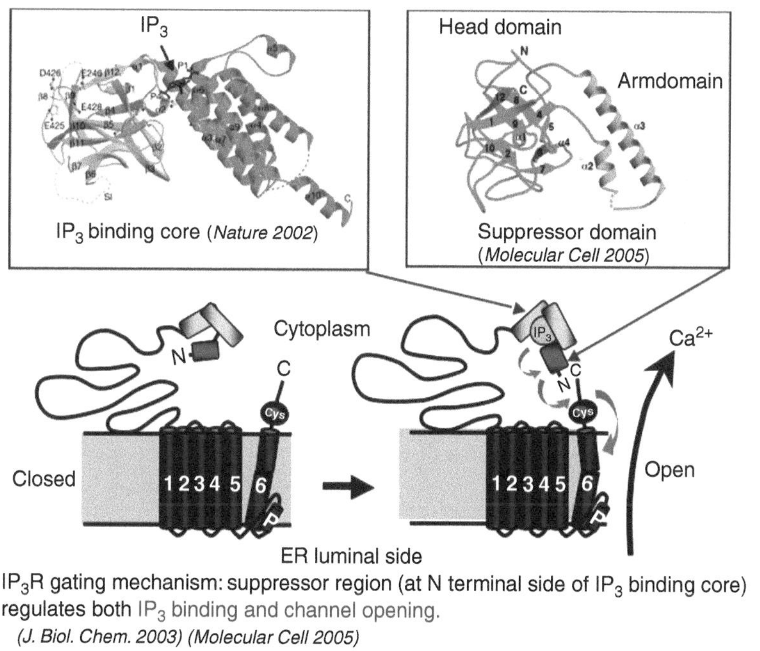

Fig. 9. Gating mechanism of IP_3 ligated channel pore opening. IP_3 binding core and N-terminal suppressor region revealed by X-ray crystallographic analysis interact together with C-terminal portion of IP_3R to regulate both IP_3 binding and channel opening. Both cryo-EM structural analysis and X-ray crystallographic analysis provided important information to understand the gating mechanism of the IP_3R channel pore (See Color Plate 34, p. 532).

unique biophysical properties. Rapid transport of ER vesicles carrying IP_3R as well as mRNA granules carrying IP_3R1 mRNA gives us important information to understand the novel mechanism of neural plasticity and how their movements are modified by various synaptic stimuli.

As IP_3R associates with many molecules, the IP_3R/Ca^{2+} channel has scaffolding protein-like properties. Ca^{2+}-related molecules may form "Calcio-signalsome" to regulate the function of IP_3R, and IP_3R works as a "signaling center" inside the cell.

References

1. Streb H, Irvine RF, Berridge MJ and Schulz I. (1983) *Nature* 306, 67–69.
2. Supattapone S, Worley PF, Baraban JM and Snyder SH. (1988) *J Biol Chem* 263, 1530–1534.
3. Sudhof TC, Newton CL, Archer BTD, Ushkaryov YA and Mignery GA. (1991) *EMBO J* 10, 3199–3206.
4. Maeda N, Niinobe M, Inoue Y and Mikoshiba K. (1989) *Dev Biol* 133, 67–76.
5. Mikoshiba K, Nagaike K, Aoki E and Tsukada Y. (1979) *Brain Res* 177, 287–299.
6. Maeda N, Niinobe M, Nakahira K and Mikoshiba K. (1988) *J Neurochem* 51, 1724–1730.
7. Maeda N, Niinobe M and Mikoshiba K. (1990) *EMBO J* 9, 61–67.
8. Furuichi T, Yoshikawa S, Miyawaki A, Wada K, Maeda N and Mikoshiba K. (1989) *Nature* 342, 32–38.
9. Maeda N, Kawasaki T, Nakade S, Yokota N, Taguchi T, Kasai M and Mikoshiba K. (1991) *J Biol Chem* 266, 1109–1116.
10. Hirota J, Michikawa T, Miyawaki A, Furuichi T, Okura I and Mikoshiba K. (1995) *J Biol Chem* 270, 19046–19051.
11. Ferris CD, Huganir RL and Snyder SH. (1990) *Proc Natl Acad Sci USA* 87, 2147–2151.
12. Miyawaki A, Furuichi T, Maeda N and Mikoshiba K. (1990) *Neuron* 5, 11–18.
13. Yamamoto A, Otsu H, Yoshimori T, Maeda N, Mikoshiba K and Tashiro Y. (1991) *Cell Struct Funct* 16, 419–432.
14. Matsumoto M, Nakagawa T, Inoue T, Nagata E, Tanaka K, Takano H, Minowa O, Kuno J, Sakakibara S, Yamada M, Yoneshima H, Miyawaki A, Fukuuchi Y, Furuichi T, Okano H, Mikoshiba K and Noda T. (1996) *Nature* 379, 168–171.
15. Ogura H, Matsumoto M and Mikoshiba K. (2001) *Behav Brain Res* 122, 215–219.
16. Ito M. (1989) *Annu Rev Neurosci* 12, 85–102.
17. Aiba A, Kano M, Chen C, Stanton ME, Fox GD, Herrup K, Zwingman TA and Tonegawa S. (1994) *Cell* 79, 377–388.
18. Conquet F, Bashir ZI, Davies CH, Daniel H, Ferraguti F, Bordi F, Franz-Bacon K, Reggiani A, Matarese V, Conde F. et al. (1994) *Nature* 372, 237–243.
19. Hartell NA. (1994) *Neuroreport* 5, 833–836.
20. Chen C, Kano M, Abeliovich A, Chen L, Bao S, Kim JJ, Hashimoto K, Thompson RF and Tonegawa S. (1995) *Cell* 83, 1233–1242.
21. Crepel F and Krupa M. (1988) *Brain Res* 458, 397–401.
22. Inoue T, Kato K, Kohda K and Mikoshiba K. (1998) *J Neurosci* 18, 5366–5373.
23. Bliss TV and Lomo T. (1973) *J Physiol (Lond)* 232, 331–356.
24. Bliss TV and Collingridge GL. (1993) *Nature* 361, 31–39.
25. Fujii S, Saito K, Miyakawa H, Ito K and Kato H. (1991) *Brain Res* 555, 112–122.
26. Bashir ZI and Collingridge GL. (1994) *Exp Brain Res* 100, 437–443.
27. Fujii S, Kuroda Y, Miura M, Furuse H, Sasaki H, Kaneko K, Ito K, Chen Z and Kato H. (1996) *Exp Brain Res* 111, 305–312.
28. Fujii S, Matsumoto M, Igarashi K, Kato H and Mikoshiba K. (2000) *Learn Mem* 7, 312–320.

29. Nishiyama, M, Hong, K, Mikoshiba, K, Poo, MM and Kato K. (2000) *Nature* 408, 584–588.
30. Suzuki H, Takano H, Yamamoto Y, Komuro T, Saito M, Kato K and Mikoshiba K. (2000) *J Physiol (Lond)* 525, 105–111.
31. Street VA, Robinson LC, Erford SK and Tempel BL. (1995) *Genomics* 29, 123–130.
32. Street VA, Bosma MM, Demas VP, Regan MR, Lin DD, Robinson LC, Agnew WS and Tempel BL. (1997) *J Neurosci* 17, 635–645.
33. Futatsugi A, Nakamura T, Yamada MK, Ebisui E, Nakamura K, Uchida K, Kitaguchi T, Takahashi-Iwanaga H, Noda T, Aruga J and Mikoshiba K. (2005) *Science* 309, 2232–2234.
34. Hasan G and Rosbash M. (1992) *Development* 116, 967–975.
35. Yoshikawa S, Tanimura T, Miyawaki A, Nakamura M, Yuzaki M, Furuichi T and Mikoshiba K. (1992) *J Biol Chem* 267, 16613–16619.
36. Raghu P and Hasan G. (1995) *Dev Biol* 171, 564–577.
37. Venkatesh K and Hasan G. (1997) *Curr Biol* 7, 500–509.
38. Acharya JK, Jalink K, Hardy RW, Hartenstein V and Zuker CS. (1997) *Neuron* 18, 881–887.
39. DalSanto P, Logan MA, Chisholm AD and Jorgensen EM. (1999) *Cell* 98, 757–767.
40. Jaffe LF. (1985) In: *Biology of Fertilization* (Metz CB and Montroy A, eds), vol. 3, 127–165.
41. Miyazaki S, Shirakawa H, Nakada K and Honda Y. (1993) *Dev Biol* 158, 62–78.
42. Miyazaki S, Yuzaki M, Nakada K, Shirakawa H, Nakanishi S, Nakade S and Mikoshiba K. (1992) *Science* 257, 251–255.
43. Whitaker M and Patel R. (1990) *Development* 108, 525–542.
44. Whitaker M and Swann K. (1993) *Development* 117, 1–12.
45. Whitaker MJ and Baker PF. (1983) *Proc R Soc Lond B Biol Sci* 218, 397–413.
46. Zimmerberg J and Whitaker M. (1985) *Nature* 315, 581–584.
47. Ciapa B, Pesando D, Wilding M and Whitaker M. (1994) *Nature* 368, 875–878.
48. Poenie M, Alderton J, Tsien RY and Steinhardt RA. (1985) *Nature* 315, 147–149.
49. Steinhardt RA and Alderton J. (1988) *Nature* 332, 364–366.
50. Twigg J, Patel R, and Whitaker, M. (1988) *Nature* 332, 366–369.
51. Iwasaki H, Chiba K, Uchiyama T, Yoshikawa F, Suzuki F, Ikeda M, Furuichi T and Mikoshiba K. (2002) *J Biol Chem* 277, 2763–2772.
52. Berridge MJ, Downes CP and Hanley MR. (1989) *Cell* 59, 411–419.
53. Kao KR, Masui Y and Elinson RP. (1986) *Nature* 322, 371–373.
54. Condie BG and Harland RM. (1987) *Development* 101, 93–105.
55. Cooke J and Smith EJ. (1988) *Development* 102, 85–99.
56. Kao KR and Elinson RP. (1989) *Dev Biol* 132, 81–90.
57. Yamaguchi Y and Shinagawa A. (1989) *Dev Growth Differ* 31, 531–541.
58. Slack J. (1991) From egg to embryo. Regional specification in early development (Slack JMW, ed.). *Developmental and Cell Biology Series*, vol. 26. Cambridge University Press.
59. Maslanski JA, Leshko L and Busa WB. (1992) *Science* 256, 243–245.
60. Kume S, Muto A, Inoue T, Suga K, Okano H and Mikoshiba K. (1997) *Science* 278, 1940–1943.
61. Klein PS and Melton DA. (1996) *Proc Natl Acad Sci USA* 93, 8455–8459.
62. Hedgepeth CM, Conrad LJ, Zhang J, Huang HC, Lee VM and Klein PS. (1997) *Dev Biol* 185, 82–91.
63. Saneyoshi T, Kume S, Amasaki Y and Mikoshiba K. (2002) *Nature* 417, 295–299.
64. Bosanac I, Alattia JR, Mal TK, Chan J, Talarico S, Tong FK, Tong KI, Yoshikawa F, Furuichi T, Iwai M, Michikawa T, Mikoshiba K and Ikura M. (2002) *Nature* 420, 696–700.
65. Bosanac I, Yamazaki H, Matsu-Ura T, Michikawa T, Mikoshiba K and Ikura M. (2005) *Mol Cell* 17, 193–203.
66. da Fonseca PC, Morris SA, Nerou EP, Taylor CW and Morris EP. (2003) *Proc Natl Acad Sci USA* 100, 3936–3941.
67. Hamada K, Miyata T, Mayanagi K, Hirota J and Mikoshiba K. (2002) *J Biol Chem*, 277, 21115–21118.

68. Hamada K, Terauchi A and Mikoshiba K. (2003) *J Biol Chem* 278, 52881–52889.
69. Jiang QX, Thrower EC, Chester DW, Ehrlich BE and Sigworth FJ. (2002) *EMBO J* 21, 3575–3581.
70. Serysheva II, Bare DJ, Ludtke SJ, Kettlun CS, Chiu W and Mignery GA. (2003) *J Biol Chem* 278, 21319–21322.
71. Sato C, Hamada K, Ogura T, Miyazawa A, Iwasaki K, Hiroaki Y, Tani K, Terauchi A, Fujiyoshi Y and Mikoshiba K. (2004) *J Mol Biol* 336, 155–164.
72. Ando H, Mizutani A, Matsuura T and Mikoshiba K. (2003) *J Biol Chem* 278, 10602–10612.
73. Ando H, Mizutani A, Kiefer H, Tsuzurugi D, Michikawa T and Mikoshiba K. (2006) *Mol Cell* 22, 795–806.
74. Shirakabe K, Priori G, Yamada H, Ando H, Horita S, Fujita T, Fujimoto I, Mizutani A, Seki G and Mikoshiba K. (2006) *Proc Natl Acad Sci USA* 103, 9542–9547.
75. Tanimura A, Nezu A, Morita T, Turner RJ and Tojyo Y. (2004) *J Biol Chem* 279, 38095–38098.
76. Sato M, Ueda Y, Shibuya M and Umezawa Y. (2005) *Anal Chem* 77, 4751–4758.
77. Remus TP, Zima AV, Bossuyt J, Bare DJ, Martin JL, Blatter LA, Bers DM and Mignery GA. (2006) *J Biol Chem* 281, 608–616.
78. Matsu-ura T, Michikawa T, Inoue T, Miyawaki A, Yoshida M and Mikoshiba K. (2006) *J Cell Biol* 173, 755–765.
79. Yoo SH, So SH, Kweon HS, Lee JS, Kang MK and Jeon CJ. (2000) *J Biol Chem* 275, 12553–12559.
80. Hirota J, Ando H, Hamada K and Mikoshiba K. (2003) *Biochem J* 372, 435–441.
81. Boehning D, Patterson RL, Sedaghat L, Glebova NO, Kurosaki T and Snyder SH. (2003) *Nat Cell Biol* 5, 1051–1061.
82. Bezprozvanny I and Hayden MR. (2004) *Biochem Biophys Res Commun* 322, 1310–1317.
83. Tang TS, Tu H, Chan EY, Maximov A, Wang Z, Wellington CL, Hayden MR and Bezprozvanny I. (2003) *Neuron* 39, 227–239.
84. DeSouza N, Reiken S, Ondrias K, Yang YM, Matkovich S and Marks AR. (2002) *J Biol Chem* 277, 39397–39400.
85. Patterson RL, vanRossum DB, Barrow RK and Snyder SH. (2004) *Proc Natl Acad Sci USA* 101, 2328–2332.
86. Bourguignon LY, Jin H, Iida N, Brandt NR and Zhang SH. (1993) *J Biol Chem* 268, 7290–7297.
87. Tu JC, Xiao B, Yuan JP, Lanahan AA, Leoffer t K, Li M, Linden DJ and Worley PF. (1998) *Neuron* 21, 717–726.
88. Higo T, Hattori M, Nakamura T, Natsume T, Michikawa T and Mikoshiba K. (2005) *Cell* 120, 85–98.
89. Maximov A, Tang TS and Bezprozvanny I. (2003) *Mol Cell Neurosci* 22, 271–283.
90. Zhang S, Mizutani A, Hisatsune C, Higo T, Bannai H, Nakayama T, Hattori M and Mikoshiba K. (2003) *J Biol Chem* 278, 4048–4056.
91. Fukatsu K, Bannai H, Zhang S, Nakamura H, Inoue T and Mikoshiba K. (2004) *J Biol Chem* 279, 48976–48982.
92. Fukatsu K, Bannai H, Inoue T and Mikoshiba K. (2006) *Biochem Biophys Res Commun* 342, 573–576.
93. Nakade, S, Maeda, N, and Mikoshiba, K. (1991) *Biochem J* 277, 125–131.
94. Aizman O, Uhlen P, Lal M, Brismar H and Aperia A. (2001) *Proc Natl Acad Sci USA* 98, 13420–13424.
95. Harwood S and Yaqoob MM. (2005) *Front Biosci* 10, 2011–2017.
96. Miyakawa-Naito A, Uhlen P, Lal M, Aizman O, Mikoshiba K, Brismar H, Zelenin S and Aperia A. (2003) *J Biol Chem* 278, 50355–50361.
97. Yuan, Z, Cai, T, Tian, J, Ivanov, AV, Giovannucci, DR and Xie, Z. (2005) *Mol Biol Cell* 16, 4034–4045.
98. Abramowitz J, Dai C, Hirschi KK, Dmitrieva RI, Doris PA, Liu L and Allen JC. (2003) *Circulation* 108(24), 3048–3053.
99. Liu L, Abramowitz J, Askari A and Allen JC. (2004) *Am J Physiol Heart Circ Physiol* 287, H2173–2182.

100. Wang XQ and Yu SP. (2005) *J Neurochem* 93, 1515–1523.
101. Barwe SP, Anilkumar G, Moon SY, Zheng Y, Whitelegge JP, Rajasekaran SA and Rajasekaran AK. (2005) *Mol Biol Cell* 16(3), 1082–1094.
102. Li J, Zelenin S, Aperia A and Aizman O. (2006) *J Am Soc Nephrol* 17, 1848–1857.
103. Nogradi, A, Jonsson N, Walker R, Caddy K, Carter N and Kelly C. (1997) *Brain Res Dev Brain Res* 98, 91–101.
104. Kato K. (1990) *FEBS Lett.* 271, 137–140.
105. Bannai H, Inoue T, Nakayama T, Hattori M and Mikoshiba K. (2004) *J Cell Sci* 117, 163–175.
106. Furuichi T, Simon-Chazottes D, Fujino I, Yamada N, Hasegawa M, Miyawaki A, Yoshikawa S, Guenet JL and Mikoshiba K. (1993) *Rec Chann* 1, 11–24.
107. Knowles RB, Sabry JH, Martone ME, Deerinck TJ, Ellisman MH, Bassell GJ and Kosik KS. (1996) *J Neurosci* 16, 7812–7820.
108. Tiedge H and Brosius J. (1996) *J Neurosci* 16, 7171–7181.
109. Torre ER and Steward O. (1996) *J Neurosci* 16, 5967–5978.
110. Feng Y, Gutekunst CA, Eberhart DE, Yi H, Warren ST and Hersch SM. (1997) *J Neurosci* 17, 1539–1547.
111. Kiebler MA, Hemraj I, Verkade P, Kohrmann M, Fortes P, Marion RM, Ortin J and Dotti CG. (1999) *J Neurosci* 19, 288–297.
112. Bannai H, Fukatsu K, Mizutani A, Natsume T, Iemura S, Ikegami T, Inoue T and Mikoshiba K. (2004) *J Biol Chem* 279, 53427–53434.
113. Feng W, Tu J, Yang T, Vernon PS, Allen PD, Worley PF and Pessah IN. (2002) *J Biol Chem* 277, 44722–44730.
114. Sandona D, Scolari A, Mikoshiba K and Volpe P. (2003) *Neurochem Res* 28, 1151–1158.
115. Mikoshiba K. (2006) J Neurochem. 97, 1627–1633.

J. Krebs and M. Michalak (Editors)
Calcium: A Matter of Life or Death

ISSN: 0167-7306/DOI: 10.1016/S0167-7306(06)41012-7

CHAPTER 12

Ryanodine receptor structure, function and pathophysiology

Spyros Zissimopoulos and F. Anthony Lai

Department of Cardiology, Wales Heart Research Institute, School of Medicine, Cardiff University, Cardiff CF14 4XN, UK, Tel.: +44 29 207 42338; Fax: +44 29 207 43500; E-mail: lait@cardiff.ac.uk, zissimopouloss@cardiff.ac.uk

Abstract

The ryanodine receptor (RyR) is an intracellular calcium release channel located on the sarco(endo)plasmic reticulum of muscle and non-muscle cells. The functional channel is composed of four identical subunits of approximately 560 kDa, which combine to form a high-conductance cation-permeable protein pore. There are three mammalian RyR isoforms that have a wide tissue expression. Their highest levels are in striated muscles where they mediate the release of stored Ca^{2+} leading to a rise in intracellular Ca^{2+} concentration and muscle contraction. Channel activity is regulated by Ca^{2+}, Mg^{2+}, ATP and post-translational modifications, i.e. oxidation/reduction and phosphorylation. In addition, the RyR is regulated by intramolecular protein–protein interactions, as well as by interacting with numerous accessory proteins including the dihydropyridine receptor (DHPR), FK506-binding protein (FKBP), calmodulin (CaM), sorcin and calsequestrin (CSQ). Inherited or acquired defective channel regulation results in abnormal Ca^{2+} handling and leads to neuromuscular disorders and arrhythmogenic cardiac disease.

Keywords: calcium release channel, calmodulin, calsequestrin, DHPR, E–C coupling, FK506BP, heart failure, neuromuscular disorder, phosphorylation, redox status, ryanodine, ryanodine receptor, RyR accessory proteins, RyR pathophysiology, sarco(endo)-plasmic reticulum, sorcin

1. Introduction

The ryanodine receptor (RyR) is a family of Ca^{2+} channels located in the membranes of internal Ca^{2+} storage organelles. It is closely related to the other family of intracellular Ca^{2+} channels, the inositol trisphosphate receptor (IP_3R), with which they provide a regulated pathway for the release of stored Ca^{2+} during Ca^{2+} signalling processes such as muscle contraction and fertilization (reviewed in [1]). The RyR and IP_3R proteins share structural similarities, high homology especially at their carboxyl-terminal sequences that form the channel pore (approximately 40% homology) and functional similarities including Ca^{2+} and ATP regulation.

RyRs were initially observed in skeletal muscle in the early 1970s, where they were visualized in electron micrographs as large electron-dense masses situated along the face of the sarcoplasmic reticulum (SR), spanning the junctional gap between SR and the plasma membrane, and they were therefore termed junctional foot proteins (reviewed in [2]). The RyR gained its present name in the late 1980s after it was found to be the protein that binds ryanodine, a plant alkaloid that enabled purification and molecular characterization of the protein.

Ryanodine is a natural product found in members of the genus *Ryania*, which grow as shrubs or slender trees in several tropical locations in Central and South America, with potent paralytic actions on skeletal and cardiac muscles (reviewed in [3]). Crude ryanodine extracts were originally used to prepare poison arrowheads, and in the 1940s, powdered *Ryania* wood was marketed as an insecticide. A compound purified from stem wood was found to have 700 times the insecticidal potency of the starting material and was designated ryanodine. Ryanodine binds to the RyR with high affinity and specificity, preferably in its open conformation, and thus, ryanodine binding is used as an index of channel activation. The RyR is the only cellular target of ryanodine reported to date.

The RyR was purified from skeletal and cardiac muscle using [^{3}H]–ryanodine as a selective marker and found to exist in a tetrameric form that sediments as a very large (30 S) complex by sucrose density gradient centrifugation [4–11]. The purified protein, incorporated into an artificial planar lipid bilayer, functions as a Ca^{2+} channel, with characteristics identical to the Ca^{2+} release channel observed upon incorporation of the crude SR fraction [6,8,10,12,13].

2. *RyR isoforms and distribution*

2.1. *Isoforms*

Biochemical, molecular or pharmacological evidence for the presence of RyRs has been found in vertebrates (mammals, birds, amphibians, reptiles and fish) as well as invertebrates including crustaceans, insects and nematodes (reviewed in [14,15]).

In mammals, three RyR isoforms encoded by distinct genes have been cloned and fully sequenced, RyR1 in skeletal muscle [16,17], RyR2 in heart [18,19] and RyR3 in brain [20], revealing subunit composition of approximately 5000 amino acids and molecular weight of approximately 560 kDa (Table 1). The currently accepted terminology is based on the abundance and timing of purification of the RyRs from various tissues. Thus, RyR1 is also known as the skeletal type, RyR2 is also known as the cardiac type, whereas RyR3 was initially but misleadingly termed the brain type. Cloned RyRs have been heterologously expressed in various mammalian cell lines including CHO, HEK293 and COS-1 [16,21–26]. Expressed proteins were indistinguishable from the native channels in immunoreactivity, molecular size, sucrose density gradient sedimentation and ryanodine-binding properties. Furthermore, recombinant channels had conductance, kinetics of opening, current–voltage relationship, cation permeability and modulation by physiological and pharmacological ligands similar to the native channels.

The overall percentage identity of the three mammalian RyRs is approximately 70% (Table 1), with the highest sequence identity present at the carboxyl-terminal

Table 1

Homology between ryanodine receptor isoforms

	RyR1 human	RyR1 pig	RyR1 rabbit	RyR1 fish	RyR1 frog	RyR2 rabbit	RyR2 human	RyR2 mouse	RyR3 human	RyR3 rabbit	RyR3 mink	RyR3 chicken	RyR3 frog	RyR sea urchin	RyR *Drosophila*	RyR *Caenorhabditis elegans*	Length(aa)	GenBank accession number
RyR1 human	100																5032	J05200
RyR1 pig[a]	96	100															5035	M91452
RyR1 rabbit	96	96	100														5037	X15750
RyR1 fish	74	74	74	100													5081	U97329
RyR1 frog	78	78	78	78	100												5037	D21070
RyR2 rabbit[b]	65	65	65	64	66	100											4969	M59743
RyR2 human	65	65	65	64	66	98	100										4967	X98330
RyR2 mouse	64	64	64	63	66	97	97	100									4968	AF295105
RyR3 human[c]	64	64	64	63	66	67	66	67	100								4866	AB001025
RyR3 rabbit	64	64	65	63	66	67	67	67	95	100							4872	X68650
RyR3 mink	64	65	65	64	67	67	67	67	95	95	100						4859	Y07749
RyR3 chicken	64	64	64	64	66	67	67	67	86	86	86	100					4869	X95267
RyR3 frog	65	64	65	64	68	68	68	67	85	85	85	85	100				4868	D21071
RyR sea urchin	44	44	44	43	44	46	45	45	34[d]	33[d]	33[d]	34[d]	34[d]	100			5317	AB051576
RyR *Drosophila*	44	44	44	42	43	45	45	45	34[d]	33[d]	34[d]	35[d]	35[d]	44	100		5126	D17389
RyR *Caenorhabditis elegans*	39	39	39	38	38	41	41	41	34[d]	35[d]	35[d]	32[d]	33[d]	40	45	100	5071	D45899

Table illustrating the percentage sequence identity of the various optimally aligned ryanodine receptors (RyRs). The sequence alignments between RyR pairs were produced with the 'pairwise BLAST' software available at the NCBI website (http://www.ncbi.nlm.nih.gov).

[a]A second pig RyR1 cDNA sequence has also been published (X62880); at the protein level, the second sequence presents a deletion of one residue and five additional mismatches.

[b]A second rabbit RyR2 cDNA sequence has also been published (U50465); at the protein level, the second sequence presents a deletion of one residue and six additional mismatches.

[c]A second human RyR3 cDNA sequence has also been published (AJ001515); at the protein level, the second sequence is four residues longer due to two alternative splicing events: the first involves replacement of 26 amino acids by a distinct 25-residue fragment and the second involves insertion of five amino acids and there are also 18 additional mismatches.

[d]Vertebrate RyR3 isoforms do not align with sea urchin, *Drosophila* and *Caenorhabditis elegans* RyRs over the entire sequence but only over fragments of low identity; the figure given is the average one.

end. Lower overall percentage identity is due to three regions of marked divergence (D1, D2 and D3) [27]. With reference to the RyR1 sequence, D1 spans amino acids 4254–4631, D2 spans amino acids 1342–1403 and D3 lies between residues 1872 and 1923. D1 is the largest and most variable region, whereas D2 is completely absent in RyR3.

Non-mammalian vertebrates (birds, amphibians, reptiles and fish) also express three subtypes, initially called the α-type, β-type and the cardiac type [28–33]. On the basis of sequence comparisons with the mammalian isoforms, the α- and β-RyRs are identified as the mammalian RyR1 and RyR3, respectively [34,35]. Only one type of RyR has been detected in invertebrates including lobster [36], fly [37] and nematode [38]. Full-length cDNA sequences are known for sea urchin (*Hemicentrotus pulcherrimus*), fly (*Drosophila melanogaster*) and nematode (*Caenorhabditis elegans*), which show 35–45% similarity with the vertebrate RyRs, the highest similarity with the RyR2 mammalian isoform (Table 1). Phylogenetic analysis suggests that the RyR2 isoform diverged from a single ancestral gene before RyR1 and RyR3 isoforms to form a distinct branch of the RyR family tree [39]. The RyR2 is also the subtype with the highest conservation across species (Table 1).

Alternative splicing that could introduce functional complexity in the RyR family has been described [18,39–44]. The resulting RyR proteins may have unique functional roles as some splice variants are expressed in a tissue- and developmental stage-specific manner. Indeed, a dominant negative RyR3 splice variant, which lacks residues 4406–4434 encompassing a putative transmembrane (TM) domain, has been described in smooth muscle tissues but not in skeletal muscle, heart or brain [44]. Heterologous expression of the splice variant did not result in a functional channel, whereas co-expression with wild-type RyR3 resulted in heteromeric channels with suppressed activity compared to wild-type homotetramers. In addition, aberrant splicing of RyR1 mRNA has been implicated in impaired Ca^{2+} homeostasis in myotonic dystrophy type 1 (DM1) muscle [45]. It was shown that the foetal RyR1 variant, which lacks residues 3481–3485, was significantly increased in DM1 skeletal muscle and exhibited reduced activity compared to the wild-type protein. Additional complexity may arise from the formation of mixed heterotetrameric RyR channels. An immunoprecipitation study of recombinant, heterologously co-expressed RyRs provided evidence that RyR2 could form heterotetramers with RyR1 and RyR3 but not RyR1 with RyR3 [46]. However, the presence of native tissue RyR heterotetramers has not so far been reported.

2.2. Tissue and cellular distribution

Biochemical, molecular or pharmacological evidence indicate that RyRs have a widespread expression including skeletal, cardiac and smooth muscles, neurons of both the central and peripheral nervous systems, liver, lung, kidney, testis, ovary, endothelial, epithelial, pancreatic and adrenal chromaffin cells, osteocytes, neutrophils and macrophages (reviewed in [14,15]). The RyR is primarily located on the SR of muscle cells and the endoplasmic reticulum of non-muscle cells. It has also been reported that the RyR is found in mitochondria, where it mediates mitochondrial Ca^{2+} uptake [47], and in secretory vesicles of a pancreatic β-cell line [48].

The highest RyR levels exist in striated muscle, RyR1 in skeletal and RyR2 in cardiac muscle, whereas all three isoforms are expressed in brain and several smooth muscles [16,18–20,28,30,31,33,49–56]. In mammals, the RyR1 isoform is the predominant form expressed in skeletal muscle and in some parts of the brain, most prominently in Purkinje cells of the cerebellum. The RyR2 isoform is the predominant form in cardiac muscle and is also the most widely and abundantly distributed isoform in the brain. RyR3 is expressed in a wide range of cells including the brain, skeletal and smooth muscles and epithelial cells, but at low levels. In contrast to mammalian skeletal muscle, amphibian and avian skeletal muscles express both RyR1 and RyR3 (the α- and β-types, respectively) at approximately equal amounts. It is now appreciated that the myocardium is the source with the purest RyR composition (RyR2), whereas mammalian skeletal muscle is the richest source of RyR1 with a minor contamination with RyR3. The highest levels of RyR3 in mammals have been reported in diaphragm muscle, but they still represent $<5\%$ of the overall RyR population [57,58].

2.3. Gene knock-out studies

The physiological role of the RyR has been addressed by gene knock-out studies. Mice deficient in RyR1 died at birth with gross abnormalities of the skeletal muscle and had abolished contractility following electrical stimulation [59]. Similarly, mice deficient in RyR2 died at the embryonic stage due to severe cardiac defects and impaired Ca^{2+} homeostasis although there was apparently normal excitation–contraction (E–C) coupling [60]. Both RyR1- and RyR2-deficient mice presented marked degeneration of skeletal and cardiac muscle, respectively, indicating the importance of these proteins in normal muscle maturation. On the other hand, embryonic lethality did not allow investigation of the role that RyR1/2 could play in other tissues. In contrast to RyR1/2 knock-out animals, mice deficient in RyR3 were viable with no gross abnormalities and exhibited functional E–C coupling in adults [61,62]. However, skeletal muscle contraction was impaired during the first weeks after birth suggesting a role for RyR3 in neonatal skeletal muscle. In addition, physiological and behavioural studies have revealed that the RyR3-deficient mice exhibit altered hippocampal synaptic plasticity and impairment of learning capabilities [63,64].

3. Channel properties

Ion conduction through the RyR, being an intracellular channel, cannot be monitored using standard whole-cell recording techniques, but it can be monitored following incorporation of the channel into artificial lipid bilayers (single channel recordings). Although RyR is impermeable to anions, the channel is permeable to a wide range of inorganic divalent and monovalent cations as well as organic monovalent cations (reviewed in [65]). The unitary channel conductance for the three mammalian RyR subtypes is similar and extremely high, approximately 10-fold higher than the plasma membrane Ca^{2+} channel [66–70]. Despite channel conductance being generally lower with divalent (e.g. 135–150 pS for Ca^{2+}) than monovalent cations (e.g. 720–800 pS for K^+) as the charge carrier, divalent cations are six to seven times more permeable than K^+. Hence, the RyR selects rather poorly between cations, and it is certainly not as selective as the plasma membrane Ca^{2+} channel. However, Ca^{2+} is essentially the only ion released from the SR, because it is the only cation with a significant concentration gradient across the SR membrane. It appears that due to its localization, the RyR has evolved into a high-conduction pore with little need for ion discrimination.

The RyR1 is permeable to relatively large compounds including choline and xylose (radius approximately 2.9 Å and approximately 3.4 Å, respectively) suggesting a rather wide pore [66,71]. Molecular sieving experiments using organic monovalent cations of increasing size have indicated that the minimal width of the RyR2 pore is approximately 3.4 Å [72], in agreement with streaming potential measurements indicating that H_2O (radius 1.5 Å) and Cs^+ (radius 2 Å) must pass in a single file fashion [73]. Electrical distance measurements using trimethylammonium derivatives of varying length indicated that the length of the voltage drop along the RyR2 channel pore is approximately 10.4 Å [74], consistent with streaming potential measurements showing that three H_2O molecules (approximately 9 Å) occupy the single file region of the pore [73]. Voltage-dependent block of ion translocation by organic compounds such as tetramethylammonium and triethylamine (radius 3.6 Å) was found to occur at a site located 90% into the voltage drop from the cytosolic face of the channel [72,75]. These investigations indicate that the RyR pore is effectively a wide tunnel with a short constriction of approximately 7 Å in diameter near its luminal exit, and a total length of approximately 10 Å. Thus, the RyR provides a pathway for cation translocation across the SR membrane (45–50 Å) through a short pore (approximately 10 Å) that is flanked by large cytosolic and/or luminal mouths or vestibules.

Several studies have indicated the presence of negatively charged residues on both the cytosolic and the luminal ends of the conductance pathway of the RyRs. Addition of carboxyl-neutralizing chemical modifiers to the luminal face of RyR2 significantly reduced channel conductance [76]. Large organic and inorganic cations, anaesthetics and peptides were shown to block ion translocation through RyR1 and RyR2 channels from the cytosolic side [75,77–79], whereas polycations such as neomycin and ruthenium red were capable of blocking the channel from both sides [80,81]. It has been proposed that the high density of negative charge at both mouths of the RyR pore serves not only to deny access to anions but to increase the local cation concentration ensuring a high-delivery rate to the pore, whereas it could also contribute to the relative permeability of divalent over monovalent cations [82].

The tetrameric RyR channel comprises a single pore although sublevel conductance states corresponding to one-fourth, one-half and three-fourth of the main conductance level for the purified channel have been reported [66]. Four subconductance states have also been detected following dissociation of the small accessory

FK506-binding protein 12 (FKBP12), which led to the proposal that the channel contains four pores with each subunit contributing an individual ion-conducting pathway [83]. However, single pore channels can also exhibit subconductance states because of uncoordinated monomers. Evidence for a single pore is provided by a study that combined two approaches: modification of the RyR2 pore by methanethiosulfonate ethylammonium ($MTSEA^{+}$) to reduce current amplitudes and the use of different-sized current carriers [84]. $MTSEA^{+}$ modification decreased the number of channel substates as the diameter of the current carrier increased, indicating that the conduction pathway is comprised of a single central pore. Further support for a single ion-conducting pathway comes from heterologous co-expression of wild-type RyR2 and a low-conductance pore mutant that produced six groups of channel with conductance ranging from wild type to that of homomeric mutant [85]. The six conductance groups were then correlated with possible arrangements of mutant and wild-type subunits, and it was concluded that the conduction pathway is a single pore created by an equal contribution from each subunit. A single pore model is also consistent with the three-dimensional structure of the RyR as visualized by single-particle electron microscopy (see below).

4. *RyR structure*

4.1. *Three-dimensional architecture*

The detailed atomic three-dimensional structure of the RyR is not known and awaits crystallization of the protein. However, gross three-dimensional reconstructions have been obtained for the three mammalian RyRs at a resolution of approximately 30 Å using image-processing techniques applied to electron micrographs of individual frozen-hydrated purified channels [58,86–88]. In general, the shapes of the three mammalian RyRs are very similar; there are however, some points of structural heterogeneity that could account for the specific properties of each isoform. Overall, the RyR resembles a mushroom with a 'cap' composed of the cytoplasmic domains and the 'stalk' made up of the membrane-spanning domains (Fig. 1). The structure has a 4-fold cyclic symmetry around an axis perpendicular to the membrane consistent with the tetrameric nature of the channel. The cytoplasmic part has an overall square prism shape (approximately 28 nm × 28 nm × 12 nm), composed of globular parts interconnected by segments. The TM assembly has a square shape in transverse cross section, with an edge measuring around 12 nm at the attachment side to the cytoplasmic region and a length of about 7 nm in the direction perpendicular to the SR membrane, more than enough to traverse a membrane bilayer. More recently, through improvements in the preparation of the frozen samples, software analysis and processing of a larger number of pictures, structures have been resolved at a higher resolution (approximately 14 Å) [89,90]. At the highest resolution achieved to date (9.6 Å), some rod-like densities indicative of α-helices have been assigned in the membrane-spanning region of the channel [91].

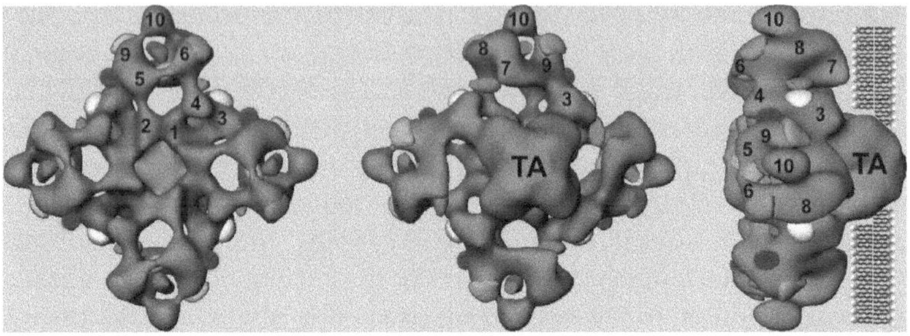

Fig. 1. Ryanodine receptor three-dimensional structure. RyR architecture displayed as a projection view from the cytoplasm to sarcoplasmic reticulum (SR) lumen (left panel), the view from SR lumen to cytoplasm (middle panel) and side view (right panel). The numerals on the cytoplasmic assembly refer to distinguishable globular domains. TA indicates the transmembrane assembly. The location of the three divergent regions D1 (yellow), D2 (orange) and D3 (purple) as well as the N-terminal (red) and central domains (light green) is shown. The location of the binding sites for FK506-binding protein 12 (FKBP12) (brown) and calmodulin (green) is also shown. Images were kindly prepared by Dr. Zheng Liu (Wadsworth Center, Albany, NY, USA) (See Color Plate 35, p. 533).

The structure of the RyR has also been determined in an open conformation, and although the gross shape is similar to the closed state, there are some differences [92–94]. The channel in the open state is slightly taller because of a vertical elongation of the TM assembly and the clamp-shaped domains at the corners of the structure. Most significant is the presence of a low central density (approximately 2 nm in diameter) in the TM assembly of the open form connecting the luminal and cytoplasmic sides, indicating a single pore within the tetrameric channel. The TM assembly is also rotated by approximately 4° with respect to the cytoplasmic region. In addition, two subdomains of the clamp region that are joined in the closed conformation of the channel are separated in the open state. These studies indicate that alterations in the functional state of the RyR involve widespread changes in the structure of both the cytoplasmic and TM domains of the channel.

The binding sites for two accessory proteins, calmodulin (CaM) and FKBP12, have been localized by determining differences in shape between free RyR and RyR protein-bound complexes [95]. Both CaM and FKBP12 are situated further than 10 nm away from the putative TM pore, suggesting that conformational changes can be transmitted over long distances (Fig. 1). Several RyR regions including the N-terminal [96,97] and central domains [98] and the three divergent regions [99–102] have also been localized using chimaeras of either RyR with insertions of green fluorescent protein (GFP) or glutathione *S*-transferase (GST) or binding of sequence-specific antibodies.

4.2. Membrane topology

Molecular cloning and hydropathy plot analysis of the predicted peptide sequence have indicated a large hydrophilic N-terminal region and a smaller hydrophobic C-terminus [16,17]. The large N-terminal region faces the cytoplasm and constitutes the 'foot' structure seen in electron micrographs of junctional SR. Studies of truncated channels have provided evidence that the C-terminal portion contains the TM sequences that form the channel pore. Limited proteolysis of the native RyR1 tetramer (30 S) yielded a smaller tetrameric complex (14 S) composed of the C-terminal 76-kDa fragment of the RyR1 monomers, which formed a functional ion channel upon incorporation into planar lipid bilayers [103]. Heterologous expression of a truncated RyR1 encompassing the N-terminal 182 residues fused with the C-terminal 1030 amino acids and yielded a protein of about 130-kDa that functioned as an ion channel with conductance comparable with that of the full-length protein [104,105]. Similarly, a truncated version of an insect RyR isoform containing the N-terminal 276 residues fused with the C-terminal 1484 amino acids was also shown to function as an ion channel [106]. Cells expressing the truncated RyR1 protein had a higher resting Ca^{2+} concentration and smaller Ca^{2+} pools [104,105] that were also observed in cells expressing RyR2 C-terminal constructs [107]. These findings suggest that the RyR C-terminus forms a Ca^{2+} conduction pore that is constitutively open.

Putative membrane-spanning domains are located in the C-terminal (10–20% of the protein, last 500–1000 amino acids), and two topology models of 4TM or 10TM segments have been proposed [16,17]. Based on their hydrophobicity and their conservation across species and isoforms, the four segments of the 4TM model corresponding to M5, M6, M8 and M10 of the 10TM model emerge as strong candidates for membrane-spanning helices. Both models are consistent with the cytoplasmic location of the amino- and carboxyl-termini shown by immunolocalization studies, indicating that an even number of sequences traverse the membrane [108,109]. The further use of site-directed antibodies discriminating between cytoplasmic and luminal locations was in support of the 4TM model although inconsistent with the assignment of M9 as a membrane-spanning sequence with the corresponding M8–M10 loop found to be luminal [39,109].

The use of C-terminal-truncated channels carrying a C-terminal GFP tag, with deletions starting near the beginning or the end of predicted TM helices M1–M10, indicated that RyR1 contains eight TM sequences organized as four hairpin loops formed from M4a–M4b, M5–M6, M7a–M7b and M8–M10 [110,111]. Hydrophobic segments M1, M2, M3 and M9 did not traverse the membrane, whereas the long (approximately 50 residues) M7 domain was shown to form a double TM hairpin. M4 was found to be membrane bound, but the mechanism of membrane association is unclear; M4 can either form a TM hairpin or associate in an unorthodox fashion with the cytosolic leaflet of the membrane. Support for the above biochemical findings is provided by the most recent three-dimensional reconstruction of the RyR1 at 9.6 Å resolution using single-particle electron microscopy of frozen purified channels [91]. Five helices were clearly resolved in the membrane-spanning portion of the channel, although additional helices may exist that could not be unambiguously identified at the resolution achieved. However, the structure obtained suggested a more complicated arrangement than a simple arrangement of membrane-spanning helices, because only helix 1 (assigned to the predicted M10 domain) had either sufficient length (approximately 45 Å) or the proper orientation to span the membrane. Helix 3 was found oriented parallel to the plane of the membrane on the

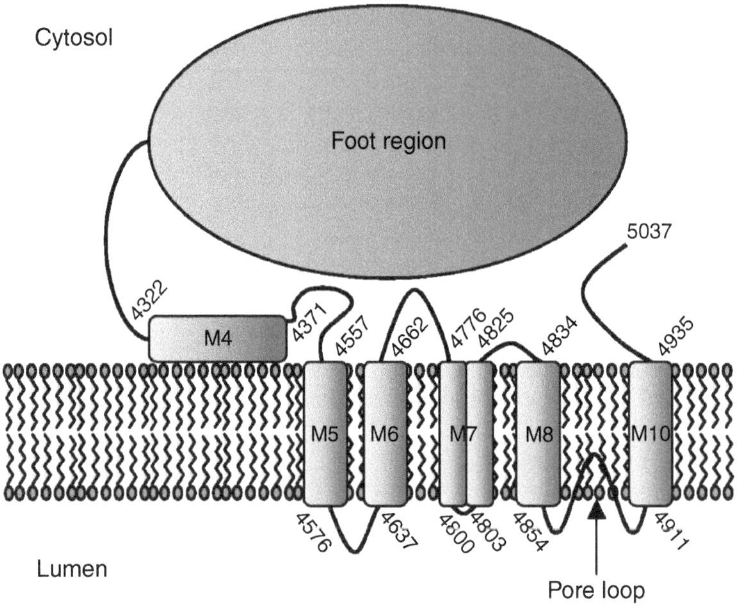

Fig. 2. Ryanodine receptor (RyR) membrane topology. Transmembrane domain assignment is based on the 10 transmembrane (10TM) model [17] and revised according to recent experimental evidence [91,110,111]. Hydrophobic sequences M5, M6, M8 and M10 are shown to span the membrane once, M7 is shown as a transmembrane hairpin and M4 to reside parallel to the membrane on the cytoplasmic side. Hydrophobic domain M9 dips inside the lipid bilayer from the luminal side and forms the pore loop and selectivity filter. The bulky amino-terminal portion forms the 'foot structure' seen in electron micrographs. Coordinates given are for rabbit RyR1.

cytoplasmic side and most probably corresponds to M4. The currently favoured RyR membrane topology is depicted in Fig. 2.

4.3. Pore structure

The determination of the structure of K^+ channels at atomic resolution [112,113] has enabled the prediction of critical sequences in RyR that participate in cation translocation and selection. The conserved GGGIG motif in RyRs (residues 4894–4898 in RyR1 within domain M9) is a sequence related to the K^+ channel TVGYG motif that comprises the selectivity filter implying that the luminal loop between the last two TM helices (M8 and M10) may fold back into the membrane in a manner similar to that found for the K^+ channel [114]. Several studies of recombinant mutant RyR1 or RyR2 channels using measurements of intracellular Ca^{2+} concentration, ryanodine binding and single channel recordings have indicated that mutations within the GGGIG motif and neighbouring residues produce profound alterations in channel activity and, most importantly, in cation conductance and selectivity [85,115–117]. Channel activation and gating were also altered by mutations in the last TM helix suggesting that it forms the inner helix of the pore [118–120]. In addition, other studies have identified the presence of rings of negative charges in the luminal (D4899 and E4900) and cytosolic (D4938 and to a lesser extent D4945) vestibules that are required for maintaining high rates of ion translocation and Ca^{2+} selectivity in RyR1 [121,122]. It has also been proposed that the conserved GLIIDA motif within M10 (residues 4934–4939 in RyR1) is analogous to the gating hinge motif (GXXXXA) identified in K^+ channels, with G4934 (G4866 for RyR2) introducing a kink in the TM helix [82,91,122]. The importance of domains M9 and M10 in channel function is also highlighted by the occurrence of natural mutations associated with skeletal and cardiac muscle diseases (see Sections 10.2. and 10.3.).

The pore structure of the closed RyR1 channel has been described at 9.6 Å resolution, the highest resolution achieved to date [91]. At this resolution, at least five helices within the membrane-spanning region could be resolved, including TM helix 1 that was noticeably kinked and short helix 2 both lining the pore. Helix 1, which is likely to correspond to M10, was assigned as the inner, pore-lining helix, whereas helix 2, likely to correspond to M9, was assigned as the pore helix associated with the selectivity filter. The arrangement of the helices and the kink in the inner, pore-lining helix suggested that the RyR structure resembles the pore structure of the *Methanobacterium thermoautrophicum* K^+ channel. A hypothetical model of the RyR ion pore is shown in Fig. 3.

5. *E–C coupling*

E–C coupling is a term used to describe the events that link plasma membrane depolarization to the release of Ca^{2+} from the SR, which in turn triggers contraction. Central to this process is the functional interaction between the RyR and the surface voltage-activated L-type Ca^{2+} channel or dihydropyridine receptor (DHPR).

In skeletal muscle, the terminal cisternae (TC) of the SR (also referred to as junctional SR) are closely apposed to either the surface membrane to form the ‘peripheral couplings’ or the transverse tubules (T-tubules), which are tubular invaginations of the plasma membrane, to yield the ‘triads’ (reviewed in [2]). The triad is so named because it is composed of two TC closely facing a central T-tubule. RyR1s are visible in electron micrographs as dense particles initially named ‘feet’ that span the whole distance between the junctional SR and the surface membrane. Electron microscopy studies have also indicated that RyR1s are concentrated on the TC and form highly ordered arrays [123,124]. The skeletal muscle DHPR isoform ($DHPR_s$) on the plasma membrane and T-tubules in particular acts as voltage sensor as well as calcium channel, and it is coupled through a direct physical association with RyR1 [125–130]. This direct mechanical coupling provides the basis of the E–C coupling mechanism in skeletal muscle (Fig. 4). Thus, T-tubule depolarization is sensed by the $DHPR_s$ that assumes a conformational change that is then transmitted to RyR1 causing the channel to open and release Ca^{2+} from the SR. Ca^{2+} influx is very small, and in fact, adult vertebrate skeletal muscles can contract vigorously even in the absence of extracellular Ca^{2+}.

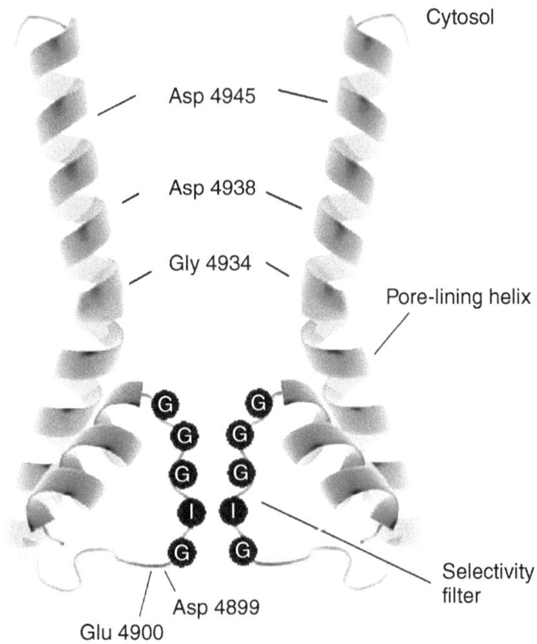

Fig. 3. Ryanodine receptor (RyR) pore structure. Hypothetical model of the RyR pore structure based on the model proposed by [122]. The inner transmembrane helix corresponds to the predicted M10 domain, whereas the pore helix corresponds to M9 with the GGGIG motif (residues 4894–4898 in RyR1) forming the selectivity filter. The putative position of negatively charged amino acids D4899 and E4900 on the luminal side and D4938 and D4945 on the cytosolic side is shown. G4934 may act as a gating hinge producing a kink in the inner transmembrane helix. Note that only two of the four pore-forming segments are shown. Image kindly prepared by Dr. Leon D'Cruz (Wales Heart Research Institute, Cardiff University, UK).

Morphological studies have also shown that $DHPR_s$ channels are clustered in groups of four ('tetrads'), which form ordered arrays that face the ordered RyR1 arrays on the SR [131–133]. The formation of $DHPR_s$ tetrads is dictated by their interaction with RyR1 in such a way that each of the four $DHPR_s$s composing the tetrad is linked to a subunit of an underlying RyR1. Interestingly, $DHPR_s$ tetrads are associated with alternate RyR1s, leaving a set of orphan RyR1s that are not directly linked to $DHPR_s$s. It is believed that uncoupled RyR1s are either activated by Ca^{2+} released from adjacent, $DHPR_s$-coupled channels or by transmission of the conformational change through physical interactions between RyR1s within the array. The functional interaction between RyR1 and $DHPR_s$ is bidirectional because $DHPR_s$ controls gating of RyR1, and the RyR1 in turn affects $DHPR_s$ channel properties [134].

In cardiac muscle, the SR has a similar but not identical arrangement to that of skeletal muscle (reviewed in [2]). Peripheral couplings are also found in cardiomyocytes, but the structures corresponding to the triads are 'diads', because TC are comparatively less numerous, occupying only one of the two faces along the

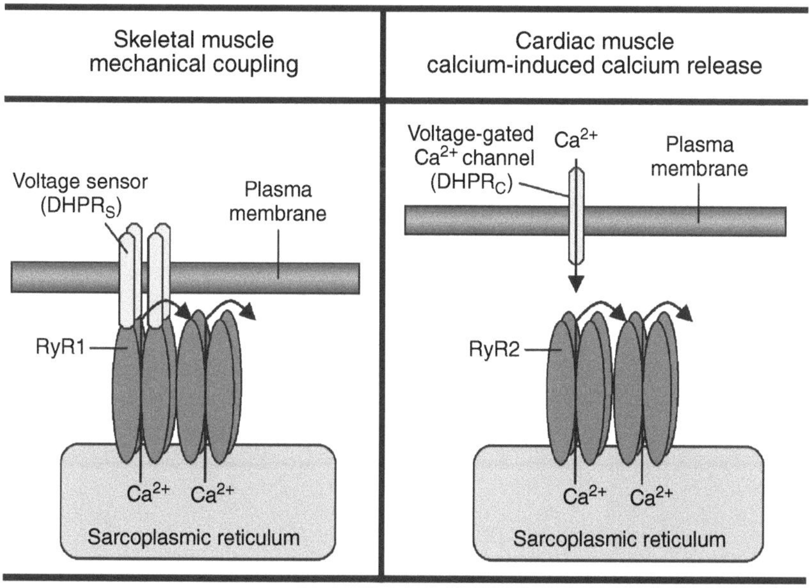

Fig. 4. Ryanodine receptor (RyR) activation in striated muscles. In skeletal muscle dihydropyridine receptor ($DHPR_s$) channels sense the plasma membrane depolarization and transmit a conformational change to the RyR1 causing the channel to open and release Ca^{2+} from the sarcoplasmic reticulum (SR) store. Ca^{2+} influx is not required. Alternate RyR1 channels that are not coupled to $DHPR_s$ may either be activated by Ca^{2+} released from a neighbouring RyR1 or by transmission of the conformational change through physical interactions between RyR1 channels within the array. In cardiac muscle, $DHPR_c$ channels are activated by plasma membrane depolarization and allow Ca^{2+} influx into the cytoplasm. The Ca^{2+} entering the cell activates RyR2 channels allowing Ca^{2+} release from the SR store. Neighbouring RyR2 channels may in turn be activated by Ca^{2+} flowing from within the SR or by conformational changes transmitted through physical interactions between RyR2 channels within the array (See Color Plate 36, p. 533).

T-tubules. Moreover, TC, termed corbular or extended SR, are also present in the interior of the cell without bearing any relationship to the plasma membrane. In cardiac muscle, although the cardiac DHPR isoform ($DHPR_c$) and RyR2 are localized in close proximity to each other, there is no evidence for direct physical interaction, and the $DHPR_c$ does not form ordered arrays of tetrads but they are randomly distributed with respect to the RyR2 arrays [135–137]. There is however, a functional $DHPR_c$–RyR2 association, and Ca^{2+} influx is an absolute requirement for the mechanism of E–C coupling in the heart [128,138–141]. Thus, plasma membrane depolarization activates the $DHPR_c$ allowing extracellular Ca^{2+} to flow into the cytoplasm, which in turn activates RyR2 channels to release Ca^{2+} from the SR (Fig. 4). It has been estimated that the opening of a single $DHPR_c$ channel triggers Ca^{2+} release from 4–6 RyR2 channels [118]. This mechanism is known as 'calcium-induced calcium release' (CICR), and in the case of the heart, this term refers to RyR2 activation by Ca^{2+} coming from both outside the cell and within the SR. The spatial organization of $DHPR_c$ with relation to RyR2 channels is very important for normal E–C coupling. Reduced efficiency of E–C coupling has been implicated in heart failure (see Section 10.1.), attributed to a reduction in the density of T-tubules and therefore $DHPR_c$ channels and/or reorganization of T-tubules resulting in a greater number of functionally uncoupled RyR2 channels [142–146].

The CICR process is intrinsically self-regenerating because the Ca^{2+} released by a RyR2 channel should feedback and further activate the channel or its neighbours, and it should therefore lead to release of all the calcium stored in the SR. However, SR Ca^{2+} release in an all-or-none fashion is not observed, but rather it is graded and tightly controlled by the magnitude and duration of the L-type Ca^{2+} current. The implication is that some sort of negative control mechanism(s) must exist to counter the inherent positive feedback of CICR and terminate calcium release. Several mechanisms have been proposed, including Ca^{2+}-dependent RyR inactivation that may also involve accessory protein(s), stochastic attrition, adaptation, coupled gating of adjacent channels and local depletion of SR calcium; however, none of these mechanisms appears sufficient by itself to terminate calcium release (reviewed in [147,148]).

6. *Modulation by pharmacological agents*

A plethora of exogenous compounds, including volatile and local anaesthetics, 4-chloro-m-cresol, polylysine, doxorubicin, dantrolene, neomycin and peptide toxins, has been shown to modulate RyR channel activity (reviewed in [149]). Of special note is caffeine, which is the most commonly used drug to activate the RyR, ruthenium red, which is the most commonly used channel blocker, and ryanodine, which is discussed Section 6.1.

6.1. *Ryanodine*

Ryanodine and related ryanoid compounds have complex effects on channel conductance and gating in all three RyR isoforms (reviewed in [3]). Initial studies using skeletal and cardiac muscle preparations demonstrated that ryanodine results in an increase as well as in a decrease in the Ca^{2+} permeability of the SR, and its effects are concentration and use dependent. In particular, ryanodine at low concentrations activates Ca^{2+} release from the SR but prolonged exposure to micromolar concentrations inhibits Ca^{2+} release.

The action of ryanodine is clearly documented by single channel recordings [66,150–152]. Several studies have shown that ryanodine at submicromolar to micromolar concentrations (up to 10 μM) results in the channel exhibiting partially conducting states. Multiple subconductance levels have been reported but the most frequently observed one is a sole, long lasting substate that is approximately 40–50% of the full conductance state. This effect is observed consistently and has become a signature for a ryanodine-modified RyR channel. Ryanodine at very low doses (approximately 10 nM) increases the frequency of channel openings to the full conductance level, whereas micromolar concentrations (>10 μM) block the channel. Further studies demonstrated that ryanodine sensitizes the RyR to Ca^{2+} activation through an allosteric mechanism of interaction [153–155]. It was also shown that the ryanodine-modified channel is sensitive to Mg^{2+} inhibition and to activation by caffeine and ATP.

The complex effects of ryanodine are most likely due to the presence of both high-affinity ($K_d \sim 10\,nM$) and low-affinity ($K_d > 1\,\mu M$) sites [8,156–158]. Results obtained for [^{3}H]–ryanodine binding to purified RyRs suggest a single high-affinity site per tetrameric RyR channel. Occupation of the low-affinity sites slows the dissociation of ryanodine from the high-affinity site. After occupation of the low-affinity sites, the channel undergoes a slow transition to a state characterized by persistent channel inactivation, which is associated with decreased ryanodine binding. It has been proposed that these effects are due to an allosteric or steric negative interaction between four initially identical binding sites, with high-affinity binding for the first ryanodine molecule locking the channel in a subconductance state and inhibiting ryanodine binding to the other sites. High-affinity ryanodine binding is observed under conditions that are associated with activation of the RyR channel, indicating that ryanodine binds to a conformation associated with the open state of the channel [156,159]. This property has proved useful, because it allows ryanodine binding to be used as an index of RyR channel activation.

The high-affinity binding site for ryanodine has been localized to the carboxyl-terminal domain of the protein. Limited proteolysis of the native RyR1 tetramer (30 S) that was covalently bound with [^{3}H]–ryanodine (with the use of a photo-activated cross-linking derivative) or pre-labelled with it yielded a smaller tetrameric complex (14 S) composed of the C-terminal 76-kDa fragment of the RyR1 monomers, which retained pre-bound [^{3}H]–ryanodine [160,161]. Further evidence comes from expression studies of truncated, C-terminal RyR protein fragments that formed functional channels upon incorporation into planar lipid bilayers and were modified to the characteristic approximately 50% subconductance state by ryanodine [104,106]. It has been suggested that ryanodine binding occurs within the channel pore, as mutation of residues within the pore region (M9 and M10 domains) alters the ability of the receptor to bind ryanodine without affecting other characteristics of channel function [117,118]. Moreover, interaction of ryanodine and related ryanoids with the RyR is voltage dependent and sensitive to the net charge of the ryanoid, suggesting that the site of interaction may be within the voltage drop across the channel pore [162,163].

7. *Modulation by endogenous effectors*

RyR channel activity is regulated by a wide variety of endogenous molecules, with Ca^{2+}, Mg^{2+} and ATP being the key regulators (reviewed in [149]).

7.1. *Cytosolic* Ca^{2+}

Ca^{2+} is thought to be the physiological channel activator because other ligands either cannot activate the channel in its absence or they require Ca^{2+} for maximum effect. SR Ca^{2+} flux measurements, single channel recordings and ryanodine-binding assays have demonstrated that cytosolic Ca^{2+} has a biphasic effect on skeletal muscle RyR channel activity [164–169]. The threshold for channel activation is approximately 100 nM with a maximum in the range of 10–100 μM, whereas millimolar Ca^{2+} almost entirely inhibits the channel. Such a bell-shaped, cytosolic Ca^{2+}-dependence curve suggests that the RyR1 contains high-affinity Ca^{2+}-binding site(s) that stimulate the channel and inhibitory low-affinity site(s). However, Ca^{2+} alone cannot fully activate RyR1. The cardiac RyR shows some marked differences to the skeletal type [170–175]. The most important ones are that RyR2 is less sensitive to inhibition by high Ca^{2+} concentrations and that almost maximum activation can be achieved by Ca^{2+} alone at approximately 100 μM. RyR3 resembles RyR2 more closely than RyR1, with lower sensitivity to Ca^{2+} inactivation and full activation by Ca^{2+} alone at approximately 100 μM [58,70,167,168,176]. It has also been reported that RyR1 and to a lesser extent RyR2 channels are functionally heterogenous with respect to Ca^{2+} regulation [177,178], which may be due to differences in the redox state between individual channels [179]. Characterization of the three mammalian, recombinant RyRs expressed in a non-myogenic cell line (HEK293) revealed cytosolic Ca^{2+}-dependence properties similar to the native channels [22,25,180].

7.2. Luminal Ca^{2+}

The rate of Ca^{2+} release from the SR is sensitive to the store Ca^{2+} content [181,182], and single channel recordings have demonstrated effective RyR regulation by luminal Ca^{2+} [183–188]. Although differences may exist between RyR1 and RyR2, increasing luminal Ca^{2+} concentration augments channel activity by increasing the channel sensitivity to agonists such as cytosolic Ca^{2+}, ATP and caffeine and by alleviating Mg^{2+} inhibition. However, increasing luminal Ca^{2+} levels above a threshold causes a decrease in channel activity. The mechanism of RyR activation by luminal Ca^{2+} is likely to involve either Ca^{2+} regulatory site(s) on the luminal side of the RyR [183,184,187] and/or the cytosolic Ca^{2+} regulatory site(s) following permeation through the pore [185,186]. It has been proposed that both mechanisms are operational, with luminal Ca^{2+} altering the affinity of cytosolic regulatory site(s) through an allosteric effect in single isolated channels, whereas in close-packed RyR1 arrays, luminal Ca^{2+} flowing through one channel interacts directly with cytosolic regulatory site(s) of neighbouring channels [188]. The luminal Ca^{2+} sensor can either be an intrinsic property of the RyR or is imparted by accessory protein(s). Evidence for the former is provided by proteolysis studies suggesting the presence of both activating and inactivating Ca^{2+} sites on the RyR luminal side [189]. In contrast, dissociation of calsequestrin (CSQ), triadin and junctin abolished RyR2 regulation by luminal Ca^{2+} [190], whereas CSQ association amplified channel responses to luminal Ca^{2+} [191].

7.3. Mg^{2+}

Mg^{2+} is a potent RyR channel inhibitor. SR Ca^{2+} flux measurements, single channel recordings and ryanodine-binding assays have demonstrated that Mg^{2+} inhibits

RyR1 in a dose-dependent manner, with millimolar concentrations resulting in complete inactivation [164–166,188,192–194]. Mg^{2+} inhibition may be a significant mechanism for maintaining the RyR1 closed as the cytoplasmic concentration is approximately 1 mM. RyR2 and RyR3 are also sensitive to Mg^{2+} inhibition but to a lesser extent [25,70,170,194–196]. It has been suggested that Mg^{2+} exerts its inhibitory effects by competing with Ca^{2+} for the high-affinity Ca^{2+}-stimulatory site(s) and by binding to the low-affinity inactivation site(s) with an affinity comparable with Ca^{2+} [193,194,196]. More recent evidence suggests that Mg^{2+} is an antagonist that inhibits RyR1 channel activity in the absence of Ca^{2+} [188].

7.4. ATP

The adenine nucleotides ATP, ADP, AMP and cyclic-AMP, as well as adenine, activate RyR channel activity [164–166]. SR Ca^{2+} flux measurements, single channel recordings and ryanodine-binding assays have demonstrated that ATP at millimolar levels is a potent agonist of skeletal muscle RyR [164–166,197,198]. RyR1 is stimulated by ATP even at very low nanomolar Ca^{2+} concentrations, whereas the combination of micromolar Ca^{2+} and millimolar ATP elicits persistent channel activation. It has been suggested that ATP increases the sensitivity of the channel to Ca^{2+} activation and decreases its sensitivity to Ca^{2+} inactivation [169]. The effect of ATP on the cardiac RyR is qualitatively similar although less marked [170,195,199]. ATP also stimulates RyR3 channel activity and overrides inhibition by millimolar Ca^{2+} [25,200].

7.5. Redox status

Pharmacological sulphydryl-reactive reagents, including thimerosal, dithiodipyridines, *N*-ethylmaleimide and diamide, have been shown to activate skeletal and cardiac muscle RyRs [201–203]. This effect was reversed by reducing reagents such as dithiothreitol, whereas prolonged exposure to oxidants, or use of high concentrations, led to irreversible loss of channel activity. Endogenous physiological redox modulators have also been shown to regulate the RyR. Glutathione in its reduced form was found to inhibit the channel (but oxidized glutathione was stimulatory) [204,205], whereas reactive oxygen species such as hydrogen peroxide and superoxide anion radical activated the channel [206–209]. Reactive oxygen species and superoxide anion radical in particular, generated by an endogenous SR NAD(P)H oxidase, have been implicated in modulation of skeletal and cardiac RyRs [209–212]. Nitric oxide has also been proposed to be an important physiological modulator of RyR function, although results obtained from different studies are contradictory [213–218]. Native RyRs are endogenously S-glutathionylated as well as S-nitrosylated [212,215,216,219], suggesting that these post-translational modifications may play a role in RyR function.

The redox potential of RyR1, estimated to be about −165 mV, was found to be sensitive to channel modulators [205]. Conditions that favour channel opening lowered RyR1 redox potential that favours the oxidation of critical thiols, whereas conditions that close down the channel have the opposite effect. There is evidence for at least three classes of functional cysteines that modulate RyRs, and channel activity has been correlated with the number of free thiols [202,215,220]. Under physiological conditions, RyR1-free thiol content is high (approximately 48 per subunit) and is associated with low channel activity. Oxidation of approximately 10 thiols has little effect on channel activity, whereas oxidation of approximately 15 extra thiols increased channel activity and was reversible. More extensive oxidation (loss of approximately 10 additional thiols) inactivated the channel irreversibly.

RyR sulphydryl modification alters Ca^{2+} dependence and increases the sensitivity of the channel to Ca^{2+} activation [179,205,220–222]. In addition, it was shown that RyR oxidation increases the sensitivity of the channel to ATP activation [221,222] and decreases its sensitivity to Mg^{2+} inhibition [217,223]. Sulphydryl modification of RyR involves altered protein–protein interactions, in particular formation of intra-subunit and intersubunit cross-links within RyR1 [224], high-molecular weight complexes with triadin [225], CaM binding [208,216,219,226] and FKBP12 binding [219]. Notably, defective regulation of RyR2 channels due to oxidative stress has been implicated in heart failure [227] (see Section 10.1.).

7.6. Phosphorylation status

In vitro phosphorylation studies have shown that both skeletal and cardiac RyRs are substrates for serine/threonine protein kinases, primarily for cAMP-dependent protein kinase (PKA) and Ca^{2+}/calmodulin-dependent protein kinase II (CaMKII) and less efficiently for cGMP-dependent protein kinase (PKG) [228–233]. Dephosphorylation is carried out primarily by protein phosphatases 1 (PP1) and 2A (PP2A) [234,235]. CaMKII is tightly bound to SR membranes and co-purifies with the RyR through a direct association [229,232,236]. RyR association with PKA, PP1 and PP2A appears to be indirect, through specific anchoring proteins [235]; this explains why endogenous CaMKII but not PKA (or PKG) activity is often reported in isolated SR membranes [229,230,232,237–239]. Although results vary, much evidence suggests that there are multiple RyR sites phosphorylated by CaMKII [230–233,240–242]. Two in vivo phosphorylation sites have been identified: RyR2 S2808[1] and the corresponding RyR1 S2843, whereas a second site, S2814[1], has been found for RyR2 only.

Studies aiming to identify the PKA site(s) have produced conflicting results over the issue of one or two sites and which one is PKA specific. Although RyR2 S2808 was initially reported to be CaMKII specific [231], it was later proposed that this residue (and the corresponding RyR1 S2843) is phosphorylated exclusively by PKA [234,241,243]. However, recent studies have identified a second PKA site, S2031[1] in addition to S2808, and both appeared to be phosphorylated in vivo [242,244]. Expression of mutant channels, where these serines were substituted by alanines (which cannot be phosphorylated) either individually or in combination, did not resolve the discrepancy. One group observed that single S2808A substitution (but not

S2031A) abolished PKA-dependent phosphorylation [245,246], whereas another group demonstrated PKA-dependent phosphorylation occurred with either individual mutant RyR2 and was absent only for the double mutant [242,244]. Determination of S2808 phosphorylation characteristics is important because it has been implicated in the pathogenesis of heart failure (see Section 10.1.). Many studies have indicated that S2808 is phosphorylated to high levels (up to 75%) even in normal hearts at rest and can be phosphorylated efficiently by several protein kinases including PKA, CaMKII and PKG [231,233,240,242,244,247,248].

The functional effects of RyR phosphorylation/dephosphorylation have been examined by SR Ca^{2+} flux measurements, single channel recordings and by ryanodine-binding assays. Addition of exogenous PKA results in increased RyR activity because of enhanced sensitivity to Ca^{2+} activation and substantially reduced Mg^{2+} inhibition [228,230,234,237,238,243,248–252]. These effects were reversed by PP. One study reported that PKA activated the RyR but only during rapid rises in local Ca^{2+} concentration, whereas under fixed Ca^{2+} levels, it promoted channel closure [249]. Another study has suggested that the PKA-stimulatory effects occur only when stoichiometric phosphorylation of S2808 is achieved; phosphorylation by up to 75%, as often is the case in normal hearts at rest, had negligible effects on channel activity [248]. Exogenous CaMKII displayed enhanced RyR channel activity [231,236–238,241,250] although one study reported inhibition [253]. Contradictory results were also obtained when the effects of endogenous CaMKII (activated by Ca^{2+} and/or CaM) were assessed [230,237–239]. The reasons for these discrepancies are unknown. Notably, PKA- and CaMKII-dependent phosphorylation produced different functional effects in parallel investigations, suggesting that they target different residues and/or a different number of residues [230,237,238,250]. Dephosphorylation of RyRs is expected to inhibit channel activity because phosphorylation results in channel activation (in the majority of studies). However, RyRs treated with PP displayed increased channel activity [248,253,254]. These apparently disparate findings are explained when the basal RyR phosphorylation level is considered. It has been suggested that phosphorylation levels above (i.e. additional phosphorylation) or below (i.e. dephosphorylation) an intrinsic homeostatic level results in channel activation [248].

In cardiomyocytes, β-adrenergic stimulation leads to cAMP elevation and PKA activation, resulting in enhancement of E–C coupling (reviewed in [255]). This is due to increased DHPR Ca^{2+} current, as well as increased SR Ca^{2+} content caused by higher SR Ca^{2+}-ATPase pump activity following relief of phospholamban inhibition. RyR activity may also be up-regulated following β-adrenergic stimulation as in vitro functional assays suggest. In support of this hypothesis, one study found enhancement of highly localized Ca^{2+} release events despite lower SR Ca^{2+} content [256]. However, other studies did not concur with an effect through PKA-mediated phosphorylation. β-Adrenergic stimulation was found to have no effect on Ca^{2+} transients and E–C coupling (when normalized for DHPR Ca^{2+} current and SR Ca^{2+} content), although it did enhance the initial rate of SR Ca^{2+} release [257]. It was further shown that direct PKA activation, upon addition of cAMP or PKA catalytic subunit in permeabilized cardiomyocytes of phospholamban knock-out mice, did not affect Ca^{2+} spark frequency (under conditions that excluded PKA effects on DHPR Ca^{2+} current and SR Ca^{2+} content) [258,259]. Interestingly, several studies have indicated that CaMKII enhances E–C coupling by directly acting on the RyR. Activation of endogenous CaMKII in cardiomyocytes augmented electrically evoked Ca^{2+} transients (when normalized for DHPR Ca^{2+} current and SR Ca^{2+} content) and increased Ca^{2+} spark frequency and duration (under conditions that excluded CaMKII effects on DHPR Ca^{2+} current and SR Ca^{2+} content), whereas CaMKII inhibition decreased Ca^{2+} spark frequency [236,259,260]. CaMKII (δC isoform) over-expression, chronically in transgenic mice or acutely in isolated cardiomyocytes through adenovirus, enhanced electrically evoked Ca^{2+} transients and Ca^{2+} spark frequency (normalized for SR Ca^{2+} content) [261,262]. One study reported disparate results, where constitutively active CaMKII reduced and CaMKII inhibition enhanced Ca^{2+} release [263]. Notably, PKA- and/or CaMKII-dependent phosphorylation of RyR2 has been implicated in heart failure (see Section 10.1.).

The effects of PP1 and PP2A in cardiomyocytes are unclear at present. Intracellular dialysis of PP1 and PP2A was found to decrease Ca^{2+} transients and E–C coupling without affecting SR Ca^{2+} content [264]. In contrast, in permeabilized cardiomyocytes PP1 and PP2A initially enhanced Ca^{2+} spark frequency followed by a subsequent decline because of depletion of the SR calcium store [254].

8. Functional interactions within the RyR

8.1. Physical coupling between channels

A characteristic feature of RyRs is their membrane organization into ordered arrays, observed in both skeletal and cardiac muscle [123,124,137]. Such a regular organization is an intrinsic property of the RyR, because recombinant RyR1 expressed in a non-myogenic (CHO) cell line formed extensive arrays [265] and purified RyR1 was shown to self-assemble into large two-dimensional crystalline arrays in a strict ionic strength-dependent manner [266,267]. In these 'checkerboard-like' lattices, each RyR1 is interlocked with four adjacent channels through a specific intermolecular domain–domain interaction that appears to involve sequences in close proximity to where the D2 region has been localized [102,267]. Biochemical evidence has implicated a sequence (residues 2540–3207) in the central region of RyR2 participating in tetramer–tetramer interactions [268]. Physical coupling between RyRs may facilitate intermolecular allosteric interactions and conformational changes transmitted from one RyR to four adjacent channels. This inter-RyR physical coupling mechanism may explain how skeletal muscle RyR channels that are not associated with DHPR voltage sensors can be regulated or how a given number of RyRs within a cluster open and close simultaneously during a Ca^{2+} spark. Ca^{2+} sparks are small localized Ca^{2+} release events generated by spontaneous openings of RyR channels (reviewed in [269]). Ca^{2+} sparks are widely used as an index of intrinsic RyR activity in cell-based assays.

8.2. Interdomain interactions within the tetrameric channel

The RyR carboxyl-terminus contains the membrane-spanning domains, is capable of self-association and functions as a Ca^{2+}-conducting pore (see Section 4.2.). The cytoplasmic C-terminal tail is important in channel oligomerization as a truncated RyR1 protein lacking the last 15 residues formed inactive channels due to impaired assembly of the tetrameric complex [270], whereas the RyR2 extreme C-terminal 100 residues form dimers and tetramers [271]. The large N-terminal cytoplasmic portion remains associated with the TM assembly in partial protease digestion studies of RyR1 [272]. This is consistent with complementation assays demonstrating that N-terminal RyR2 fragments can combine with overlapping C-terminal fragments encompassing the pore-forming TM domains to form functional channels [107,273].

An important role for the central domain in channel regulation and a putative interaction with the N-terminal region has been proposed [274–279]. SR Ca^{2+} flux measurements, single channel recordings and ryanodine-binding assays showed that a synthetic peptide termed DP4 (residues 2442–2477 of RyR1), as well as anti-DP4 antibody, enhances RyR1 channel activity and potentiates agonist-induced Ca^{2+} release. DP4 also enhanced Ca^{2+} release and contraction in skeletal muscle fibres, overriding Mg^{2+} inhibition, and increased the frequency of Ca^{2+} sparks. DP4 was shown to bind the approximately 150-kDa N-terminal RyR1 calpain fragment suggesting that the entral domain interacts with the N-terminal region. Interestingly, when DP4 had a single residue change to mimic a malignant hyperthermia (MH) mutation (R2458C) (see Section 10.3.), the resulting peptide lacked any stimulatory effects. Similarly, an RyR2 central domain peptide (DPc10, residues 2460–2495) was found to activate RyR2, an effect that was abolished by a single residue change (R2475S) mimicking a mutation involved in catecholaminergic polymorphic ventricular tachycardia (CPVT) (see Section 10.2.) [280]. Furthermore, DPc10 decreased Ca^{2+} transients and cell shortening due to Ca^{2+} leak from the SR calcium store while transiently increasing Ca^{2+} spark frequency in cardiomyocytes [281,282]. Like DP4, DPc10 was also shown to bind the approximately 120-kDa N-terminal calpain fragment of RyR2. Hence, stimulatory effects of central DPcs may be due to disruption of the normal interdomain interactions that stabilize the closed state of the channel, with DP4 or DPc10 competing with the corresponding domain of the native RyR for a binding site on the N-terminal region. However, other studies appear to be inconsistent with this hypothesis. A RyR1 N-terminal fragment (residues 281–620) interacted with two native RyR1 fragments (residues 799–1172 and 2937–3225) that do not encompass the DP4 sequence [283]. A RyR2 N-terminal peptide termed DP1-2 (residues 601–639) and the corresponding RyR1 peptide (residues 590–628) both stimulated channel activity of the cardiac RyR but had no effect on skeletal RyR function [284]. Further peptide-probe studies demonstrated that dantrolene, which inhibits Ca^{2+} release from skeletal but not from cardiac muscle, binds native RyR1 at residues 590–609 (DP1 peptide) and inhibits the stimulatory effect of DP4. However, dantrolene does not bind native RyR2 although it does bind the identical cardiac DP1 peptide (residues 601–620) [285–287]. Similarly, anti-DP1 antibody recognizes native RyR1 but not native RyR2. In additon, DP4 was reported to have no effect on the type 3 RyR [288]. These latter studies suggest that the postulated N-terminal interaction with the central domain may be isoform specific.

The primary CaM-binding site (see Section 9.3.) has also been implicated in interdomain interactions [289–292]. A synthetic peptide comprising the CaM site (amino acids 3614–3643) activated RyR1 channel activity in SR Ca^{2+} flux measurements, single channel recordings and ryanodine-binding assays and increased Ca^{2+} spark frequency in skeletal muscle fibres, effects which were CaM independent. Furthermore, RyR1 mutant channels lacking this region displayed a severely compromised agonist-induced Ca^{2+} release. Peptide 3614–3643 were able to bind to a region (amino acids 4064–4210) with structural similarities to CaM in a Ca^{2+}- and CaM-dependent way. Thus, it was proposed that an interdomain, CaM-sensitive interaction occurs between the CaM-binding site and the CaM-like domain, which stabilizes the channel in a closed conformation.

Evidence also exists for intersubunit contacts between cytoplasmic domains of RyR1, from a combination of protease-protection and cysteine-modification assays [202,224,293,294]. These studies demonstrated that an intersubunit disulphide bond could be formed by treating RyR1 with oxidizing reagents, involving C3635 within the primary CaM-binding site and a second cysteine located within residues 2000–2401. An additional intrasubunit disulphide bond was also detected, but the cysteines involved remain to be identified.

9. *RyR accessory proteins*

The RyR is a dynamic macromolecular complex interacting directly or indirectly with numerous proteins that affect its channel function, including protein kinases and phosphatases docked through anchoring proteins [234,235], S100 [295], calexcitin [296], homer [297] and snapin [298]. The DHPR, FKBP, CaM, sorcin, CSQ, triadin and junctin are the best characterized RyR accessory proteins and are discussed Section 9.1.–9.5.

9.1. *DHPR*

The DHPR is the L-type plasma membrane Ca^{2+} channel that senses surface membrane depolarization and initiates the process of E–C coupling in striated muscle (see Section 5.). The DHPR is composed of α_1-, α_2-δ-, β- and γ-subunits, with α_1 constituting the channel-forming, voltage-sensing and dihydropyridine-sensitive portion of the receptor. The skeletal muscle α_{1s}-subunit is composed of 1873 amino acids, with a predicted topology consisting of four repeats of six TM segments linked by cytoplasmic loops of variable length bounded by cytoplasmic N- and C-termini [299]. The cardiac muscle α_{1c}-subunit displays 66% homology with the skeletal isoform, with similar TM topology but longer N- and C-terminal domains [300].

Expression of DHPR isoforms in skeletal myocytes derived from the dysgenic mouse, a naturally occurring mutant that lacks functional DHPR α_1-subunit, demonstrated that α_{1s} restores depolarization-induced intracellular Ca^{2+} transients in the absence of Ca^{2+} influx, i.e. skeletal type E–C coupling, whereas expression of α_{1c} results in cardiac type E–C coupling (dependent on Ca^{2+} influx) [127,139]. Analysis of skeletal and cardiac isoform chimaeras of the α_1-subunit indicated that the major determinant of skeletal E–C coupling is the II–III cytoplasmic loop of the α_{1s}-subunit [128]. Further studies of additional chimaeras, or mutant α_{1s}-subunits with scrambled sequences, deletions or point substitutions, identified residues 720–764 within the II–III loop (also known as peptide C) to be crucial for skeletal-type E–C coupling [301–305]. The minimal region required was found within residues 725–742, with four individual residues being particularly important (A739, F741, P742 and D744).

The recombinant α_{1s}-subunit II–III loop activated RyR1 channels in ryanodine-binding assays and single channel recordings [306,307]. Further studies to identify the minimal region within the II–III loop affecting RyR1 channel activity have produced conflicting results. Synthetic peptide C (or a larger fragment containing this region) was reported to have no effect, inhibit or activate the RyR1, depending on experimental conditions (e.g. peptide concentration, incubation time and effectors such as Ca^{2+}, Mg^{2+} or ATP) [279,308–315]. A second peptide (termed peptide A) corresponding to residues 671–690 (or a larger fragment containing this region) activated RyR1 in SR Ca^{2+} flux measurements, single channel recordings and ryanodine-binding assays [308,309,311,312,316–319]. These studies showed that a basic sequence RKRRK is essential for the action of peptide A and that phosphorylation of S687 could also be involved in RyR1 activation. Peptide A activation of RyR1 was also dependent on the presence of FKBP12. Interestingly, peptide C suppressed the stimulatory effects of peptide A [309,310,314]. These studies suggest that peptide A is a key region of the II–III loop in RyR1 activation. However, additional experiments in dysgenic myocytes showed that skeletal E–C coupling was not altered when this region was scrambled, deleted or replaced [302–304].

DHPR α_{1s} regions other than the II–III loop may also be involved in skeletal-type E–C coupling. The I–II loop was proposed to support weak E–C coupling [128], a synthetic C-terminal peptide (amino acids 1487–1506) was able to inhibit RyRs [320] and the recombinant III–IV loop competed with the II–III loop for binding to a RyR1 fragment [321]. Furthermore, expression of α_1-subunit chimaeras in dysgenic myocytes suggested that regions distinct from the II–III loop are involved in skeletal-type E–C coupling and the tetrad organization of DHPRs [322,323]. In addition, the β-subunit has recently been shown to interact directly with RyR1 and strengthen E–C coupling [324].

The type of E–C coupling depends not only on the DHPR isoform but also on the RyR isoform. Expression of RyR isoforms in skeletal myocytes derived from a knock-out mouse of RyR1 (also known as dyspedic mouse) demonstrated that RyR1 restores skeletal-type E–C coupling, whereas expression of RyR2 or RyR3 results in the cardiac type [325,326]. Mapping of the DHPR interaction site(s) on RyR1 has produced conflicting results. Screening of recombinant fragments encompassing the large cytoplasmic N-terminal region of RyR1 identified residues 922–1112 to be capable of binding the II–III loop of the DHPR α_{1s}-subunit

[321,327]. The binding site was further localized to a 37-residue stretch (amino acids 1076–1112) by use of RyR1/2 chimaeric fragments, and this region was also found to bind the α_{1s} III–IV loop. However, the involvement of this RyR1 sequence in skeletal-type E–C coupling remains in question because expression of a chimaeric RyR1, where this sequence was replaced with the equivalent RyR2 sequence, restored normal skeletal-type E–C coupling in dyspedic myocytes [328]. Further chimaeras between type 1 and type 2 RyRs identified two regions of RyR1, residues 1635–2559 and 2659–3720, as essential for skeletal-type E–C coupling and formation of DHPR tetrads [328,329]. These results were also verified by RyR1/RyR3 chimaeras and these studies were able to further refine the DHPR interaction sites to residues 1924–2446 and 2644–3223 [330].

The use of RyR1/RyR3 chimaeras also suggested that a large N-terminal region (amino acids 1–1680) might participate in skeletal-type E–C coupling [331]. Deletion of divergent domain D2 within the RyR1 N-terminus abolished skeletal-type E–C coupling; however, chimaeras between RyR1 and RyR2 or RyR3 indicated that this domain by itself is not sufficient to promote skeletal-type E–C coupling and most probably plays a structural role in RyR1 conformation [331,332]. Peptide A of the α_{1s} DHPR subunit was reported to bind to two discrete RyR1 fragments, residues 1021–1631 and 3201–3661, although deletion of only the second region abolished peptide A interaction with RyR1 [333]. Yeast two-hybrid analysis showed that the peptide C region of the α_{1s}-DHPR subunit interacted with RyR1 amino acids 1837–2168, but expression of RyR1/2 chimaeras in dyspedic myocytes indicated that this region is only partially involved in skeletal-type E–C coupling [334].

Distinct RyR1 regions have also been implicated in DHPR interaction that do not involve the II–III loop of the α_{1s}-subunit. The C-terminal tail of the α_{1s}-subunit (or shorter peptides within this region) interacts with RyR1 residues 3609–3643 and 4064–4210 in a CaM-sensitive manner [291,335]. Also the DHPR β subunit binds to RyR1 residues 3201–3661 through a cluster of positively charged amino acids [324]. Thus, all the above studies taken together suggest that the DHPR-RyR association involves multiple regions on both proteins and/or is conformation sensitive.

9.2. FKBP

The cytosolic FKBP, FKBP12 and the closely related FKBP12.6 isoform are classified as immunophilins as they are the receptor proteins for the immunosuppressant drugs, FK506 and rapamycin, and are the smallest members of an extensive and ubiquitous FKBP family (reviewed in [336]). Both isoforms have peptidyl-prolyl *cis-trans* isomerase activity that is inhibited by FK506 and rapamycin. Both FKBP12 and FKBP12.6 consist of 108 amino acids including a cleaved N-terminal methionine residue, and they share approximately 85% homology. The structures of both isoforms, determined by X-ray crystallography, were found to be nearly identical to each other and unaltered in their drug complexes [337–339]. Northern blot analysis has revealed transcripts for both isoforms in all mammalian tissues but significantly higher levels for FKBP12 [340].

RyR1 binds both FKBP12/12.6 with similar affinities, as shown by co-immunoprecipitation and co-purification assays and by GST-FKBP affinity chromatography [341–344]. However, native RyR1 from skeletal muscle is isolated as a complex with FKBP12 because of relative abundance of FKBP12 over FKBP12.6. The RyR1-FKBP12 association in skeletal muscle is common to each of the five classes of vertebrates, i.e. mammals (rabbit), birds (chicken), reptiles (turtle), fish (salmon and rainbow trout) and amphibians (frog). The stoichiometry is four molecules of either FKBP12 or FKBP12.6 per RyR1 tetramer, i.e. one FKBP per RyR1 subunit. Mutation analysis identified three important residues (Q3, R18 and M49) in the FKBP12 interaction with RyR1 [345] and revealed that FKBP isomerase activity is not required [342]. Surface plasmon resonance analysis of RyR1–FKBP12 binding characteristics revealed a conformation-sensitive, high-affinity interaction; strong FKBP12 binding occurs when the channel is in its open state ($K_d \sim 1$ nM), but it is further enhanced (by approximately 4–5 orders of magnitude) when the channel is in its closed configuration [346]. Using GST-FKBP12 affinity chromatography, the interaction was found to be largely unaffected by temperature and pH changes, but it was ionic strength dependent (optimal at 100–200 mM NaCl) [347]. FKBP12-bound RyR1 (in its closed state) has been visualized by cryo-electron microscopy and found to be localized along the edge of the square-shaped cytoplasmic assembly, approximately 12 nm away from the putative channel pore (Fig. 1) [95].

RyR2 associates specifically with FKBP12.6 also with a stoichiometry of one FKBP12.6 per RyR2 subunit [340,343,348,349]. The RyR2-FKBP12.6 association in the heart is common in mammals and other vertebrates including chicken, fish and frog. Although canine RyR2 binds FKBP12.6 exclusively, RyR2 from other species is also able to bind FKBP12 but with a 7- to 8-fold lower affinity, and therefore native RyR2 from cardiac muscle is isolated as a complex with FKBP12.6. Mutation analysis has identified three residues (Q31, N32 and F59) critical for selective binding of FKBP12.6 to RyR2 [350] and revealed that enzymatic isomerase activity is not required [351]. The location of the FKBP12.6-binding site on RyR2 (in its open state) visualized by cryo-electron microscopy appears similar, although slightly displaced (by approximately 2 nm) from the FKBP12 site of RyR1 [352]. RyR3 is capable of binding both FKBP12/12.6 with similar affinities [353]. The immunosuppressant drugs FK506 and rapamycin disrupt the RyR-FKBP interactions [343,354–356].

Initial reports using yeast two-hybrid assay or GST-FKBP affinity chromatography combined with limited RyR proteolysis have implicated the central RyR region (amino acids 2407–2520 in RyR1) in the interaction with FKBP12/12.6 [234,353,357,358]. Furthermore, mutation analysis has identified V2461 in RyR1 (and the corresponding V2322 in RyR3) as an important residue for FKBP binding as assessed by co-immunoprecipitation, GST-FKBP-affinity chromatography and functional assays [353,359–361]. However, other studies, using very similar experimental approaches and RyRs with mutations or deletions in the central region, have questioned the involvement of this domain in FKBP binding [273,360,362]. Two alternative, very large domains have been proposed to form the FKBP12.6-binding site for the cardiac RyR isoform: the amino-terminus (residues 307–1855) [273,363] or the carboxyl-terminus (residues 3788–4967) [364]. The reason for the conflicting results is unclear but may be due to isoform and/or species differences. Hence, there may be multiple FKBP-binding sites in the RyR (at the N-terminus, central domain and C-terminus) with different binding properties and different sensitivities, depending on channel state. Notably, the FKBP12-binding site visualized on RyR1 by cryo-electron microscopy at approximately 16 Å resolution does not prove or disprove any of the three proposed sites, because it appears to be in close proximity, but it does not overlap with any of them [365].

Functional effects of FKBP12 on skeletal muscle RyR activity have been studied in isolated channels or SR preparations following FKBP12 dissociation with FK506 or rapamycin. SR Ca^{2+} flux measurements, single channel recordings and ryanodine-binding assays have demonstrated that FKBP12 removal results in enhanced channel activation. In particular, FKBP12-deficient RyR1 channels have an increased sensitivity to Ca^{2+} or caffeine activation and reduced sensitivity to millimolar Ca^{2+} or Mg^{2+} inhibition, whereas control values are restored upon reconstitution with FKBP12/12.6 [342,344,354,356,366–369]. Single channel recordings of RyR1 have revealed pronounced subconductance activity upon FKBP12 dissociation. FKBP12-deficient RyR1 channels, generated either by FK506 or rapamycin treatment (with or without subsequent removal of FKBP12-drug complexes), or recombinant channels expressed in insect cells devoid of FKBP12/12.6, or native channels isolated from FKBP12 knock-out mice, displayed openings to approximately 25, 50 and 75% of the maximum conductance [356,366,367,370]; however, subconductance states are not always observed [368,369]. The effects of FKBP12.6 on cardiac RyR are similar to that of FKBP12 on RyR1. Single channel recordings of FKBP-deficient RyR2 channels, either by application of FK506 or rapamycin or by gene-targeted deletion of FKBP12 or FKBP12.6 in mice, also indicated subconductance behaviour [355,370–372]. However, other studies found no effect on RyR2 channel activity or conductance upon FKBP12.6 removal or addition [343,369]. FKBP12/12.6 has also been proposed to enhance the functional coupling between adjacent RyR channels (termed 'coupled gating') in both skeletal and cardiac muscle [373,374], although FKBPs are not involved in their physical coupling [267]. Single channel recording studies have shown that addition of FKBP induced simultaneous openings and closings of two channels incorporated into the planar lipid bilayer, whereas rapamycin treatment resulted in uncoupled channels. However, FKBP binding occurs at the opposite site of the RyR–RyR contact region (which is close to the D2 region) indicating that FKBP is not a direct intermediate protein but may allosterically modulate tetramer–tetramer interactions [267].

FKBP12/12.6 functional effects have also been studied in isolated myocytes or muscle fibres. FK506 and rapamycin-enhanced caffeine- and halothane-induced skeletal muscle contractions [367,375] that are consistent with FKBP12-promoting channel closure. These studies also suggested an involvement of FKBP12 dissociation in MH (see Section 10.3.); however, MH responses were not detected in in vitro caffeine contracture tests with FKBP12-deficient mice [376]. Depolarization-induced Ca^{2+} release studies revealed a role for an intact RyR1-FKBP12 association in skeletal-type E–C coupling. FKBP12-deficiency resulted in reduced depolarization-induced Ca^{2+} release and contraction [360,376–378]. As SR Ca^{2+} content was normal, these findings suggest that FKBP12 dissociation suppresses the intrinsic ability of the DHPR voltage sensors to efficiently activate Ca^{2+} release. Contrary to skeletal muscle, FKBP12.6 deficiency in cardiomyocytes, induced by application of FK506 or rapamycin or by gene-targeted deletion of FKBP12.6, resulted in increased intracellular Ca^{2+} transients and contractions and also in increased Ca^{2+} spark frequency and/or amplitude and duration [371,379,380]. The effects of FKBP12.6 have been investigated in heterologous systems expressing recombinant RyR2 channels, where co-expression of FKBP12.6 (but not FKBP12) suppressed spontaneous or agonist-induced Ca^{2+} release [381,382]. Adenovirus-mediated FKBP12.6 over-expression in isolated cardiomyocytes resulted in increased electrically evoked Ca^{2+} transients and SR Ca^{2+} content, whereas it decreased Ca^{2+} spark frequency, amplitude and duration [383–385]. Although increased Ca^{2+} transients upon FKBP12.6 over-expression appears to contradict the results obtained by FK506 or rapamycin treatment (where increased Ca^{2+} transients were also observed), the former were due to increased SR Ca^{2+} content, most likely because of suppressed RyR2 spontaneous activity. FKBP12.6 over-expression in transgenic mice resulted in enhanced contractility and cardiac output, although detailed analysis of potential adaptive changes was not carried out [351]. These studies taken together indicate that FKBP12.6 has an inhibitory effect on RyR2 activity and promotes channel closure.

The physiological role of FKBP12/12.6 has been addressed by gene knock-out studies in mice. FKBP12-deficient mice have normal skeletal muscle but severe cardiac defects and die at embryonic stage or shortly after birth [370]. Mice with a skeletal muscle-specific deletion of FKBP12 live normally, but certain types of skeletal muscle have compromised contractile properties [376]. The generation of mice deficient in FKBP12.6 by two independent groups has produced two different phenotypes [372,380]. In both cases, mice had a normal phenotype; however, in one laboratory mild cardiac hypertrophy was observed in male but not in female mice. This was discrepant with the second group, where both sexes presented with exercise-induced cardiac arrhythmias. The reason for disparate phenotypes is unclear, but it could be due to different mouse strains.

In summary, FKBP12/12.6 stabilizes channel conductance, promotes channel closure, supports coupled gating between adjacent channels and enhances skeletal-type E–C coupling. In contrast, FKBP dissociation results in enhanced sensitivity of Ca^{2+} activation and 'leaky' channels. Because of their profound effects on RyR channel properties, FKBPs are considered essential components of the RyR complex in both skeletal and cardiac muscle. It should be noted that defective regulation of the RyR2-FKBP12.6 association has been implicated in the pathogenesis of heart failure and cardiac arrhythmias (see Sections 10.1. and 10.2.).

9.3. CaM

CaM is a ubiquitous, highly conserved 17-kDa protein that has a high affinity for Ca^{2+} and participates in numerous Ca^{2+} signalling processes (reviewed in [386]). CaM has a dumbbell-shaped structure consisting of N-terminal and C-terminal globular domains connected by a flexible linker, each containing two EF-hand Ca^{2+}-binding motifs. CaM binds all three mammalian RyRs with nanomolar affinity irrespective of whether it is in a Ca^{2+}-free or Ca^{2+}-bound state, although affinity is slightly higher at micromolar calcium [387–391]. There are four CaM molecules per RyR channel, i.e. one CaM per RyR subunit, at both low and high Ca^{2+} levels. This 1:1 stoichiometry replaces earlier findings based on chemically labelled CaM (either [^{125}I] or fluorescent label) or mapping studies with RyR fragments, suggesting multiple binding sites for CaM per RyR subunit [392–395]. The RyR1–CaM binding site has been visualized by cryo-electron microscopy; CaM was localized at the RyR1 periphery, approximately 10 nm away from the putative channel pore (Fig. 1) [95].

The primary binding site for both Ca^{2+}-free and Ca^{2+}-bound CaM lies within residues 3614–3643 of RyR1 from assessment of protease protection assays, cysteine modification protection assays, peptide-binding analysis and mutation analysis [226,387,396,397]. CaM is suggested to bind at a site of intersubunit contact and protects RyR1 from oxidation, with identified RyR1 residues, W3620 and L3624, being critical for the interaction. The presence of Ca^{2+} shifts the position of CaM several amino acids N-terminal of the Ca^{2+}-free site, because a smaller C-terminal-truncated peptide (amino acids 3614–3634) retained the interaction with the Ca^{2+}-bound, but not Ca^{2+}-free, CaM [396]. Further studies indicate that the CaM C-terminal globular domain binds RyR1 at 3614–3643, whereas the N-terminal domain binds to an adjacent RyR1 subunit at residues 1975–1999 [294,398]. The primary CaM site is also conserved in RyR2 and RyR3, because mutations or deletion of RyR2 residues 3583–3603 or RyR3 residues 3469–3489 abolished CaM binding [391,399]. Mutation analysis also suggests that additional, isoform-specific regions are involved in the differential modulation of the three RyRs by CaM [391,400].

CaM has a biphasic, Ca^{2+}-dependent effect on RyR1 activity from SR Ca^{2+} flux measurements, single channel recordings and ryanodine binding assays [388,393,401–404]. At low nanomolar Ca^{2+} levels, CaM activates RyR1, but at higher Ca^{2+} concentrations ($\geq 1\ \mu M$), it inhibits the channel. However, CaM is not essential for E–C coupling, because mutation of its binding site on the RyR1 did not affect depolarization-induced Ca^{2+} release [405]. Similar to RyR1, RyR3 activity is also stimulated by CaM at low Ca^{2+} levels but inhibited at high Ca^{2+} [25,391,406]. In contrast, RyR2 is inhibited by CaM at all Ca^{2+} concentrations, although inhibition is more pronounced at high Ca^{2+} [170,390,402,404]. CaM has been speculated to be involved in Ca^{2+} release termination, as it has an inhibitory effect for all RyRs at micromolar Ca^{2+}. However, CaM was found to associate and dissociate from RyR2 on a time scale of seconds to minutes [390], which suggests that CaM occupancy of the channel remains unchanged during the cardiac cycle, and the kinetics of CaM-regulated RyR2 activity suggested only a facilitatory role in Ca^{2+} release termination [407]. In skeletal muscle, CaM was shown to increase Ca^{2+} spark frequency, whereas other spark properties were unaffected, suggesting a role for CaM in initiation, but not termination, of Ca^{2+} release [408].

CaM may also affect SR Ca^{2+} release by acting on other proteins including the DHPR Ca^{2+} channel, CaMKII and calcineurin, a CaM-activated protein phosphatase (reviewed in [409]).

9.4. Sorcin

Sorcin is a 22-kDa member of the penta-EF-hand Ca^{2+} binding protein family, which binds Ca^{2+} with high affinity (reviewed in [410]). It has a widespread expression including the heart, where it localizes along the Z-lines, in particular the dyadic junctions of T-tubules and SR TC [411–413]. Sorcin is implicated in modulation of cardiac E–C coupling, because it associates with RyR2 and DHPR channels [411,414] and translocates from cytosol to membrane-embedded targets upon increasing intracellular Ca^{2+} (to approximately 200 μM) [413]. Sorcin interacts with DHPR through the cytoplasmic C-terminal tail of the α_1-subunit [414], whereas the sorcin-binding site on RyR2 remains to be determined. Single channel recordings and ryanodine-binding assays have demonstrated isoform-specific effects; submicromolar sorcin levels result in near-complete RyR2 inhibition that is Ca^{2+}-independent,whereas it has a stimulatory effect on RyR1 [415]. Kinetics analysis indicates that RyR2 inhibition occurs within milliseconds, suggesting that sorcin may be involved in termination of Ca^{2+} release [413].

Sorcin studies in vivo and in isolated cardiomyocytes have yielded disparate results. Evidence of a role for sorcin in suppression of E–C coupling consistent with RyR2 inhibition is provided by some studies but not all. Sorcin depressed intracellular Ca^{2+} transients and contraction and decreased Ca^{2+} spark frequency, when dialysed into cells through a patch-pipette, following adenovirus-mediated over-expression, in transgenic mice with cardiac-specific over-expression [412,413,416]. However, there were also some differences. In isolated rabbit cardiomyocytes infected with an adenoviral sorcin construct, depressed Ca^{2+} transients were attributed to a reduced SR Ca^{2+} content resulting from enhanced activity of the Na^+–Ca^{2+} exchanger, although decreased Ca^{2+} spark frequency was indeed due to RyR2 inhibition. In transgenic animals, sorcin over-expression accelerated the rate of L-type Ca^{2+} current inactivation, whereas L-type Ca^{2+} current density and kinetics were unaffected in the other two studies. More significant differences were reported by three other studies where an enhancement of E–C coupling upon sorcin over-expression was observed [417–419]. Adenoviral transfer of sorcin in isolated cardiomyocytes or in vivo in mouse and rat hearts resulted in enhanced Ca^{2+} transients and contractility. One study also reported stimulation of SR Ca^{2+} uptake activity by sorcin and enhanced SR Ca^{2+} content as the underlying mechanism [419]. The reason(s) for the above discrepancies remain unclear. It could be due to species differences, although most of the studies used rat cardiomyocytes, and/or due to different sorcin levels. The above studies also suggest that sorcin has other targets in addition to the RyR and DHPR channels, although a direct interaction has not been demonstrated for the Na^+–Ca^{2+}exchanger or the SR Ca^{2+} pump. Sorcin is also known to form dimers and tetramers, a process that is Ca^{2+} and H^+ dependent [420], and it is possible that its target binding is differentially affected by its oligomeric state.

9.5. CSQ, triadin and junctin

CSQ is a highly acidic glycoprotein that comprises the main Ca^{2+} storage protein in the SR lumen. Isolated first from striated muscle, where it is primarily expressed, CSQ is a low-affinity, high-capacity Ca^{2+}-binding protein (approximately 40 mol Ca^{2+} per mol protein) that undergoes conformational changes upon binding Ca^{2+} [421,422]. CSQ forms dimers, tetramers and eventually polymers as the Ca^{2+} concentration increases [423]. CSQ polymers are visible in electron micrographs of junctional muscle preparations, as electron-dense material exclusively in the TC of the SR, and they appear to be anchored near the luminal face of RyRs [2].

Two isoforms of CSQ are known, comprising approximately 390 residues and sharing 86% homology, termed the skeletal and cardiac type due to their relative abundance in skeletal and cardiac muscle, respectively [424,425]. Their sequences reveal no TM domains and no apparent SR retention signal. CSQs are rich in acidic residues (approximately 30%), most of which are involved in Ca^{2+} binding at the C-terminal region, with pairs of acidic residues creating a net surface charge to which Ca^{2+} binds [423]. Ca^{2+} causes conformational changes in the monomer and induces CSQ polymerisaton [423]. Micromolar Ca^{2+} (10 μM) causes compaction of the CSQ monomer, whereas further increases (10–100 μM) lead to dimerization and then formation of linear polymers at millimolar Ca^{2+}. Hence, CSQ exists as a stable polymer at physiological free luminal Ca^{2+} concentration of approximately 1 mM [426]. Both skeletal and cardiac CSQs bind more Ca^{2+} ions than predicted by their net negative charges (–80 and –69, respectively), with approximately 50 and 36 mol of Ca^{2+} per protein molecule, respectively, at $\geq$5 mM Ca^{2+}, as they form their polymeric states [427]. Skeletal and cardiac CSQ have almost identical three-dimensional structures revealed by X-ray crystallography, each with three similar domains and a topology reminiscent of *Escherichia coli* thioredoxin [423,427].

CSQ is anchored to the junctional face of the SR membrane through interactions with two SR integral proteins, triadin and junctin, and together with the RyR, the four proteins form a macromolecular complex [428,429]. Junctin and triadin binding are mediated through an aspartate-rich region (amino acids 354–367) at the C-terminus of CSQ [430]. This region is also implicated in Ca^{2+} binding, suggesting that Ca^{2+} and triadin/junctin compete for binding to the same site. CSQ can be dissociated from the junctional face membrane by either low (<10 μM) or high Ca^{2+} concentrations (10 mM) or high ionic strength [429,431]. The interaction between CSQ and RyR is either direct (although the two main RyR luminal loops do not bind CSQ [432]) or indirect through triadin and junctin [429]. Recent evidence suggests that CSQ is dissociated from RyR at Ca^{2+} concentrations >4 mM [191].

Triadin was isolated from skeletal muscle as a 95-kDa membrane glycoprotein from junctional SR [433]. It is now known to be encoded by a single gene but to exist in multiple forms in both skeletal and cardiac muscle because of alternative splicing

[434–438]. Four subtypes have been found in skeletal muscle, termed Trisk 95, 51, 49 and 32 based on their molecular weights, with the first two being the predominant ones. In the heart, three subtypes have been identified, named CT1, 2 and 3 with molecular weights of 32, 35 and 75 kDa, with CT1 being the predominant one. All the isoforms share a common N-terminal cytoplasmic segment, a single TM domain (residues 47–67) and luminal parts of variable length, with unique sequences at the C-terminal end. A single TM segment is the accepted membrane topology based on hydropathy plot analysis and immunolocalization studies [434,439]. Triadin forms oligomers through disulphide bridges [440]. Interaction of triadin with RyR is mediated through both cytoplasmic and luminal domains [428,441,442]. Triadin binding to the SR membrane cytoplasmic face involves residues 18–46 and is Ca^{2+} dependent, with an optimum at $<10\,\mu M$. Binding to the SR luminal side involves a triadin sequence of alternating positive and negative residues (amino acids 210–224, known as the KEKE motif) interacting with the most C-terminal RyR luminal loop (M8–M10 loop) where three negatively charged residues (D4878, D4907 and E4908) are important for the interaction. Triadin interaction with CSQ is also mediated through the KEKE motif [443]. Interestingly, Trisk 49 and 32 do not bind RyR, although they contain both the N-terminal cytoplasmic domain and the KEKE motif [436].

Junctin, a protein of 26 kDa encoded by a single gene, is also an integral membrane protein localized in the junctional SR of both skeletal and cardiac muscle [444]. It has a similar predicted structure to triadin with a short, 21-residue cytoplasmic N-terminus, a single TM domain with the highly charged remainder residing in the SR lumen. Unlike triadin, junctin is not glycosylated, and it does not form disulphide-linked oligomers. Through its luminal domain, junctin binds RyR and triadin in a Ca^{2+}-independent manner, whereas binding to CSQ is disrupted by either low- or high Ca^{2+} concentrations [429].

It is generally assumed that the primary role of triadin and junctin is to dock CSQ to the RyR, though their individual roles remain to be investigated in detail. In single channel recordings, triadin was reported to inhibit RyR1 but only when added to the cytoplasmic side of the channel [445], whereas luminal addition of triadin and junctin to purified RyR2 led to activation [190]. These differences could be due to the use of native versus purified RyRs (see below). The functions of triadin have been studied in transgenic animals with cardiac-specific over-expression [446,447]. Transgenic animals developed cardiac hypertrophy, had altered muscle contraction and SR Ca^{2+} handling and enhanced SR Ca^{2+} content. Over-expression of triadin (CT1) was accompanied by reduced expression of junctin and RyR2, although the number of functional RyR2s was unchanged. Similar results were obtained with transgenic animals over-expressing junctin, although SR Ca^{2+} content was reduced [448,449]. Although these studies suggest that triadin and junctin play an important role in the Ca^{2+} release process, interpretation of the results is complicated by the various adaptive changes in these animals. To overcome these problems, the role of triadin was studied following acute, adenovirus-mediated over-expression in primary myocytes. Over-expression of CT1 in cardiomyocytes enhanced E–C coupling and Ca^{2+} spark frequency and increased the activity of RyR2 channels suggesting a stimulatory role for triadin [450]. The role of triadin in skeletal muscle is not clear, because acute over-expression of Trisk 95 (but not Trisk 51) in skeletal muscle myotubes did not affect pharmacological activation of RyR1 or Ca^{2+}-induced Ca^{2+} release but abolished depolarization-induced Ca^{2+} release, i.e. skeletal-type E–C coupling [451].

Reports of functional effects of CSQ on RyR have been contradictory. It appears however, that the disparate results are due to the use of native or purified channels that lack the complement of accessory proteins, in particular triadin and junctin [431]. Initial single channel recording studies using either purified channels or channels, where the association with triadin and/or junctin is likely to be disrupted, reported that CSQ addition activates the RyR [431,445,452,453]. However, studies with native channels or channels, where the macromolecular complex with triadin and junctin was reconstituted, demonstrated that CSQ inhibits both skeletal and cardiac RyRs [190,191,431]. The latter studies also revealed the need for an intact macromolecular complex for luminal Ca^{2+} regulation of the RyR (see Section 7.2.).

An inhibitory effect of CSQ on RyR activity is supported by CSQ over-expressing transgenic mice, because these animals had higher SR Ca^{2+} content, but Ca^{2+} transients and Ca^{2+} spark frequency were reduced [454–456]. Notably, the mutant mice developed severe hypertrophy and heart failure and/or had altered expression of RyR, triadin and junctin complicating the interpretation of results. Acute, adenovirus-mediated over-expression in primary myocytes produced conflicting results. Limited over-expression of CSQ (approximately 1.6-fold) in cardiomyocytes resulted in increased SR Ca^{2+} content but reduced E–C coupling, in agreement with an inhibitory role for CSQ [457]. In contrast, a higher over-expression of CSQ (approximately 4-fold) resulted in increased E–C coupling in cardiomyocytes [458], as well as in skeletal muscle cells [459]. Thus, it seems that CSQ functional effects depend on the stoichiometry of association with the RyR/triadin/junctin/CSQ macromolecular complex.

The importance of CSQ in control of SR Ca^{2+} release was demonstrated by the identification of homozygous mutations in patients with CPVT (see Section 10.2.). Functional characterization of two mutations (R33Q and D307H) revealed two distinct mechanisms of action, although they both resulted in arrhythmogenic membrane potentials following β-adrenergic stimulation in cardiomyocytes over-expressing the mutant proteins [460,461]. Adenovirus-mediated over-expression of R33Q CSQ enhanced E–C coupling and Ca^{2+} spark frequency despite normal (total) SR Ca^{2+} content (versus wild type) [461]. In single channel recordings, the R33Q CSQ mutation abolished its inhibitory effect on RyR2, and it was therefore proposed that this mutation results in enhanced RyR2 channel sensitivity to luminal Ca^{2+} activation. In contrast, over-expression of D307H CSQ resulted in decreased E–C coupling and Ca^{2+} spark frequency consistent with the concomitant reduction in SR Ca^{2+} store content [460]. In vitro characterization of this mutant revealed that it had a profoundly altered conformation and reduced affinity for Ca^{2+}, triadin and junctin [462]. It is unclear how the two CSQ mutations result in cardiac arrhythmias while they produce opposing effects on RyR2-mediated SR Ca^{2+} release.

10. RyR pathophysiology

10.1. Heart failure

Heart failure resulting from different forms of cardiomyopathy is defined as the inability of the heart to pump sufficient blood to meet the body's metabolic demands (reviewed in [463]). It is a leading cause of disability and sudden death because of ventricular arrhythmias. In the early stages of heart failure, compensatory neurohormonal mechanisms are triggered, including the sympathetic nervous system, and catecholamine-induced stimulation of β-adrenergic receptors leads to production of cAMP and activation of PKA. PKA-mediated phosphorylation enhances L-type Ca^{2+} current, whereas phosphorylation of phospholamban relieves its inhibition of the SR Ca^{2+} ATPase pump, resulting in higher SR Ca^{2+} content. The net effect is enhancement of E–C coupling and improvement of cardiac function. However, in later stages of heart failure, calcium homeostasis is impaired, and contractile dysfunction is caused primarily by reduced intracellular Ca^{2+} transients and by concomitant decrease in the SR Ca^{2+} content (reviewed in [464]). There is now considerable evidence that diastolic SR Ca^{2+} leak, through RyR2, contributes to reduced SR Ca^{2+} content and systolic Ca^{2+} transients in heart failure [227,234,259,465,466]. However, there is controversy with regard to the mechanism that leads to RyR2-mediated Ca^{2+} leak.

An extensive series of studies from one laboratory have provided compelling evidence that in heart failure, chronic β-adrenergic stimulation results in RyR2 'hyper-phosphorylation' at S2808 by PKA, inducing dissociation of the stabilizing FKBP12.6 co-protein, thereby enhancing channel activity [234,241,245,246,467–471]. This hypothesis was based on the following observations: (i) failing cardiomyocytes displayed higher phosphorylation levels exclusively at S2808, (ii) S2808 was uniquely and exclusively phosphorylated by PKA, whereas CaMKII phosphorylated a different, unique residue, S2814, (iii) RyR2 hyper-phosphorylation was due to decreased local protein phosphatase and cAMP hydrolysis activities (reduced levels of PP1, PP2A and phosphodiesterase 4D3 associated with RyR2), (iv) FKBP12.6 levels associated with RyR2 were severely reduced in failing cardiomyocytes, PKA-treated channels or S2808D mutant channels (mimicking constitutive phosphorylation) but were normal in S2808A channels (that cannot be phosphorylated), (v) RyR2 channels isolated from failing cardiomyocytes or treated with PKA or S2808D (but not S2808A) mutant channels displayed enhanced channel activity and subconductance behaviour characteristic of FKBP12.6 dissociation and consistent with SR Ca^{2+} leak in cells, (vi) β-adrenergic receptor blockers reversed RyR2 hyper-phosphorylation, FKBP12.6 dissociation and channel activation in heart failure, (vii) RyR2 S2808A transgenic mice were resistant to heart failure, (viii) a benzothiazepine, JTV519, was found to restore the RyR2-FKBP12.6 association in failing cardiomyocytes, PKA-treated channels or S2808D mutant channels and (ix) JTV519 protected mice from heart failure, but it was without effect in FKBP12.6-deficient animals.

An independent group has supported some of the above findings, including increased RyR2 phosphorylation, FKBP12.6 dissociation and the beneficial effects of beta-blockers and JTV519 in heart failure [472–475]. However, it also reported some important differences, namely that the initial critical step leading to RyR2 dysregulation is a defective interdomain interaction that is induced by oxidative stress [227,281]. These conclusions are based on the finding that disruption of the RyR N-terminal–central domain interaction by the DPc10 peptide or by oxidizing reagents resulted in Ca^{2+} leak through RyR2 without removing FKBP12.6 although it facilitated PKA-mediated phosphorylation and FKBP12.6 dissociation. Importantly, RyR2 was shown to be oxidized, and the interdomain interaction was disrupted in an animal model of heart failure but was restored by antioxidant administration. In isolated failing cardiomyocytes, JTV519 also restored the interdomain interaction and enhanced Ca^{2+} transients without reversing RyR2 phosphorylation state or reassociation of FKBP12.6. These results suggest that oxidative stress-induced disruption of interdomain interactions within RyR2 is the primary cause of Ca^{2+} leak, whereas phosphorylation and FKBP12.6 dissociation are the consequence, rather than the cause, of RyR2 leaky channels.

In addition, certain aspects of the hyper-phosphorylation hypothesis have been questioned by results from several other laboratories that used very similar experimental approaches. First, S2808 was found to be highly phosphorylated in normal hearts even at rest, and to be a substrate for CaMKII and PKG in addition to PKA, and that it is not a unique PKA site (see Section 7.6.). Second, PKA-mediated RyR2 phosphorylation levels and channel activity were similar in failing and normal cardiomyocytes [242,476], whereas the increased S2808 phosphorylation that was found in some animal models of heart failure was unrelated to β-adrenergic stimulation and RyR2 dysfunction [244,477]. In line with these studies, PKA activation in permeabilized cardiomyocytes obtained from phospholamban knock-out mice did not affect Ca^{2+} spark frequency (under conditions that excluded PKA effects on DHPR Ca^{2+} current and SR Ca^{2+} content) [258,259]. Third, no cardiomyopathies or heart failures were reported in FKBP12.6 knock-out mice (although mild hypertrophy for male mice was reported in one of the two models) [372,380]. Fourth, PKA-dependent phosphorylation or S2808D substitution did not cause FKBP12.6 dissociation, and S2808D mutant channel activity was identical to wild type [242,247,363,382,476]. A recent study detected a reduced association between RyR2 and FKBP12.6 (by approximately 40%) in failing cardiomyocytes but also reported a decrease in cellular FKBP12.6 levels, which at the mRNA level was reduced by approximately 50% [478]. This finding suggests that the reduced RyR2-FKBP12.6 association often seen in heart failure could simply be a result of lower FKBP12.6 expression levels.

An alternative mechanism in the pathogenesis of heart failure involves CaMKII-dependent phosphorylation of RyR2. It has been shown that CaMKII activity is increased upon sustained β-adrenergic stimulation and in heart failure [478–481]. In an animal model of heart failure, RyR2 had increased phosphorylation levels due to higher levels of associated CaMKII, and failing cardiomyocytes had an increased RyR2-mediated SR Ca^{2+} leak that was inhibited by CaMKII inhibitors [478]. Chronic over-expression of CaMKIIδC in transgenic mice resulted in contractile dysfunction and heart failure, with decreased Ca^{2+} transients and SR Ca^{2+} content, as well as changes in the expression profile of proteins involved in E–C coupling [261,479]. Although these adaptive changes make it difficult to pinpoint the primary defect, there was increased CaMKII-dependent phosphorylation of RyR2 that was associated with increased Ca^{2+} spark frequency. In contrast, transgenic mice over-expressing a CaMKII inhibitory peptide were resistant to maladaptive remodelling associated with heart failure and had preserved contractility and Ca^{2+} homeostasis [481]. It should be noted that CaMKII-mediated phosphorylation of RyR2 did not result in FKBP12.6 dissociation [259,262].

Increased SR Ca^{2+} leak may also be due to increased luminal Ca^{2+} activation of RyR2. It has been reported in an animal model of heart failure that there was an increase in Ca^{2+} spark frequency despite lower SR Ca^{2+} content, and the RyR2 studied by single channel recordings was found to have an enhanced sensitivity to luminal Ca^{2+} activation [466].

10.2. Arrhythmogenic cardiac diseases

CPVT is an inherited arrhythmogenic disease characterized by adrenergically mediated bidirectional or polymorphic ventricular tachycardia leading to syncope and/or sudden cardiac death (reviewed in [482,483]). Although it is relatively rare, CPVT is a highly malignant disease (mortality rates of 30–35%) with incomplete penetrance, manifesting in childhood and adolescence. Patients with CPVT have structurally normal hearts and typically present with ventricular arrhythmias because of physical or emotional stress, and they can also be induced reproducibly by the infusion of catecholamines. Administration of β-adrenergic receptor blockers is the standard treatment; however, it is not always effective. CPVT has been linked to two cardiac SR proteins, RyR2 and CSQ. Mutations in RyR2 cause a dominant form of CPVT, whereas mutations in CSQ are linked to an autosomal recessive form. Mutations in RyR2 are also associated with arrhythmogenic right ventricular dysplasia type 2 (ARVD2). Patients with ARVD2 exhibit progressive fibro-fatty replacement of the right ventricular myocardium, in addition to stress-induced polymorphic ventricular tachycardia. However, the structural abnormalities are relatively mild and ARVD2 mimics the CPVT phenotype, suggesting an overlap between the two diseases. To date, 68 point mutations and a 2-residue insertion have been identified in RyR2 that tend to segregate in three distinct regions: the N-terminus, the central domain and the C-terminus containing the membrane-spanning pore-forming helices (Fig. 5).

Several studies have characterized some RyR2 mutations following expression of recombinant channels in cardiac (HL-1) or other (HEK293) cell lines [372,484–491]. Most studies suggest that mutations do not affect channel state in resting conditions, although enhanced basal activity for at least some mutants (N4104K, R4497C[1] and N4895D) has been reported [484,486]. All studies indicate that RyR2 mutations result in increased SR Ca^{2+} release after channel activation; however, more detailed examination to reveal the molecular mechanism(s) underlying channel dysfunction has produced contradictory results. In a cell-based assay, some mutant channels were

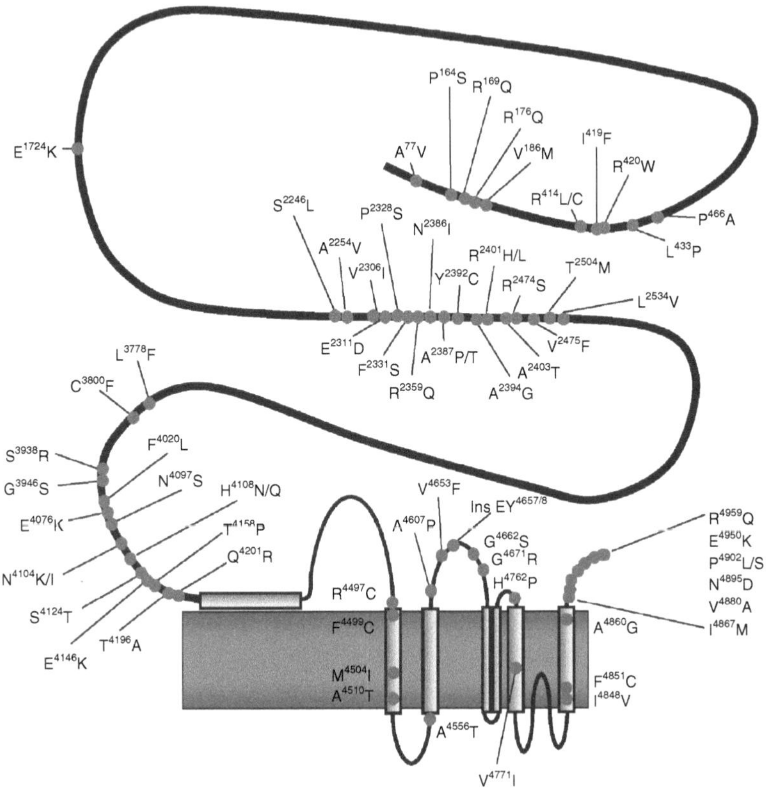

Fig. 5. Ryanodine receptor 2 (RyR2) mutations associated with arrhythmogenic diseases. Schematic to illustrate catecholaminergic polymorphic ventricular tachycardia/arrhythmogenic right ventricular dysplasia type 2 (CPVT/ARVD2) mutations (red circles) clustered primarily in three discrete regions of the RyR2 polypeptide: the N-terminal (residues 77–466), the central (residues 2246–2534) and the C-terminal domain (residues 3778–4959). Image kindly prepared by Dr. N. Lowri Thomas (Wales Heart Research Institute, Cardiff University, UK) (See Color Plate 37, p. 534).

found to have altered sensitivity to caffeine or cytoplasmic Ca^{2+} activation, whereas other mutants (R176Q/T2504M and L433P) had markedly reduced sensitivity to cytoplasmic Ca^{2+} inactivation [487,489]. The latter observation is in line with reduced Mg^{2+} inhibition of other mutants (P2328S, Q4201R and V4653F) when studied by single channel recordings [488]. It has also been proposed that PKA-mediated phosphorylation, with the ensuing FKBP12.6 dissociation, results in abnormal SR Ca^{2+} release during β-adrenergic stimulation, in common with the mechanism proposed in heart failure [372,470,488,492]. This proposal was based on the observations that mutant RyR2 channels (S2246L, P2328S, R2474S, Q4201R, R4497C and V4653F) had reduced affinity for FKBP12.6, mice deficient (or haploinsufficient) in FKBP12.6 presented with exercise-induced ventricular arrhythmias similar to CPVT patients and JTV519 prevented triggered arrhythmias by restoring the RyR2-FKBP12.6 association. However, FKBP12.6 dissociation upon

PKA-dependent phosphorylation has been heavily contested, and JTV519 was shown to act independently of FKBP12.6 (see Section 10.1.). In addition, other studies found normal FKBP12.6 interaction with channels harbouring the same mutations [485,490,493], whereas β-blockers or JTV519 could not prevent cardiac arrhythmias in a mouse knock-in RyR2 mutant model for R4497C [493,494].

An alternative hypothesis proposes that RyR2 mutations exclusively enhance the sensitivity of the channel to luminal Ca^{2+} activation [486,490]. It was found that spontaneous Ca^{2+} release occurred at a lower Ca^{2+} store threshold in cells expressing mutant channels (R176Q/T2504M, L433P, S2246L, R2474S, N4104K, Q4201R, R4497C, I4867M and N4895D), and single channel recordings indicated enhanced luminal Ca^{2+} activation but unaltered cytoplasmic Ca^{2+} sensitivity. This hypothesis is consistent with involvement of CSQ mutations in CPVT (see Section 9.5.). Enhanced sensitivity to luminal calcium is likely to involve long range, conformational changes and allosteric interactions, at least for the N-terminal and central domain mutations. The disruption of regulatory interactions between the N-terminus and central domain by CPVT-linked mutations has been proposed to underlie the defective regulation of RyR2 mutant channels (see Section 8.2.). In addition, RyR2 mutations (N4104K and R4497C) were shown by a protein fragment complementation assay to result in altered interactions within a region (amino acids 3722–4610) suggested to transduce cytoplasmic signals to the pore-forming TM assembly [491].

The generation of two mouse knock-in models of R4497C and R176Q linked to CPVT and ARVD2, respectively, has yielded important information about the mechanism of inherited, RyR2-linked, triggered arrhythmias [493–495]. These animal models demonstrated that RyR2 mutations pre-dispose the mouse heart to the development of ventricular arrhythmias upon exercise or infusion of β-adrenergic agonists. About 30% of R4497C heterozygous mice developed ventricular tachycardia, which was unrelated to the relative expression levels of the wild-type and mutant alleles [494], consistent with incomplete penetrance of the CPVT disease, where some family members carrying RyR2 mutations are asymptomatic. Electrophysiological analysis indicated that triggered arrhythmias are mediated by delayed after-depolarizations [493–495], in common with cardiac arrhythmias in heart failure (reviewed in [496,497]). The likely mechanism is that RyR2-mediated diastolic SR Ca^{2+} leak activates the plasma membrane Na^{+}–Ca^{2+} exchanger (which exchanges one Ca^{2+} for three Na^{+}) thereby generating a net inward current that could induce diastolic depolarizations (called delayed after-depolarizations). Delayed after-depolarizations could in turn elicit pre-mature action potentials and initiation of arrhythmias. Interestingly, R4497C knock-in mice occasionally demonstrated delayed after-depolarizations at rest, which were enhanced upon β-adrenergic stimulation, suggesting that mutant RyR2 has a basal defect that is exacerbated by β-adrenergic stimulation [493]. This phenotype is consistent with the clinical observations that β-blockers provide only partial protection in CPVT patients. Mice heterozygous for R176Q exhibited catecholamine-induced ventricular tachycardia and had structurally normal hearts (although right ventricular end-diastolic volume was decreased), suggesting that ARVD2 and CPVT may represent subtle but distinct variations of a single disorder [495].

The use of animal knock-in models should help to identify the molecular mechanism(s) underlying RyR2 dysfunction. In vitro functional assays have suggested several mechanisms including altered sensitivity to luminal Ca^{2+} activation, cytoplasmic Ca^{2+} and Mg^{2+} inhibition, inter-domain interactions, phosphorylation status and FKBP12.6 interaction. The reason for the disparate results is not clear. It is notable that expression of recombinant channels in HEK293 cells results in homotetramers, whereas in HL-1 (which contain endogenous wild type RyR2), this results in heterotetramers, mimicking the heterozygous CPVT (or ARVD2) patients. HL-1 cells also have an appropriate genetic background, expressing the complement of cardiac-specific accessory proteins. However, different findings were reported for the same RyR2 mutation, even when expressed in the same cell line. At present, the important unresolved issue is to determine the primary causative defect. For example, altered inter-domain interactions within the RyR2 structure may be the primary cause, resulting in a channel that is more prone to open and/or remain open. This could be manifested as enhanced sensitivity to cytoplasmic and/or luminal Ca^{2+} activation and/or reduced sensitivity to Ca^{2+} and Mg^{2+} inhibition. However, given the dispersion of mutations across the RyR2 sequence, it may be unlikely that there is a common mechanism to account for channel dysregulation.

10.3. Neuromuscular disorders

Three clinically distinct hereditary skeletal muscle disorders are known to be associated with mutations in the gene encoding RyR1: MH, central core disease (CCD) and multi-minicore disease (MmD) (reviewed in [498]). MH is an autosomal dominant pharmacogenetic disorder of skeletal muscle, triggered in pre-disposed individuals by inhalation of anaesthetics (e.g. halothane) or by depolarising muscle relaxants (e.g. succinylcholine). Susceptible individuals respond with skeletal muscle rigidity, hypermetabolism, tachycardia, unstable and rising blood pressure and eventually dramatic hyperthermia. MH is a life-threatening disease and is treated with the RyR1 antagonist dantrolene.

CCD is a congenital myopathy characterized by hypotonia during infancy, proximal muscle weakness, delayed motor development and reduced muscle bulk. It is usually inherited as an autosomal-dominant disease although recessive forms have also been described. Diagnosis of CCD is determined through histological identification of large amorphous areas (cores) that lack mitochondria and thus oxidative enzyme activity and cover a considerable area of primarily type 1 muscle fibres. MmD is an autosomal recessive congenital myopathy that is characterized histologically by the presence of multiple cores devoid of mitochondria and with disorganized sarcomeric structures. Multiple small cores can occur in both type 1 and 2 fibres that do not run the entire length of the muscle fibre. Patients affected by MmD have symptoms similar to CCD, and there may be some overlap between the two diseases. More than 80 different mutations in the RyR1 gene have so far been associated with MH, CCD and MmD. They are mostly point mutations although small, in-frame deletions or out-of-frame insertions have also been found. RyR1 mutations cluster in three separate regions: in the N-terminus, the central domain and the C-terminus containing the channel pore, with very few mutations outside these regions. These three regions correspond to the ones identified in RyR2-associated arrhythmogenic diseases (see Section 10.2.), which might suggest a common mechanism of action. RyR1 is the only gene so far linked to CCD, whereas mutations in RyR1 account for the majority of MH cases and only a minority of MmD cases. Although both MH- and CCD-associated mutations are found in all three regions, MH-linked mutations are mostly in the N-terminal and central regions, whereas CCD-linked mutations predominate in the C-terminus. Interestingly, some RyR1 mutations cause both MH and CCD.

The functional effects of MH-/CCD-linked mutations have been extensively studied by a combination of methods. These include single-channel recordings and ryanodine-binding assays of isolated channels, as well as Ca^{2+} release measurements in heterologous (HEK293 and COS-7) cell lines, or in primary dyspedic or normal myocytes expressing recombinant mutant RyR1, or of endogenous mutant channels in skeletal myocytes or lymphocytes [isolated from affected individuals, pigs with a naturally occurring MH mutation (R614C[1]) or a mouse knock-in MH model, Y522S]. These studies have indicated that RyR1 mutations in the N-terminal and central regions and most mutations in the C-terminus result in higher sensitivity to pharmacological agonists such as caffeine and halothane, as well as to activation by plasma membrane depolarization, and also in reduced sensitivity to inactivation by Mg^{2+} and Ca^{2+} [499–510]. Although differences exist between studies, it was found that RyR1 mutant channels had enhanced basal activity at rest but to a variable extent, resulting in higher resting intracellular Ca^{2+} concentration and lower Ca^{2+} store content and consequently reduced magnitude of Ca^{2+} release [505,507,510–516]. CCD-linked mutations had the most profound effect on calcium store depletion irrespective of their location on the RyR1 sequence, whereas MH-linked mutations had smaller or no effects. Contradictory findings have been reported for a subset of C-terminal CCD-associated mutations located within the pore-forming luminal loop (e.g. I4898T[1]). Although some studies have shown that these mutations also produce 'leaky' channels with increased resting intracellular Ca^{2+} concentration and lower Ca^{2+} store content [512,516–518], others found normal cytoplasmic and SR Ca^{2+} levels but severely reduced sensitivity to Ca^{2+} activation and depolarization- or agonist-induced Ca^{2+} release [513,519,520].

The above investigations suggest that although both MH- and CCD-linked mutations result in 'hyper-sensitive' channels and increased basal activity, only CCD-linked mutations cause a substantial Ca^{2+} leak from the SR that cannot be compensated for by increased Ca^{2+} uptake. This leads to a partially depleted SR Ca^{2+} store and in turn decreased Ca^{2+} release during E–C coupling and therefore muscle weakness. Alternatively, compromised E–C coupling is due to RyR1 mutations within the pore-forming sequences that result in reduced Ca^{2+} permeability. In MH, RyR1 mutations result in reduced inactivation of channels by Ca^{2+} and Mg^{2+} and therefore prolonged elevation in intracellular Ca^{2+} concentration and muscle rigidity. Disruption of inter-domain interactions involved in stabilization of the channel in a closed state has also been implicated in MH (see Section 8.2.).

11. Conclusions

Since its purification and molecular cloning in the late 1980s, the RyR has been extensively studied using a variety of approaches that range from single channel analysis to animal models. These studies have contributed to our understanding of the RyR properties at the molecular level and also in the context of the intact cell and whole animal. We now know a lot about the biophysical properties of the channel, its regulation by endogenous and exogenous effectors and its physiological mechanism of activation in skeletal and cardiac muscle. There are, however, aspects that we still do not understand fully. Of paramount importance is the lack of information about the three-dimensional structure of the protein and of mechanism(s) that satisfactorily explain termination of SR Ca^{2+} release despite the inherent positive feedback of CICR.

Defective RyR2 regulation, either induced (in heart failure) or inherited (in CPVT and ARVD2), is directly involved in cardiac disease and sudden death. Extensive research is currently focused on the delineation of the underlying pathway(s) of RyR2-mediated pathology, but thus far, conflicting reports have only produced a confusing picture. Clearly much more research is needed in order to clarify this field and identify the causative molecular defect(s). It is only then that with the combined efforts of basic scientists and clinicians, treatments can be developed aimed at restoring RyR2 function and therefore cellular Ca^{2+} homeostasis and cardiac output.

Acknowledgements

We are grateful to Dr. Zheng Liu, Dr. Leon D'Cruz and Dr. N. Lowri Thomas for their kind assistance in preparing the images used in Fig. 1, 3 and 5, respectively. Our research is funded by grants from the British Heart Foundation, The Wellcome Trust, European Union (CONTICA programme) and Medical Research Council.

Note

1. Numbering given for the human sequence.

References

1. Berridge, M., Lipp, P. and Bootman, M. (2000) *Nat Rev Mol Cell Biol* **1**, 11–21.
2. Franzini-Armstrong, C. (1999) *FASEB J* **13**, S266–S270.
3. Sutko, J., Airey, J., Welch, W. and Ruest, L. (1997) *Pharmacol Rev* **49**, 53–98.
4. Fleischer, S., Ogunbunm, I., Dixon, M. and Fleer, E. (1985) *Proc Natl Acad Sci USA* **82**, 7256–7259.
5. Pessah, I., Francini, A., Scales, D., Waterhouse, A. and Casida, J. (1986) *J Biol Chem* **261**, 8643–8648.
6. Campbell, K., Knudson, C., Imagawa, T., Leung, A., Sutko, J., Kahl, S., Raab, C. and Madson, L. (1987) *J Biol Chem* **262**, 6460–6463.
7. Inui, M., Saito, A. and Fleischer, S. (1987) *J Biol Chem* **262**, 1740–1747.

8. Lai, F., Erickson, H., Rousseau, E., Liu, Q. and Meissner, G. (1988) *Nature* **331**, 315–319.
9. Inui, M., Saito, A. and Fleischer, S. (1987) *J Biol Chem* **262**, 15637–15642.
10. Anderson, K., Lai, F., Liu, Q., Rousseau, E., Erickson, H. and Meissner, G. (1989) *J Biol Chem* **264**, 1329–1335.
11. Lai, F., Misra, M., Xu, L., Smith, H. and Meissner, G. (1989) *J Biol Chem* **264**, 16776–16785.
12. Imagawa, T., Smith, J., Coronado, R. and Campbell, K. (1987) *J Biol Chem* **262**, 16636–16643.
13. Hymel, L., Inui, M., Fleischer, S. and Schindler, H. (1988) *Proc Natl Acad Sci USA* **85**, 441–445.
14. Sutko, J. and Airey, J. (1996) *Physiol Rev* **76**, 1027–1071.
15. Franzini-Armstrong, C. and Protasi, F. (1997) *Physiol Rev* **77**, 699–729.
16. Takeshima, H., Nishimura, S., Matsumoto, T., Ishida, H., Kangawa, K., Minamino, N., Matsuo, H., Ueda, M., Hanaoka, M., Hirose, T. and Numa, S. (1989) *Nature* **339**, 439–445.
17. Zorzato, F., Fujii, J., Otsu, K., Phillips, M., Green, N., Lai, F., Meissner, G. and MacLennan, D. (1990) *J Biol Chem* **265**, 2244–2256.
18. Nakai, J., Imagawa, T., Hakamata, Y., Shigekawa, M., Takeshima, H. and Numa, S. (1990) *FEBS Lett* **271**, 169–177.
19. Otsu, K., Willard, H., Khanna, V., Zorzato, F., Green, N. and MacLennan, D. (1990) *J Biol Chem* **265**, 13472–13483.
20. Hakamata, Y., Nakai, J., Takeshima, H. and Imoto, K. (1992) *FEBS Lett* **312**, 229–235.
21. Chen, S., Vaughan, D., Airey, J., Coronado, R. and MacLennan, D. (1993) *Biochemistry* **32**, 3743–3753.
22. Chen, S., Leong, P., Imredy, J., Bartlett, C., Zhang, L. and MacLennan, D. (1997) *Biophys J* **73**, 1904–1912.
23. Imagawa, T., Nakai, J., Takeshima, H., Nakasaki, Y. and Shigekawa, M. (1992) J *Biochemistry* **112**, 508–513.
24. Bhat, M., Hayek, S., Zhao, J., Zang, W., Takeshima, H., Wier, W. and Ma, J. (1999) *Biophys J* **77**, 808–816.
25. Chen, S., Li, X., Ebisawa, K. and Zhang, L. (1997) *J Biol Chem* **272**, 24234–24246.
26. Saeki, K., Obi, I., Ogiku, N., Hakamata, Y. and Matsumoto, T. (1998) *Life Sci* **63**, 575–588.
27. Sorrentino, V. and Volpe, P. (1993) *Trends Pharmacol Sci* **14**, 98–103.
28. Airey, J., Beck, C., Murakami, K., Tanksley, S., Deerinck, T., Ellisman, M. and Sutko, J. (1990) *J Biol Chem* **265**, 14187–14194.
29. Olivares, E., Tanksley, S., Airey, J., Beck, C., Ouyang, Y., Deerinck, T., Ellisman, M. and Sutko, J. (1991) *Biophys J* **59**, 1153–1163.
30. Lai, F., Liu, Q., Xu, L., El-Hashem, A., Kramarcy, N., Sealock, R. and Meissner, G. (1992) *Am J Physiol* **263**, C365–C372.
31. Murayama, T. and Ogawa, Y. (1992) *J Biochem* **112**, 514–522.
32. O'Brien, J., Meissner, G. and Block, B. (1993) *Biophys J* **65**, 2418–2427.
33. Airey, J., Grinsell, M., Jones, L., Sutko, J. and Witcher, D. (1993) *Biochemistry* **32**, 5739–5745.
34. Oyamada, H., Murayama, T., Takagi, T., Iino, M., Iwabe, N., Miyata, T., Ogawa, Y. and Endo, M. (1994) *J Biol Chem* **269**, 17206–17214.
35. Ottini, L., Marziali, G., Conti, A., Charlesworth, A. and Sorrentino, V. (1996) *Biochem J* **315**, 207–216.
36. Seok, J., Xu, L., Kramarcy, N., Sealock, R. and Meissner, G. (1992) *J Biol Chem* **267**, 15893–15901.
37. Takeshima, H., Nishi, M., Iwabe, N., Miyata, T., Hosoya, T., Masai, I. and Hotta, Y. (1994) *FEBS Lett* **337**, 81–87.
38. Sakube, Y., Ando, H. and Kagawa, H. (1993) *Ann NY Acad Sci* **707**, 540–545.
39. Tunwell, R., Wickenden, C., Bertrand, B., Shevchenko, V., Walsh, M., Allen, P. and Lai, F. (1996) *Biochem J* **318**, 477–487.
40. Zorzato, F., Sacchetto, R. and Margreth, A. (1994) *Biochem Biophys Res Commun* **203**, 1725–1730.
41. Futatsugi, A., Kuwajima, G. and Mikoshiba, K. (1995) *Biochem J* **305**, 373–378.

42. Marziali, G., Rossi, D., Giannini, G., Charlesworth, A. and Sorrentino, V. (1996) *FEBS Lett* **394**, 76–82.
43. Miyatake, R., Furukawa, A., Masayuki, M., Iwahashi, K., Nakamura, K., Ichikawa, Y. and Suwaki, H. (1996) *FEBS Lett* **395**, 123–126.
44. Jiang, D., Xiao, B., Li, X. and Chen, S. (2003) *J Biol Chem* **278**, 4763–4769.
45. Kimura, T., Nakamori, M., Lueck, J., Pouliquin, P., Aoike, F., Fujimura, H., Dirksen, R., Takahashi, M., Dulhunty, A. and Sakoda, S. (2005) *Hum Mol Genet* **14**, 2189–2200.
46. Xiao, B., Masumiya, H., Jiang, D., Wang, R., Sei, Y., Zhang, L., Murayama, T., Ogawa, Y., Lai, F., Wagenknecht, T. and Chen, S. (2002) *J Biol Chem* **277**, 41778–41785.
47. Beutner, G., Sharma, V., Giovannucci, D., Yule, D. and Sheu, S. (2001) *J Biol Chem* **276**, 21482–21488.
48. Mitchell, K., Pinton, P., Varadi, A., Tacchetti, C., Ainscow, E., Pozzan, T., Rizzuto, R. and Rutter, G. (2001) *J Cell Biol* **155**, 41–51.
49. Kuwajima, G., Futatsugi, A., Niinobe, M., Nakanishi, N. and Mikoshiba, K. (1992) *Neuron* **9**, 1133–1142.
50. Lai, F., Dent, M., Wickenden, C., Xu, L., Kumari, G., Misra, M., Lee, H., Sar, M. and Meissner, G. (1992) *Biochem J* **288**, 553–564.
51. Ouyang, Y., Deerinck, T., Walton, P., Airey, J., Sutko, J. and Ellisman, M. (1993) *Brain Res* **620**, 269–280.
52. Ledbetter, M., Preiner, J., Louis, C. and Mickelson, J. (1994) *J Biol Chem* **269**, 31544–31551.
53. Furuichi, T., Furutama, D., Hakamata, Y., Nakai, J., Takeshima, H. and Mikoshiba, K. (1994) *J Neurosci* **14**, 4794–4805.
54. Giannini, G., Conti, A., Mammarella, S., Scrobogna, S. and Sorrentino, V. (1995) *J Cell Biol* **128**, 893–904.
55. Neylon, C., Richards, S., Larsen, M., Agrotis, A. and Bobik, A. (1995) *Biochem Biophys Res Commun* **215**, 814–821.
56. Martin, C., Chapman, K., Seckl, J. and Ashley, R. (1998) *Neuroscience* **85**, 205–216.
57. Murayama, T. and Ogawa, Y. (1997) *J Biol Chem* **272**, 24030–24037.
58. Jeyakumar, L., Copello, J., O'Malley, A., Wu, G.-M., Grassucci, R., Wagenknecht, T. and Fleischer, S. (1998) *J Biol Chem* **273**, 16011–16020.
59. Takeshima, H., Iino, M., Takekura, H., Nishi, M., Kuno, J., Minowa, O., Takano, H. and Noda, T. (1994) *Nature* **369**, 556–559.
60. Takeshima, H., Komazaki, S., Hirose, K., Nishi, M., Noda, T. and Iino, M. (1998) *EMBO J* **17**, 3309–3316.
61. Takeshima, H., Ikemoto, T., Nishi, M., Nishiyama, N., Shimuta, M., Sugitani, Y., Kuno, J., Saito, I., Saito, H., Endo, M., Iino, M. and Noda, T. (1996) *J Biol Chem* **271**, 19649–19652.
62. Bertocchini, F., Ovitt, C., Conti, A., Barone, V., Scholer, H., Bottinelli, R., Reggiani, C. and Sorrentino, V. (1997) *EMBO J* **16**, 6956–6963.
63. Balschun, D., Wolfer, D., Bertocchini, F., Barone, V., Conti, A., Zuschratter, W., Missiaen, L., Lipp, H., Frey, U. and Sorrentino, V. (1999) *EMBO J* **18**, 5264–5273.
64. Futatsugi, A., Kato, K., Ogura, H., Li, S., Nagata, E., Kuwajuma, G., Tanaka, K., Itohara, S. and Mikoshiba, K. (1999) *Neuron* **24**, 701–713.
65. Williams, A., West, D. and Sitsapesan, R. (2001) *Q Rev Biophys* **34**, 61–104.
66. Smith, J., Imagawa, T., Ma, J., Fill, M., Campbell, K. and Coronado, R. (1988) *J Gen Physiol* **92**, 1–26.
67. Liu, Q., Lai, F., Rousseau, E., Jones, R. and Meissner, G. (1989) *Biophys J* **55**, 415–424.
68. Lindsay, A., Manning, S. and Williams, A. (1991) *J Physiol* **439**, 463–480.
69. Tinker, A. and Williams, A. (1992) *J Gen Physiol* **100**, 479–493.
70. Murayama, T., Oba, T., Katayama, E., Oyamada, H., Oguchi, K., Kobayashi, M., Otsuka, K. and Ogawa, Y. (1999) *J Biol Chem* **274**, 17297–17308.
71. Kasai, M., Kawasaki, T. and Yamamoto, K. (1992) *J Biochem* **112**, 197–203.

72. Tinker, A. and Williams, A. (1993) *J Gen Physiol* **102**, 1107–1129.
73. Tu, Q., Velez, P., Brodwick, M. and Fill, M. (1994) *Biophys J* **67**, 2280–2285.
74. Tinker, A. and Williams, A. (1995) *Biophys J* **68**, 111–120.
75. Tinker, A., Lindsay, A. and Williams, A. (1992) *J Membr Biol* **127**, 149–159.
76. Tu, Q., Velez, P., Cortes-Gutierrez, M. and Fill, M. (1994) *J Gen Physiol* **103**, 853–867.
77. Tinker, A. and Williams, A. (1993) *Biophys J* **65**, 852–864.
78. Xu, L., Jones, R. and Meissner, G. (1993) *J Gen Physiol* **101**, 207–233.
79. Mead, F., Sullivan, D. and Williams, A. (1998) *J Membr Biol* **163**, 225–234.
80. Xu, L., Tripathy, A., Pasek, D. and Meissner, G. (1999) *J Biol Chem* **274**, 32680–32691.
81. Mead, F. and Williams, A. (2004) *Biophys J* **87**, 3814–3825.
82. Welch, W., Rheault, S., West, D. and Williams, A. (2004) *Biophys J* **87**, 2335–2351.
83. Marks, A. (1996) *Physiol Rev* **76**, 631–649.
84. Anyatonwu, G., Buck, E. and Ehrlich, B. (2003) *J Biol Chem* **278**, 45528–45538.
85. Zhao, M., Li, P., Li, X., Zhang, L., Winkfein, R. and Chen, S. (1999) *J Biol Chem* **274**, 25971–25974.
86. Radermacher, M., Rao, V., Grassucci, R., Frank, J., Timerman, A., Fleischer, S. and Wagenknecht, T. (1994) *J Cell Biol* **127**, 411–423.
87. Serysheva, I., Orlova, E., Chiu, W., Sherman, M., Hamilton, S. and Van Heel, M. (1995) *Nat Struct Biol* **2**, 18–24.
88. Sharma, M., Penczek, P., Grassucci, R., Xin, H., Fleischer, S. and Wagenknecht, T. (1998) *J Biol Chem* **273**, 18429–18434.
89. Serysheva, I., Hamilton, S., Chiu, W. and Ludtke, S. (2005) *J Mol Biol* **345**, 427–431.
90. Samso, M., Wagenknecht, T. and Allen, P. (2005) *Nat Struct Mol Biol* **12**, 539–544.
91. Ludtke, S., Serysheva, I., Hamilton, S. and Chiu, W. (2005) *Structure* **13**, 1203–1211.
92. Orlova, E., Serysheva, I., van Heel, M., Hamilton, S. and Chiu, W. (1996) *Nat Struct Biol* **3**, 547–552.
93. Serysheva, I., Schatz, M., Van Heel, M., Chiu, W. and Hamilton, S. (1999) *Biophys J* **77**, 1936–1944.
94. Sharma, M., Jeyakumar, L., Fleischer, S. and Wagenknecht, T. (2000) *J Biol Chem* **275**, 9485–9491.
95. Wagenknecht, T., Radermacher, M., Grassucci, R., Berkowitz, J., Xin, H. and Fleisher, S. (1997) *J Biol Chem* **272**, 32463–32471.
96. Liu, Z., Zhang, J., Sharma, M., Li, P., Chen, S. and Wagenknecht, T. (2001) *Proc Natl Acad Sci USA* **98**, 6104–6109.
97. Baker, M., Serysheva, I., Sencer, S., Wu, Y., Ludtke, S., Jiang, W., Hamilton, S. and Chiu, W. (2002) *Proc Natl Acad Sci USA* **99**, 12155–12160.
98. Liu, Z., Wang, R., Zhang, J., Chen, S. and Wagenknecht, T. (2005) *J Biol Chem* **280**, 37941–37947.
99. Benacquista, B., Sharma, M., Samso, M., Zorzato, F., Treves, S. and Wagenknecht, T. (2000) *Biophys J* **78**, 1349–1358.
100. Liu, Z., Zhang, J., Li, P., Chen, S. and Wagenknecht, T. (2002) *J Biol Chem* **277**, 46712–46719.
101. Zhang, J., Liu, Z., Masumiya, H., Wang, R., Jiang, D., Li, F., Wagenknecht, T. and Chen, S. (2003) *J Biol Chem* **278**, 14211–14218.
102. Liu, Z., Zhang, J., Wang, R., Chen, S. and Wagenknecht, T. (2004) *J Mol Biol* **338**, 533–545.
103. Wang, J., Needleman, D., Seryshev, A., Aghdasi, B., Slavik, K., Liu, S.-Q., Pedersen, S. and Hamilton, S. (1996) *J Biol Chem* **271**, 8387–8393.
104. Bhat, M., Zhao, J., Takeshima, H. and Ma, J. (1997) *Biophys J* **73**, 1329–1336.
105. Bhat, M., Zhao, J., Zang, W., Balke, C., Takeshima, H., Wier, W. and Ma, J. (1997) *J Gen Physiol* **110**, 749–762.
106. Xu, X., Bhat, M., Nishi, M., Takeshima, H. and Ma, J. (2000) *Biophys J* **78**, 1270–1281.
107. George, C., Jundi, H., Thomas, N., Scoote, M., Walters, N., Williams, A. and Lai, F. (2004) *Mol Biol Cell* **15**, 2627–2638.
108. Marty, I., Villaz, M., Arlaud, G., Bally, I. and Ronjat, M. (1994) *Biochem J* **298**, 743–749.
109. Grunwald, R. and Meissner, G. (1995) *J Biol Chem* **270**, 11338–11347.

110. Du, G., Sandhu, B., Khanna, V., Guo, X. and MacLennan, D. (2002) *Proc Natl Acad Sci USA* **99**, 16725–16730.
111. Du, G., Avila, G., Sharma, P., Khanna, V., Dirksen, R. and MacLennan, D. (2004) *J Biol Chem* **279**, 37566–37574.
112. Doyle, D., Morais Cabral, J., Pfuetzner, R., Kuo, A., Gulbis, J., Cohen, S., Chait, B. and MacKinnon, R. (1998) *Science* **280**, 69–77.
113. Jiang, Y., Lee, A., Chen, J., Cadene, M., Chait, B. and MacKinnon, R. (2002) *Nature* **417**, 523–526.
114. Balshaw, D., Gao, L. and Meissner, G. (1999) *Proc Natl Acad Sci USA* **96**, 3345–3347.
115. Gao, L., Balshaw, D., Xu, L., Tripathy, A., Xin, C. and Meissner, G. (2000) *Biophys J* **79**, 828–840.
116. Du, G., Guo, X., Khanna, V. and MacLennan, D. (2001) *J Biol Chem* **276**, 31760–31771.
117. Chen, S., Li, P., Zhao, M., Li, X. and Zhang, L. (2002) *Biophys J* **82**, 2436–2447.
118. Wang, R., Zhang, L., Bolstad, J., Diao, N., Brown, C., Ruest, L., Welch, W., Williams, A. and Chen, S. (2003) *J Biol Chem* **278**, 51557–51565.
119. Wang, R., Bolstad, J., Kong, H., Zhang, L., Brown, C. and Chen, S. (2004) *J Biol Chem* **279**, 3635–3642.
120. Du, G. and MacLennan, D. (1998) *J Biol Chem* **273**, 31867–31872.
121. Wang, Y., Xu, L., Pasek, D., Gillespie, D. and Meissner, G. (2005) *Biophys J* **89**, 256–265.
122. Xu, L., Wang, Y., Gillespie, D. and Meissner, G. (2006) *Biophys J* **90**, 443–453.
123. Ferguson, D., Schwartz, H. and Franini-Armstrong, C. (1984) *J Cell Biol* **99**, 1735–1742.
124. Saito, A., Inui, M., Radermacher, M., Frank, J. and Fleischer, S. (1988) *J Cell Biol* **107**, 211–219.
125. Schneider, M. and Chandler, W. (1973) *Nature* **242**, 244–246.
126. Rios, E. and Brum, G. (1987) *Nature* **325**, 717–720.
127. Tanabe, T., Beam, K., Powell, J. and Numa, S. (1988) *Nature* **336**, 134–139.
128. Tanabe, T., Beam, K., Adams, B. and Niidome, T. (1990) *Nature* **346**, 567–569.
129. Jacquemond, V., Csernoch, L., Klein, M. and Schneider, M. (1991) *Biophys J* **60**, 867–873.
130. Marty, I., Robert, M., Villaz, M., De Jongh, K., Lai, Y., Catterall, W. and Ronjat, M. (1994) *Proc Natl Acad Sci USA* **91**, 2270–2274.
131. Block, B., Leung, A., Campbell, K. and Franzini-Armstrong, C. (1988) *J Cell Biol* **107**, 2587–2600.
132. Takekura, H., Bennett, L., Tanabe, T., Beam, K. and Franzini-Armstrong, C. (1994) *Biophys J* **67**, 793–804.
133. Franzini-Armstrong, C. and Kish, J. (1995) *J Muscle Res Cell Motil* **16**, 319–324.
134. Nakai, J., Dirksen, R., Nguyen, H., Pessah, I., Beam, K. and Allen, P. (1996) *Nature* **380**, 72–75.
135. Sun, X., Protasi, F., Takahashi, M., Takeshima, H., Ferguson, D. and Franzini-Armstrong, C. (1995) *J Cell Biol* **129**, 659–667.
136. Carl, S., Felix, K., Caswell, A., Brandt, N., Ball, W., Vaghy, P., Meissner, G. and Ferguson, D. (1995) *J Cell Biol* **129**, 672–682.
137. Protasi, F., Sun, X. and Franzini-Armstrong, C. (1996) *Dev Biol* **173**, 265–278.
138. Reuter, H. and Beeler, G.J. (1969) *Science* **163**, 399–401.
139. Tanabe, T., Mikami, A., Numa, S. and Beam, K. (1990) *Nature* **344**, 451–453.
140. Cleemann, L. and Morad, M. (1991) *J Physiol* **432**, 283–312.
141. Sham, J., Cleeman, L. and Morad, M. (1995) *Proc Natl Acad Sci USA* **92**, 121–125.
142. Gomez, A., Valdivia, H., Cheng, H., Lederer, M., Santana, L., Cannel, M., McCune, S., Altschuld, R. and Lederer, W. (1997) *Science* **276**, 800–806.
143. Gomez, A., Guatimosim, S., Dilly, K., Vassort, G. and Lederer, W. (2001) *Circulation* **104**, 688–693.
144. He, J.-Q., Conklin, M., Foell, J., Wolff, M., Haworth, R., Coronado, R. and Kamp, T. (2001) *Cardiovasc Res* **49**, 298–307.
145. Balijepalli, R., Lokuta, A., Maertz, N., Buck, J., Haworth, R., Valdivia, H. and Kamp, T. (2003) *Cardiovasc Res* **59**, 67–77.
146. Song, L.-S., Sobie, E., McCulle, S., Lederer, W., Balke, C. and Cheng, H. (2006) *Proc Natl Acad Sci USA* **103**, 4305–4310.

147. Fill, M. and Copello, J. (2002) *Physiol Rev* **82**, 893–922.
148. Stern, M. and Cheng, H. (2004) *Cell Calcium* **35**, 591–601.
149. Zucchi, R. and Ronca-Testoni, S. (1997) *Pharmacol Rev* **49**, 1–51.
150. Rousseau, E., Smith, J. and Meissner, G. (1987) *Am J Physiol* **253**, C364–C368.
151. Buck, E., Zimanyi, I., Abramson, J. and Pessah, I. (1992) *J Biol Chem* **267**, 23560–23567.
152. Zimanyi, I., Buck, E., Abramson, J., Mack, M. and Pessah, I. (1992) *Mol Pharmacol* **42**, 1049–1057.
153. Fessenden, J., Chen, L., Wang, Y., Paolini, C., Franzini-Armstrong, C., Allen, P. and Pessah, I. (2001) *Proc Natl Acad Sci USA* **98**, 2865–2870.
154. Masumiya, H., Li, P., Zhang, L. and Chen, S. (2001) *J Biol Chem* **276**, 39727–39735.
155. Du, G., Guo, X., Khanna, V. and MacLennan, D. (2001) *Proc Natl Acad Sci USA* **98**, 13625–13630.
156. Chiu, A., Diaz-Munoz, M., Hawkes, M., Brush, K. and Hamilton, S. (1990) *Mol Pharmacol* **37**, 735–741.
157. Pessah, I. and Zimanyi, I. (1991) *Mol Pharmacol* **39**, 679–689.
158. Wang, J., Needleman, D. and Hamilton, S. (1993) *J Biol Chem* **268**, 20974–20982.
159. Holmberg, S. and Williams, A. (1990) *Biochim Biophys Acta* **1022**, 187–193.
160. Callaway, C., Seryshev, A., Wang, J., Slavik, K., Needleman, D., Cantu, C., Wu, Y., Jayaraman, T., Marks, A. and Hamilton, S. (1994) *J Biol Chem* **269**, 15876–15884.
161. Witcher, D., McPherson, P., Kahl, S., Lewis, T., Bentley, P., Mullinnix, M., Windass, J. and Campell, K. (1994) *J Biol Chem* **269**, 13076–13079.
162. Tanna, B., Welch, W., Ruest, L., Sutko, J. and Williams, A. (2000) *J Gen Physiol* **116**, 1–9.
163. Tanna, B., Welch, W., Ruest, L., Sutko, J. and Williams, A. (2003) *J Gen Physiol* **121**, 551–561.
164. Meissner, G., Darling, E. and Eveleth, J. (1986) *Biochemistry* **25**, 236–244.
165. Smith, J., Coronado, R. and Meissner, G. (1986) *J Gen Physiol* **88**, 573–588.
166. Pessah, I., Stambuk, R. and Casida, J. (1987) *Mol Pharmacol* **31**, 232–238.
167. Bull, R. and Marengo, J. (1993) *FEBS Lett* **331**, 223–227.
168. Percival, A., Williams, A., Kenyon, J., Grinsell, M., Airey, J. and Sutko, J. (1994) *Biophys J* **67**, 1834–1850.
169. Meissner, G., Rios, E., Tripathy, A. and Pasek, D. (1997) *J Biol Chem* **272**, 1628–1638.
170. Meissner, G. and Henderson, J. (1987) *J Biol Chem* **262**, 3065–3073.
171. Ashley, R. and Williams, A. (1990) *J Gen Physiol* **95**, 981–1005.
172. Chu, A., Fill, M., Stefani, E. and Entman, M. (1993) *J Membr Biol* **135**, 49–59.
173. Sitsapesan, R. and Williams, A. (1994) *Biophys J* **67**, 1484–1494.
174. Laver, D., Roden, L., Ahern, G., Eager, K., Junankar, P. and Dulhunty, A. (1995) *J Membr Biol* **147**, 7–22.
175. Schiefer, A., Meissner, G. and Isenberg, G. (1995) *J Physiol* **489**, 337–348.
176. Sonnleitner, A., Conti, A., Bertocchini, F., Schindler, H. and Sorrentino, V. (1998) *EMBO J* **17**, 2790–2798.
177. Ma, J. (1995) *Biophys J* **68**, 893–899.
178. Copello, J., Barg, S., Onoue, H. and Fleiscer, S. (1997) *Biophys J* **73**, 141–156.
179. Marengo, J., Hidalgo, C. and Bull, R. (1998) *Biophys J* **74**, 1263–1277.
180. Li, P. and Chen, S. (2001) *J Gen Physiol* **118**, 33–44.
181. Donoso, P., Prieto, H. and Hidalgo, C. (1995) *Biophys J* **68**, 507–515.
182. Lamb, G., Cellini, M. and Stephenson, D. (2001) *J Physiol* **531**, 715–728.
183. Sitsapesan, R. and Williams, A. (1994) *J Membr Biol* **137**, 215–226.
184. Sitsapesan, R. and Williams, A. (1995) *J Membr Biol* **146**, 133–144.
185. Tripathy, A. and Meissner, G. (1996) *Biophys J* **70**, 2600–2615.
186. Xu, L. and Meissner, G. (1998) *Biophys J* **75**, 2302–2312.
187. Gyorke, I. and Gyorke, S. (1998) *Biophys J* **75**, 2801–2810.
188. Laver, D., O'Neill, E. and Lamb, G. (2004) *J Gen Physiol* **124**, 741–758.

189. Ching, L., Williams, A. and Sitsapesan, R. (2000) *Circ Res* **87**, 201–206.
190. Gyorke, I., Hester, N., Jones, L. and Gyorke, S. (2004) *Biophys J* **86**, 2121–2128.
191. Beard, N., Casarotto, M., Wei, L., Varsanyi, M., Laver, D. and Dulhunty, A. (2005) *Biophys J* **88**, 3444–3454.
192. Lamb, G. and Stephenson, D. (1994) *J Physiol* **478**, 331–339.
193. Laver, D., Baynes, T. and Dulhunty, A. (1997) *J Membr Biol* **156**, 213–229.
194. Copello, J., Barg, S., Sonnleitner, A., Porta, M., Diaz-Sylvester, P., Fill, M., Schindler, H. and Fleischer, S. (2002) *J Membr Biol* **187**, 51–64.
195. Xu, L., Mann, G. and Meissner, G. (1996) *Circ Res* **79**, 1100–1109.
196. Liu, W., Pasek, D. and Meissner, G. (1998) *Am J Physiol* **274**, C120–C128.
197. Meissner, G. (1984) *J Biol Chem* **259**, 2365–2374.
198. Laver, D., Lenz, G. and Lamb, G. (2001) *J Physiol* **537**, 763–778.
199. Kermode, H., Williams, A. and Sitsapesan, R. (1998) *Biophys J* **74**, 1296–1304.
200. Manunta, M., Rossi, D., Simeoni, I., Butelli, E., Romanin, C., Sorrentino, V. and Schindler, H. (2000) *FEBS Lett* **471**, 256–260.
201. Abramson, J., Zable, A., Favero, T. and Salama, G. (1995) *J Biol Chem* **270**, 29644–29647.
202. Aghdasi, B., Zhang, J.-Z., Wu, Y., Reid, M. and Hamilton, S. (1997) *J Biol Chem* **272**, 3739–3748.
203. Eager, K., Roden, L. and Dulhunty, A. (1997) *Am J Physiol* **272**, C1908–C1918.
204. Zable, A., Favero, T. and Abramson, J. (1997) *J Biol Chem* **272**, 7069–7077.
205. Xia, R., Stangler, T. and Abramson, J. (2000) *J Biol Chem* **275**, 36556–36561.
206. Boraso, A. and Williams, A. (1994) *Am J Physiol* **267**, H1010–H1016.
207. Favero, T., Zable, A. and Abramson, J. (1995) *J Biol Chem* **270**, 25557–25563.
208. Kawakami, M. and Okabe, E. (1998) *Mol Pharmacol* **53**, 497–503.
209. Zima, A., Copello, J. and Blatter, L. (2004) *J Physiol* **555**, 727–741.
210. Xia, R., Webb, J., Gnall, L., Cutler, K. and Abramson, J. (2003) *Am J Physiol* **285**, C215–C221.
211. Cherednichenko, G., Zima, A., Feng, W., Schaefer, S., Blatter, L. and Pessah, I. (2004) *Circ Res* **94**, 478–486.
212. Sanchez, G., Pedroso, Z., Domenech, R., Hidalgo, C. and Donoso, P. (2005) *J Mol Cell Cardiol* **39**, 982–991.
213. Meszaros, L., Minarovic, I. and Zahradnikova, A. (1996) *FEBS Lett* **380**, 49–52.
214. Stoyanovsky, D., Murphy, T., Anno, P., Kim, Y.-M. and Salama, G. (1997) *Cell Calcium* **21**, 19–29.
215. Xu, L., Eu, J., Meissner, G. and Stamler, J. (1998) *Science* **279**, 234–237.
216. Eu, J., Sun, J., Xu, L., Stamler, J. and Meissner, G. (2000) *Cell* **102**, 499–509.
217. Aracena, P., Sanchez, G., Donoso, P., Hamilton, S. and Hidalgo, C. (2003) *J Biol Chem* **278**, 42927–42935.
218. Cheong, E., Tumbev, V., Stoyanovsky, D. and Salama, G. (2005) *Cell Calcium* **38**, 481–488.
219. Aracena, P., Tang, W., Hamilton, S. and Hidalgo, C. (2005) *Antioxid Redox Signal* **7**, 870–881.
220. Sun, J., Xu, L., Eu, J., Stamler, J. and Meissner, G. (2001) *J Biol Chem* **276**, 15625–15630.
221. Oba, T., Murayama, T. and Ogawa, Y. (2002) *Am J Physiol* **282**, C684–C692.
222. Bull, R., Marengo, J., Finkelstein, J., Behrens, M. and Alvarez, O. (2003) *Am J Physiol* **285**, C119–C128.
223. Donoso, P., Aracena, P. and Hidalgo, C. (2000) *Biophys J* **79**, 279–286.
224. Wu, Y., Aghdasi, B., Dou, S., Zhang, J., Liu, S. and Hamilton, S. (1997) *J Biol Chem* **272**, 25051–25061.
225. Liu, G., Abramson, J., Zable, A. and Pessah, I. (1994) *Mol Pharmacol* **45**, 189–200.
226. Zhang, J., Wu, Y., Williams, B., Rodney, G., Mandel, F., Strasburg, G.M. and Hamilton, S. (1999) *Am J Physiol* **276**, C46–C53.
227. Yano, M., Okuda, S., Oda, T., Tokuhisa, T., Tateishi, H., Mochizuki, M., Noma, T., Doi, M., Kobayashi, S., Yamamoto, T., Ikeda, Y., Ohkusa, T., Ikemoto, N. and Matsuzaki, M. (2005) *Circulation* **112**, 3633–3643.

228. Takasago, T., Imagawa, T. and Shikegawa, M. (1989) *J Biochem* **106**, 872–877.
229. Chu, A., Sumbilla, C., Inesi, G., Jay, S. and Campbell, K. (1990) *Biochemistry* **29**, 5899–5905.
230. Takasago, T., Imagawa, T., Furukawa, K., Ogurusu, T. and Shikegawa, M. (1991) *J Biochem* **109**, 163–170.
231. Witcher, D., Kovacs, R., Schulman, H., Cefali, D. and Jones, L. (1991) *J Biol Chem* **266**, 11144–11152.
232. Hohenegger, M. and Suko, J. (1993) *Biochem J* **296**, 303–308.
233. Suko, J., Maurer-Fogy, I., Plank, B., Bertel, O., Wyskowsky, W., Hohenegger, M. and Hellmann, G. (1993) *Biochim Biophys Acta* **1175**, 193–206.
234. Marx, S., Reiken, S., Hisamatsu, Y., Jayaraman, T., Burkhoff, D., Rosemblit, N. and Marks, A. (2000) *Cell* **101**, 365–376.
235. Marx, S., Reiken, S., Hisamatsu, Y., Gaburjakova, M., Gaburjakova, J., Yang, Y.-M., Rosemblit, N. and Marks, A. (2001) *J Cell Biol* **153**, 699–708.
236. Currie, S., Loughrey, C., Craig, M.-A. and Smith, G. (2004) *Biochem J* **377**, 357–366.
237. Hain, J., Nath, S., Mayrleitner, M., Fleischer, S. and Schindler, H. (1994) *Biophys J* **67**, 1823–1833.
238. Hain, J., Onoue, H., Mayrleitner, M., Fleischer, S. and Schindler, H. (1995) *J Biol Chem* **270**, 2074–2081.
239. Dulhunty, A., Laver, D., Curtis, S., Pace, S., Haarmann, C. and Gallant, E. (2001) *Biophys J* **81**, 3240–3252.
240. Rodriguez, P., Bhogal, M. and Colyer, J. (2003) *J Biol Chem* **278**, 38593–38600.
241. Wehrens, X., Lehnart, S., Reiken, S. and Marks, A. (2004) *Circ Res* **94**, e61–e70.
242. Xiao, B., Jiang, M., Zhao, M., Yang, D., Sutherland, C., Lai, F., Walsh, M., Warltier, D., Cheng, H. and Chen, S. (2005) *Circ Res* **96**, 847–855.
243. Reiken, S., Lacampagne, A., Zhou, H., Kherani, A., Lehnart, S., Ward, C., Huang, F., Gaburjakova, M., Gaburjakova, J., Rosemblit, N., Warren, M., He, K., Yi, G., Wang, J., Burkhoff, D., Vassort, G. and Marks, A. (2003) *J Cell Biol* **160**, 919–928.
244. Xiao, B., Zhong, G., Obayashi, M., Yang, D., Chen, K., Walsh, M., Shimoni, Y., Cheng, H., ter Keurs, H. and Chen, S. (2006) *Biochem J* **396**, 7–16.
245. Lehnart, S., Wehrens, X., Reiken, S., Warrier, S., Belevych, A., Harvey, R., Richter, W., Jin, S., Conti, M. and Marks, A. (2005) *Cell* **123**, 25–35.
246. Wehrens, X., Lehnart, S., Reiken, S., Vest, J., Wronska, A. and Marks, A. (2006) *Proc Natl Acad Sci USA* **103**, 511–518.
247. Stange, M., Xu, L., Balshaw, D., Yamaguchi, N. and Meissner, G. (2003) *J Biol Chem* **278**, 51693–51702.
248. Carter, S., Colyer, J. and Sitsapesan, R. (2006) *Circ Res* **98**, 1506–1513.
249. Valdivia, H., Kaplan, J., Ellis-Davies, G. and Lederer, W. (1995) *Science* **267**, 1997–2000.
250. Mayrleitner, M., Chandler, R., Schindler, H. and Fleischer, S. (1995) *Cell Calcium* **18**, 197–206.
251. Patel, J., Coronado, R. and Moss, R. (1995) *Circ Res* **77**, 943–949.
252. Uehara, A., Yasukochi, M., Mejia-Alvarez, R., Fill, M. and Imanaga, I. (2002) *Pflugers Arch* **444**, 202–212.
253. Lokuta, A., Rogers, T., Lederer, W. and Valdivia, H. (1995) *J Physiol* **487**, 609–622.
254. Terentyev, D., Viatchenko-Karpinski, S., Gyorke, I., Terentyeva, R. and Gyorke, S. (2003) *J Physiol* **552**, 109–118.
255. Bers, D. (2002) *Nature* **415**, 198–205.
256. Lindegger, N. and Niggli, E. (2005) *J Physiol* **565**, 801–813.
257. Ginsburg, K. and Bers, D. (2004) *J Physiol* **556**, 463–480.
258. Li, Y., Kranias, E., Mignary, G. and Bers, D. (2002) *Circ Res* **90**, 309–316.
259. Guo, T., Zhang, T., Mestril, R. and Bers, D. (2006) *Circ Res.* **99**, 398–406.
260. Li, L., Satoh, H., Ginsburg, K. and Bers, D. (1997) *J Physiol* **501**, 17–31.
261. Maier, L., Zhang, T., Chen, L., DeSantiago, J., Brown, J. and Bers, D. (2003) *Circ Res* **92**, 904–911.

262. Kohlhaas, M., Zhang, T., Seidler, T., Zibrova, D., Dybkova, N., Steen, A., Wagner, S., Chen, L., Brown, J., Bers, D. and Maier, L. (2006) *Circ Res* **98**, 235–244.
263. Wu, Y., Colbran, R. and Anderson, M. (2001) *Proc Natl Acad Sci USA* **98**, 2877–2881.
264. duBell, W., Lederer, W. and Rogers, T. (1996) *J Physiol* **493**, 793–800.
265. Takekura, H., Takeshima, H., Nishimura, S., Takahashi, M., Tanabe, T., Flockerzi, V., Hofmann, F. and Franzini-Armstrong, C. (1995) *J Muscle Res Cell Motil* **16**, 465–480.
266. Yin, C. and Lai, F. (2000) *Nat Cell Biol* **2**, 669–671.
267. Yin, C., Blayney, L. and Lai, F. (2005) *J Mol Biol* **349**, 538–546.
268. Blayney, L., Zissimopoulos, S., Ralph, E., Abbot, E., Matthews, L. and Lai, F. (2004) *J Biol Chem* **279**, 14639–14648.
269. Niggli, E. (1999) *Annu Rev Physiol* **61**, 311–335.
270. Gao, L., Tripathy, A., Lu, X. and Meissner, G. (1997) *FEBS Lett* **412**, 223–226.
271. Stewart, R., Zissimopoulos, S. and Lai, F. (2003) *Biochem J* **376**, 795–799.
272. Chen, S., Airey, J. and MacLennan, D. (1993) *J Biol Chem* **268**, 22642–22649.
273. Masumiya, H., Wang, R., Zhang, J., Xiao, B. and Chen, S. (2003) *J Biol Chem* **278**, 3786–3792.
274. Yamamoto, T., El-Hayek, R. and Ikemoto, N. (2000) *J Biol Chem* **275**, 11618–11625.
275. Lamb, G., Posterino, G., Yamamoto, T. and Ikemoto, N. (2001) *Am J Physiol* **281**, C207–C214.
276. Shtifman, A., Ward, C., Yamamoto, T., Wang, J., Olbinski, B., Valdivia, H., Ikemoto, N. and Schneider, M. (2002) *J Gen Physiol* **119**, 15–32.
277. Yamamoto, T. and Ikemoto, N. (2002) *Biochemistry* **41**, 1492–1501.
278. Kobayashi, S., Yamamoto, T., Parness, J. and Ikemoto, N. (2004) *Biochem J* **380**, 561–569.
279. Bannister, M. and Ikemoto, M. (2006) *Biochem J* **394**, 145–152.
280. Yamamoto, T. and Ikemoto, N. (2002) *Biochem Biophys Res Commun* **291**, 1102–1108.
281. Oda, T., Yano, M., Yamamoto, T., Tokuhisa, T., Okuda, S., Doi, M., Ohkusa, T., Ikeda, Y., Kobayashi, S., Ikemoto, N. and Matsuzaki, M. (2005) *Circulation* **111**, 3400–3410.
282. Yang, Z., Ikemoto, N., Lamb, G. and Steele, D. (2006) *Cardiovasc Res* **70**, 475–485.
283. Zorzato, F., Menegazzi, P., Treves, S. and Ronjat, M. (1996) *J Biol Chem* **271**, 22759–22763.
284. El-Hayek, R., Saiki, Y., Yamamoto, T. and Ikemoto, N. (1999) *J Biol Chem* **274**, 33341–33347.
285. Paul-Pletzer, K., Yamamoto, T., Bhat, M., Ma, J., Ikemoto, N., Jimenez, L., Morimoto, H., Williams, P. and Parness, J. (2002) *J Biol Chem* **277**, 34918–34923.
286. Kobayashi, S., Bannister, M., Gangopadhyay, J., Hamada, T., Parness, J. and Ikemoto, N. (2005) *J Biol Chem* **280**, 6580–6587.
287. Paul-Pletzer, K., Yamamoto, T., Ikemoto, N., Jimenez, L., Morimoto, H., Williams, P., Ma, J. and Parness, J. (2005) *Biochem J* **387**, 905–909.
288. Murayama, T., Oba, T., Kobayashi, S., Ikemoto, N. and Ogawa, Y. (2005) *Am J Physiol* **288**, C1222–C1230.
289. Zhu, X., Ghanta, J., Walker, J., Allen, P. and Valdivia, H. (2004) *Cell Calcium* **35**, 165–177.
290. Rodney, G., Wilson, G. and Schneider, M. (2005) *J Biol Chem* **280**, 11713–11722.
291. Xiong, L., Zhang, J.-Z., He, R. and Hamilton, S. (2006) *Biophys J* **90**, 173–182.
292. Gangopadhyay, J. and Ikemoto, N. (2006) *Biophys J* **90**, 2015–2026.
293. Moore, C., Zhang, J.-Z. and Hamilton, S. (1999) *J Biol Chem* **274**, 36831–36834.
294. Zhang, H., Zhang, J.-Z., Danila, C. and Hamilton, S. (2003) *J Biol Chem* **278**, 8348–8355.
295. Treves, S., Scutari, E., Robert, M., Groh, S., Ottolia, M., Prestipino, G., Ronjat, M. and Zorzato, F. (1997) *Biochemistry* **36**, 11496–11503.
296. Nelson, T., Zhao, W., Yuan, S., Favit, A., Pozzo-Miller, L. and Alkon, D. (1999) *Biochem J* **341**, 423–433.
297. Feng, W., Tu, J., Yang, T., Vernon, P., Allen, P., Worley, P. and Pessah, I. (2002) *J Biol Chem* **277**, 44722–44730.
298. Zissimopoulos, S., West, D., Williams, A. and Lai, F. (2006) *J Cell Sci* **119**, 2386–2397.

299. Tanabe, T., Takeshima, H., Mikami, A., Flockerzi, V., Takahashi, H., Kangawa, K., Kojima, M., Matsuo, H., Hirose, T. and Numa, S. (1987) *Nature* **328**, 313–318.
300. Mikami, A., Imoto, K., Tanabe, T., Niidome, T., Mori, Y., Takeshima, H., Narumiya, S. and Numa, S. (1989) *Nature* **340**, 230–233.
301. Nakai, J., Tanabe, T., Konno, T., Adams, B. and Beam, K. (1998) *J Biol Chem* **273**, 24983–24986.
302. Proenza, C., Wilkens, C. and Beam, K. (2000) *J Biol Chem* **275**, 29935–29937.
303. Ahern, C., Bhattacharya, D., Mortenson, L. and Coronado, R. (2001) *Biophys J* **81**, 3294–3307.
304. Wilkens, C., Kasielke, N., Flucher, B., Beam, K. and Grabner, M. (2001) *Proc Natl Acad Sci USA* **98**, 5892–5897.
305. Kugler, G., Weiss, R., Flucher, B. and Grabner, M. (2004) *J Biol Chem* **279**, 4721–4728.
306. Lu, X., Xu, L. and Meissner, G. (1994) *J Biol Chem* **269**, 6511–6516.
307. O'Reilly, F., Robert, M., Jona, I., Szegedi, C., Albrieux, M., Geib, S., De Waard, M., Villaz, M. and Ronjat, M. (2002) *Biophys J* **82**, 145–155.
308. Lu, X., Xu, L. and Meissner, G. (1995) *J Biol Chem* **270**, 18459–18464.
309. El-Hayek, R., Antoniu, B., Wang, J., Hamilton, S. and Ikemoto, N. (1995) *J Biol Chem* **270**, 22116–22118.
310. Saiki, Y., El-Hayek, R. and Ikemoto, N. (1999) *J Biol Chem* **274**, 7825–7832.
311. Lamb, G., El-Hayek, R., Ikemoto, N. and Stephenson, D. (2000) *Am J Physiol* **279**, C891–C905.
312. Stange, M., Tripathy, A. and Meissner, G. (2001) *Biophys J* **81**, 1419–1429.
313. Yamamoto, T., Rodriguez, J. and Ikemoto, N. (2002) *J Biol Chem* **277**, 993–1001.
314. Haarmann, C., Green, D., Casarotto, M., Laver, D. and Dulhunty, A. (2003) *Biochem J* **372**, 305–316.
315. Haarmann, C., Dulhunty, A. and Laver, D. (2005) *Biochem J* **387**, 429–436.
316. El-Hayek, R. and Ikemoto, N. (1998) *Biochemistry* **37**, 7015–7020.
317. Dulhunty, A., Laver, D., Gallant, E., Casarotto, M., Pace, S. and Curtis, S. (1999) *Biophys J* **77**, 189–203.
318. Gurrola, G., Arevalo, C., Sreekumar, R., Lokuta, A., Walker, J. and Valdivia, H. (1999) *J Biol Chem* **274**, 7879–7886.
319. Casarotto, M., Green, D., Pace, S., Curtis, S. and Dulhunty, A. (2001) *Biophys J* **80**, 2715–2726.
320. Slavik, K., Wang, J., Aghdasi, B., Zhang, J., Mandel, F., Malouf, N. and Hamilton, S. (1997) *Am J Physiol* **272**, C1475–C1481.
321. Leong, P. and MacLennan, D. (1998) *J Biol Chem* **273**, 29958–29964.
322. Takekura, H., Paolini, C., Franzini-Armstrong, C., Kugler, G., Grabner, M. and Flucher, B. (2004) *Mol Biol Cell* **15**, 5408–5419.
323. Carbonneau, L., Bhattacharya, D., Sheridan, D. and Coronado, R. (2005) *Biophys J* **89**, 243–255.
324. Cheng, W., Altafaj, X., Ronjat, M. and Coronado, R. (2005) *Proc Natl Acad Sci USA* **102**, 19225–19230.
325. Nakai, J., Ogura, T., Protasi, F., Franzini-Armstrong, C., Allen, P. and Beam, K. (1997) *Proc Natl Acad Sci USA* **94**, 1019–1022.
326. Fessenden, J., Wang, Y., Moore, R., Chen, S., Allen, P. and Pessah, I. (2000) *Biophys J* **79**, 2509–2525.
327. Leong, P. and MacLennan, D. (1998) *J Biol Chem* **273**, 7791–7794.
328. Nakai, J., Sekiguchi, N., Rando, T., Allen, P. and Beam, K. (1998) *J Biol Chem* **273**, 13403–13406.
329. Protasi, F., Paolini, C., Nakai, J., Beam, K., Franzini-Armstrong, C. and Allen, P. (2002) *Biophys J* **83**, 3230–3244.
330. Perez, C., Voss, A., Pessah, I. and Allen, P. (2003) *Biophys J* **84**, 2655–2663.
331. Perez, C., Mukherjee, S. and Allen, P. (2003) *J Biol Chem* **278**, 39644–39652.
332. Yamazawa, T., Takeshima, H., Shimuta, M. and Iino, M. (1997) *J Biol Chem* **272**, 8161–8164.
333. Altafaj, X., Cheng, W., Esteve, E., Urbani, J., Grunwald, D., Sabatier, J., Coronado, R., De Waard, M. and Ronjat, M. (2005) *J Biol Chem* **280**, 4013–4016.

334. Proenza, C., O'Brien, J., Nakai, J., Mukherjee, S., Allen, P. and Beam, K. (2002) *J Biol Chem* **277**, 6530–6535.
335. Sencer, S., Papineni, R., Halling, D., Pate, P., Krol, J., Zhang, J. and Hamilton, S. (2001) *J Biol Chem* **276**, 38237–38241.
336. Kay, J. (1996) *Biochem J* **314**, 361–385.
337. Michnick, S., Rosen, M., Wandless, T., Karplus, M. and Schreiber, S. (1991) *Science* **252**, 836–839.
338. Van Duyne, G., Standaert, R., Schreiber, S. and Clardy, J. (1991) *J Am Chem Soc* **113**, 7433–7434.
339. Deivanayagam, C., Carson, M., Thotakura, A., Narayama, S. and Chodavarapu, R. (2000) *Acta Crystallogr D* **56**, 266–271.
340. Lam, E., Martin, M., Timerman, A., Sabers, C., Fleischer, S., Lukas, T., Abraham, R., O'Keefe, S., O'Neill, E. and Wiederrecht, G. (1995) *J Biol Chem* **270**, 26511–26522.
341. Jayaraman, T., Brillantes, A., Timerman, A., Fleischer, S., Erdjument-Bromage, H., Tempst, P. and Marks, A. (1992) *J Biol Chem* **267**, 9474–9477.
342. Timerman, A., Wiederrecht, G., Marcy, A. and Fleischer, S. (1995) *J Biol Chem* **270**, 2451–2459.
343. Timerman, A., Onoue, H., Xin, H., Barg, S., Copello, J., Wiederrecht, G. and Fleischer, S. (1996) *J Biol Chem* **271**, 20385–20391.
344. Qi, Y., Ogunbunmi, E., Freund, E., Timerman, A. and Fleischer, S. (1998) *J Biol Chem* **273**, 34813–34819.
345. Lee, E., Rho, S., Kwon, S., Eom, S., Allen, P. and Kim, D. (2004) *J Biol Chem* **279**, 26481–26488.
346. Jones, J., Reynolds, D., Lai, F. and Blayney, L. (2005) *J Cell Sci* **118**, 4613–4619.
347. Mackrill, J., O'Driscoll, S., Lai, F. and McCarthy, T. (2001) *Biochem Biophys Res Commun* **285**, 52–57.
348. Timerman, A., Jayaraman, T., Wiederrecht, G., Onoue, H., Marks, A. and Fleischer, S. (1994) *Biochem Biophys Res Commun* **198**, 701–706.
349. Jeyakumar, L., Ballester, L., Cheng, D., McIntyre, J., Chang, P., Olivey, H., Rollins-Smith, L., Barnett, J., Murray, K., Xin, H.-B. and Fleischer, S. (2001) *Biochem Biophys Res Commun* **281**, 979–986.
350. Xin, H., Rogers, K., Qi, Y., Kanematsu, T. and Fleischer, S. (1999) *J Biol Chem* **274**, 15315–15319.
351. Huang, F., Shan, J., Reiken, S., Wehrens, X. and Marks, A. (2006) *Proc Natl Acad Sci USA* **103**, 3456–3461.
352. Sharma, M., Jeyakumar, L., Fleischer, S. and Wagenknecht, T. (2006) *Biophys J* **90**, 164–172.
353. Bultynck, G., Rossi, D., Callewaert, G., Missiaen, L., Sorrentino, V., Parys, J. and De Smet, H. (2001) *J Biol Chem* **276**, 47715–47724.
354. Timerman, A., Ogunbumni, E., Freund, E., Wiederrecht, G., Marks, A. and Fleischer, S. (1993) *J Biol Chem* **268**, 22992–22999.
355. Kaftan, E., Marks, A. and Ehrlich, B. (1996) *Circ Res* **78**, 990–997.
356. Ahern, G., Junankar, P. and Dulhunty, A. (1997) *Biophys J* **72**, 146–162.
357. Cameron, A., Nucifora, F., Fung, E., Livingston, D., Aldape, R., Ross, C. and Snyder, S. (1997) *J Biol Chem* **272**, 27582–27588.
358. Bultynck, G., De Smet, P., Rossi, D., Callewaert, G., Missiaen, L., Sorrentino, V., De Smet, H. and Parys, J. (2001) *Biochem J* **354**, 413–422.
359. Gaburjakova, M., Gaburjakova, J., Reiken, S., Huang, F., Marx, S., Rosemblit, N. and Marks, A. (2001) *J Biol Chem* **276**, 16931–16935.
360. Avila, G., Lee, E., Perez, C., Allen, P. and Dirksen, R. (2003) *J Biol Chem* **278**, 22600–22608.
361. Van Acker, K., Bultynck, G., Rossi, D., Sorrentino, V., Boens, N., Missiaen, L., De Smedt, H., Parys, J. and Callewaert, G. (2004) *J Cell Sci* **117**, 1129–1137.
362. Zissimopoulos, S. and Lai, F. (2005) *Cell Biochem Biophys* **43**, 203–220.
363. Xiao, B., Sutherland, C., Walsh, M. and Chen, S. (2004) *Circ Res* **94**, 487–495.
364. Zissimopoulos, S. and Lai, F. (2005) *J Biol Chem* **280**, 5475–5485.
365. Samso, M., Shen, X. and Allen, P. (2006) *J Mol Biol* **356**, 917–927.

366. Ahern, G., Junankar, P. and Dulhunty, A. (1994) *FEBS Lett* **352**, 369–374.
367. Brillantes, A., Ondrias, K., Scott, A., Kobrinsky, E., Ondriasova, E., Moschella, M., Jayaraman, T., Landers, M., Ehrlich, B. and Marks, A. (1994) *Cell* **77**, 513–523.
368. Mayrleitner, M., Timerman, A., Wiederrecht, G. and Fleischer, S. (1994) *Cell Calcium* **15**, 99–108.
369. Barg, S., Copello, J. and Fleisher, S. (1997) *Am J Physiol* **272**, C1726–C1733.
370. Shou, W., Aghdasi, B., Armstrong, D., Guo, Q., Bao, S., Charng, M., Mathews, L., Schneider, M., Hamilton, S. and Matzuk, M. (1998) *Nature* **391**, 489–492.
371. Xiao, R., Valdivia, H., Bogdanov, K., Valdivia, C., Lakatta, E. and Cheng, H. (1997) *J Physiol* **500**, 343–354.
372. Wehrens, X., Lenhart, S., Huang, F., Vest, J., Reiken, S., Mohler, P., Sun, J., Guatimosim, S., Song, L.-S., Rosemblit, N., D'Armiento, J., Napolitano, C., Memmi, M., Priori, S., Lederer, W. and Marks, A. (2003) *Cell* **113**, 829–840.
373. Marx, S., Ondrias, K. and Marks, A. (1998) *Science* **281**, 818–821.
374. Marx, S., Gaburjakova, J., Gaburjakova, M., Henrikson, C., Ondrias, K. and Marks, A. (2001) *Circ Res* **88**, 1151–1158.
375. Brooksbank, R., Badenhorst, M., Isaacs, H. and Savage, N. (1998) *Anesthesiology* **89**, 693–698.
376. Tang, W., Ingalls, C., Durham, W., Snider, J., Reid, M., Wu, G., Matzuk, M. and Hamilton, S. (2004) *FASEB J* **18**, 1597–1599.
377. Lamb, G. and Stephenson, D. (1996) *J Physiol* **494**, 569–576.
378. Avila, G. and Dirksen, R. (2005) *Cell Calcium* **38**, 35–44.
379. McCall, E., Li, L., Satoh, H., Shannon, T., Blatter, L. and Bers, D. (1996) *Circ Res* **79**, 1110–1121.
380. Xin, H.-B., Senbonmatsu, T., Cheng, D.-S., Wang, Y.-X., Copello, J., Ji, G.-J., Collier, M., Deng, K.-Y., Jeyakumar, L., Magnuson, M., Inagami, T., Kotlikoff, M. and Fleischer, S. (2002) *Nature* **416**, 334–337.
381. George, C., Sorathia, R., Bertrand, B. and Lai, F. (2003) *Biochem J* **370**, 579–589.
382. Goonasekera, S., Chen, S. and Dirksen, R. (2005) *Am J Physiol* **289**, C1476–C1484.
383. Prestle, J., Janssen, P., Janssen, A., Zeitz, O., Lehnart, S., Bruce, L., Smith, G. and Hasenfuss, G. (2001) *Circ Res* **88**, 188–194.
384. Loughrey, C., Seidler, T., Miller, S., Prestle, J., MacEachern, K., Reynolds, D., Hasenfuss, G. and Smith, G. (2004) *J Physiol* **556**, 919–934.
385. Gomez, A., Schuster, I., Fauconnier, J., Prestle, J., Hasenfuss, G. and Richard, S. (2004) *Am J Physiol* **287**, H1987–H1993.
386. Chin, D. and Means, A. (2000) *Trends Cell Biol* **10**, 322–328.
387. Moore, C., Rodney, G., Zhang, J.-Z., Santacruz-Toloza, L., Strasburg, G. and Hamilton, S. (1999) *Biochemistry* **38**, 8532–8537.
388. Rodney, G., Williams, B., Strasburg, G., Beckingham, K. and Hamilton, S. (2000) *Biochemistry* **39**, 7807–7812.
389. Fruen, B., Bardy, J., Byrem, T., Strasburg, G. and Louis, C. (2000) *Am J Physiol* **279**, C724–C733.
390. Balshaw, D., Xu, L., Yamaguchi, N., Pasek, D. and Meissner, G. (2001) *J Biol Chem* **276**, 20144–20153.
391. Yamaguchi, N., Xu, L., Pasek, D., Evans, K., Chen, S. and Meissner, G. (2005) *Biochemistry* **44**, 15074–15081.
392. Yang, H., Reedy, M., Burke, C. and Strasburg, G. (1994) *Biochemistry* **33**, 518–525.
393. Tripathy, A., Xu, L., Mann, G. and Meissner, G. (1995) *Biophys J* **69**, 106–119.
394. Chen, S. and MacLennan, D. (1994) *J Biol Chem* **269**, 22698–22704.
395. Menegazzi, P., Larini, F., Treves, S., Guerrini, R., Quadroni, M. and Zorzato, F. (1994) *Biochemistry* **33**, 9078–9084.
396. Rodney, G., Moore, C., Williams, B., Zhang, J., Krol, J., Pedersen, S. and Hamilton, S. (2001) *J Biol Chem* **276**, 2069–2074.
397. Yamaguchi, N., Xin, C. and Meissner, G. (2001) *J Biol Chem* **276**, 22579–22585.

398. Xiong, L., Newman, R., Rodney, G., Thomas, O., Zhang, J., Persechini, A., Shea, M. and Hamilton, S. (2002) *J Biol Chem* **277**, 40862–40870.
399. Yamaguchi, N., Xu, L., Pasek, D., Evans, K. and Meissner, G. (2003) *J Biol Chem* **278**, 23480–23486.
400. Yamaguchi, N., Xu, L., Evans, K., Pasek, D. and Meissner, G. (2004) *J Biol Chem* **279**, 36433–36439.
401. Meissner, G. (1986) *Biochemistry* **25**, 244–251.
402. Smith, J., Rousseau, E. and Meissner, G. (1989) *Circ Res* **64**, 352–359.
403. Fuentes, O., Valdivia, C., Vaughan, D., Coronado, R. and Valdivia, H. (1994) *Cell Calcium* **15**, 305–316.
404. Fruen, B., Black, D., Bloomquist, R., Bardy, J., Johnson, J., Louis, C. and Balog, E. (2003) *Biochemistry* **42**, 2740–2747.
405. O'Connell, K., Yamaguchi, N., Meissner, G. and Dirksen, R. (2002) *J Gen Physiol* **120**, 337–347.
406. Ikemoto, T., Takeshima, H., Iino, M. and Endo, M. (1998) *Pflugers Arch* **437**, 43–48.
407. Xu, L. and Meissner, G. (2004) *Biophys J* **86**, 797–804.
408. Rodney, G. and Schneider, M. (2003) *Biophys J* **85**, 921–932.
409. Anderson, M. (2002) *J Cardiovasc Electrophysiol* **13**, 195–197.
410. Maki, M., Kitaura, Y., Satoh, H., Ohkouchi, S. and Shibata, H. (2002) *Biochim Biophys Acta* **1600**, 51–60.
411. Meyers, M., Pickel, V., Sheu, S., Sharma, V., Scotto, K. and Fishman, G. (1995) *J Biol Chem* **270**, 26411–26418.
412. Meyers, M., Fischer, A., Sun, Y., Lopes, C., Rohacs, T., Nakamura, T., Zhou, Y., Lee, P., Altschuld, R., McCune, S., Coetzee, W. and Fishman, G. (2003) *J Biol Chem* **278**, 28865–28871.
413. Farrell, E., Antaramian, A., Rueda, A., Gomez, A. and Valdivia, H. (2003) *J Biol Chem* **278**, 34660–34666.
414. Meyers, M., Puri, T., Chien, A., Gao, T., Hus, P.-H., Hosey, M. and Fishman, G. (1998) *J Biol Chem* **273**, 18930–18935.
415. Lokuta, A., Meyers, M., Sander, P., Fishman, G. and Valdivia, H. (1997) *J Biol Chem* **272**, 25333–25338.
416. Seidler, T., Miller, S., Loughrey, C., Kania, A., Burow, A., Kettlewell, S., Teucher, N., Wagner, S., Kogler, H., Meyers, M., Hasenfuss, G. and Smith, G. (2003) *Circ Res* **93**, 132–139.
417. Suarez, J., Belke, D., Gloss, B., Dieterle, T., McDonough, P., Kim, Y., Brunton, L. and Dillmann, D. (2004) *Am J Physiol* **286**, H68-H75.
418. Frank, K., Block, B., Ding, Z., Krause, D., Hattebuhr, N., Malik, A., Brixius, K., Hajjar, R., Schrader, J. and Schwinger, R. (2005) *J Mol Cell Cardiol* **38**, 607–615.
419. Matsumoto, T., Hisamatsu, Y., Ohkusa, T., Inoue, N., Sato, T., Suzuki, S., Ikeda, Y. and Matsuzaki, M. (2005) *Basic Res Cardiol* **100**, 250–262.
420. Ilari, A., Johnson, K., Nastopoulos, V., Verzili, D., Zamparelli, C., Colotti, G., Tsernoglou, D. and Chiancone, E. (2002) *J Mol Biol* **317**, 447–458.
421. MacLennan, D. and Wong, P. (1971) *Proc Natl Acad Sci USA* **68**, 1231–1235.
422. Ikemoto, N., Bhatnagar, G., Nagy, B. and Gergely, J. (1972) *J Biol Chem* **247**, 7835–7837.
423. Wang, S., Trumble, W., Liao, H., Wesson, C., Dunker, A. and Kang, C. (1998) *Nat Struct Biol* **5**, 476–483.
424. Fliegel, L., Ohnishi, M., Carpenter, M., Khanna, V., Reithmeier, R. and MacLennan, D. (1987) *Proc Natl Acad Sci USA* **84**, 1167–1171.
425. Scott, B., Simmerman, H., Collins, J., Nadal-Ginard, B. and Jones, L. (1988) *J Biol Chem* **263**, 8958–8964.
426. Fryer, M. and Stephenson, D. (1996) *J Physiol* **493**, 357–370.
427. Park, H., Park, I., Kim, E., Youn, B., Fields, K., Dunker, A. and Kang, C. (2004) *J Biol Chem* **279**, 18026–18033.
428. Guo, W. and Campbell, K. (1995) *J Biol Chem* **270**, 9027–9030.

429. Zhang, L., Kelley, J., Schmeisser, G., Kobayashi, Y. and Jones, L. (1997) *J Biol Chem* **272**, 23389–23397.

430. Shin, D., Ma, J. and Kim, D. (2000) *FEBS Lett* **486**, 178–182.

431. Beard, N., Sakowska, M., Dulhunty, A. and Laver, D. (2002) *Biophys J* **82**, 310–320.

432. Herzog, A., Szegedi, C., Jona, I., Herberg, F. and Varsanyi, M. (2000) *FEBS Lett* **472**, 73–77.

433. Brandt, N., Caswell, A., Wen, S. and Talvenheimo, J. (1990) *J Membr Biol* **113**, 237–251.

434. Knudson, C., Stang, K., Moomaw, C., Slaughter, C. and Campbell, K. (1993) *J Biol Chem* **268**, 12646–12654.

435. Marty, I., Thevenon, D., Scotto, C., Groh, S., Sainnier, S., Robert, M., Grunwald, D. and Villaz, M. (2000) *J Biol Chem* **275**, 8206–8212.

436. Vassilopoulos, S., Thevenon, D., Rezgui, S., Brocard, J., Chapel, A., Lacampagne, A., Lunardi, J., De Waard, M. and Marty, I. (2005) *J Biol Chem* **280**, 28601–28609.

437. Guo, W., Jorgensen, A., Jones, L. and Campbell, K. (1996) *J Biol Chem* **271**, 458–465.

438. Kobayashi, Y. and Jones, L. (1999) *J Biol Chem* **274**, 28660–28668.

439. Marty, I., Robert, M., Ronjat, M., Bally, I., Arlaud, G. and Villaz, M. (1995) *Biochem J* **307**, 769–774.

440. Caswell, A., Brandt, N., Brunschwig, J. and Purkerson, S. (1991) *Biochemistry* **30**, 7507–7513.

441. Groh, S., Marty, I., Ottolia, M., Prestipino, G., Chapel, A., Villaz, M. and Ronjat, M. (1999) *J Biol Chem* **274**, 12278–12283.

442. Lee, J., Rho, S.-H., Shin, D., Cho, C., Park, W., Eom, S., Ma, J. and Kim, D. (2004) *J Biol Chem* **279**, 6994–7000.

443. Kobayashi, Y., Alseikhan, B. and Jones, L. (2000) *J Biol Chem* **275**, 17639–17646.

444. Jones, L., Zhang, L., Sanborn, K., Jorgensen, A. and Kelley, J. (1995) *J Biol Chem* **270**, 30787–30796.

445. Ohkura, M., Furukawa, K., Fujimori, H., Kuruma, A., Kawano, S., Hiraoka, M., Kuniyasu, A., Nakayama, H. and Ohizumi, Y. (1998) *Biochemistry* **37**, 12987–12993.

446. Kirchhefer, U., Neumann, J., Baba, H., Begrow, F., Kobayashi, Y., Reinke, U., Schmitz, W. and Jones, L. (2001) *J Biol Chem* **276**, 4142–4149.

447. Kirchhefer, U., Jones, L., Begrow, F., Boknik, P., Hein, L., Lohse, M., Riemann, B., Schmitz, W., Stypmann, J. and Neumann, J. (2004) *Cardiovasc Res* **62**, 122–134.

448. Hong, C., Cho, M., Kwak, Y., Song, C., Lee, Y., Lim, J., Kwon, Y., Chae, S. and Kim, D. (2002) *FASEB J* **16**, 1310–1312.

449. Kirchhefer, U., Neumann, J., Bers, D., Buchwalow, I., Fabritz, L., Hanske, G., Justus, I., Riemann, B., Schmitz, W. and Jones, L. (2003) *Cardiovasc Res* **59**, 369–379.

450. Terentyev, D., Cala, S., Houle, T., Viatchenko-Karpinski, S., Gyorke, I., Terentyeva, R., Williams, S. and Gyorke, S. (2005) *Circ Res* **96**, 651–658.

451. Rezgui, S., Vassilopoulos, S., Brocard, J., Platel, J., Bouron, A., Arnoult, C., Oddoux, S., Garcia, L., De Waard, M. and Marty, I. (2005) *J Biol Chem* **280**, 39302–39308.

452. Kawasaki, T. and Kasai, M. (1994) *Biochem Biophys Res Commun* **199**, 1120–1127.

453. Szegedi, C., Sarkozi, S., Herzog, A., Jona, I. and Varsanyi, M. (1999) *Biochem J* **337**, 19–22.

454. Jones, L., Suzuki, Y., Wang, W., Kobayashi, Y., Ramesh, V., Franzini-Armstrong, C., Cleemann, L. and Morad, M. (1998) *J Clin Invest* **101**, 1385–1393.

455. Sato, Y., Ferguson, D., Sako, H., Dorn, G., Kadambi, V., Yatani, A., Hoit, B., Walsh, R. and Kranias, E. (1998) *J Biol Chem* **273**, 28470–28477.

456. Wang, W., Cleemann, L., Jones, L. and Morad, M. (2000) *J Physiol* **524**, 399–414.

457. Miller, S., Currie, S., Loughrey, C., Kettlewell, S., Seidler, T., Reynolds, D., Hasenfuss, G. and Smith, G. (2005) *Cardiovasc Res* **67**, 667–677.

458. Terentyev, D., Viatchenko-Karpinski, S., Gyorke, I., Volpe, P., Williams, S. and Gyorke, S. (2003) *Proc Natl Acad Sci USA* **100**, 11759–11764.

459. Shin, D., Pan, Z., Kim, E., Lee, J., Bhat, M., Parness, J., Kim, D. and Ma, J. (2003) *J Biol Chem* **278**, 3286–3292.
460. Viatchenko-Karpinski, S., Terentyev, D., Gyorke, I., Terentyeva, R., Volpe, P., Priori, S., Napolitano, C., Nori, A., Williams, S. and Gyorke, S. (2004) *Circ Res* **94**, 471–477.
461. Terentyev, D., Nori, A., Santoro, M., Viatchenko-Karpinski, S., Kubalova, Z., Gyorke, I., Terentyeva, R., Vedamoorthyrao, S., Blom, N., Valle, G., Napolitano, C., Williams, S., Volpe, P., Priori, S. and Gyorke, S. (2006) *Circ Res* **98**, 1151–1158.
462. Houle, T., Ram, M. and Cala, S. (2004) *Cardiovasc Res* **64**, 227–233.
463. Liew, C.-C. and Dzau, V. (2004) *Nat Rev Genet* **5**, 811–825.
464. Bers, D., Eisner, D. and Valdivia, H. (2003) *Circ Res* **93**, 487–490.
465. Shannon, T., Pogwizd, S. and Bers, D. (2003) *Circ Res* **93**, 592–594.
466. Kubalova, Z., Terentyev, D., Viatchenko-Karpinski, S., Nishijima, Y., Gyorke, I., Terentyeva, R., da Cunha, D., Sridhar, A., Feldman, D., Hamlin, R., Carnes, C. and Gyorke, S. (2005) *Proc Natl Acad Sci USA* **102**, 14104–14109.
467. Reiken, S., Gaburjakova, M., Gaburjakova, J., He, K., Prieto, A., Becker, E., Yi, G., Wang, J., Burkhoff, D. and Marks, A. (2001) *Circulation* **104**, 2843–2848.
468. Reiken, S., Gaburjakova, M., Guatimosim, S., Gomez, A., D'Armiento, J., Burkhoff, D., Wang, J., Vassort, G., Lederer, W. and Marks, A. (2003) *J Biol Chem* **278**, 444–453.
469. Reiken, S., Wehrens, X., Vest, J., Barbone, A., Klotz, S., Mancini, D., Burkhoff, D. and Marks, A. (2003) *Circulation* **107**, 2459–2466.
470. Wehrens, X., Lehnart, S., Reiken, S., Deng, S., Vest, J., Cervantes, D., Coromilas, J., Landry, D. and Marks, A. (2004) *Science* **304**, 292–296.
471. Wehrens, X., Lehnart, S., Reiken, S., Van der Nagel, R., Morales, R., Sun, J., Cheng, Z., Deng, S., De Windt, L., Landry, D. and Marks, A. (2005) *Proc Natl Acad Sci USA* **102**, 9607–9612.
472. Yano, M., Ono, K., Ohkusa, T., Suetsugu, M., Kohno, M., Hisaoka, T., Kobayashi, S., Hisamatsu, Y., Yamamoto, T., Kohno, M., Noguchi, N., Takasawa, S., Okamoto, H. and Matsuzaki, M. (2000) *Circulation* **102**, 2131–2136.
473. Ono, K., Yano, M., Ohkusa, T., Kohno, M., Hisaoka, T., Tanigawa, T., Kobayashi, S., Kohno, M. and Matsuzaki, M. (2000) *Cardiovasc Res* **48**, 323–331.
474. Doi, M., Yano, M., Kobayashi, S., Kohno, M., Tokuhisa, T., Okuda, S., Suetsugu, M., Hisamatsu, Y., Ohkusa, T., Kohno, M. and Matsuzaki, M. (2002) *Circulation* **105**, 1374–1379.
475. Yano, M., Kobayashi, S., Kohno, M., Doi, M., Tokuhisa, T., Okuda, S., Suetsugu, M., Hisaoka, T., Obayashi, M., Ohkusa, T., Kohno, M. and Matsuzaki, M. (2003) *Circulation* **107**, 477–484.
476. Jiang, M., Lokuta, A., Farrell, E., Wolff, M., Haworth, R. and Valdivia, H. (2002) *Circ Res* **91**, 1015–1022.
477. Obayashi, M., Xiao, B., Stuyvers, B., Davidoff, A., Mei, J., Chen, S. and ter Keurs, H. (2006) *Cardiovasc Res* **69**, 140–151.
478. Ai, X., Curran, J., Shannon, T., Bers, D. and Pogwizd, S. (2005) *Circ Res* **97**, 1314–1322.
479. Zhang, T., Maier, L., Dalton, N., Miyamoto, S., Ross, J.J., Bers, D. and Brown, J. (2003) *Circ Res* **92**, 912–919.
480. Wang, W., Zhu, W., Wang, S., Yang, D., Crow, M., Xiao, R.-P. and Cheng, H. (2004) *Circ Res* **95**, 798–806.
481. Zhang, R., Khoo, M., Wu, Y., Yang, Y., Grueter, C., Ni, G., Price, E.J., Thiel, W., Guatimosim, S., Song, L.-S., Madu, E., Shah, A., Vishnivetskaya, T., Atkinson, J., Gurevich, V., Salama, G., Lederer, W., Colbran, R. and Anderson, M. (2005) *Nat Med* **11**, 409–417.
482. Francis, J., Sankar, V., Nair, V. and Priori, S. (2005) *Heart Rhythm* **2**, 550–554.
483. Kontula, K., Laitinen, P., Lehtonen, A., Toivonen, L., Viitasalo, M. and Swan, H. (2005) *Cardiovasc Res* **67**, 379–387.
484. Jiang, D., Xiao, B., Zhang, L. and Chen, S. (2002) *Circ Res* **91**, 218–225.
485. George, C., Higgs, G. and Lai, F. (2003) *Circ Res* **93**, 531–540.

486. Jiang, D., Xiao, B., Yang, D., Wang, R., Choi, P., Zhang, L., Cheng, H. and Chen, S. (2004) *Proc Natl Acad Sci USA* **101**, 13062–13067.
487. Thomas, L., George, C. and Lai, F. (2004) *Cardiovasc Res* **64**, 52–60.
488. Lehnart, S., Wehrens, X., Laitinen, P., Reiken, S., Deng, S., Cheng, Z., Landry, D., Kontula, K., Swan, H. and Marks, A. (2004) *Circulation* **109**, 3208–3214.
489. Thomas, L., Lai, F. and George, C. (2005) *Biochem Biophys Res Commun* **331**, 231–238.
490. Jiang, D., Wang, R., Xiao, B., Kong, H., Hunt, D., Choi, P., Zhang, L. and Chen, S. (2005) *Circ Res* **97**, 1173–1181.
491. George, C., Jundi, H., Walters, N., Thomas, N., West, R. and Lai, F. (2006) *Circ Res* **98**, 88–97.
492. Lehnart, S., Terrenoire, C., Reiken, S., Wehrens, X., Song, L., Tillman, E., Mancarella, S., Coromillas, J., Lederer, W., Kass, R. and Marks, A. (2006) *Proc Natl Acad Sci USA* **103**, 7906–7910.
493. Liu, N., Colombi, B., Memmi, M., Zissimopoulos, S., Rizzi, N., Negri, S., Imbriani, M., Napolitano, C., Lai, F. and Priori, S. (2006) *Circ Res* **99**, 292–298.
494. Cerrone, M., Colombi, B., Santoro, M., Raffaele di Barletta, M., Scelsi, M., Villani, L., Napolitano, C. and Priori, S. (2005) *Circ Res* **96**, e77–e82.
495. Kannankeril, P., Mitchell, B., Goonasekera, S., Chelu, M., Zhang, W., Sood, S., Kearney, D., Danila, C., De Biasi, M., Wehrens, X., Pautler, R., Roden, D., Taffet, G., Dirksen, R., Anderson, M. and Hamilton, S. (2006) *Proc Natl Acad Sci USA* **103**, 12179–12184.
496. Janse, M. (2004) *Cardiovasc Res* **61**, 208–217.
497. Pogwizd, S. and Bers, D. (2004) *Trends Cardiovasc Med* **14**, 61–66.
498. Treves, S., Anderson, A., Ducreux, S., Divet, A., Bleunven, C., Grasso, C., Paesante, S. and Zorzato, F. (2005) *Neuromuscul Disord* **15**, 577–587.
499. Gallant, E. and Lentz, L. (1992) *Am J Physiol* **262**, C422–C426.
500. Otsu, K., Nishida, K., Kimura, Y., Kuzuya, T., Hori, M., Kamada, T. and Tada, M. (1994) *J Biol Chem* **269**, 9413–9415.
501. Treves, S., Larini, F., Menegazzi, P., Steinberg, T., Koval, M., Vilsen, B., Andersen, J. and Zorzato, F. (1994) *Biochem J* **301**, 661–665.
502. Owen, V., Taske, N. and Lamb, G. (1997) *Am J Physiol* **272**, C203–C211.
503. Laver, D., Owen, V., Junankar, P., Taske, N., Dulhunty, A. and Lamb, G. (1997) *Biophys J* **73**, 1913–1924.
504. Tong, J., Oyamada, H., Demaurex, N., Grinstein, S., McCarthy, T. and MacLennan, D. (1997) *J Biol Chem* **272**, 26332–26339.
505. Censier, K., Urwyler, A., Zorzato, F. and Treves, S. (1998) *J Clin Invest* **101**, 1233–1242.
506. Dietze, B., Henke, J., Eichinger, H., Lehmann-Horn, F. and Melzer, W. (2000) *J Physiol* **526**, 507–514.
507. Avila, G. and Dirksen, G. (2001) *J Gen Physiol* **118**, 277–290.
508. Yang, T., Ta, T., Pessah, I. and Allen, P. (2003) *J Biol Chem* **278**, 25722–25730.
509. Wehner, M., Rueffert, H., Koenig, F. and Olthoff, D. (2004) *Neuromuscul Disord* **14**, 429–437.
510. Chelu, M., Goonasekera, S., Durham, W., Tang, W., Lueck, J., Riehl, J., Pessah, I., Zhang, P., Bhattacharjee, M., Dirksen, R. and Hamilton, S. (2005) *FASEB J* **20**, 329–330.
511. Tong, J., McCarthy, T. and MacLennan, D. (1999) *J Biol Chem* **274**, 693–702.
512. Tilgen, N., Zorzarto, F., Halliger-Keller, B., Muntoni, F., Sewry, C., Palmucci, L., Schneider, C., Hauser, E., Lehmann-Horn, F., Muller, C. and Treves, S. (2001) *Hum Mol Genet* **10**, 2879–2887.
513. Avila, G., O'Connell, K. and Dirksen, R. (2003) *J Gen Physiol* **121**, 277–286.
514. Zorzato, F., Yamaguchi, N., Xu, L., Meissner, G., Muller, C., Pouliquin, P., Muntoni, F., Sewry, C., Girard, T. and Treves, S. (2003) *Hum Mol Genet* **12**, 379–388.
515. Dirksen, R. and Avila, G. (2004) *Biophys J* **87**, 3193–3204.
516. Brini, M., Manni, S., Pierobon, N., Du, G., Sharma, P., MacLennan, D. and Carafoli, E. (2005) *J Biol Chem* **280**, 15380–15389.

517. Lynch, P., Tong, J., Lehane, M., Mallet, A., Giblin, L., Heffron, J., Vaughan, P., Zafra, G., MacLennan, D. and McCarthy, T. (1999) *Proc Natl Acad Sci USA* **96**, 4164–4169.
518. Ducreux, S., Zorzato, F., Muller, C., Sewry, C., Muntoni, F., Quinlivan, R., Restagno, G., Girard, T. and Treves, S. (2004) *J Biol Chem* **279**, 43838–43846.
519. Avila, G., O'Brien, J. and Dirksen, R. (2001) *Proc Natl Acad Sci USA* **98**, 4215–4220.
520. Du, G., Khanna, V., Guo, X. and MacLennan, D. (2004) *Biochem J* **382**, 557–564.

Modulation of Calcium Functions

J. Krebs and M. Michalak (Editors)
Calcium: A Matter of Life or Death

ISSN: 0167-7306/DOI: 10.1016/S0167-7306(06)41013-9

CHAPTER 13

The calcium/calmodulin-dependent protein kinase cascades

Felice A. Chow[1,2] and Anthony R. Means[2]

[1]*Department of Cell Biology, FibroGen, Inc., South San Francisco, CA 94080, USA;*
[2]*Department of Pharmacology and Cancer Biology, Duke University Medical Center, Box 3813, Durham, NC 27710-3813, USA, Tel.: +1 919 681 6209; Fax: +1 919 681 7767;E-mail: means001@mc.duke.edu*

Abstract

Calcium/calmodulin-dependent kinase kinase (CaMKK), CaMKI, and CaMKIV constitute CaMK cascades that have gained increased recognition during the past 10 years for their roles in varied biological functions, including learning and memory, T-cell function, and cell survival. This chapter summarizes the tremendous work, performed in the last decade or so, that has advanced our understanding of these kinase cascades. We examine the progress that has been made toward elucidating the mechanisms regulating CaMKK, CaMKI, and CaMKIV; highlight the various approaches that have been used to study these kinases; discuss identification of the physiological roles of these kinases; and consider some of the issues that will keep the field occupied well into the future.

Keywords: Ca^{2+}/CaM, Ca^{2+}/CaM-dependent, calcium, calcium/calmodulin, calcium/calmodulin-dependent, calmodulin, CaMK, CaMKI, CaMKIV, CaMKK, cascade, kinase

Calcium is one of the most frequently utilized signaling molecules in the cell, and it plays roles in nearly every physiological response. One of the most intriguing characteristics of calcium is that, although it is essential for life, excessively high levels of intracellular Ca^{2+} are toxic to the cell [1,2]. For this reason, intracellular Ca^{2+} levels are tightly controlled and kept relatively low (approximately 100 nM) by the actions of plasma membrane ion channels, ion exchangers, and ATPase pumps that transport Ca^{2+} out of the cell or sequester it in intracellular organelles. Cells are exquisitely sensitive to changes in intracellular Ca^{2+} levels, and they are able to respond differentially to Ca^{2+} signals based on their source, duration, amplitude, and subcellular localization [3,4]. Intracellular Ca^{2+} ions that function as second messengers may originate from the influx of extracellular Ca^{2+} into the cell through activated ligand-gated or voltage-gated ion channels or from the release of Ca^{2+} from intracellular stores within the endoplasmic reticulum [5]. Calcium release from the endoplasmic reticulum may be triggered by the binding of IP_3 to IP_3-gated ion channel receptors on the endoplasmic reticulum membrane or by the activation of Ca^{2+}-induced Ca^{2+} channels (ryanodine receptors), which can be directly activated by increases in intracellular Ca^{2+} [5]. The cellular response to

Ca^{2+} is mediated by a multitude of Ca^{2+}-binding proteins in the cell, which act as Ca^{2+} buffers, Ca^{2+} transducers, and Ca^{2+} effectors.

One of the major Ca^{2+}-binding proteins in the cell is calmodulin (CaM), a 17-kDa protein conserved among all members of the animal, plant, fungal, and protozoa kingdoms. It consists of four EF-hand Ca^{2+}-binding motifs, two in its N-terminal domain and two in its C-terminal domain [6,7]. The binding of Ca^{2+} to CaM results in conformational changes that modulate CaM's ability to interact with its targets. CaM may bind to, and regulate, some targets in either its Ca^{2+}-free state (apo-CaM) or its Ca^{2+}-bound state (Ca^{2+}/CaM), although the effect of CaM on such targets is often determined by CaM's state of Ca^{2+} occupancy. Other proteins are targets only of calcium-bound CaM. An important group of cellular targets positively regulated by Ca^{2+}/CaM is the family of Ca^{2+}–CaM-dependent protein kinases.

Calcium/CaM-dependent protein kinases (CaMKs) can be divided into two main classes: dedicated kinases (which are dedicated to a single substrate) and multifunctional kinases (which can phosphorylate multiple targets). The dedicated kinases include phosphorylase kinase, myosin light chain kinase, and eukaryotic elongation factor-2 kinase [also known as Ca^{2+}/CaM-dependent kinase III (CaMKIII)]; the multifunctional kinases include CaMKI, CaMKII, CaMKIV, and CaMKK. The multifunctional CaMKs share considerable sequence homology (20–40% global identity) and have similar domain structures, with each kinase possessing a well-conserved N-terminal kinase domain as well as a central regulatory domain containing overlapping autoinhibitory and Ca^{2+}/CaM-binding regions. Much of the current understanding of the structural mechanisms that regulate each of the multifunctional CaMKs has been achieved through mutagenesis studies and through the resolution of the crystal structures for CaMKI and CaMKII [8–14]. In the absence of bound Ca^{2+}/CaM, each CaMK is inhibited by intramolecular interactions between residues within the regulatory domain and residues within the catalytic domain. The binding of Ca^{2+}/CaM leads to conformational changes that relieve the steric blockade of the catalytic site and generates an active enzyme. Despite their structural similarities and their common regulation by Ca^{2+}/CaM, the CaMKs are highly distinct functionally. They vary in their tissue distributions, subcellular localizations, protein–protein interactions, and substrate specificities and are subject to different regulatory mechanisms.

Of the multifunctional CaMKs, CaMKII may be the most divergent in its enzymatic regulation, existing as a 12-subunit multimer that is subject to distinctive intraholoenzyme regulation. The binding of Ca^{2+}/CaM to CaMKII leads to the relief of autoinhibition and CaMKII kinase activity [8,15]. Intraholenzyme, intersubunit autophosphorylation on T^{286}, can occur between proximal Ca^{2+}/CaM-bound CaMKII subunits, leading to increased affinity of the phosphorylated subunit for CaM (the so-called CaM trapping), as well as Ca^{2+}/CaM independence (often referred to as autonomy), such that full kinase activity is retained upon the loss of Ca^{2+}/CaM binding [16–19]. Ca^{2+}/CaM dissociation from phospho-T^{286} CaMKII is followed by additional autophosphorylation events, which inhibit Ca^{2+}/CaM from rebinding (CaM capping) [20]. The unique regulatory properties of CaMKII allow it to act as a Ca^{2+} frequency decoder and as a form of "molecular memory" [21]. During low-frequency Ca^{2+} signals, Ca^{2+}/CaM dissociates from CaMKII between

Ca^{2+} spikes, whereas during high-frequency Ca^{2+} signals, there is insufficient time for such dissociation between spikes. Thus, upon each subsequent Ca^{2+} spike, more and more subunits of the multimeric holoenzyme are bound by Ca^{2+}/CaM and become active. As the number of Ca^{2+}/CaM-bound subunits increases, the likelihood of autophosphorylation between proximal Ca^{2+}/CaM-bound subunits also increases. Thus, the responsiveness of the CaMKII holoenzyme to the frequency of Ca^{2+} signals allows it to act as a frequency detector, whereas its ability to maintain its activation status over long durations, even after a Ca^{2+} signal has dissipated, allows it to act as a form of molecular memory [22–24].

The remarkable regulatory properties and important biological roles of CaMKII have been the subject of many excellent reviews [21,25–31]. This review will instead focus on the other multifunctional CaMKs, CaMKI, CaMKIV, and CaMKK, which constitute CaMK cascades that have garnered increased attention during the past 10 years.

1. *The CaMK cascade*

CaMKI and CaMKIV were first isolated from the brain where they are particularly abundant, and it was initially theorized that, similar to CaMKII, these two kinases might be regulated by autophosphorylation [32–35]. Early studies found that CaMKI and CaMKIV that had been isolated from brain extracts became phosphorylated in vitro upon incubation with Ca^{2+} and CaM. In the case of CaMKIV, this phosphorylation was associated with a dramatic increase in kinase activity and the development of some Ca^{2+}/CaM-independent activity [36,37]. Although originally ascribed to autophosphorylation, the activation of brain-isolated CaMKIV in these early studies was later attributed to the presence of a contaminating kinase because subsequent studies, using recombinant CaMKIV, found that autophosphorylation resulted in only slight increases in kinase activity, whereas the addition of brain extract considerably activated CaMKIV [11,38]. Similarly, it was found that recombinant CaMKI could be activated in vitro by a factor present in brain extract [39]. A CaMKK was later identified and cloned and found to be an activator of both CaMKI and CaMKIV [12,38–45]. Numerous studies have since validated the existence of physiological CaMK cascades (consisting of the upstream activator CaMKK and its two target CaMKs, CaMKI and CaMKIV) in numerous organisms and cell types [46–48]. This review will summarize what is currently known about the regulation of CaMKK, CaMKI, and CaMKIV; highlight some of the inroads that have been made into identifying and understanding the physiological roles of these kinases; and discuss some of the issues that still remain to be resolved.

2. *Tools of the trade*

One of the most popular approaches for studying CaMK function and mechanism of action has been the use of pharmacological inhibitors. Two compounds, KN62 and KN93, are frequently used as inhibitors of CaMKI, CaMKII, and CaMKIV, whereas another compound, STO-609, has been useful as a selective CaMKK inhibitor. KN62 and KN93 were originally described as specific CaMKII inhibitors but actually inhibit CaMKI and CaMKIV with similar K_i values to CaMKII [49–52]. Although their mechanism of action has not been thoroughly studied, it has been suggested that KN62 and KN93 bind close to the regulatory region of target CaMKs, thus preventing Ca^{2+}/CaM binding and subsequent relief of enzyme autoinhibition [49,50]. As KN62 and KN93 cannot effectively differentiate between CaMKI, CaMKII, and CaMKIV, great care must be taken when interpreting experiments in which they are used, because effects may be due to the inhibition of one or more CaMKs that exhibit similar and overlapping substrate specificities. In addition to their potency against CaMKI, CaMKII, and CaMKIV, KN62 and/or KN93 also reportedly inhibit eukaryotic elongation factor-2 kinase, IP_3 receptors, and a variety of voltage-gated K^+ and Ca^{2+} channels [53–64]. Because of their pleiotropic effects, KN62 and KN93 should not be used as the sole means for verifying the involvement of any particular CaMK in biological processes. Another recent addition to the CaMK toolbox is STO-609, which potently inhibits CaMKK, but not CaMKI, CaMKII, or CaMKIV [65]. Although STO-609 is not a specific CaMKK inhibitor [it has been found to inhibit several other kinases, including AMP-activated protein kinase (AMPK), casein kinase 2, and dual-specificity tyrosine-regulated protein kinase], its selectivity among the CaMKs makes it a particularly helpful tool for studying the CaMK cascades [66].

Another oft-used strategy in the study of CaMKs involves the overexpression in cells of wild-type, constitutively active, or dominant-negative proteins in the attempt to drive or inhibit functions normally regulated by the endogenous CaMK. These approaches have contributed greatly to our knowledge of CaMK function. However, because overexpression of proteins frequently creates cellular conditions that differ markedly from normal physiological states, the results of such experiments must be interpreted cautiously.

Constitutively active CaMKs may be generated by mutating autoinhibitory residues within the regulatory domain or by truncating the enzyme at the C-terminal end such that the autoinhibitory domain is removed. Constitutively active mutants generated through point mutations are probably less anomalous in their function than constitutively active truncation mutants and more likely to reflect the function of wild-type, endogenous protein. The behavior of truncation mutants must be interpreted with particular scrutiny. For example, although a constitutively active form of CaMKI (CaMKIα 1-295) has been shown to drive transcription in reporter assays in several studies, this behavior is probably not reflective of normal wild-type CaMKI function; wild-type CaMKIα is excluded from the nucleus, whereas CaMKIα 1-295 localizes to the nucleus because of removal of a C-terminal nuclear export sequence [67–69]. On the contrary, constitutively active CaMKI mutants created through point mutations within the regulatory domain are expected to localize appropriately in the cytoplasm. The functions of the C-terminal domains of CaMKIV and CaMKK are unknown, but truncation of these regions is also likely to result in unintended effects.

Catalytically inactive CaMK mutants are often used as "dominant-negative" proteins. They are believed to inhibit endogenous, wild-type CaMK either by multimerizing with it and impairing activating cross-phosphorylation (as in the case of CaMKII) or by squelching factors necessary for the activation or function of wild-type CaMK (as in the case of CaMKI and CaMKIV). Because of this second mechanism, dominant-negative mutants may have deleterious effects in addition to inhibition of the target of interest. For example, dominant-negative CaMKI and CaMKIV mutants are believed to act by squelching CaMKK and thus might be expected to inhibit activation of multiple CaMKK targets, not just the CaMK of interest. Attempts have been made to address these concerns. For example, catalytically inactive CaMKIV mutants are generally incapable of localizing appropriately to the nucleus and instead accumulate in the cytoplasm where they sequester cytoplasmic CaMKK [70]. One study found that targeting a catalytically inactive CaMKIV mutant to the nucleus using a nuclear localization signal maintained the ability of the mutant to inhibit functions normally attributed to wild-type, endogenous CaMKIV but eliminated other deleterious effects that ostensibly may have been due to the inhibition of another CaMKK substrate, CaMKI [71,72].

Pharmacological inhibitors and exogenously expressed CaMKs can be valuable tools for probing CaMK function. However, the limitations and complications of each approach should always be considered when interpreting experiments in which inhibitors or overexpressed proteins are used.

3. *CaMKK*

Two CaMKK isoforms have been identified (α and β) that are encoded by distinct genes and are approximately 50% identical at the amino acid level [41,42,44,73,74]. They have a similar gross tissue distribution, with high expression levels in the brain, moderate levels in the testis, thymus, and spleen, and low levels in many other tissues [41,74]. However, the two isoforms are differentially expressed in tissue subregions, at least in the case of the brain. CaMKKα is most highly expressed in the hippocampus and outer cortex and is expressed to a low degree in the cerebellum, whereas CaMKKβ is most highly expressed in the cerebellum and outer cortex and is expressed to a low degree in the hippocampus [41]. Both isoforms are predominantly cytoplasmic although CaMKKβ has been observed in the nucleus on occasion [75–78].

As with all CaMKs, CaMKK is activated by Ca^{2+}/CaM binding that relieves intramolecular autoinhibition, thus generating kinase activity. In the case of CaMKKα, Ca^{2+}/CaM binding is absolutely required for kinase activity [79]. However, CaMKKβ differs from CaMKKα in that it displays 50–70% Ca^{2+}/CaM-independent activity [41]. This property may be dependent on a 23-amino acid region in the CaMKKβ N-terminus because deletion of this region results in increased Ca^{2+}/CaM dependence of CaMKKβ [80]. Both CaMKK isoforms are subject to autophosphorylation, although this appears to have minimal effect on enzyme activity [41,81]. CaMKK activity is negatively regulated by phosphorylation on T^{108}, S^{458}, and S^{74} by protein kinase A [82,83]. Phosphorylation of T^{108} and S^{458} has direct inhibitory effects on CaMKK activity, whereas phosphorylation of S^{74}

leads to the recruitment of the 14-3-3 protein, which both directly inhibits CaMKK activity and blocks dephosphorylation of residue T^{108} by the protein phosphatase PP2A [82].

In addition to phosphorylating CaMKI and CaMKIV, CaMKK has been identified as an upstream activator for both protein kinase B (a protein involved in cell proliferation, apoptosis, angiogenesis, and glucose homeostasis) and AMPK (a central player in sensing and maintaining intracellular energy balance) (Fig. 1) [66,84–86]. The role of CaMKK in regulating these additional pathways may provide an answer to questions regarding the physiological significance of a kinase cascade where both upstream kinase (CaMKK) and target (CaMKI or CaMKIV) are regulated by Ca^{2+}/CaM. CaMKK appears to have a greater affinity for Ca^{2+}/CaM than either CaMKI or CaMKIV [83]. Thus, CaMKK may be activated under circumstances where neither CaMKI nor CaMKIV can be activated but where CaMKK is important as an upstream kinase for protein kinase B or AMPK. The role of CaMKK in potentiating these two pathways has been discussed elsewhere [84,87].

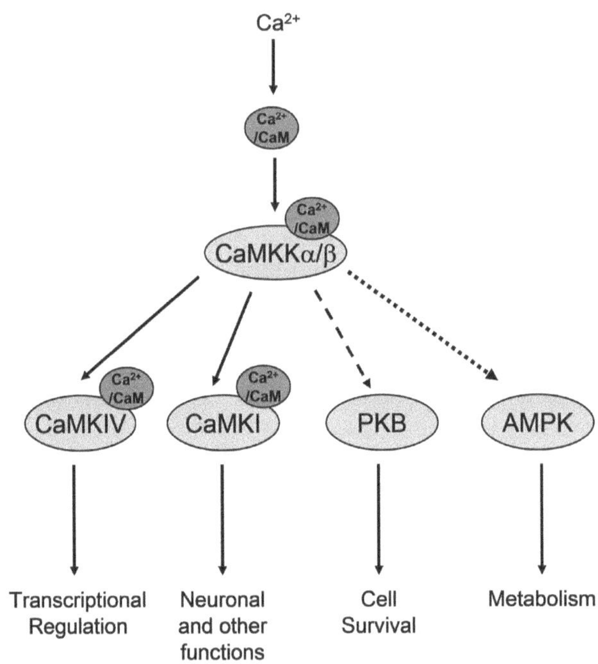

Fig. 1. Multiple pathways are regulated by calcium/calmodulin-dependent kinase kinase (CaMKK). CaMKK is an upstream activator for CaMKI, CaMKIV, PKB (Protein Kinase B), and AMP-activated protein kinase (AMPK). CaMKKα and CaMKKβ can activate CaMKI and CaMKIV (indicated by solid arrows); CaMKKα has been shown to activate PKB (indicated by dashed arrow); CaMKKβ appears to be the major CaMKK isoform contributing to AMPK activation (indicated by dotted arrow). Ca^{2+}/CaM regulates both the upstream kinase (CaMKK) and target kinase (CaMKI or CaMKIV) of the CaMKK–CaMKI and CaMKK–CaMKIV cascades (See Color Plate 38, p. 535).

4. The CaMKK–CaMKI cascade

Four isoforms of CaMKI, encoded by distinct genes, have been identified in mammals: CaMKIα, CaMKIβ (PNCK), CaMKIγ (CLICK III), and CaMKIδ (CKLiK) [12,36,88–90]. CaMKIα, the first isoform to be identified (often referred to simply as CaMKI), is the best studied. It is broadly expressed throughout all the tissues of the body and has an exclusively cytoplasmic subcellular distribution. CaMKI is completely dependent on Ca^{2+}/CaM for its activity. The binding of Ca^{2+}/CaM to CaMKI has two primary effects—it relieves the steric autoinhibition of CaMKI, thus generating an active enzyme, and it exposes the activation loop threonine of CaMKI (T177), which can then be phosphorylated by CaMKK [9,43].

In addition to increasing CaMKI activity, phosphorylation of CaMKI by CaMKK can also result in a shift in CaMKI peptide substrate preference [12,91]. Although some CaMKI substrates (such as the ADR1 peptide) are good substrates for the dephospho form of CaMKI, other targets require phosphorylation of CaMKI on residue T^{177}. For example, the synapsin (4–13) peptide is efficiently phosphorylated by CaMKI only after the kinase has been activated by CaMKK. This activation increases the V_{max}/K_m ratio toward the peptide by over 100-fold, primarily due to a 44-fold decrease in K_m [91]. This suggests that activation of CaMKI may serve to broaden the substrate pool of the enzyme. But, despite the fact that CaMKI has been shown to phosphorylate a variety of proteins in vitro, few physiological targets of CaMKI have been identified, and it is not fully understood how T^{177} phosphorylation regulates CaMKI's activity toward its genuine substrates. Interestingly, CaMKK is among the proteins identified as an in vitro substrate for CaMKI. A truncated, constitutively active form of CaMKI was found to phosphorylate CaMKK in vitro on T^{108}, a residue that has been shown to inhibit CaMKK activity [82,83]. Although it is unknown whether CaMKI phosphorylates CaMKK in vivo, it is interesting to speculate that activated CaMKI might inhibit CaMKK in a negative feedback mechanism.

Two kinases other than CaMKK—AMPKK and CaMKIβ—are capable of activating CaMKIα in vitro [92,93]. Although there is no evidence that either AMPKK or CaMKIβ can activate CaMKIα in vivo (indeed CaMKIα is a relatively poor substrate for AMPKK compared with the physiological AMPKK target, AMPK), the fact that kinases other than CaMKK are *capable* of activating CaMKIα does call attention to the possibility that additional CaMKKs may exist for CaMKI (or CaMKIV).

It is unknown how CaMKI is deactivated. In hippocampal neurons, CaMKI activation is fairly persistent: CaMKI activity remains maximally elevated up to 30 min after glutamate-induced activation [94]. Although a phosphatase capable of dephosphorylating CaMKI in vitro has been identified and cloned, it is not clear whether this phosphatase has any role in regulating CaMKI in vivo [95].

The biological roles of CaMKI have not been thoroughly explored, but in recent years, a role for a CaMKK–CaMKI cascade in the brain has been elucidated. CaMKK and CaMKI appear to be important in neuronal motility; inhibition of either kinase (using pharmacological inhibitors or dominant-negative mutant proteins) significantly decreases neuronal outgrowth and branching in hippocampal and cerebellar neurons [71]. It was shown that overexpression of a full-length constitutively active CaMKI mutant was able to rescue the effects of CaMKK inhibition on neuronal morphology, suggesting that CaMKK functioned upstream of CaMKI in neuronal branching [71]. A role for CaMKK and CaMKI in neuronal activation of the mitogen-activated protein kinases (MAPKs), ERK1 and ERK2, has also been identified. Studies utilizing pharmacological inhibitors and dominant-negative CaMKs demonstrated that inhibition of either CaMKK or CaMKI perturbed depolarization-induced ERK1/2 activation in neuroblastoma NG108 cells. Additionally, expression of constitutively active forms of CaMKK or CaMKI in NG108 cells led to ERK1/2 activation in the absence of other stimuli [96]. The Ras/MEK/ERK pathway has been implicated in a variety of neuronal processes including neurite branching, and it is possible that CaMKK and CaMKI might regulate neuron outgrowth and differentiation partly through their effects on ERK activation. CaMKI also appears to be involved in neuronal long-term potentiation (LTP) through a mechanism that involves activation of ERK by the CaMKK–CaMKI cascade. Inhibition of either CaMKK or CaMKI using pharmacological inhibitors or dominant-negative CaMKs was found to block NMDA-induced ERK activation in primary hippocampal neurons. Additionally, the CaMKK inhibitor STO-609 inhibited CaMKI activation, ERK activation, and early LTP induction in mouse hippocampal slices. The inhibition of early LTP by STO-609 was similar to that observed with the ERK inhibitor U0126. The effects of STO-609 and U0126 were not additive suggesting that the inhibitors were acting on the same pathway [72]. Taken together, these data suggest that a CaMKK–CaMKI cascade contributes to ERK activation in early LTP induction.

In addition to its roles in neuronal processes, CaMKI has been implicated in cell-cycle events. It was found that in WI-38 human fibroblasts, both the CaMK inhibitor KN93 and the dominant-negative CaMKI mutant led to the inhibition of cyclin D/cdk4 activation and arrest of the cell cycle in G1 [97]. A dominant-negative CaMKII mutant did not have this inhibitory effect, and WI-38 cells do not express CaMKIV (which can also be inhibited by KN93), supporting the conclusion that CaMKI plays a role in cdk4 activation during the G1 to S transition. In *Aspergillus nidulans*, it has been demonstrated that two kinases highly similar to mammalian CaMKI/IV and CaMKK (CMKB and CMKC, respectively) are important for entry into the cell cycle and for proper G1 to S transition [48]. These two studies indicate that CaMKK and CaMKI are both important in regulating the cell cycle, and at least in the case of *A. nidulans*, genetic analysis suggests that these kinases are acting together in a kinase cascade to regulate cell-cycle events.

The other mammalian CaMKI isoforms—CaMKIβ, CaMKIγ, and CaMKIδ—are only beginning to be characterized. They can all be activated by CaMKK in vitro, and it seems likely that each might be capable of participating in CaMKK–CaMKI cascades in vivo [88,90,93,98,99]. However, the biological roles mediated by CaMKIβ, CaMKIγ, and CaMKIδ are probably very different from those mediated by CaMKIα, because each isoform displays different tissue expression patterns and subcellular localizations. CaMKIβ is broadly expressed throughout the tissues of the body and can be found both in the nucleus and in the cytoplasm [89,93,100]. CaMKIγ appears to be more restricted in its expression; it is most highly expressed in certain subregions of the brain and expressed to a lesser degree in skeletal muscle, kidney, spleen, and liver [90]. CaMKIγ has been observed localized in the cytoplasm and enriched at membrane compartments such as the Golgi and plasma membrane [90,99]. A CAAX motif in the C-terminus of CaMKIγ, which directs the addition of a prenyl lipid group to the motif's cysteine residue, appears to be critical for membrane targeting [90]. CaMKIδ has so far been shown to be highly expressed in the brain and in granulocytes [88,101]. In hippocampal neurons, CaMKIδ is localized in the cytoplasm but has been observed to translocate to the nucleus upon KCl-induced depolarization. Activation of CaMKIδ appeared to be critical for this phenomenon, as the CaMKK inhibitor STO-609 inhibited CaMKIδ nuclear translocation, and mutation of the activation loop threonine prevented the depolarization-induced translocation of CaMKIδ as well. In cellular assays, wild-type CaMKIδ (but not an activation loop threonine CaMKIδ mutant) was capable of driving transcription in cellular assays, through a mechanism that was potentiated by the co-transfection of CaMKKα. These results support the conjecture that CaMKI isoforms other than CaMKIα may function in CaMKK–CaMKI cascades, in roles that are distinct from those of CaMKIα.

5. The CaMKK–CaMKIV cascade

Two isoforms of CaMKIV (α and β) have been described; they are identical except that the latter isoform carries 28–30 extra amino acids at its N-terminus [102–104]. They are encoded by a single gene, which also encodes calspermin, a testis-specific protein of unknown function, which consists of the 169 C-terminal amino acids of CaMKIV [103,105,106]. The three proteins are expressed from the single *CaMKIV* gene through the use of alternate transcriptional initiation sites [103]. The majority of studies involving CaMKIV have employed the α-isoform of CaMKIV; it is unknown whether the regulation or function of the two CaMKIV isoforms differ significantly, although one study found that the two isoforms had different patterns of expression in the brain [107]. CaMKIV is expressed only in specific tissues; it is most highly expressed in distinct regions of the brain and in the thymus but is also found in the testis, spleen, ovary, and in certain bone marrow-derived cells [108,109]. CaMKIV exhibits predominantly nuclear subcellular localization but has been found in the cytoplasm in some instances [110,111]. Although CaMKIV appears to lack a classical nuclear localization sequence, the enzyme is transported into the nucleus through an importin α-dependent (but importin β- and Ran-independent) mechanism that is dependent on CaMKIV ATP binding [70,112].

As with the other CaMKs, CaMKIV activity is restrained by autoinhibitory interactions, which are relieved upon Ca^{2+}/CaM binding, leading to basal CaMKIV activity [11]. The conformational changes elicited by Ca^{2+}/CaM binding also expose the CaMKIV activation loop threonine (T^{200} in human CaMKIV; T^{196} in mouse CaMKIV), which can then be phosphorylated by CaMKK [79,113]. Phosphorylation of the CaMKIV activation loop results in increased CaMKIV activity toward a variety of protein and peptide substrates [79,113]. One of the few validated physiological substrates for CaMKIV is the transcriptional co-activator, cAMP response element-binding protein (CREB). Examination of the kinetics of CREB phosphorylation by CaMKIV found that activation of CaMKIV had little effect on the enzyme's affinity for CREB (the K_m actually increased from 3.6 to 5.5 μM) but increased the V_{max} of CREB phosphorylation more than 30-fold [114]. This consequence is quite different from the effect seen on synapsin (4–13) peptide phosphorylation following CaMKI activation, wherein CaMKI's increased kinase activity is due primarily to a dramatic increase in the affinity of CaMKI for the peptide substrate rather than a large increase in V_{max} [91]. However, it remains to be determined how activation affects the kinetics of CaMKIV activity toward physiological substrates other than CREB.

In addition to increased kinase activity, the activation of CaMKIV results in the generation of Ca^{2+}/CaM-independent activity. Most studies assessing CaMKIV activation have reported the development of only partial Ca^{2+}/CaM independence upon phosphorylation of CaMKIV by CaMKK [45,115]. However, a recent study demonstrated that maximal CaMKIV activation (achieved only by incubating CaMKIV with at least equimolar amounts of CaMKK in the presence of Ca^{2+}/CaM) resulted in the generation of a fully Ca^{2+}/CaM-independent enzyme [14]. This finding suggested that in many circumstances where CaMKIV activation had previously been examined, limiting amounts of CaMKK may have precluded activation of all CaMKIV molecules. The fact that complete activation of CaMKIV was observed only in the presence of similar molar amounts of CaMKK led to the suggestion that CaMKIV and CaMKK do not share a typical Michaelis–Menton enzyme–substrate relationship, but rather, activation of CaMKIV by CaMKK occurs within a stable stoichiometric complex [14]. This conjecture is supported by the following observations: (i) CaMKIV and CaMKK stably interact in co-immunoprecipitation experiments and (ii) disruption of CaMKK–CaMKIV binding has been found to prevent CaMKIV activation [an RP-domain mutant of CaMKKα that is incapable of stably binding CaMKIV fails to activate CaMKIV even though it is unimpaired in its phosphorylation of other substrates (such as protein kinase B)] [14,70,81].

Despite biochemical and biological evidence supporting a physiological CaMKK–CaMKIV cascade, it is unknown where in the cell CaMKIV might be activated by CaMKK, as these two kinases predominantly occupy separate cellular compartments (CaMKK is mostly cytoplasmic and CaMKIV is mostly nuclear) [75,77,78,111]. It is possible that some small fraction of cellular CaMKK normally resides in the nucleus or translocates to the nucleus upon a Ca^{2+} stimulus. Alternatively, some portion of CaMKIV may cycle in and out of the nucleus allowing CaMKIV to be activated in the cytoplasm before returning to the nucleus.

While CaMKIV is phosphorylated and activated by CaMKK, CaMKIV can also reportedly phosphorylate CaMKK in vitro, although it is not known whether this has any functional importance [43]. However, phosphorylation on T^{108} inhibits CaMKK, and both PKA and CaMKI can phosphorylate this residue; an interesting and untested possibility is that activated CaMKIV might also phosphorylate CaMKK on this site and negatively regulate CaMKK in a feedback mechanism (Fig. 2B) [82].

CaMKIV is negatively regulated by the protein phosphatase PP2A that dephosphorylates its activation loop threonine [116]. It has been suggested that a substantial fraction of CaMKIV in the cell may normally exist in a complex with PP2A. One study reported that up to 43% of cellular CaMKIV immunoprecipitated from cells was associated with PP2A in a 1:1 stoichiometric complex [116]. Additionally, disruption of the CaMKIV–PP2A interaction in cells (by overexpressing the small T

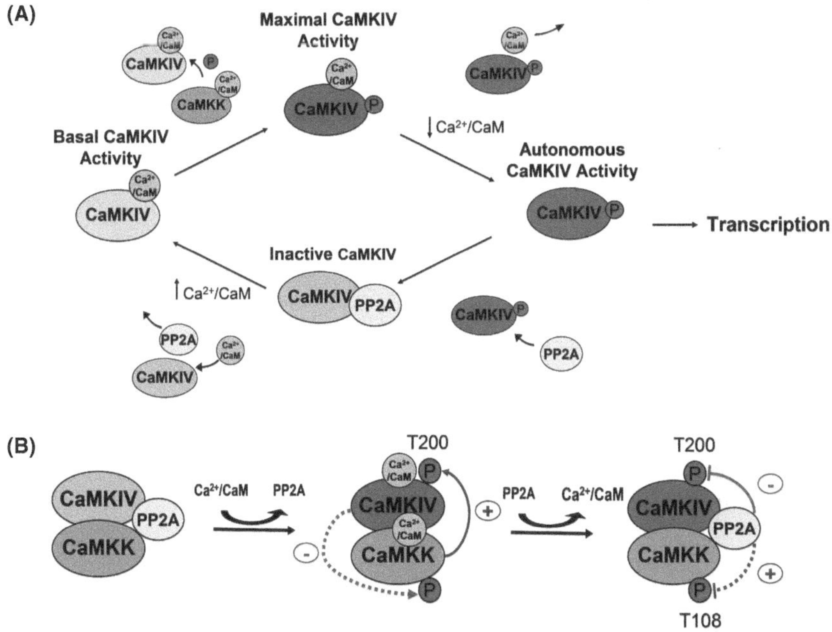

Fig. 2. (**A**) Regulation of calcium/calmodulin-dependent kinase IV (CaMKIV) activity. CaMKIV interacts with PP2A and Ca^{2+}/CaM in a mutually exclusive manner; CaMKIV/PP2A complexes may be disrupted upon a Ca^{2+} stimulus. CaMKIV becomes basally active upon Ca^{2+}/CaM binding but only achieves autonomous activity (which is required for CaMKIV transcriptional function) after phosphorylation on residue T^{200} by CaMKK. PP2A negatively regulates CaMKIV by dephosphorylating residue T^{200} of CaMKIV. (**B**) Potential regulation of CaMKK–CaMKIV cascade components within a complex. CaMKIV stably interacts with both CaMKK and PP2A, and it is possible that a complex consisting of all three components might exist under some cellular conditions. In such a complex, CaMKIV would be readily accessible to phosphorylation by CaMKK, such that CaMKIV activation could occur almost immediately upon Ca^{2+}/CaM binding. CaMKIV can phosphorylate CaMKK in vitro, and an intriguing possibility is that activated CaMKIV might phosphorylate CaMKK on inhibitory residues such as T^{108}, thus regulating CaMKK in a negative feedback fashion. PP2A negatively regulates CaMKIV by dephosphorylating residue T^{200} and positively regulates CaMKK by dephosphorylating residue T^{108}, although it is unknown whether these two PP2A functions can occur within the same complex. Positive regulatory events are depicted in blue, whereas negative regulatory events are depicted in red. Confirmed regulatory events are represented with solid lines, whereas speculative events are represented by dotted lines (See Color Plate 39, p. 536).

antigen or by mutagenizing CaMKIV residues important for PP2A binding) results in elevated levels of activated CaMKIV in the absence of a Ca^{2+} stimulus [14,116,117]. This suggests that normal cellular Ca^{2+} transients are capable of initiating CaMKIV activation and that persistent association of CaMKIV and PP2A is important for maintaining CaMKIV in a deactivated state in the absence of appropriate stimuli.

In co-immunoprecipitation experiments, CaMKIV binds PP2A and Ca^{2+}/CaM in a mutually exclusive manner, suggesting that CaMKIV–PP2A complexes in cells might be disrupted following a Ca^{2+} stimulus [117]. This temporary reprieve from PP2A's dephosphorylating activity might be necessary to allow CaMKIV to become activated and remain activated long enough to carry out its functions. Indeed, PP2A appears to keep tight reins over CaMKIV activation levels. In T cells, hippocampal neurons, and 293A human embryonic kidney cells, CaMKIV achieves maximal activation within minutes of a Ca^{2+} stimulus but is rapidly dephosphorylated such that CaMKIV activity returns to basal levels within 30 min [94,117,118]. The inactivation of CaMKIV following a Ca^{2+} stimulus can be slowed by inhibiting PP2A, a finding that further indicates a role for PP2A as a negative regulator of CaMKIV activity [117].

The ability of CaMKIV to complex with both CaMKK and PP2A raises the possibility that some fraction of CaMKIV in the cell might exist in a complex with both of its regulators (Fig. 2B). (It is not known whether CaMKIV can bind PP2A and CaMKK simultaneously or whether the interactions occur in a mutually exclusive manner.) Although the interaction between CaMKKα and CaMKIV has been reported to be Ca^{2+}/CaM dependent, the interaction between CaMKKβ and CaMKIV is reportedly independent of Ca^{2+}/CaM, and thus it is conceivable that, in the absence of bound Ca^{2+}/CaM, CaMKIV might exist in a complex with both CaMKKβ and PP2A [14,81]. In such a complex, CaMKIV would be readily accessible to phosphorylation by CaMKKβ, such that CaMKIV activation could occur almost immediately upon Ca^{2+}/CaM binding. Additionally, the complex might also subject CaMKKβ to regulation by PP2A, as PP2A can positively regulate CaMKK by dephosphorylating its T^{108} residue [82].

The most widely recognized function of CaMKIV has been its role as a transcriptional regulator. In cellular assays, CaMKIV has been shown to activate transcription mediated by a number of transcription factors, including serum response factor, retinoid-related orphan receptor α/γ, MADS box transcription enhancer factor 2 (MEF2), and chicken ovalbumin upstream promoter-transcription factor 1 [119–121]. One mechanism of CaMKIV function in transcription is phosphoregulation of specific transcription-related components such as histone deacetylase 4 (HDAC4), calcineurin-binding protein 1 (Cabin1), CREB, and CREB-binding protein (CBP) [51,110,122–126]. Among identified CaMKIV functions, the role of CaMKIV in regulating CREB-dependent transcription has been the best characterized.

One of the ways in which CaMKIV potentiates CREB-dependent transcription is direct phosphorylation of CREB on S^{133}, which leads to the recruitment of CBP and subsequent CREB/CBP-dependent transcription [110,122,127]. In CREB-mediated transcriptional assays, the ability of CaMKIV to potentiate transcription is dependent on CaMKIV activation by CaMKK. Overexpression of CaMKK enhances CaMKIV-mediated CREB transcription, whereas mutation of the CaMKIV activation loop threonine to a non-phosphorylatable alanine abrogates the ability of CaMKIV to drive CREB transcription in cellular assays [45,74,113]. Although activation has been shown to increase the kinase activity of CaMKIV toward CREB, the primary role of activation loop phosphorylation may be the generation of CaMKIV autonomy (Ca^{2+}/CaM independence). A recent study demonstrated that although wild-type CaMKIV achieved autonomous activity (and transcriptional competence) only after Ca^{2+}/CaM binding and subsequent T^{200} phosphorylation, full-length constitutively active CaMKIV mutants could drive transcription in the absence of bound Ca^{2+}/CaM or activation loop phosphorylation. In the case of each CaMKIV protein tested, transcriptional activity correlated with the generation, or levels, of autonomous activity in cells, but not with levels of Ca^{2+}/CaM-dependent activity in vitro [14]. These data strongly argue that autonomous activity is the primary requirement for CaMKIV potentiation of CREB-dependent transcription. Thus, the most important role of activation loop phosphorylation may be the generation and/or maintenance of autonomous activity (Fig. 2A). This would be quite similar to the role of T^{286} phosphorylation in the related kinase, CaMKII; phosphorylation of T^{286} prevents a return of CaMKII to its autoinhibited state upon Ca^{2+}/CaM removal and confers enzyme autonomy [21,29,128].

Activation of CaMKIV generates an autonomous enzyme that can function for a period of time that outlasts the duration of the Ca^{2+} stimulus (levels of activated/autonomous CaMKIV remain substantially elevated 5–10 min after a Ca^{2+} stimulus). The importance of autonomy to CaMKIV function makes sense when one considers the requirements for CREB-dependent transcription. Several studies have demonstrated that in hippocampal neurons, sustained (rather than transient) CREB S^{133} phosphorylation is required for CREB-dependent transcription and depolarization-induced LTP; persistent activity of CREB kinases is required to sustain CREB S^{133} phosphorylation [129–131]. In hippocampal neurons, both CaMKIV and MAPK are activated by depolarizing stimuli and are the dominant pathways contributing to the CREB phosphorylation that is required for subsequent CREB-dependent transcription and LTP [132,133]. CaMKIV contributes to the rapid, early phase of CREB phosphorylation (approximately 0–20 min following stimulation), whereas MAPK contributes to the slower, later phase (approximately 15-60 min following stimulation) [133]. Inhibition of the MAPK or CaMK pathway using pharmacological inhibitors, or more specific inhibition of CaMKIV using antisense oligonucleotides or dominant-negative mutant proteins, results in repression of stimulus-induced CREB phosphorylation, CREB-dependent transcription, and LTP in hippocampal neurons [125,130,131,133,134]. In addition to the direct role of CaMKIV in neuronal depolarization-induced CREB phosphorylation, an unidentified CaMK (or CaMKs) contributes indirectly to CREB phosphorylation by mediating activation of the MAPK pathway. In the neuroblastoma cell line, NG108, both CaMKI and CaMKIV have been implicated in activation of MAPKs. Overexpression of a constitutively active CaMKIV mutant leads to the activation of

ERK2, JNK1, and p38, whereas overexpression of a constitutively active CaMKI mutant results in increased activation of ERK1 and ERK2 [96,135]. In hippocampal neurons, the effects of pharmacological CaMK inhibitors on CREB phosphorylation are due partly to the inhibition of CaMKIV and partly to the inhibition of CaMK-mediated activation of the MAPK pathway [53].

In addition to its contribution to hippocampal LTP, CaMKIV has also been implicated in CREB-dependent events essential for neuronal survival, T-cell function, and long-term depression in Purkinje cells [55,136–140]. The importance of CaMKIV in CREB-dependent processes is supported by phenotypes of CaMKIV-deficient mice. Basal levels of phospho-CREB are decreased in CaMKIV knockout mice, and many defects displayed by these mice (such as impaired learning and memory and defective T-cell responses) appear related to a failure to adequately induce CREB phosphorylation in response to stimuli [141–145].

Apart from its well-documented role in CREB phosphorylation, CaMKIV also contributes to CREB/CBP-dependent transcription through a second mechanism. Several studies have shown that recruitment of CBP to CREB is insufficient to drive CREB/CBP-dependent transcription; a second event mediated by CaMKs or protein kinase A (but not MAPKs) is required [125,146,147]. CaMKIV can fulfill this second requirement by phosphorylating CBP; CaMKIV specifically phosphorylates CBP on S^{301} in vitro, and mutation of S^{301} of CBP to a non-phosphorylatable alanine reduces the ability of CaMKIV (but not protein kinase A) to drive CBP-dependent transcription in hippocampal neurons [125]. These data indicate that CaMKIV participates in CREB/CBP-dependent transcription through phosphorylation of both CREB and CBP and support the critical role of CaMKIV in hippocampal neurons. In hippocampal neurons, both CaMKIV and MAPK are activated in response to stimuli and lead to CREB phosphorylation, but only CaMKIV can mediate the second required step of CBP phosphorylation for CREB/CBP-dependent transcription [125,146,147]. It will be interesting to explore whether regulation of CBP contributes to CaMKIV's role in regulating gene expression by other transcription factors (in addition to CREB) that utilize CBP as a co-activator.

In addition to phosphorylation of CREB and CBP, it has been proposed that CaMKIV can regulate gene expression through its actions on two transcriptional repressors, Cabin1 and HDAC4. In T cells, MEF2 release from its transcriptional repressor, Cabin1, is promoted by phosphorylation of Cabin1 on S^{2126}, an event that leads to the interaction of Cabin1 with 14-3-3 and subsequent Cabin1 nuclear export. CaMKIV can phosphorylate this critical serine residue of Cabin1 in vitro and activated CaMKIV augments Cabin1 phosphorylation, Cabin1/14-3-3 association, Cabin1 nuclear export, and MEF2 transcriptional activity in T cells [126]. Thus, CaMKIV may regulate MEF2 transcriptional activity through Cabin1. It has also been suggested that CaMKIV regulates MEF2 transcription by phosphorylating another MEF2 transcriptional repressor, HDAC4. In the U2OS osteosarcoma cell line, a truncated, constitutively active CaMKIV mutant was found to drive the nuclear export of HDAC4, ostensibly by phosphorylating HDAC4/14-3-3 complexes present in the nucleus and promoting their nuclear export [124]. Whether endogenous CaMKIV is also capable of regulating HDAC4 in this manner is unknown.

6. Conclusions

The activation of CaMKI and CaMKIV by their upstream activator, CaMKK, has been fairly well characterized in vitro, but much still remains obscure about the regulation and function of the CaMK cascades in vivo. For instance, the extent to which activation is required for the biological functions of CaMKI and CaMKIV is unknown. It has been shown that CaMKI can phosphorylate some peptide substrates independently of activation loop phosphorylation [91]. A question that naturally arises from this finding is whether CaMKI, or CaMKIV for that matter, might have physiological functions that are independent of CaMKK. This seems an especially relevant query for CaMKIV, given that CaMKIV and CaMKK reside mostly in separate cellular compartments; can CaMKIV act in certain capacities without the aid of CaMKK?

Another conundrum has been the question of where in the cell CaMKIV is activated, as CaMKIV resides mostly in the nucleus, whereas CaMKK is predominantly cytoplasmic. One possibility is that some fraction of CaMKIV cycles in and out of the nucleus and can be activated by CaMKK while in the cytoplasm. Conversely, CaMKIV may be activated in the nucleus by some fraction of CaMKK that either normally exists in the nucleus or translocates to the nucleus in response to an appropriate stimulus. A variation on these first two possibilities is that some fraction of CaMKIV exists in a stable complex with CaMKK, and perhaps, this complex may cycle between subcellular compartments. The stability of such a complex between CaMKIV and CaMKK may depend on which CaMKK isoform is involved; one group has reported (as data not shown) that the interaction between CaMKKα and CaMKIV is Ca^{2+}/CaM dependent [81], whereas our laboratory has observed that the interaction between CaMKKβ and CaMKIV is Ca^{2+}/CaM independent.

How the two identified CaMKK isoforms differ in their biological roles is another area that is ripe for further study. One of the most obvious distinctions between CaMKKα and CaMKKβ is the latter's Ca^{2+}/CaM-independent activity. It is not clear how the Ca^{2+}/CaM independence of CaMKKβ might influence its function toward CaMK targets, because neither CaMKI nor CaMKIV can be activated in the absence of Ca^{2+}/CaM, given their own requirement for Ca^{2+}/CaM binding prior to activation. The differential expression patterns of the CaMKK isoforms (in the brain, CaMKKβ expression tends to be expressed in regions also expressing CaMKIV, whereas CaMKKα expression more closely mirrors CaMKIα) also point to distinct functions of the two enzymes [148,149]. However, differences between CaMKKα and CaMKKβ in regulation of the CaMK cascades in vivo have not been explicitly explored. Several groups *have* shown significant differences in the behavior of CaMKKβ and CaMKKα toward another of their shared substrates, AMPK. In vitro, CaMKKβ phosphorylates AMPK much more efficiently than CaMKKα, and although there is not as yet complete consensus, studies seem to indicate that, in cells, CaMKKβ plays a much greater role than CaMKKα in AMPK activation [87]. It will be interesting to determine whether there is also functional divergence between CaMKKα and CaMKKβ in regulation of the CaMK cascades.

The past 10 years have yielded a flurry of studies uncovering biological functions of the CaMKs. But as the CaMKs have been implicated in an ever greater array of activities, it has become an increasing challenge to ascribe credit for a specific cellular function to any particular kinase. Addressing this challenge will require more sensitive tools than the relatively non-specific CaMK inhibitors and non-physiological overexpression-type approaches that the field has relied upon in the past. To untangle the functions of the various CaMKs and probe the differences between CaMK isoforms, it will be necessary to take greater advantage of more selective approaches—such as RNA interference, inducible knockouts, and genetically engineered analog-sensitive kinases. These powerful tools hold great promise for the field and will undoubtedly bring greater understanding of, and deeper appreciation for, the CaMKs.

References

1. Hanson, C. J., Bootman, M. D., and Roderick, H. L. (2004) *Curr Biol* **14**, R933–R935.
2. Orrenius, S., Zhivotovsky, B., and Nicotera, P. (2003) *Nat Rev Mol Cell Biol* **4**, 552–565.
3. Macrez, N., and Mironneau, J. (2004) *Curr Mol Med* **4**, 263–275.
4. Thomas, A. P., Bird, G. S., Hajnoczky, G., Robb-Gaspers, L. D., and Putney, J. W., Jr. (1996) *FASEB J* **10**, 1505–1517.
5. Ghosh, A., and Greenberg, M. E. (1995) *Science* **268**, 239–247.
6. Chin, D., and Means, A. R. (2000) *Trends Cell Biol* **10**, 322–328.
7. Babu, Y. S., Bugg, C. E., and Cook, W. J. (1988) *J Mol Biol* **204**, 191–204.
8. Cruzalegui, F. H., Kapiloff, M. S., Morfin, J. P., Kemp, B. E., Rosenfeld, M. G., and Means, A. R. (1992) *Proc Natl Acad Sci USA* **89**, 12127–12131.
9. Goldberg, J., Nairn, A. C., and Kuriyan, J. (1996) *Cell* **84**, 875–887.
10. Rosenberg, O. S., Deindl, S., Sung, R. J., Nairn, A. C., and Kuriyan, J. (2005) *Cell* **123**, 849–860.
11. Cruzalegui, F. H., and Means, A. R. (1993) *J Biol Chem* **268**, 26171–26178.
12. Haribabu, B., Hook, S. S., Selbert, M. A., Goldstein, E. G., Tomhave, E. D., Edelman, A. M., Snyderman, R., and Means, A. R. (1995) *EMBO J* **14**, 3679–3686.
13. Tokumitsu, H., Wayman, G. A., Muramatsu, M., and Soderling, T. R. (1997) *Biochemistry* **36**, 12823–12827.
14. Chow, F. A., Anderson, K. A., Noeldner, P. K., and Means, A. R. (2005) *J Biol Chem* **280**, 20530–20538.
15. Payne, M. E., Fong, Y. L., Ono, T., Colbran, R. J., Kemp, B. E., Soderling, T. R., and Means, A. R. (1988) *J Biol Chem* **263**, 7190–7195.
16. Chin, D., and Means, A. R. (2002) *Biochemistry* **41**, 14001–14009.
17. Meyer, T., Hanson, P. I., Stryer, L., and Schulman, H. (1992) *Science* **256**, 1199–1202.
18. Rich, R. C., and Schulman, H. (1998) *J Biol Chem* **273**, 28424–28429.
19. Thiel, G., Czernik, A. J., Gorelick, F., Nairn, A. C., and Greengard, P. (1988) *Proc Natl Acad Sci USA* **85**, 6337–6341.
20. Hanson, P. I., and Schulman, H. (1992) *Annu Rev Biochem* **61**, 559–601.
21. Hudmon, A., and Schulman, H. (2002) *Biochem J* **364**, 593–611.
22. Miller, P., Zhabotinsky, A. M., Lisman, J. E., and Wang, X. J. (2005) *PLoS Biol* **3**, e107.
23. Lisman, J. E., and McIntyre, C. C. (2001) *Curr Biol* **11**, R788–R791.
24. Lisman, J. E., and Zhabotinsky, A. M. (2001) *Neuron* **31**, 191–201.
25. Hudmon, A., and Schulman, H. (2002) *Annu Rev Biochem* **71**, 473–510.

26. Griffith, L. C., Lu, C. S., and Sun, X. X. (2003) *Mol Interv* **3**, 386–403.
27. Schulman, H. (2004) *J Neurosci* **24**, 8399–8403.
28. Lisman, J., Schulman, H., and Cline, H. (2002) *Nat Rev Neurosci* **3**, 175–190.
29. Colbran, R. J. (1992) *Neurochem Int* **21**, 469–497.
30. Elgersma, Y., Sweatt, J. D., and Giese, K. P. (2004) *J Neurosci* **24**, 8410–8415.
31. Fink, C. C., and Meyer, T. (2002) *Curr Opin Neurobiol* **12**, 293–299.
32. Huttner, W. B., DeGennaro, L. J., and Greengard, P. (1981) *J Biol Chem* **256**, 1482–1488.
33. Huttner, W. B., and Greengard, P. (1979) *Proc Natl Acad Sci USA* **76**, 5402–5406.
34. Ohmstede, C. A., Jensen, K. F., and Sahyoun, N. E. (1989) *J Biol Chem* **264**, 5866–5875.
35. Nairn, A. C., and Greengard, P. (1987) *J Biol Chem* **262**, 7273–7281.
36. Picciotto, M. R., Czernik, A. J., and Nairn, A. C. (1993) *J Biol Chem* **268**, 26512–26521.
37. Frangakis, M. V., Ohmstede, C. A., and Sahyoun, N. (1991) *J Biol Chem* **266**, 11309–11316.
38. Tokumitsu, H., Brickey, D. A., Glod, J., Hidaka, H., Sikela, J., and Soderling, T. R. (1994) *J Biol Chem* **269**, 28640–28647.
39. Lee, J. C., and Edelman, A. M. (1994) *J Biol Chem* **269**, 2158–2164.
40. Okuno, S., and Fujisawa, H. (1993) *J Biochem (Tokyo)* **114**, 167–170.
41. Anderson, K. A., Means, R. L., Huang, Q. H., Kemp, B. E., Goldstein, E. G., Selbert, M. A., Edelman, A. M., Fremeau, R. T., and Means, A. R. (1998) *J Biol Chem* **273**, 31880–31889.
42. Hsu, L. S., Tsou, A. P., Chi, C. W., Lee, C. H., and Chen, J. Y. (1998) *J Biomed Sci* **5**, 141–149.
43. Matsushita, M., and Nairn, A. C. (1998) *J Biol Chem* **273**, 21473–21481.
44. Edelman, A. M., Mitchelhill, K. I., Selbert, M. A., Anderson, K. A., Hook, S. S., Stapleton, D., Goldstein, E. G., Means, A. R., and Kemp, B. E. (1996) *J Biol Chem* **271**, 10806–10810.
45. Selbert, M. A., Anderson, K. A., Huang, Q. H., Goldstein, E. G., Means, A. R., and Edelman, A. M. (1995) *J Biol Chem* **270**, 17616–17621.
46. Eto, K., Takahashi, N., Kimura, Y., Masuho, Y., Arai, K., Muramatsu, M. A., and Tokumitsu, H. (1999) *J Biol Chem* **274**, 22556–22562.
47. Kimura, Y., Corcoran, E. E., Eto, K., Gengyo-Ando, K., Muramatsu, M. A., Kobayashi, R., Freedman, J. H., Mitani, S., Hagiwara, M., Means, A. R., and Tokumitsu, H. (2002) *EMBO Rep* **3**, 962–966.
48. Joseph, J. D., and Means, A. R. (2000) *J Biol Chem* **275**, 38230–38238.
49. Tokumitsu, H., Chijiwa, T., Hagiwara, M., Mizutani, A., Terasawa, M., and Hidaka, H. (1990) *J Biol Chem* **265**, 4315–4320.
50. Sumi, M., Kiuchi, K., Ishikawa, T., Ishii, A., Hagiwara, M., Nagatsu, T., and Hidaka, H. (1991) *Biochem Biophys Res Commun* **181**, 968–975.
51. Enslen, H., Sun, P., Brickey, D., Soderling, S. H., Klamo, E., and Soderling, T. R. (1994) *J Biol Chem* **269**, 15520–15527.
52. Mochizuki, H., Ito, T., and Hidaka, H. (1993) *J Biol Chem* **268**, 9143–9147.
53. Wu, G. Y., Deisseroth, K., and Tsien, R. W. (2001) *Nat Neurosci* **4**, 151–158.
54. Sihra, T. S., and Pearson, H. A. (1995) *Neuropharmacology* **34**, 731–741.
55. Ahn, S., Ginty, D. D., and Linden, D. J. (1999) *Neuron* **23**, 559–568.
56. Li, G., Hidaka, H., and Wollheim, C. B. (1992) *Mol Pharmacol* **42**, 489–498.
57. Cui, Z. J., Hidaka, H., and Dannies, P. S. (1996) *Biochim Biophys Acta* **1310**, 343–347.
58. Marley, P. D., and Thomson, K. A. (1996) *Biochem Biophys Res Commun* **221**, 15–18.
59. Tornquist, K., and Ekokoski, E. (1996) *J Endocrinol* **148**, 131–138.
60. Tsutsui, M., Yanagihara, N., Fukunaga, K., Minami, K., Nakashima, Y., Kuroiwa, A., Miyamoto, E., and Izumi, F. (1996) *J Neurochem* **66**, 2517–2522.
61. Puhl, H. L., Raman, P. S., Williams, C. L., and Aronstam, R. S. (1997) *Biochem Pharmacol* **53**, 1107–1114.
62. Smyth, J. T., Abbott, A. L., Lee, B., Sienaert, I., Kasri, N. N., De Smedt, H., Ducibella, T., Missiaen, L., Parys, J. B., and Fissore, R. A. (2002) *J Biol Chem* **277**, 35061–35070.

63. Ledoux, J., Chartier, D., and Leblanc, N. (1999) *J Pharmacol Exp Ther* **290**, 1165–1174.
64. Li, P. M., Fukazawa, H., Mizuno, S., and Uehara, Y. (1993) *Anticancer Res* **13**, 1957–1964.
65. Tokumitsu, H., Inuzuka, H., Ishikawa, Y., Ikeda, M., Saji, I., and Kobayashi, R. (2002) *J Biol Chem* **277**, 15813–15818.
66. Hawley, S. A., Pan, D. A., Mustard, K. J., Ross, L., Bain, J., Edelman, A. M., Frenguelli, B. G., and Hardie, D. G. (2005) *Cell Metab* **2**, 9–19.
67. Condon, J. C., Pezzi, V., Drummond, B. M., Yin, S., and Rainey, W. E. (2002) *Endocrinology* **143**, 3651–3657.
68. Sun, P., Lou, L., and Maurer, R. A. (1996) *J Biol Chem* **271**, 3066–3073.
69. Stedman, D. R., Uboha, N. V., Stedman, T. T., Nairn, A. C., and Picciotto, M. R. (2004) *FEBS Lett.* **566**, 275–280.
70. Lemrow, S. M., Anderson, K. A., Joseph, J. D., Ribar, T. J., Noeldner, P. K., and Means, A. R. (2004) *J Biol Chem* **279**, 11664–11671.
71. Wayman, G. A., Kaech, S., Grant, W. F., Davare, M., Impey, S., Tokumitsu, H., Nozaki, N., Banker, G., and Soderling, T. R. (2004) *J Neurosci* **24**, 3786–3794.
72. Schmitt, J. M., Guire, E. S., Saneyoshi, T., and Soderling, T. R. (2005) *J Neurosci* **25**, 1281–1290.
73. Kitani, T., Okuno, S., and Fujisawa, H. (1997) *J Biochem (Tokyo)* **122**, 243–250.
74. Tokumitsu, H., Enslen, H., and Soderling, T. R. (1995) *J Biol Chem* **270**, 19320–19324.
75. Sakagami, H., Umemiya, M., Saito, S., and Kondo, H. (2000) *Eur J Neurosci* **12**, 89–99.
76. Nakamura, Y., Okuno, S., Kitani, T., Otake, K., Sato, F., and Fujisawa, H. (1996) *Neurosci Lett* **204**, 61–64.
77. Nakamura, Y., Okuno, S., Kitani, T., Otake, K., Sato, F., and Fujisawa, H. (2001) *Neurosci Res* **39**, 175–188.
78. Alliance for Cellular Signaling Data Center, http://www.signaling-gateway.org.
79. Tokumitsu, H., and Soderling, T. R. (1996) *J Biol Chem* **271**, 5617–5622.
80. Tokumitsu, H., Iwabu, M., Ishikawa, Y., and Kobayashi, R. (2001) *Biochemistry* **40**, 13925–13932.
81. Tokumitsu, H., Takahashi, N., Eto, K., Yano, S., Soderling, T. R., and Muramatsu, M. (1999) *J Biol Chem* **274**, 15803–15810.
82. Davare, M. A., Saneyoshi, T., Guire, E. S., Nygaard, S. C., and Soderling, T. R. (2004) *J Biol Chem* **279**, 52191–52199.
83. Matsushita, M., and Nairn, A. C. (1999) *J Biol Chem* **274**, 10086–10093.
84. Yano, S., Tokumitsu, H., and Soderling, T. R. (1998) *Nature* **396**, 584–587.
85. Hamilton, S. R., O'Donnell, J. B., Jr., Hammet, A., Stapleton, D., Habinowski, S. A., Means, A. R., Kemp, B. E., and Witters, L. A. (2002) *Biochem Biophys Res Commun* **293**, 892–898.
86. Hong, S. P., Momcilovic, M., and Carlson, M. (2005) *J Biol Chem* **280**, 21804–21809.
87. Witters, L. A., Kemp, B. E., and Means, A. R. (2006) *Trends Biochem Sci* **31**, 13–16.
88. Verploegen, S., Lammers, J. W., Koenderman, L., and Coffer, P. J. (2000) *Blood* **96**, 3215–3223.
89. Gardner, H. P., Rajan, J. V., Ha, S. I., Copeland, N. G., Gilbert, D. J., Jenkins, N. A., Marquis, S. T., and Chodosh, L. A. (2000) *Genomics* **63**, 279–288.
90. Takemoto-Kimura, S., Terai, H., Takamoto, M., Ohmae, S., Kikumura, S., Segi, E., Arakawa, Y., Furuyashiki, T., Narumiya, S., and Bito, H. (2003) *J Biol Chem* **278**, 18597–18605.
91. Hook, S. S., Kemp, B. E., and Means, A. R. (1999) *J Biol Chem* **274**, 20215–20222.
92. Hawley, S. A., Selbert, M. A., Goldstein, E. G., Edelman, A. M., Carling, D., and Hardie, D. G. (1995) *J Biol Chem* **270**, 27186–27191.
93. Naito, Y., Watanabe, Y., Yokokura, H., Sugita, R., Nishio, M., and Hidaka, H. (1997) *J Biol Chem* **272**, 32704–32708.
94. Uezu, A., Fukunaga, K., Kasahara, J., and Miyamoto, E. (2002) *J Neurochem* **82**, 585–593.
95. Kitani, T., Ishida, A., Okuno, S., Takeuchi, M., Kameshita, I., and Fujisawa, H. (1999) *J Biochem (Tokyo)* **125**, 1022–1028.

96. Schmitt, J. M., Wayman, G. A., Nozaki, N., and Soderling, T. R. (2004) *J Biol Chem* **279**, 24064–24072.

97. Kahl, C. R., and Means, A. R. (2004) *J Biol Chem* **279**, 15411–15419.

98. Ishikawa, Y., Tokumitsu, H., Inuzuka, H., Murata-Hori, M., Hosoya, H., and Kobayashi, R. (2003) *FEBS Lett* **550**, 57–63.

99. Nishimura, H., Sakagami, H., Uezu, A., Fukunaga, K., Watanabe, M., and Kondo, H. (2003) *J Neurochem* **85**, 1216–1227.

100. Ueda, T., Sakagami, H., Abe, K., Oishi, I., Maruo, A., Kondo, H., Terashima, T., Ichihashi, M., Yamamura, H., and Minami, Y. (1999) *J Neurochem* **73**, 2119–2129.

101. Sakagami, H., Kamata, A., Nishimura, H., Kasahara, J., Owada, Y., Takeuchi, Y., Watanabe, M., Fukunaga, K., and Kondo, H. (2005) *Eur J Neurosci* **22**, 2697–2707.

102. Sakagami, H., and Kondo, H. (1993) *Brain Res Mol Brain Res* **19**, 215–218.

103. Sun, Z., Means, R. L., LeMagueresse, B., and Means, A. R. (1995) *J Biol Chem* **270**, 29507–29514.

104. Strausberg, R. L., Feingold, E. A., Grouse, L. H., Derge, J. G., Klausner, R. D., Collins, F. S., Wagner, L., Shenmen, C. M., Schuler, G. D., Altschul, S. F., Zeeberg, B., Buetow, K. H., Schaefer, C. F., Bhat, N. K., Hopkins, R. F., Jordan, H., Moore, T., Max, S. I., Wang, J., Hsieh, F., Diatchenko, L., Marusina, K., Farmer, A. A., Rubin, G. M., Hong, L., Stapleton, M., Soares, M. B., Bonaldo, M. F., Casavant, T. L., Scheetz, T. E., Brownstein, M. J., Usdin, T. B., Toshiyuki, S., Carninci, P., Prange, C., Raha, S. S., Loquellano, N. A., Peters, G. J., Abramson, R. D., Mullahy, S. J., Bosak, S. A., McEwan, P. J., McKernan, K. J., Malek, J. A., Gunaratne, P. H., Richards, S., Worley, K. C., Hale, S., Garcia, A. M., Gay, L. J., Hulyk, S. W., Villalon, D. K., Muzny, D. M., Sodergren, E. J., Lu, X., Gibbs, R. A., Fahey, J., Helton, E., Ketteman, M., Madan, A., Rodrigues, S., Sanchez, A., Whiting, M., Young, A. C., Shevchenko, Y., Bouffard, G. G., Blakesley, R. W., Touchman, J. W., Green, E. D., Dickson, M. C., Rodriguez, A. C., Grimwood, J., Schmutz, J., Myers, R. M., Butterfield, Y. S., Krzywinski, M. I., Skalska, U., Smailus, D. E., Schnerch, A., Schein, J. E., Jones, S. J., and Marra, M. A. (2002) *Proc Natl Acad Sci USA* **99**, 16899–16903.

105. Means, A. R., Cruzalegui, F., LeMagueresse, B., Needleman, D. S., Slaughter, G. R., and Ono, T. (1991) *Mol Cell Biol* **11**, 3960–3971.

106. Ohmstede, C. A., Bland, M. M., Merrill, B. M., and Sahyoun, N. (1991) *Proc Natl Acad Sci USA* **88**, 5784–5788.

107. Sakagami, H., Umemiya, M., Kobayashi, T., Saito, S., and Kondo, H. (1999) *Eur J Neurosci* **11**, 2531–2536.

108. Means, A. R., Ribar, T. J., Kane, C. D., Hook, S. S., and Anderson, K. A. (1997) *Recent Prog Horm Res* **52**, 389–406; discussion 406–407.

109. Wang, S. L., Ribar, T. J., and Means, A. R. (2001) *Cell Growth Differ* **12**, 351–361.

110. Matthews, R. P., Guthrie, C. R., Wailes, L. M., Zhao, X., Means, A. R., and McKnight, G. S. (1994) *Mol Cell Biol* **14**, 6107–6116.

111. Wu, J. Y., Gonzalez-Robayna, I. J., Richards, J. S., and Means, A. R. (2000) *Endocrinology* **141**, 4777–4783.

112. Kotera, I., Sekimoto, T., Miyamoto, Y., Saiwaki, T., Nagoshi, E., Sakagami, H., Kondo, H., and Yoneda, Y. (2005) *EMBO J* **24**, 942–951.

113. Chatila, T., Anderson, K. A., Ho, N., and Means, A. R. (1996) *J Biol Chem* **271**, 21542–21548.

114. Enslen, H., Tokumitsu, H., and Soderling, T. R. (1995) *Biochem Biophys Res Commun* **207**, 1038–1043.

115. Park, I. K., and Soderling, T. R. (1995) *J Biol Chem* **270**, 30464–30469.

116. Westphal, R. S., Anderson, K. A., Means, A. R., and Wadzinski, B. E. (1998) *Science* **280**, 1258–1261.

117. Anderson, K. A., Noeldner, P. K., Reece, K., Wadzinski, B. E., and Means, A. R. (2004) *J Biol Chem* **279**, 31708–31716.

118. Nakai, J., Ohkura, M., and Imoto, K. (2001) *Nat Biotechnol* **19**, 137–141.

119. Miranti, C. K., Ginty, D. D., Huang, G., Chatila, T., and Greenberg, M. E. (1995) *Mol Cell Biol* **15**, 3672–3684.
120. Kane, C. D., and Means, A. R. (2000) *EMBO J* **19**, 691–701.
121. Blaeser, F., Ho, N., Prywes, R., and Chatila, T. A. (2000) *J Biol Chem* **275**, 197–209.
122. Sun, P., Enslen, H., Myung, P. S., and Maurer, R. A. (1994) *Genes Dev* **8**, 2527–2539.
123. Miska, E. A., Langley, E., Wolf, D., Karlsson, C., Pines, J., and Kouzarides, T. (2001) *Nucleic Acids Res* **29**, 3439–3447.
124. Zhao, X., Ito, A., Kane, C. D., Liao, T. S., Bolger, T. A., Lemrow, S. M., Means, A. R., and Yao, T. P. (2001) *J Biol Chem* **276**, 35042–35048.
125. Impey, S., Fong, A. L., Wang, Y., Cardinaux, J. R., Fass, D. M., Obrietan, K., Wayman, G. A., Storm, D. R., Soderling, T. R., and Goodman, R. H. (2002) *Neuron* **34**, 235–244.
126. Pan, F., Means, A. R., and Liu, J. O. (2005) *EMBO J* **24**, 2104–2113.
127. Parker, D., Ferreri, K., Nakajima, T., LaMorte, V. J., Evans, R., Koerber, S. C., Hoeger, C., and Montminy, M. R. (1996) *Mol Cell Biol* **16**, 694–703.
128. Soderling, T. R., Fukunaga, K., Rich, D. P., Fong, Y. L., Smith, K., and Colbran, R. J. (1990) *Adv Second Messenger Phosphoprotein Res* **24**, 206–211.
129. Liu, F. C., and Graybiel, A. M. (1996) *Neuron* **17**, 1133–1144.
130. Bito, H., Deisseroth, K., and Tsien, R. W. (1996) *Cell* **87**, 1203–1214.
131. Hardingham, G. E., Chawla, S., Cruzalegui, F. H., and Bading, H. (1999) *Neuron* **22**, 789–798.
132. Kasahara, J., Fukunaga, K., and Miyamoto, E. (2001) *J Biol Chem* **276**, 24044–24050.
133. Wu, G. Y., Deisseroth, K., and Tsien, R. W. (2001) *Proc Natl Acad Sci USA* **98**, 2808–2813.
134. Kasahara, J., Fukunaga, K., and Miyamoto, E. (2000) *J Neurosci Res* **59**, 594–600.
135. Enslen, H., Tokumitsu, H., Stork, P. J., Davis, R. J., and Soderling, T. R. (1996) *Proc Natl Acad Sci USA* **93**, 10803–10808.
136. See, V., Boutillier, A. L., Bito, H., and Loeffler, J. P. (2001) *FASEB J* **15**, 134–144.
137. Tremper-Wells, B., Mathur, A., Beaman-Hall, C. M., and Vallano, M. L. (2002) *J Neurochem* **81**, 314–324.
138. Shieh, P. B., and Ghosh, A. (1999) *J Neurobiol* **41**, 127–134.
139. Shieh, P. B., Hu, S. C., Bobb, K., Timmusk, T., and Ghosh, A. (1998) *Neuron* **20**, 727–740.
140. Yu, C. T., Shih, H. M., and Lai, M. Z. (2001) *J Immunol* **166**, 284–292.
141. Ho, N., Liauw, J. A., Blaeser, F., Wei, F., Hanissian, S., Muglia, L. M., Wozniak, D. F., Nardi, A., Arvin, K. L., Holtzman, D. M., Linden, D. J., Zhuo, M., Muglia, L. J., and Chatila, T. A. (2000) *J Neurosci* **20**, 6459–6472.
142. Anderson, K. A., and Means, A. R. (2002) *Mol Cell Biol* **22**, 23–29.
143. Ribar, T. J., Rodriguiz, R. M., Khiroug, L., Wetsel, W. C., Augustine, G. J., and Means, A. R. (2000) *J Neurosci* **20**, RC107.
144. Wei, F., Qiu, C. S., Liauw, J., Robinson, D. A., Ho, N., Chatila, T., and Zhuo, M. (2002) *Nat Neurosci* **5**, 573–579.
145. Kang, H., Sun, L. D., Atkins, C. M., Soderling, T. R., Wilson, M. A., and Tonegawa, S. (2001) *Cell* **106**, 771–783.
146. Hu, S. C., Chrivia, J., and Ghosh, A. (1999) *Neuron* **22**, 799–808.
147. Chawla, S., Hardingham, G. E., Quinn, D. R., and Bading, H. (1998) *Science* **281**, 1505–1509.
148. Anderson, K. A., and Kane, C. D. (1998) *Biometals* **11**, 331–343.
149. Sawamura, Y., Sakagami, H., and Kondo, H. (1996) *Brain Res* **706**, 259–266.

J. Krebs and M. Michalak (Editors)
Calcium: A Matter of Life or Death

ISSN: 0167-7306/DOI: 10.1016/S0167-7306(06)41014-0

CHAPTER 14

The Ca^{2+}–calcineurin–NFAT signalling pathway

Stefan Feske, Anjana Rao, and Patrick G. Hogan

Harvard Medical School and the CBR Institute for Biomedical Research, Harvard Medical School, 200 Longwood Avenue, Boston, MA 02115, USA, Tel.: +1 617 278 3260; Fax: +1 617 278 3280; E-mail: arao@cbr.med.harvard.edu

Abstract

Most if not all eukaryotic cells utilize Ca^{2+} as a second messenger but the mechanisms employed to control Ca^{2+} levels and the downstream molecules used to transmit the Ca^{2+} signal vary between different cell types and tissues. A major pathway for Ca^{2+} entry in non-excitable cells is store-operated Ca^{2+} entry (SOCE) via the opening of Ca^{2+} release activated Ca^{2+} (CRAC) channels. Subsequent elevation of intracellular Ca^{2+} levels results in the activation of a large number of calmodulin (CaM)-dependent enzymes including the serine-threonine phosphatase calcineurin. The transcription factor NFAT (nuclear factor of activated T cells) is dephosphorylated and activated by calcineurin and in turn controls expression of numerous genes. Sustained Ca^{2+} influx and NFAT activation are crucial for lymphocyte activation and many other cellular processes including protein secretion, cell metabolism, cell differentiation as well as opposite processes like activation induced cell death or T cell unresponsiveness. In this chapter we will describe the Ca^{2+}-calcineurin-NFAT pathway using the immune system, where it was first described and is best characterized, as a model.

Keywords: Anergy, AP-1 (activator protein-1), Calcineurin, Calcium, CK1 (casein kinase 1), CRAC (calcium release activated calcium) channel, Cyclosporin A, DYRK (Dual Specificity Tyrosine Phosphorylation Regulated Kinase), FK506, FOXP3 (forkhead box P3), IL-2 (interleukin-2), NFAT (nuclear factor of activated T cells), Orai1, PVIVIT, SCID (severe combined immunodeficiency), SOCE (store-operated calcium entry), STIM (stromal interaction molecule), TRP (transient receptor potential) channel

Ca^{2+} is a universal second messenger used by practically all cell types throughout evolution. Intracellular Ca^{2+} levels in resting cells are low (on the order of 50–100 nM) compared with concentrations found in the extracellular space or intracellular Ca^{2+} storage compartments, for instance, the endoplasmic reticulum (ER). The resulting concentration gradient of approximately 10 000:1 can be utilized for fast changes in intracellular Ca^{2+} concentrations ($[Ca^{2+}]_i$) through the opening of Ca^{2+} channels. Cells have developed a plethora of different mechanisms to control Ca^{2+} levels. A variety of Ca^{2+} and other ion channels, transporters and exchangers are used to increase Ca^{2+} levels; their action is counteracted by mechanisms limiting the Ca^{2+} signal such as extruding Ca^{2+} from the cell through Ca^{2+} pumps and exchangers, re-uptake into Ca^{2+} stores or absorption by Ca^{2+}-binding proteins.

The mechanisms employed to control Ca^{2+} levels may vary in different cell types. Store-operated Ca^{2+} entry (SOCE) through Ca^{2+} release-activated Ca^{2+} (CRAC) channels is arguably the most important Ca^{2+} influx pathway in cells of the immune system and will be discussed in detail in Chapter 1 (Fig. 1). In lymphocytes, CRAC channels are required for sustained Ca^{2+} influx and consequently lymphocyte proliferation and activation [1]. Increases in $[Ca^{2+}]_i$ result in activation of a large number of calmodulin (CaM)-dependent enzymes including the serine–threonine phosphatase calcineurin (CN). The main substrate of CN is the transcription factor, nuclear factor of activated T cells (NFAT), which is dephosphorylated by CN and subsequently translocates to the nucleus, where it controls expression of numerous genes. Ca^{2+} signals and activated NFAT control many cellular processes including gene transcription, protein secretion, cell metabolism and differentiation, but they also mediate opposite processes such as activation-induced cell death or T-cell unresponsiveness.

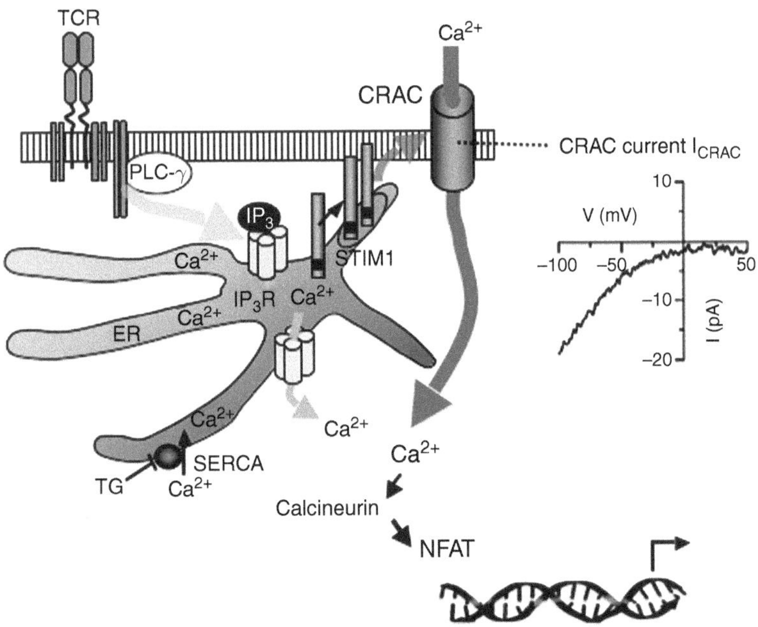

Fig. 1. Overview of the Ca^{2+}–calcineurin–nuclear factor of activated T cells (Ca^{2+}–CN–NFAT) pathway. Stimulation of T cells through the T-cell receptor (TCR) induces activation of phospholipase C-γ (PLC-γ), which generates the second messenger inositol-1,4,5-trisphosphate (IP_3), which in turn binds to the IP_3 receptor (IP_3R), releasing Ca^{2+} from endoplasmic reticulum (ER) Ca^{2+} stores. Ca^{2+} store depletion activates an inside-out signalling pathway leading to the opening of store-operated Ca^{2+} channels like the Ca^{2+} release-activated Ca^{2+} (CRAC) channel, which gives rise to well-defined CRAC currents (I_{CRAC}). The process of CRAC channel activation is thought to involve relocation of stromal interaction molecule 1 (STIM1). Ca^{2+} influx through CRAC channels results in activation of the phosphatase CN, dephosphorylation of NFAT and gene expression. TG, thapsigargin; SERCA, sarcoplasmic–ER Ca^{2+}-ATPase (See Color Plate 40, p. 537).

1. Ca^{2+} influx pathways

1.1. Ca^{2+} channels

1.1.1. Ca^{2+} channel overview

Ca^{2+} can enter cells through a variety of Ca^{2+} channels such as voltage-gated, intracellular ligand-gated (or second messenger-gated), extracellular ligand-gated (or receptor-operated) and SOC channels. Numerous Ca^{2+} entry mechanisms and channels have been studied extensively in electrically excitable tissues such as neurons, cardiomyocytes or muscle cells. Voltage-gated Ca^{2+} channels directly sense a change in membrane potential through a voltage-sensor domain integral to the channel pore forming protein resulting in opening of the channel. These channels control processes such as neurotransmission, vascular tonus and the release of Ca^{2+} from the sarcoplasmic reticulum necessary for the contraction of heart and skeletal muscles. In neurons, ionotropic glutamate receptors like the NMDA receptor represent extracellular ligand-gated ion channels permeable to Na^{+}, K^{+} and also Ca^{2+}, which bind neurotransmitters through an extracellular ligand-binding domain and ultimately modulate processes such as learning and memory. Lymphoctyes, like many other electrically nonexcitable cells (i.e. cells that do not create an action potential), use different mechanisms to regulate intracellular Ca^{2+} levels.

1.1.2. SOCE

In cells of haematopoietic lineages, several immunoreceptors such as the T-cell receptor (TCR), B-cell receptor (BCR) and receptor tyrosine kinases are able to activate phospholipase C-γ (PLC-γ), whereas chemokine receptors are coupled to heterotrimeric G proteins that activate PLC-β. Activated PLC proteins hydrolyze phosphatidylinositol-4,5-bisphosphate at the plasma membrane resulting in the synthesis of two second messengers, inositol-1,4,5-trisphosphate (IP_3) and diacylglycerol (DAG). Although DAG has a role in the activation of, for instance, some protein kinase C (PKC) family members and presumably several transient receptor potential (TRP) ion channels, IP_3 binds to IP_3 receptors (IP_3R) localized in the membrane of the ER. Binding of IP_3 to the IP_3R, which is a nonselective Ca^{2+}-permeable ion channel itself, results in the release of Ca^{2+} from ER Ca^{2+} stores (Fig. 1). In lymphocytes, Ca^{2+} stores are small, and the contribution of ER Ca^{2+} to overall increases in cytoplasmic Ca^{2+} concentrations is limited. Store depletion also triggers an 'inside out' signalling mechanism, which results in the opening of Ca^{2+} channels in the plasma membrane, a process originally termed capacitative Ca^{2+} entry but now commonly referred to as store-operated Ca^{2+} entry (SOCE) [1,5]. This process ensures prolonged increases in $[Ca^{2+}]_i$ and, therefore, facilitates the transcriptional reprogramming of cells. Although several different SOCE channels may exist, the predominant and best-characterized store-operated Ca^{2+} channel in lymphocytes is the CRAC channel. Two main mysteries prevail with regard to SOCE: the nature of the signal mediating opening of the SOC channel in the plasma membrane upon depletion of Ca^{2+} stores and the molecular nature of the SOC, and particularly the CRAC, channel itself.

1.2. CRAC channels are SOC channels

Ca^{2+} release-activated Ca^{2+} channel activation is the main source of Ca^{2+} influx in T cells, mast cells and presumably other cells of the immune system. CRAC channels are SOC channels activated by depletion of Ca^{2+} stores following cell-surface receptor stimulation.

In T cells, physiological activation of CRAC channels is achieved through TCR stimulation, which results in 'active' depletion of Ca^{2+} stores mediated by IP_3 binding to IP_3R. Passive depletion of Ca^{2+} stores by thapsigargin, a selective inhibitor of the sarcoplasmic–ER Ca^{2+}-ATPase (SERCA) pump, which transports Ca^{2+} back into ER Ca^{2+} stores, activates CRAC channels to the same degree as TCR stimulation does. This indicates that depletion of Ca^{2+} stores is necessary and sufficient for activation of CRAC channels. CRAC channels are the primary source of Ca^{2+} influx following TCR engagement, a notion which is supported by the complete absence of Ca^{2+} influx in patients with a severe T-cell immune deficiency who lack functional CRAC channels [6–9].

A CRAC current (I_{CRAC}) was first described in mast cells [10] and in T cells [11,12] as a highly Ca^{2+} selective current with unique biophysical properties. I_{CRAC} has since then been carefully defined based on these properties using, for the most part, whole-cell patch-clamp recordings, while the molecular identity of the channel remained elusive. Key biophysical hallmarks of the CRAC channel include a lack of significant voltage-dependent gating, an extremely high selectivity for Ca^{2+} over monovalent cations, extremely low single-channel conductance (<1 pS), inwardly rectifying I–V relationship, rapid Ca^{2+}-dependent inactivation, blockade by submicromolar concentrations of La^{3+} and modulation by a drug called 2-aminoethoxydiphenyl borate (for detailed reviews, see [1,13,14]).

1.3. Mechanisms regulating SOCE

1.3.1. Models of SOCE

The mechanisms, which control SOCE, have long been controversial although mobilization of Ca^{2+} from Ca^{2+} stores is undoubtedly sufficient to activate the process. The original (first) model termed capacitative Ca^{2+} entry [1,5] evolved into a conformational coupling hypothesis, postulating a store depletion-induced conformational change of the IP_3R, which would allow it to interact with and activate the SOC channel in the plasma membrane. Binding of the IP_3R to TRP channels seemed to support this model [15,16], but it is not universally accepted that TRP channels are indeed gated by store depletion [17–19]. Genetic ablation of all three IP_3R isoforms in DT40 B cells abrogated Ca^{2+} store-depletion but did not impair Ca^{2+} influx [20,21]. Alternative hypotheses explaining SOCE have been formulated. A second model proposes a diffusible Ca^{2+} influx factor (CIF), which is believed to be released upon store-depletion to activate SOC channels [22,23]. The main evidence for this model comes from experiments in which cytoplasm isolated from thapsigargin-stimulated Jurkat T cells when injected into oocytes activates Ca^{2+} influx and a CRAC-like current [24]. More recently, CIF was proposed to displace inhibitory CaM from the Ca^{2+}-independent phospholipase A_2 (PLA_2) resulting in the generation of lysophospholipids, which in turn activate the CRAC channel [23,25]. A third model suggested that upon depletion of Ca^{2+} stores, SOC channels are inserted into the plasma membrane through exocytosis of SOC-containing vesicles. Supporting this model were experiments showing that inhibition of vesicle secretion by botulinus toxin A or dominant-negative forms of synaptosomal-associated protein 25 (SNAP25) inhibited Ca^{2+} influx and SOC channel activation [26]. SNAP25 is part of soluble *N*-ethylmaleimide sensitive factor attachment protein receptor (SNARE) complexes necessary for regulated exocytosis in neuroendocrine and endocrine cells [27]. It is currently unclear how the different mechanisms proposed to regulate SOCE can be unified into one model.

1.4. The ER Ca^{2+} sensor STIM

1.4.1. Stromal interaction molecule (STIM) acts as an ER Ca^{2+} sensor and is a key regulator of SOCE

An exciting step forward in understanding the molecular mechanisms of SOCE was the discovery of stromal interaction molecule (STIM) as an essential regulator of SOC influx in RNA interference (RNAi) screens in *Drosophila* and mammalian cells [28,29]. *Drosophila* Stim and its two human homologues, STIM1 and STIM2, are single-transmembrane proteins localized in the membrane of the ER and the plasma membrane [30]. Early studies had identified STIM1 as a protein binding to and promoting survival of pre-B cells [31] and located in a region on human chromosome 11p15 that is associated with rhabdomyosarcoma development [32]. Although STIM1 was shown to homo- and heteromultimerize with itself and STIM2, and a conserved EF-hand domain located in the N-terminus of the protein was recognized early on (Fig. 2) [33], a clear functional role of STIM1 could not be established. The 're-discovery' of dStim and STIM1 in RNAi screens showed that ablation of dStim/STIM1 expression in several cell types strongly suppressed SOC influx and CRAC channel currents [28,29]. *Drosophila* S2 cells display a SOC current that is very similar to CRAC in mammalian cells [34]. It was soon recognized that the N-terminal EF-hand, located in the ER lumen, is of crucial importance for the ability of dStim/STIM1 to activate SOCE. Mutation of one or several Ca^{2+}-binding glutamate residues within the EF-hand constitutively activates SOCE and I_{CRAC} (Fig. 2). Thus, the EF-hand is thought to act as a sensor of ER Ca^{2+} levels. In support of this hypothesis, the K_D of Ca^{2+} binding to the EF-hand domain (200–600 μM) [35] is ideally matched to detect decreasing concentrations of free Ca^{2+} in the ER upon store depletion from resting levels estimated to be in the 100–800 μM range [36–39].

How exactly dSTIM/STIM1 transmit the depletion signal to the Ca^{2+} channel is not understood at this point, but upon store depletion, STIM1 undergoes a relocalization into clusters or 'puncta' [28]. A similar phenomenon is observed in cells expressing the EF-hand mutant of STIM1 [28]. Confocal and total internal reflection fluorescence (TIRF) microscopy showed that these 'puncta' are located in or near the plasma membrane suggesting that STIM1 may interact, directly or indirectly, with the CRAC channel and activate it [28,40,41]. Early studies on STIM1 have shown that a fraction of STIM1, in addition to being located in the ER, is present in the

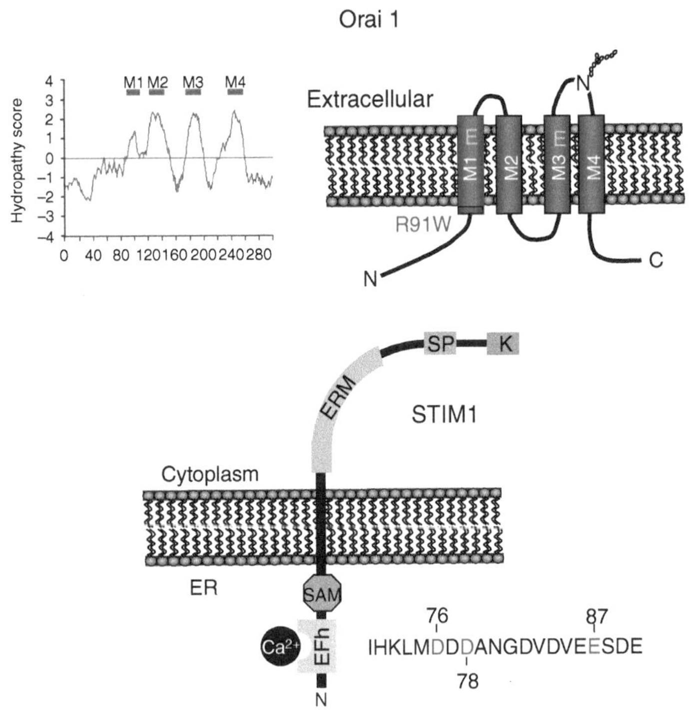

Fig. 2. Domain architecture of Orai1 and stromal interaction molecule 1 (STIM1). Hydropathy analysis of Orai1 using the Kyte–Doolittle algorithm predicts a protein with four transmembrane segments (M1–M4) [59–61]. The predicted membrane topology of Orai1 is derived from the hydropathy plot and experiments suggesting localization of Orai1 in the plasma membrane and extension of the N- and C-termini of the protein in the cytoplasm [48,60]. The site of the R91W mutation in immunodeficient patients is indicated by a red bar [60], and two conserved glutamate (E) residues important for Ca^{2+} permeability of the Ca^{2+} release-activated Ca^{2+} (CRAC) channel are shown in yellow. STIM1 is a single-pass transmembrane protein, which is predominantly localized in the endoplasmic reticulum (ER). The N-terminus contains a Ca^{2+} binding EF-hand (EFh) motif, which senses ER Ca^{2+} levels. Several negatively charged residues critical for Ca^{2+} sensing are indicated in red (D76, D78 and E87); [28,40,239]. STIM1 also contains a sterile alpha motif (SAM), a coiled coil (CC)/ezrin–radixin–moesin (ERM) domain, a serine–proline-rich region (SP) and a lysine-rich region (K), which are likely to mediate protein–protein interactions (See Color Plate 41, p. 538).

plasma membrane [30]. Although clustering of STIM1 is thought to be required for CRAC channel activation, it remains unclear whether STIM1 actually has to translocate to the plasma membrane to achieve activation of I_{CRAC} or whether clustering of STIM1 in parts of the ER juxtaposed to the plasma membrane is sufficient.

1.5. CRAC channel candidate genes

1.5.1. TRP channels

The molecular nature of the CRAC channel has long remained a mystery. Among the most thoroughly investigated candidate genes are proteins of the TRP family [42,43]. *Drosophila* TRP was first identified in the fly retina as a light-sensitive cation channel

[44,45]. More than 30 mammalian TRP proteins have been identified to date, which all share a six-transmembrane domain structure and are thought to homo- and heteromultimerize to form cation channels. Among all TRP channels, TRPV5 and TRPV6 (previously termed ECaC1/CaT2 and ECaC2/CaT1 respectively) are the most Ca^{2+}-selective channels when ectopically expressed [42,46,47]. As some other biophysical properties of TRPV6 are also similar to those of I_{CRAC} (e.g. sequence of divalent conductivity [46]), it had been suggested that TRPV6 is part of the CRAC channel pore [46]. The unitary monovalent conductance of TRPV6 (approximately 45 pS) and CRAC (<1 pS) differ significantly, however, as do their relative permeabilities to Na^+ and Cs^+: $P_{Cs}/P_{Na} \approx 0.48$ for TRPV6 and $P_{Cs}/P_{Na} \approx 0.1$ for I_{CRAC} [46–48]. Several other TRPV6 properties, also, differ from those of endogenous CRAC channels; most importantly, store-dependent activation of TRPV6 could not unambiguously be shown [47,49,50]. Furthermore, RNAi-mediated 'knock-down' of TRPV6 failed to affect CRAC channel function in mast cells [51], thus arguing against a role for TRPV6 in SOCE.

Ca^{2+} currents emanating from few other TRP channels have been demonstrated in lymphocytes, but their biophysical properties differ from those of the CRAC channel. TRPM7 (also called TRP-PLIK, MIC or MagNuM), for instance, is a Ca^{2+}- and Mg^{2+}-permeable ion channel activated by the washout of intracellular Mg^{2+} and ATP during whole-cell patch-clamp recordings [50,52,53] and was initially confused with CRAC channels because of its slow-induction kinetics in whole-cell recordings using Ca^{2+} chelators in the patch pipette [49]. TRPM7 and CRAC have distinct biophysical properties [50,54], and TRPM7 currents are normal in patients lacking functional CRAC channels [6]. The contribution of other TRP channels – for instance TRPC1 [55–57] or TRPC3 [58] – to SOCE is still controversial.

1.5.2. Orai: an essential component of the CRAC channel

As none of the TRP channels discussed above possess biophysical properties fully compatible with I_{CRAC}, it was suggested that heteromultimers of two or more TRP channels or coexpression of a TRP channel protein with a regulatory subunit could constitute the CRAC channel. More recent evidence suggests, however, that the CRAC channel belongs to an entirely different class of genes than previously expected. Several groups identified the *Drosophila* gene *olf-186F* and one of its human homologues, FLJ14466, as a molecule essential for SOCE and CRAC channel function [59–61]. At the time of their discovery by genome-wide searches for regulators of the Ca^{2+}–CN–NFAT pathway, olf-186F and FLJ14466 were hypothetical proteins without known function. Genome-wide RNAi screens in *Drosophila* cells found that depletion of olf-186F abrogated SOCE [59–61]. Similarly, RNAi-mediated depletion of the human gene product FLJ14466 strongly interfered with SOCE and I_{CRAC} [61]. FLJ14466 is located on human chromosome 12q24, a region linked to a particular form of severe combined immunodeficiency (SCID) disease characterized by the absence of functional CRAC channels [6,60]. These SCID patients are homozygous for a missense mutation in FLJ14466, resulting in an arginine-to-tryptophan amino acid exchange at position 91 (R91W) of the protein. Complementation of SCID patients' T cells with FLJ14466 reconstitutes SOC influx and I_{CRAC} with properties indistinguishable from bona fide CRAC channels. FLJ14466 was renamed Orai1 (after the 'Hours', the keepers of the gates of heaven in Greek mythology) [60], CRAC modulator 1 [61] and transmembrane protein 142A (HUGO Gene Nomenclature Committee symbol), respectively. The requirement of Orai1 for CRAC channel activity was almost instantaneously accepted, but given the fact that the protein was not related structurally to any other known ion channel, its function as either the CRAC channel itself or a regulatory molecule needed to be addressed.

Hydrophobicity analysis of the protein sequence predicted a transmembrane protein with four membrane domains and intracellular N- and C-termini. This topology, experimentally confirmed by immunocytochemical studies [48,60,61], could be expected for an ion channel activated by an intracellular gating mechanism like store depletion. The main question then remained whether Orai1 is part of the CRAC channel pore itself. Two strongly conserved glutamate residues in two of the four transmembrane domains of Orai1 are potential Ca^{2+}-binding sites analogous to binding sites in other highly selective Ca^{2+} channels, e.g. voltage-gated Ca^{2+} channels or TRPV6 [62,63]. Site-directed mutagenesis of glutamate 106 in human Orai1 and the corresponding glutamate 178 in *Drosophila* Orai and whole-cell patch-clamp recordings showed that this residue is critically important for ion conductance and Ca^{2+} selectivity of the CRAC channel [48,64]. (Note that glutamate 178 [48] in *Drosophila* Orai is identical with glutamate 180 described in another study [64].) Mutation of E106 in Orai1 and E178 in dOrai results in greatly decreased Ca^{2+} selectivity and increased Na^+ and Cs^+ conductance of the CRAC channel. Together with the plasma membrane localization and topology of the protein, Orai1 very likely is part of the CRAC channel pore. The function of two proteins structurally related to Orai1, Orai2 (encoded by TMEM142B located on chromosome 7) and Orai3 (encoded by TMEM142C located on chromosome 16) is less well understood despite some evidence that overexpression of Orai2 and Orai3 can also mediate SOCE [65] [240].

1.5.3. STIM1 and Orai1 both function in the SOCE–I_{CRAC} pathway

As discussed in chapter 1.4.1 and 1.5.2, both STIM1 and Orai1 are necessary for SOCE and CRAC channel activation. Overexpression of both dStim and dOrai in *Drosophila* S2 cells [64] or Orai1 and STIM1 in 293 cells [66,67] and RBL cells [66] resulted in a dramatic increase in CRAC currents, suggesting that the proteins are sufficient for the activation of CRAC channels or are the only limiting components of the pathway. In the simplest model derived from these studies, STIM1 would function as a direct regulator of the CRAC channel (i.e. Orai1) in the plasma membrane, but additional proteins may be involved in the regulation of SOCE.

1.6. Decoding the Ca^{2+} signal

Ca^{2+} signals activate cellular programmes as opposite as cell differentiation and apoptotic cell death. Increases in $[Ca^{2+}]_i$ can have many different causes, but the spatio-temporal pattern of the Ca^{2+} signal contains information that determines its downstream effect. This pattern in turn is influenced by the expression and distribution of proteins and cellular organelles that bind, take up or are activated by Ca^{2+} such as Ca^{2+}-binding proteins, kinases and phosphatases, Ca^{2+} channels and pumps as well as mitochondria or the ER.

A *local* increase in $[Ca^{2+}]_i$ mediated by voltage-gated Ca^{2+} channels in neurons promotes specific phosphorylation of the cAMP response element binding (CREB) transcription factor CREB [68,69]. Micromolar Ca^{2+} concentrations in 'nanodomains' around Ca^{2+} channels are 100- to 1000-fold higher than 'bulk' cytoplasmic concentrations and are thought to allow Ca^{2+} binding to Calmodulin (CaM) and subsequent activation of CaM kinase IV (CaMKIV) and mitogen-activated protein kinases (MAPK), which in turn activate CREB and CREB-dependent gene transcription [70–72]. Similar microdomains of Ca^{2+} influx around clusters of Orai1 and STIM1 proteins were recently described in lymphocytes [73].

The *duration* of the Ca^{2+} signal also determines which downstream events are triggered by Ca^{2+}. Brief activation of Ca^{2+} and calcineurin suffices for acute secretory processes such as mast cell degranulation, which do not require de novo gene transcription [74–76]. In contrast, T- and B-cell activation and differentiation are long-term events, which are strongly dependent on sustained Ca^{2+}–CN–NFAT signalling because they require several sequential rounds of immediate–early, early and late gene transcription. It was shown that interactions of T cells with antigen-pulsed B cells are accompanied by production of phosphatidylinositol-3,4,5-triphosphate and an increase in $[Ca^{2+}]_i$ levels for at least 10 h [77]. These long durations of signalling are required for maximal proliferation and cytokine production by T cells, as shown by the use of antibodies that rapidly disrupt T- to B-cell contact [77] or by addition of the cell-permeant agents cyclosporin A (CsA) or FK506 that inhibit CN activity within minutes of addition to the cells [78,79]. Many mechanisms exist to terminate or dampen the rise in $[Ca^{2+}]_i$ and thus limit the exposure of the cell to Ca^{2+} signals: Ca^{2+} from the cytoplasm is quickly taken up into the ER and other intracellular Ca^{2+} stores or is extruded from the cytoplasm by plasma membrane Ca^{2+}-ATPase pumps or bound by Ca^{2+}-sequestering proteins. How the temporal pattern of the Ca^{2+} signal shapes the transcriptional responses mediated by transcription factors like NFAT will be discussed in Section 3.4.

2. CN

2.1. Biological function in T cells

A principal mediator of calcium-initiated signalling in T cells is the protein phosphatase CN [80–82]. The critical role of CN in activation of T cells was first established through studies of the immunosuppressive compounds CsA and FK506. The complexes of CsA with the abundant cytoplasmic protein cyclophilin A or with cyclophilin C, and the complex of FK506 with the cytoplasmic protein FKBP12, were shown to bind tightly to CN and to inhibit CN activity against its peptide and protein substrates [83–86]. This inhibition correlated with the biological effects of CsA and FK506 on NFAT transcriptional signalling and interleukin-2 (IL-2) production

[84,87–89]. It was later recognized that CsA and FK506 block induction of a battery of cytokine genes and other activation markers (references in [80]) and that there is extensive overlap in T cells between the set of genes activated by Ca^{2+} mobilization and the set of genes activated by CN [8].

2.2. Structure, activation and catalysis

Calcineurin in resting cells is an AB heterodimer (Fig. 3) [8,90–93]. On activation, CaM binds as a third tightly associated subunit. X-ray crystallography has yielded a model delineating the CNA catalytic domain, the interaction of CNA with CNB and the positioning of the autoinhibitory peptide of CNA over the catalytic site [94]. Important structural questions remain, however, because a large section of the regulatory region of CNA was apparently unstructured in the crystals and because CaM has not been cocrystallized with intact CN.

The key to activation of the phosphatase is binding of Ca^{2+} to CNB and to CaM [95–97]. Each of these proteins is composed largely of four tandem EF-hands (Fig. 4). The EF-hand is a widely used Ca^{2+}-binding motif, usually occurring as pairs of EF-hands, that changes conformation on Ca^{2+} binding to expose a surface that interacts with a protein target [98–101]. Two of the four Ca^{2+}-binding sites of CNB (sites III and IV) are occupied even at the very low Ca^{2+} concentrations of resting cells [97,102], and presumably for this reason, a part of the interaction surface of CNB is exposed, and CNB is constitutively associated with CNA. All four Ca^{2+}-binding sites of free CaM are, for practical purposes, unoccupied at the Ca^{2+} concentrations prevailing in resting cells [103,104].

Two sites in CNB (sites I and II), with K_d around 1 μM, are occupied or partially occupied by Ca^{2+} only when cytoplasmic Ca^{2+} is elevated by physiological signalling [97,102]. Presumably, the major conformational rearrangement induced by Ca signalling is in sites I and II and their associated binding surfaces in the N-terminal half of CNB.

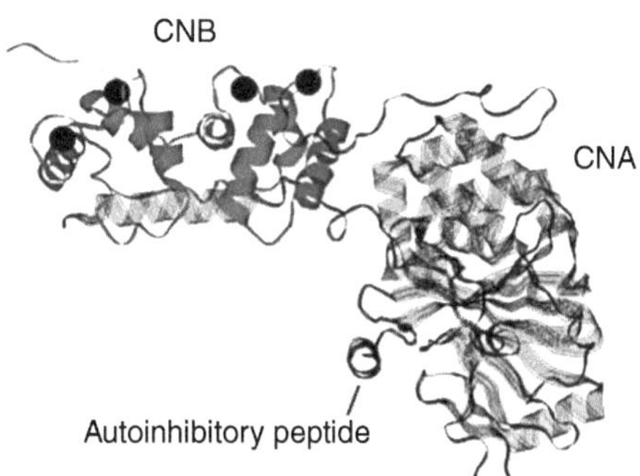

Fig. 3. Overview of the human calcineurin (CN) heterodimer from PDB entry 1AUI. The individual subunits CNAα (approximately 59 kDa) and CNB (approximately 19 kDa) are indicated. The part of the CNAα regulatory region from residues 374 to 521 is not visible in the model, except for the autoinhibitory peptide CNAα (469–486). Four Ca^{2+} ions bound to CNB are depicted as dark spheres. Figs 3, 4, 6 were prepared using RasMol.

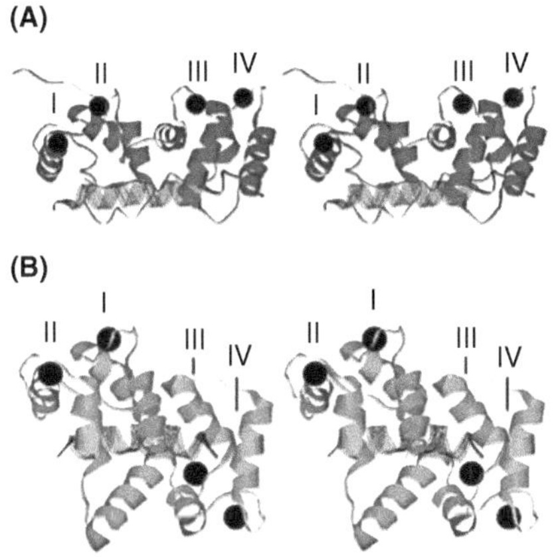

Fig. 4. Stereo views of the EF-hand proteins calcineurin B (CNB) and calmodulin (CaM) bound to specific target peptides. (**A**) CNB draped over the CNB-binding helix of CNA. The four Ca^{2+} sites are numbered, and the successive EF-hands are clearly recognizable. The orientation is the same as in Fig. 3. (**B**) CaM interacting with a helical peptide from chicken smooth muscle myosin light chain kinase, as an approximation of the interaction with the corresponding segment of CNA. Again, the four Ca^{2+} sites are numbered. Coordinates are from PDB entry 1CDL, chains A and E.

However, the nature of this rearrangement and the conformation of CNB in resting cells are unknown, because the crystal structure represents the conformation with all four Ca^{2+}-binding sites occupied. Significant Ca^{2+} binding to free CaM requires elevated cytoplasmic Ca^{2+}, in the low micromolar range [97,103,104], but the subsequent interaction with CN or other target protein stabilizes the binding of Ca^{2+} [97,105].

Biochemical studies [79,91,97,102,106–108] indicate that the probable sequence of steps in CN activation is (i) binding of Ca^{2+} to the two moderate-affinity sites in CNB, (ii) triggering a conformational change in the regulatory region of CNA that exposes the CaM-binding segment, (iii) binding of CaM and (iv) a further conformational change that removes the CN autoinhibitory peptide from the active site (Fig. 5).

The enzymatic function of active CN is hydrolysis of serine or threonine phosphoester bonds. Catalysis depends on bound Fe^{2+} and Zn^{2+} at the CN active site and on a set of active site residues conserved between CN and other phosphatases in its family (reviewed in [92]). Proposed mechanisms and the underlying evidence have been discussed for CN and the related phosphatases PP1 and bacteriophage λ protein phosphatase [92,109–111]. There is agreement that hydrolysis involves an enzyme-assisted direct attack by water on the phosphorus, displacing the protein serine or threonine leaving group, without formation of a phosphoenzyme intermediate. Typical K_m values for protein substrates are 2–20 μM, and the maximum catalytic rate of native CN is 40 s^{-1} [91].

2.3. CN-substrate targeting

Although CN has many substrates, its involvement with the transcription factor NFAT is of special interest in T cells [80,81]. The next few paragraphs cover aspects

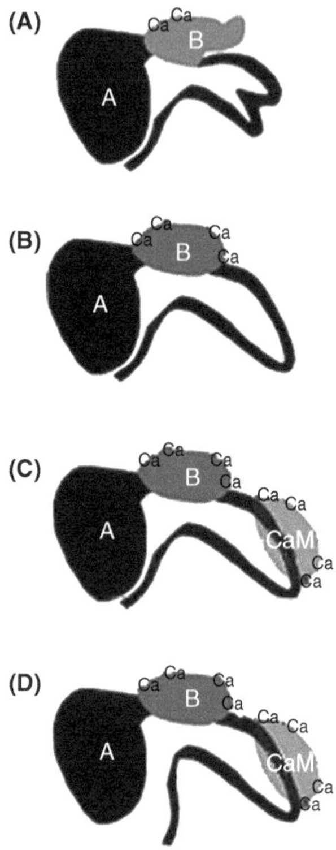

Fig. 5. Steps in the activation of calcineurin (CN). (**A**) CNA and CNB are associated at rest. Ca^{2+} ions are bound to the two high-affinity sites of CNB, and the calmodulin (CaM)-binding segment is unavailable for interaction with CaM. (**B**) Ca^{2+} binds to the low-affinity sites of CNB, promoting a conformational change in the regulatory region of CNA that exposes the CaM-binding peptide. (**C**) Ca^{2+}–CaM associates with CN. (**D**) A further conformational change displaces the autoinhibitory peptide from the catalytic site.

of CN–NFAT signalling primarily related to CN or its enzymatic activity. NFAT itself is discussed in detail below.

Calcineurin is directed to NFAT as a substrate by the conserved sequence PxIxIT in NFAT proteins [112–115]. T > A substitution in this recognition sequence in NFAT1 effectively abrogates CN–NFAT signalling in T cells [112]. Conversely, substitution of the higher affinity PVIVIT sequence into NFAT1 results in a protein that is largely dephosphorylated even in resting cells [116]. The dephosphorylation still depends on CN, because it is blocked by CsA. Thus, effective recognition and signalling in the CN–NFAT pathway is not synonymous with high-affinity recognition. In fact, this example may hold a lesson for the study of many transient docking interactions that characterize intracellular signalling.

The NFAT docking site on CN has been mapped by a combination of protein–peptide crosslinking, in silico docking and introduction of complementary mutations and, in finer detail, by x-ray crystallography of a CN–PVIVIT peptide complex (Fig. 6) [117,118]. The core PxIxIT motif binds as a β strand alongside β strand 14 of CNA,

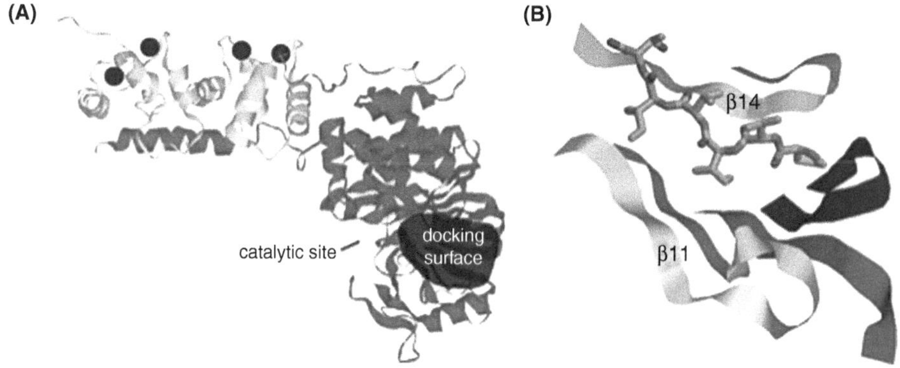

Fig. 6. Complex of PVIVIT peptide with calcineurin Aα (CNAα). (**A**) The docking surface mapped experimentally is distinct from the catalytic site of CN. (**B**) The PxIxIT recognition sequence aligns in a parallel β-sheet interaction with β-strand 14 of CNA. Coordinates are from reference [117].

extending a β sheet of CN, with each of the conserved PxIxIT residues making close contact with a complementary portion of the CN surface. Mutations of the CN surface in contact with the peptide, like mutations of the PxIxIT peptide itself, interfere with the ability of CN to dephosphorylate NFAT [117].

Calcineurin long predates the evolutionary appearance of NFAT proteins, and the substrate docking site utilized by NFAT is conserved throughout the evolution of CN. Thus, it is not surprising that the same docking site is utilized by some other substrates in yeast and mammals [119–122]. However, it is clearly established that not all CN substrates use the site. It is of particular interest that the CN–NFκB pathway in T cells and the CN–MEF2 pathway in skeletal muscle are not affected by blockade with PVIVIT peptide, which completely prevents CN–NFAT signalling [116,123]. This finding may imply that other interactions targeting CN to specific substrates remain to be discovered.

2.4. Relating biochemistry to cellular signalling

The K_d for CN binding to SPRIEIT peptide, the native recognition sequence of NFAT1, is approximately 20 μM. The K_d for CN–NFAT binding has not been measured directly, but the biochemical observations are consistent with a value in the low micromolar range. This weak interaction is reminiscent of the reported interactions of CN with other protein substrates [91]. There is no indication that NFAT in cells exists in a stable complex with CN. Nevertheless, under physiological conditions, CN has access to all the NFAT in a T or B cell in the course of 1–2 min, because a strong Ca^{2+} mobilization results in dephosphorylation of the entire NFAT pool in this time [124,125]. A similar situation has been described for the related phosphatase PP1 and its glycogen particle substrates [126].

Calcineurin is kinetically competent to carry out the observed rapid dephosphorylation of NFAT in cells, even assuming that the protein–protein binding is not tighter than CN–SPRIEIT peptide binding. With a simple Michaelis–Menten model that sets K_m to 20 μM, CN turnover number to 40 s^{-1}, and taking a plausible CN concentration between 10^{-7} and 10^{-6} M and NFAT concentration 1 μM or less, the time constant for the dephosphorylation reaction ranges from 0.5–5 s, varying inversely with the concentration of CN. This oversimplified calculation ignores the possibilities, though, that CN is sequestered by other substrates or by regulatory proteins, which would decrease the rate, or that some phosphorylated sites on NFAT are not readily accessible to CN at the outset of the reaction. In fact, the experimental findings, cited below, are that CN availability does limit signalling in T cells.

Another factor ignored in the simple Michaelis–Menten analysis is that activation of NFAT does not depend solely on its dephosphorylation by CN. The net dephosphorylation of NFAT in the face of competing phosphorylation or rephosphorylation by kinases is the true determinant of activation. Mathematical modelling provides some help in appreciating the interplay between CN and competing kinases and provides a possible explanation for the dependence of NFAT activation on the amplitude and pattern of Ca^{2+} oscillations [127]. NFAT kinases are discussed further in section 3.2.2.

It is an open question how the enzymatic hydrolysis of more than a dozen serine phosphates in NFAT1 [128], and comparable numbers in other NFAT proteins, proceeds. Docking at the PxIxIT-binding site is a prerequisite for any dephosphorylation of NFAT1 under physiological conditions [112,116], but a continuing requirement for docking throughout the process of dephosphorylation has not been demonstrated. Thus, there are two possibilities: that docking is essential only for an initial dephosphorylation, e.g. of the SRR1 region of NFAT described below, after which the completion of dephosphorylation is not limited by a need for docking, or that docking, in a single encounter or sequential encounters of NFAT with active CN, affords sufficient dwell time for serine phosphates to be presented successively to the catalytic site.

The utilization of additional docking sites to facilitate later steps in the dephosphorylation has not been excluded. A proposed second docking site is the LxVP peptide, a CN-interacting peptide identified in NFAT2 and NFAT4 that is only partially conserved in NFAT1 [129–132]. The case for its involvement in dephosphorylation is weakened by the finding that NFAT4(1-351), a fragment lacking the LxVP site, is activated in cells with normal kinetics [115]. Furthermore, synthetic LxVP peptide blocks dephosphorylation of RII phosphopeptide [132], a CN substrate that does not require docking. It is conceivable that the observed blockade of CN–NFAT signalling in cells by LxVP peptide reflects a partial or complete inhibition of CN enzyme activity, independent of any effect on docking. Finally, the contribution of the LxVP site to protein–protein interaction has been analysed only with bacterially expressed fragments of NFAT [129–132]. There is evidence that bacterially expressed NFAT is unstructured [132], whereas NFAT in mammalian cells assumes detectably different conformations in unstimulated and stimulated cells [128]. The question will be resolved only by examining whether substitutions in the LxVP site, in the context of the full-length protein, alter NFAT activation in mammalian cells.

The sensitivity of CN–NFAT transcriptional signalling to the kinetics and pattern of Ca^{2+} signals has been noted above. Ca^{2+} binding to CaM is rapid compared with alterations of cytoplasmic Ca^{2+} in T cells [105], the rate of CaM–CN interaction measured with the purified proteins [133] indicates that this step will be rapid unless the concentration of free CaM is severely limited, and CaM dissociates rapidly when the free Ca^{2+} concentration falls [133,134]. Thus, the earliest steps where it is likely that decoding of a pattern of Ca^{2+} signals could take place are the movements of the regulatory domain and the autoinhibitory peptide of CN, which have not been examined directly. It is perhaps more likely that decoding occurs in the dephosphorylation and rephosphorylation of NFAT or at later steps involved in nuclear import and export. This issue will have to be examined experimentally.

2.5. *Inhibitors and regulators of CN–NFAT signalling*

Several endogenous protein inhibitors or regulators of CN have been identified. The RCANs, or regulators of CN [also termed calcipressins, MCIPs or Down syndrome critical region 1 (DSCR1) and DSCR1L proteins], were first identified as inhibitors [135–138]. It is now evident that some of these proteins have a dual function as activators and inhibitors [139] and that their binding to and functional effect on CN may be regulated by kinases that phosphorylate RCANs [140,141]. Mice lacking one or more of these proteins can display either enhanced CN–NFAT signalling or reduced CN–NFAT signalling [142–144]. DSCR1 is encoded in the genomic region implicated in Down syndrome, and it has been suggested that many features of Down syndrome may arise from a *DSCR1* gene dosage effect [145,146]. Furthermore, expression of some *RCAN* transcripts is upregulated by CN–NFAT signalling [107,108,136,145], adding a further layer of complexity to the effects on CN pathways. Another endogenous protein, Cabin/cain, can inhibit CN and is sometimes used experimentally as a blocker of CN signalling [147,148]. Its primary physiological role is unclear. Targeting proteins, such as AKAP79/150 and calsarcins, are critical to organizing CN signalling in some other cells [149–153] but are not known to have effects on CN–NFAT signalling in T cells.

The viral protein A238L, from African swine fever virus, inhibits CN and, independently, NFκB signalling [154–156]. These are two of several molecular strategies employed by the virus to limit the immune response to infection [157].

Inhibition by the immunosuppressive compounds CsA and FK506, as explained above, results from binding of CsA–cyclophlin A or FK506–FKBP12 complexes to CN. Structural models of CN–FK506–FKBP12 and CN–CsA–cyclophilin A complexes, derived from x-ray crystallography, show a partial occlusion of the CN catalytic site [94,158,159]. These structures largely explain how the complexes can inhibit CN activity against protein substrates without inhibiting activity against the small substrate pNPP [85,86,160]. The paradoxical ability of some, but not all, such complexes to augment activity against pNPP has been tentatively attributed to their causing subtle alterations in the geometry of the active site [160]. Two remaining unanswered questions are whether binding of the inhibitory complexes also causes a reorganization of the regulatory region of CNA and whether this contributes to their physiological effects.

The docking inhibitors, PVIVIT peptide and INCA compounds, demonstrate that effective inhibition of the CN–NFAT pathway can be achieved by preventing CN-substrate recognition [114,116,123,161–163]. Their action is clearly independent of an effect on the catalytic site and thus differs from the action of CsA and FK506. Inhibition of CN–NFAT docking could have fewer undesired side effects than broad inhibition of enzymatic activity, because pathways such as CN–NFκB and CN–MEF2 that are sensitive to CsA are unaffected by PVIVIT peptide [116,123]. The set of substrates that depend on docking at the PxIxIT-binding site, and that therefore will be sensitive to docking inhibitors, remains to be fully defined.

Calcineurin is at limiting levels in T cells: an increase or a decrease in CN activity is reflected directly in increased or decreased physiological signalling [84,88,89, 164–168]. This fact has practical implications for therapeutic inhibition by CsA or FK506. In fact, circulating levels of CsA that are clinically effective in patients produce only partial inhibition of CN in T cells [165] and by implication even less inhibition in cells where CN is expressed at higher concentrations. The same advantage of blocking signalling in T cells with relatively low levels of inhibitor would be expected to hold for PVIVIT and other docking site inhibitors, in addition to their possible advantage of increased specificity.

3. *NFAT*

3.1. *General introduction*

Nuclear factor of activated T cells denotes a small family of four Ca^{2+}-regulated transcription factors (NFAT1-4, also known as NFATc1-c4) that, contrary to their name, are expressed in almost all vertebrate cell types. NFAT has a critical role in many developmental programmes, including T-cell differentiation, osteoclast differentiation, cardiac valve development, differentiation of slow-twitch skeletal muscle fibres, and neuronal and vascular patterning (reviewed in [2,3,80,169,170]). As a result, NFAT is implicated in many pathological processes, among them osteoporosis, myocardial hypertrophy, transplant rejection and autoimmune/inflammatory disorders including asthma, rheumatoid arthritis, type I diabetes and multiple sclerosis [2,3,80,169,170].

3.2. *Regulation of NFAT activation*

3.2.1. *The NFAT regulatory domain*

Unlike many other transcription factors that are activated through phosphorylation, NFAT proteins are activated by dephosphorylation. In resting cells, NFAT proteins are heavily phosphorylated on multiple serine residues and reside in the cytoplasm; upon stimulation, they are dephosphorylated by CN and translocate to the nucleus [2,3,80,128,169–171]. A majority of the phosphoserine residues known to be involved in NFAT activation are present in the NFAT regulatory domain, a conserved 300 amino acid region located immediately N-terminal to the DNA-binding domain (Fig. 7). Studies using GFP fusion proteins show that the regulatory domain is both necessary and sufficient for regulated nuclear transport of NFAT [128,172].

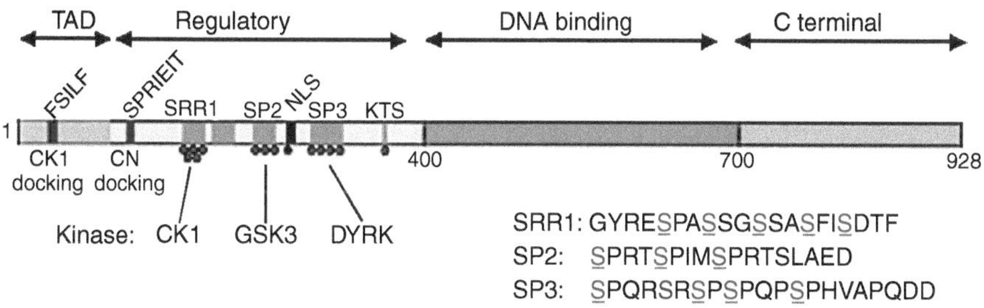

Fig. 7. Schematic structure of the nuclear factor of activated T cells 1 (NFAT1) regulatory domain. Conserved phosphorylation sites are indicated by red circles and serine-rich motifs (SRR1, SP1, SP2, SP3 and KTS) by cyan bars. The black circle represents a conserved phosphorylation site near the nuclear localization sequence (NLS) that is not dephosphorylated by calcineurin (CN). Kinases (CK1, GSK3 and DYRK for the SRR1, SP2 and SP3 motifs, respectively) are listed below. At lower right are the sequences of the SRR1, SP2 and SP3 motifs, with the phosphorylated serines shown in red (See Color Plate 42, p. 538).

The regulatory domain is encoded in a single exon in all four NFAT proteins. It bears a large number of phosphates distributed among several classes of conserved serine-rich sequence motifs. In NFAT1, the family member that is the most abundant in resting T cells and so has been most thoroughly studied, mass spectrometric analysis has identified five conserved sequence motifs – SRR1, SP2, SP3, SRR2 and KTS – which together are stoichiometrically phosphorylated at 13 phosphoserine residues in resting cells (Fig. 7) [128]. Only particular serine residues in each motif are phosphorylated (Fig. 7) [6]. CN dephosphorylates four of the five motifs (red circles in Color Plate 40), targeting 12 of the 13 phosphoserine residues in the NFAT regulatory domain, thus triggering NFAT nuclear accumulation and increasing the affinity of NFAT for its target sites in DNA [125,128,173,174]; a phosphoserine residue immediately N-terminal to the nuclear localization signal (NLS) is apparently spared [128]. A handful of other phosphoserine residues has also been identified, both in the C-terminus of the proteins and in the regulatory domain; these residues are not conserved among the four NFAT family members and tend to show substoichiometric phosphorylation and, therefore, have not yet been investigated in any detail. There are no phosphorylated sites in the DNA-binding domain under resting conditions [128].

Analysis of serine-to-alanine mutations in NFAT1 has provided clues to the functions of the phosphorylated motifs. The N-terminal SRR1 motif is a critical 'gatekeeper' region whose dephosphorylation controls exposure of an NLS in the regulatory domain (Fig. 7). Dephosphorylation of the SRR1 motif may also control the accessibility of the SP2 and SP3 motifs to CN: NFAT mutants with S > A substitutions or small deletions in the SRR1 region are more susceptible than wild-type NFAT to dephosphorylation of the SP2 and SP3 motifs by CN [115,116]. In contrast, the SP2 and SP3 motifs may control DNA-binding affinity, because NFAT mutants with S > A substitutions in these two motifs show increased affinity for DNA [173]. Fully phosphorylated NFAT binds the exportin Crm 1 [128,175], either because it exposes an intrinsic nuclear export sequence (NES) [128,171] or because it binds a partner that bears an NES [176].

3.2.2. NFAT kinases

The activation status of NFAT is controlled by the balance between CN and NFAT kinases. The kinases may be classified as 'maintenance' kinases that act in the cytoplasm of resting cells to keep NFAT in its phosphorylated state and 'export' kinases that rephosphorylate NFAT in the nucleus and promote its nuclear export [177]. The 13 serine residues that control NFAT1 nuclear localization are located in diverse sequence contexts that are unlikely to be recognized by a single kinase (Fig. 7), implying strongly that multiple kinases act in concert to maintain the phosphorylation status of NFAT. Indeed, it is now clear that three different constitutive kinases, DYRK, CK1 and GSK3, are required to inactivate NFAT by rephosphorylating 12 serine residues in the regulatory domain that are dephosphorylated by CN (Fig. 7) [128,177,178]. An integrated picture of NFAT phosphorylation has emerged in which these three kinases cooperate to deactivate NFAT and promote its nuclear export [178]. CK1 phosphorylates five equally spaced serine residues in the SRR1 motif, in a manner consistent with its known sequence preference for acidic or phosphorylated residues at position –3 (Fig. 7). DYRK phosphorylates the SP3 motif, thereby priming for GSK3-mediated phosphorylation of the SP2 motif.

Like the NFAT–CN interaction discussed in section 2.3, the interaction between NFAT and CK1 is mediated through a conserved docking site located near the N-terminus of the protein. The motif has the consensus docking sequence FxxxF (FSILF in NFAT1 and FEFLF in NFAT2) [177]. Resting cells contain a high-molecular weight complex of CK1 and NFAT that dissociates upon activation [177]. Mutation of the FxxxF motif (ASILA in NFAT1) eliminates the interaction between NFAT and CK1 in vitro, resulting in aberrant nuclear translocation of NFAT1 in resting cells [177].

The ability to mutate the CN and CK1 docking sites independently, to high and low affinity, respectively, means that it should theoretically be possible to generate hyperactivable versions of NFAT that are more readily dephosphorylated by CN and less readily rephosphorylated by CK1. These hyperactivable proteins would be expected to be more susceptible to activation at low input levels of stimulus, but the activation should remain reversible when the stimulus is withdrawn. In this behaviour, the hyperactivable NFATs would be expected to differ from the constitutively active NFATs used previously [128], in which the phosphorylated serine residues were mutated to alanines. The constitutively active proteins are constitutively localized to the nucleus and constitutively and irreversibly activated; they tend to be expressed at very low levels in cells, presumably because constitutive NFAT activation has deleterious consequences for the expressing cells. Nevertheless, a mouse in which constitutively active NFAT2 was expressed in osteoblasts showed a striking phenotype of increased bone density, even though expression of the constitutively active protein was so low as to be undetectable in standard cell lysates [179]. It will be interesting to use hyperactivable NFATs in similar experiments.

3.3. Nuclear import and export

The phosphorylation status of the NFAT regulatory domain determines whether the NLSs or NESs are exposed [125,128,171,173,174]. There are two potential NLSs in NFAT proteins, located in the regulatory and the DNA-binding domains, respectively, with the core of each comprising a short stretch of basic amino acids. Mutational analysis shows that the regulatory domain NLS has the major role in NFAT1 [180], whereas both the regulatory domain and the DNA-binding domain NLSs are functional in NFAT2 [181]. The identity of the NES is less clear: an intrinsic NES has been identified in NFAT2 [175], whereas for NFAT3, NES function has been suggested to be conferred by proteins such as 14.3.3 that bind to the phosphorylated form [176]. However, the putative NES of 14.3.3 functions in ligand binding rather than directly mediating nuclear transport [182], thus further analysis is needed.

An interesting question is why dephosphorylation of so many sites is needed for full activation of NFAT. One suggestion, which has not yet been confirmed at a structural level, is that different conserved phosphorylated motifs on NFAT interact with different basic regions on NFAT itself or on partner proteins and that dephosphorylation interrupts these interactions and causes a global conformational switch from an inactive to an active configuration [128,171,173,174]. In a mathematical model of the NFAT conformational switch, Salazar and Höfer [127] calculated how the sensitivity of NFAT to changes in CN activity would be affected by the number of phosphoserine residues that needed to be dephosphorylated during activation. If dephosphorylation of 12 residues was assumed to be required to induce the active conformation, as is actually observed [128], a steep and highly cooperative dose-response curve was obtained, in which low CN activity induced almost no NFAT activation, whereas CN activity above a sharp threshold led to efficient NFAT activation. In contrast, if the model postulated that dephosphorylation of only 1–2 sites was sufficient for full activation, a very shallow dose-response curve with no threshold effect was obtained: i.e., even a large increase in CN activity was capable of inducing only a small increase in the activation status of NFAT [127].

Recently, there have been hints that the nuclear import of NFAT is regulated by a cytoplasmic scaffold complex that may contain a variety of novel components, both RNA and protein. Endogenous NFAT1 was shown to exist in a high-molecular-weight complex that contains CK1, but not GSK3 or CN [177]. Upon stimulation, CK1 dissociates from this complex, although the overall size of the complex is only slightly diminished [177]. In a quest for the biological functions of noncoding RNAs conserved between humans and mice, a noncoding RNA termed noncoding (RNA) repressor of NFAT (NRON) was identified as a repressor of NFAT nuclear import [183]. RNAi-mediated depletion of NRON resulted in an increase of basal and stimulated levels of nuclear NFAT–GFP and an increase in transcriptional activity of endogenous NFAT as measured in reporter assays [183]. NRON may serve as the core of an RNA–protein complex that brings together multiple regulators of NFAT: it binds the scaffold protein IQGAP as well as several proteins involved in nuclear transport [183]. In turn, IQGAP binds a variety of cytoskeletal proteins, Rho-family GTPases and CaM [184] and could potentially provide the missing bridge between NFAT and its upstream regulators such as CN and NFAT kinases.

3.4. Kinetics and pattern of Ca^{2+} signals regulating NFAT activation

Because of its strong dependence on rises in $[Ca^{2+}]_i$ and consecutive dephosphorylation by CN, NFAT is a sensitive indicator of dynamic changes in $[Ca^{2+}]_i$ levels. NFAT-based reporter systems are frequently used to assess Ca^{2+}-mediated signalling processes or, more recently, to identify modulators of the Ca^{2+}–CN–NFAT pathway in large-scale screens [60,178]. NFAT is only poorly activated in response to a single pulse of high $[Ca^{2+}]_i$ but requires prolonged elevations of $[Ca^{2+}]_i$ with relatively low $[Ca^{2+}]_i$ being sufficient to achieve dephosphorylation and nuclear import [124]. There are fundamental differences in how the amplitude and duration of $[Ca^{2+}]_i$ changes elicit optimal activation of NFAT versus other signalling pathways such as NFκB and JNK1/ATF2 [124]. Transient high Ca^{2+} spikes evoked sustained activation of NFκB and JNK1/ATF2 but not of NFAT, whereas conversely, NFAT activation could be elicited by prolonged low increases in $[Ca^{2+}]_i$ that were not sufficient to induce NFκB or JNK1/ATF2 [124]. Another level of regulation exists in the transcriptional response to oscillations of $[Ca^{2+}]_i$: low-frequency oscillations induced NFκB alone, whereas high-frequency oscillations activated both NFAT and NFκB [185]. In addition, oscillations enhanced signalling efficiency specifically at low levels of stimulation. The differential responsiveness of NFAT versus NFκB activation to $[Ca^{2+}]_i$ changes most likely reflects the fact that CN is quickly and reversibly activated and deactivated in response to changes in $[Ca^{2+}]_i$, whereas NFκB activation requires irreversible degradation of IκB and its inactivation the resynthesis of IκB [186].

3.5. Signal integration in the immune response: NFAT and its transcriptional partners on DNA

NFAT is a member of the NFκB/Rel family of transcription factors [3,187]. The DNA-binding regions of all members of this family comprise about 300 amino acids and consist of two separate domains, both of which contact DNA. However, only the N-terminal domain makes base-specific contacts with DNA, through a conserved loop whose amino acid sequence is RxRYxCEG in NFκB and RAHYETEG in NFAT [3,187]. The C-terminal domain is also involved in dimer formation both in NFAT and in NFκB. Although NFκB/Rel proteins are obligate dimers in solution and on DNA, NFAT can bind DNA either as a monomer or as a dimer [188,189]. This flexibility is largely due to the presence of an unstructured linker region between the N- and C-terminal DNA-binding domains of NFAT, which allows the two domains to adopt widely different relative orientations, as illustrated by the very diverse conformations attained by an NFAT monomer on DNA [190]. In part because of this lack of structural constraint, NFAT possesses an unusual ability to pair with diverse classes of partner proteins on DNA.

Complexes of NFAT with other transcription factors represent a remarkably efficient means of integrating signals through distinct signalling pathways: the Ca^{2+}–CN pathway that activates NFAT and other intracellular pathways that

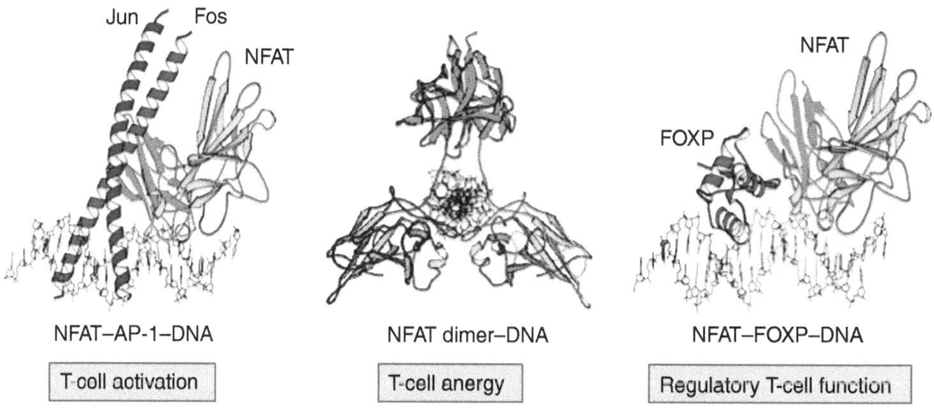

Fig. 8. Nuclear factor of activated T cells (NFAT) can bind DNA in multiple configurations, each eliciting a distinct transcriptional programme characteristic of a distinct cellular response. Left, NFAT–DNA binding in cooperation with AP-1 (Fos-Jun) activates high-level transcription of cytokine and chemokine genes leading to productive T-cell immune responses; middle, DNA binding by NFAT in the absence of AP-1 induces T-cell anergy, in part through formation of NFAT dimers on DNA; and right, NFAT binding to NFAT–AP-1 composite DNA elements with a different transcriptional partner, FOXP3, induces 'dominant' peripheral tolerance by promoting the suppressive function of regulatory T cells (See Color Plate 43, p. 539).

activate specific but unrelated transcriptional partners of NFAT. There is evidence in T cells that different types of NFAT–DNA complexes turn on different transcriptional programmes with very distinct biological consequences (Fig. 8).

3.5.1. NFAT–AP-1 complexes mediate T-cell activation

The best-characterized partners of NFAT are AP-1 (Fos-Jun) transcription factors, which are expressed in essentially all cell types and bind DNA as Jun dimers and Fos-Jun heterodimers. NFAT and AP-1 bind cooperatively to form strong complexes on 'composite' NFAT–AP-1 DNA elements present in gene regulatory regions [3,80,191]. Because Ca^{2+}–CN signalling activates NFAT, whereas signalling through Ras/MAPK pathways results in de novo synthesis and post-transcriptional activation of AP-1 proteins, formation of cooperative NFAT–AP-1 complexes in the nucleus informs the cell that both signalling pathways have been successfully activated. This results in transcriptional activation of a large number of genes associated with the productive immune response, among them the many cytokines and chemokines produced by activated T cells, NK cells and mast cells that have been stimulated through their antigen and Fc receptors (immunoreceptors) [3,80,82,169]. Composite elements capable of binding NFAT–AP-1 complexes are found in the regulatory regions of many of these genes, and the role of the Ca^{2+}–CN–NFAT pathway in their induction has been validated in several ways: through investigation of the pronounced cytokine defect observed in T cells from the SCID patients with a primary defect in CRAC channel function, through the use of the CN inhibitors CsA and FK506 and finally by showing by chromatin immunoprecipitation that NFAT actually binds to the regulatory regions in intact cells [3,80,82,169].

3.5.2. NFAT–FOXP3 complexes control the function of regulatory T cells
The obverse of immune activation and effector function is immune tolerance, a system of checks and balances aimed at preventing autoimmune destruction of an organism's own cells and tissues. One mechanism to achieve immune tolerance is represented by regulatory T cells that suppress the effector functions of self-reactive T cells [192–194]. Various types of regulatory T cells, with potentially overlapping properties, have been described [195–201]. $CD4^+$ $CD25^+$ T-regulatory cells express the forkhead transcription factor FOXP3, which controls differentiation of thymic-derived T-regulatory cells (for a review, see [202]). It was recently shown that NFAT–FOXP3 complexes control the function of regulatory T cells [41] by forming DNA-bound complexes that are analogous to the NFAT–AP-1 complexes that control induction of activation-associated genes (Fig. 8). Reporter assays showed that FOXP3 represses transcription driven by the IL-2 promoter and that this repressive activity is specifically targeted to NFAT–AP-1 composite sites. Based on the crystal structure of an NFAT–FOXP2–DNA complex, structure-guided mutations were made in FOXP3, which were predicted to disrupt progressively its interaction with NFAT [41]. These mutations interfered in a graded manner with the ability of FOXP3 to repress IL-2 expression, upregulate CTLA4 and CD25 expression and confer suppressor function in the murine model of autoimmune diabetes. The results indicate that by switching transcriptional partners, NFAT converts the acute T-cell activation programme into the suppressive programme of regulatory T cells.

4. *Biology of the Ca^{2+}–CN–NFAT pathway*

4.1. *Gene expression controlled by the Ca^{2+}–CN–NFAT pathway*

Gene expression in many cell types is controlled by the Ca^{2+}–CN–NFAT pathway. In lymphocytes, transcription of cytokine genes such as *IL2, IL4, IL10, IFNG* or *TNF* is tightly regulated [80,203]. To assess the contributions of Ca^{2+} signals to global gene expression in T cells, researchers have used DNA microarrays and found that hundreds of genes are expressed or repressed following stimulation of human or murine T cells with staphylococcal enterotoxin B, a superantigen [204], anti-CD3 + PMA [205] or PMA + the Ca^{2+} ionophore ionomycin [8,206]. In addition, lack of Ca^{2+} signals following stimulation of CRAC deficient T cells resulted in drastically altered transcriptional profiles with reduced expression of approximately 60% and enhanced expression of approximately 40% of all activation-dependent genes [6,8]. These findings suggest that Ca^{2+} signals are crucially important for gene expression and have both activating and repressive effects on gene transcription.

Several studies have evaluated the contribution of CN to overall gene expression in T cells by inhibiting its activity with the immune suppressants CsA and FK506 [8,205–207]. CN was found to exert a substantial influence on the transcriptional programme of activated T cells [205]. Depending on the study, stimulation of T cells in the presence of CsA and FK506, respectively, resulted in altered expression of approximately 15–65% of all stimulation-dependent genes [8,205–207]. CN signals, like Ca^{2+} signals, not only exert a positive influence on transcription in activated T cells but also, to an unexpectedly large degree, pronounced negative effects. Thus, Ca^{2+} and CN signals have a role in both promoting and repressing gene transcription, but the mechanisms through which repression is mediated are not well defined. Interestingly, inhibition of Ca^{2+} and CN signals results in largely overlapping changes in global gene transcription, as approximately 75% of all genes depending on Ca^{2+} signals for proper expression are similarly dependent on the CsA-inhibitable function of CN [8]. Hence, in T cells, a majority of Ca^{2+} transcriptional signals are transmitted by the phosphatase CN, the major substrate of which is NFAT.

4.2. Ca^{2+} and NFAT in T- and B-cell anergy

As discussed in Sections 3.5.1 and 4.1., the Ca^{2+}–CN–NFAT pathway and NFAT complexed with AP-1 are important for the productive activation of T cells, their transcriptional reprogramming and effector functions such as cytokine production and cytolytic activity. Conversely, NFAT is also a primary effector of T- and B-cell anergy under conditions where AP-1 proteins are poorly activated [206,208–211]. Anergy is the name given to a cell-intrinsic programme of negative signalling in which T cells downmodulate their responsiveness to stimulation through the TCR, while remaining responsive to stimulation with cytokines such as IL-2. This phenomenon has been termed clonal anergy in experiments performed in vitro using T-cell clones; the analogous process in vivo has been termed adaptive tolerance [192].

As expected, T cells stimulated with a combination of the phorbol ester PMA and the Ca^{2+} ionophore ionomycin, or T cells stimulated through both the TCR and the costimulatory ligands, show optimal activation of NFAT, AP-1 and NFκB and altered expression of a large number (>1000) of activation-associated genes, many of which were confirmed to be dependent on NFAT and AP-1 [206]. In contrast, T cells exposed to Ca^{2+} ionophores without PMA or stimulated through the TCR without costimulation show sustained activation of the Ca^{2+}–CN–NFAT signalling pathway without concomitant activation of AP-1 or NFκB. These T cells are said to receive a partial signal (signal 1, the TCR signal, without signal 2, the costimulatory signal), in response to which they turn on an alternate transcriptional programme, which involves a smaller number of genes (approximately 200). At the same time, the partially stimulated T cells become anergic and refractory to subsequent full stimulation through the TCR and costimulatory receptors.

It is interesting that the 'anergy-associated' genes show almost no overlap with the much larger number of activation-associated genes: rather than encoding cytokines and proteins involved in positive signalling, they encode obvious negative regulators of signalling and transcription, including inhibitory cell-surface receptors; tyrosine phosphatases expected to counter the effects of tyrosine kinases downstream of the TCR; DAG kinase which by metabolizing DAG would attenuate the PKC/Ras/MAPK arm of the signal transduction pathway; proteases and E3 ubiquitin ligases that degrade key signalling proteins in the TCR signalling pathway; and chromatin-level transcriptional repressors that could potentially act to downregulate cytokine gene expression [206,210,211].

In one model of T-cell unresponsiveness, the E3 ubiquitin ligases Itch, Cbl-b and GRAIL are upregulated along with Tsg101, the ubiquitin-binding component of the ESCRT-1 endosomal sorting complex, and together target two key signalling proteins, PKC-θ and PLC-γ1 for degradation, thus imparing the ability of T cells to respond to subsequent stimulation [211]. NFAT is also involved in B-cell tolerance, although the underlying mechanism is not well understood [208,209]. Low-amplitude Ca^{2+} oscillations and activated NFAT were observed in B cells from mice bearing a transgenic BCR and which were tolerized in vivo with the antigen recognized by that receptor [212], indicating that low-level Ca^{2+} signals may be necessary for anergy induction in B cells.

The majority of the anergy-associated genes in T cells are transcriptionally activated, directly or indirectly, by NFAT. These genes are not upregulated in T cells lacking a predominant NFAT family member, NFAT1, and conversely, NFAT1-deficient T cells were resistant to development of the unresponsive state [206]. Many of the anergy-associated genes were also upregulated in T cells from orally tolerized mice, suggesting that the same mechanism of NFAT-induced unresponsiveness is operative in vivo as well. The implication is that a single transcription factor, NFAT, is capable of regulating two contrasting aspects of T-cell function: T-cell activation that requires NFAT–AP-1 cooperation and a negative regulatory programme that opposes T-cell activation and is predominant whenever NFAT–AP-1 cooperation does not occur (Fig. 8) [206].

4.3. *In vivo models of Ca^{2+}–CN–NFAT function*

As discussed in Sections 3.5.1, 3.5.2, 4.1, and 4.2, the Ca^{2+}–CN–NFAT pathway is critically important for the function of lymphocytes. Genetic disruption of genes encoding CN and NFAT isoforms has furthermore revealed the importance of this pathway for the development of lymphocytes and confirmed the notion that CN is limiting in T cells. It had previously been recognized that signalling through the pre-TCR in immature thymocytes results in Ca^{2+} signals and subsequent NFAT activation [213] and that inhibition of CN with CsA results in impaired lymphocyte development [214,215].

4.3.1. *CN-deficient mice*

Two of the three catalytic CNA isoforms (CnAα and CnAβ) and one regulatory CNB isoform (CNB1) are expressed in T cells [216,217]. Mice carrying targeted deletions of the *CnAα* or *CnAβ* have been generated and are viable. T cells develop normally in $CNA\alpha^{-/-}$ mice and respond normally to stimulation in vitro with concanavalin A, anti-CD3ε or PMA and ionomycin [218]. They display reduced proliferation and reduced secretion of IL-2 and IFN-γ under certain conditions of antigen stimulation [218,219]. In $CNA\beta^{-/-}$ mice, T-cell maturation is somewhat impaired, and peripheral T cells show reduced proliferation and IL-2 production even in response to stimulation with anti-CD3 or with PMA and ionomycin [168]. Correspondingly, CNAβ$^{-/-}$ mice are more tolerant than wildtype mice of allogeneic cell transplants [168]. Mice lacking the regulatory CNB1 subunit are embryonically lethal around E11 [220], but conditional, T-cell specific deletion showed a selective involvement of CNB1 in positive but not negative selection of T cells in the thymus [217].

4.3.2. NFAT-deficient mice

Of the four Ca^{2+}-responsive NFAT isoforms, three (NFAT1, NFAT2 and NFAT4) are expressed and functional in T cells [221,222]. All NFAT isoforms have been genetically targeted, either alone or in combination, and the results reveal a complex role of the transcription factor in T-cell function [3,170]. Mice lacking only one NFAT isoform show only relatively mild immune dysfunction of T cells suggesting a partially redundant role for NFAT1 [223–225], NFAT2 [226,227] and NFAT4 [228]. NFAT4 expression in thymocytes is higher than that of NFAT1 and NFAT2, and consequently, NFAT4-deficient mice show a defect in thymocytes differentiation with reduced numbers of CD4 and CD8 single positive thymocytes and increased apoptosis of CD4, CD8 double positive thymnocytes [228]. Combined ablation of either NFAT1 and NFAT2 or NFAT1 and NFAT4 results in more striking alterations in the function of mature T cells [229,230]. NFAT1 and NFAT2 double-deficient T cells show impairment of effector functions such as cytokine production or cytolytic activity [230], whereas T cells lacking NFAT1 and NFAT4 display increased expression of Th2 cytokines and hyperproliferation [229]. Together, these studies indicate that NFAT transcription factors play both positive and negative roles in T-cell development and function.

4.4. The Ca^{2+}–CN–NFAT pathway and disease

Abnormal regulation of Ca^{2+} channels and Ca^{2+} signalling pathways in excitable cells is linked to a variety of diseases of the central nervous system, cardiovascular system and skeletal muscle [231–233]. Several Ca^{2+}-permeable TRP channels have been implicated in the pathophysiology of asthma (TRPC1, TRPC3 and TRPV4) or prostate cancer (TRPV6 and TRPM8) [234]. In the immune system, the importance of Ca^{2+} influx for T-cell activation and CN-mediated gene expression has been highlighted by the existence of a small number of patients with hereditary SCID, who lack SOC influx and CRAC channel function [6–9,60]. T cells from these patients show strongly impaired NFAT activity, cytokine expression and adaptive immune responses. The genetic defect in some of these patients is a missense mutation located in the first transmembrane region of Orai1 [60]. Abrogating Orai1 and CRAC channel function predominantly impairs T-cell function, although the differentiation of myocytes and ectodermally derived skin appendages also seems affected (S.F., unpublished data).

Monogenic diseases caused by mutations in CN or NFAT isoforms are not known, but NFAT dysregulation has recently been suggested to underlie Down syndrome (trisomy 21) because the chromosome 21 region amplified in this disease contains two regulators of NFAT, DSCR1 and DYRK1A [146]. Repression of

NFAT4 by integration of the murine lymphomagenic retrovirus SL3-3 has been shown to be associated with induction of T-cell lymphomas in mice implicating NFAT4 as a putative tumour suppressor [235], whereas increased NFAT1 and NFAT2 activity in human B-cell lymphomas suggests that these NFAT isoforms may play a role in tumour survival [236,237].

4.4.1. Therapeutic potential

Pharmacological inhibition of the Ca^{2+}–CN–NFAT pathway has been proven to be beneficial in a clinical context in preventing rejection of allogeneic organ transplants by using the CN inhibitors CsA or FK506. Specific peptide inhibitors of CN–NFAT interaction have been tested in an experimental model of islet cell transplantation and found to prolong graft survival [112,238]. CsA is also used for the treatment of some autoimmune diseases including rheumatoid arthritis, psoriasis or ulcerative colitis, which are associated with increased T-cell activity. The development of novel inhibitors of the Ca^{2+}–CN–NFAT pathway that do not display the nephro- and neurotoxic side effects of CsA, therefore, promises alternative means to treat T-cell-mediated diseases in the future.

Acknowledgments

This work was supported by grants from the National Institutes of Health to A.R. and S.F. and from the Charles H. Hood Foundation to S.F. The authors thank Drs F. Macian and M. Prakriya for critical reading of the manuscript.

References

1. Lewis, R. S. 2001. Calcium signaling mechanisms in T lymphocytes. *Annu Rev Immunol 19:497.*
2. Crabtree, G. R., and E. N. Olson. 2002. NFAT signaling: choreographing the social lives of cells. *Cell 109 Suppl:S67.*
3. Hogan, P. G., L. Chen, J. Nardone, and A. Rao. 2003. Transcriptional regulation by calcium, calcineurin, and NFAT. *Genes Dev 17:2205.*
4. Graef, I. A., F. Chen, and G. R. Crabtree. 2001. NFAT signaling in vertebrate development. *Curr Opin Genet Dev 11:505.*
5. Putney, J. W., Jr. 1986. A model for receptor-regulated calcium entry. *Cell Calcium 7:1.*
6. Feske, S., M. Prakriya, A. Rao, and R. S. Lewis. 2005. A severe defect in CRAC Ca2+ channel activation and altered K+ channel gating in T cells from immunodeficient patients. *J Exp Med 202:651.*
7. Le Deist, F., C. Hivroz, M. Partiseti, C. Thomas, H. A. Buc, M. Oleastro, B. Belohradsky, D. Choquet, and A. Fischer. 1995. A primary T-cell immunodeficiency associated with defective transmembrane calcium influx. *Blood 85:1053.*
8. Feske, S., J. Giltnane, R. Dolmetsch, L. Staudt, and A. Rao. 2001. Gene regulation by calcium influx in T lymphocytes. *Nat Immunol 2:316.*
9. Partiseti, M., F. Le Deist, C. Hivroz, A. Fischer, H. Korn, and D. Choquet. 1994. The calcium current activated by T cell receptor and store depletion in human lymphocytes is absent in a primary immunodeficiency. *J Biol Chem 269:32327.*

10. Hoth, M., and R. Penner. 1992. Depletion of intracellular calcium stores activates a calcium current in mast cells. *Nature 355:353.*
11. Zweifach, A., and R. S. Lewis. 1993. Mitogen-regulated Ca2+ current of T lymphocytes is activated by depletion of intracellular Ca2+ stores. *Proc Natl Acad Sci USA 90:6295.*
12. Lewis, R. S., and M. D. Cahalan. 1989. Mitogen-induced oscillations of cytosolic Ca2+ and transmembrane Ca2+ current in human leukemic T cells. *Cell Regul 1:99.*
13. Parekh, A. B., and J. W. Putney, Jr. 2005. Store-operated calcium channels. *Physiol Rev 85:757.*
14. Prakriya, M., and R. S. Lewis. 2003. CRAC channels: activation, permeation, and the search for a molecular identity. *Cell Calcium 33:311.*
15. Boulay, G., D. M. Brown, N. Qin, M. Jiang, A. Dietrich, M. X. Zhu, Z. Chen, M. Birnbaumer, K. Mikoshiba, and L. Birnbaumer. 1999. Modulation of Ca(2+) entry by polypeptides of the inositol 1,4, 5-trisphosphate receptor (IP3R) that bind transient receptor potential (TRP): evidence for roles of TRP and IP3R in store depletion-activated Ca(2+) entry. *Proc Natl Acad Sci USA 96:14955.*
16. Yuan, J. P., K. Kiselyov, D. M. Shin, J. Chen, N. Shcheynikov, S. H. Kang, M. H. Dehoff, M. K. Schwarz, P. H. Seeburg, S. Muallem, and P. F. Worley. 2003. Homer binds TRPC family channels and is required for gating of TRPC1 by IP3 receptors. *Cell 114:777.*
17. Zitt, C., A. G. Obukhov, C. Strubing, A. Zobel, F. Kalkbrenner, A. Luckhoff, and G. Schultz. 1997. Expression of TRPC3 in Chinese hamster ovary cells results in calcium-activated cation currents not related to store depletion. *J Cell Biol 138:1333.*
18. Putney, J. W., Jr., M. Trebak, G. Vazquez, B. Wedel, and G. S. Bird. 2004. Signalling mechanisms for TRPC3 channels. *Novartis Found Symp 258:123.*
19. Lintschinger, B., M. Balzer-Geldsetzer, T. Baskaran, W. F. Graier, C. Romanin, M. X. Zhu, and K. Groschner. 2000. Coassembly of Trp1 and Trp3 proteins generates diacylglycerol- and Ca2+-sensitive cation channels. *J Biol Chem 275:27799.*
20. Sugawara, H., M. Kurosaki, M. Takata, and T. Kurosaki. 1997. Genetic evidence for involvement of type 1, type 2 and type 3 inositol 1,4,5-trisphosphate receptors in signal transduction through the B-cell antigen receptor. *EMBO J 16:3078.*
21. Prakriya, M., and R. S. Lewis. 2001. Potentiation and inhibition of Ca(2+) release-activated Ca(2+) channels by 2-aminoethyldiphenyl borate (2-APB) occurs independently of IP(3) receptors. *J Physiol 536:3.*
22. Randriamampita, C., and R. Y. Tsien. 1993. Emptying of intracellular Ca2+ stores releases a novel small messenger that stimulates Ca2+ influx. *Nature 364:809.*
23. Smani, T., S. I. Zakharov, P. Csutora, E. Leno, E. S. Trepakova, and V. M. Bolotina. 2004. A novel mechanism for the store-operated calcium influx pathway. *Nat Cell Biol 6:113.*
24. Csutora, P., Z. Su, H. Y. Kim, A. Bugrim, K. W. Cunningham, R. Nuccitelli, J. E. Keizer, M. R. Hanley, J. E. Blalock, and R. B. Marchase. 1999. Calcium influx factor is synthesized by yeast and mammalian cells depleted of organellar calcium stores. *Proc Natl Acad Sci USA 96:121.*
25. Smani, T., S. I. Zakharov, E. Leno, P. Csutora, E. S. Trepakova, and V. M. Bolotina. 2003. Ca2+-independent phospholipase A2 is a novel determinant of store-operated Ca2+ entry. *J Biol Chem 278:11909.*
26. Yao, Y., A. V. Ferrer-Montiel, M. Montal, and R. Y. Tsien. 1999. Activation of store-operated Ca2+ current in Xenopus oocytes requires SNAP-25 but not a diffusible messenger. *Cell 98:475.*
27. Stojilkovic, S. S. 2005. Ca2+-regulated exocytosis and SNARE function. *Trends Endocrinol Metab 16:81.*
28. Liou, J., M. L. Kim, W. D. Heo, J. T. Jones, J. W. Myers, J. E. Ferrell, Jr., and T. Meyer. 2005. STIM is a Ca2+ sensor essential for Ca2+-store-depletion-triggered Ca2+ influx. *Curr Biol 15:1235.*
29. Roos, J., P. J. DiGregorio, A. V. Yeromin, K. Ohlsen, M. Lioudyno, S. Zhang, O. Safrina, J. A. Kozak, S. L. Wagner, M. D. Cahalan, G. Velicelebi, and K. A. Stauderman. 2005. STIM1, an essential and conserved component of store-operated Ca2+ channel function. *J Cell Biol 169:435.*
30. Manji, S. S., N. J. Parker, R. T. Williams, L. van Stekelenburg, R. B. Pearson, M. Dziadek, and P. J. Smith. 2000. STIM1: a novel phosphoprotein located at the cell surface. *Biochim Biophys Acta 1481:147.*
31. Oritani, K., and P. W. Kincade. 1996. Identification of stromal cell products that interact with pre-B cells. *J Cell Biol 134:771.*

32. Sabbioni, S., G. Barbanti-Brodano, C. M. Croce, and M. Negrini. 1997. GOK: a gene at 11p15 involved in rhabdomyosarcoma and rhabdoid tumor development. *Cancer Res 57:4493.*
33. Williams, R. T., S. S. Manji, N. J. Parker, M. S. Hancock, L. Van Stekelenburg, J. P. Eid, P. V. Senior, J. S. Kazenwadel, T. Shandala, R. Saint, P. J. Smith, and M. A. Dziadek. 2001. Identification and characterization of the STIM (stromal interaction molecule) gene family: coding for a novel class of transmembrane proteins. *Biochem J 357:673.*
34. Yeromin, A. V., J. Roos, K. A. Stauderman, and M. D. Cahalan. 2004. A store-operated calcium channel in Drosophila S2 cells. *J Gen Physiol 123:167.*
35. Stathopulos, P. B., G. V. Li, M. J. Plevin, J. B. Ames, and M. Ikura. 2006. Stored Ca^{2+}-depletion induced oligomerization of stromal interaction molecule 1 (STIM1) via the EF-SAM region: An initiation mechanism for capacitive Ca^{2+} entry. *J Biol Chem 281:35855.*
36. Meldolesi, J., and T. Pozzan. 1998. The endoplasmic reticulum Ca2+ store: a view from the lumen. *Trends Biochem Sci 23:10.*
37. Montero, M., M. J. Barrero, and J. Alvarez. 1997. [Ca2+] microdomains control agonist-induced Ca2+ release in intact HeLa cells. *FASEB J 11:881.*
38. Miyawaki, A., J. Llopis, R. Heim, J. M. McCaffery, J. A. Adams, M. Ikura, and R. Y. Tsien. 1997. Fluorescent indicators for Ca2+ based on green fluorescent proteins and calmodulin. *Nature 388:882.*
39. Burdakov, D., O. H. Petersen, and A. Verkhratsky. 2005. Intraluminal calcium as a primary regulator of endoplasmic reticulum function. *Cell Calcium 38:303.*
40. Zhang, S. L., Y. Yu, J. Roos, J. A. Kozak, T. J. Deerinck, M. H. Ellisman, K. A. Stauderman, and M. D. Cahalan. 2005. STIM1 is a Ca2+ sensor that activates CRAC channels and migrates from the Ca2+ store to the plasma membrane. *Nature 437:902.*
41. Wu, M. M., J. Buchanan, R. M. Luik, and R. S. Lewis. 2006. Ca^{2+} store depletion causes STIM1 to accumulate in ER regions closely associated with the plasma membrane. *J Cell Biol 174:803–13.*
42. Clapham, D. E. 2003. TRP channels as cellular sensors. *Nature 426:517.*
43. Ramsey, I. S., M. Delling, and D. E. Clapham. 2006. An introduction to TRP channels. *Annu Rev Physiol 68:619.*
44. Montell, C., K. Jones, E. Hafen, and G. Rubin. 1985. Rescue of the Drosophila phototransduction mutation trp by germline transformation. *Science 230:1040.*
45. Minke, B. 1977. Drosophila mutant with a transducer defect. *Biophys Struct Mech 3:59.*
46. Yue, L., J. B. Peng, M. A. Hediger, and D. E. Clapham. 2001. CaT1 manifests the pore properties of the calcium-release-activated calcium channel. *Nature 410:705.*
47. Voets, T., J. Prenen, A. Fleig, R. Vennekens, H. Watanabe, J. G. Hoenderop, R. J. Bindels, G. Droogmans, R. Penner, and B. Nilius. 2001. CaT1 and the calcium release-activated calcium channel manifest distinct pore properties. *J Biol Chem 276:47767.*
48. Prakriya, M., S. Feske, Y. Gwack, S. Srikanth, A. Rao, and P. G. Hogan. 2006. Orai1 is an essential pore subunit of the CRAC channel. *Nature 443:230.*
49. Clapham, D. E. 2002. Sorting out MIC, TRP, and CRAC ion channels. *J Gen Physiol 120:217.*
50. Prakriya, M., and R. S. Lewis. 2002. Separation and characterization of currents through store-operated CRAC channels and Mg(2+)-inhibited cation (MIC) channels. *J Gen Physiol 119:487.*
51. Kahr, H., R. Schindl, R. Fritsch, B. Heinze, M. Hofbauer, M. E. Hack, M. A. Mortelmaier, K. Groschner, J. B. Peng, H. Takanaga, M. A. Hediger, and C. Romanin. 2004. CaT1 knock-down strategies fail to affect CRAC channels in mucosal-type mast cells. *J Physiol 557:121.*
52. Runnels, L. W., L. Yue, and D. E. Clapham. 2001. TRP-PLIK, a bifunctional protein with kinase and ion channel activities. *Science 291:1043.*
53. Nadler, M. J., M. C. Hermosura, K. Inabe, A. L. Perraud, Q. Zhu, A. J. Stokes, T. Kurosaki, J. P. Kinet, R. Penner, A. M. Scharenberg, and A. Fleig. 2001. LTRPC7 is a Mg.ATP-regulated divalent cation channel required for cell viability. *Nature 411:590.*
54. Hermosura, M. C., M. K. Monteilh-Zoller, A. M. Scharenberg, R. Penner, and A. Fleig. 2002. Dissociation of the store-operated calcium current I(CRAC) and the Mg-nucleotide-regulated metal ion current MagNuM. *J Physiol 539:445.*

55. Zitt, C., A. Zobel, A. G. Obukhov, C. Harteneck, F. Kalkbrenner, A. Luckhoff, and G. Schultz. 1996. Cloning and functional expression of a human Ca2+-permeable cation channel activated by calcium store depletion. *Neuron 16:1189.*
56. Zhu, X., M. S. Jiang, M. Peyton, G. Boulay, R. Hurst, E. Stefani, and L. Birnbaumer. 1996. Trp, a novel mammalian gene family essential for agonist-activated capacitative Ca2+ entry. *Cell 85:661.*
57. Mori, Y., M. Wakamori, T. Miyakawa, M. Hermosura, Y. Hara, M. Nishida, K. Hirose, A. Mizushima, M. Kurosaki, E. Mori, K. Gotoh, T. Okada, A. Fleig, R. Penner, M. Iino, and T. Kurosaki. 2002. Transient receptor potential 1 regulates capacitative Ca(2+) entry and Ca(2+) release from endoplasmic reticulum in B lymphocytes. *J Exp Med 195:673.*
58. Philipp, S., B. Strauss, D. Hirnet, U. Wissenbach, L. Mery, V. Flockerzi, and M. Hoth. 2003. TRPC3 mediates T-cell receptor-dependent calcium entry in human T-lymphocytes. *J Biol Chem 278:26629.*
59. Zhang, S. L., A. V. Yeromin, X. H. Zhang, Y. Yu, O. Safrina, A. Penna, J. Roos, K. A. Stauderman, and M. D. Cahalan. 2006. Genome-wide RNAi screen of Ca2+ influx identifies genes that regulate Ca2+ release-activated Ca2+ channel activity. *Proc Natl Acad Sci USA 103:9357.*
60. Feske, S., Y. Gwack, M. Prakriya, S. Srikanth, S. H. Puppel, B. Tanasa, P. G. Hogan, R. S. Lewis, M. Daly, and A. Rao. 2006. A mutation in Orai1 causes immune deficiency by abrogating CRAC channel function. *Nature 441:179.*
61. Vig, M., C. Peinelt, A. Beck, D. L. Koomoa, D. Rabah, M. Koblan-Huberson, S. Kraft, H. Turner, A. Fleig, R. Penner, and J. P. Kinet. 2006. CRACM1 is a plasma membrane protein essential for store-operated Ca2+ entry. *Science 312:1220.*
62. Owsianik, G., D. D'Hoedt, T. Voets, and B. Nilius. 2006. Structure-function relationship of the TRP channel superfamily. *Rev Physiol Biochem Pharmacol 156:61.*
63. Sather, W. A., and E. W. McCleskey. 2003. Permeation and selectivity in calcium channels. *Annu Rev Physiol 65:133.*
64. Yeromin, A. V., S. L. Zhang, W. Jiang, Y. Yu, O. Safrina, and M. D. Cahalan. 2006. Molecular identification of the CRAC channel by altered ion selectivity in a mutant of Orai. *Nature 443:226.*
65. Mercer, J. C., W. I. Dehaven, J. T. Smyth, B. Wedel, R. R. Boyles, G. S. Bird, and J. W. Putney, Jr. 2006. Large store-operated calcium selective currents due to co-expression of Orai1 or Orai2 with the intracellular calcium sensor, Stim1. *J Biol Chem 281:24979.*
66. Soboloff, J., M. A. Spassova, X. D. Tang, T. Hewavitharana, W. Xu, and D. L. Gill. 2006. Orai1 and STIM reconstitute store-operated calcium channel function. *J Biol Chem 281:20661.*
67. Peinelt, C., M. Vig, D. L. Koomoa, A. Beck, M. J. Nadler, M. Koblan-Huberson, A. Lis, A. Fleig, R. Penner, and J. P. Kinet. 2006. Amplification of CRAC current by STIM1 and CRACM1 (Orai1). *Nat Cell Biol 8:771.*
68. Weick, J. P., R. D. Groth, A. L. Isaksen, and P. G. Mermelstein. 2003. Interactions with PDZ proteins are required for L-type calcium channels to activate cAMP response element-binding protein-dependent gene expression. *J Neurosci 23:3446.*
69. Deisseroth, K., H. Bito, and R. W. Tsien. 1996. Signaling from synapse to nucleus: postsynaptic CREB phosphorylation during multiple forms of hippocampal synaptic plasticity. *Neuron 16:89.*
70. Naraghi, M., and E. Neher. 1997. Linearized buffered Ca2+ diffusion in microdomains and its implications for calculation of [Ca2+] at the mouth of a calcium channel. *J Neurosci 17:6961.*
71. Augustine, G. J., F. Santamaria, and K. Tanaka. 2003. Local calcium signaling in neurons. *Neuron 40:331.*
72. Yamada, W. M., and R. S. Zucker. 1992. Time course of transmitter release calculated from simulations of a calcium diffusion model. *Biophys J 61:671.*
73. Luik, R., M. Wu, J. Buchanan, and R. S. Lewis. 2006. The elementary unit of store-operated Ca2+ entry: local activation of CRAC channels by STIM1 at ER-plasma membrane junctions. *J Cell Biol 174:815.*
74. Ozawa, K., Z. Szallasi, M. G. Kazanietz, P. M. Blumberg, H. Mischak, J. F. Mushinski, and M. A. Beaven. 1993. Ca(2+)-dependent and Ca(2+)-independent isozymes of protein kinase C mediate exocytosis in antigen-stimulated rat basophilic RBL-2H3 cells. Reconstitution of secretory responses with Ca2+ and purified isozymes in washed permeabilized cells. *J Biol Chem 268:1749.*

75. Nadler, M. J., and J. P. Kinet. 2002. Uncovering new complexities in mast cell signaling. *Nat Immunol 3:707.*
76. Penner, R., and E. Neher. 1988. Secretory responses of rat peritoneal mast cells to high intracellular calcium. *FEBS Lett 226:307.*
77. Huppa, J. B., M. Gleimer, C. Sumen, and M. M. Davis. 2003. Continuous T cell receptor signaling required for synapse maintenance and full effector potential. *Nat Immunol 4:749.*
78. Timmerman, L. A., N. A. Clipstone, S. N. Ho, J. P. Northrop, and G. R. Crabtree. 1996. Rapid shuttling of NF-AT in discrimination of Ca2+ signals and immunosuppression. *Nature 383:837.*
79. Loh, C., K. T. Shaw, J. Carew, J. P. Viola, C. Luo, B. A. Perrino, and A. Rao. 1996. Calcineurin binds the transcription factor NFAT1 and reversibly regulates its activity. *J Biol Chem 271:10884.*
80. Rao, A., C. Luo, and P. G. Hogan. 1997. Transcription factors of the NFAT family: regulation and function. *Annu Rev Immunol 15:707.*
81. Serfling, E., F. Berberich-Siebelt, S. Chuvpilo, E. Jankevics, S. Klein-Hessling, T. Twardzik, and A. Avots. 2000. The role of NF-AT transcription factors in T cell activation and differentiation. *Biochim Biophys Acta 1498:1.*
82. Feske, S., H. Okamura, P. G. Hogan, and A. Rao. 2003. Ca2+/calcineurin signalling in cells of the immune system. *Biochem Biophys Res Commun 311:1117.*
83. Friedman, J., and I. Weissman. 1991. Two cytoplasmic candidates for immunophilin action are revealed by affinity for a new cyclophilin: one in the presence and one in the absence of CsA. *Cell 66:799.*
84. Fruman, D. A., C. B. Klee, B. E. Bierer, and S. J. Burakoff. 1992. Calcineurin phosphatase activity in T lymphocytes is inhibited by FK 506 and cyclosporin A. *Proc Natl Acad Sci USA 89:3686.*
85. Liu, J., J. D. Farmer, Jr., W. S. Lane, J. Friedman, I. Weissman, and S. L. Schreiber. 1991. Calcineurin is a common target of cyclophilin-cyclosporin A and FKBP-FK506 complexes. *Cell 66:807.*
86. Swanson, S. K., T. Born, L. D. Zydowsky, H. Cho, H. Y. Chang, C. T. Walsh, and F. Rusnak. 1992. Cyclosporin-mediated inhibition of bovine calcineurin by cyclophilins A and B. *Proc Natl Acad Sci USA 89:3741.*
87. Liu, J., M. W. Albers, T. J. Wandless, S. Luan, D. G. Alberg, P. J. Belshaw, P. Cohen, C. MacKintosh, C. B. Klee, and S. L. Schreiber. 1992. Inhibition of T cell signaling by immunophilin-ligand complexes correlates with loss of calcineurin phosphatase activity. *Biochemistry 31:3896.*
88. O'Keefe, S. J., J. Tamura, R. L. Kincaid, M. J. Tocci, and E. A. O'Neill. 1992. FK-506- and CsA-sensitive activation of the interleukin-2 promoter by calcineurin. *Nature 357:692.*
89. Clipstone, N. A., and G. R. Crabtree. 1992. Identification of calcineurin as a key signalling enzyme in T-lymphocyte activation. *Nature 357:695.*
90. Klee, C. B., T. H. Crouch, and M. H. Krinks. 1979. Calcineurin: a calcium- and calmodulin-binding protein of the nervous system. *Proc Natl Acad Sci USA 76:6270.*
91. Klee, C. B., H. Ren, and X. Wang. 1998. Regulation of the calmodulin-stimulated protein phosphatase, calcineurin. *J Biol Chem 273:13367.*
92. Rusnak, F., and P. Mertz. 2000. Calcineurin: form and function. *Physiol Rev 80:1483.*
93. Aramburu, J., A. Rao, and C. B. Klee. 2000. Calcineurin: from structure to function. *Curr Top Cell Regul 36:237.*
94. Kissinger, C. R., H. E. Parge, D. R. Knighton, C. T. Lewis, L. A. Pelletier, A. Tempczyk, V. J. Kalish, K. D. Tucker, R. E. Showalter, and E. W. Moomaw, et al. 1995. Crystal structures of human calcineurin and the human FKBP12-FK506-calcineurin complex. *Nature 378:641.*
95. Perrino, B. A., L. Y. Ng, and T. R. Soderling. 1995. Calcium regulation of calcineurin phosphatase activity by its B subunit and calmodulin. Role of the autoinhibitory domain. *J Biol Chem 270:340.*
96. Stewart, A. A., T. S. Ingebritsen, A. Manalan, C. B. Klee, and P. Cohen. 1982. Discovery of a Ca2+-and calmodulin-dependent protein phosphatase: probable identity with calcineurin (CaM-BP80). *FEBS Lett 137:80.*

97. Stemmer, P. M., and C. B. Klee. 1994. Dual calcium ion regulation of calcineurin by calmodulin and calcineurin B. *Biochemistry 33:6859.*
98. Kretsinger, R. H., and C. E. Nockolds. 1973. Carp muscle calcium-binding protein. II. Structure determination and general description. *J Biol Chem 248:3313.*
99. Strynadka, N. C., and M. N. James. 1989. Crystal structures of the helix-loop-helix calcium-binding proteins. *Annu Rev Biochem 58:951.*
100. Ikura, M., and J. B. Ames. 2006. Genetic polymorphism and protein conformational plasticity in the calmodulin superfamily: two ways to promote multifunctionality. *Proc Natl Acad Sci USA 103:1159.*
101. Grabarek, Z. 2006. Structural basis for diversity of the EF-hand calcium-binding proteins. *J Mol Biol 359:509.*
102. Feng, B., and P. M. Stemmer. 1999. Interactions of calcineurin A, calcineurin B, and Ca2+. *Biochemistry 38:12481.*
103. Linse, S., A. Helmersson, and S. Forsen. 1991. Calcium binding to calmodulin and its globular domains. *J Biol Chem 266:8050.*
104. Haiech, J., C. B. Klee, and J. G. Demaille. 1981. Effects of cations on affinity of calmodulin for calcium: ordered binding of calcium ions allows the specific activation of calmodulin-stimulated enzymes. *Biochemistry 20:3890.*
105. Peersen, O. B., T. S. Madsen, and J. J. Falke. 1997. Intermolecular tuning of calmodulin by target peptides and proteins: differential effects on Ca2+ binding and implications for kinase activation. *Protein Sci 6:794.*
106. Feng, B., and P. M. Stemmer. 2001. Ca2+ binding site 2 in calcineurin-B modulates calmodulin-dependent calcineurin phosphatase activity. *Biochemistry 40:8808.*
107. Yang, J., B. Rothermel, R. B. Vega, N. Frey, T. A. McKinsey, E. N. Olson, R. Bassel-Duby, and R. S. Williams. 2000. Independent signals control expression of the calcineurin inhibitory proteins MCIP1 and MCIP2 in striated muscles. *Circ Res 87:E61.*
108. Yang, S. A., and C. B. Klee. 2000. Low affinity Ca2+-binding sites of calcineurin B mediate conformational changes in calcineurin A. *Biochemistry 39:16147.*
109. Egloff, M. P., P. T. Cohen, P. Reinemer, and D. Barford. 1995. Crystal structure of the catalytic subunit of human protein phosphatase 1 and its complex with tungstate. *J Mol Biol 254:942.*
110. Barford, D. 1996. Molecular mechanisms of the protein serine/threonine phosphatases. *Trends Biochem Sci 21:407.*
111. Voegtli, W. C., D. J. White, N. J. Reiter, F. Rusnak, and A. C. Rosenzweig. 2000. Structure of the bacteriophage lambda Ser/Thr protein phosphatase with sulfate ion bound in two coordination modes. *Biochemistry 39:15365.*
112. Aramburu, J., F. Garcia-Cozar, A. Raghavan, H. Okamura, A. Rao, and P. G. Hogan. 1998. Selective inhibition of NFAT activation by a peptide spanning the calcineurin targeting site of NFAT. *Mol Cell 1:627.*
113. Garcia-Cozar, F. J., H. Okamura, J. F. Aramburu, K. T. Shaw, L. Pelletier, R. Showalter, E. Villafranca, and A. Rao. 1998. Two-site interaction of nuclear factor of activated T cells with activated calcineurin. *J Biol Chem 273:23877.*
114. Chow, C. W., M. Rincon, and R. J. Davis. 1999. Requirement for transcription factor NFAT in interleukin-2 expression. *Mol Cell Biol 19:2300.*
115. Zhu, J., F. Shibasaki, R. Price, J. C. Guillemot, T. Yano, V. Dotsch, G. Wagner, P. Ferrara, and F. McKeon. 1998. Intramolecular masking of nuclear import signal on NF-AT4 by casein kinase I and MEKK1. *Cell 93:851.*
116. Aramburu, J., M. B. Yaffe, C. Lopez-Rodriguez, L. C. Cantley, P. G. Hogan, and A. Rao. 1999. Affinity-driven peptide selection of an NFAT inhibitor more selective than cyclosporin A. *Science 285:2129.*
117. Li, H., A. Rao, and P. G. Hogan. 2004. Structural delineation of the calcineurin-NFAT interaction and its parallels to PP1 targeting interactions. *J Mol Biol 342:1659.*
118. Li, H., L. Zhang, A. Rao, S. C. Harrison, and P. G. Hogan. 2007. Structure of calcineurin in complex with PVIVIT peptide: portrait of a low-affinity signaling interaction. Manuscript submitted.

119. Boustany, L. M., and M. S. Cyert. 2002. Calcineurin-dependent regulation of Crz1p nuclear export requires Msn5p and a conserved calcineurin docking site. *Genes Dev 16:608.*
120. Bultynck, G., V. L. Heath, A. P. Majeed, J. M. Galan, R. Haguenauer-Tsapis, and M. S. Cyert. 2006. Slm1 and slm2 are novel substrates of the calcineurin phosphatase required for heat stress-induced endocytosis of the yeast uracil permease. *Mol Cell Biol 26:4729.*
121. Czirjak, G., and P. Enyedi. 2006. Targeting of calcineurin to an NFAT-like docking site is required for the calcium-dependent activation of the background K+ channel, TRESK. *J Biol Chem 281:14677.*
122. Heath, V. L., S. L. Shaw, S. Roy, and M. S. Cyert. 2004. Hph1p and Hph2p, novel components of calcineurin-mediated stress responses in Saccharomyces cerevisiae. *Eukaryot Cell 3:695.*
123. McCullagh, K. J., E. Calabria, G. Pallafacchina, S. Ciciliot, A. L. Serrano, C. Argentini, J. M. Kalhovde, T. Lomo, and S. Schiaffino. 2004. NFAT is a nerve activity sensor in skeletal muscle and controls activity-dependent myosin switching. *Proc Natl Acad Sci USA 101:10590.*
124. Dolmetsch, R. E., R. S. Lewis, C. C. Goodnow, and J. I. Healy. 1997. Differential activation of transcription factors induced by Ca2+ response amplitude and duration. *Nature 386:855.*
125. Shaw, K. T., A. M. Ho, A. Raghavan, J. Kim, J. Jain, J. Park, S. Sharma, A. Rao, and P. G. Hogan. 1995. Immunosuppressive drugs prevent a rapid dephosphorylation of transcription factor NFAT1 in stimulated immune cells. *Proc Natl Acad Sci USA 92:11205.*
126. Hubbard, M. J., and P. Cohen. 1989. Regulation of protein phosphatase-1G from rabbit skeletal muscle. 1. Phosphorylation by cAMP-dependent protein kinase at site 2 releases catalytic subunit from the glycogen-bound holoenzyme. *Eur J Biochem 186:701.*
127. Salazar, C., and T. Höfer. 2003. Allosteric regulation of the transcription factor NFAT1 by multiple phosphorylation sites: a mathematical analysis. *J Mol Biol 327:31.*
128. Okamura, H., J. Aramburu, C. Garcia-Rodriguez, J. P. Viola, A. Raghavan, M. Tahiliani, X. Zhang, J. Qin, P. G. Hogan, and A. Rao. 2000. Concerted dephosphorylation of the transcription factor NFAT1 induces a conformational switch that regulates transcriptional activity. *Mol Cell 6:539.*
129. Liu, J., E. S. Masuda, L. Tsuruta, N. Arai, and K. Arai. 1999. Two independent calcineurin-binding regions in the N-terminal domain of murine NF-ATx1 recruit calcineurin to murine NF-ATx1. *J Immunol 162:4755.*
130. Liu, J., K. Arai, and N. Arai. 2001. Inhibition of NFATx activation by an oligopeptide: disrupting the interaction of NFATx with calcineurin. *J Immunol 167:2677.*
131. Park, S., M. Uesugi, and G. L. Verdine. 2000. A second calcineurin binding site on the NFAT regulatory domain. *Proc Natl Acad Sci USA 97:7130.*
132. Martinez-Martinez, S., A. Rodriguez, M. D. Lopez-Maderuelo, I. Ortega-Perez, J. Vazquez, and J. M. Redondo. 2006. Blockade of NFAT activation by the second calcineurin binding site. *J Biol Chem 281:6227.*
133. Quintana, A. R., D. Wang, J. E. Forbes, and M. N. Waxham. 2005. Kinetics of calmodulin binding to calcineurin. *Biochem Biophys Res Commun 334:674.*
134. Perrino, B. A., A. J. Wilson, P. Ellison, and L. H. Clapp. 2002. Substrate selectivity and sensitivity to inhibition by FK506 and cyclosporin A of calcineurin heterodimers composed of the alpha or beta catalytic subunit. *Eur J Biochem 269:3540.*
135. Fuentes, J. J., L. Genesca, T. J. Kingsbury, K. W. Cunningham, M. Perez-Riba, X. Estivill, and S. de la Luna. 2000. DSCR1, overexpressed in Down syndrome, is an inhibitor of calcineurin-mediated signaling pathways. *Hum Mol Genet 9:1681.*
136. Kingsbury, T. J., and K. W. Cunningham. 2000. A conserved family of calcineurin regulators. *Genes Dev 14:1595.*
137. Gorlach, J., D. S. Fox, N. S. Cutler, G. M. Cox, J. R. Perfect, and J. Heitman. 2000. Identification and characterization of a highly conserved calcineurin binding protein, CBP1/calcipressin, in Cryptococcus neoformans. *EMBO J 19:3618.*
138. Rothermel, B., R. B. Vega, J. Yang, H. Wu, R. Bassel-Duby, and R. S. Williams. 2000. A protein encoded within the Down syndrome critical region is enriched in striated muscles and inhibits calcineurin signaling. *J Biol Chem 275:8719.*

139. Hilioti, Z., and K. W. Cunningham. 2003. The RCN family of calcineurin regulators. *Biochem Biophys Res Commun 311:1089.*
140. Hilioti, Z., D. A. Gallagher, S. T. Low-Nam, P. Ramaswamy, P. Gajer, T. J. Kingsbury, C. J. Birchwood, A. Levchenko, and K. W. Cunningham. 2004. GSK-3 kinases enhance calcineurin signaling by phosphorylation of RCNs. *Genes Dev 18:35.*
141. Abbasi, S., J. D. Lee, B. Su, X. Chen, J. L. Alcon, J. Yang, R. E. Kellems, and Y. Xia. 2006. Protein kinase-mediated regulation of calcineurin through the phosphorylation of modulatory calcineurin-interacting protein 1. *J Biol Chem 281:7717.*
142. Vega, R. B., B. A. Rothermel, C. J. Weinheimer, A. Kovacs, R. H. Naseem, R. Bassel-Duby, R. S. Williams, and E. N. Olson. 2003. Dual roles of modulatory calcineurin-interacting protein 1 in cardiac hypertrophy. *Proc Natl Acad Sci USA 100:669.*
143. Ryeom, S., R. J. Greenwald, A. H. Sharpe, and F. McKeon. 2003. The threshold pattern of calcineurin-dependent gene expression is altered by loss of the endogenous inhibitor calcipressin. *Nat Immunol 4:874.*
144. Sanna, B., E. B. Brandt, R. A. Kaiser, P. Pfluger, S. A. Witt, T. R. Kimball, E. van Rooij, L. J. De Windt, M. E. Rothenberg, M. H. Tschop, S. C. Benoit, and J. D. Molkentin. 2006. Modulatory calcineurin-interacting proteins 1 and 2 function as calcineurin facilitators in vivo. *Proc Natl Acad Sci USA 103:7327.*
145. Lange, A. W., J. D. Molkentin, and K. E. Yutzey. 2004. DSCR1 gene expression is dependent on NFATc1 during cardiac valve formation and colocalizes with anomalous organ development in trisomy 16 mice. *Dev Biol 266:346.*
146. Arron, J. R., M. M. Winslow, A. Polleri, C. P. Chang, H. Wu, X. Gao, J. R. Neilson, L. Chen, J. J. Heit, S. K. Kim, N. Yamasaki, T. Miyakawa, U. Francke, I. A. Graef, and G. R. Crabtree. 2006. NFAT dysregulation by increased dosage of DSCR1 and DYRK1A on chromosome 21. *Nature 441:595.*
147. Sun, L., H. D. Youn, C. Loh, M. Stolow, W. He, and J. O. Liu. 1998. Cabin 1, a negative regulator for calcineurin signaling in T lymphocytes. *Immunity 8:703.*
148. Lai, M. M., P. E. Burnett, H. Wolosker, S. Blackshaw, and S. H. Snyder. 1998. Cain, a novel physiologic protein inhibitor of calcineurin. *J Biol Chem 273:18325.*
149. Dell'Acqua, M. L., K. L. Dodge, S. J. Tavalin, and J. D. Scott. 2002. Mapping the protein phosphatase-2B anchoring site on AKAP79. Binding and inhibition of phosphatase activity are mediated by residues 315-360. *J Biol Chem 277:48796.*
150. Frey, N., J. A. Richardson, and E. N. Olson. 2000. Calsarcins, a novel family of sarcomeric calcineurin-binding proteins. *Proc Natl Acad Sci USA 97:14632.*
151. Frey, N., and E. N. Olson. 2002. Calsarcin-3, a novel skeletal muscle-specific member of the calsarcin family, interacts with multiple Z-disc proteins. *J Biol Chem 277:13998.*
152. Oliveria, S. F., L. L. Gomez, and M. L. Dell'Acqua. 2003. Imaging kinase–AKAP79–phosphatase scaffold complexes at the plasma membrane in living cells using FRET microscopy. *J Cell Biol 160:101.*
153. Frey, N., T. Barrientos, J. M. Shelton, D. Frank, H. Rutten, D. Gehring, C. Kuhn, M. Lutz, B. Rothermel, R. Bassel-Duby, J. A. Richardson, H. A. Katus, J. A. Hill, and E. N. Olson. 2004. Mice lacking calsarcin-1 are sensitized to calcineurin signaling and show accelerated cardiomyopathy in response to pathological biomechanical stress. *Nat Med 10:1336.*
154. Miskin, J. E., C. C. Abrams, L. C. Goatley, and L. K. Dixon. 1998. A viral mechanism for inhibition of the cellular phosphatase calcineurin. *Science 281:562.*
155. Miskin, J. E., C. C. Abrams, and L. K. Dixon. 2000. African swine fever virus protein A238L interacts with the cellular phosphatase calcineurin via a binding domain similar to that of NFAT. *J Virol 74:9412.*
156. Tait, S. W., E. B. Reid, D. R. Greaves, T. E. Wileman, and P. P. Powell. 2000. Mechanism of inactivation of NF-kappa B by a viral homologue of I kappa b alpha. Signal-induced release of i kappa b alpha results in binding of the viral homologue to NF-kappa B. *J Biol Chem 275:34656.*
157. Dixon, L. K., C. C. Abrams, G. Bowick, L. C. Goatley, P. C. Kay-Jackson, D. Chapman, E. Liverani, R. Nix, R. Silk, and F. Zhang. 2004. African swine fever virus proteins involved in evading host defence systems. *Vet Immunol Immunopathol 100:117.*

158. Huai, Q., H. Y. Kim, Y. Liu, Y. Zhao, A. Mondragon, J. O. Liu, and H. Ke. 2002. Crystal structure of calcineurin-cyclophilin-cyclosporin shows common but distinct recognition of immunophilin-drug complexes. *Proc Natl Acad Sci USA 99:12037.*

159. Jin, L., and S. C. Harrison. 2002. Crystal structure of human calcineurin complexed with cyclosporin A and human cyclophilin. *Proc Natl Acad Sci USA 99:13522.*

160. Etzkorn, F. A., Z. Y. Chang, L. A. Stolz, and C. T. Walsh. 1994. Cyclophilin residues that affect noncompetitive inhibition of the protein serine phosphatase activity of calcineurin by the cyclophilin.cyclosporin A complex. *Biochemistry 33:2380.*

161. Kang, S., H. Li, A. Rao, and P. G. Hogan. 2005. Inhibition of the calcineurin-NFAT interaction by small organic molecules reflects binding at an allosteric site. *J Biol Chem 280:37698.*

162. Roehrl, M. H., S. Kang, J. Aramburu, G. Wagner, A. Rao, and P. G. Hogan. 2004. Selective inhibition of calcineurin-NFAT signaling by blocking protein-protein interaction with small organic molecules. *Proc Natl Acad Sci USA 101:7554.*

163. Zoeteweij, J. P., A. V. Moses, A. S. Rinderknecht, D. A. Davis, W. W. Overwijk, R. Yarchoan, J. M. Orenstein, and A. Blauvelt. 2001. Targeted inhibition of calcineurin signaling blocks calcium-dependent reactivation of Kaposi sarcoma-associated herpesvirus. *Blood 97:2374.*

164. Fruman, D. A., S. Y. Pai, S. J. Burakoff, and B. E. Bierer. 1995. Characterization of a mutant calcineurin A alpha gene expressed by EL4 lymphoma cells. *Mol Cell Biol 15:3857.*

165. Batiuk, T. D., F. Pazderka, J. Enns, L. DeCastro, and P. F. Halloran. 1995. Cyclosporine inhibition of calcineurin activity in human leukocytes in vivo is rapidly reversible. *J Clin Invest 96:1254.*

166. Tsuboi, A., E. S. Masuda, Y. Naito, H. Tokumitsu, K. Arai, and N. Arai. 1994. Calcineurin potentiates activation of the granulocyte-macrophage colony-stimulating factor gene in T cells: involvement of the conserved lymphokine element 0. *Mol Biol Cell 5:119.*

167. Batiuk, T. D., L. Kung, and P. F. Halloran. 1997. Evidence that calcineurin is rate-limiting for primary human lymphocyte activation. *J Clin Invest 100:1894.*

168. Bueno, O. F., E. B. Brandt, M. E. Rothenberg, and J. D. Molkentin. 2002. Defective T cell development and function in calcineurin A beta-deficient mice. *Proc Natl Acad Sci USA 99:9398.*

169. Macian, F., C. Lopez-Rodriguez, and A. Rao. 2001. Partners in transcription: NFAT and AP-1. *Oncogene 20:2476.*

170. Macian, F. 2005. NFAT proteins: key regulators of T-cell development and function. *Nat Rev Immunol 5:472.*

171. Beals, C. R., C. M. Sheridan, C. W. Turck, P. Gardner, and G. R. Crabtree. 1997. Nuclear export of NF-ATc enhanced by glycogen synthase kinase-3. *Science 275:1930.*

172. Shibasaki, F., E. R. Price, D. Milan, and F. McKeon. 1996. Role of kinases and the phosphatase calcineurin in the nuclear shuttling of transcription factor NF-AT4. *Nature 382:370.*

173. Neal, J. W., and N. A. Clipstone. 2001. Glycogen synthase kinase-3 inhibits the DNA binding activity of NFATc. *J Biol Chem 276:3666.*

174. Porter, C. M., M. A. Havens, and N. A. Clipstone. 2000. Identification of amino acid residues and protein kinases involved in the regulation of NFATc subcellular localization. *J Biol Chem 275:3543.*

175. Klemm, J. D., C. R. Beals, and G. R. Crabtree. 1997. Rapid targeting of nuclear proteins to the cytoplasm. *Curr Biol 7:638.*

176. Chow, C. W., and R. J. Davis. 2000. Integration of calcium and cyclic AMP signaling pathways by 14-3-3. *Mol Cell Biol 20:702.*

177. Okamura, H., C. Garcia-Rodriguez, H. Martinson, J. Qin, D. M. Virshup, and A. Rao. 2004. A conserved docking motif for CK1 binding controls the nuclear localization of NFAT1. *Mol Cell Biol 24:4184.*

178. Gwack, Y., S. Sharma, J. Nardone, B. Tanasa, A. Iuga, S. Srikanth, H. Okamura, D. Bolton, S. Feske, P. G. Hogan, and A. Rao. 2006. A genome-wide Drosophila RNAi screen identifies DYRK-family kinases as regulators of NFAT. *Nature 441:646.*

179. Winslow, M. M., M. Pan, M. Starbuck, E. M. Gallo, L. Deng, G. Karsenty, and G. R. Crabtree. 2006. Calcineurin/NFAT signaling in osteoblasts regulates bone mass. *Dev Cell 10:771.*

180. Luo, C., K. T. Shaw, A. Raghavan, J. Aramburu, F. Garcia-Cozar, B. A. Perrino, P. G. Hogan, and A. Rao. 1996. Interaction of calcineurin with a domain of the transcription factor NFAT1 that controls nuclear import. *Proc Natl Acad Sci USA 93:8907.*
181. Beals, C. R., N. A. Clipstone, S. N. Ho, and G. R. Crabtree. 1997. Nuclear localization of NF-ATc by a calcineurin-dependent, cyclosporin-sensitive intramolecular interaction. *Genes Dev 11:824.*
182. Brunet, A., F. Kanai, J. Stehn, J. Xu, D. Sarbassova, J. V. Frangioni, S. N. Dalal, J. A. DeCaprio, M. E. Greenberg, and M. B. Yaffe. 2002. 14-3-3 transits to the nucleus and participates in dynamic nucleocytoplasmic transport. *J Cell Biol 156:817.*
183. Willingham, A. T., A. P. Orth, S. Batalov, E. C. Peters, B. G. Wen, P. Aza-Blanc, J. B. Hogenesch, and P. G. Schultz. 2005. A strategy for probing the function of noncoding RNAs finds a repressor of NFAT. *Science 309:1570.*
184. Briggs, M. W., and D. B. Sacks. 2003. IQGAP1 as signal integrator: Ca2+, calmodulin, Cdc42 and the cytoskeleton. *FEBS Lett 542:7.*
185. Dolmetsch, R. E., K. Xu, and R. S. Lewis. 1998. Calcium oscillations increase the efficiency and specificity of gene expression. *Nature 392:933.*
186. Hoffmann, A., A. Levchenko, M. L. Scott, and D. Baltimore. 2002. The IkappaB-NF-kappaB signaling module: temporal control and selective gene activation. *Science 298:1241.*
187. Chen, L., J. N. Glover, P. G. Hogan, A. Rao, and S. C. Harrison. 1998. Structure of the DNA-binding domains from NFAT, Fos and Jun bound specifically to DNA. *Nature 392:42.*
188. Jin, L., P. Sliz, L. Chen, F. Macian, A. Rao, P. G. Hogan, and S. C. Harrison. 2003. An asymmetric NFAT1 dimer on a pseudo-palindromic kappa B-like DNA site. *Nat Struct Biol 10:807.*
189. Giffin, M. J., J. C. Stroud, D. L. Bates, K. D. von Koenig, J. Hardin, and L. Chen. 2003. Structure of NFAT1 bound as a dimer to the HIV-1 LTR kappa B element. *Nat Struct Biol 10:800.*
190. Stroud, J. C., and L. Chen. 2003. Structure of NFAT bound to DNA as a monomer. *J Mol Biol 334:1009.*
191. Jain, J., P. G. McCaffrey, Z. Miner, T. K. Kerppola, J. N. Lambert, G. L. Verdine, T. Curran, and A. Rao. 1993. The T-cell transcription factor NFATp is a substrate for calcineurin and interacts with Fos and Jun. *Nature 365:352.*
192. Schwartz, R. H. 2003. T cell anergy. *Annu Rev Immunol 21:305.*
193. Kamradt, T., and N. A. Mitchison. 2001. Tolerance and autoimmunity. *N Engl J Med 344:655.*
194. Goodnow, C. C., J. Sprent, B. Fazekas de St Groth, and C. G. Vinuesa. 2005. Cellular and genetic mechanisms of self tolerance and autoimmunity. *Nature 435:590.*
195. Roncarolo, M. G., and M. K. Levings. 2000. The role of different subsets of T regulatory cells in controlling autoimmunity. *Curr Opin Immunol 12:676.*
196. O'Garra, A., and P. Vieira. 2004. Regulatory T cells and mechanisms of immune system control. *Nat Med 10:801.*
197. Bluestone, J. A., and A. K. Abbas. 2003. Natural versus adaptive regulatory T cells. *Nat Rev Immunol 3:253.*
198. von Boehmer, H. 2005. Mechanisms of suppression by suppressor T cells. *Nat Immunol 6:338.*
199. Sakaguchi, S. 2004. Naturally arising CD4+ regulatory t cells for immunologic self-tolerance and negative control of immune responses. *Annu Rev Immunol 22:531.*
200. Kronenberg, M., and A. Rudensky. 2005. Regulation of immunity by self-reactive T cells. *Nature 435:598.*
201. Piccirillo, C. A., and E. M. Shevach. 2004. Naturally-occurring CD4+CD25+ immunoregulatory T cells: central players in the arena of peripheral tolerance. *Semin Immunol 16:81.*
202. Ziegler, S. F. 2006. FOXP3: of mice and men. *Annu Rev Immunol 24:209.*
203. Feske, S., R. Draeger, H. H. Peter, K. Eichmann, and A. Rao. 2000. The duration of nuclear residence of NFAT determines the pattern of cytokine expression in human SCID T cells. *J Immunol 165:297.*
204. Teague, T. K., D. Hildeman, R. M. Kedl, T. Mitchell, W. Rees, B. C. Schaefer, J. Bender, J. Kappler, and P. Marrack. 1999. Activation changes the spectrum but not the diversity of genes expressed by T cells. *Proc Natl Acad Sci USA 96:12691.*

205. Cristillo, A. D., and B. E. Bierer. 2002. Identification of novel targets of immunosuppressive agents by cDNA-based microarray analysis. *J Biol Chem 277:4465.*

206. Macian, F., F. Garcia-Cozar, S. H. Im, H. F. Horton, M. C. Byrne, and A. Rao. 2002. Transcriptional mechanisms underlying lymphocyte tolerance. *Cell 109:719.*

207. Diehn, M., A. A. Alizadeh, O. J. Rando, C. L. Liu, K. Stankunas, D. Botstein, G. R. Crabtree, and P. O. Brown. 2002. Genomic expression programs and the integration of the CD28 costimulatory signal in T cell activation. *Proc Natl Acad Sci USA 99:11796.*

208. Barrington, R. A., M. Borde, A. Rao, and M. C. Carroll. 2006. Involvement of NFAT1 in B cell self-tolerance. *J Immunol 177:1510.*

209. Borde, M., R. A. Barrington, V. Heissmeyer, M. C. Carroll, and A. Rao. 2006. Transcriptional basis of lymphocyte tolerance. *Immunol Rev 210:105.*

210. Heissmeyer, V., and A. Rao. 2004. E3 ligases in T cell anergy–turning immune responses into tolerance. *Sci STKE 2004:pe29.*

211. Heissmeyer, V., F. Macian, S. H. Im, R. Varma, S. Feske, K. Venuprasad, H. Gu, Y. C. Liu, M. L. Dustin, and A. Rao. 2004. Calcineurin imposes T cell unresponsiveness through targeted proteolysis of signaling proteins. *Nat Immunol 5:255.*

212. Healy, J. I., R. E. Dolmetsch, L. A. Timmerman, J. G. Cyster, M. L. Thomas, G. R. Crabtree, R. S. Lewis, and C. C. Goodnow. 1997. Different nuclear signals are activated by the B cell receptor during positive versus negative signaling. *Immunity 6:419.*

213. Aifantis, I., F. Gounari, L. Scorrano, C. Borowski, and H. von Boehmer. 2001. Constitutive pre-TCR signaling promotes differentiation through Ca2+ mobilization and activation of NF-kappaB and NFAT. *Nat Immunol 2:403.*

214. Jenkins, M. K., R. H. Schwartz, and D. M. Pardoll. 1988. Effects of cyclosporine A on T cell development and clonal deletion. *Science 241:1655.*

215. Gao, E. K., D. Lo, R. Cheney, O. Kanagawa, and J. Sprent. 1988. Abnormal differentiation of thymocytes in mice treated with cyclosporin A. *Nature 336:176.*

216. Jiang, H., F. Xiong, S. Kong, T. Ogawa, M. Kobayashi, and J. O. Liu. 1997. Distinct tissue and cellular distribution of two major isoforms of calcineurin. *Mol Immunol 34:663.*

217. Neilson, J. R., M. M. Winslow, E. M. Hur, and G. R. Crabtree. 2004. Calcineurin B1 is essential for positive but not negative selection during thymocyte development. *Immunity 20:255.*

218. Zhang, B. W., G. Zimmer, J. Chen, D. Ladd, E. Li, F. W. Alt, G. Wiederrecht, J. Cryan, E. A. O'Neill, C. E. Seidman, A. K. Abbas, and J. G. Seidman. 1996. T cell responses in calcineurin A alpha-deficient mice. *J Exp Med 183:413.*

219. Chan, V. S., C. Wong, and P. S. Ohashi. 2002. Calcineurin Aalpha plays an exclusive role in TCR signaling in mature but not in immature T cells. *Eur J Immunol 32:1223.*

220. Graef, I. A., F. Chen, L. Chen, A. Kuo, and G. R. Crabtree. 2001. Signals transduced by Ca(2+)/calcineurin and NFATc3/c4 pattern the developing vasculature. *Cell 105:863.*

221. Lyakh, L., P. Ghosh, and N. R. Rice. 1997. Expression of NFAT-family proteins in normal human T cells. *Mol Cell Biol 17:2475.*

222. Timmerman, L. A., J. I. Healy, S. N. Ho, L. Chen, C. C. Goodnow, and G. R. Crabtree. 1997. Redundant expression but selective utilization of nuclear factor of activated T cells family members. *J Immunol 159:2735.*

223. Xanthoudakis, S., J. P. Viola, K. T. Shaw, C. Luo, J. D. Wallace, P. T. Bozza, D. C. Luk, T. Curran, and A. Rao. 1996. An enhanced immune response in mice lacking the transcription factor NFAT1. *Science 272:892.*

224. Schuh, K., B. Kneitz, J. Heyer, U. Bommhardt, E. Jankevics, F. Berberich-Siebelt, K. Pfeffer, H. K. Muller-Hermelink, A. Schimpl, and E. Serfling. 1998. Retarded thymic involution and massive germinal center formation in NF- ATp-deficient mice. *Eur J Immunol 28:2456.*

225. Kiani, A., J. P. Viola, A. H. Lichtman, and A. Rao. 1997. Down-regulation of IL-4 gene transcription and control of Th2 cell differentiation by a mechanism involving NFAT1. *Immunity 7:849.*

226. Yoshida, H., H. Nishina, H. Takimoto, L. E. Marengere, A. C. Wakeham, D. Bouchard, Y. Y. Kong, T. Ohteki, A. Shahinian, M. Bachmann, P. S. Ohashi, J. M. Penninger, G. R. Crabtree, and T. W. Mak. 1998. The transcription factor NF-ATc1 regulates lymphocyte proliferation and Th2 cytokine production. *Immunity 8:115.*

227. Ranger, A. M., M. R. Hodge, E. M. Gravallese, M. Oukka, L. Davidson, F. W. Alt, F. C. de la Brousse, T. Hoey, M. Grusby, and L. H. Glimcher. 1998. Delayed lymphoid repopulation with defects in IL-4-driven responses produced by inactivation of NF-ATc. *Immunity 8:125.*

228. Oukka, M., I. C. Ho, F. C. de la Brousse, T. Hoey, M. J. Grusby, and L. H. Glimcher. 1998. The transcription factor NFAT4 is involved in the generation and survival of T cells. *Immunity 9:295.*

229. Ranger, A. M., M. Oukka, J. Rengarajan, and L. H. Glimcher. 1998. Inhibitory function of two NFAT family members in lymphoid homeostasis and Th2 development. *Immunity 9:627.*

230. Peng, S. L., A. J. Gerth, A. M. Ranger, and L. H. Glimcher. 2001. NFATc1 and NFATc2 together control both T and B cell activation and differentiation. *Immunity 14:13.*

231. Striessnig, J., J. C. Hoda, A. Koschak, F. Zaghetto, C. Mullner, M. J. Sinnegger-Brauns, C. Wild, K. Watschinger, A. Trockenbacher, and G. Pelster. 2004. L-type Ca2+ channels in Ca2+ channelopathies. *Biochem Biophys Res Commun 322:1341.*

232. Benkusky, N. A., E. F. Farrell, and H. H. Valdivia. 2004. Ryanodine receptor channelopathies. *Biochem Biophys Res Commun 322:1280.*

233. Felix, R. 2006. Calcium channelopathies. *Neuromolecular Med 8:307.*

234. Nilius, B., T. Voets, and J. Peters. 2005. TRP channels in disease. *Sci STKE 2005:re8.*

235. Glud, S. Z., A. B. Sorensen, M. Andrulis, B. Wang, E. Kondo, R. Jessen, L. Krenacs, E. Stelkovics, M. Wabl, E. Serfling, A. Palmetshofer, and F. S. Pedersen. 2005. A tumor-suppressor function for NFATc3 in T-cell lymphomagenesis by murine leukemia virus. *Blood 106:3546.*

236. Fu, L., Y. C. Lin-Lee, L. V. Pham, A. Tamayo, L. Yoshimura, and R. J. Ford. 2006. Constitutive NF-kappaB and NFAT activation leads to stimulation of the BLyS survival pathway in aggressive B-cell lymphomas. *Blood 107:4540.*

237. Pham, L. V., A. T. Tamayo, L. C. Yoshimura, Y. C. Lin-Lee, and R. J. Ford. 2005. Constitutive NF-kappaB and NFAT activation in aggressive B-cell lymphomas synergistically activates the CD154 gene and maintains lymphoma cell survival. *Blood 106:3940.*

238. Noguchi, H., M. Matsushita, T. Okitsu, A. Moriwaki, K. Tomizawa, S. Kang, S. T. Li, N. Kobayashi, S. Matsumoto, K. Tanaka, N. Tanaka, and H. Matsui. 2004. A new cell-permeable peptide allows successful allogeneic islet transplantation in mice. *Nat Med 10:305.*

239. Spassova, M. A., J. Soboloff, L. P. He, W. Xu, M. A. Dziadek, and D. L. Gill. 2006. STIM1 has a plasma membrane role in the activation of store-operated Ca(2+) channels. *Proc Natl Acad Sci USA 103:4040.*

240. Gwack, Y., S. Srikanth, S. Feske, F. Crue-Guilloty, M. OH-Hora, D. S. Neems, P. G. Hogan, and A. Rao. 2007. Biochemical and functional characterization of Orai family proteins. *J Biol Chem.* [Epub ahead of print].

J. Krebs and M. Michalak (Editors)
Calcium: A Matter of Life or Death

ISSN: 0167-7306/DOI: 10.1016/S0167-7306(06)41015-2

CHAPTER 15

Temporal and spatial regulation of calcium-dependent transcription

Jacob Brenner, Natalia Gomez-Ospina, and Ricardo Dolmetsch

Department of Neurobiology, Stanford University School of Medicine, 299 Campus Drive, Fairchild Building Rm. 227, Stanford, CA 94305, USA, Tel.: +1 650 723 9812; Fax: +1 650 725 3958; E-mail: ricardo.dolmetsch@stansford.edu

Abstract

Intracellular calcium ($[Ca^{2+}]_i$) signals play an important role in regulating gene expression in cells. Activation of receptors and ion channels at the cell membrane generates a wide diversity of $[Ca^{2+}]_i$ signals that range from microdomains of elevated Ca^{2+} around ion channels to $[Ca^{2+}]_i$ oscillations and waves. In this review, we discuss how the biochemical machinery that regulates transcription interprets spatially and temporally varying $[Ca^{2+}]_i$ signals in cells. We focus on how distinct sets of Ca^{2+} channels can differentially regulate the activation of the transcription factor calcium-response-element binding protein (CREB) and on how $[Ca^{2+}]_i$ oscillations regulate the activation of the transcription factors nuclear factor of activated T cells (NFAT) and nuclear factor κB (NFκB). Finally, we discuss how the transcription factor myocyte enhancer factor 2 (MEF2) and the associated histone deacetylase 4 (HDAC4) are regulated by dynamic $[Ca^{2+}]_i$ signals in muscle cells. These examples illustrate how cells use spatially and temporally varying $[Ca^{2+}]_i$ signals to selectively activate distinct sets of genes, thereby increasing the versatility of Ca^{2+} as a second messenger.

Keywords: Ca^{2+}-channel, Ca^{2+}-dependent gene transcription, CREB, HDAC4, MEF2, NFAT, NKκB, NMDA receptor, transcription factors

$[Ca^{2+}]_i$-regulated transcription plays a key role in converting signals at the membrane into long-lasting biochemical and structural changes in cells. One important way in which $[Ca^{2+}]_i$ regulates cellular phenotypes is by activating signaling cascades that lead to changes in gene expression [1,2]. At least 30 Ca^{2+}-regulated transcription factors have been identified, and thousands of genes are regulated by changes in $[Ca^{2+}]_i$ [2]. Because Ca^{2+} regulates the expression of such a large number of genes, cells have developed mechanisms for using the temporal and spatial features of $[Ca^{2+}]_i$ signals to selectively activate distinct sets of calcium-regulated genes. These mechanisms increase the versatility of Ca^{2+} as a signaling molecule and allow Ca^{2+} to participate in a large number of cellular responses. In this review, we will discuss the mechanisms by which the spatial and temporal features of a calcium signal are decoded by the machinery that regulates gene expression in cells.

Electrical, mechanical, and chemical stimulation of cells leads to a wide diversity of intracellular $[Ca^{2+}]_i$ signals [3]. Distinct types of stimuli activate different sets of ion channels and therefore produce spatially and temporally distinct $[Ca^{2+}]_i$ signals. An example of spatial Ca^{2+} patterns arising from the activation of distinct types of receptors occurs in cortical neurons. Cortical neurons express NMDA receptors and nine types of voltage-gated Ca^{2+} channels (VGCCs), grouped into L, N, P/Q, T, and R families (encoded by the Ca_V 1, 2, and 3 family of genes) [4]. VGCCs all have the common ability to elevate $[Ca^{2+}]_i$ in response to changes in membrane potential, but they are activated by different stimuli and differ in their cellular localization and in their activation kinetics [5]. L-Type calcium channels are localized in the dendrites and cell bodies of neurons [6,7] and inactivate relatively slowly [8]. In contrast, N- and P/Q-type channels are found mainly in synaptic boutons [7] and inactivate rapidly [8]. Finally, NMDA receptors are localized in the cell body, and postsynaptic spines [9] are gated by extracellular glutamate and do not inactivate at all [10,11]. Activation of each of these types of Ca^{2+} channel therefore leads to a $[Ca^{2+}]_i$ rise that originates in a distinct part of the cell, that is triggered by specific extracellular events, and that has a characteristic temporal signature.

The subcellular localization of a Ca^{2+} channel is particularly important for signaling because activation of a Ca^{2+} channel generates a calcium gradient that is high close to the channel and decays rapidly as a function of distance. These "microdomains" of Ca^{2+} that are found close to channels activate signaling pathways very rapidly in response to the opening of a specific type of Ca^{2+} channel [12,13]. Channels can confer specificity to a $[Ca^{2+}]_i$ signal because activation of a specific type of Ca^{2+} channel generates a region of high $[Ca^{2+}]_i$ close to the channel that will activate Ca^{2+}-binding proteins that are localized in this domain and not Ca^{2+}-binding proteins in other parts of the cell. The binding of Ca^{2+}-binding proteins that regulate signaling proteins close to the mouth of Ca^{2+} channels probably accounts for the general observation that Ca^{2+} influx through distinct types of Ca^{2+} channels selectively activates different cellular responses. For example, L-type channels produce dendritic or somatic $[Ca^{2+}]_i$ elevations and regulate dendritic arborization [14], whereas N- and P/Q-type channels cause $[Ca^{2+}]_i$ elevations in presynaptic boutons and cause synaptic vesicle release [15]. Therefore, the spatial features of $[Ca^{2+}]_i$ contain information about the stimuli impinging on a cell.

In addition to spatial $[Ca^{2+}]_i$ patterns that arise from the activation of distinct types of calcium channels, different types of stimuli also lead to the generation of $[Ca^{2+}]_i$ signals with characteristic kinetic features. In cells such as T lymphocytes, mast cells, hepatocytes, and pancreatic acinar cells, stimulation of PLC-coupled receptors leads to $[Ca^{2+}]_i$ oscillations and waves [3]. Oscillations are generated by a variety of mechanisms that have in common the interplay between periodic Ca^{2+} release from internal stores and either oscillatory or sustained Ca^{2+} influx through cell-surface channels [16,17]. In hepatocytes, the frequency of $[Ca^{2+}]_i$ oscillations reflects the concentration of the extracellular stimulus [18], whereas in T cells [19,20], oscillations occur only at a narrow stimulus concentration range. In both cases, the temporal pattern of $[Ca^{2+}]_i$ reflects features of the extracellular stimulus impinging on the cell and thus provides information to the signaling machinery in the nucleus. In cells that fire action potentials such as neurons and muscle cells, patterns of action potentials that reflect the magnitude and frequency of stimulation produce oscillatory or sustained $[Ca^{2+}]_i$ signals that are a filtered reflection of the electrical activity of the cell [21]. Because $[Ca^{2+}]_i$ entry is fast, but the mechanisms that remove Ca^{2+} from the intracellular space (buffers and pumps) are slow relative to changes in membrane potential, low-frequency action potentials produce oscillatory $[Ca^{2+}]_i$ elevations, but very fast action potential trains lead to a sustained $[Ca^{2+}]_i$ rise whose amplitude is proportional to the frequency of the action potentials [22]. In both cases, the features of the $[Ca^{2+}]_i$ signal contain information about the firing frequency of the cell.

The observation that distinct stimuli lead to distinct $[Ca^{2+}]_i$ signals has led to the idea that cells can interpret differences in $[Ca^{2+}]_i$ to generate specific cellular responses. This idea has been particularly well studied for a few transcription factors (see sections 1–4), but it is very likely that the temporal and spatial kinetics of $[Ca^{2+}]_i$ affect all Ca^{2+}-activated signaling cascades. This is because most Ca^{2+}-binding proteins are not distributed uniformly in cells but are localized in specific subcellular compartments such as the presynaptic or postsynaptic membranes or the T tubules in muscle cells. Therefore, spatial calcium gradients, which tend to be very steep around ion channels, will preferentially activate Ca^{2+}-regulated proteins that are close to those Ca^{2+} channels. Similarly, the temporal features of $[Ca^{2+}]_i$ will also have different effects on different Ca^{2+}-activated signaling cascades depending on kinetics of activation and inactivation of the signaling cascade. Ca^{2+}-regulated signaling cascades that deactivate quickly after a decrease in $[Ca^{2+}]_i$ will be activated best by sustained $[Ca^{2+}]_i$ elevations or by high-frequency $[Ca^{2+}]_i$ oscillations. In contrast, cascades that inactivate slowly after $[Ca^{2+}]_i$ decreases in the cell will require less frequent calcium elevations. Other kinetic aspects of the response of a Ca^{2+}-regulated transcriptional pathway are also important for determining the response of a transcription factor to a particular pattern of $[Ca^{2+}]_i$. For example, a signaling cascade that is first activated and then inactivated by $[Ca^{2+}]_i$ will respond best to periodic $[Ca^{2+}]_i$ elevations and not to $[Ca^{2+}]_i$ calcium rises. This is because a sustained $[Ca^{2+}]_i$ elevation will inactivate the signaling system, whereas periodic elevations of the right frequency will cause activation but not trigger inactivation. By tuning the rate of activation and inactivation of a Ca^{2+}-regulated signaling cascade, it is possible to generate signaling systems that are exquisitely sensitive to the frequency of $[Ca^{2+}]_i$ oscillations in cells. In the following sections, we will explore the effects of $[Ca^{2+}]_i$ dynamics on Ca^{2+}-dependent transcriptional activation pathways, focusing on several well-studied transcription factors, CREB, NFAT, NFκB, MEF2.

1. CREB regulation by $[Ca^{2+}]_i$

The transcription factor CREB is one of the most extensively studied targets of $[Ca^{2+}]_i$ in the nucleus of cells. It mediates expression of target genes by binding as a dimer to the palindromic CRE, TGACGTCA [23], and its activity is important for neuronal survival, glucose homeostasis, and synaptic plasticity [24]. CREB was originally identified as a target of cAMP and protein kinase A (PKA), but it is also a major mediator of the effects of calcium on transcription [25]. Several thousand

CREB-binding sites have been identified in the human genome that regulate the expression of a wide variety of genes including BDNF, c-Fos, and a large number of micro-RNAs [26]. Mice lacking CREB and mice that express various dominant-negative CREB proteins show defects in neuronal survival, in neuronal arborization, and in various aspects of neuronal behavior [27]. CREB therefore plays an important role in regulating experience-dependent plasticity in the mammalian brain.

CREB is regulated by phosphorylation of at least three serines: Ser133 [28], Ser142, and Ser143 [29–31]. Phosphorylation of CREB Ser133 is required for the activation of CREB-dependent transcription, whereas Ser142 and Ser143 play more complex modulatory roles. Phosphorylation of Ser133 recruits CREB to a transcriptional activation complex [32] that includes transcriptional coregulators such as CREB-binding protein (CBP [33]) and TORC [34]. CBP and its homolog P300 are adaptor proteins that bind to the kinase-inducible domain of CREB and link it to other transcription factors and to transcriptional modulators such as histone deacetylases (HDACs) and basal transcription factors; TORC is a newly discovered family of transcriptional regulators that regulates glucose homeostasis [34]. Both CBP and TORC are themselves regulated by $[Ca^{2+}]_i$. CBP is phosphorylated by a $[Ca^{2+}]_i$-activated kinase [35], and TORC translocates into the nucleus in response to dephosphorylation by the Ca^{2+}–calomodulin-dependent phosphatase calcineurin [36]. In addition to regulation by CBP and TORC, CREB is also regulated by a Ca^{2+}-dependent activation of nitric oxide synthase that triggers the nitrosylation of nuclear target proteins [32]. Thus, CREB-dependent transcription involves multiple Ca^{2+}-dependent steps and signal transduction pathways.

Because Ser133 phosphorylation is essential for activation of CREB, the mechanisms that lead to this event have been studied extensively. Many kinases can phosphorylate CREB Ser133 including PKA [37], the Ca^{2+}–CaM-activated kinases I [38], II, and IV [31,39], the kinases Rsk I and II [40,41], Msk I and Msk II [42–44], and Akt [45,46]. Which kinase cascade is the major CREB regulator seems to depend on the stimulus and cell type. CamK I and IV and Rsk I and II are thought to be the major players in regulating CREB in response to $[Ca^{2+}]_i$ elevations in hippocampal and cortical neurons (reviewed in [24]). The Rsk kinases are targets of the MAP kinase cascade that includes, Ras, Raf, MapKK, and Erk. $[Ca^{2+}]_i$ regulates this pathway through the kinase Pyk-2 [47] and the Ras guanine nucleotide exchange factors like RAS-GRF [48], RAS-GRP [49], CAPRI [50], and RASL [51], which activate Ras by promoting the exchange of GDP for GTP on Ras and other small G proteins. In neurons, both of these kinase signaling cascades play a key role in regulating activity-dependent gene expression.

In contrast to the phosphorylation of CREB, much less is known about CREB dephosphorylation following decreases in intracellular $[Ca^{2+}]_i$. The phosphatases PP1 and PP2 have both been proposed to directly dephosphorylate CREB [52–55]. PP1 in particular has been suggested to play a role in the transient CREB phosphorylation in response to brief electrical stimuli; however, precisely how CREB phosphatases are regulated and how they regulate CREB in response to increases in $[Ca^{2+}]_i$ have not been fully determined.

1.1. CREB regulation by L-type channels and NMDA receptors

Because CREB is activated by $[Ca^{2+}]_i$ signals in cells like neurons with complex intracellular $[Ca^{2+}]_i$ signaling patterns, a great deal is known about how it responds to particular types of $[Ca^{2+}]_i$ signals. Three features of $[Ca^{2+}]_i$ signals play an important role in how CREB-dependent transcription is regulated, the source of $[Ca^{2+}]_i$, the kinetics of the $[Ca^{2+}]_i$ rise, and the subcellular localization of the $[Ca^{2+}]_i$ elevation.

The source of Ca^{2+} influx into a cell seems to play a key role in determining to what extent $[Ca^{2+}]_i$ signals can activate CREB-dependent transcription. In neurons, Ca^{2+} can enter cells through both VGCCs and NMDA receptors. Ca^{2+} influx through VGCCs leads to rapid CREB phosphorylation that is maintained over the course of several hours and that leads to robust activation of CREB-dependent transcription [56]. Of the VGCCs, L-type Ca^{2+} channels (LTCs) are particularly effective at activating CREB [57]. Blockers of LTCs like nimodipine strongly inhibit CREB phosphorylation and activation in response to bath depolarization of neuronal cultures. In contrast, blockers of N-, P/Q- and R-type Ca^{2+} channels have no effect on CREB activation suggesting that LTCs are privileged in their ability to activate CREB. The ability of LTCs to activate CREB does not seem to arise because LTCs elevate $[Ca^{2+}]_i$ more effectively than other channels [58]. Even when the nuclear Ca^{2+} levels are matched by altering extracellular Ca^{2+} levels, LTCs continue to be more effective at activating CREB than other channels. Thus, LTC signaling to CREB seems to be largely independent of their ability to elevate global Ca^{2+} levels.

It is not known what features of LTCs are necessary for their ability to activate CREB in neurons. The binding of CaM to an IQ motif on the C-terminus of the LTC is important for activating a signaling pathway that leads from LTCs to the nucleus because channels lacking this motif can elevate $[Ca^{2+}]_i$ but are inefficient at activating CREB [58]. The binding of the CaM to the IQ motif seems to be important for activating the MAP kinase cascade downstream of the LTC suggesting that this signaling pathway is closely associated with the channel. In addition to the binding of CaM to the IQ motif of the LTC, the last four amino acids in the C-terminus of the LTC have also been implicated in connecting the LTCs to activation of CREB [59]. This domain is a putative binding site for proteins containing PDZ motifs suggesting that PDZ motif proteins might be involved in connecting Ca^{2+} channels to signaling proteins. The importance of the PDZ domain for activation of signaling to CREB by the LTC, however, has not been replicated by all groups suggesting that the PDZ-binding domain of the LTC might not be universally required for activating signaling leading to CREB. It has also been suggested that of the four LTCs in the genome, only one CaV 1.3 (Alpha1D) is specialized for signaling to the nucleus [60]. In summary, although it is clear that LTCs are specialized for activating CREB at least in response to sustained depolarizing stimuli, it is not known precisely which types of L-type channels are important or how they are linked to signaling pathways that lead to the nucleus.

In addition to VGCCs, Ca^{2+} also enters neurons through NMDA receptors. Activation of NMDA receptors by bath application of NMDA or glutamate leads to transient CREB phosphorylation in a manner that is dependent on the age of the neurons [61]. Because CREB phoshorylation must be sustained for activation of CREB-dependent transcription, bath application of NMDA receptors is generally inefficient at activating CREB-dependent transcription. Why NMDA receptor activation leads to transient CREB activation is not fully understood. Activation of NMDA receptors after activation of LTCs causes a decline in CREB phosphorylation suggesting that NMDA receptors activate a CREB phosphatase. The identity of this phosphatase is not known although it has been suggested that under some conditions of elevated $[Ca^{2+}]_i$, calcineurin might be involved in negatively regulating CREB either by regulating PP1 or by inactivating CamKIV [53]. Therefore, global activation of NMDA receptors leads to transient CREB phosphorylation and to a weak activation of CREB-dependent transcription.

1.2. CREB regulation by action potentials and synaptic stimulation

In an effort to study the effects of more physiological stimuli on CREB activation, several groups have examined how action potentials and synaptic stimulation regulate CREB. Trains of action potentials activate CREB presumably by producing a backpropagating depolarization that leads to the activation of VGCCs. Fields and colleagues [62] studied CREB activation in response to different patterns of electrical stimulation. The authors used 10 Hz bursts of action potential of different durations (1.8, 3, 5.4, and 9 s) spaced in such a way that all stimuli resulted in the same total number of action potentials over the 30-min duration of the experiment. Surprisingly, they observed that CREB was most effectively phosphorylated when 1.8-s bursts were spaced 1-min apart or when 9-s bursts were spaced 5 min apart but poorly phosphorylated by the other stimuli. The authors suggested that CREB is regulated by Ca^{2+}-dependent activators and inhibitors and that high $[Ca^{2+}]_i$ elevations produced by 9-s bursts spaced every 5 min or by 1.8-s bursts every minute overcome a CREB dephosphorylation. In contrast, infrequent bursts of short duration do not overcome CREB inhibition and thus result in lower levels of CREB activation. These results provide evidence that the temporal pattern of $[Ca^{2+}]_i$ elevations plays a role in regulating CREB activity perhaps by maintaining CREB phosphorylation in the face of an active pathway that dephosphorylates CREB.

The response of CREB to another type of physiological stimulation, synaptic activity, has also been studied extensively. Unlike bath stimulation with glutamate which activates both synaptic and nonsynaptic glutamate receptors but fails to activate CREB, synaptic stimulation triggered either chemically by increasing the electrical activity of a dish of cells or electrically by stimulating presynaptic cells leads to sustained CREB phosphorylation and to the activation of transcription [63–65]. Why synaptic activation of CREB is different from the activation of CREB in response to bath application of glutamate is still controversial. One idea is that synaptic stimulation leads to the indirect activation of postsynaptic LTCs that then activate an MAP kinase signaling pathways leading to the nucleus [63]. An alternative hypothesis is that synaptic NMDA receptors containing the NR2B subunit are uniquely able to activate the signaling pathways that lead to the nucleus, whereas nonsynaptic NMDA receptors cannot activate CREB [64]. In the latter view, Ca^{2+} influx through synaptic NMDA receptors activates a local signaling pathway that can promote survival and plasticity by activating transcription, whereas nonsynaptic NMDA receptors lead to only transient CREB phosphorylation and to cell death.

1.3. Nuclear versus local $[Ca^{2+}]_i$ signals

Both voltage-gated and synaptic NMDA receptors seem to generate $[Ca^{2+}]_i$ signals that lead to efficient CREB phosphorylation and to the activation of transcription. This implies that there must exist a signaling molecule that transfers information from channels to the machinery in the nucleus that regulates transcription. One candidate for this signaling molecule is Ca^{2+} itself, which can diffuse to the nucleus and trigger activation of Ca^{2+}-dependent CREB kinases such as CaMKIV [66]. To test the idea that nuclear Ca^{2+} is necessary and sufficient for CREB activation, Hardingham and colleagues [67] injected BAPTA dextran, a nondifusable Ca^{2+} chelator, into the nucleus of neurons. Injection of BAPTA dextran into the nucleus prevents activation of CREB-dependent transcription suggesting that nuclear Ca^{2+} is indeed necessary for CREB-dependent transcription. It is unclear how the BAPTA dextran experiment works in these experiments because the ability of BAPTA to chelate Ca^{2+} is likely to be exhausted in a few seconds given the large amounts of Ca^{2+} that enter the nucleus and stimulation of cells to study CREB are generally minutes or hours long. Therefore, the nuclear effects of the BAPTA dextran may be unrelated to its ability to buffer nuclear Ca^{2+}. A second line of evidence for the idea that nuclear Ca^{2+} elevation is sufficient to lead to CREB phosphorylation is that this event occurs even in isolated nuclei suggesting that all the machinery for activation of CREB is in the nucleus of cells [68]. Finally, injecting cells with wheat germ agglutinin to prevent nuclear translocation of proteins has no effect on the activation of CREB suggesting that at least under some circumstances CREB can be activated independently of the nuclear translocation of any large proteins [68].

One difficulty with Ca^{2+} as the main signaling molecule carrying signals from NMDA receptors and LTCs to the nucleus is that under many circumstances Ca^{2+} influx into the nucleus does not cause prolonged CREB phosphorylation or activate transcription. Activation of N- and P/Q-type VGCCs and activation of extrasynaptic NMDA receptors elevates nuclear Ca^{2+} but does not activate CREB-dependent transcription efficiently [57,58]. In addition, LTCs that have mutations in the IQ domain can elevate intracellular Ca^{2+} but do not lead to CREB-dependent transcription. Conversely, loading cells with the Ca^{2+} buffer EGTA to prevent elevation of nuclear Ca^{2+} has no effect on nuclear CREB phosphorylation suggesting that nuclear Ca^{2+} is not always necessary to activate CREB-dependent transcription. This has lead to the idea that local elevation of Ca^{2+} around Ca^{2+} channels can lead to the activation of signaling cascades that propagate to the nucleus and lead to CREB activation.

The identity of the molecule that transmits signals from the cell membrane to the nucleus is not known. One candidate is CaM, a ubiquitous Ca^{2+}-binding protein that binds to many proteins and confers Ca^{2+}-dependent regulation [57]. In response to increases in $[Ca^{2+}]_i$, CaM translocates to the nucleus of neurons and other cells. Because CREB activation depends on CaM-binding proteins like CaMKK and CaMKIV, CaM has been proposed to be a signal that propagates from ion channels that can regulate CREB activation. CaM however is not likely to be the main signal from LTCs or NMDA receptors to the nucleus. First, CaM levels are already high in the nucleus of most cells, and translocation is only apparent in cells that have been significantly starved of electrical activity. Furthermore, Ca^{2+} binding to CaM is necessary for its translocation, and this is unlikely to persist under conditions of high cytoplasmic EGTA where CREB phosphorylation is known to occur. Therefore, changes in nuclear CaM while contributing to the amplification of nuclear signaling are not likely to be the sole mode of signaling from ion channels to the nucleus [69].

A second possibility is that local activation of the MAP kinase cascade around Ca^{2+} channels generates a signal that travels to the nucleus [58,70]. Activation of LTCs and NMDA receptors can activate Ras at the membrane through Ca^{2+}-regulated exchange factors like Ras-GRS and Ras-GRP. In addition, mutations of the IQ domain of the LTC that reduce CREB phosphorylation reduce the ability of the LTC to activate Erk. However, it is presently not clear how binding of CaM to the LTC would regulate Ras signaling leaving room for other possibilities.

One way of reconciling the need for both local and global $[Ca^{2+}]_i$ signals is to suggest that both nuclear and local $[Ca^{2+}]_i$ signals are necessary for Ca^{2+} activation of CREB. One idea is that MAP kinases and CaMKs respond to different aspects of the Ca^{2+} signal and mediate different phases of CREB activation. According to this idea, CaMKs are activated by nuclear Ca^{2+} elevations and mediate rapid activation of CREB in the seconds or minutes following stimulation. The MAP kinase enzymatic cascade responds to local Ca^{2+} elevations and mediates long-lasting $[Ca^{2+}]_i$ signals in neurons. This scheme would allow neurons to activate transcription in response to both changes in local $[Ca^{2+}]_i$ elevations such as synaptic stimulation and large $[Ca^{2+}]_i$ elevations such as backpropagating action potentials.

In summary, CREB activation depends on both the spatial and the kinetic features of a $[Ca^{2+}]_i$ rise. Ca^{2+} influx through particular types of Ca^{2+} channels seems to activate a local signaling cascade that propagates local $[Ca^{2+}]_i$ signals to the nucleus. At the same time, global $[Ca^{2+}]_i$ elevations can regulate CREB activity by regulating nuclear targets of $[Ca^{2+}]_i$.

2. *$[Ca^{2+}]_i$ regulation of NFAT*

In terms of its response to temporal $[Ca^{2+}]_i$ signals, NFAT is the most thoroughly studied of the Ca^{2+}-regulated transcription factors. NFAT is activated in response to a very broad range of $[Ca^{2+}]_i$ patterns, from the >3 Hz contraction cycle in cardiac muscle to the <0.01 Hz $[Ca^{2+}]_i$ oscillations of T lymphocytes. A great deal of phenomenological information has been collected regarding which temporal patterns of $[Ca^{2+}]_i$ activate NFAT in these cells. More recently, a combination of cell biological and computational modeling approaches has begun to elucidate the mechanisms by which NFAT and its regulatory proteins finely discriminate between patterns of $[Ca^{2+}]_i$.

The NFAT family of transcription factors contains five mammalian genes (NFAT1–5). The primordial member of the family, NFAT5, is the only NFAT-like gene in *Drosophila* and does not share the Ca^{2+} responsiveness that distinguish NFAT1–4 [71]. NFAT1–4 have significant homology at the amino acid level and share common modes of regulation. Indeed, although each of the NFAT1–4 isoforms have unique features, their relative redundancy is evidenced by the mild phenotypes of mice lacking single NFAT isoforms (reviewed in [72]). However, mice simultaneously deficient in two or three NFAT isoforms show severe phenotypes, such as failure of T-cell activation [71] and lethal vascular patterning defects [73], illustrating the cellular processes controlled by NFAT.

NFAT's activity is primarily determined by its phosphorylation state. In unstimulated cells, NFAT resides primarily in the cytoplasm, phosphorylated on 13 serines by a variety of kinases, including DYRK1A, DYRK2, GSK3, and CS1. After reception of a variety of stimuli that cause a sufficient rise in $[Ca^{2+}]_i$, the Ca^{2+}–CaM-dependent phosphatase calcineurin dephosphorylates NFAT. Dephosphorylated NFAT translocates to the nucleus, where it regulates gene expression, mostly through transactivation. NFAT's phosphorylation state determines not just subcellular distribution but also DNA binding affinity and, consequently, transcriptional activity [74–79].

2.1. *NFAT activation by* $[Ca^{2+}]_i$ *oscillations*

NFAT's $[Ca^{2+}]_i$-response dynamics have been most extensively studied in the system in which NFAT was first identified, T-lymphocyte activation [80]. When a T cell encounters an antigen-presenting cell bearing a peptide/myosin heavy chain (MHC) complex of sufficient strength, the T-cell receptor (TCR) activates a signaling cascade that leads to the transcription of more than 4000 genes, including many genes controlled by NFAT, such as IL-2 and IL-4 [81]. TCR stimulation induces Ca^{2+} influx through the Ca^{2+} release-activated Ca^{2+} channel (CRAC) and a consequent rise in global $[Ca^{2+}]_i$ that is critical for NFAT transactivation. Both in vitro and in intact ex vivo thymus, TCR stimulation produces a rise not only in average $[Ca^{2+}]_i$ but also in $[Ca^{2+}]_i$ oscillations [19,82,83].

The effect of $[Ca^{2+}]_i$ oscillation frequency and amplitude on the activation of NFAT in T lymphocytes has been examined extensively. Initial experiments by Negulescu and colleagues [84] suggested that $[Ca^{2+}]_i$ oscillations led to increased activation of the IL-2 promoter in cells stimulated through the TCR. However, because TCR stimulation also activates Ca^{2+}-independent signaling pathways and because the response of T cells to antigen stimulation is highly variable, it was not possible to determine whether any particular parameter of the $[Ca^{2+}]_i$ response was either necessary or sufficient to cause NFAT activation.

A more careful analysis of the effects of $[Ca^{2+}]_i$ oscillation amplitude and frequency on the activation of NFAT required the development of the Ca^{2+} clamp, a method for coordinating the $[Ca^{2+}]_i$ signal in large population of cells by rapidly changing the extracellular Ca^{2+} concentration. Dolmetsch et al. [85] used the Ca^{2+} clamp to determine whether NFAT is more sensitive to $[Ca^{2+}]_i$ oscillations than to a constant rise in $[Ca^{2+}]_i$. The NFAT response was compared for constant $[Ca^{2+}]_i$ rises versus $[Ca^{2+}]_i$ oscillations of the same average $[Ca^{2+}]_i$. When the average $[Ca^{2+}]_i$ was greater than approximately 250 nM, the NFAT response to oscillations was nearly identical to that of a constant $[Ca^{2+}]_i$ rise, suggesting that oscillations did not confer additional sensitivity for $[Ca^{2+}]_i$ signals of large average amplitude. As the average $[Ca^{2+}]_i$ input decreased below 250 nM, however, the NFAT responses produced by oscillations became increasingly stronger than those produced by constant $[Ca^{2+}]_i$ elevation. Thus, NFAT is more sensitive to oscillations than to a constant $[Ca^{2+}]_i$ rise at low $[Ca^{2+}]_i$ levels. These observations provide insight into how NFAT acts as a signal processor of the $[Ca^{2+}]_i$ pattern. First, these results show that NFAT does not simply integrate the $[Ca^{2+}]_i$ over time. Second, this feature of NFAT signal processing supports the hypothesis that oscillations increase the "efficiency" of Ca^{2+}-induced signaling, in that oscillations require less total $[Ca^{2+}]_i$ (integrated over time) than constant $[Ca^{2+}]_i$ elevation to achieve the same level of NFAT activation. As high levels of $[Ca^{2+}]_i$ can be deleterious to cells, efficient oscillations may be beneficial in that they minimize the amount of time $[Ca^{2+}]_i$ remains high.

Why are $[Ca^{2+}]_i$ oscillations more effective than sustained $[Ca^{2+}]_i$ rises at activating NFAT and what is the function of $[Ca^{2+}]_i$ oscillations? A potential explanation for NFAT's heightened response to $[Ca^{2+}]_i$ oscillations is that oscillations allow $[Ca^{2+}]_i$ levels to rise above a threshold required for NFAT activation, whereas a constant $[Ca^{2+}]_i$ rise that results in equal average $[Ca^{2+}]_i$ never surpasses such a threshold. NFAT activation does in fact have a pronounced threshold. For T cells stimulated with a constant $[Ca^{2+}]_i$ rise, the dose-response curve of $[Ca^{2+}]_i$ versus NFAT reporter gene activity is very steep, with an effective Hill coefficient of $n_H = 4.7$ [85]. For comparison, most single phosphorylation events have much shallower dose-response curves, with $n_H = 1$. The unique property of having an effective $n_H > 1$ is termed "ultrasensitivity" and serves as a type of switch in numerous signaling events [86]. One potential explanation for NFAT's ultrasensitivity is that NFAT requires phosphorylation of 13 serine residues for full activation [76]. To understand why multisite phosphorylation might affect NFAT's activation threshold, Salazar and Hofer [87] used a thermodynamic model to explore NFAT's switching between active and inactive conformations based on phosphorylation state. The model found that the dose-response curve between the fraction of activated calcineurin and activated NFAT became increasingly steep as the number of NFAT phosphorylation sites increased. With only one phosphorylation site, the dose-response curve was graded, with $n_H = 1$. But with 13 phosphorylation sites, the n_H was of order 10. The appearance of a threshold when a protein has a large number of phosphorylation sites may be seen as a consequence of linking the probability of multiple phosphorylation events. Thus, multisite phosphorylation may explain part or all of the observed threshold in Ca^{2+} activation of NFAT.

However, other mechanisms, such as the cooperative binding of NFAT to certain promoter elements and the cooperative binding of $[Ca^{2+}]_i$ ions to CaM, are also likely to contribute to the steep threshold observed in NFAT activation.

Increased "efficiency" is not the only possible function of $[Ca^{2+}]_i$ oscillations. One of the earliest questions regarding $[Ca^{2+}]_i$ oscillations was whether the frequency of oscillations could specify which signaling pathways were activated [88]. This question was addressed for TCR-induced transcription by comparing the $[Ca^{2+}]_i$ frequency response for NFAT with that of another Ca^{2+}-regulated transcription factor, NFκB [85]. Using the $[Ca^{2+}]_i$-clamp protocol, T cells were stimulated with $[Ca^{2+}]_i$ oscillations of 1 μM amplitude and periods of 1–30 min for 3 h, after which reporter gene activities for the two transcription factors were measured. For both NFAT and NFκB, reporter gene activity decreased with increasing period. However, for periods greater than 6 min, NFAT was completely inactive, while NFκB showed substantial activity even at periods as long as 30 min. The different $[Ca^{2+}]_i$ frequency responses of these two Ca^{2+}-regulated transcription factors suggests that the frequency of TCR-stimulated $[Ca^{2+}]_i$ oscillations may specify which transcription factors are activated and thus which downstream genes are upregulated. Indeed, an enhancer element of the *IL-2* gene, which contains NFAT consensus sequences, had a frequency response similar to that of NFAT, while the IL-8 enhancer, possessing NFκB-binding sites, behaved like the NFκB reporter. NFκB is discussed further in section 3.

NFAT's intriguing frequency response, with a requirement for $[Ca^{2+}]_i$ oscillations with periods < 6 min, has been explained by detailed measurements of the kinetics of a number of steps of NFAT regulation. Tomida et al. [89] used the Ca^{2+}-clamp technique to control $[Ca^{2+}]_i$ and subsequently measured NFAT phosphorylation and subcellular distribution by western blot of cytosolic and nuclear protein extracts. After a sustained high-amplitude $[Ca^{2+}]_i$ rise, NFAT was dephosphorylated very rapidly ($t_{1/2}$ ~2 min), whereas NFAT translocation to the nucleus was significantly slower ($t_{1/2}$ ~12 min). When calcineurin activity declined to baseline, NFAT was rapidly phosphorylated but then translocated from the nucleus on a much slower time scale (with a $t_{1/2}$ ~15 min) [90]. Tomida et al. then directly compared the *net* "on-rate" of NFAT activation (dephosphorylation + nuclear import) and net "off-rate" (phosphorylation + nuclear export). Using the Ca^{2+} clamp and video microscopy to track a transfected green fluorescent protein (GFP)-tagged NFAT (NFAT-GFP), it was found that the rate of NFAT entry into the nucleus during a $[Ca^{2+}]_i$ rise is greater than the rate of exit out of nucleus upon return to baseline $[Ca^{2+}]_i$, consistent with the biochemical data. Thus, the "off-rate" of NFAT activation is slower than the "on-rate," thereby allowing NFAT to build up if successive $[Ca^{2+}]_i$ pulses are delivered with a period not much greater than the kinetics of the off-rate. Given that a threshold concentration of NFAT must reside in the nucleus for a sufficient amount of time to activate gene expression [84], the time scale of the on- and off-rates (minutes) is consistent with the requirement of $[Ca^{2+}]_i$ oscillations with a period < 6 min to NFAT-activated gene expression [85]. Interestingly, it seems that the kinetics governing NFAT's frequency response might be tuned to read the $[Ca^{2+}]_i$ oscillations elicited by TCR stimulation.

NFAT's apparent tuning to the $[Ca^{2+}]_i$ signals in T cells might be a liability in other cell types. Therefore, it is possible that the NFAT regulatory network has different kinetics in different cell types. The need for cell-type-specific regulation of NFAT dynamics is suggested by the observation that $[Ca^{2+}]_i$ signals vary tremendously between different cell types. As discussed above, stimulated T cells typically have $[Ca^{2+}]_i$ oscillations of around 200 nM in amplitude with a duration approximately 70 s, and T-cell NFAT may be able to discriminate between frequencies of 4 and 1×10^{-3} Hz. But in the relatively similar cell-type B lymphocytes, $[Ca^{2+}]_i$ signals are completely different, with binding of antigen causing $[Ca^{2+}]_i$ to rise from a baseline of approximately 75 nM to a peak of approximately 1.4 μM and then decline to a plateau of approximately 225 nM, with the plateau alone sufficient to activate NFAT [91]. In cultured neonatal cardiac myocytes, NFAT is activated by increasing the beat frequency from 1 to 3 Hz, with each beat typically bringing the $[Ca^{2+}]_i$ from a baseline of ~100 to ~600 nM, with a decay constant of $t_{1/2}$ ~250 ms [92–94]. This small sampling of the different $[Ca^{2+}]_i$ patterns that regulate NFAT activity shows that NFAT can be activated by modulation of very slow $[Ca^{2+}]_i$ oscillations (T cells) or very fast oscillations (cardiac myocytes) or by no oscillations at all (B cells).

It is currently unknown how the NFAT regulatory network can be tuned to read all the widely varying $[Ca^{2+}]_i$ patterns that exist among the numerous cell types in which NFAT regulates transcription. One leading hypothesis is that the presence and concentrations of the numerous proteins that regulate NFAT differ dramatically between cell types. Indeed, the expression of calcineurin is approximately 10-fold higher in cardiac myocytes than in immune cells [72], and the numerous NFAT kinases may also differ in tissue distribution [79]. Future research should thoroughly investigate which parameters of the NFAT regulatory network vary between cell types thus serving as the "loci of flexibility" [95] that permit NFAT to vary its dynamic range so dramatically between tissues.

Another future direction will be to examine how NFAT's response changes after long-term stimulation. There is reason to believe that NFAT's Ca^{2+} response changes over time, as NFAT's activity is modulated by transcriptionally mediated feedback loops. Both NFAT2 and NFAT3 positively regulate their own transcription through NFAT-binding sites in their regulatory regions [96–98]. Additionally, NFAT upregulates the expression of DSCR1, a protein which inhibits calcineurin and therefore NFAT activity [99]. These positive and negative feedback loops should make NFAT's Ca^{2+} response change over time. The importance of these feedback regulations has been examined with a dynamic model of NFAT activation in the context of Down's syndrome (DS) [98]. The trisomy 21 characteristic of DS patients results in 1.5 times the copy number of both DSCR1 and DYRK1A, an NFAT kinase. Dynamic modeling that included the transcriptional feedback loops shows that the increased copy number of DSCR1 and DYRK1A can prevent NFAT activation in many contexts, possibly producing some of the characteristic phenotypes of DS. Future work may aim to investigate how these transcriptional feedbacks affect NFAT's Ca^{2+}-response dynamics in both health and disease.

3. Regulation of NFκB by $[Ca^{2+}]_i$ oscillations

The NFκB family of transcription factors contains five members: RelA, RelB, c-Rel, NFκB1, and NFκB2 [100]. All NFκB family members posses a conserved Rel domain that mediates dimerization, nuclear localization, and a DNA-binding domain. RelA, RelB, and c-Rel have transactivation domains able to bind NFκB consensus sites, whereas NFκB1 and NFκB2 lack these domains and therefore function as transcriptional repressors [100]. Before stimulation, NFκB dimers are held in a latent state in the cytoplasm, bound to inhibitory proteins called inhibitors of NFκB (IkBs). NFκB-inducing stimuli cause the IkB proteins to be phosphorylated, ubiquitinated, and finally proteolyzed by the proteosome. The destruction of IkB allows NFκB to translocate to the nucleus, where it binds to NFκB consensus sites to regulate gene expression (reviewed in [101]).

At least one isoform of NFκB is found in most tissues [101]. In most cell types, NFκB has been primarily studied during regulation by Ca^{2+}-independent signaling pathways. However, in both lymphocytes and neurons, NFκB has been shown to respond to $[Ca^{2+}]_i$ changes with intriguing dynamics [85,91,102]. In lymphocytes, NFκB regulates the expression of numerous genes involved in the immune response, such as the cytokine IL-8 [103]. In B lymphocytes, binding of the expressed immunoglobulin by cognate antigen immediately produces a biphasic $[Ca^{2+}]_i$ rise, characterized by a brief spike of large amplitude (approximately 1 μM) followed by a sustained plateau (approximately 100 nM) lasting tens of minutes [91]. This biphasic $[Ca^{2+}]_i$ rise leads to activation of at least two Ca^{2+}-regulated transcription factors, NFAT and NFκB, and subsequent activation of gene expression.

An interesting question is which phase of the $[Ca^{2+}]_i$ rise is most effective in activating NFκB, the spike or the sustained plateau? Using the Ca^{2+}-clamp technique, Dolmetsch et al. [91] produced a single $[Ca^{2+}]_i$ spike that resembled the antigen-induced spike in terms of amplitude and duration. This single spike activated NFκB but failed to activate NFAT. The Ca^{2+} clamp was then used to produce a sustained low-amplitude $[Ca^{2+}]_i$ plateau with amplitude equal to that of the $[Ca^{2+}]_i$ plateau of antigen stimulation. Intriguingly, the low-amplitude $[Ca^{2+}]_i$ plateau activated NFAT but not NFκB. Thus, NFκB and NFAT seem to be tuned to respond to different parameters of the $[Ca^{2+}]_i$ pattern produced by a natural stimulus.

The different Ca^{2+}-response dynamics of NFκB and NFAT can be partly explained by the kinetics of their regulatory networks. As discussed in section 2, NFAT is activated by Ca^{2+}-mediated phosphorylation and then rephosphorylated and deactivated relatively quickly (time scale of minutes) after the decline of $[Ca^{2+}]_i$. Thus, only a relatively sustained $[Ca^{2+}]_i$ elevation can keep NFAT dephosphorylated for more than a few minutes. In contrast, NFκB is activated by the Ca^{2+}-mediated degradation of IkB, and then NFκB is deactivated comparatively slowly (time scale of at least tens of minutes), as new IkB must be synthesized. Thus, if a $[Ca^{2+}]_i$ signal is large enough in amplitude to activate NFκB, the $[Ca^{2+}]_i$ signal can be withdrawn and NFκB will remain active for tens of minutes. Thus, a spike of Ca^{2+} is capable of activating NFκB, whereas a low-amplitude plateau is not. As was the case for

NFAT, the off-rate of NFκB activation (synthesis of IkB + nuclear export) plays a major role in determining the $[Ca^{2+}]_i$-response dynamics.

NFκB's slow inactivation rate also largely determines the transcription factor's response to the $[Ca^{2+}]_i$ oscillations observed in T cells. In Ca^{2+}-clamped T cells, NFκB was activated by $[Ca^{2+}]_i$ oscillations with periods of 30 min, whereas NFAT could not be activated by oscillations with periods of more than 6 min [85]. This probably reflects the fact that a single pulse of $[Ca^{2+}]$ can activate NFκB for tens of minutes, so that NFκB activity can remain above baseline for much of the 30 min between $[Ca^{2+}]_i$ pulses.

4. *MEF2, HDAC4, and the dynamics of Ca^{2+}-regulated chromatin packing*

Most Ca^{2+}-regulated transcription factors have been characterized as proteins which recruit the basal transcription machinery (BTM) to promoter sequences. However, $[Ca^{2+}]_i$ signaling also regulates transcription by controlling the packing state of chromatin, through control of histone-modifying enzymes. The Ca^{2+}-sensitive MEF2 family of transcription factors interact with both the BTM and a variety of histone-modifying enzymes, regulating histone methylation and acetylation. One group of histone-modifying enzymes that interact with MEF2, the class II HDACs, are themselves regulated directly by Ca^{2+}-dependent phosphorylation by CaMKII. As chromatin modification has different kinetics than BTM recruitment, MEF2 and HDAC potentially provide cells with Ca^{2+}-response dynamics distinct from other Ca^{2+}-regulated transcription factors.

Genomic DNA is compacted by its association with histone proteins. Histones are regulated by a variety of posttranslational modifications that affect histone charge and conformation, which in turn control DNA conformation and access of transcriptional machinery to DNA [104]. One type of histone modification, acetylation, has been shown to be sensitive to $[Ca^{2+}]_i$ signaling, thus providing a link between $[Ca^{2+}]_i$ and chromatin packing state [105]. At least two types of Ca^{2+}-sensitive proteins, MEF2 and HDAC, have been shown to regulate histone acetylation-dependent transcription. The four vertebrate MEF2 genes (*MEF2A–D*) are expressed in many tissues, but MEF2 protein is most abundant in lymphocytes, myocytes, and neurons [106]. MEF2 binds to both the BTM and also numerous histone-modifying enzymes to regulate gene expression. When a cell receives a transactivating stimulus, MEF2 binds to histone acetyltransferases, such as p300 and CBP, to increase lysine acetylation of histones bound to MEF2-responsive DNA sequences [107]. The acetylation destabilizes histone–DNA interactions, thereby permitting access of transcription factors and induction of gene expression. Under other conditions, MEF2 binds to HDACs, which generally decrease expression of MEF2-responsive genes.

MEF2 and HDAC are regulated by $[Ca^{2+}]_i$ signaling in a variety of ways. Calcineurin has been shown to dephosphorylate MEF2 in response to $[Ca^{2+}]_i$ signals such as those caused by depolarization in neurons [108], Ca^{2+} ionophore in skeletal myoblasts, and electrically stimulated mouse muscle in vivo [109]. Dephosphorylated MEF2 binds DNA with higher affinity, thereby increasing MEF2 transactivation [108]. MEF2 transactivation is also augmented by $[Ca^{2+}]_i$ signals that cause CaMKII to phosphorylate HDAC (specifically HDAC4), thereby removing HDAC's repressive effect on MEF2 [110]. These multiple modes of Ca^{2+}-dependent regulation of MEF2 and HDAC may confer complex $[Ca^{2+}]_i$-response dynamics to chromatin packing.

4.1. MEF2 and HDAC response dynamics in muscle fiber-type switching

From a dynamic perspective, the most extensively studied phenotype employing MEF2 and HDAC is muscle fiber-type switching. Mammalian skeletal muscle is grouped into two types, fast twitch (FT) and slow twitch (ST) fibers, based on their contraction and relaxation speeds [111]. The two fiber types express different isoforms and concentrations of many myofibrillar proteins, such as the MHC. In live animals, muscle fibers can be experimentally induced to switch fiber type by de-enervating the muscle and then stimulating the muscle for weeks with a pattern of electrical activity characteristic of the opposite fiber type [112]. Similarly, in vitro-cultured adult mouse FT fibers display increased ST-type MHC expression after 5 days of ST-type stimulation (5-s trains of 5 Hz delivered once per minute) [113].

These in vivo and in vitro models have been employed to study the dynamic properties of the signaling pathways, including MEF2 and class II HDACs, leading to fiber-type switching. MEF2 was first identified as a regulator of fiber type by the finding that MEF2 consensus sites are abundant in myofibrillar gene regulatory regions [109]. MEF2 transactivation was then shown to increase upon slow-type stimulation of FT muscle in vivo [109]. The mechanisms by which MEF2 can discriminate between the $[Ca^{2+}]_i$ patterns experienced by ST and FT fibers can be partly understood by examining the dynamics of the network of molecules that interact with MEF2.

The first point of control that aids MEF2-dependent pattern discrimination is that activated MEF2 has a differential effect on the transcriptional regulatory regions of ST- versus FT-specific genes. ST-specific enhancer elements are more stimulated by active MEF2 than are FT-specific enhancers. This difference in enhancer activity is due in part to the fact that the MEF2 consensus sites in ST-specific enhancers exhibit higher affinity binding to MEF2 protein than do the MEF2 consensus sites in FT enhancers [109]. Thus, different promoter-binding affinity causes ST-specific genes to increase their expression relative to FT genes when an ST-type electrical stimulus activates MEF2.

MEF2's ability to discriminate ST- from FT-type patterns of electrical stimulation is also determined by the dynamics of Ca^{2+}-dependent phosphorylation of MEF2. Calcineurin activity is increased by ST-type electrical stimulation of muscle through changes in $[Ca^{2+}]_i$. Active calcineurin then dephosphorylates and thereby activates MEF2 in skeletal fibers [109]. Thus, the relative kinetics of phosphorylation and de-phosphorylation of MEF2 help determine which patterns of electrical activity stimulate MEF2 transactivation.

The point of control of MEF2 in which $[Ca^{2+}]_i$ dynamics have been best studied is that of regulation of HDAC4 phosphorylation and subsequent nuclear export of HDAC4. In cultured FT fibers, HDAC4-GFP translocates out of the nucleus in response to slow-type electrical stimulation [114]. This HDAC4-GFP translocation is abolished by the CaMK inhibitor KN-62. The HDAC4-GFP translocation was accompanied by an increase in activity of an expression reporter construct containing MEF2 consensus sites [114].

Interestingly, the stimulation patterns that activate HDAC4-GFP translocation are very different from the patterns that activate NFAT-GFP translocation. HDAC4 moved out of the nucleus in response to both 10 and 1 Hz stimulation, whereas NFAT only entered the nucleus in response to 10 Hz stimulation [115,116]. By employing Ca^{2+}-regulated transcription factors that have different frequency responses, skeletal muscle fibers may be able to more reliably convert electrical stimulation patterns into the appropriate expression levels of myofibrillar proteins.

The mechanism of HDAC4's particular signal processing of $[Ca^{2+}]_i$ patterns has not yet been fully elucidated. However, because phosphorylation state and nuclear export of HDAC4 are strongly regulated by CaMKII [110], it is likely that understanding CaMKII's Ca^{2+}-response dynamics will shed light on HDAC4/MEF2 activation. CaMKII responds to increased $[Ca^{2+}]_i$ by phosphorylating numerous downstream targets, including CaMKII itself. The CaMKII holoenzyme is composed of 12 CaMKII subunits. In low $[Ca^{2+}]_i$, a subunit's kinase domain is covered by its autoinhibitory domain, resulting in an inactive subunit. When $[Ca^{2+}]_i$ rises, Ca^{2+}–CaM binds to a CaMKII subunit, inducing a conformational change that relieves autoinhibition and allows the subunit to phosphorylate its substrates. An active subunit can then phosphorylate a neighboring subunit, as long as that subunit is also bound to Ca^{2+}–CaM. Autophosphorylation shifts a subunit into a Ca^{2+}-independent (autonomous) state.

To measure CaMKII's $[Ca^{2+}]_i$-response dynamics, De Koninck and Schulman [117] exposed purified CaMKII protein to a variety of time-varying patterns of $[Ca^{2+}]_i$ concentrations in vitro. As a measure of CaMKII activity, these investigators measured the fraction of CaMKII in the autonomous state (% autonomy). CaMKII autonomy was found to be very sensitive to the frequency of $[Ca^{2+}]_i$ oscillations, with a monotonically increasing frequency response over the tested range. $[Ca^{2+}]_i$ pulses of 4 Hz produced over 6 times the % autonomy as 1 Hz pulses, despite the fact that the two protocols exposed CaMKII to elevated Ca^{2+} for the same length of time by adjusting the number of pulses. Similarly, increased pulse duration or amplitude of the Ca^{2+}–CaM stimulus also increased % autonomy. Interestingly, increased amplitude and pulse duration broadened the frequency-response curve; a sharper threshold, and therefore perhaps increased signal-to-noise ratio, was observed in the frequency response at lower amplitudes and pulse durations. This broadening of the frequency-response curve may explain why some cells are much more sensitive than others to the frequency of $[Ca^{2+}]_i$ oscillations.

Structural and biochemical data suggest an explanation to why CaMKII activity does not function as a simple integrator of $[Ca^{2+}]_i$. When $[Ca^{2+}]_i$ declines, CaM dissociates from CaMKII subunits, decreasing subunit activity and providing an effective off-rate for the creation of autonomous CaMKII. At high-frequency stimulation (4 Hz), there is little dissociation of CaM between $[Ca^{2+}]_i$ pulses. After a few seconds of 4 Hz stimulation, however, Ca^{2+}–CaM-bound subunits have accumulated to a high enough concentration that there is a high probability of neighboring subunits being Ca^{2+}–CaM bound. Therefore, autophosphorylation can occur (measured as % autonomy). Furthermore, the rate of autophosphorylation continues to increase because phosphorylated subunits remain active between $[Ca^{2+}]_i$ pulses. During low-frequency stimulation, however, Ca^{2+}–CaM has largely dissociated from CaMKII during the interval between $[Ca^{2+}]_i$ pulses. Consequently, the probability of neighboring subunits both being Ca^{2+}–CaM bound is low, so the autophosphorylation rate is low. As with NFAT and NFκB, the frequency dependence is largely a result of a relatively fast off-rate of Ca^{2+} activation and the existence of a threshold concentration of activated subunits for the output (% autonomy) to be augmented.

At this point, it is not possible to map CaMKII's in vitro $[Ca^{2+}]_i$-response dynamics onto the in vivo response of MEF2/HDAC-responsive genes. The network of other proteins involved in regulation of these transcription factors should have a significant effect on the dynamics of MEF2-regulated genes. For instance, although CaMKII autophosphorylation alone in vitro displays a steep dependence on $[Ca^{2+}]_i$ ($n_H \approx 5$), simply adding in the endogenous phosphatase PP1 causes CaMKII activation to become even more ultrasensitive ($n_H \approx 8$) [118]. Thus, fully understanding the $[Ca^{2+}]_i$-response dynamics of MEF2/HDAC-responsive genes in vivo may require elucidation of the entire regulatory network. Nonetheless, key principles, such as the effects of off-rates and various mechanisms of ultrasensitivity, can be derived from studying the individual components of the system.

5. *Conclusions*

In conclusion, the temporal and spatial features of $[Ca^{2+}]_i$ signals play a key role in the activation of transcription factors such as CREB, NFAT, NFκB, and MEF2. Cells have developed complex feedback and feed-forward loops that allow these transcription factors to extract information about the spatial and temporal patterns of intracellular calcium and to convert them into changes in the global pattern of gene expression. For each Ca^{2+}-regulated transcription factor, there will be an optimal spatial and temporal pattern of $[Ca^{2+}]_i$ that will result in its optimal activation. Therefore, each Ca^{2+}-regulated transcription factor effectively filters the $[Ca^{2+}]_i$ signal and activates transcription in proportion to how well the $[Ca^{2+}]_i$ signal approximates its optimal stimulus pattern. Thus, every $[Ca^{2+}]_i$ elevation generates a coordinated response of Ca^{2+}-regulated transcription factors that by leads to unique patterns of Ca^{2+}-dependent gene expression. The challenge for the future will be to characterize the response properties of all the Ca^{2+}-regulated transcription factors and to understand how specific calcium patterns regulate cellular phenotypes.

References

1. Crabtree, G.R. (2001) *J Biol Chem* 276, 2313–2316.
2. West, A.E., Chen, W.G., Dalva, M.B., Dolmetsch, R.E., Kornhauser, J.M., Shaywitz, A.J., Takasu, M.A., Tao, X., and Greenberg, M.E. (2001) *Proc Natl Acad Sci USA* 98, 11024–11031.

3. Berridge, M.J., Lipp, P., and Bootman, M.D. (2000) *Nat Rev Mol Cell Biol* 1, 11–21.
4. Tsien, R.W., Lipscombe, D., Madison, D., Bley, K., and Fox, A. (1995) *Trends Neurosci* 18, 52–54.
5. Yu, F.H., Yarov-Yarovoy, V., Gutman, G.A., and Catterall, W.A. (2005) *Pharmacol Rev* 57, 387–395.
6. Hell, J.W., Westenbroek, R.E., Warner, C., Ahlijanian, M.K., Prystay, W., Gilbert, M.M., Snutch, T.P., and Catterall, W.A. (1993) *J Cell Biol* 123, 949–962.
7. Timmermann, D.B., Westenbroek, R.E., Schousboe, A., and Catterall, W.A. (2002) *J Neurosci Res* 67, 48–61.
8. Fox, A.P., Nowycky, M.C., and Tsien, R.W. (1987) *J Physiol* 394, 149–172.
9. Conti, F., Minelli, A., DeBiasi, S., and Melone, M. (1997) *Mol Neurobiol* 14, 1–18.
10. Nakanishi, S., Nakajima, Y., Masu, M., Ueda, Y., Nakahara, K., Watanabe, D., Yamaguchi, S., Kawabata, S., and Okada, M. (1998) *Brain Res Brain Res Rev* 26, 230–235.
11. Moriyoshi, K., Masu, M., Ishii, T., Shigemoto, R., Mizuno, N., and Nakanishi, S. (1991) *Nature* 354, 31–37.
12. Wang, S.Q., Wei, C., Zhao, G., Brochet, D.X., Shen, J., Song, L.S., Wang, W., Yang, D., and Cheng, H. (2004) *Circ Res* 94, 1011–1022.
13. Rios, E. and Stern, M.D. (1997) *Annu Rev Biophys Biomol Struct* 26, 47–82.
14. McAllister, A.K., Katz, L.C., and Lo, D.C. (1996) *Neuron* 17, 1057–1064.
15. Hirning, L.D., Fox, A.P., McCleskey, E.W., Olivera, B.M., Thayer, S.A., Miller, R.J., and Tsien, R.W. (1988) *Science* 239, 57–61.
16. Thomas, A.P., Bird, G.S., Hajnoczky, G., Robb-Gaspers, L.D., and Putney, J.W. Jr. (1996) *FASEB J* 10, 1505–1517.
17. Fewtrell, C. (1993) *Annu Rev Physiol* 55, 427–454.
18. Thomas, A.P., Renard-Rooney, D.C., Hajnoczky, G., Robb-Gaspers, L.D., Lin, C., and Rooney, T.A. (1995) *Ciba Found Symp* 188, 18–35; discussion 35–49.
19. Lewis, R.S. and Cahalan, M.D. (1989) *Cell Regul* 1, 99–112.
20. Dolmetsch, R.E. and Lewis, R.S. (1994) *J Gen Physiol* 103, 365–388.
21. Sheng, H.Z., Fields, R.D., and Nelson, P.G. (1993) *J Neurosci Res* 35, 459–467.
22. Eshete, F. and Fields, R.D. (2001) *J Neurosci* 21, 6694–6705.
23. Montminy, M.R. and Bilezikjian, L.M. (1987) *Nature* 328, 175–178.
24. Lonze, B.E. and Ginty, D.D. (2002) *Neuron* 35, 605–623.
25. Sheng, M., McFadden, G., and Greenberg, M.E. (1990) *Neuron* 4, 571–582.
26. Impey, S., McCorkle, S.R., Cha-Molstad, H., Dwyer, J.M., Yochum, G.S., Boss, J.M., McWeeney, S., Dunn, J.J., Mandel, G., and Goodman, R.H. (2004) *Cell* 119, 1041–1054.
27. Mantamadiotis, T., Lemberger, T., Bleckmann, S.C., Kern, H., Kretz, O., Martin Villalba, A., Tronche, F., Kellendonk, C., Gau, D., Kapfhammer, J., Otto, C., Schmid, W., and Schutz, G. (2002) *Nat Genet* 31, 47–54.
28. Gonzalez, G.A., Yamamoto, K.K., Fischer, W.H., Karr, D., Menzel, P., Biggs, W. 3rd, Vale, W.W., and Montminy, M.R. (1989) *Nature* 337, 749–752.
29. Gau, D., Lemberger, T., von Gall, C., Kretz, O., Le Minh, N., Gass, P., Schmid, W., Schibler, U., Korf, H.W., and Schutz, G. (2002) *Neuron* 34, 245–253.
30. Kornhauser, J.M., Cowan, C.W., Shaywitz, A.J., Dolmetsch, R.E., Griffith, E.C., Hu, L.S., Haddad, C., Xia, Z., and Greenberg, M.E. (2002) *Neuron* 34, 221–233.
31. Sun, P., Enslen, H., Myung, P.S., and Maurer, R.A. (1994) *Genes Dev* 8, 2527–2539.
32. Riccio, A., Alvania, R.S., Lonze, B.E., Ramanan, N., Kim, T., Huang, Y., Dawson, T.M., Snyder, S.H., and Ginty, D.D. (2006) *Mol Cell* 21, 283–294.
33. Chrivia, J.C., Kwok, R.P., Lamb, N., Hagiwara, M., Montminy, M.R., and Goodman, R.H. (1993) *Nature* 365, 855–859.
34. Conkright, M.D., Canettieri, G., Screaton, R., Guzman, E., Miraglia, L., Hogenesch, J.B., and Montminy, M. (2003) *Mol Cell* 12, 413–423.

35. Impey, S., Fong, A.L., Wang, Y., Cardinaux, J.R., Fass, D.M., Obrietan, K., Wayman, G.A., Storm, D.R., Soderling, T.R., and Goodman, R.H. (2002) *Neuron* 34, 235–244.
36. Bittinger, M.A., McWhinnie, E., Meltzer, J., Iourgenko, V., Latario, B., Liu, X., Chen, C.H., Song, C., Garza, D., and Labow, M. (2004) *Curr Biol* 14, 2156–2161.
37. Gonzalez, G.A. and Montminy, M.R. (1989) *Cell* 59, 675–680.
38. Wayman, G.A., Impey, S., Marks, D., Saneyoshi, T., Grant, W.F., Derkach, V., and Soderling, T.R. (2006) *Neuron* 50, 897–909.
39. Matthews, R.P., Guthrie, C.R., Wailes, L.M., Zhao, X., Means, A.R., and McKnight, G.S. (1994) *Mol Cell Biol* 14, 6107–6116.
40. Xing, J., Kornhauser, J.M., Xia, Z., Thiele, E.A., and Greenberg, M.E. (1998) *Mol Cell Biol* 18, 1946–1955.
41. Xing, J., Ginty, D.D., and Greenberg, M.E. (1996) *Science* 273, 959–963.
42. Deak, M., Clifton, A.D., Lucocq, L.M., and Alessi, D.R. (1998) *EMBO J* 17, 4426–4441.
43. Darragh, J., Soloaga, A., Beardmore, V.A., Wingate, A.D., Wiggin, G.R., Peggie, M., and Arthur, J.S. (2005) *Biochem J* 390, 749–759.
44. Wiggin, G.R., Soloaga, A., Foster, J.M., Murray-Tait, V., Cohen, P., and Arthur, J.S. (2002) *Mol Cell Biol* 22, 2871–2881.
45. Du, K. and Montminy, M. (1998) *J Biol Chem* 273, 32377–32379.
46. Lin, C.H., Yeh, S.H., Lin, C.H., Lu, K.T., Leu, T.H., Chang, W.C., and Gean, P.W. (2001) *Neuron* 31, 841–851.
47. Lev, S., Moreno, H., Martinez, R., Canoll, P., Peles, E., Musacchio, J.M., Plowman, G.D., Rudy, B., and Schlessinger, J. (1995) *Nature* 376, 737–745.
48. Farnsworth, C.L., Freshney, N.W., Rosen, L.B., Ghosh, A., Greenberg, M.E., and Feig, L.A. (1995) *Nature* 376, 524–527.
49. Ebinu, J.O., Bottorff, D.A., Chan, E.Y., Stang, S.L., Dunn, R.J., and Stone, J.C. (1998) *Science* 280, 1082–1086.
50. Lockyer, P.J., Kupzig, S., and Cullen, P.J. (2001) *Curr Biol* 11, 981–986.
51. Liu, Q., Walker, S.A., Gao, D., Taylor, J.A., Dai, Y.F., Arkell, R.S., Bootman, M.D., Roderick, H.L., Cullen, P.J., and Lockyer, P.J. (2005) *J Cell Biol* 170, 183–190.
52. Hagiwara, M., Alberts, A., Brindle, P., Meinkoth, J., Feramisco, J., Deng, T., Karin, M., Shenolikar, S., and Montminy, M. (1992) *Cell* 70, 105–113.
53. Bito, H., Deisseroth, K., and Tsien, R.W. (1996) *Cell* 87, 1203–1214.
54. Genoux, D., Haditsch, U., Knobloch, M., Michalon, A., Storm, D., and Mansuy, I.M. (2002) *Nature* 418, 970–975.
55. Alberts, A.S., Montminy, M., Shenolikar, S., and Feramisco, J.R. (1994) *Mol Cell Biol* 14, 4398–4407.
56. Sheng, M., Dougan, S.T., McFadden, G., and Greenberg, M.E. (1988) *Mol Cell Biol* 8, 2787–2796.
57. Deisseroth, K., Heist, E.K., and Tsien, R.W. (1998) *Nature* 392, 198–202.
58. Dolmetsch, R.E., Pajvani, U., Fife, K., Spotts, J.M., and Greenberg, M.E. (2001) *Science* 294, 333–339.
59. Weick, J.P., Groth, R.D., Isaksen, A.L., and Mermelstein, P.G. (2003) *J Neurosci* 23, 3446–3456.
60. Zhang, H., Fu, Y., Altier, C., Platzer, J., Surmeier, D.J., and Bezprozvanny, I. (2006) *Eur J Neurosci* 23, 2297–2310.
61. Sala, C., Rudolph-Correia, S., and Sheng, M. (2000) *J Neurosci* 20, 3529–3536.
62. Fields, R.D., Eshete, F., Dudek, S., Ozsarac, N., and Stevens, B. (2001) *Novartis Found Symp* 239, 160–172; discussion 172–166, 234–140.
63. Mermelstein, P.G., Bito, H., Deisseroth, K., and Tsien, R.W. (2000) *J Neurosci* 20, 266–273.
64. Hardingham, G.E., Fukunaga, Y., and Bading, H. (2002) *Nat Neurosci* 5, 405–414.
65. Pokorska, A., Vanhoutte, P., Arnold, F.J., Silvagno, F., Hardingham, G.E., and Bading, H. (2003) *J Neurochem* 84, 447–452.

66. Hardingham, G.E. and Bading, H. (1998) *Biometals* 11, 345–358.
67. Hardingham, G.E., Chawla, S., Johnson, C.M., and Bading, H. (1997) *Nature* 385, 260–265.
68. Hardingham, G.E., Arnold, F.J., and Bading, H. (2001) *Nat Neurosci* 4, 261–267.
69. Mermelstein, P.G., Deisseroth, K., Dasgupta, N., Isaksen, A.L., and Tsien, R.W. (2001) *Proc Natl Acad Sci USA* 98, 15342–15347.
70. Hardingham, G.E., Arnold, F.J., and Bading, H. (2001) *Nat Neurosci* 4, 565–566.
71. Graef, I.A., Chen, F., Chen, L., Kuo, A., and Crabtree, G.R. (2001) *Cell* 105, 863–875.
72. Crabtree, G.R. and Olson, E.N. (2002) *Cell* 109 (Suppl), S67–S79.
73. Peng, S.L., Gerth, A.J., Ranger, A.M., and Glimcher, L.H. (2001) *Immunity* 14, 13–20.
74. Shaw, K.T., Ho, A.M., Raghavan, A., Kim, J., Jain, J., Park, J., Sharma, S., Rao, A., and Hogan, P.G. (1995) *Proc Natl Acad Sci USA* 92, 11205–11209.
75. Beals, C.R., Clipstone, N.A., Ho, S.N., and Crabtree, G.R. (1997) *Genes Dev* 11, 824–834.
76. Okamura, H., Aramburu, J., Garcia-Rodriguez, C., Viola, J.P., Raghavan, A., Tahiliani, M., Zhang, X., Qin, J., Hogan, P.G., and Rao, A. (2000) *Mol Cell* 6, 539–550.
77. Porter, C.M., Havens, M.A., and Clipstone, N.A. (2000) *J Biol Chem* 275, 3543–3551.
78. Neal, J.W. and Clipstone, N.A. (2001) *J Biol Chem* 276, 3666–3673.
79. Gwack, Y., Sharma, S., Nardone, J., Tanasa, B., Iuga, A., Srikanth, S., Okamura, H., Bolton, D., Feske, S., Hogan, P.G., and Rao, A. (2006) *Nature* 441, 646–650.
80. Shaw, J.P., Utz, P.J., Durand, D.B., Toole, J.J., Emmel, E.A., and Crabtree, G.R. (1988) *Science* 241, 202–205.
81. Diehn, M., Alizadeh, A.A., Rando, O.J., Liu, C.L., Stankunas, K., Botstein, D., Crabtree, G.R., and Brown, P.O. (2002) *Proc Natl Acad Sci USA* 99, 11796–11801.
82. Bhakta, N.R. and Lewis, R.S. (2005) *Semin Immunol* 17, 411–420.
83. Bhakta, N.R., Oh, D.Y., and Lewis, R.S. (2005) *Nat Immunol* 6, 143–151.
84. Negulescu, P.A., Shastri, N., and Cahalan, M.D. (1994) *Proc Natl Acad Sci USA* 91, 2873–2877.
85. Dolmetsch, R.E., Xua, K., and Lewis, R.S. (1998) *Nature* 392, 933–936.
86. Goldbeter, A. and Koshland, D.E., Jr. (1982) *Q Rev Biophys* 15, 555–591.
87. Salazar, C. and Hofer, T. (2003) *J Mol Biol* 327, 31–45.
88. Berridge, M.J. and Galione, A. (1988) *FASEB J* 2, 3074–3082.
89. Tomida, T., Hirose, K., Takizawa, A., Shibasaki, F., and Iino, M. (2003) *EMBO J* 22, 3825–3832.
90. Loh, C., Carew, J.A., Kim, J., Hogan, P.G., and Rao, A. (1996) *Mol Cell Biol* 16, 3945–3954.
91. Dolmetsch, R.E., Lewis, R.S., Goodnow, C.C., and Healy, J.I. (1997) *Nature* 386, 855–858.
92. Andrews, C., Ho, P.D., Dillmann, W.H., Glembotski, C.C., and McDonough, P.M. (2003) *Cardiovasc Res* 59, 46–56.
93. McDonough, P.M. and Glembotski, C.C. (1992) *J Biol Chem* 267, 11665–11668.
94. Xia, Y., McMillin, J.B., Lewis, A., Moore, M., Zhu, W.G., Williams, R.S., and Kellems, R.E. (2000) *J Biol Chem* 275, 1855–1863.
95. Bhalla, U.S., Ram, P.T., and Iyengar, R. (2002) *Science* 297, 1018–1023.
96. Northrop, J.P., Ho, S.N., Chen, L., Thomas, D.J., Timmerman, L.A., Nolan, G.P., Admon, A., and Crabtree, G.R. (1994) *Nature* 369, 497–502.
97. Zhou, B., Cron, R.Q., Wu, B., Genin, A., Wang, Z., Liu, S., Robson, P., and Baldwin, H.S. (2002) *J Biol Chem* 277, 10704–10711.
98. Arron, J.R., Winslow, M.M., Polleri, A., Chang, C.P., Wu, H., Gao, X., Neilson, J.R., Chen, L., Heit, J.J., Kim, S.K., Yamasaki, N., Miyakawa, T., Francke, U., Graef, I.A., and Crabtree, G.R. (2006) *Nature* 441, 595–600.
99. Yang, J., Rothermel, B., Vega, R.B., Frey, N., McKinsey, T.A., Olson, E.N., Bassel-Duby, R., and Williams, R.S. (2000) *Circ Res* 87, E61–E68.
100. Verma, I.M., Stevenson, J.K., Schwarz, E.M., Van Antwerp, D., and Miyamoto, S. (1995) *Genes Dev* 9, 2723–2735.
101. Li, Q. and Verma, I.M. (2002) *Nat Rev Immunol* 2, 725–734.

102. Meffert, M.K., Chang, J.M., Wiltgen, B.J., Fanselow, M.S., and Baltimore, D. (2003) *Nat Neurosci* 6, 1072–1078.
103. Okamoto, S., Mukaida, N., Yasumoto, K., Rice, N., Ishikawa, Y., Horiguchi, H., Murakami, S., and Matsushima, K. (1994) *J Biol Chem* 269, 8582–8589.
104. Jenuwein, T. and Allis, C.D. (2001) *Science* 293, 1074–1080.
105. Backs, J. and Olson, E.N. (2006) *Circ Res* 98, 15–24.
106. Dodou, E., Sparrow, D.B., Mohun, T., and Treisman, R. (1995) *Nucleic Acids Res* 23, 4267–4274.
107. McKinsey, T.A., Zhang, C.L., and Olson, E.N. (2002) *Trends Biochem Sci* 27, 40–47.
108. Mao, Z. and Wiedmann, M. (1999) *J Biol Chem* 274, 31102–31107.
109. Wu, H., Naya, F.J., McKinsey, T.A., Mercer, B., Shelton, J.M., Chin, E.R., Simard, A.R., Michel, R.N., Bassel-Duby, R., Olson, E.N., and Williams, R.S. (2000) *EMBO J* 19, 1963–1973.
110. Backs, J., Song, K., Bezprozvannaya, S., Chang, S., and Olson, E.N. (2006) *J Clin Invest* 116, 1853–1864.
111 Zierath, J.R. and Hawley, J.A. (2004) *PLoS Biol* 2, e348.
112. Windisch, A., Gundersen, K., Szabolcs, M.J., Gruber, H., and Lomo, T. (1998) *J Physiol* 510Pt 2, 623–632.
113. Liu, Y. and Schneider, M.F. (1998) *J Physiol* 512 (Pt 2), 337–344.
114. Liu, Y., Randall, W.R., and Schneider, M.F. (2005) *J Cell Biol* 168, 887–897.
115. Liu, Y., Cseresnyes, Z., Randall, W.R., and Schneider, M.F. (2001) *J Cell Biol* 155, 27–39.
116. Liu, Y., Shen, T., Randall, W.R., and Schneider, M.F. (2005) *J Muscle Res Cell Motil* 26, 13–21.
117. De Koninck, P. and Schulman, H. (1998) *Science* 279, 227–230.
118. Bradshaw, J.M., Kubota, Y., Meyer, T., and Schulman, H. (2003) *Proc Natl Acad Sci USA* 100, 10512–10517.

J. Krebs and M. Michalak (Editors)
Calcium: A Matter of Life or Death

ISSN: 0167-7306/DOI: 10.1016/S0167-7306(06)41016-4

CHAPTER 16

Calcium and fertilization

Jong Tai Chun and Luigia Santella

Cell Signaling Laboratory, Stazione Zoologica "Anton Dohrn" Villa Comunale, I-80121, Naples, Italy, Tel.: +39 0815833289; Fax: +39 0817641355; E-mail: santella@szn.it

Abstract

The beginning of new cell life is dependent on the signaling events that occur inside the sperm and the egg. From the moment when the sperm acquires the ability to swim to the egg to its attachment to the plasma membrane of the latter, Ca^{2+} acts as the key messenger that allows essential information exchange between the two cells. A large body of evidence is now available on the importance of Ca^{2+} in the activation of sperm prior to the fertilization events. On the egg side, the signaling pathways that trigger the Ca^{2+} release at fertilization have been extensively studied for at least 30 years. Even if several lines of evidence indicate that the increase of Ca^{2+} in fertilized mammalian eggs demands the activation of inositol 1,4,5-trisphosphate receptors ($InsP_3Rs$), the molecular mechanisms responsible for the Ca^{2+} increase in the fertilization of other animal species have not been conclusively established.

Keywords: calcium signal, fertilization, $InsP_3Rs$, maturation, MPF, NAADP, starfish oocyte

The fertilization process occurs when the fertilizing sperm and the egg find and activate each other [1]. The activation process involves large changes in the metabolism of the two gametes that are initiated by the signals that trigger changes in the intracellular concentrations of free calcium [2]. That calcium is important in the regulation of egg physiology at fertilization has been known for a long time. By contrast, only recently information has begun to accumulate on the importance of Ca^{2+} in the activation of sperm prior to fertilization [3]. It is generally accepted that in mammals the elevation of calcium in the fertilized egg is triggered by a process of Ca^{2+} release through activated inositol 1,4,5-trisphosphate receptors ($InsP_3Rs$). The situation may be different in invertebrates (e.g., echinoderms) in which the novel Ca^{2+} messenger nicotinic acid adenine dinucleotide phosphate (NAADP) appears to initiate the Ca^{2+} response at fertilization.

1. Ca^{2+} and sperm activation

1.1. Chemotaxis

Fertilization depends on the specific recognition and binding of spermatozoa by the egg. Thus, germ cells are endowed with surface structures that allow the interaction of gametes belonging to the opposite sexes. The chemotaxis of sperm is an important step toward fertilization. The ability of sperm (from marine to nonmarine species) to swim to the egg and to fertilize it is modulated by calcium changes induced in the sperm mainly by soluble components released by the egg's outer layers. During chemotaxis, the swimming trajectory of sperm is controlled by Ca^{2+} spikes in the flagellum. *Speract*, the first sperm-activating peptide ever purified [4], binds to a guanylyl cyclase receptor on the sperm flagellum activating two separate signaling pathways that are necessary to regulate its motility. A membrane potential hyperpolarization caused by K^+ efflux through cGMP-activated K^+ channels would enhance the Ca^{2+} extrusion activity of a K^+-dependent Na^+/Ca^{2+} exchanger (NCKX) as a mechanism to decrease the intracellular Ca^{2+} that is observed in the initial phase of the sperm response. Subsequently, a voltage-gated Ca^{2+} channel as well as a cAMP-regulated Ca^{2+} transporter may be involved in the intracellular Ca^{2+} increase that promotes asymmetry of flagellar movement [5,6]. Stimulation of sperm by another chemoattractant, *resact*, gives rise to a rapid and transient rise in the concentration of cGMP, which elicits Ca^{2+} spikes in the flagellum [7–9]. The onset of the Ca^{2+} rise in the flagellum precedes the rise in the head, and the short duration of the Ca^{2+} signals is regulated by two Ca^{2+} extruding proteins, an NCKX and a plasma membrane Ca^{2+}-ATPase (PMCA) pump. The PMCA pump (isoform 4) is almost exclusively located in the tail of the sperm, and its Ca^{2+} extruding action is essential to its hyperactivated motility. Knocking out one of the pump genes in mice completely blocks sperm motility [10,11]. The role of NCKX is less clearly established, but inhibition of both NCKX and PMCA elevates intracellular Ca^{2+} and stops motility also in sea urchin [12,13]. Previous studies had proposed that the influx of Ca^{2+} was initiated by the rise in intracellular pH. According to this proposal, a cGMP-induced hyperpolarization would activate a Na^+/H^+ exchanger that expels H^+ from the cell. However, re-examination of the role of protons in sperm swimming behavior has recently shown that a rise in intracellular pH is not required for chemoattractant-induced Ca^{2+} entry in sea urchin sperm [14]. The events that occur during chemotaxis of sea urchin and starfish gametes are similar even if, at variance with sea urchin, the starfish chemoattractant *asterosap* does not noticeably increase the cAMP concentration [15]. The similarities of the Ca^{2+} fluctuations and the swimming behavior between starfish and sea urchin sperms indicate that the chemotactic cGMP signaling pathway has been conserved in species that diverged approximately 500 million years ago [16]. By contrast, in ascidians, the activation of sperm motility is due to the increased cAMP that follows the influx of Ca^{2+} promoted by a chemoattractant that activates a calmodulin-dependent kinase pathway, causing membrane hyperpolarization [17]. Mammalian sperm show hyperactivated motility just before fertilization. An increase in flagellar bend amplitude and beat asymmetry is required for successful fertilization as it enhances the ability of sperm to detach from the wall of the oviduct, to move around its labyrinthine lumen, to penetrate mucous substances, and finally, to penetrate the zona pellucida (ZP), the egg extracellular coat. The intracellular Ca^{2+} increase in the flagella of the sperm is crucial for the initiation and maintenance of hyperactivation, which is a change in movement pattern. Recent data suggest that hyperactivated motility may be regulated by the emptying of an $InsP_3$-dependent intracellular Ca^{2+} store [18]. Whole-cell patch-clamp measurements of spermatozoa have described a sperm cation channel, CatSper, in the principal piece of the sperm tail that is strongly potentiated by intracellular alkalinization [19]. The amino acid sequence of CatSper closely resembles that of a single six-transmembrane spanning repeat of the voltage-dependent Ca^{2+}-channel alpha subunit. Distributed in the principal piece of the sperm tail, CatSper is a component of the flagellar calcium channel mediating Ca^{2+} influx and sperm hyperactivation [20].

1.2. Sperm capacitation and acrosome reaction in mammals

Capacitation is a complex physiological event that is accompanied by an increase in mammalian sperm respiration and motility. Although numerous events occur during the process including the removal of cholesterol from the plasma membrane that increases membrane fluidity [21], the intracellular pH increase, and the cAMP-dependent tyrosine phosphorylation of specific sperm proteins [22–24], it appears that the regulation of intracellular Ca^{2+} is a crucial event. The intracellular pH increase that could be induced by several external factors such as HCO_3-, heparin and serum albumin could be responsible for the influx of Ca^{2+} as it modulates the voltage-gated Ca^{2+} channels [25]. Ca^{2+} increase could also result from the decline of the sarco(endo)plasmic reticulum Ca^{2+} ATPase (SERCA) pump activity that is responsible for the filling of the internal Ca^{2+} stores, possibly located in the acrosome [26]. Capacitated spermatozoa undergo the acrosome reaction (AR), a Ca^{2+}-dependent process that enables the outer acrosomal membrane and the overlying plasma membrane to come into proximity, to fuse, to secrete the acrosomal content, and to trigger the opening of store-operated Ca^{2+} channels (SOCs) in the plasma membrane. Thapsigargin, which increases intracellular Ca^{2+} by depressing SERCA pump activity, indeed stimulates the acrosomal reaction [27]. Various members of the transient receptor potential channel (TRPC) superfamily (TRPC 1, 2, 3, and 6) have been recently known to be involved in the regulation of sperm flagellar motility in mice and in humans [28]. A glycoprotein, ZP3, which is also a component of the egg extracellular matrix, has a role in the AR. It mediates an initial Ca^{2+} influx into the sperm through activated T-type channels but also activates a phospholipase C (PLC), eventually depleting the sperm intracellular Ca^{2+} stores through newly formed $InsP_3$. This in turn activates TRPC2 inducing sustained Ca^{2+} influx and eventually the AR [29]. A novel protein, termed enkurin, has been identified as an adapter that interacts with TRPCs tethering signal transduction proteins to TRPCs [30]. The impairment of the ZP-induced AR in PLCδ4–/– sperm emphasizes the importance of the intracellular Ca^{2+} mobilization for the AR [31]. In mammals, after binding to the ZP of the egg [32], only capacitated spermatozoa can undergo the AR. Prior to its occurrence, gelsolin, an actin severing protein, and F-actin associate to the sperm membranes acting as a barrier to inhibit their fusion. The binding of the capacitated sperm to ZP induces a fast increase in sperm intracellular Ca^{2+} that activates gelsolin that breaks down the actin fibers, allowing membrane fusion and AR to take place [33,34].

1.3. Acrosome reaction in echinoderms

Most of the work on the AR has been carried out on sea urchin sperm, the first comprehensive study dating back to the 1950s [35]. It was found that exposure to the so-called "egg water" led to an extension of a process at the head of several sea urchins sperms, which was termed the "AR." The fucose sulfate polymer (FSP) of the egg jelly has specie-specificity in inducing the exposure of the globular actin of the acrosomal granule that polymerizes to form the acrosomal filament, a process that is required for egg coat penetration and for fusion with the egg plasma membrane [36]. The acrosomal filament exposes bindin, a protein that mediates sea urchin sperm–egg attachment and membrane fusion and is thus important in species recognition and speciation [37]. The requirement for Ca^{2+} in the AR is absolute; indeed, it may be the only role for extracellular Ca^{2+} in fertilization, because sea urchin sperm that have already undergone the AR in the presence of Ca^{2+} can fertilize eggs in Ca^{2+}-free sea water [38]. It was shown that by the time the acrosomal filament had extended, the Ca^{2+} concentration in the sperm reached values several orders of magnitude higher than in the egg. Some of this Ca^{2+} was likely to be associated with calmodulin that was found to be present in high levels in the sperm head [39,40]. Recently, experiments in which Ca^{2+}-sensitive fluorescent dyes were coupled to high-sensitivity imaging techniques have enabled the monitoring of the intracellular Ca^{2+} changes that occur during echinoderm sperm activation. The measurements have shown the involvement of two different Ca^{2+} channels that are involved in the exocytosis of the sea urchin sperm acrosomal vesicle. The first is sensitive to the voltage-gated Ca^{2+} channel blockers, and the second is blocked by SKF96365, an inhibitor of SOC [41]. Production of $InsP_3$ can also be stimulated by the FSP. The FSP-induced $InsP_3$ and cAMP increase is dependent on external Ca^{2+} as it is inhibited by verapamil and nifedipine [42]. During acrosomal exocytosis, $InsP_3$ could act on $InsP_3Rs$ that empty the Ca^{2+} store and thus mediate the sustained uptake of Ca^{2+} through the SOC channels. A Ca^{2+} store has been reported in human sperm, and it has been shown that calcium efflux from this compartment through $InsP_3$-sensitive channels is a necessary event for the triggering of acrosomal exocytosis [43].

In sea urchin sperm, a receptor for egg jelly-3 (suREJ3) has been identified on the plasma membrane covering the sperm acrosomal vesicle as the homolog of polycystin-1, together with the sea urchin sperm homolog of polycystin-2 (suPC2). Coimmunoprecipitation has shown that suREJ3 and suPC2 were associated within the membrane, indicating that the two proteins operate as receptor–calcium channel complexes in the regulation of AR during fertilization [44,45]. In starfish, the AR is induced by the cooperation of three components of the outer surface of the egg coat, namely AR-inducing substance (ARIS), co-ARIS, and *asterosap*. A sustained elevation of intracellular Ca^{2+} and the AR occur if sperms are simultaneously treated with ARIS and *asterosap*. They are inhibited by SKF96365 and Ni^{2+}, specific blockers of SOC channels, and by a slight intracellular pH reduction. Thus, the *asterosap*-induced intracellular pH elevation is required for triggering the ARIS-induced sustained Ca^{2+} elevation and consequent AR [46,47]. Once formed, the long thin acrosomal process (20 μm) passes through the jelly layer and establishes contact with the oocyte surface [48 and Fig. 4].

2. *Adaptation of the egg for the development of the Ca^{2+} response at fertilization*

2.1. *Endoplasmic reticulum dynamics during the maturation process of the oocytes*

In a few animals such as echiuran worms and some bivalve mollusks, fertilization occurs when the immature oocyte still contains a large nucleus [germinal vesicle (GV)] and is thus still arrested at the prophase I stage of meiosis [49]. However, before becoming fertilizable, the early oocytes of mollusk undergo a maturation process or meiosis, ending with a halving of the number of chromosomes and with the ability of the egg to produce a normal Ca^{2+} response at fertilization. Thus, in most animal species, meiosis is arrested at different cycle stages and is completed only upon shedding or fertilization of the egg. Sea urchins are one of the few groups of animals in which the female gamete goes through the entire maturation process within the ovary and is shed as a haploid cell [50].

In the starfish, fully grown oocytes are arrested at the first prophase of the meiotic division and contain a very large GV in the animal hemisphere [51]. Re-initiation of the maturation process is induced in vitro by the hormone 1-methyladenine (1-MA). The breakdown of the envelope of the GV coincides with the initiation of a series of physiological changes that are connected with the ability of the egg cytoplasm to respond to activating stimuli (Fig. 1). Research reported in the literature has shown

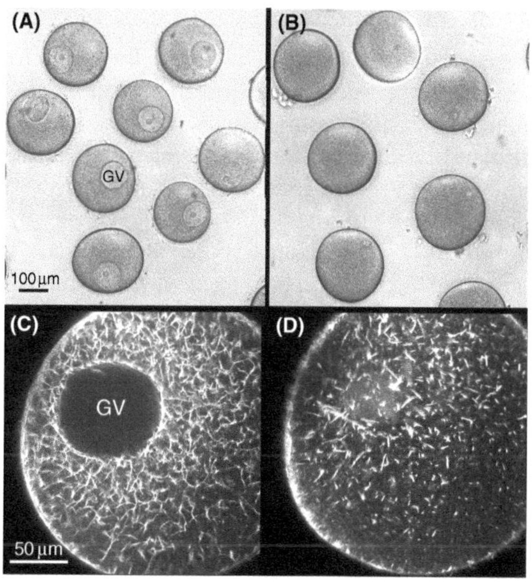

Fig. 1. The maturation process of starfish oocytes involves profound changes in the actin cytoskeleton. (**A**) Light microscopy images of immature oocytes arrested at the prophase of the first meiotic division. The large nucleus [germinal vesicle (GV)] is well evident. (**B**) Oocytes matured by the exposure of the hormone 1-methyladenine (1-MA) for 60 min. The GV is no longer visible. (**C**) An oocyte of panel **A** stained with phalloidin to visualize the actin cytoskeleton. (**D**) A mature oocyte of panel **B** in which the phalloidin staining shows very evident changes in the organization of the actin cytoskeleton.

marked differences in the spatiotemporal pattern of the Ca^{2+} transients that are promoted when eggs are inseminated at different stages of the maturation process. Thus, the ability of eggs to produce a proper Ca^{2+} response and to undergo normal cortical granule (CG) exocytosis is a consequence of the structural reorganization of the endoplasmic reticulum (ER), the major internal store of calcium during egg maturation. By monitoring ER structure in vivo, alterations that occur following maturation have been assessed in most species [52]. The first in vivo analysis of ER reorganization was performed in starfish oocytes using the ER-specific probe 1,1′-dioctadecyl-3,3,3′,3′-tetramethylindocarbocyanine perchlorate (DiI) [53]. It was found that the ER in fully grown immature oocytes is composed of interconnected membrane sheets that appear to become associated with the yolk platelets and to form special shells following the triggering of the oocyte maturation by 1-MA. Changes in ER structure were also demonstrated during meiotic maturation of mammalian oocytes. In mouse, the DiI signal shifted from a uniform cytoplasmic distribution with no ER accumulations in the cortex of I-prophase-arrested mouse oocytes to a more cortical localization with ER accumulations concentrated in the vegetal hemisphere at the end of mouse oocyte maturation [54]. Alternatively, in hamster, ER masses visible in immature oocytes disaggregated to form an organized network in the cortex [55]. A correlation between ER dynamics and the fertilization-linked Ca^{2+} response in several species clearly showed that ER clusters begin to form in maturing oocytes at the time when they become competent for a normal Ca^{2+} signal when inseminated [52]. The disappearance of the developed ER microdomains after fertilization is coincident with the termination of the Ca^{2+} oscillations [56]. Research evaluating maturation-associated Ca^{2+} signaling modifications in the oocyte has shown changes in the distribution and expression of the $InsP_3R$ following oocyte maturation. A common consequence of $InsP_3R$ reorganization is the formation of $InsP_3R$ clusters that could be essential for the enhancement of the Ca^{2+} response at fertilization in mammalian eggs. Very recently, it has been shown that the concentration, cellular distribution, and sensitivity of the $InsP_3R$ type 1 isoform ($InsP_3R$-1) increase during maturation [57]. An increased sensitivity, that is, the ability of the receptor to release more calcium when challenged with the same amount of ligand, is a phenomenon dependent on the maturation stage of the oocyte. The process has been analyzed in great detail in starfish. In starfish, at variance with mammalian oocytes [58,59] in starfish, the increased release of Ca^{2+} through the $InsP_3R$ was not linked to an increased loading of the ER with Ca^{2+} nor to the increased level of expression of the $InsP_3R$-1 during maturation [60,61]. In addition, the increased sensitivity of the Ca^{2+} stores to $InsP_3$ begins at the animal hemisphere of the cell (Fig. 2) and then propagates to the opposite vegetal hemisphere [62]. Various lines of evidence have implicated the maturation promoting factor (MPF) in the regulation of the increased sensitivity of the $InsP_3Rs$, a finding that supports recent biochemical studies describing the presence of the cyclin-dependent kinase/MPF consensus site in the $InsP_3R$-1 [63,64]. High MPF activity can prolong fertilization-induced Ca^{2+} oscillations in several species, and the inhibition of the kinase activity induces a rapid disassembly of ER clusters together with termination of Ca^{2+} oscillations [52]. In starfish, the modifications of the $InsP_3R$ sensitivity correlates nicely with the MPF activity that increases in maturing oocytes.

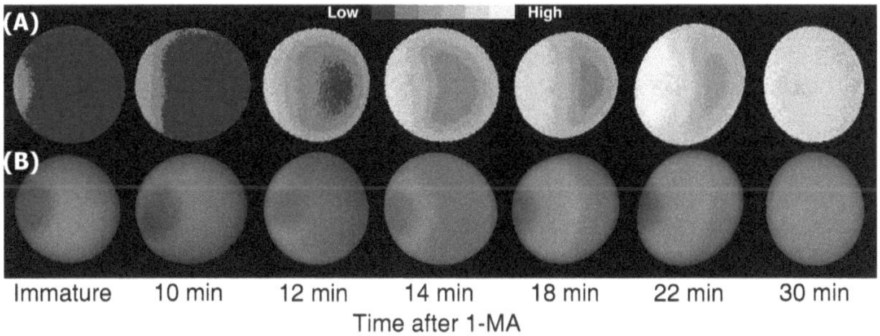

Fig. 2. Increased sensitivity of the inositol 1,4,5-triphosphate receptors ($InsP_3Rs$) to $InsP_3$ during the starfish oocyte maturation process induced by 1-methyladenine (1-MA). (**A**) Relative fluorescence images showing the initiation of the Ca^{2+} signal in the animal hemisphere of seven oocytes at different maturation stages. Ca^{2+} release was induced by the uncaging of pre-injected caged $InsP_3$. The uncaging was induced after the injected $InsP_3$ had diffused globally to the entire oocyte. (**B**) Overlay of fluorescence and transmitted light microscopy of the same seven oocytes shown in panel **A** (See Color Plate 44, p. 539).

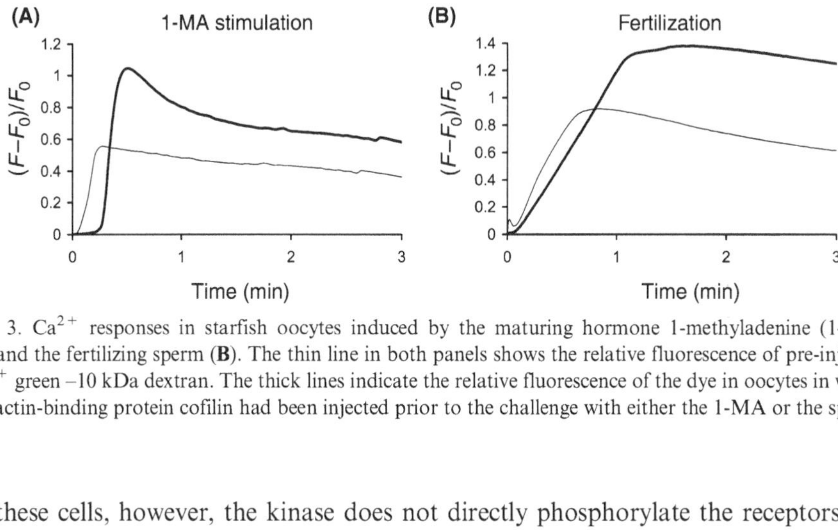

Fig. 3. Ca^{2+} responses in starfish oocytes induced by the maturing hormone 1-methyladenine (1-MA) (**A**) and the fertilizing sperm (**B**). The thin line in both panels shows the relative fluorescence of pre-injected Ca^{2+} green –10 kDa dextran. The thick lines indicate the relative fluorescence of the dye in oocytes in which the actin-binding protein cofilin had been injected prior to the challenge with either the 1-MA or the sperm.

In these cells, however, the kinase does not directly phosphorylate the receptors but modulates the channel activity through F-actin rearrangement [62]. Very recently, the actin binding protein cofilin has been identified as the F-actin-associated protein that increases the Ca^{2+} signaling induced by the uncaging of $InsP_3$ [65]. Moreover, cofilin modulates the Ca^{2+} release that occurs naturally during starfish oocyte maturation and fertilization (Fig. 3).

3. How does a spermatozoon activate an egg?

At fertilization, the first change in the egg is a transient increase in Ca^{2+} that results from its release from intracellular stores and that sets in motion the entire program of development. The finding that Ca^{2+} increases at fertilization dates back to the early 1920s when it was discovered that the activation of the egg could be induced by promoting Ca^{2+} entry from the medium by pricking them with a needle [2,66]. The morphological changes occurring in eggs at activation include the cortical reaction, which is a Ca^{2+}-dependent extrusion of the cortical granule's (CG) content to the extracellular space [67,68]. The 1-μm granules located around the point of sperm–egg interaction are the first to exocytose, the wave of exocytosis then spreading around the egg surface [69,70]. One of the strongest pieces of evidence for the role of intracellular Ca^{2+} release in mediating the triggering of CG exocytosis was provided by the usage of the Ca^{2+} ionophore A23187 that induced the cortical reaction [71,72]. Further support for a role of Ca^{2+} in regulating the cortical exocytosis was obtained by inducing cortical granules–plasma membrane fusion in isolated preparations of cortices by Ca^{2+} addition to CG isolated in Ca^{2+}-free sea water [73].

Direct evidence that sperm interaction induces an intracellular Ca^{2+} elevation at the sperm entry site that then propagates to the opposite pole of the egg was originally provided in medaka fish by experiments using injected aequorin [74]. The increase in intracellular Ca^{2+} is now recognized as a general feature of the fertilization process [75,76]. Depending on the animal species, the increase may occur as a single Ca^{2+} transient or in the form of repetitive oscillations [49].

Sea urchins have been a useful model to correlate the sequence of ultrastructural events at fertilization with changes in the electrical properties of the egg plasma membrane. A few seconds following the attachment of the fertilizing sperm, a small step-like depolarization of 1–2 mV occurs across the egg plasma membrane [77]. The initial electrical event, which occurs nearly simultaneously with sperm–egg fusion, has been suggested to result from the incorporation of ion channels of the plasma membrane of the sperm to the egg plasma membrane [78,79]. The depolarization step is followed by a transient increase in the voltage noise [fertilization potential (FP)] driven by voltage-gated Ca^{2+} channels, resulting in a transient ring of fluorescence of Ca^{2+} indicators beneath the plasma membrane (cortical flash). The prolonged plateau after the FP is due to the entry of Na^{+}, which is dependent on an intracellular Ca^{2+} wave in echinoderms and nemerteans, whereas the second series of oscillations in the membrane potential in the ascidian *Ciona intestinalis* requires the contribution of a current activated by Ca^{2+} release [80]. In golden hamster eggs, the FP consists of a series of periodic hyperpolarizations because of a Ca^{2+}-activated K^{+} conductance increase accompanying the repetitive Ca^{2+} spikes that mark calcium signaling in mammalian fertilization [81]. The mechanism responsible for the initial depolarization has been identified in the ascidian *C. intestinalis* as a cationic current activated by ADP ribose (ADPr), but information on the trigger of the initial depolarization in other invertebrates is scarce [82]. The positive shift in membrane potential at fertilization is responsible for the block to polyspermy, protecting the egg from the entry of supernumerary sperms during the first few minutes after insemination. The fast electrical block hypothesis has been applied to frog, starfish, worm, and mammalian oocytes [83–87]. In mammals, the sperm-induced Ca^{2+} oscillations give rise to the ZP block to polyspermy. Recently, on examining the relationship between sperm-induced Ca^{2+} signaling and block to polyspermy, it has been found that different concentrations of the Ca^{2+} chelator 1,2-bis(2-aminophenoxy)-N,N,N′,N′-tetraacetic acid tetrakis(acetoxymethyl ester) completely abolish Ca^{2+} spiking following fertilization, inducing an increased rate of polyspermy [88].

As mentioned above (see Section 1.2), in the starfish, the AR is triggered by the contact of the spermatozoon with the outer layer of the jelly coat, which extends for about 20 μm outside the outer border of the oocyte plasma membrane. This was first indicated by the isolation from the starfish oocyte of a jelly coat fraction that had the ability to induce the AR (Fig. 4) [89]. A thin acrosomal filament (20 μm in length) is formed following polymerization of the acrosomal globular actin (panel C). Recent experiments in our laboratory have exploited the length of the acrosomal filament of the fertilizing sperm to show that the Ca^{2+} wave in the fertilized eggs initiates well before the sperm is incorporated into them, that is, when it is still far away from

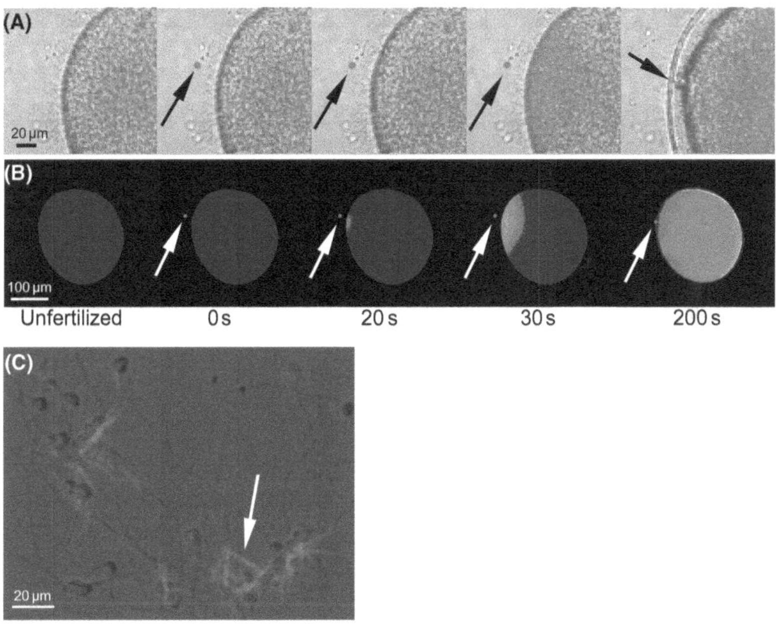

Fig. 4. Ca^{2+} signaling induced by the sperm in starfish oocytes. Panel **A** shows an overlay of fluorescence images and transmitted light of a number of spermatozoa surrounding the oocyte. One of them (arrow), indicated in red, is going to extend the long acrosomal process that will initiate the Ca^{2+} response in the oocyte (third image from the left and panel **C**). The circumscribed Ca^{2+} elevation will spread centripetally while the spermatozoon is still at some distance from the surface of the oocyte (fourth image from the left). Eventually, the head of the spermatozoon makes contact with the oocyte at the time when the Ca^{2+} wave has become global (fifth image from the left). At this point, the fertilization membrane becomes completely elevated. At a later stage, the spermatozoon will be engulfed by the oocyte. Panel **B** shows relative fluorescence images of the same sequence of events shown in panel **A**. Panel **C** shows phalloidin-stained F-actin on the long acrosomal process of reacted spermatozoa (arrow) (See Color Plate 45, p. 540).

the egg plasma membrane: the distance corresponds to the 20-μm length of the acrosomal process (panels A and B). Whether the activation of the egg requires fusion with the spermatozoon or whether the processes leading to egg activation begin before penetration has been the matter of continued debate in the past [90]. Early findings supporting the latter view were provided by Lillie [91] who took advantage of the long delay between sperm attachment and penetration into the egg of *Nereis* to attempt removing the "activating" spermatozoon by centrifugation. He found that the removal of the sperm still allowed the cortical reaction to occur. In line with this suggestion, cytochalasin B, which prevents the incorporation of the attached spermatozoon, failed to inhibit egg activation in several species [92–95]. This led to the hypothesis that fusion, as opposed to incorporation, of the sperm introduced a factor in the egg that was responsible for triggering the Ca^{2+} elevation. It was indeed suggested that one reason for the accumulation of Ca^{2+} in the sperm during the AR might have been to deposit it in the egg to trigger the cortical reaction [96,97].

The hypothesis that the fertilization wave in eggs is propagated by calcium-stimulated calcium release from some internal sources was based on the finding that the fertilization wave was nearly independent of extracellular Ca^{2+} as it was only slightly slowed (by perhaps 15%) in a medium containing 5 mM EGTA [98]. Thus, the search for the intracellular Ca^{2+} store begun. Ca^{2+} was suggested to be sequestered in cytoplasmic organelles because electro-cytochemical studies showed its presence in the ER, in the CG, in the mitochondria, and in large, clear acidic vesicles of unknown function [99]. The ER was then identified as the source of the intracellular calcium released upon fertilization in organelle-stratified centrifuged egg pre-injected with the Ca^{2+} photoprotein aequorin [100]. Thus, at fertilization, intracellular Ca^{2+} was assumed to be released from the egg's ER and to propagate in the form of a wave that initiated at the site of sperm binding and spread to the opposite pole.

The mechanism(s) by which the sperm-induced Ca^{2+} signals are produced to initiate the development of the early embryo has been the subject of a vigorous debate. Two accepted ideas attempt to explain how the spermatozoon triggers the dormant egg into metabolic activity. One is that the interaction of the spermatozoon with the egg receptors transduces signals from the egg surface to the interior of the cell producing the synthesis of Ca^{2+}-linked second messengers and Ca^{2+} increase. In sea urchin eggs, $InsP_3$ injection was shown to induce a very rapid fertilization membrane elevation consistent with a massive release of intracellular Ca^{2+} [101,102], and it was suggested that the sperm activated the phosphoinositide turnover to produce $InsP_3$ and Ca^{2+} mobilization, as proposed for pancreatic acinar cells [103].

The other hypothesis surmises that the spermatozoon contains a factor, soluble in the egg cytoplasm, that is released into the egg following sperm fusion and that directly triggers the mobilization of second messengers. This idea is not new: actually goes back to almost 100 years ago when it was suggested that the spermatozoon injects a lysine that breaks down a calcium–lipid complex in the cortex, thus releasing a catalyst that activates the egg [104].

3.1. What is the signaling pathway leading to the intracellular Ca^{2+} elevation at fertilization?

After the observations described in the preceding paragraph of Section 3, the mechanism that links the egg–sperm interaction to the Ca^{2+} increase rapidly became a hot topic. According to the receptor hypothesis, generation of the Ca^{2+} signal at fertilization might occur through several routes, all eventually leading to the activation of PLC [105]. Irrespective of the signal transduction pathway, the action of a PLC on phosphatidylinositol 4,5-bisphosphate (PIP_2) would result in the rapid formation of $InsP_3$, which became accepted as the main actor in the Ca^{2+} mobilization at fertilization [106,107]. However, the interest on conventional G protein/PLC/$InsP_3$ involvement has decreased considerably based on ambiguous results with a number of G-protein inhibitors and activators [108]. In echinoderms, it has recently become evident that a receptor could have a role, but it would be a tyrosine kinase receptor that initiates Ca^{2+} release through a PLCγ-mediated pathway [109]. Indeed, immunoprecipitation studies have demonstrated that a significant fraction of PLCγ is translocated from the cytosol to the membrane compartment of the egg at fertilization [110]. As PLCγ is activated when its two Src homology-2 (SH2) domains bind to an activated tyrosine kinase, the injection of a PLCγ–SH2 domain fusion protein strongly delayed the Ca^{2+} signal at fertilization in starfish eggs, indicating that the enzyme was activated at fertilization to stimulate the production of $InsP_3$ and Ca^{2+} mobilization [111]. Recently, an oocyte cDNA-encoding PLCγ has been isolated from starfish confirming that an endogenous PLCγ regulates Ca^{2+} release at fertilization through a PLCγ–SH2 domain-mediated mechanism [112].

The Fyn kinase has also been suggested to participate in the signaling pathway, producing the Ca^{2+} wave at fertilization of sea urchin eggs [113]. Even if it appears that the pathway leading to release of Ca^{2+} from the ER in echinoderms includes activation of a Src homolog, followed by PLCγ activation and formation of $InsP_3$, the upstream activators or modulators of this signaling pathway are not known. Recently, the role of a canonical G-protein signaling pathway at fertilization in sea urchin has been re-evaluated by suggesting that the activation of Src homologs could be dependent on the activation of G-proteins [114].

The sperm factor hypothesis is based on the action of a diffusible messenger in the cytoplasm of the spermatozoa that enters the oocyte cytoplasm after sperm–egg interaction. This hypothesis was based on pioneering experiments in which microinjection of soluble components of crushed spermatozoa into sea urchin eggs induced the exocytosis of CG consistent with an intracellular Ca^{2+} release [115]. Sperm extracts of ascidians injected into oocytes also induce repetitive intracellular Ca^{2+} increases with kinetics consistent with those at fertilization and cause re-initiation and progression of meiosis as in fertilized oocytes, by the same pathway activated at fertilization [116,117]. In mammals, several lines of evidence suggest that a sperm factor injected into the oocytes induces Ca^{2+} oscillations that mimic those induced by the sperm at fertilization [118]. Very recently, an isoform of PLC (PLCζ) has been proposed as an egg-activating factor that diffuses from the sperm to the egg of mammalian oocytes and produces $InsP_3$ by hydrolyzing PIP_2 [119].

3.2. $InsP_3$ is a key actor in the Ca^{2+} signal at fertilization

A large body of evidence in a number of laboratories has indicated that the release of Ca^{2+} from the ER stores at fertilization is mediated by the $InsP_3Rs$. Early experiments in sea urchin eggs had shown a substantial increase of polyphosphoinositide content at fertilization [120], and generation of $InsP_3$ was found to coincide with the calcium wave [121]. Perhaps, the strongest support for a critical role of $InsP_3$ in the development of the Ca^{2+} spikings at fertilization in mammalian eggs in which the Ca^{2+} signal occurs in the form of repetitive spikes was the abolishment of the Ca^{2+} response mouse eggs activated by sperm by blocking the $InsP_3Rs$ with specific antibodies [122]. The role of $InsP_3$ has been most clearly shown by blockade of Ca^{2+} oscillations in mammalian eggs upon ablation of PLCζ, which recent work has shown to be the $InsP_3$-generating factor in the sperm extract used to challenge the egg [123]. One could also quote experiments in which the expression of recombinant PLCζ in mouse eggs mimicked the effect of the sperm extract [124].

At variance with the mammalian fertilization process, the $InsP_3$ role is not as conclusively established in echinoderms. Experiments more recent than those on sea urchin eggs quoted in the preceding paragraph [121] have failed to see a temporal coincidence between the $InsP_3$ increase and the generation of the Ca^{2+} wave: in fact, the development of the Ca^{2+} signal coincided rather with the elevation of the second messenger cGMP [125]. The suggestion has thus been put forward that cGMP could possibly increase the level of the ryanodine receptor (RyR) agonist cyclic ADPr (cADPr) through the free radical nitrogen oxide (NO) [126]. That $InsP_3$ may not be the exclusive Ca^{2+} mobilizer in echinoderms has also been indicated by experiments in which the inhibitor of the $InsP_3Rs$, heparin, even if injected in massive concentrations, failed to completely abolish the Ca^{2+} response at fertilization in sea urchin and starfish eggs but only delayed its onset and diminished its amplitude [127,128]. Experiments of microinjection of a recombinant portion of the murine $InsP_3R$ that sequesters $InsP_3$ (the $InsP_3$ sponge) have been widely claimed to provide conclusive support for the exclusive role of $InsP_3$ during fertilization of starfish oocytes. These experiments have been performed on immature oocytes (Fig. 1) in which the injected $InsP_3$ sponge showed normal maturation upon addition of 1-MA, and the failure of the fertilization envelope to properly elevate following the sperm challenge [61]. These are certainly interesting experiments: however, the $InsP_3$ sponge evidently failed to inhibit the generation of the Ca^{2+} wave that is essential for maturation [129,130]. It would also be necessary to establish whether in blocking the elevation of the fertilization membrane the $InsP_3$ sponge also blocks the Ca^{2+} wave generated by the sperm contact. At the present state of knowledge, it thus appears premature to subscribe to the $InsP_3$ only idea to interpret the generation of the Ca^{2+} wave at fertilization in starfish oocytes. This seems to be a particularly plausible conclusion for the case of echinoderms but may be valid for mammals as well. It is particularly revealing that very recent work on Ca^{2+} oscillations induced in mouse eggs by the delivery of PLCζ has failed to reveal significant swings in $InsP_3$ prior to the initiation of the Ca^{2+} signal. As the technology used in these experiments was extremely sensitive, the authors conclude that the $InsP_3$ oscillation was even smaller than the detection level [131]. Although permissible, the suggestion clearly demands additional work.

3.3. Ca^{2+}-linked second messengers different from $InsP_3$ may initiate the Ca^{2+} signal at fertilization

Although general consensus points toward a major role for $InsP_3$-induced Ca^{2+} release at fertilization, other second messengers such as cGMP, cADPr, and NAADP have also been shown to trigger Ca^{2+} elevation when injected in whole eggs. The RyR system (RyRs) has been detected in the eggs of most of the species, but its behavior at fertilization is not clear. Its contribution to the process has only been established in sea urchin eggs where the cADPr is generated by the activating sperm after the initiation of the Ca^{2+} wave to regulate the duration of the Ca^{2+} transient [126]. In starfish, the involvement of cADPr/RyRs in the sperm-induced Ca^{2+} response is still uncertain. The issue has been investigated by injecting oocytes with 8-NH_2cADPr, which had been shown to inhibit the cADPr-sensitive Ca^{2+} release in these cells [132]. The pre-treatment with the inhibitor neither prevented nor attenuated the Ca^{2+} response to the sperm. In addition, ruthenium red or inhibitory concentrations of ryanodine [133], both known to suppress the Ca^{2+} oscillations induced by the action of caffeine on RyRs (unpublished results), did not impair the development of the Ca^{2+} wave [134]. In hamster eggs, the spatiotemporal development of the Ca^{2+} signal at fertilization is only linked to the $InsP_3$-mediated Ca^{2+} release [122], because a feedback system underlying calcium-induced calcium release mediated by RyRs [135] is lacking. Very recently, the RyRs have been shown to directly associate with GC in bovine oocytes (the presence of a ryanodine-sensitive calcium release channel has been previously shown to be located within the sea urchin egg cortex as well [136]) from metaphase I stage through fertilization, suggesting that these granules may have a physiological role in the cortical exocytosis process [137].

The second messenger NAADP, in which the nicotinamide ring in the molecule of NADP is replaced by nicotinic acid, has also been found to release Ca^{2+} from intact eggs [138]. Ever since the discovery of its action in sea urchin egg homogenates, NAADP has always exhibited a distinct profile among Ca^{2+}-linked second messengers as it acts on a Ca^{2+} store different from the ER [139]. NAADP- and $InsP_3$-activated receptors have recently been explored in detail in starfish oocytes as determinants of the spatiotemporal dynamics of Ca^{2+} signaling at fertilization. The global Ca^{2+} wave that follows egg activation by the sperm is preceded by a cortical Ca^{2+} flash during the depolarization generated by the influx of Ca^{2+} through voltage-dependent channels [140,141]. Global photoactivation of pre-injected caged NAADP generates a Ca^{2+} signal that resembles closely that induced by the sperm: it is uniformly distributed in the cortical region of the oocytes and then spreads centripetally as a wave. Influx of Ca^{2+} from the external sea water through a channel sensitive to NAADP seems to be involved in the effect of the latter, because the Ca^{2+} response is completely inhibited by L-type Ca^{2+} channel blockers. The finding that the response is also blocked by SOC inhibitors suggests that the channel gated by NAADP may be a TRP-like channel. Analysis of the NAADP-activated channel has revealed that the membrane current is mainly carried by extracellular Ca^{2+}, displaying an inwardly rectifying I–V relationship. The current is not blocked by inhibitors of $InsP_3$ and RyRs receptors [142]. In line with the idea that NAADP may initiate the Ca^{2+} response at fertilization, the NAADP-activated current triggers the membrane depolarization leading to the initial cortical Ca^{2+} increase as the sperm does. Thus, the following chain of events may be envisaged: i) Ca^{2+} entry during the NAADP-triggered FP would initiate the Ca^{2+} response, ii) the $InsP_3$-mediated Ca^{2+} would subsequently interact with the NAADP-evoked depolarization to amplify the Ca^{2+} pulse [143]. The idea is reinforced by the similarities between the biophysical and pharmacological properties of the sperm-elicited depolarization and those displayed by the NAADP-activated Ca^{2+}-mediated current [144]. NAADP has also been shown to activate a cortical flash dependent on Ca^{2+} influx in sea urchin eggs, and the involvement of NAADP in initiating the Ca^{2+} response in those eggs has been corroborated by the finding that NAADP, which is present at high concentrations in sea urchin spermatozoa [145], could be delivered by them into the eggs to induce the cortical Ca^{2+} flash [146]. The finding that NAADP desensitization (auto-inactivation by sub-threshold concentrations of NAADP) prevented the onset of the FP in starfish eggs [143] and strongly reduced the cortical flash, and the Ca^{2+} wave in sea urchin eggs [146] supports the suggestion. The nature of the Ca^{2+} pool mobilized by NAADP remains to be clarified. In sea urchin, it has been suggested that the NAADP-sensitive store activated by the sperm is located on lysosomal-related acidic organelles [147]. However, neither drugs interfering with acidic granules nor agents known to block the action of NAADP on RyRs [148] failed to affect the NAADP-induced membrane depolarization in starfish oocytes [149].

4. Conclusions

Many of the molecular aspects of the fertilization process are still unknown, or at best controversial. One aspect that is undisputed and that has been established decades ago is the necessity of calcium for the successful completion of the process. Both fertilization partners in the operation are involved in the Ca^{2+} effect: the sperm must be activated by a sequel of Ca^{2+}-regulated processes prior to challenging the eggs. The egg, in turn, must experience a significant elevation of Ca^{2+} in the cytoplasm, which in different species may take the form of a single wave sweeping across the cell, or of repetitive transients. Early research on the mechanisms of Ca^{2+} elevation in the eggs had discovered a role for $InsP_3$, which in the course of the years became firmly entrenched in the field as the $InsP_3$-only hypothesis. The early $InsP_3$ evidence and the more novel findings pertain essentially to mammals: other organisms, for example, echinoderms, have begun to produce evidence showing that other Ca^{2+}-linked second messengers, particularly the newest of them all, NAADP, may also have a significant role. In particular, it is now becoming clear that $InsP_3$, which is clearly responsible for the globalization of the Ca^{2+} response, is not for its initial generation. This seems instead to be a role for NAADP, as recently convincingly shown for echinoderms. In fact, NAADP is emerging as a multifaceted second messenger that may have more than one role in the fertilization process. Finally, it is appropriate to mention one particular exciting novel finding that promises to become more and more important in the fertilization field in the future: indications for a role of a new isoform of PLC, PLCζ. This isoform may turn out to be, at least in mammals, the key actor in the generation of the repetitive Ca^{2+} signals at fertilization.

Acknowledgments

The authors are indebted to their colleagues Dr Keiichiro Kyozuka, Asamushi Marine Biological Station, Japan, for the donation of Fig. 4C and to Mr Gianni Gragnaniello, Stazione Zoologica, Naples, for the analysis of the data presented in the figures. Special thanks are due to Dr Ernesto Carafoli (Padova) for many discussions and for his helpful criticisms.

References

1. Monroy, A. (1986) *Adv. Exp. Med. Biol.* 207, 25–35.
2. Jaffe, L.F. (1985) In: C. B. Metz and A. Monroy (Eds), *Biology of fertilization. The fertilization response of the egg,* Vol. 3, Academic Press, pp. 128–157.
3. Darszon, A., Nishigaki, T., Wood, C., Treviño, C.L., Felix, R. and Beltrán, C. (2005) *Intern. Rev. Cytol.* 243, 79–172.
4. Suzuki, N., Nomura, K. and Isaka, S. (1981) *Biochem. Biophys. Res. Commun.* 99, 1238–1244.
5. Cook, S.P., Brokaw, C.J., Muller, C.H. and Babcock, D.F. (1994) *Dev. Biol.* 165, 10–19.
6. Nishigaki, T., Wood, C.D., Tatsu, Y., Yumoto, N., Furuta, T., Elias, D., Shiba, K., Baba, S.A. and Darszon, A. (2004) *Dev. Biol.* 272, 376–388.
7. Kaupp, U.B., Solzin, J., Hildebrand, E., Brown, J.E., Helbig, A., Hagen, V., Beyerman, M., Pampaloni, F. and Weyand, I. (2003) *Nat. Cell Biol.* 5, 109–117.
8. Wood, C.D., Darszon, A. and Whitaker, M. (2003) *J. Cell Biol.* 161, 89–101.
9. Wood, C.D., Nishigaki, T., Furuta, T., Baba, S.A. and Darszon, A. (2005) *J. Cell Biol.* 169, 725–731.
10. Schuh, K., Cartwright, E.J., Jankevics, E., Bundschu, K., Liebermann, J., Williams, J.C., Armesilla, A.L., Emerson, M., Oceandy, D., Knobeloch, K.P. and Neyses, L. (2004) *J. Biol. Chem.* 279, 28220–28226.
11. Okunade, G.W., Millerr, M.L., Pyne, G.J., Sutliff, R.L., O'Connor, K.T., Neumann, J.C., Andringa, A., Miller, D.A., Prasad, V., Doetschman, T., Paul, R.J. and Shull, G.E. (2004) *J. Biol. Chem.* 279, 33742–33750.
12. Su, Y.H. and Vacquier, V.D. (2002) *Proc. Natl. Acad. Sci. U. S. A.* 99, 6743–6748.
13. Gunaratne, H.J., Neill, A.T. and Vacquier, V.D. (2006) *J. Cell Physiol.* 207, 413–419.
14. Solzin, J., Helbig, A., Van, Q., Brown, J.E., Hildebrand, E., Weyand, I. and Kaupp, U.B. (2004) *J. Gen. Physiol.* 124, 115–124.
15. Matsumoto, M., Solzin, J., Helbig, A., Hagen, V., Ueno, S., Kawase, O., Maruyama, Y., Ogiso, M., Godde, M., Minakata, H., Kaupp, U.B., Hoshi, M. and Weyand, I. (2003) *Dev. Biol.* 260, 314–324.
16. Böhmer, M., Van, Q., Weyand, I., Hagen, V., Beyermann, M., Matsumoto, M., Hoshi, M., Hildebrand, E. and Kaupp, U.B. (2005) *EMBO J.* 24, 2741–2752.
17. Nomura, M., Yoshida, M. and Morisawa, M. (2004) *Cell Motil. Cytoskeleton* 59, 28–37.
18. Ho, H.C. and Suarez, S.S. (2001) *Biol. Reprod.* 65, 1606–1615.
19. Ren, D., Navarro, B., Perez, G., Jackson, A.C., Hsu, S., Shi, Q., Tilly, J.L. and Clapham, D.E. (2001) *Nature* 413, 603–609.
20. Kirichok, Y., Navarro, B. and Clapham, D.E. (2006) *Nature* 439, 737–740.

21. Cross, N.L. (2004) *Biol. Reprod.* 71, 1367–1373.
22. Visconti, P.E., Galantino-Homer, H., Ning, X., Moore, G.D., Valenzuela, J.P., Jorgez, C.J., Alvarez, J.G. and Kopf, G.S. (1999) *J. Biol. Chem.* 274, 3235–3242.
23. Dorval, V., Dufour, M. and Leclerc, P. (2002) *Biol. Reprod.* 67, 1538–1545.
24. Galantino-Homer, H.L., Florman, H.M., Storey, B.T., Dobrinski, I. and Kopf, G.S. (2004) *Mol. Reprod. Dev.* 67, 487–500.
25. Neri-Vidaurri Pdel, C., Torres-Flores, V. and Gonzalez-Martinez, M.T. (2006) *Biochem. Biophys. Res. Commun.* 343, 105–109.
26. Adeoya-Osiguwa, S.A. and Fraser, L.R. (1996) *Mol. Reprod. Dev.* 44, 111–120.
27. Dragileva, E., Rubinstein, S. and Breitbart, H. (1999) *Biol. Reprod.* 61, 1226–1234.
28. Castellano, L.E., Trevino, C.L., Rodriguez, D., Serrano, C.J., Pacheco, J., Tsutsumi, V., Felix, R. and Darszon, A. (2003) *FEBS Lett.* 541, 69–74.
29. Jungnickel, M.K., Marrero, H., Birnbaumer, L., Lemos, J.R. and Florman, H.M. (2001) *Nat. Cell Biol.* 3, 499–502.
30. Sutton, K.A., Jungnickel, M.K., Wang, Y., Cullen, K., Lambert, S. and Florman, H.M. (2004) *Dev. Biol.* 274, 426–435.
31. Fukami, K., Yoshida, M., Inoue, T., Kurokawa, M., Fissore, R.A., Yoshida, N., Mikoshiba, K. and Takenawa, T. (2003) *J. Cell Biol.* 161, 79–88.
32. Bleil, J.D. and Wassarman, P.M. (1983) *Dev. Biol.* 95, 317–324.
33. Cabello-Agueros, J.F., Hernandez-Gonzalez, E.O. and Mujica, A. (2003) *Cell Motil. Cytoskeleton* 56, 94–108.
34. Breitbart, H., Cohen, G. and Rubinstein, S. (2005) *Reproduction* 129, 263–268.
35. Dan, J.C. (1952) *Biol. Bull.* 103, 54–66.
36. Tilney, L.G., Kiehart, D.P., Sardet, C. and Tilney, M. (1978) *J. Cell Biol.* 77, 536–550.
37. Vacquier, V.D. and Moy, G.W. (1977) *Proc. Natl. Acad. Sci. U. S. A.* 74, 2456–2460.
38. Takahashi, Y.M. and Sugiyama, M. (1973) *Dev. Growth Differ.* 15, 261–267.
39. Schackmann, R.W., Eddy, E.M. and Shapiro, B.M. (1978) *Dev. Biol.* 65, 483–495.
40. Jones, H.P., Bradford, M.M., McRorie, R.A. and Cormier, M.J. (1978) *Biochem. Biophys. Res. Commun.* 82, 1264–1272.
41. Hirohashi, N. and Vacquier, V.D. (2003) *Biochem. Biophys. Res. Commun.* 304, 285–292.
42. Domino, S.E. and Garbers, D.L. (1988) *J. Biol. Chem.* 263, 690–695.
43. De Blas, G., Michaut, M., Trevino, C.L., Tomes, C.N., Yunes, R., Darszon, A. and Mayorga, L.S. (2002) *J. Biol. Chem.* 277, 49326–49331.
44. Mengerink, K.J., Moy, G.W. and Vacquier, V.D. (2002) *J. Biol. Chem.* 277, 943–948.
45. Watnick, T.J., Jin, Y., Matunis, E., Kernan, M.J. and Montell, C. (2003) *Curr. Biol.* 13, 2179–2184.
46. Hoshi, M., Amano, T., Okita, Y., Okinaga, T. and Matsui, T. (1990) *J. Reprod. Fertil. Suppl.* 42, 23–31.
47. Kawase, O., Minakata, H., Hoshi, M. and Matsumoto, M. (2005) *Zygote* 13, 63–71.
48. Dale, B., Dan-Sohkawa, M., De Santis, A. and Hoshi, M. (1981) *Exp. Cell Res.* 132, 505–510.
49. Stricker, S.A. (1999) *Dev. Biol.* 211, 157–176.
50. Whitaker, M. (2006) *Physiol. Rev.* 86, 25–88.
51. Meijer, L. and Guerrier, P. (1984) *Int. Rev. Cytol.* 86, 129–196.
52. Stricker, S.A. (2006) *Semin. Cell Dev. Biol.* 17, 303–313.
53. Jaffe, L.A. and Terasaki, M. (1994) *Dev. Biol.* 164, 579–587.
54. Mehlmann, L.M., Mikoshiba, K. and Kline, D. (1996) *Dev. Biol.* 180, 489–498.
55. Shiraishi, K., Okada, A., Shirakawa, H., Nakanishi, S., Mikoshiba, K. and Miyazaki, S. (1995) *Dev. Biol.* 170, 594–606.
56. Stricker, S.A., Silva, R. and Smythe, T. (1998) *Dev. Biol.* 203, 305–322.
57. Lee, B., Yoon, S.Y. and Fissore, R.A. (2006) *Semin. Cell Dev. Biol.* 17, 274–284.
58. Jones, K.T., Carroll, J. and Whittingham, D.G. (1995) *J. Biol. Chem.* 270, 6671–6677.
59. Vincent, C., Cheek, T.R. and Johnson, M.H. (1992) *J. Cell Sci.* 103, 389–396.

60. Chiba, K., Kado, R.T. and Jaffe, L.A. (1990) *Dev. Biol.* 140, 300–306.
61. Iwasaki, H., Chiba, K., Uchiyama, T., Yoshikawa, F., Suzuki, F., Ikeda, M., Furuichi, T. and Mikoshiba, K. (2002) *J. Biol. Chem.* 277, 2763–2772.
62. Lim, D., Ercolano, E., Kyozuka, K., Nusco, G.A., Moccia, F., Lange, K. and Santella, L. (2003) *J. Biol. Chem.* 278, 42505–42514.
63. Malathi, K., Kohyama, S., Ho, M., Soghoian, D., Li, X., Silane, M., Berenstein, A. and Jayaraman, T. (2003) *J. Cell Biochem.* 90, 1186–1196.
64. Jellerette, T., Kurokawa, M., Lee, B., Malcuit, C., Yoon, S.Y., Smyth, J., Vermassen, E., De Smedt, H., Parys, J.B. and Fissore, R.A. (2004) *Dev. Biol.* 274, 94–109.
65. Nusco, G.A., Chun, J.T., Ercolano, E., Lim, D., Gragnaniello, G., Kyozuka, K. and Santella, L. (2006) *Biochem. Biophys. Res. Commun.* 348, 109–114.
66. Loeb, J. (1921) *J. Gen. Physiol.* 3, 539–545.
67. Schuel, H. (1985) In: C. H. Metz and A. Monroy (Eds), *Biology of fertilization. The fertilization response of the egg*, Vol. 3, Academic Press, pp. 1–43.
68. Jaffe, L.A., Giusti, A.F., Carroll, D.J. and Foltz, K.R. (2001) *Semin. Cell Dev. Biol.* 12, 45–51.
69. Schuel, H. (1978) *Gamete Res.* 1, 299–382.
70. Vacquier, V.D. (1981) *Dev. Biol.* 84, 1–26.
71. Steinhardt, R.A. and Epel, D. (1974) *Proc. Natl. Acad. Sci. U. S. A.* 71, 1915–1919.
72. Steinhardt, R. (2006) *Semin. Cell Dev. Biol.* 17, 226–228.
73. Vacquier, V.D. (1976) *J. Supramol. Struct.* 5, 27–35.
74. Ridgway, E.B., Gilkey, J.C. and Jaffe, L.F. (1977) *Proc. Natl. Acad. Sci. U. S. A.* 74, 623–627.
75. Carafoli, E., Santella, L., Branca, D. and Brini, M. (2001) *Crit. Rev. Biochem. Mol. Biol.* 36, 107–260.
76. Santella, L., Lim, D. and Moccia, F. (2004) *Trends Biochem. Sci.* 29, 400–408.
77. Dale, B., De Belice, L.J. and Taglietti, V. (1978) *Nature* 275, 217–219.
78. Longo, F.J., Lynn, J.W., McCulloh, D.H. and Chambers, E.L. (1986) *Dev. Biol.* 118, 155–166.
79. Hinkley, R.E., Wright, B.D. and Lynn, J.W. (1986) *Dev. Biol.* 118, 148–154.
80. Arnoult, C., Grunwald, D. and Villaz, M. (1996) *Dev. Biol.* 174, 322–334.
81. Miyazaki, S., Hashimoto, N., Yoshimoto, Y., Kishimoto, T., Igusa, Y. and Hiramoto, Y. (1986) *Dev. Biol.* 118, 259–267.
82. Wilding, M., Russo, G.L., Galione, A., Marino, M. and Dale, B. (1998) *Am. J. Physiol.* 275, C1277–C1283.
83. Jaffe, L.A. (1976) *Nature* 261, 68–71.
84. Gould, M.C. and Stephano, J.L. (2003) *Microsc. Res. Tech.* 61, 379–388.
85. Dale, B. (1987) *Nature* 325, 762–763.
86. Cross, N.L. and Elinson, R.P. (1980) *Dev. Biol.* 75, 187–198.
87. Miyazaki, S. and Igusa, Y. (1981) *Nature* 290, 702–704.
88. McAvey, B.A., Wortzman, G.B., Williams, C.J. and Evans, J.P. (2002) *Biol. Reprod.* 67, 1342–1352.
89. Uno, Y. and Hoshi, M. (1978) *Science* 200, 58–59.
90. Shapiro, B.M. and Eddy, E.M. (1980) *Int. Rev. Cytol.* 66, 257–302.
91. Lillie, F.R. (1911) *J. Morphol.* 22, 361–393.
92. Gould-Somero, M., Holland, L. and Paul, M. (1997) *Dev. Biol.* 58, 11–22.
93. Longo, F.J. (1978) *Dev. Biol.* 67, 249–265.
94. Byrd, W. and Perry, G. (1980) *Exp. Cell Res.* 126, 333–342.
95. Kyozuka, K. and Osanai, K. (1994) *Zygote* 2, 103–109.
96. Jaffe, L.F. (1980) *Ann. N. Y. Acad. Sci.* 339, 86–101.
97. Jaffe, L.F. (1983) *Dev. Biol.* 99, 265–276.
98. Gilkey, J.C., Jaffe, L.F., Ridgway, E.B. and Reynolds, G.T. (1978) *J. Cell Biol.* 76, 448–466.
99. Cardasis, C.A., Schuel, H. and Herman, L. (1978) *J. Cell Sci.* 31, 101–115.
100. Eisen, A. and Reynolds, G.T. (1985) *J. Cell Biol.* 100, 1522–1527.
101. Swann, K. and Whitaker, M. (1986) *J. Cell Biol.* 103, 2333–2342.

102. Whitaker, M. (2006) *Semin. Cell Dev. Biol.* 17, 230–232.
103. Streb, H., Irvine, R.F., Berridge, M.J. and Schulz, I. (1983) *Nature* 306, 67–69.
104. Loeb, J.(1913). *Artificial parthenogenesis and fertilization,* Chicago University Press.
105. Ciapa, B. and Chiri, S. (2000) *Biol. Cell* 92, 215–233.
106. Ciapa, B. and Whitaker, M. (1986) *FEBS Lett.* 195, 347–351.
107. Nuccitelli, R., Yim, D.L. and Smart, T. (1993) *Dev. Biol.* 158, 200–212.
108. Runft, L.L., Jaffe, L.A. and Mehlmann, L.M. (2002) *Dev. Biol.* 245, 237–254.
109. Townley, I.K., Roux, M.M. and Foltz, K.R. (2006) *Semin. Cell Dev. Biol.* 17, 293–302.
110. Rongish, B.J., Wu, W. and Kinsey, W.H. (1999) *Dev. Biol.* 215, 147–154.
111. Carroll, D.J., Ramarao, C.S., Mehlmann, L.M., Roche, S., Terasaki, M. and Jaffe, L.A. (1997) *J. Cell Biol.* 138, 1303–1311.
112. Runft, L.L., Carroll, D.J., Gillett, J., Giusti, A.F., O'Neill, F.J. and Foltz, K.R. (2004) *Dev. Biol.* 269, 220–236.
113. Kinsey, W.H. and Shen, S.S. (2000) *Dev. Biol.* 225, 253–264.
114. Voronina, E. and Wessel, G.M. (2004) *J. Cell Sci.* 117, 5995–6005.
115. Dale, B., DeFelice, L.J. and Ehrenstein, G. (1985) *Experientia* 41, 1068–1070.
116. Kyozuka, K., Deguchi, R., Mohri, T. and Miyazaki, S. (1998) *Development* 125, 4099–4105.
117. Runft, L.L. and Jaffe, L.A. (2000) *Development* 127, 3227–3236.
118. Swann, K. (1990) *Development* 110, 1295–1302.
119. Swann, K., Larman, M.G., Saunders, C.M. and Lai, F.A. (2004) *Reproduction* 127, 431–439.
120. Turner, P.R., Sheetz, M.P. and Jaffe, L.A. (1984) *Nature* 310, 414–415.
121. Ciapa, B., Borg, B. and Whitaker, M. (1992) *Development* 115, 187–195.
122. Miyazaki, S., Yuzaki, M., Nakada, K., Shirakawa, H., Nakanishi, S., Nakade, S. and Mikoshiba, K. (1992) *Science* 257, 251–255.
123. Nomikos, M., Blayney, L.M., Larman, M.G., Campbell, K., Rossbach, A., Saunders, C.M., Swann, K. and Lai, F.A. (2005) *J. Biol. Chem.* 280, 31011–31018.
124. Swann, K., Saunders, C.M., Rogers, N.T. and Lai, F.A. (2006) *Semin. Cell Dev. Biol.* 17, 264–273.
125. Kuroda, R., Kontani, K., Kanda, Y., Katada, T., Nakano, T., Satoh, Y., Suzuki, N. and Kuroda, H. (2001) *Development* 128, 4405–4414.
126. Leckie, C., Empson, R., Becchetti, A., Thomas, J., Galione, A. and Whitaker, M. (2003) *J. Biol. Chem.* 278, 12247–12254.
127. Mohri, T., Ivonnet, P.I. and Chambers, E.L. (1995) *Dev. Biol.* 172, 139–157.
128. Santella, L., De Riso, L., Gragnaniello, G. and Kyozuka, K. (1999) *Exp. Cell Res.* 248, 567–574.
129. Santella, L. and Kyozuka, K. (1994) *Biochem. Biophys. Res. Commun.* 203, 674–680.
130. Santella, L., Ercolano, E., Lim, D., Nusco, G.A. and Moccia, F. (2003) *Biochem. Soc. Trans.* 31, 79–82.
131. Shirakawa, H., Ito, M., Sato, M., Umezawa, Y. and Miyazaki, S. (2006) *Biochem. Biophys. Res. Commun.* 345, 781–788.
132. Nusco, G.A., Lim, D., Sabala, P. and Santella, L. (2002) *Biochem. Biophys. Res. Commun.* 290, 1015–1021.
133. Gerasimenko, O. and Gerasimenko, J. (2004) *J. Cell Sci.* 117, 3087–3094.
134. Moccia, F., Nusco, G.A., Lim, D., Kyozuka, K. and Santella, L. (2006) *Dev. Biol.* 294, 24–38.
135. Roderick, H.L., Berridge, M.J. and Bootman, M.D. (2003) *Curr. Biol.* 27, R425.
136. McPherson, S.M., McPherson, P.S., Mathews, L., Campbell, K.P. and Longo, F.J. (1992) *J. Cell Biol.* 116, 1111–1121.
137. Wang, L., White, K.L., Reed, W.A. and Campbell, K.D. (2005) *Cloning Stem Cells* 7, 306–320.
138. Lee, H.C. (2002) In: H. C. Lee (Ed.), *Cyclic ADP-ribose and NAADP, structures, metabolism and functions,* Kulver, pp. 1–21.
139. Lee, H.C. and Aarhus, R.A. (1995) *J. Biol. Chem.* 270, 2152–2157.
140. Lim, D., Kyozuka, K., Gragnaniello, G., Carafoli, E. and Santella, L. (2001) *FASEB J.* 15, 2257–2267.

141. Santella, L. (2005) *Mol. Interv.* 5, 70–72.
142. Moccia, F., Lim, D., Nusco, G.A., Ercolano, E. and Santella, L. (2003) *FASEB J.* 17, 1907–1909.
143. Moccia, F., Nusco, G.A., Lim, D., Kyozuka, K. and Santella, L. (2006) *Dev. Biol.* 294, 24–38.
144. Moccia, F., Lim, D., Kyozuka, K. and Santella, L. (2004) *Cell Calcium* 36, 515–524.
145. Billington, R.A., Ho, A. and Genazzani, A.A. (2002) *J. Physiol.* 544, 107–112.
146. Churchill, G.C., O'Neill, J.S., Masgrau, R., Patel, S., Thomas, J.M., Genazzani, A.A. and Galione, A. (2003) *Curr. Biol.* 13, 125–128.
147. Churchill, G.C., Okada, Y., Thomas, J.M., Genazzani, A.A., Patel, S. and Galione, A. (2002) *Cell* 111, 703–708.
148. Galione, A. and Petersen, O. (2005) *Mol. Interv.* 5, 73–79.
149. Moccia, F., Billington, R.A. and Santella, L. (2006) *Biochem. Biophys. Res. Commun.* 348, 329–336.

J. Krebs and M. Michalak (Editors)
Calcium: A Matter of Life or Death

ISSN: 0167-7306/DOI: 10.1016/S0167-7306(06)41017-6

CHAPTER 17

Ca^{2+} signaling during embryonic cytokinesis in animal systems

Sarah E. Webb and Andrew L. Miller

Department of Biology, The Hong Kong University of Science and Technology, Clear Water Bay, Hong Kong, People's Republic of China, Tel.: +852 2358 8631; Fax: +852 2358 1559; E-mail: almiller@ust.hk

Abstract

In this chapter, we provide a review of the literature that describes the role of Ca^{2+} signaling during embryonic cytokinesis in animal systems. We begin with a historic overview (starting with the earliest reports published in the latter part of the nineteenth century) that gives some of the first descriptions linking Ca^{2+} with cell division in embryos. This introductory overview also outlines the different techniques that were used and developed over time, from the pre-Ca^{2+} imaging days through to the sophisticated approaches that are available today for directly visualizing Ca^{2+} signals in living cells. In the remainder of the chapter, we describe the more recent advances in cytokinetic Ca^{2+}-signaling research, starting where the introduction finished, in the 1990s. Reports of the various Ca^{2+} signals visualized during cytokinesis in fish, amphibian, echinoderm, and insect embryos using both fluorescent and luminescent Ca^{2+} probes are described, as well as the investigations carried out to determine both the requirement of elevated Ca^{2+} during cytokinesis and the source of the Ca^{2+} involved in this process. The current hypotheses regarding the possible roles and targets of the different cytokinetic Ca^{2+} signals observed are also briefly discussed.

Keywords: actomyosin, apposition, Ca^{2+}, cleavage furrow, cytokinesis, deepening, positioning, propagation

1. Introduction: a historical perspective

There is a vast literature concerning cytokinesis in animal cells, and many hypotheses and ideas have been proposed over the years attempting to explain this complex process in a variety of different cell types. These have been nicely reviewed by numerous authors, reflecting the current views that prevailed [1–12]. No one, however, perhaps does so with such insight and eloquence than Rappaport in his excellent book on animal cell cytokinesis [13]. In this chapter, however, we have attempted to sift through both the past and current literature and focus specifically on reports concerning the possible roles played by Ca^{2+} in initiating and orchestrating the cytokinesis process. We have narrowed the scope of this chapter even further, by concentrating on the role of Ca^{2+} signaling during embryonic cytokinesis in animal systems, as this is our major research interest. By doing so, we hope to present the reader with a more in-depth review of what we propose to be a distinct form of cytokinesis and one that has been particularly amenable for visualizing cytokinesis-related intracellular Ca^{2+} transients. There are other previous reviews where the possible roles played by Ca^{2+} during mitosis and the cell cycle have been considered. Although these do not specifically focus on Ca^{2+} and cytokinesis, we would like to draw the readers' attention to four of these in particular for further reference, for the dual reason that they help to place cytokinetic Ca^{2+} signaling into an overall cell cycle signaling context and the fact that their authors have made significant contributions to our understanding of the Ca^{2+} signaling field. The first is by Berridge [14], where he discusses the role played by Ca^{2+} and cyclic nucleotides in the cell division process; the second is by Steinhardt [15], where he considers Ca^{2+} regulation of the first cell cycle of the sea urchin embryo; the third by Hepler [16] mainly focuses on Ca^{2+} and mitosis but also contains a section on Ca^{2+} and cell division; and finally the fourth by Whitaker [17], where he considers the regulation of the cell division cycle by inositol trisphosphate (IP_3) and the Ca^{2+}-signaling pathway.

The study of cytokinesis in embryonic systems has a significant experimental history that allows one to follow the link between Ca^{2+} and cytokinesis from pre-Ca^{2+} imaging days, through to the sophisticated techniques we currently possess for visualizing Ca^{2+} signals in living cells. As pointed out by Rappaport [13] with regards to cytokinesis in general – but even more so concerning the possible roles played by Ca^{2+} in this process – it is quite easy when searching through the early literature to miss important Ca^{2+}-related observations, as they often do not appear prominently in the title of papers but are consigned to footnotes, figure legends, and discussion sections. Thus, it is hard to present a comprehensive review of the literature on this subject. The strategy we adopted, therefore, was to identify what we felt were key publications, where an investigation of the role played by Ca^{2+} during embryonic cytokinesis was the focus of the study, rather than an adjunct observation. We do apologize in advance, however, to those who have published excellent work concerning Ca^{2+} signaling during cytokinesis in non-embryonic situations (such as tissue culture cells and dividing unicellular systems). We suggest that this would make an excellent focus for a separate review. We also believe that the study of embryonic cytokinesis in particular is most relevant to the title of this Volume – "*Calcium, A Matter of Life or Death*" – for without the successful completion of the post-fertilization phase of rapid embryonic cell division, a multicellular Blastula Stage would never be attained, resulting in certain embryonic death.

We suggest that our current ideas regarding the possible roles played by Ca^{2+} during embryonic cytokinesis can be traced back to the early mechanistic hypothesis relating to increases in equatorial tension proposed in 1876 by Bütschli [18]. He suggested that cells divide because the tension at the equatorial surface increases and that the increase somehow results from the action of the mitotic apparatus. The unknowns that lay ahead at this time were the nature of the equatorial contractile apparatus and how it got positioned, assembled, and modulated, as well as the identity of the cellular messengers that orchestrated these processes. In 1937, Mazia [19] was one of the first to suggest that Ca^{2+} might be responsible for triggering cell division after fertilization. However, a direct mechanistic connection between cytokinesis and Ca^{2+} perhaps has its origins in the observations of physiologists in the 1950s, who saw similarities in the contractile nature of muscle and the movements of, and within, cells – including cell division [20–22]. At the time, these observations were regarded as being one of the most exciting developments with respect to understanding the biochemical processes involved in animal cell cleavage [2]. Following on from this general view, Tilney and Marsland [23] proposed that the deposition of contractile filaments in the furrow cortex of dividing sea urchin (*Arbacia punctulata*) eggs might be triggered by the mobilization of Ca^{2+} from intracellular stores. They proposed that this store was the interior of the nucleus and that Ca^{2+} was released on nuclear envelope breakdown (NEB). In addition, they also suggested that polar relaxation (thought at the time to be an essential pre-requisite for successful cleavage) was achieved via localized Ca^{2+} sequestration activity of "relaxing factors" [23]. Thus, they were among the first to suggest that both the release and the re-sequestration of Ca^{2+} from intracellular stores might play important roles in the cytokinesis process.

Following on from these Ca^{2+}-related observations, Bluemink [24] reported the presence of a diastema during cleavage in the salamander (*Ambystoma mexicanum*) egg. The most prominent features of this cytokinetic structure (seen in the large yolky eggs of certain primitive fish and amphibians) were elongated, closed profiles of double-stranded membranes, which he termed "cisternae." Associated with and presumably originating from the cisternae, Bluemink [24] also described "light vesicles." He proposed that these helped to polarize and organize the filaments in the contractile ring and also served as membrane-bound compartments for Ca^{2+} transport in order to stimulate the contraction processes during cytokinesis. Another significant report from this time was that of Gingell [25], who demonstrated that directly injecting Ca^{2+} into the cortex of eggs of the frog (*Xenopus laevis*) resulted in a strong cortical contraction. He also observed that wound-induced surface wrinkling displayed properties similar to early cleavage furrow formation. Furthermore, he demonstrated an absolute requirement for Ca^{2+} in the extracellular medium, which would induce a cortical contraction around a tear in the plasma membrane in order to seal the opening and that the contraction process involved microfilaments. These observations formed the basis of the idea that elevated intracellular Ca^{2+} in some way stimulated microfilaments in the cortex to contract and suggested that there might be mechanistic similarities between the contraction processes involved in surface wound healing and cytokinesis.

Soon after, Baker and Warner [26] reported that Ca^{2+} was essential for early embryonic cleavage in *Xenopus* embryos. They could not, however, determine whether changes in Ca^{2+} regulated cytokinesis, as opposed to cytokinesis merely having a requirement for some intracellular free Ca^{2+}. This is an especially significant paper, as Baker and Warner were also among the first investigators to use the Ca^{2+}-sensitive bioluminescent reporter, aequorin, to attempt to directly record temporal Ca^{2+} dynamics in developing *Xenopus* embryos during cytokinesis. They reported that five of eight aequorin-injected embryos showed an increase in aequorin-generated light output at the time of both the first and second embryonic cleavages. These transient changes were seen in eggs bathed in Ca^{2+}-free solution containing EGTA, which they suggested indicated that the increases in intracellular Ca^{2+} could not have resulted from Ca^{2+} entry across the plasma membrane during cleavage but resulted from release from intracellular stores. They did, however, suggest that the results from their aequorin-based experiments should be treated with considerable caution. As we will see in subsequent sections, Baker and Warner [26] were being perhaps overcautious with regard to their seminal observations. However, the early days of recording Ca^{2+} dynamics from living cells were fraught with technical difficulties when gathering data, and an understandable caution was displayed when interpreting results.

A second report from the same year by Timourian et al. [27], utilizing the Ca^{2+} buffer EDTA, clearly demonstrated that in sea urchin (*Lytechinus pictus*) embryos, Ca^{2+} appeared to play a role in specifically determining the site of the cleavage furrow. They proposed that Ca^{2+} was released from the mitotic centers and, via diffusion, attained its highest concentration at the cortex equidistant from the two centers. Furthermore, using frozen longitudinal sections of *L. pictus* eggs prepared during metaphase and anaphase of the first cell division cycle, in conjunction with electron-probe microanalysis, the same group was able to demonstrate that Ca^{2+} was more concentrated at the furrow region during metaphase, but not anaphase [28].

Durham [29] also considered the possible role of Ca^{2+} in positioning and stimulating the embryonic cleavage furrow in an attempt to unify the then current ideas and suggestions regarding the control of actin and myosin in non-muscle movement. He proposed that what he described as the "ion permeability barrier" would be reduced in the maximally stretched region of the egg surface (due to the expanding mitotic apparatus) and that a resulting local Ca^{2+} influx would thus enhance actomyosin contraction. In this fashion, the contractile ring would be assembled and stimulated to contract. Thus, the idea that a localized elevation in intracellular Ca^{2+} might play a crucial role in positioning the cleavage furrow became an area of considerable experimental interest. To further explore this possibility, Schroeder and Strickland [30], using the recently developed Ca^{2+} ionophore A23187, reported that furrow-like deformations resulted from ionophore-induced increases in intracellular Ca^{2+} concentrations in frog (*Rana pipiens*) eggs and as a result suggested that A23187 might be a good experimental tool for investigating the role of Ca^{2+} during cytokinesis. Subsequently, Arnold [31] demonstrated that the application of A23187 to the surface of squid (*Loligo pealei*) embryos could precociously induce the appearance of furrows. Furthermore, application of the ionophore, while cleavage furrow progression was underway, speeded up this process and resulted in an extension of the furrows beyond their normal meroblastic boundaries. Both these effects were reported to occur in Ca^{2+}-free medium, suggesting that the ionophore was releasing Ca^{2+} from internal stores. Following up on the observations of Gingell [25], Hollinger and Schuetz [32] demonstrated that microinjection of Ca^{2+} into frog (*R. pipiens*) oocytes after NEB resulted in a cleavage-like constriction and the formation of an embryonic structure that resembled a two-cell embryo. Furthermore, the location of the microinjected Ca^{2+} determined the orientation of the resulting cleavage furrow. They suggested that Ca^{2+} diffused away from the injectate bolus toward the cortex, and at some specific, critical concentration of Ca^{2+}, it activated some contractile elements (which they proposed might be microfilaments) in the embryonic cortex. The following year, Conrad and Davis [33] reported that the microiontophoretic injection of Ca^{2+} resulted in rapid shape changes in fertilized eggs of the gastropod *Ilyanassa obsoleta* that resembled, in some cases, cleavage furrows. The same group subsequently demonstrated, via the use of Ca^{2+} and calmodulin antagonists, that Ca^{2+} may be necessary for cytokinesis (and polar lobe formation) in *I. obsoleta* and that the Ca^{2+} involved was released from intracellular, sequestered stores rather than being derived from exogenous sources [34].

The year 1977 also saw the first report suggesting that the early cleavages in teleost embryos may also be accompanied by transient elevations in intracellular Ca^{2+}. Although mainly focused on the large Ca^{2+} transient associated with egg activation, Ridgway et al. [35] – using aequorin-loaded medaka (*Oryzias latipes*) eggs and a photomultiplier tube (PMT) – reported that in one experimental embryo, small transient increases in intracellular Ca^{2+} appeared to correlate with the first two cell divisions. When considered along with Baker and Warner's 1972 [26] report from dividing *Xenopus* embryos, the data presented from these first attempts to directly measure changes in intracellular Ca^{2+} associated with cell division in live embryos are somewhat equivocal. Subsequent experiments, however, carried out on both of these embryonic systems [36,37] using improved intracellular Ca^{2+} reporters in combination with better imaging technology (and discussed in following sections) clearly support and vindicate these early reports. The investigators responsible for these early experiments thus deserve appropriate acknowledgement for their pioneering attempts to advance our understanding of the role played by Ca^{2+} in the cytokinetic process.

The 1980s saw a dramatic step forward in our understanding of cytokinetic Ca^{2+} signaling, through the development and use of reliable fluorescent-based intracellular Ca^{2+} reporters such as fura-2, which had the advantage of improved sensitivity as well as easier handling and loading into cells [38]. These were used to describe, for the first time, temporal changes in intracellular Ca^{2+} during cell cycle progression in individual sea urchin (*L. pictus*) eggs, where Ca^{2+} transients recorded after mitotic events were reported to be associated with cytokinesis [39]. In addition, through the use of inserted Ca^{2+}-sensitive microelectrodes, Shantz [40] reported that in one experimental medaka embryo, during two successive cleavages, Ca^{2+} rose transiently fourfold above the original resting level in synchrony with each cell division. During this period, however, contradictory results with regard to the role played by Ca^{2+} were also reported. For example, Yoshimoto et al. [41], like Ridgway et al. [35], used a PMT to measure the light output of single aequorin-injected medaka eggs. They, however, reported that the intracellular concentration of Ca^{2+} (judged by the luminescent output from the whole egg) appeared to be lowest at the time of furrowing. They also reported similar results from the eggs of several echinoderm species, including those of the sea urchin, *Hemicentrotus pulcherrimus*. However, in the case of the echinoderms, because they could not detect any post-fertilization light emitted from single aequorin-loaded eggs, they devised a method where they suspended a population of eggs in a solution of Ca^{2+}-free sea water (plus EGTA), and then both aequorin and the Ca^{2+} ionophore A23187 were added to the suspension medium to measure the total amount of Ca^{2+} that leaked out of the eggs during successive cell division cycles. They reasoned that the total Ca^{2+} efflux induced by the ionophore would reflect, at least qualitatively, changes in the intracellular Ca^{2+}

concentration within a population of synchronously diving echinoderm eggs. Eggs were removed from the measuring chamber at regular intervals in order to check on cytokinetic progression. Like the temporal data they collected from individual medaka eggs, their echinoderm-based results suggested that the times of furrowing always corresponded with temporal troughs in a cyclic pattern of luminescent output [41]. The indirect method used to try and measure subtle intracellular Ca^{2+} dynamics, combined with the assumptions that were made with respect to interpreting the data, serve as a good example to illustrate the difficulties involved in attempting to measure changes in intracellular Ca^{2+} in dividing cells. One of the keys to addressing this problem would be the development of Ca^{2+} reporters and imaging devices with the resolution and sensitivity to visualize small, localized changes within individual, healthy, dividing cells. In addition to these attempts to directly (or indirectly) measure Ca^{2+} changes during cytokinesis, investigators continued to report experimental results demonstrating that the direct injection of Ca^{2+} into eggs such as those of *Xenopus* could initiate precocious pseudo-cleavage formation [42].

Thus, there is a considerable body of evidence from the older literature to support the suggestion that positioning, assembly, and stimulation of the cell division machinery may indeed involve localized regions of elevated intracellular Ca^{2+}. The technical challenges associated with measuring changes in localized intracellular Ca^{2+} levels during cytokinesis have in the past, and still continue to, generate uncertainties with regard to the timing, location, and even the direction of Ca^{2+} changes. The development, however, of non-disturbing, intracellular Ca^{2+} reporters has greatly improved our understanding of the Ca^{2+}-signaling events that occur throughout the cytokinetic process, and these will be discussed in the following sections of this chapter. It must be stressed and acknowledged, however, that much of what we currently know about Ca^{2+} signaling during cytokinesis has its origins in the series of elegant experiments and careful observations made by individuals over the past few decades, who did not have access to the techniques and equipment we sometimes take for granted today.

2. *The sequential stages of embryonic cytokinesis*

In animal cells, cytokinesis begins with the invagination of existing plasma membrane as a result of the contraction of an actomyosin ring or arc [6,13] and is complete with the permanent separation of the daughter cells. How this cleavage process is achieved varies depending on both the size and geometry of the cells involved. In what we would classify as "large" embryos (e.g., zebrafish, with a diameter of approximately 600 μm, and *Xenopus*, with a diameter of approximately 1200 μm), we suggest that there are several distinct basic processes that contribute to cytokinesis. These include an initial positioning of the furrow within the cell cortex and the propagation (without significant deepening) of the furrow across the surface; this is then followed by furrow deepening (i.e., furrow ingression), which results in two daughter cells that are separated by a distinct groove, and finally furrow apposition, where the daughter cells "zip" up together. In small embryos (e.g., echinoderms, with a diameter of approximately 60 μm), the first two components (i.e., furrow positioning and propagation) may be indistinguishable as separate entities, as the furrow appears almost simultaneously in the equatorial cortex of the late anaphase cell. In addition, in small somatic and tissue culture cells, as well as dividing unicellular organisms, the final stage of cytokinesis may involve the abscission of the final intercellular connection between the prospective daughter cells, which results in complete cell separation [13,43–45]. During embryonic cell division, however, the cells are faced with two opposing requirements: daughter cells must be separated from one another, while at the same time the embryonic blastoderm must be held together. Thus, following the process of furrow deepening that separates the daughter nuclei, the respective daughter cell plasma membranes do not separate from one another but undergo a process of apposition that holds the cells together. Cleavage furrow apposition, therefore, represents the final step in this type of distinct embryonic cytokinesis [36,46]. Due to the large size of their embryos, and thus the fact that the different cytokinetic components (positioning, propagation, deepening, and apposition) are separated both temporally and spatially, the first few meroblastic cleavages of teleost embryos and the early holoblastic cleavages of amphibian embryos thus provide a unique opportunity to explore both the Ca^{2+}-signaling pathways and the mechanisms specific to each stage in the embryonic cell cleavage process.

3. *Recent advances in cytokinetic Ca^{2+}-signaling research*

The following sections review more recent reports regarding cytokinetic Ca^{2+} signaling, starting from the early 1990s (i.e., where the Introduction finished). The first Ca^{2+} imaging studies were reported at this time, using both luminescent (i.e., aequorin) and fluorescent (e.g. calcium green-1) Ca^{2+} reporters, in combination with custom-built photon imaging microscopy [36] and time-lapse confocal microscopy [47], respectively. Furthermore, a combination of these improved Ca^{2+} imaging techniques, along with the intracellular Ca^{2+} modulation procedures that were developed previously (e.g., using Ca^{2+} ionophores and chelators, as well as Ca^{2+} channel agonists and antagonists), provided researchers with the tools required to explore the significance and function of Ca^{2+} signaling in the cytokinetic process.

3.1. *Cytokinetic Ca^{2+} signaling in fish embryos*

3.1.1. *Visualization of Ca^{2+} transients that accompany cytokinesis*

Using the bioluminescent reporter aequorin, Fluck et al. [36] were the first to show that slow Ca^{2+} waves accompanied the progression of furrows across blastomere surfaces in medaka embryos, during the first few meroblastic cell divisions. Two successive Ca^{2+} waves, of approximately 0.5 μm/s, were observed; the first, relatively narrow wave, was reported to accompany furrow extension (i.e., propagation) while the second, which was approximately three to seven times wider than the first, accompanied the subsequent furrow deepening then apposition of the daughter cells.

The first reports describing Ca^{2+} transients during the early embryonic cleavages in zebrafish (*Danio rerio*) yielded conflicting results. Reinhard et al. [48] reported no Ca^{2+} activity during the early cleavage cycles, whereas Chang and Meng [49] observed that localized elevations in free Ca^{2+} were associated with cytokinesis. Furthermore, Chang and Meng [49] clearly showed that intracellular Ca^{2+} was elevated "not only in the 'right' place, but also at the 'right' time" and suggested that such a signal played a role in determining the position of the furrowing plane. Both groups used the fluorescent Ca^{2+} reporter, calcium green-1 dextran, the only apparent difference between the two investigations being the size of the dextran that the reporter was conjugated to. These conflicting reports from the same embryonic system only served to illustrate the challenges involved in visualizing cytokinetic Ca^{2+} transients.

This apparent contradiction in dividing zebrafish embryos was subsequently addressed using bioluminescent aequorin-based imaging [50]. Using this method, several distinct Ca^{2+} signals were observed to accompany the sequential stages of cytokinesis. The first was a clear localized elevation of intracellular Ca^{2+} (subsequently given the name the "furrow positioning signal"), which was observed to precede the first appearance of the furrow arc at the blastodisc surface. As the leading edges of the arc progressed outward toward the margins of the blastodisc, they were accompanied by two sub-surface slow Ca^{2+} waves (called the "furrow propagation signal") moving at approximately 0.5 μm/s. Fig. 1A shows a representative animal pole view of an *f*-aequorin-loaded zebrafish embryo demonstrating the changes in intracellular free Ca^{2+} during positioning and propagation of the first cell division cycle. As these propagation wave fronts approached the edge of the blastodisc, another Ca^{2+} signal appeared in the central region where the positioning signal had originally appeared. Like the propagation signal, it likewise extended outward to the margins of the blastodisc at approximately 0.5 μm/s, but in this case it also moved downward (at ~0.1 μm/s), accompanying the deepening process that separates the daughter cells. This Ca^{2+} signal was given the name the "furrow deepening signal." The region of localized elevated Ca^{2+} persisted throughout the furrow apposition process and only returned to the resting level once the cleavage furrow was fully apposed [46]. This signaling sequence was also observed during the second cell division cycle [50]. The results obtained in zebrafish loaded with aequorin thus both supported and also extended further the observations of Chang and Meng [49], by demonstrating that localized increases in Ca^{2+} were associated with furrow positioning, propagation, deepening, and apposition during the first two cleavages in these embryos. Subsequently, other reports confirmed that such localized increases in Ca^{2+} do occur in zebrafish embryos during the different phases of cytokinesis using both aequorin [51] and fluorescent Ca^{2+} reporters [52]. The accumulated data suggest that the positioning, propagation, and deepening transients are generated exclusively by Ca^{2+} released from an IP_3-sensitive store, i.e., the ER [46,49,52–54]. The apposition transient, however, may rely on Ca^{2+} derived from this same common store as well as a contribution from extracellular sources, due to the fact that furrow apposition fails to proceed normally when embryos are bathed in Ca^{2+}-free medium [36,50,55].

Apart from the reports from medaka and zebrafish described above, little else has been published regarding the Ca^{2+}-signaling events during cytokinesis in other fish species. However, here we present new data, which suggest that a sequential series of Ca^{2+} transients might be a conserved feature of embryonic cytokinesis in fish. Fig. 2 (panels A and B) illustrates the Ca^{2+} transients that accompany cytokinesis of the first cell division cycle in embryos of the rosy barb (*Puntius conchonius*; Fig. 2A) and mummichog (*Fundulus heteroclitus*; Fig. 2B). It appears, therefore, that localized Ca^{2+} transients are a conserved feature during embryonic cytokinesis within at least three orders of teleosts: Cypriniformes (zebrafish and rosy barb), Beloniformes (medaka), and Atheriniformes (mummichog). Beloniformes and Atheriniformes are acanthopterygians, whereas Cypriniformes are ostariophysans [56]. As both of these groups are within the Eutelostei, cytokinetic Ca^{2+} signaling during development is likely an ancestral trait at this taxonomic level and may well be present in more basal teleosts and other actinopterygians.

3.1.2. Determination of the requirement of elevated Ca^{2+} for cytokinesis

In addition to direct visualization, there is also good indirect evidence (e.g., from injecting Ca^{2+} chelators such as BAPTA-type buffers [57] at various times during the cell cycle) to indicate that Ca^{2+} signaling plays a required role in cleavage of both medaka [58] and zebrafish [46,49–51,54] embryos.

In order to understand more clearly what effect Ca^{2+} chelators (such as BAPTA buffers) have on the generation of a particular transient, and the developmental significance of blocking or modulating that transient, the timing of their introduction as well as their rate of spread within an early embryo is crucial. Taking this into account, Webb et al. [50] focused on the propagation transient and waited until the furrow had been positioned on the blastodisc surface (by observing either the appearance of the furrow on the surface or the Ca^{2+} transient associated with this event) before introducing the buffer. These experiments clearly indicated a Ca^{2+} requirement for furrow propagation in cleaving zebrafish embryos. Subsequently, and again by careful timing of the introduction of the Ca^{2+} buffer, it has more recently been shown that a localized elevation of Ca^{2+} is also essential for both furrow deepening [46] and furrow positioning [54] in zebrafish embryos.

3.1.3. Determination of the source of the Ca^{2+} generating the various cytokinetic transients

Zebrafish embryos have also been treated with antagonists of the various Ca^{2+} release channels in order to explore the Ca^{2+} stores responsible for the generation of the cytokinetic Ca^{2+} transients. Chang and Meng [49] demonstrated that the cytokinetic Ca^{2+} signal that they observed using calcium green-1 dextran could be blocked via the introduction of heparin, an antagonist of IP_3 receptors (IP_3Rs), but was not affected by ryanodine (a ryanodine receptor antagonist), nifedipine and La^{3+} (inhibitors of plasma membrane Ca^{2+} channels), or the

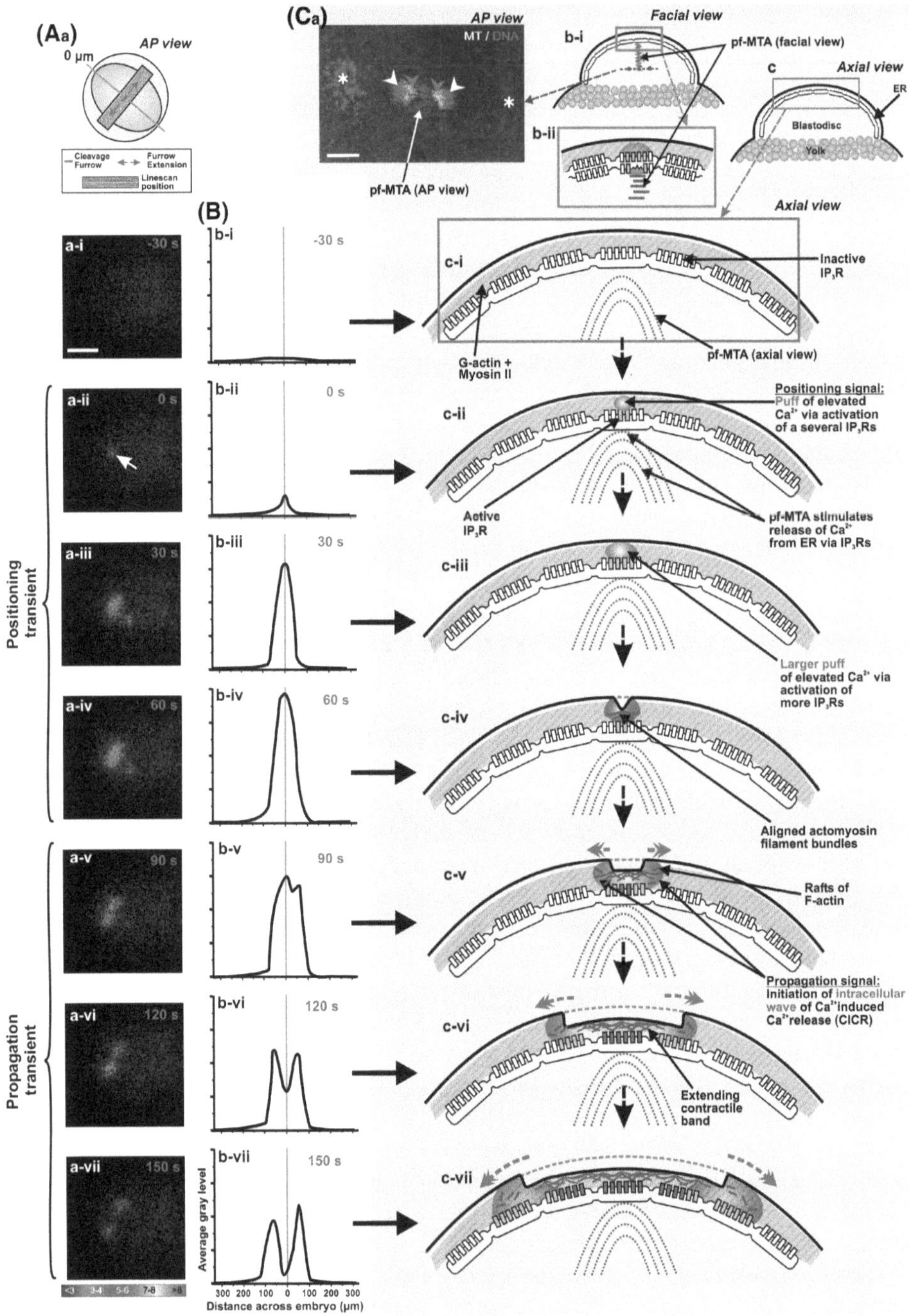
(Aa)
AP view
0 µm
Cleavage Furrow
Furrow Extension
Linescan position
(B)
(Ca)
AP view
MT / DNA
pf-MTA (AP view)
Facial view
b-i
b-ii
pf-MTA (facial view)
c
Axial view
ER
Blastodisc
Yolk
Axial view
a-i -30 s
a-ii 0 s
a-iii 30 s
a-iv 60 s
a-v 90 s
a-vi 120 s
a-vii 150 s
b-i -30 s
b-ii 0 s
b-iii 30 s
b-iv 60 s
b-v 90 s
b-vi 120 s
b-vii 150 s
Positioning transient
Propagation transient
Average gray level
300 200 100 0 100 200 300
Distance across embryo (µm)
c-i
c-ii
c-iii
c-iv
c-v
c-vi
c-vii
Inactive IP3R
G-actin + Myosin II
pf-MTA (axial view)
Positioning signal:
Puff of elevated Ca2+ via activation of a several IP3Rs
Active IP3R
pf-MTA stimulates release of Ca2+ from ER via IP3Rs
Larger puff of elevated Ca2+ via activation of more IP3Rs
Aligned actomyosin filament bundles
Rafts of F-actin
Propagation signal:
Initiation of intracellular wave of Ca2+induced Ca2+release (CICR)
Extending contractile band

Fig. 1. Cleavage furrow positioning and propagation in zebrafish zygotes. **(A)** Representative animal pole (AP) view of an *f*-aequorin-loaded zebrafish embryo demonstrating the changes in intracellular free Ca^{2+} during positioning and propagation of the first cell division cycle. **(Aa)** A schematic illustration of the embryo to show the location and extension of the cleavage furrow as well as the position of a Linescan analysis (the results of which are shown in **(B)**). The luminescent images (pseudo-color panels, labeled a-i to a-vii) represent 10 s of accumulated luminescence and were obtained at the following time intervals: (a-i) 30 s before the appearance of the positioning transient; (a-ii) during the initial appearance of the positioning transient (see arrow); then (a-iii) 30 s, (a-iv) 60 s, (a-v) 90 s, (a-vi) 120 s, and (a-vii) 150 s after the initiation of the positioning transient. **(B)** Analysis of luminescence along the first cleavage furrow, before (b-i), and then during the positioning (b-ii to b-iv) and propagation (b-v to b-vii) transients. The profiles represent the average gray level of light emission from the region of the blastodisc shown in **(Aa)** and were measured using the Linescan function in the Metamorph (v.6.1) image analysis software. 0 μm represents the location on the blastodisc where the positioning signal was first detected and corresponds to the mid-point of the furrow. **(Ca)** A pre-furrowing microtubule array (pf-MTA) appears to arise from the mid-zone spindle and grows upward and outward to the blastodisc cortex where it precedes the appearance of the furrow on the surface of the blastodisc. This representative embryo was fixed during the metaphase/anaphase transition of the first division cycle and the microtubules (MTs; red) and DNA (green) labeled. In this single optical section through the blastodisc, the pf-MTA is located between the separating chromosomes (green arrowheads) and originates from the remnants of the mitotic spindle (white arrowheads). The MTOCs (asterisks) at the spindle poles are also shown. **(Cb-i)** A schematic representation of the blastodisc from a facial view to show the approximate location of the single confocal section in **(Ca)**. **(Cb-ii)** We propose that the pf-MTA transmits positional information to the blastodisc cortex, which results in the localized release of Ca^{2+} via the activation of IP_3Rs in the ER, and subsequently the correct positioning of the cytokinetic contractile apparatus. **(Cc)** Hypothetical model showing an axial view of a zebrafish blastodisc to illustrate how Ca^{2+} released via the activation of IP_3Rs in the ER might generate the furrow positioning and propagation transients during the first cell division cycle via a blip/puff/wave Ca^{2+}-signaling cascade. Scale bars represent 200 μm (in **A**) and 20 μm (in **Ca**). Reproduced with kind permission from *Zygote* [53,54] (See Color Plate 46, p. 541).

removal of Ca^{2+} from the external medium. The authors concluded that the cytokinetic Ca^{2+} transient arose from internal stores through the release of Ca^{2+} via IP_3Rs [49].

Webb et al. [50] confirmed with aequorin that zebrafish embryos could generate a regular series of cytokinetic Ca^{2+} transients and divide normally (for at least the first few cell division cycles) in Ca^{2+}-free medium, thus supporting Chang and Meng's [49] observation that extracellular Ca^{2+} is not involved in generating these transients. More recently, Lee et al. [46,54] investigated the source of cytokinetic Ca^{2+} in further detail by carefully timing the introduction of the various antagonists in order to focus specifically on the deepening and positioning Ca^{2+} transients. They showed that the introduction of heparin or another IP_3R antagonist, 2-aminoethoxydiphenylborate (2-APB), at the appropriate time to challenge only the deepening transient, blocked the Ca^{2+} signal and resulted in an inhibition of furrow deepening. However, antagonists of the ryanodine receptor and NAADP-sensitive channel had no effect on either furrow deepening or on the deepening Ca^{2+} transient [46]. They also demonstrated that the endoplasmic reticulum (ER) and IP_3R are both localized on either side of the cleavage furrow during the deepening process and thus provided additional evidence for the possible intracellular Ca^{2+} store and release mechanism for the deepening Ca^{2+} transient. Most recently, Lee et al. [54] demonstrated that the positioning Ca^{2+} transient is also generated by Ca^{2+} release via IP_3Rs. They also

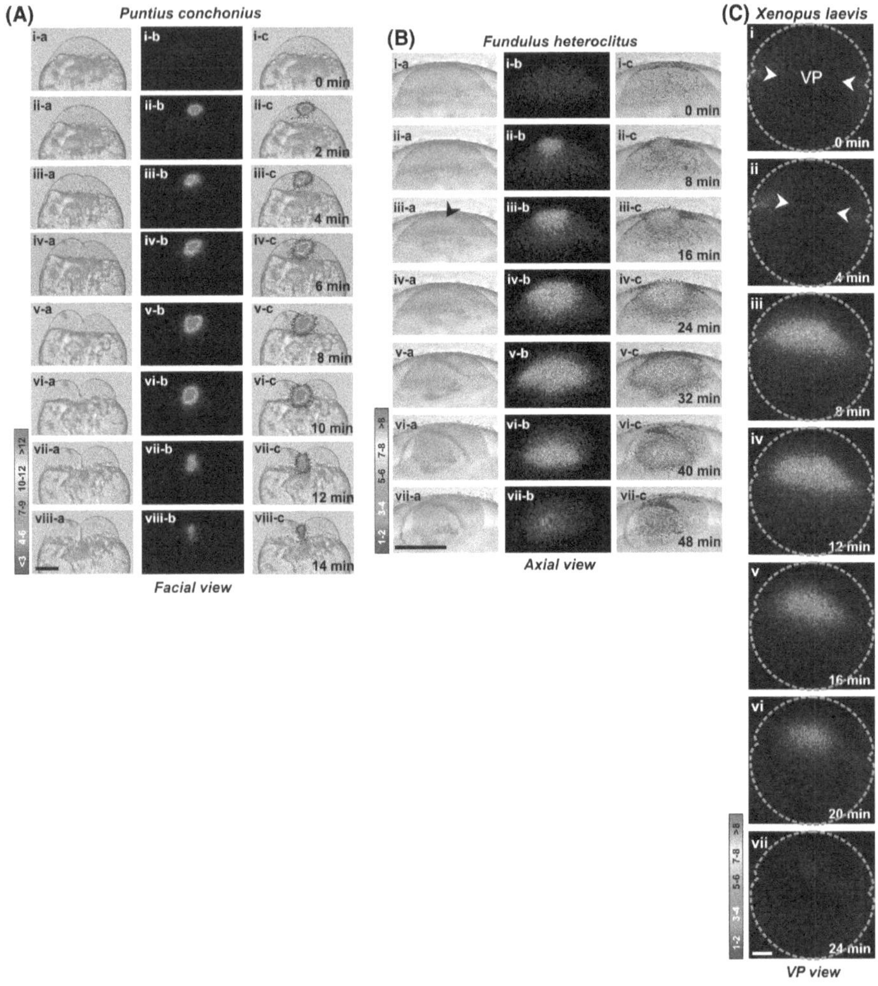

Fig. 2. Representative examples of the changes in intracellular free Ca^{2+} that occur during the first cell division cycle in *f*-aequorin-loaded **(A)** *Puntius conchonius*, **(B)** *Fundulus heteroclitus*, and **(C)** *Xenopus laevis* embryos. In **(A)**, the luminescent images (pseudo-colored panels, labeled i-b to viii-b) represent 30 s of accumulated luminescence with a 1.5-min gap between each image, in **(B)** the luminescent images (i-b to vii-b) represent 30 s of accumulated luminescence with an 7.5-min gap between each image and in **(C)** the luminescent images (i to vii) represent 120 s of accumulated luminescence with a 2-min gap between each image. In **(A)** and **(B)**, corresponding bright-field images (in **A** labeled i-a to viii-a and in **B** labeled i-a to vii-a) were grabbed immediately after the respective luminescent images. The superimposed luminescent and bright-field images (shown in **A**, panels i-c to viii-c and in **B**, panels i-c to vii-c) show the location of the Ca^{2+} signals more clearly. Color scale indicates luminescent flux in photons/pixel. Scale bars represent 200 μm. **(A)**, **(B)**, and **(C)** are Webb et al., Chan et al., and Lee et al., unpublished results, respectively (See Color Plate 47, p. 543).

showed that this transient is a required component in positioning the cleavage furrow at the blastodisc surface and has a distinct rising phase, which clearly distinguishes it from the subsequent propagation transient. Fig. 1B shows a series of line-graphs to illustrate the dynamics of the positioning (panels b-ii to b-iv) and propagation (panels b-v to b-vii) Ca^{2+} transients that were obtained from the embryo shown in Fig. 1A.

With regards to exploring the upstream events that might be involved in organizing the furrow positioning Ca^{2+} transient, evidence has recently been presented to demonstrate that a dynamic array of microtubules may be involved. This array, which originates from the mid-zone of the mitotic spindle, and then expands both upward and outward toward the surface of the blastodisc, may play a role in mechanistically linking (in both a spatial and temporal manner) mitotic and cytokinetic events [53]. Lee et al. [53] reported that this "pre-furrowing microtubule array" (or pf-MTA) localized together with a zone of corticular ER and IP_3Rs in the blastoderm cortex just prior to the morphological appearance of the cleavage furrow at the blastodisc surface. The authors suggested that the pf-MTA might be involved in organizing the ER and IP_3Rs, required to generate the Ca^{2+} signals that are essential for cleavage furrow formation in zebrafish embryos. Fig. 1Ca shows a single confocal section taken from an animal pole view through the middle of the blastodisc of a representative zebrafish embryo during the metaphase/anaphase transition of the first cell division cycle. The microtubules (red) were labeled by immunohistochemistry and the DNA (green) labeled with SYTOX Green. This image illustrates that the pf-MTA appears to arise from the mid-zone spindle. Fig. 1Cb (panels i and ii) are schematic representations of a blastodisc from a facial view to show (i) the approximate location of the confocal scan in Fig. 1Ca and (ii) the co-localization of the pf-MTA, ER, IP_3Rs, and Ca^{2+}. In addition, representative examples of the localization patterns of these four elements before the morphological appearance of the cleavage furrow at the start of the second cell division cycle are shown in Fig. 3. Fig. 1Cc is a hypothetical model to illustrate how Ca^{2+} released via the activation of IP_3Rs in the ER might generate the furrow positioning and propagation transients during the first cell division cycle via a blip/puff/wave Ca^{2+}-signaling cascade [59–61].

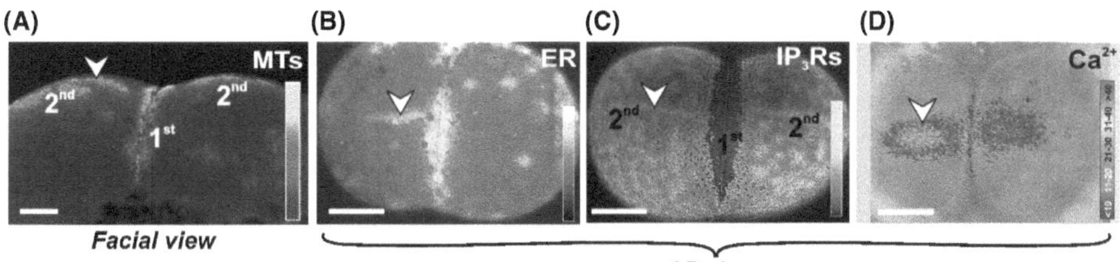

Fig. 3. Localization of microtubules (MTs), endoplasmic reticulum (ER), IP_3 receptors (IP_3R_s), and Ca^{2+} before the first morphological appearance of the second cleavage furrow. These representative examples of embryos viewed from **(A)** a facial and **(B–D)** animal pole views show the co-localization (see arrowheads) of **(A)** the MTs, **(B)** the ER, **(C)** IP_3Rs, and **(D)** Ca^{2+} at the site of the future furrows of the second cell division cycle. **(D)** The aequorin-generated image (in pseudo-color) is superimposed on the appropriate bright-field image of the embryo (acquired just prior to the aequorin image) to show the position of the Ca^{2+} signal more clearly. This panel represents 60 s of accumulated luminescence. Color scale indicates luminescent flux in photons/pixel. Scale bars represent 100 μm. The first and second cleavage furrows are labeled in panels **(A)** and **(C)**. Reproduced with kind permission from *Zygote* [53] (See Color Plate 48, p. 544).

3.2. *Cytokinetic Ca^{2+} signaling in amphibian embryos*

3.2.1. *Visualization of Ca^{2+} transients that accompany cytokinesis*

In amphibians, much of the research on Ca^{2+} signaling during cytokinesis has been done using *Xenopus* embryos, although there are a few reports suggesting a possible role of Ca^{2+} during cleavage in other amphibian species. These include the older reports on the salamander and other frog species that are described in the Introduction, as well as more recent publications from the newt, *Cynops pyrrhogaster* [62,63]. The role of Ca^{2+} in amphibian cytokinesis has always been a somewhat controversial issue. As mentioned earlier (see the Introduction), Baker and Warner [26], using aequorin in conjunction with a PMT, reported that in *Xenopus* embryos, transient changes in Ca^{2+} were sometimes detected during the first and second cell cleavages. On the contrary, Rink et al. [64], who measured the free Ca^{2+} in *Xenopus* embryos with inserted Ca^{2+}-selective microelectrodes, concluded that cell division is not accompanied by a change in the level of free Ca^{2+}. More recent reports were also contradictory regarding the role of Ca^{2+} in cleavage in these embryos. In 1991, for example, Grandin and Charbonneau [65] again using Ca^{2+}-selective microelectrodes demonstrated that periodic Ca^{2+} oscillations do accompany cell division in *Xenopus* embryos. They showed that these oscillations reached a peak just a few minutes after the start of a membrane hyperpolarization [65], which is reported to correspond to the onset of cleavage and the production of new plasma membrane in the forming blastomeres [66]. These observations during embryonic cleavage in *Xenopus* were supported by those of Aimar and Grant [67], who reported that the injection of Ca^{2+} into enucleated newt (*Pleurodeles waltl*) embryos stimulated the formation of pseudo-furrows on the egg surface. However, two other independent reports that were published in the early 1990s appeared to contradict these findings. Using semi-synthetic aequorin in conjunction with a PMT, Kubota et al. [68] and Keating et al. [69] reported that whereas the Ca^{2+} oscillations that occurred had the same frequency as cytokinesis, cleavage actually began when Ca^{2+} was at a minimum, with the level of Ca^{2+} reaching maximum just before the onset of mitosis. Furthermore, both these groups demonstrated that the Ca^{2+} oscillations continued even when cleavage was blocked with microtubule disrupting drugs, such as colchicine or nocodazole. A potential problem that might result from measuring the total light output from an embryo during cytokinesis using a non-spatial photon counting system such as a PMT is that it does not enable one to distinguish Ca^{2+} dynamics within localized regions (i.e., in the cleavage furrow) of the dividing embryo. Once again, direct visualization using a spatially sensitive detection system was required to help resolve this controversy in amphibian systems.

Thus, using a combination of calcium green-1 (or calcium green-1-dextran, 10 kDa) and confocal microscopy, Muto et al. [37] reported that a series of highly localized Ca^{2+} waves propagated along the cleavage furrows in *Xenopus* embryos. Furthermore, they reported that these localized Ca^{2+} waves could be distinguished from the sinusoidal oscillating Ca^{2+} waves identified previously by other groups [68,69]. These cleavage furrow waves were reported to begin a few minutes after the completion of the furrow propagation process (i.e., after furrows had spread from the animal to the vegetal pole) and they then propagated along the deepening furrows at a velocity of approximately 3 μm/s. The waves lasted for approximately 10 min, after which the newly formed blastomeres, which at this stage were separated from each other by a distinct groove, "zipped" back up together. These data thus suggested that localized Ca^{2+} transients might be associated with furrow deepening and apposition in *Xenopus* embryos.

Most recently, Noguchi and Mabuchi [70], again using calcium green-1-dextran (10 kDa) and confocal microscopy, reported the presence of not one but two distinct Ca^{2+} waves during the first cleavage in *Xenopus* embryos. Like Muto et al. [37], Noguchi and Mabuchi did not detect any type of Ca^{2+} signal associated with either furrow positioning (i.e., as reported in zebrafish) or furrow propagation (as observed in zebrafish and medaka). They suggested that the first Ca^{2+} wave might be the counterpart of the furrow deepening transient observed in zebrafish and medaka and that the second might be involved in the apposition of the two daughter blastomeres after furrow deepening has been completed. However, as the waves (that lasted for only approximately 5 min) did not propagate into the vegetal region of embryos and were not localized in the region of the contractile band, the authors concluded that Ca^{2+} was not specifically required for either the formation of the cleavage furrow or its deepening during cytokinesis in *Xenopus* embryos [70].

To add to the controversy, Fig. 2 (panel C) illustrates some preliminary imaging data to show a Ca^{2+} transient that accompanies cytokinesis of the first cell division cycle in an aequorin-injected *Xenopus* embryo (Lee et al., unpublished results). In this vegetal pole view of the embryo, two waves of Ca^{2+}, which last for approximately 20 min, can be seen to propagate from the equator of the embryo (possibly arising at the animal pole) and then converge on the vegetal pole. These preliminary data, therefore, suggest that in *Xenopus* embryos, propagating waves of intracellular Ca^{2+} release perhaps accompany the progression of the leading edges of the holoblastic cleavage furrow. When considering all the reported data, however, it would appear that we are still far from fully understanding the possible roles played by Ca^{2+} during cytokinesis in amphibian embryos.

3.2.2. Determination of the requirement of elevated Ca^{2+} for cytokinesis

The function and developmental significance of the Ca^{2+} transients observed during cytokinesis in *Xenopus* embryos has been explored once again by introducing Ca^{2+} chelators such as BAPTA and EGTA. As described earlier (see Section 3.1.2.), the timing of the injection of the buffer is critical. For example, it has been shown by injecting BAPTA-type buffers into dividing *Xenopus* embryos that if the buffer is introduced too early then it blocks karyokinesis rather than cytokinesis, and the embryos fail to divide [71,72]. If, however, it is introduced immediately after the completion of karyokinesis, but before the onset of cytokinesis, then furrow positioning is affected [71]. Indeed, Miller et al. [71] demonstrated that the position of a buffer-induced ectopic furrow always lay on a meridian passing through the animal pole, and they suggested that this might be due to the disruptive action of the buffer on a central spindle (carrying positional information to the cell cortex) rather than on the mitotic asters. Thus, they did not suggest that Ca^{2+} itself was the primary positioning signal but that it might play a key role in the formation and/or subsequent stability of the central spindle. If, on the other hand, buffer was injected after furrow propagation had begun, the buffer quickly arrested furrow extension and eventually resulted in its regression [71]. These results thus suggest that Ca^{2+} is required to position, extend, and maintain the cleavage furrow in *Xenopus* embryos.

As mentioned previously, Noguchi and Mabuchi [70] challenged the proposal that elevated Ca^{2+} was required for furrow formation and deepening. This was supported by data demonstrating that the use of dibromo-BAPTA (or EGTA) at concentrations sufficient to suppress the two Ca^{2+} waves they visualized had no effect on the cleavage process in *Xenopus* embryos. This once again clearly indicates that further work is required in order to settle these seemingly contradictory findings with regard to the specific roles, if any, played by Ca^{2+} transients during embryonic cleavage in amphibians.

3.2.3. Determination of the source of the Ca^{2+} generating the various cytokinetic transients

Evidence suggests that in *Xenopus* (as for zebrafish and medaka), the cytokinetic Ca^{2+} signals arise from internal stores through the release of Ca^{2+} via IP_3Rs. For example, Han et al. [73] demonstrated that when one blastomere of a two-cell stage *Xenopus* was injected with heparin or an antibody to PIP_2 (an upstream component of the phosphatidylinositol (PI) pathway), when the first cleavage furrow had formed, then the second cleavage of this blastomere was blocked. In addition, Muto et al. [37] reported that both the progression of the Ca^{2+} wave that they observed and also the morphological process of furrow deepening could be blocked by the injection of heparin and thus concluded that IP_3-mediated Ca^{2+} mobilization played a role in the propagation of the furrow deepening Ca^{2+} wave.

It has also been reported that the microinjection of Ca^{2+} store-enriched microsomal fractions (derived from mouse cerebella and CHO cells) into dividing newt (*C. pyrrhogaster*) eggs can induce extra cleavage furrows at the site of injection [62]. When these microsomal fractions were co-injected with heparin or an IP_3R antibody, however, cleavage furrow induction was suppressed. These authors concluded that in the newt, the cleavage furrow is induced via IP_3-mediated Ca^{2+} release. Thus, Ca^{2+} release via IP_3Rs might also play a key role in positioning as well as deepening these embryonic cleavage furrows. There is also evidence from amphibian embryos that, like in zebrafish [53], microtubules are involved in the initiation of cleavage. For example, Mitsuyama and Sawai [63] demonstrated that when Ca^{2+} store-enriched microsomal fractions that expressed IP_3Rs were injected into newt embryos, these stores moved toward the next cleavage furrow, in a microtubule-dependent manner.

Thus, in spite of the various contradictory reports that have been published, evidence is slowly accumulating to suggest that localized, elevated levels of intracellular Ca^{2+} released via IP_3Rs in the ER (which are in some way organized in the cleavage furrow via microtubules) are a conserved feature of cytokinesis both in amphibians and in fish.

3.3. Cytokinetic Ca^{2+} signaling in echinoderm embryos

As mentioned in the Introduction, Poenie et al. [39] were the first to demonstrate that a Ca^{2+} transient was generated during cleavage in embryos of the sea urchin *L. pictus*. They reported that the Ca^{2+} transient, which rises to approximately threefold above the resting level, is initiated just before the furrow appears and then accompanies the developing furrow during cleavage. More recently, Ciapa et al. [74] demonstrated that in the fura-2-dextran-loaded embryos of another sea urchin (*Paracentrotus lividus*), a broad, 200-nM Ca^{2+} transient accompanies cleavage. In addition, they showed that this transient coincides with an increase in IP_3 levels. Suzuki et al. [75] subsequently presented indirect evidence to suggest that Ca^{2+} might play a role in cytokinesis in sea urchin embryos by demonstrating that when *Lytechinus variegatus* embryos were treated with the Ca^{2+} ionophore, A23187, shortly after the furrow appeared, then progression of the furrow stopped followed by regression. The authors suggested that the ionophore might inhibit the furrowing process in these embryos by causing a global elevation of Ca^{2+} and thus disrupt the localized rise in Ca^{2+} required for successful cleavage furrow formation [75]. A nice aspect of these experiments was the fact that the effects of the A23187 were reversible. Thus, when the ionophore was washed away, cytokinesis resumed.

A broad peak of Ca^{2+} was also shown to occur in sand dollar (*Echinaracnius parma*) embryos during cytokinesis [76]. When one of the two blastomeres in two-cell stage embryos was injected with aequorin, and then the Ca^{2+} dependent luminescence measured with a microscope photometer, a broad peak of luminescence was observed in the injected blastomere, that correlated with the 10–12 min duration of cytokinesis [76].

Starfish embryos have also been shown to undergo transient rises in their intracellular free Ca^{2+} during cleavage [47]. Using calcium green dextran, loaded into *Asterina miniata* and *Pisaster ochraceus* oocytes, in conjunction with time-lapse confocal microscopy, a series of repetitive Ca^{2+} oscillations were imaged, which began within approximately 90 min following fertilization. Some of these Ca^{2+} transients were reported to precede the formation of the cleavage furrow. Furthermore, treatment with heparin both blocked the Ca^{2+} oscillations and inhibited cleavage, thus suggesting that the Ca^{2+} release was once again mediated by IP_3Rs [47].

There are a variety of technical challenges associated with attempting to visualize cytokinetic Ca^{2+} signals from echinoderm embryos. These are currently being overcome [47] and thus this classic animal group, historically used for studying embryonic cytokinesis, will hopefully provide much additional new data in the future.

3.4. Cytokinetic Ca^{2+} signaling in insect embryos

By imaging *Drosophila melanogaster* embryos that had been loaded with aequorin, Créton et al. [77] showed that Ca^{2+} is elevated in the ventral region of the embryo during early development (stages 1–4) and oscillates with the cell cycle. However, as the earliest stages of *Drosophila* development are characterized by nuclear division without cytokinesis, these Ca^{2+} signals are more likely to be associated with some aspect of the nuclear cycle and not with embryonic cell cleavage. Ca^{2+} levels were also shown to increase (this time largely on the dorsal side of the embryo), during cell formation, which starts at stage 5 and lasts for approximately 60 min. This increase in intracellular Ca^{2+} on the dorsal side resulted in a dorsal-to-ventral Ca^{2+} gradient, which was suggested to play a role in dorsal–ventral axis formation [77]. This, to date, is the only report suggesting that elevated Ca^{2+} might play some role in the cell cycle and cellularization process in *Drosophila* embryos, but no data were presented to link either of these elevated Ca^{2+} events to a specific cytokinetic process.

In *Drosophila* spermatocytes, however, Ca^{2+} has been shown to play a key role in cytokinesis [78]. Using indirect methods, for example, by treating the spermatocytes with BAPTA-AM or the Ca^{2+} ionophores A23187 and ionomycin, these investigators showed that Ca^{2+} is required to stabilize the cleavage furrow during deepening but is not required for the initiation of cleavage. They also demonstrated that PIP_2 is localized in the plasma membrane and in the cleavage furrows of dividing *Drosophila* spermatocytes and, like Ca^{2+}, is not needed for the initiation of cytokinesis but is required for continued furrow deepening. They concluded, therefore, that in these particular cells, Ca^{2+} released from intracellular stores via the PI pathway is required for the normal progression of cytokinesis.

In spite of the paucity of Ca^{2+} imaging data reported from *Drosophila* embryos, this system – with its powerful genetics – represents a potent model for exploring the interaction between Ca^{2+} signaling, cell division, and development.

4. *Possible targets of the cytokinetic Ca^{2+} signals*

It has been well documented from studies in both embryonic (see Introduction) and tissue culture cells that cleavage involves the contraction of an actomyosin band (or ring) that lies at the base of the cleavage furrow, via a mechanism comparable with the contraction of smooth muscle (reviewed by Satterwhite and Pollard, [6]). In addition, it was suggested that an increase in Ca^{2+} in the furrow region might (i) induce the recruitment of microfilaments into the contractile band [36], (ii) regulate the binding of myosin to the actin filaments [75,79], and/or (iii) trigger actomyosin contraction via the activation of the Ca^{2+}-sensitive myosin light chain kinase (MLCK; [80,81]). Indeed, a current proposition that was discussed at length at the 2004 ASCB Summer Symposium on cytokinesis (see conference report [82]) was that furrow positioning is closely associated with myosin regulation via the activation of MLCK. This might mechanistically link the localized region of elevated Ca^{2+} with a cascade of molecular events that results in the assembly of the contractile apparatus, followed by its activation (see Fig. 4). In newt (*C. pyrrhogaster*) embryos, for example, bundles of actin filaments were identified in surgically isolated cleavage furrows. These isolated furrows were induced to contract by the addition of ATP and Ca^{2+} [83]. In addition, it has been demonstrated that in embryos of the sand dollar (*Clypeaster japonicus*) and

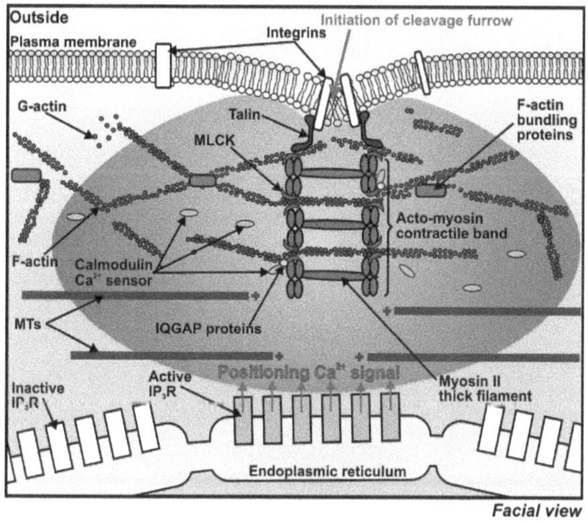

Fig. 4. The possible cytoskeletal targets of the localized release of Ca^{2+}, as well as other molecular elements reported to be involved in cytokinesis. Modified from Figure 2 in [9]. The main Ca^{2+}-sensitive target is the calmodulin Ca^{2+} sensor, which acts via MLCK to organize the actomyosin-based contractile band (See Color Plate 49, p. 544).

several sea urchin species (i.e., *Pseudocentrotus*, *Strongylocentrotus*, and *Hemicentrotus* sps), the phosphorylation of myosin by MLCK is required for the formation of the contractile ring [84]. In addition, a more recent report suggests that in *Drosophila* syncytial blastoderm embryos, F-actin and myosin II are transported along microtubules and co-localize in areas where pseudo-cleavage furrows form [85]. Taken together, these results thus suggest that many of the components of the contractile mechanism required for successful cytokinesis are a conserved feature of many animal cells, as might be the suggested modulatory role played by Ca^{2+}. Furthermore, more recently, Field et al. [86] have demonstrated that PIP_2 is required for the adhesion of the contractile ring to the plasma membrane in tissue culture cells, thus providing evidence for another possible role of cytokinetic Ca^{2+} signaling in the formation of the cleavage furrow.

In addition to its assembly and contractile-related activities, another suggested downstream function of the furrow deepening and apposition Ca^{2+} transients, is the recruitment followed by the exocytosis of vesicles at the ingressing furrow membrane [46]. Several lines of evidence indicate that membrane trafficking plays a key role in membrane remodeling during cytokinesis in animal embryos [87,88]. For example, the addition of new membrane has been found to be a common feature of cytokinesis in many species including *Xenopus*, *Drosophila*, zebrafish, and various sea urchin species [87,89–91]. In addition, an inhibition of cytokinesis has in some cases been shown to be associated with the lack of new membrane formation [92]. It has been suggested that membrane addition may be due to a mobilization of membranes from internal stores [55,93] with the Golgi being the prime candidate [93]. How the furrow membrane is restructured, as well as the identity and precise function of the

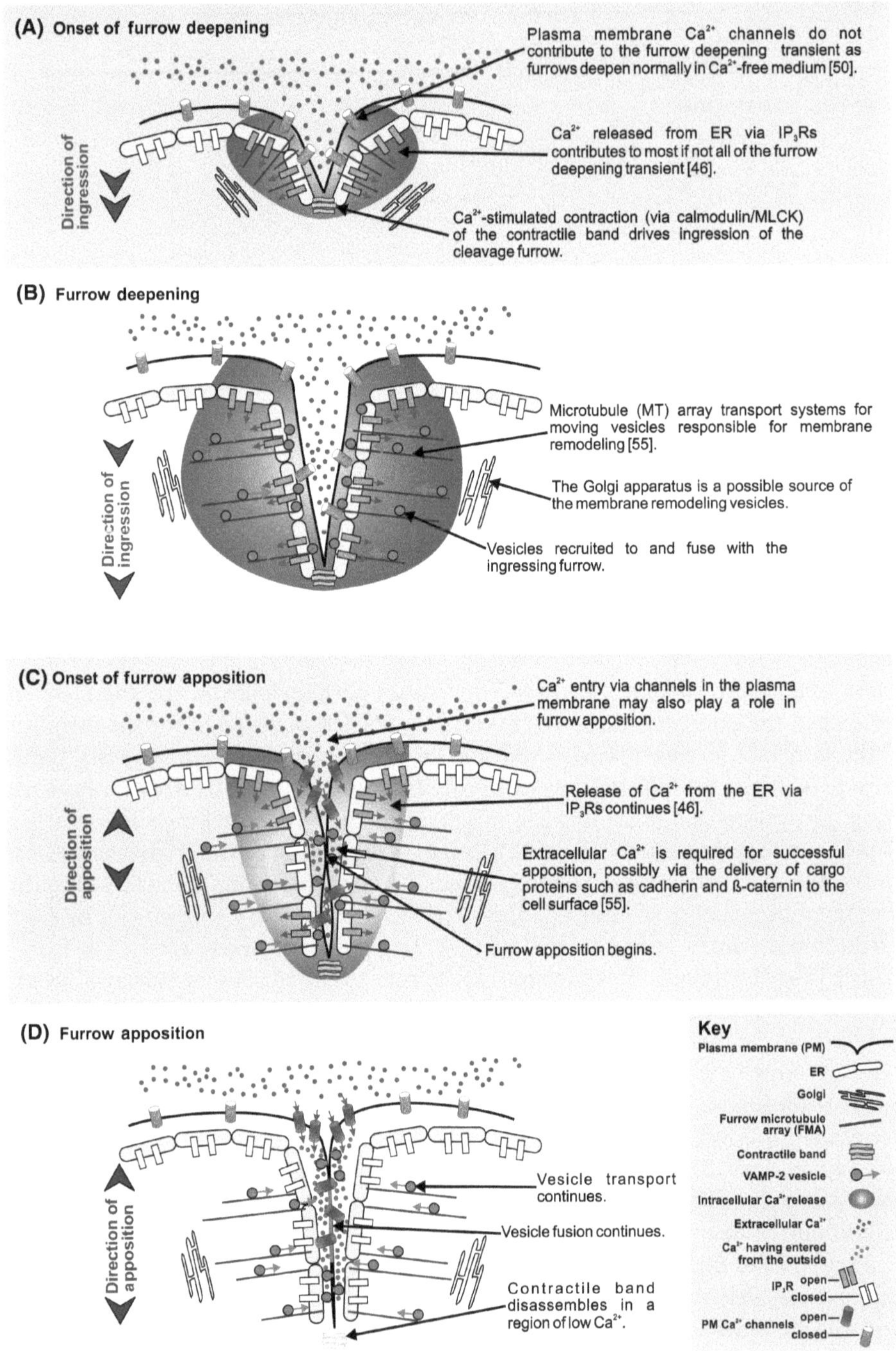
(A) Onset of furrow deepening
Plasma membrane Ca2+ channels do not contribute to the furrow deepening transient as furrows deepen normally in Ca2+-free medium [50].
Ca2+ released from ER via IP3Rs contributes to most if not all of the furrow deepening transient [46].
Ca2+-stimulated contraction (via calmodulin/MLCK) of the contractile band drives ingression of the cleavage furrow.
Direction of ingression
(B) Furrow deepening
Microtubule (MT) array transport systems for moving vesicles responsible for membrane remodeling [55].
The Golgi apparatus is a possible source of the membrane remodeling vesicles.
Vesicles recruited to and fuse with the ingressing furrow.
Direction of ingression
(C) Onset of furrow apposition
Ca2+ entry via channels in the plasma membrane may also play a role in furrow apposition.
Release of Ca2+ from the ER via IP3Rs continues [46].
Extracellular Ca2+ is required for successful apposition, possibly via the delivery of cargo proteins such as cadherin and ß-catenin to the cell surface [55].
Furrow apposition begins.
Direction of apposition
(D) Furrow apposition
Vesicle transport continues.
Vesicle fusion continues.
Contractile band disassembles in a region of low Ca2+.
Direction of apposition
Key
Plasma membrane (PM)
ER
Golgi
Furrow microtubule array (FMA)
Contractile band
VAMP-2 vesicle
Intracellular Ca2+ release
Extracellular Ca2+
Ca2+ having entered from the outside
IP3R open closed
PM Ca2+ channels open closed

Fig. 5. Possible roles of Ca^{2+} signaling during cleavage furrow deepening and apposition in zebrafish zygotes. Hypothetical model showing a facial view of a zebrafish blastodisc to illustrate how Ca^{2+} released via the activation of IP_3Rs in the ER might generate the furrow deepening transient and how Ca^{2+} entering from the outside of the embryo through plasma membrane Ca^{2+} channels might generate the furrow apposition Ca^{2+} transient. The model also suggests that one of the possible downstream functions of these Ca^{2+} transients is to regulate vesicle recruitment and fusion during ingression and apposition. It is suggested that vesicle recruitment might be mediated by cognate v- and t-SNARE partners located on the vesicles and ingressing furrow membrane, respectively. It is also proposed that vesicles are transported to the deepening furrow by an array of perpendicular microtubules [55], which may have developed from the pf-MTA (See Color Plate 50, p. 545).

trafficking molecules involved in this process, is currently an area of intense interest. We propose that membrane-bound t-SNARE complexes, that possess a Ca^{2+}-sensitive element, may be one of the targets of the deepening and apposition Ca^{2+} transients. Thus, vesicles, decorated with the appropriate v-SNARE, could be moved to the ingressing furrow membrane along microtubule arrays [53,55], where docking and fusion might be stimulated via these Ca^{2+} transients (see Fig. 5).

5. *Conclusions*

In all the embryonic systems studied so far, even though a close relationship has been shown to exist between cytokinetic Ca^{2+} transients and various morphological events during the cytokinesis process – be it furrow positioning, propagation, deepening, or apposition (see Tables 1 and 2) – the precise roles of these transients still remain obscure. Several hypothetical models have been proposed over the years, however, that attempt to link these Ca^{2+} transients to specific cytokinetic events [36,46,50,53,54]. We have also attempted to do so in this chapter, and this is illustrated in Figs 1C, 4, and 5. It is obvious that a great deal of additional effort needs to be spent in order to fully understand the specific, and perhaps multiple, functions of each of the cytokinetic Ca^{2+} transients. We suggest that this work will be greatly aided through the development of even more sensitive imaging devices and intracellular Ca^{2+} reporters, as well as the extension of these studies into different animal systems, especially those of invertebrate species that are particularly suited to imaging studies.

Acknowledgements

We acknowledge financial support from Hong Kong RGC grant HKUST6214/02M. This chapter was prepared while A.L.M. was the recipient of a Croucher Senior Research Fellowship. Special thanks to Dr Osamu Shimomura for his generous support of aequorin-based imaging over the years. Thanks also to Prof. Robert Baker and Dr Edwin Gilland for their helpful discussion with regard to embryonic Ca^{2+} signaling in teleost species.

Table 1

Summary of the reports that describe Ca^{2+} transients during cytokinesis, indicating the system under examination and Ca^{2+} reporter used

Animal	**Embryonic system**	**Ca^{2+} reporter**	**Cytokinetic events where Ca^{2+} signals were observed**	**Reference**
Fish	Medaka (*Oryzias latipes*)	Aequorin	Furrow extension, deepening, and apposition	[36]
	Zebrafish (*Danio rerio*)	Calcium-green dextran (10 kDa)	Determination of the plane of furrow formation	[49]
		Aequorin	Furrow Positioning, propagation, and deepening	[50]
		Aequorin	Furrow propagation and deepening	[51]
		Calcium-green dextran	Furrow positioning, propagation, and deepening	[52]
		Aequorin	Furrow positioning, propagation, deepening, and zipping	[46,54]
	Rosy barb (*Puntius conchonius*)	Aequorin	Furrow positioning, propagation, and deepening	Webb et al., unpublished data
	Mummichog (*Fundulus heteroclitus*)	Aequorin	Furrow positioning, propagation, and deepening	Chan et al., unpublished data
Amphibians	African Clawed Frog (*Xenopus laevis*)	Calcium green-1	Furrow deepening	[37]
		Calcium green-1 dextran (10 kDa)	Not involved in cytokinesis	[70]
		Aequorin	Furrow propagation and deepening	Lee et al., unpublished data
Echinoderms	Starfish (*Asterina miniata* and *Pisaster ochraceus*)	Calcium-green dextran (10 kDa)	Cleavage furrow formation	[47]
	Sand dollar (*Echinaracnius parma*)	Aequorin	Cytokinesis	[76]

Table 1
(Continued)

Animal	Embryonic system	Ca^{2+} reporter	Cytokinetic events where Ca^{2+} signals were observed	Reference
Echinoderms (contd.)	Sea urchin (*Lytechinus pictus*)	Fura-2	Starts just before and accompanies the developing furrow during cleavage	[39]
	Sea urchin (*Paracentrotus lividus*)	Fura-2-dextran	Accompanies cleavage	[74]

Table 2
Summary of the reports outlining possible roles played by Ca^{2+} in the cytokinesis process determined via modulation of the cytokinetic Ca^{2+} transient

Animal	Embryonic system	Ca^{2+} chelator/ antagonist/agonist	Ca^{2+}-dependent cytokinetic event(s) affected	Reference
Fish	Medaka (*Oryzias latipes*)	Dibromo-BAPTA	Cytokinesis blocked	[58]
	Zebrafish (*Danio rerio*)	BAPTA, Heparin	Cytokinesis blocked	[49]
		Dibromo-BAPTA	Furrow propagation blocked	[50]
		BAPTA	Cleavage blocked	[51]
		Dibromo-BAPTA, Heparin, 2-APB	Furrow deepening blocked	[46]
		Dibromo-BAPTA, 2-APB	Furrow positioning blocked	[54]
		Dibromo-BAPTA, heparin	Furrow positioning blocked (DBB) and normal apposition prevented (heparin)	[52]
Amphibians	Frog (*Xenopus laevis*)	EGTA	Onset of cleavage slowed/blocked	[26]
		Heparin, Anti-PIP_2 antibody	Cleavage inhibited	[73]
		BAPTA buffers	Furrow propagation (blocked) and apposition (delayed)	[71]
		BAPTA buffers	Cleavage delayed	[72]

(*Continued*)

Table 2
(Continued)

Animal	Embryonic system	Ca^{2+} chelator/ antagonist/agonist	Ca^{2+}-dependent cytokinetic event(s) affected	Reference
Amphibians (contd.)	Frog (*xenopus laevis*) (contd.)	Heparin	Onset and extension of cleavage furrow delayed	[37]
		Dibromo-BAPTA	No effect on cleavage	[70]
Echinoderms	Starfish (*Asterina miniata* and *Pisaster ochraceus*)	Heparin	Cleavage furrow formation blocked	[47]
	Sea urchin (*Lytechinus pictus*)	EDTA	Furrow positioning arrested	[27]
	Sea urchin (*Lytechinus variegatus*)	A23187	Blocked furrow progression	[75]
Mollusks	Squid (*Loligo pealei*)	A23187	Promoted cleavage formation	[31]
Insects	Fruit fly (*Drosophila melanogaster*) spermatocytes	BAPTA-AM, 2-APB, U73122	Furrow deepening blocked	[78]

The system under examination and Ca^{2+}-modulating agent used are indicated.

References

1. Swann, M.M. and Mitchison, J.M. (1958) *Biol. Rev. Camb. Philos. Soc.* 33, 103–135.
2. Wolpert, L. (1960) *Int. Rev. Cytol.* 10, 163–216.
3. Mabuchi, I. (1986) *Int. Rev. Cytol.* 101, 175–213.
4. Salmon, E.D. (1989) *Curr. Opin. Cell Biol.* 1, 541–547.
5. Schroeder, T.E. (1990) *Ann. N.Y. Acad. Sci.* 582, 78–87.
6. Satterwhite, L.L. and Pollard, T.D. (1992) *Curr. Opin. Cell Biol.* 4, 43–52.
7. Fishkind, D.J. and Wang, Y.L. (1995) *Curr. Opin. Cell Biol.* 7, 23–31.
8. Field, C., Li, R. and Oegema, K. (1999) *Curr. Opin. Cell Biol.* 11, 68–80.
9. Robinson, D.N. and Spudich, J.A. (2000) *Trends Cell Biol.* 10, 228–237.
10. Wang, Y.-L. (2001) *Cell Struct. Funct.* 26, 633–638.
11. Glotzer, M. (2003) *Curr. Opin. Cell Biol.* 15, 684–690.
12. Strickland, L.I. and Burgess, D.R. (2004) *Trends Cell Biol.* 14, 115–118.
13. Rappaport, R. (1996) Cytokinesis in animal cells. Cambridge University Press, Cambridge, UK.
14. Berridge, M.J. (1976) *Symp. Soc. Exp. Biol.* 30, 219–231.
15. Steinhardt, R.A. (1990) *Ann. N.Y. Acad. Sci.* 582, 199–206.

16. Hepler, P.K. (1992) *Int. Rev. Cytol.* 138, 239–268.
17. Whitaker, M. (1995) *Adv. Second Messenger Phosphoprotein Res.* 30, 299–310.
18. Bütschli, O. (1876) *Abhandlungen, Herausgegeben von der Senckenbergischen Naturforschenden Gesellschaft* 10, 213–464.
19. Mazia, D. (1937) *J. Cell Comp. Physiol.* 10, 291–304.
20. Hoffmann-Berling, H. and Weber, H.H. (1953) *Biochim. Biophys. Acta* 10, 629–630.
21. Hayashi, T. (1953) *Am. Nat.* 87, 209–227.
22. Weber, H.H. (1955) *Symp. Soc. Exp. Biol.* 9, 271–281.
23. Tilney, L.G. and Marsland, D. (1969) *J. Cell Biol.* 42, 170–184.
24. Bluemink, J.G. (1970) *J. Ultrastruct. Res.* 32, 142–166.
25. Gingell, D. (1970) *J. Embryol. Exp. Morph.* 23, 583–609.
26. Baker, P.F. and Warner, A.E. (1972) *J. Cell Biol.* 53, 579–581.
27. Timourian, H., Clothier, G. and Watchmaker, G. (1972) *Exp. Cell Res.* 25, 296–298.
28. Timourian, H., Jotz, M.M. and Clothier, G.E. (1974) *Exp. Cell Res.* 83, 380–386.
29. Durham, A.C.H. (1974) *Cell* 2, 123–135.
30. Schroeder, T.E. and Strickland, D.L. (1974) *Exp. Cell Res.* 83, 139–142.
31. Arnold, J. (1975) *Cytobiologie* 11, 1–9.
32. Hollinger, T.G. and Schuetz, A.W. (1976) *J. Cell Biol.* 71, 395–401.
33. Conrad, G.W. and Davis, S.E. (1977) *Dev. Biol.* 61, 184–201.
34. Conrad, G.W., Glackin, P.V., Hay, R.A. and Patron, R.R. (1987) *J. Exp. Zool.* 243, 245–258.
35. Ridgway, E.B., Gilkey, J.G. and Jaffe, L.F. (1977) *Proc. Natl. Acad. Sci. USA* 74, 623–627.
36. Fluck, R.A., Miller, A.L. and Jaffe, L.F. (1991) *J. Cell Biol.* 115, 1259–1265.
37. Muto, A., Kume, S., Inoue, T., Okano, H. and Mikoshiba, K. (1996) *J. Cell Biol.* 135, 181–190.
38. Grynkiewicz, G., Poenie, M. and Tsien, R.Y. (1985) *J. Biol. Chem.* 260, 3440–3450.
39. Poenie, M., Alderton, J., Tsien, R.Y. and Steinhardt, R.A. (1985) *Nature* 315, 147–149.
40. Shantz, A.R. (1985) *J. Cell Biol.* 100, 947–954.
41. Yoshimoto, Y., Iwamatsu, T. and Hiramoto, Y. (1985) *Biomed. Res.* 6, 387–394.
42. Ezzell, R.M., Cande, W.D. and Brothers, A.J. (1985) *Roux's Arch. Dev. Biol.* 194, 140–147.
43. Low, S.H., Li, X., Miura, M., Kudo, N., Quinones, B. and Weimbs, T. (2003) *Dev. Cell* 4, 753–759.
44. Gromley, A., Yeaman, C., Rosa, J., Redick, S., Chen, C.T., Mirabelle, S., Guha, M., Sillibourne, J. and Doxsey, S.J. (2005) *Cell* 123, 75–87.
45. Glotzer, M. (2005) *Science* 307, 1735–1739.
46. Lee, K.W., Webb, S.E. and Miller, A.L. (2003) *Int. J. Dev. Biol.* 47, 411–421.
47. Stricker, S.A. (1995) *Dev. Biol.* 170, 496–518.
48. Reinhard, E., Yokoe, H., Niebling, K.R., Allbritton, N.L., Kuhn, M.A. and Meyer, T. (1995) *Dev. Biol.* 170, 50–61.
49. Chang, D.C. and Meng, C. (1995) *J. Cell Biol.* 131, 1539–1545.
50. Webb, S.E., Lee, K.W., Karplus, E. and Miller, A.L. (1997) *Dev. Biol.* 192, 78–92.
51. Créton, R., Speksnijder, J.E. and Jaffe, L.F. (1998) *J. Cell Sci.* 111, 1613–1622.
52. Chang, D.C. and Lu, P. (2000) *Microsc. Res. Tech.* 49, 111–122.
53. Lee, K.W., Ho, S.M., Wong, C.H., Webb, S.E. and Miller, A.L. (2004) *Zygote* 12, 221–230.
54. Lee, K.W., Webb, S.E. and Miller, A.L. (2006) *Zygote* 14, 143–155.
55. Jesuthasan, S. (1998) *J. Cell Sci.* 111, 3695–3703.
56. Le Comber, S.C. and Smith, C. (2004) *Biol. J. Linn. Soc.* 82, 431–442.
57. Pethig, R., Kuhn, M., Payne, R., Alder, E., Chen, T.-H. and Jaffe, L.F. (1989) *Cell Calcium* 10, 491–498.
58. Fluck, R.A., Miller, A.L., Abraham, V.C. and Jaffe, L.F. (1994) *Biol. Bull.* 186, 254–262.
59. Yao, Y., Choi, J. and Parker, I. (1995) *J. Physiol.* 482, 533–553.
60. Bootman, M.D. and Berridge, M.J. (1996) *Curr. Biol.* 6, 855–865.

61. Bootman, M., Niggli, E., Berridge, M. and Lipp, P. (1997) *J. Physiol.* 499, 307–314.
62. Mitsuyama, F., Sawai, T., Carafoli, E., Furuichi, T. and Mikoshiba, K. (1999) *Dev. Biol.* 214, 160–167.
63. Mitsuyama, F. and Sawai, T. (2001) *Int. J. Dev. Biol.* 45, 861–868.
64. Rink, T.J., Tsien, R.Y. and Warner, A.E. (1980) *Nature* 283, 658–660.
65. Grandin, N. and Charbonneau, M. (1991) *J. Cell Biol.* 112, 711–718.
66. De Laat, S.W. and Bluemink, J.G. (1974) *J. Cell Biol.* 60, 529–540.
67. Aimar, C. and Grant, N. (1992) *Biol. Cell* 76, 23–31.
68. Kubota, H.Y., Yoshimoto, Y. and Hiramoto, Y. (1993) *Dev. Biol.* 160, 512–518.
69. Keating, T.J., Cork, R.J. and Robinson, K.R. (1994) *J. Cell Sci.* 107, 2229–2237.
70. Noguchi, T. and Mabuchi, I. (2002) *Mol. Biol. Cell* 13, 1263–1273.
71. Miller, A.L., Fluck, R.A., McLaughlin, J.A. and Jaffe, L.F. (1993) *J. Cell Sci.* 106, 523–534.
72. Snow, P. and Nuccitelli, R. (1993) *J. Cell Biol.* 122, 387–394.
73. Han, J.K., Fukami, K. and Nuccitelli, R. (1992) *J. Cell Biol.* 116, 147–156.
74. Ciapa, B., Pesando, D., Wilding, M. and Whitaker, M. (1994) *Nature* 368, 875–878.
75. Suzuki, K., Roegiers, F., Tran, P. and Inoué, S. (1995) *Biol. Bull.* 189, 201–202.
76. Silver, R.B. (1996) *Cell Calcium* 20, 161–179.
77. Créton, R., Kreiling, J.A. and Jaffe, L.F. (2000) *Dev. Biol.* 217, 375–385.
78. Wong, R., Hadjiyanni, I., Wei, H.C., Polevoy, G., McBride, R., Sem, K.P. and Brill, J.A. (2005) *Curr. Biol.* 15, 1401–1406.
79. Burgess, D. (2005) *Curr. Biol.* 15, 310–311.
80. Yamakita, Y., Yamashiro, S. and Matsumura, F. (1994) *J. Cell Biol.* 124, 129–137.
81. Murthy, K. and Wadsworth, P. (2005) *Curr. Biol.* 15, 724–731.
82. Canman, J.C. and Wells, W.A. (2004) *J. Cell Biol.* 166, 943–948.
83. Mabuchi, I., Tsukita, S., Tsukita, S. and Sawai, T. (1988) *Proc. Natl. Acad. Sci. USA* 85, 5966–5970.
84. Mabuchi, I. and Takano-Ohmuro, H. (1990) *Dev. Growth Differ.* 32, 549–556.
85. Foe, V.E., Field, C.M. and Odell, G.M. (2000) *Development* 127, 1767–1787.
86. Field, S.J., Madson, N., Kerr, M.L., Galbraith, K.A., Kennedy, C.E., Tahiliani, M., Wilkins, A. and Cantley, L.C. (2005) *Curr. Biol.* 15, 1407–1412.
87. Shuster, C.B. and Burgess, D.R. (2002) *Proc. Natl. Acad. Sci. USA* 99, 3633–3638.
88. Albertson, R., Riggs, B. and Sullivan, W. (2005) *Trends Cell Biol.* 15, 92–101.
89. Bluemink, J.G. and de Laat, S.W. (1973) *J. Cell Biol.* 59, 89–108.
90. Lecuit, T. and Wieschaus, E. (2000) *J. Cell Biol.* 150, 849–860.
91. Feng, B., Schwarz, H. and Jesuthasan, S. (2002) *Exp. Cell Res.* 279, 14–20.
92. Danilchik, M.V., Funk, W.C., Brown, E.E. and Larkin, K. (1998) *Dev. Biol.* 194, 47–60.
93. Skop, A.R., Bergmann, D., Mohler, W.A. and White, J.G. (2001) *Curr. Biol.* 11, 735–746.

J. Krebs and M. Michalak (Editors)
Calcium: A Matter of Life or Death

ISSN: 0167-7306/DOI: 10.1016/S0167-7306(06)41018-8

CHAPTER 18

Mitochondrial Ca^{2+} and cell death

Tullio Pozzan[1,2] and Rosario Rizzuto[3]

[1]*Department of Biomedical Sciences and CNR Institute of Neuroscience, University of Padova, Viale Colombo 3, 35121 Padova, Italy;*
[2]*Venetian Institute of Molecular Medicine and Interdipartimental Centre for the Study of Cell Signals, University of Padova, Via Orus 2, 1-35121 Padova, Italy;*
[3]*Department of Experimental and Diagnostic Medicine, Section of General Pathology, ER-GenTech laboratory and Interdisciplinary Centre for the Study of Inflammation (ICSI), University of Ferrara, Via Borsari 46, 44100 Ferrara, Italy, Tel.: +39 0532 291361; Fax: +39 0532 247278; E-mail: r.rizzuto@unife.it*

Abstract

It has been known for a long time that cellular Ca^{2+} overload, occurring in excitotoxicity and various genetic or acquired diseases, leads to energy wasting and cell death, mostly by necrosis. The analysis of the effects of the oncogene Bcl-2 on cellular Ca^{2+} handling has led to a series of discoveries that suggest another, more specific, role of Ca^{2+} in cell death, i.e. that of modulating cellular sensitivity to apoptosis. Indeed, mitochondrial Ca^{2+} loading synergizes with toxic or signalling inputs in favouring the opening of a large-conductance mitochondrial channel, the permeability transition pore (PTP), thus altering mitochondrial structure and causing the release of caspase cofactors, such as cytochrome c, into the cytosol. The antiapoptotic protein Bcl-2 counteracts this effect by partially depleting the endoplasmic reticulum (ER) Ca^{2+} store (thus reducing mitochondrial Ca^{2+} loading), whereas the proapoptotic gene products Bax and Bak exert the opposite effect. In this contribution, we review (i) the data on the effect of pro- and antiapoptotic gene products of the Bcl-2 family on cellular Ca^{2+} homeostasis, (ii) the functional significance of the Ca^{2+} signalling alteration and (iii) the recent insights into the mechanism of this effect that may account for the apparent discrepancies between the data obtained in different studies.

Keywords: apoptosis, calcium, cell death, organelle, signal transduction

It is now firmly established that variations in cellular Ca^{2+} concentration are pivotal in the control of a variety of cellular functions, from contraction of myofilaments to secretion of hormones to modulation of metabolism. Not only, but Ca^{2+} is also central in triggering the cascade pathways controlling the mitotic stimulation is several cellular types (best known is the case of T lymphocytes and of oocytes) and strong evidence has been also accumulated supporting a central role of Ca^{2+} in the regulation of cell death (for a recent review, see [1–3]). As to the latter process, the toxicity of major increases in cytoplasmic Ca^{2+} concentration, $[Ca^{2+}]_c$, had been known for a long time, but it has been discovered only more recently that less dramatic, yet crucial increases in $[Ca^{2+}]_c$, can be the key events in triggering the cell death programme of apoptosis (for reviews, see [4,5]). The question obviously arises as to how a single signalling molecule can trigger, often in the same cell type, so vastly different functions. The key to solving this puzzle is presumably in the unique physicochemical characteristics of Ca^{2+} and its capacity to establish local concentrations within cells that in turn form a variety of recognizable signatures corresponding to specific functional effects. Indeed, a unique characteristic of Ca^{2+} within the cell cytoplasm is its very slow rate of diffusion (compared with the other classical second messengers, cAMP, cGMP and inositol trisphosphate (IP_3), Ca^{2+} is over 100-fold slower at diffusing), due to the presence of a variety of mobile or immobile binding sites [6]. Also, in distinction to the other messengers, several organelles diffused throughout the cell can sequester Ca^{2+} and, in response to appropriate signals, release it back into the cytoplasm. These conditions are ideal for the generation of subcellular heterogeneities in Ca^{2+} concentration, e.g. in the proximity of plasma membrane or organelle Ca^{2+} channels. The existence of subcellular domains of Ca^{2+} where its concentration can be significantly different from that in other cellular location has been suggested many years ago, but only in the last decade these values have been directly measured and their primary role in controlling some cell functions, in addition to the classical role in neurosecretion at the presynaptic membrane, has become evident [3].

Typical example of how differences in Ca^{2+} concentration in terms of amplitude, timing and location can lead to different outcome in a cell is represented by the process of cell death: in particular, while a controlled rise in Ca^{2+} concentration can lead to the activation of metabolism or to the entrance in G1 phase, if too large and/or too prolonged, it can cause necrotic cell death, while a limited rise, but combined with other toxic insults can initiate apoptosis.

This chapter is focused on some aspects of the Ca^{2+} signalling related to apoptosis. It stems from the authors' own work on the topic but tries to cover, at least in part, the rapidly expanding literature on the topic. We apologize in advance for the bias and the omissions and refer to more extended reviews and other chapters in this book for the aspects that are overlooked.

1. Bcl-2 and Ca^{2+}: the early link between Ca^{2+} and cell death

The knowledge that an excess Ca^{2+} within cells is highly toxic, causing massive activation of proteases and phospholipases was known to cell biologists since the late 1960s. Electron micrographs of clearly damaged cells with swollen mitochondria full of Ca^{2+} phosphate precipitates have been published several times in the 1960s and 1970s, and the toxicity of Ca^{2+} ionophores in cultured cells was one of the first effects of these molecules to be discovered. However, classically, this toxic role of Ca^{2+} has been associated to necrosis, i.e. the catastrophic derangement of cell integrity and function following exposure to different types of cell injury and leading to the activation of Ca^{2+}-responsive hydrolysing enzymes. Typical examples are complement-induced cell death and excitotoxicity, in which glutamate-dependent hyperstimulation leads neurons to necrotic death [4,7]. The link to apoptosis was appreciated only more recently, and in this field, the observation that turned the attention of many scientists to Ca^{2+} was the discovery that a classical antiapoptotic protein such as Bcl-2 was potentially capable of somehow affecting Ca^{2+} signalling.

The first of such observations dates back to 1993 and consisted in the finding that Bcl-2 overexpression was able to prevent the reduction in the Ca^{2+} concentration of the endoplasmic reticulum (ER), $[Ca^{2+}]_{er}$, that was observed upon withdrawal of IL-3 in hematopoietic cell lines [8]. Although this finding could be ascribed to the antiapoptotic effect of Bcl-2 (by preventing the loss of energy and thus of ER Ca^{2+} during apoptosis), surprisingly Bcl-2 overexpression was also reported to decrease, by itself, the size of the ER Ca^{2+} pool (reviewed in [9]). A second totally independent hint suggesting a potential role of Bcl-2 in regulating Ca^{2+} signals came from the observation that the three-dimensional structure of the Bcl-2 homologue, Bcl-X_L, bears a strong resemblance to some pore-forming bacterial toxins [10]. Bcl-2 itself was shown to form cation channels of low selectivity in artificial lipid bilayers [11], and later studies showed that these channels can conduct Ca^{2+} ions [12]. These findings, however, remained at the level of simple correlations until a few novel discoveries were made that, although complex and hard to reconcile into a unifying mechanism, unquestionably associated Bcl-2 and Ca^{2+} homeostasis. (i) Korsmeyer and colleagues [13] demonstrated in a seminal study the association of Bcl-2 with cellular membranes and, in particular, with mitochondrial membranes. This observation was pursued by a number of further studies, and it was concluded that Bcl-2 shows a heterogeneous distribution to various other cell compartments including in particular the ER. In other words, Bcl-2 tends to accumulate on the membranes of the two organelles, the ER and mitochondria, playing a key role in Ca^{2+} homeostasis [3]; (ii) it was demonstrated that the mitochondrial enzyme cytochrome c is released into the cytosol in response to several apoptotic stimuli and results in the activation of caspases. Because antiapoptotic members of the Bcl-2 family prevented this release and protected cells from various death insults, it has been assumed that Bcl-2 family members primarily regulate mitochondrial integrity, although the precise nature of this protection is still partially unclear [14].

Other studies carried out in parallel, and apparently unrelated to apoptosis, were of great importance in establishing the link between Ca^{2+} and cell death. The first of these studies were concerned with a re-evaluation of the role of mitochondria in Ca^{2+} signalling. Using luminescent [15] or green fluorescent protein (GFP)-based fluorescent probes [16], in virtually all cell types, rapid and large mitochondrial matrix Ca^{2+} concentration, $[Ca^{2+}]_m$, increases were shown to parallel the cytosolic increases occurring upon cell stimulation. In this process, the ER (and its interactions with mitochondria) was shown to play a key role. Indeed, at least in cells in which rapid Ca^{2+} responses depend on the opening of ER Ca^{2+} channels, the strategic location of mitochondria close to the source of the $[Ca^{2+}]_c$ rise (the ER) allows them to be exposed to microdomains of high $[Ca^{2+}]$ that meet the low affinity of their transporters and is the key event for the rapid accumulation of the cation in the matrix [17,18]. Ca^{2+} accumulation within the matrix, in turn, participates in functionally decoding the Ca^{2+}-mediated signals by stimulating Ca^{2+}-dependent enzymes [19–21] or metabolite carriers [22,23] and enhances ATP production [24].

The other line of research was carried out primarily in isolated mitochondria. In a series of elegant studies, the group of Bernardi and coworkers [56] rejuvenated a relatively old observation, i.e. the activation by excess Ca^{2+} accumulation into mitochondria of a large conductance pore, the so-called permeability transition pore (PTP). They showed that PTP that can be opened by an increase of [Ca^{2+}] in the matrix above a given threshold and by a series of other toxic insults, including oxidative stress, could lead to mitochondrial swelling, breaking of the outer membrane and loss into the outer medium of proteins, such as cytochrome c contained in the intermembrane space.

Thus in a few years, independent lines of research led to the following seminal observations: (i) mitochondria and ER are intimately involved in Ca^{2+} signalling; (ii) Bcl-2 is located on the membranes of both organelles and can thus potentially modulate Ca^{2+} signalling; and (iii) excess Ca^{2+} accumulation by mitochondria could lead to the release from mitochondria of proapoptotic signalling molecules.

2. *Direct measurements of [Ca^{2+}]$_{er}$ in cells overexpressing pro- and antiapoptotic proteins of the Bcl-2 family: a controversial issue*

Only in the year 2000, however, the involvement of Ca^{2+} in Bcl-2 action was directly addressed by two groups, our own and that of Krause [4,25,26]. We utilized an ER-targeted aequorin chimera [27] and transiently co-expressed it with Bcl-2 in HeLa cells. No toxicity because of Bcl-2 transient overexpression was observed, in distinction to experiments in which a Bcl-2-GFP chimera was used [28]. Rather, as expected, the cells overexpressing Bcl-2 displayed an enhanced survival upon ceramide treatment [29], in agreement with previous reports [30,31]. Most strikingly, compared with controls, Bcl-2 overexpressing cells showed an approximately 30% reduction in the Ca^{2+} levels within both the ER and the Golgi apparatus. As a consequence, the [Ca^{2+}] increases elicited in these cells by stimuli coupled to IP_3 generation were reduced both in the cytoplasm and in the mitochondria [25]. The explanation for the reduced level of ER Ca^{2+}, that in fact confirmed directly the initial observation of 1993, was an increased leak of Ca^{2+} through the ER membrane. The results of Krause and coworkers [26], using an ER-targeted 'cameleon', confirmed that overexpression of Bcl-2 results in a reduction of the steady state [Ca^{2+}]$_{er}$ most likely due to an increased Ca^{2+} leak [26].

Overall, these initial studies provide a coherent picture: despite the use of radically different probes, with different detection and calibration procedures (which may account for the significant discrepancy in the absolute [Ca^{2+}] values), the direct measurement of [Ca^{2+}]$_{er}$ reveals in all cases a 20–40% decrease of the resting values of [Ca^{2+}] within the ER lumen.

The capacity of antiapoptotic proteins to reduce [Ca^{2+}]$_{er}$ described in these initial studies was later confirmed and expanded in studies that employed other approaches and genetically altered cell models. Scorrano et al. [32] demonstrated not only that embryonic fibroblasts from mice in which the genes for Bax and Bak (proapoptotic proteins that associate with mitochondria and initiate organelle dysfunction but are also localized to the ER) had been deleted show a dramatic reduction in [Ca^{2+}]$_{er}$ but also that siRNA silencing of Bcl-2 in Bax/Bak knockout cells partially restored

$[Ca^{2+}]_{er}$ values to control levels [32]. These data suggest that these proteins directly counteract the effect of Bcl-2 on Ca^{2+} signalling (and thus the latter protein, unopposed in the Bax/Bak knockouts, allows lower $[Ca^{2+}]_{er}$ levels to be attained). In support of this possibility, we then showed that early after overexpression of Bax in HeLa cells $[Ca^{2+}]_{er}$ levels are higher than in controls, whereas at later stages (during progression into apoptosis), the difference with control cells becomes virtually undetectable [33]. Finally, recently Tsien and coworkers [34], by using a novel family of genetically engineered, GFP-based Ca^{2+} probes targeted to the ER lumen, confirmed that Bcl-2 overexpression leads to decreased ER Ca^{2+} level. They not only confirmed this fact but also showed that the green tea compound epigallocatechin gallate, known to bind and inactivate Bcl-2, reduced Ca^{2+} leakage to control levels and at the same time restored $[Ca^{2+}]_{er}$ values to that of normal cells [34].

Although the above results all consistently point to the same conclusion, i.e. that proapoptotic members of the Bcl-2 family cause an increase in the ER Ca^{2+} level, while the opposite is true for antiapoptotic proteins, no consensus has been reached yet. In particular, a few reports appeared which did not support the notion that Bcl-2 reduces $[Ca^{2+}]_{er}$. The largest body of work was carried out by the group of Distelhorst [12,28,35–37], which argued that the size of ionomycin-releasable intracellular Ca^{2+} pools is the same in controls and Bcl-2-expressing cells [38]. Other groups concur with this observation [39], and in a murine hypothalamic cell line [40], an even greater increase of $[Ca^{2+}]_c$ after thapsigargin inhibition of sarcoplasmic–ER Ca^{2+}-ATPase (SERCA) pump was observed in Bcl-2 overexpressing compared with control cells, presumably due to an up-regulation of SERCA gene expression [41,42]. Explaining these experimental discrepancies is not straightforward. One obvious possibility would be cell-specific regulatory mechanisms; a second possibility would be that in all the studies showing no effect of Bcl-2 overexpression on $[Ca^{2+}]_{er}$ (or an increase), the $[Ca^{2+}]_{er}$ was measured indirectly (i.e. by the cytosolic peak triggered by physiological or pharmacological emptying of the ER). However, this approach has been employed also in the other studies in conjunction with direct monitoring of $[Ca^{2+}]_{er}$. A third possibility is that in most of the studies where no effect of Bcl-2 on ER Ca^{2+} level is detected had been carried out on cell clones, and the conclusion may thus depend on clone-specific characteristics (or adaptive phenomena) unrelated to Bcl-2 overexpression. Last, but probably most interesting, is that, the effect on ER Ca^{2+} reflects a modulation of the activity of some specific isoform of the IP_3 receptor (IP_3R). In this case, it is reasonable to assume that the expression profile of IP_3Rs or modulatory proteins can tune the responsiveness of cells to the signalling effects of Bcl-2. Recent evidence by two independent laboratories indicates that indeed the last mechanism may be the most relevant one. The first article demonstrated that Bcl-2 controls the phosphorylation state of type I IP_3R, and this, in turn, regulates the Ca^{2+} leak through the channel [43]. In particular, the authors suggest that Bax/Bak knockout results in a reduction of calcineurin (a Ca^{2+}-dependent phosphatase) associated to the ER membranes and thus in the hyperphosphorylation of the type I isoform of the IP_3R. In addition, a siRNA specific for the type I IP_3R counteracted the reduction of $[Ca^{2+}]_{er}$ in cells from Bax/Bak knockout animals [43], whereas the siRNA for type 3 IP_3R had no effect. This observation well fits with the finding that Bcl-2 alters Ca^{2+} oscillations triggered by agonist acting by IP_3R stimulation [34] and with studies reporting a correlation between low expression of IP_3R and inhibition of apoptosis [44].

A functionally similar effect on $[Ca^{2+}]_{er}$, although by a different mechanism, was described for the other antiapoptotic protein Bcl-X_L by the groups of Foskett and Thompson [45]. They showed that Bcl-X_L binds to the IP_3R and sensitises the IP_3R to low IP_3. Recombinant expression (or addition to the patch pipette) of Bax prevented the effect of Bcl-X_L, in terms of both its binding to the IP_3R and its capacity of modifying the sensitivity to IP_3. The results are opposite of those by the Distelhorst group [35] (that showed that Bcl-2 reduces the opening probability of IP_3Rs inserted into lipid bilayers). The Bcl-X_L effect according to Foskett and Thompson is a direct modulatory role on IP_3R (through protein–protein interaction). It needs stressing however that, although by different mechanism, the end result of Bcl-2 or Bcl-X_L overexpression is the same, i.e. a reduced Ca^{2+} level within the ER. Obviously, more work needs to be done to verify whether these different results reflect true experimental discrepancies or functional complexity. Indeed, not only the Bcl-2 family members could differ in their mechanism, but the variability between cell lines in IP_3R levels, isoform expression and basal levels of IP_3 could account for thc apparcntly conflicting rcsults.

2.1. Ca^{2+} *signalling and sensitivity to apoptosis*

The data reported in Section 2 strongly support (despite the matter is not completely clarified) the notion that a reduction of $[Ca^{2+}]_{er}$ is correlated with the overexpression of antiapoptotic proteins or knockout of proapoptotic ones, while overexpression of proapoptotic genes leads to an increase of $[Ca^{2+}]_{er}$, but is this relevant for apoptosis or is it a side effect of these proteins? And what is the role of mitochondria in this scenario? We will first briefly discuss the evidence supporting the conclusion that ER Ca^{2+} level is correlated to the sensitivity to some apoptotic stimuli, and then we will discuss the role of mitochondria.

An indication that the reduction of $[Ca^{2+}]_{er}$ is a component of the antiapoptotic mechanism by Bcl-2 comes from the work of our group, published in 2001. We showed that mimicking the Bcl-2 effect on $[Ca^{2+}]_{er}$ by different pharmacological and molecular approaches (but in the absence of the oncoprotein), the cells were protected from the apoptotic stimulus ceramide, also sensitive to Bcl-2 overexpression, while treatments that increased $[Ca^{2+}]_{er}$ had the opposite effect [29]. Most important, an apoptotic stimulus, such as Fas ligand, that in the HeLa cell model is not protected by Bcl-2, was similarly unaffected by modifications of ER Ca^{2+} level (Rizzuto and Pozzan, unpublished data). A similar picture emerged from the study with the Bax/Bak knockouts. In those cells, when the ER Ca^{2+} levels were restored by recombinantly overexpressing the SERCA, not only mitochondrial Ca^{2+} uptake in response to stimulation was re-established, but the cells regained sensitivity to apoptotic stimuli such as arachidonic acid, C_2-ceramide and oxidative stress [32]. Scorrano et al. [46] also showed that other apoptotic stimuli, such as the so called 'BH_3-only' stimuli, on the contrary are unaffected by the ER Ca^{2+} level. These results are in keeping with previous work by Ma and coworkers, demonstrating that SERCA overexpression in

Cos cells causes ER Ca^{2+} overload and increases spontaneous apoptosis [47]. Last but not least, in calreticulin overexpressing cells, in which the amplitude and duration of Ca^{2+} signals from ER lumen towards cytosol are enhanced [48] without changing $[Ca^{2+}]_{er}$ [49], cell viability is drastically reduced upon ceramide treatment [29], whereas cell lines derived from calreticulin knockouts, which show a marked decrease in ER Ca^{2+} release upon cell stimulation, are more resistant to apoptosis [50].

Along the same lines of reasoning, Foskett and coworkers [45] showed that DT40 cells, in which all three IP_3R isoforms had been deleted, are more sensitive to apoptotic challenges than the wild-type cells while re-expression of IP_3R not only partially depletes the Ca^{2+} store but also provides a 'survival' signal per se [45].

Two final aspects require attention. Firstly, the idea that a $[Ca^{2+}]_{er}$ decrease protects cells from the effect of apoptotic agents may appear in contrast with the established notion that treatment of cells with SERCA inhibitors (thapsigargin, tBuBHQ and cyclopiazonic acid) is followed by apoptosis [51]. It should be remembered, however, that the effect on $[Ca^{2+}]_{er}$ is radically different in the two conditions. Upon SERCA blockade, the Ca^{2+} depletion is complete and rapid, while in Bcl-2 transfected cells, the drop in $[Ca^{2+}]_{er}$ is modest and develops slowly. A drastic reduction in the level of $[Ca^{2+}]_{er}$ might interfere with the basic ER functions (e.g. the regulation of ER protein folding and chaperone interactions) and thus cause a stress response that leads to cell death. The second aspect that needs to be remembered is that Ca^{2+} is not only associated to cell death but mediates cell-specific activities and responsiveness to mitogenic stimuli. Thus, a modification of Ca^{2+} signalling patterns may have pleiotropic effect, if both types of signalling are affected. Such a possibility was directly investigated by the Distelhorst group, which analysed in the very cell system (T cells) the functional consequences of Bcl-2-dependent reduction of Ca^{2+}-mediated signals. Interestingly, they could show that Ca^{2+} signals elicited by maximal T-cell receptor stimulation (such as those triggered by high concentrations of anti-CD3 antibody) were significantly reduced in Bcl-2-expressing cells, whereas prosurvival Ca^{2+} signals induced by weak anti-CD3 stimulation were unaffected, possibly because of different requirement (or regulation) of the IP_3R [38].

2.2. The role of mitochondria

There is no doubt that a reduction in cytosolic Ca^{2+} signals will affect the activity of enzymes that have been shown to modulate the apoptotic process. To name a few, Ca^{2+}-dependent proteases, such as calpains, have been shown to be active in cell death pathways, and the ablation of their genes associated to muscle degenerative disorders [52]. As to kinases and phosphatases, a broad literature associates calcineurin and the different isoforms of PKC to the modulation of apoptosis (with specific, even opposite effect of different isoforms) [53]. As mentioned in Section 1, it has also been shown that, in response to different apoptotic stimuli, mitochondria release a few proteins, including cytochrome c, Smac/DIABLO and AIF, i.e. potent activators of effector caspases [54]. The mechanism of this release is probably complex (including the oligomerization on the outer mitochondrial membrane of Bax and Bak that form a channel permeable to cytochrome c), but a key role is played by the PTP. The PTP in fact is activated by a variety of toxic challenges and cell signals (including increases in $[Ca^{2+}]_m$) and causes mitochondrial swelling and fragmentation. PTP opening thus appears a likely candidate for Ca^{2+}-dependent effector of apoptosis [55,56]. We and others thus verified whether apoptotic challenges led to mitochondrial alterations and whether Ca^{2+} was involved in the process. In HeLa cells, we [29] showed that the apoptotic stimulus ceramide induced the release of Ca^{2+} from the ER and its uptake by mitochondria (at least in the first phase of the process). Eventually, mitochondria swell and fragment and in parallel release cytochrome c. These changes were prevented by Bcl-2 expression as well as by experimental conditions that reduced $[Ca^{2+}]_{er}$ or prevented the rise in cytosolic Ca^{2+} [29].

The observation that ceramide-induced ER Ca^{2+} release and mitochondrial Ca^{2+} uptake results in a dramatic activation of the apoptotic process contrasts the fact that much larger Ca^{2+} releases and mitochondrial accumulations, elicited by a number of stimuli, do not trigger cell death but rather are beneficial, e.g. by increasing the cellular ATP levels (for review, see [3]). This apparent contradiction is explained by the hypothesis of the double-hit model proposed by Pinton et al. [29]. Accordingly, apoptotic stimuli such as ceramide have a dual target, on the one hand cause the release of Ca^{2+} from the ER, on the other make the mitochondria more sensitive to the potential Ca^{2+}-damaging effects. An elegant study of Hajnoczky and coworkers [57] nicely confirms the model. In particular, the authors showed that ceramide facilitates PTP opening, thus transforming physiological IP_3-mediated Ca^{2+} signals into inducers of apoptosis [57]. These results are entirely consistent with the data reported by Pinton et al. [29], i.e. ceramide-induced apoptosis is prevented by inhibiting the Ca^{2+} rises in the cytosol (e.g. by loading the cells with the Ca^{2+} chelator BAPTA or by reducing the ER Ca^{2+} content). A similar picture emerged with the other apoptotic inducers. For example, the apoptotic expression of HBx caused, both in HeLa and hepatic cell lines, a dramatic fragmentation and swelling of the mitochondrial network, which was inhibited (together with the apoptotic efficacy) by a PTP blocker [58]. Last, unpublished results by our group show that activators of mitochondrial Ca^{2+} uptake potently synergize with subthreshold doses of apoptotic agents.

This model is consistent also with the effect of proapoptotic stimuli such as Bax: its primary effect on Ca^{2+} homeostasis is, early upon expression, an increased ER Ca^{2+} loading, with consequent potentiation of mitochondrial Ca^{2+} responses; the cells then progress into an apoptotic phenotype (that included drastic reductions of the cytosolic and mitochondrial Ca^{2+} responses upon challenging of cells with stimuli causing ER Ca^{2+} release). This latter decline in ER Ca^{2+} is thus a consequence of the apoptotic process, rather than a cause, possibly linked to a collapse of the ATP synthesizing capacity of mitochondria and/or to cleavage of the IP_3R by caspases. Indeed, it has been demonstrated that the IP_3R includes a consensus site for caspase-3 cleavage, and the cleaved form of the receptor exhibits increased leakiness and reduced sensitivity to IP_3 [59,60].

In summary, Bcl-2 and other antiapoptotic proteins reduce ER Ca^{2+} levels and consequently moderate the efficacy of apoptotic mediators that use Ca^{2+} signals (and the involvement of mitochondria as downstream effectors) as a potentiation/commitment factor. Conversely, Bax enhances the loading of the ER Ca^{2+} store and thus boosts the Ca^{2+} load to which the apoptotic effector systems (including mitochondria) are exposed upon physiological and/or pathological challenges. This effect of Bax coincides with gross perturbation of mitochondrial structure and function, and finally, later in apoptotic progression, into the development of an altered signalling phenotype, which includes impaired ER Ca^{2+} release upon cell stimulation, and thus reduction of cellular Ca^{2+} signals.

3. Conclusions

Data obtained by various laboratories using different experimental approaches support the notion that Bcl-2 expression modifies the patterns of agonist-evoked intracellular Ca^{2+} signals, most likely by acting on the IP_3R and thus modifying Ca^{2+} leakage in resting conditions (and thus the steady-state $[Ca^{2+}]_{er}$) and its release kinetics upon stimulation. The other proapoptotic protein of the same family Bcl-X_L has a similar effect, although apparently through a different mechanism. On the contrary, proapoptotic molecules such as Bax and Bak have an opposite effect on $[Ca^{2+}]_{er}$, e.g. they increase it, probably by counteracting the effect of the proapoptotic gene products. As Ca^{2+} acts as a potentiation/commitment signal in some apoptotic routes (mostly by increasing the sensitivity of the mitochondrial checkpoint), this alteration tunes the efficacy of apoptotic challenges. Much still remains to be understood, including the final assessment of the molecular mechanism of the Bcl-2 effect and the significance of this process for apoptosis in vivo, but the notion that also apoptotic cell death belongs to the numerous cell functions modulated by Ca^{2+} ions adds a well-known pharmaceutical target for addressing degenerative processes.

Acknowledgments

The original work of TP and RR was supported by Telethon (grant GGP05284), the Italian Association for Cancer Research (AIRC), the Italian University Ministry (MIUR and FIRB), the EU ('Fondi Strutturali Obiettivo 2'), the PRRIITT program of the Emilia Romagna Region, the Biotech. project of the Veneto Region and the Italian Space Agency (ASI).

References

1. Berridge, M.J., Bootman, M.D., and Roderick, H.L. (2003) *Nat. Rev. Mol. Cell Biol.* **4,** 517–529.
2. Rizzuto, R., Duchen, M.R., and Pozzan, T. (2004) *Sci. STKE* **2004,** re1.
3. Rizzuto, R. and Pozzan, T. (2006) *Physiol. Rev.* **86,** 369–408.
4. Nicotera, P. and Orrenius, S. (1998) Cell Calcium **23,** 173–180.
5. Pinton, P. and Rizzuto, R. (2006) *Cell Death Differ* **13,** 1409–1418.
6. Zaccolo, M., Magalhaes, P., and Pozzan, T. (2002) *Curr. Opin. Cell Biol.* **14,** 160–166.
7. Budd, S.L. and Nicholls, D.G. (1996) *J. Neurochem.* **67,** 2282–2291.

8. Baffy, G., Miyashita, T., Williamson, J.R., and Reed, J.C. (1993) *J. Biol. Chem.* **268,** 6511–6519.
9. Distelhorst, C.W. and Shore, G.C. (2004) *Oncogene* **23,** 2875–2880.
10. Schendel, S.L., Xie, Z., Montal, M.O., Matsuyama, S., Montal, M., and Reed, J.C. (1997) *Proc. Natl. Acad. Sci. U. S. A.* **94,** 5113–5118.
11. Minn, A.J., Velez, P., Schendel, S.L., Liang, H., Muchmore, S.W., Fesik, S.W., Fill, M., and Thompson, C.B. (1997) *Nature* **385,** 353–357.
12. Lam, M., Dubyak, G., Chen, L., Nunez, G., Miesfeld, R.L., and Distelhorst, C.W. (1994) *Proc. Natl. Acad. Sci. U. S. A.* **91,** 6569–6573.
13. Hockenbery, D., Nunez, G., Milliman, C., Schreiber, R.D., and Korsmeyer, S.J. (1990) *Nature* **348,** 334–336.
14. Lucken-Ardjomande, S. and Martinou, J.C. (2005) *C. R. Biol.* **328,** 616–631.
15. Rizzuto, R., Simpson, A.W., Brini, M., and Pozzan, T. (1992) *Nature* **358,** 325–327.
16. Filippin, L., Abad, M.C., Gastaldello, S., Magalhaes, P.J., Sandona, D., and Pozzan, T. (2005) *Cell Calcium* **37,** 129–136.
17. Rizzuto, R., Pinton, P., Carrington, W., Fay, F.S., Fogarty, K.E., Lifshitz, L.M., Tuft, R.A., and Pozzan, T. (1998) *Science* **280,** 1763–1766.
18. Hajnoczky, G., Csordas, G., Krishnamurthy, R., and Szalai, G. (2000) *J. Bioenerg. Biomembr.* **32,** 15–25.
19. Denton, R.M., McCormack, J.G., and Edgell, N.J. (1980) *Biochem. J.* **190,** 107–117.
20. McCormack, J.G. and Denton, R.M. (1990) *Biochim. Biophys. Acta* **1018,** 287–291.
21. Robb-Gaspers, L.D., Burnett, P., Rutter, G.A., Denton, R.M., Rizzuto, R., and Thomas, A.P. (1998) *EMBO J.* **17,** 4987–5000.
22. Lasorsa, F.M., Pinton, P., Palmieri, L., Fiermonte, G., Rizzuto, R., and Palmieri, F. (2003) *J. Biol. Chem.* **278,** 38686–38692.
23. Hajnoczky, G., Robb-Gaspers, L.D., Seitz, M.B., and Thomas, A.P. (1995) *Cell* **82,** 415–424.
24. Jouaville, L.S., Pinton, P., Bastianutto, C., Rutter, G.A., and Rizzuto, R. (1999) *Proc. Natl. Acad. Sci. U. S. A.* **96,** 13807–13812.
25. Pinton, P., Ferrari, D., Magalhaes, P., Schulze-Osthoff, K., Di Virgilio, F., Pozzan, T., and Rizzuto, R. (2000) *J. Cell Biol.* **148,** 857–862.
26. Foyouzi-Youssefi, R., Arnaudeau, S., Borner, C., Kelley, W.L., Tschopp, J., Lew, D.P., Demaurex, N., and Krause, K.H. (2000) *Proc. Natl. Acad. Sci. U. S. A.* **97,** 5723–5728.
27. Chiesa, A., Rapizzi, E., Tosello, V., Pinton, P., de Virgilio, M., Fogarty, K.E., and Rizzuto, R. (2001) *Biochem. J.* **355,** 1–12.
28. Wang, N.S., Unkila, M.T., Reineks, E.Z., and Distelhorst, C.W. (2001) *J. Biol. Chem.* **276,** 44117–44128.
29. Pinton, P., Ferrari, D., Rapizzi, E., Di Virgilio, F.D., Pozzan, T., and Rizzuto, R. (2001) *EMBO J.* **20,** 2690–2701.
30. Zhang, J., Alter, N., Reed, J.C., Borner, C., Obeid, L.M., and Hannun, Y.A. (1996) *Proc. Natl. Acad. Sci. U. S. A.* **93,** 5325–5328.
31. Rippo, M.R., Malisan, F., Ravagnan, L., Tomassini, B., Condo, I., Costantini, P., Susin, S.A., Rufini, A., Todaro, M., Kroemer, G., and Testi, R. (2000) *FASEB J.* **14,** 2047–2054.
32. Scorrano, L., Oakes, S.A., Opferman, J.T., Cheng, E.H., Sorcinelli, M.D., Pozzan, T., and Korsmeyer, S.J. (2003) *Science* **300,** 135–139.
33. Chami, M., Prandini, A., Campanella, M., Pinton, P., Szabadkai, G., Reed, J.C., and Rizzuto, R. (2004) *J. Biol. Chem.* **279,** 54581–54589.
34. Palmer, A.E., Jin, C., Reed, J.C., and Tsien, R.Y. (2004) *Proc. Natl. Acad. Sci. U. S. A.* **101,** 17404–17409.
35. Chen, R., Valencia, I., Zhong, F., McColl, K.S., Roderick, H.L., Bootman, M.D., Berridge, M.J., Conway, S.J., Holmes, A.B., Mignery, G.A., Velez, P., and Distelhorst, C.W. (2004) *J. Cell Biol.* **166,** 193–203.
36. Distelhorst, C.W., Lam, M., and McCormick, T.S. (1996) *Oncogene* **12,** 2051–2055.

37. He, H., Lam, M., McCormick, T.S., and Distelhorst, C.W. (1997) *J. Cell Biol.* **138,** 1219–1228.
38. Zhong, F., Davis, M.C., McColl, K.S., and Distelhorst, C.W. (2006) *J. Cell Biol.* **172,** 127–137.
39. Ichimiya, M., Chang, S.H., Liu, H., Berezesky, I.K., Trump, B.F., and Amstad, P.A. (1998) *Am. J. Physiol.* **275,** C832–C839.
40. Wei, H., Wei, W., Bredesen, D.E., and Perry, D.C. (1998) *J. Neurochem.* **70,** 2305–2314.
41. Kuo, T.H., Kim, H.R., Zhu, L., Yu, Y., Lin, H.M., and Tsang, W. (1998) *Oncogene* **17,** 1903–1910.
42. Zhu, L., Ling, S., Yu, X.D., Venkatesh, L.K., Subramanian, T., Chinnadurai, G., and Kuo, T.H. (1999) *J. Biol. Chem.* **274,** 33267–33273.
43. Oakes, S.A., Scorrano, L., Opferman, J.T., Bassik, M.C., Nishino, M., Pozzan, T., and Korsmeyer, S.J. (2005) *Proc. Natl. Acad. Sci. U. S. A.* **102,** 105–110.
44. Hanson, C.J., Bootman, M.D., and Roderick, H.L. (2004) *Curr. Biol.* **14,** R933–R935.
45. White, C., Li, C., Yang, J., Petrenko, N.B., Madesh, M., Thompson, C.B., and Foskett, J.K. (2005) *Nat. Cell Biol.* **7,** 1021–1028.
46. Scorrano, L., Oakes, S.A., Opferman, J.T., Cheng, E.H., Sorcinelli, M.D., Pozzan, T., and Korsmeyer, S.J. (2003) *Science* **300,** 135–139.
47. Ma, T.S., Mann, D.L., Lee, J.H., and Gallinghouse, G.J. (1999) *Cell Calcium* **26,** 25–36.
48. Bastianutto, C., Clementi, E., Codazzi, F., Podini, P., De Giorgi, F., Rizzuto, R., Meldolesi, J., and Pozzan, T. (1995) *J. Cell Biol.* **130,** 847–855.
49. Xu, X.Z., Moebius, F., Gill, D.L., and Montell, C. (2001) *Proc. Natl. Acad. Sci. U. S. A.* **98,** 10692–10697.
50. Nakamura, K., Bossy-Wetzel, E., Burns, K., Fadel, M.P., Lozyk, M., Goping, I.S., Opas, M., Bleackley, R.C., Green, D.R., and Michalak, M. (2000) *J. Cell Biol.* **150,** 731–740.
51. Bian, X., Hughes, F.M., Jr., Huang, Y., Cidlowski, J.A., and Putney, J.W., Jr. (1997) *Am. J. Physiol.* **272,** C1241–C1249.
52. Whiteman, M., Armstrong, J.S., Cheung, N.S., Siau, J.L., Rose, P., Schantz, J.T., Jones, D.P., and Halliwell, B. (2004) *FASEB J.* **18,** 1395–1397.
53. Pinton, P., Ferrari, D., Di Virgilio, F., Pozzan, T., and Rizzuto, R. (2001) Molecular machinery and signalling events in apoptosis. *Drug Dev. Res.* **52,** 558–570.
54. Kroemer, G. and Reed, J.C. (2000) *Nat. Med.* **6,** 513–519.
55. Halestrap, A.P., Doran, E., Gillespie, J.P., and O'Toole, A. (2000) *Biochem. Soc. Trans.* **28,** 170–177.
56. Petronilli, V., Penzo, D., Scorrano, L., Bernardi, P., and Di Lisa, F. (2001) *J. Biol. Chem.* **276,** 12030–12034.
57. Szalai, G., Krishnamurthy, R., and Hajnoczky, G. (1999) *EMBO J.* **18,** 6349–6361.
58. Chami, M., Ferrari, D., Nicotera, P., Paterlini-Brechot, P., and Rizzuto, R. (2003) *J. Biol. Chem.* **278,** 31745–31755.
59. Assefa, Z., Bultynck, G., Szlufcik, K., Nadif, K.N., Vermassen, E., Goris, J., Missiaen, L., Callewaert, G., Parys, J.B., and De Smedt, H. (2004) *J. Biol. Chem.* **279,** 43227–43236.
60. Nakayama, T., Hattori, M., Uchida, K., Nakamura, T., Tateishi, Y., Bannai, H., Iwai, M., Michikawa, T., Inoue, T., and Mikoshiba, K. (2004) *Biochem. J.* **377,** 299–307.

Calcium, a Signal in Time and Space

J. Krebs and M. Michalak (Editors)
Calcium: A Matter of Life or Death

ISSN: 0167-7306/DOI: 10.1016/S0167-7306(06)41019-X

CHAPTER 19

Calcium signalling, a spatiotemporal phenomenon

Michael John Berridge

The Babraham Institute, Babraham, Cambridge CB2 4AT, UK, Tel.: +44 (0) 1223 496621; Fax: +44 (0) 1223 496033; E-mail: michael.berridge@bbsrc.ac.uk

Abstract

A large number of cellular processes are controlled by Ca^{2+} signalling. The versatility of this signalling system depends on the existence of an extensive Ca^{2+} signalling toolkit from which cells can assemble cell-specific Ca^{2+} signalsomes that are precisely suited to deliver the signals to control their particular functions. The spatial and temporal properties of such cell-specific Ca^{2+} signals are a particularly important feature of this diversity. Ca^{2+} signalling systems are adapted to control cellular processes over a wide time domain for processes such as exocytosis (microseconds), muscle contraction (milliseconds), metabolism and gene transcription (seconds to minutes) and fertilization and cell proliferation (hours).

Keywords: calcium, CRAC, cytoskeleton, endoplasmic reticulum, inositol 1,4,5-trisphosphate, PMCA, puffs, ryanodine receptors, SERCA, sparklets, sparks, STIM, store-operated channels, TRP

One of the major control mechanisms used by cells is Ca^{2+} signalling [1–4]. It triggers new life at the time of fertilization and continues to control many processes during development. Once cells have differentiated into specialized cell types, it governs the activity of most cellular processes such as secretion, contraction, metabolism, gene transcription, cell proliferation and the synaptic plasticity responsible for learning and memory. There also is a darker side to its action because larger than normal elevations can cause cell death either in the controlled manner of programmed cell death (apoptosis) or in the more catastrophic necrotic change that occurs during processes such as stroke or cardiac ischemia.

1. Ca^{2+} *signalling toolkit*

The basic mechanism of Ca^{2+} signalling is relatively simple in that it depends on an increase in the intracellular concentration of this ion. The Ca^{2+} concentration is low when cells are at rest, but when an appropriate stimulus arrives there is a sudden elevation, which is responsible for a change in cellular activity. However, there are multiple variations of this relatively simple theme. The versatility of Ca^{2+} signalling is achieved by having an extensive Ca^{2+} signalling toolkit (Fig. 1). Each cell type has

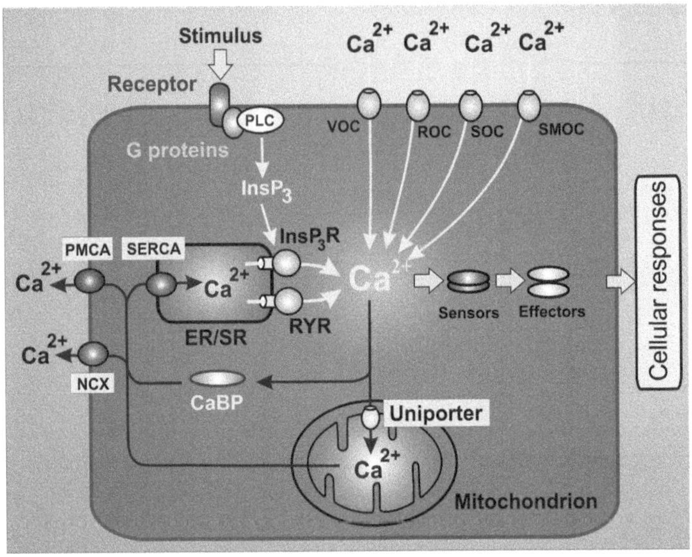

Fig. 1. Summary of the major components of the Ca^{2+} signalling toolkit. Many of the individual components are present in multiple isoforms that further enhance the diversity of the Ca^{2+} signalling systems. The yellow arrows illustrate the ON reactions that introduce Ca^{2+} into the cell, and the blue arrows depict the OFF reactions that pump Ca^{2+} either out of the cell or back into the endoplasmic reticulum (ER). During its passage through the cytoplasm, Ca^{2+} resides temporarily on the Ca^{2+}-binding proteins (CaBPs) that function as buffers, or it can pass through the mitochondria. To carry out its signalling function, Ca^{2+} binds to sensors that then employ a range of effectors to stimulate many different cellular processes. $InsP_3R$, inositol 1,4,5-trisphosphate receptor; PLC, phospholipase C; NCX, Na^+/Ca^{2+} exchanger; PMCA, plasma membrane Ca^{2+}-ATPases; ROCs, receptor-operated channels; RYR, ryanodine receptor; SOCs, store OCs; SERCA, sarcoplasmic–ER Ca^{2+}-ATPase; SMOCs, second-messenger OCs; VOCs, voltage OCs (See Color Plate 51, p. 546).

a clearly defined subset of toolkit components that will be referred to as a Ca^{2+} signalling signalsome [4]. These cell-specific signalsomes are put in place during development when a process of signalsome expression enables each differentiating cell to select out those signalling components it will require to control its particular functions. There are an enormous number of cell-specific Ca^{2+} signalsomes. The important point is that each signalsome generates a cell-specific Ca^{2+} signal with characteristic spatial and temporal properties.

This large toolkit contains many different components that can be mixed and matched to create these different Ca^{2+} signalsomes. There are Ca^{2+} channels in the plasma membrane (PM), which control the entry of Ca^{2+} from the outside. There are Ca^{2+} release (CR) channels, which control the release of Ca^{2+} from internal stores (Fig. 1). The Ca^{2+} buffers ensure that the concentration of Ca^{2+} remains within its operation range and does not rise to levels that can induce cell death. Once Ca^{2+} has carried out its signalling function, there are Ca^{2+} pumps and exchangers that remove it from the cytoplasm by either extruding it from the cell or returning it to the internal stores. Ca^{2+} signalling functions are carried out by various Ca^{2+} sensors and Ca^{2+} effectors that are responsible for translating Ca^{2+} signals into a change in cellular activity.

2. *Basic mechanism of Ca^{2+} signalling*

Cells at rest maintain a low intracellular concentration of Ca^{2+} (approximately 100 nM), but this increases rapidly into the micromolar range when cells are stimulated. This increase in intracellular Ca^{2+} can operate over a very wide time domain (e.g. microseconds to hours) to regulate many different cellular processes. An important feature of Ca^{2+} signalling is its dynamic nature as exemplified by the fact that Ca^{2+} signals invariably appear as a brief transient. The rising phase of each transient is produced by the ON reactions that introduce Ca^{2+} into the cytoplasm, whereas the falling phase depends on the OFF reactions that pump Ca^{2+} out of the cell or back into the internal stores (Fig. 1). External stimuli activate the ON reactions by stimulating both the entry and the release of Ca^{2+}. Most cells make use of both sources, but there are examples of cells using either external or internal sources to control specific processes and this will depend on the type of signalsome they are using. Most of the Ca^{2+} that enters the cytoplasm is adsorbed onto buffers while a much smaller proportion activates the effectors to stimulate cellular processes. The OFF reactions remove Ca^{2+} from the cytoplasm using a combination of mitochondria and different pumping mechanisms. When cells are at rest, these OFF reactions keep the concentration low, but these are temporarily overwhelmed when external stimuli activate the ON reactions.

Sequential activation of the ON and OFF reactions gives rise to Ca^{2+} transients, which are such a characteristic feature of Ca^{2+} signalling. At any moment in time, the level of Ca^{2+} is determined by the balance between these ON and OFF reactions. An important spatial aspect of the ON reactions is that they are often closely associated with the effector systems that respond to Ca^{2+}. For example, voltage-operated channels (VOCs) in presynaptic endings are associated with the synaptic vesicles thus producing a highly localized puff of Ca^{2+} to trigger exocytosis. Similarly, the type 2 ryanodine receptors (RYR2s) of cardiac cells are lined up close to the contractile filaments to ensure that Ca^{2+} will rapidly stimulate contraction. In cases where cells need to be stimulated over a long time, these transients are repeated at set intervals to set up Ca^{2+} oscillations. These oscillations are part of the spatiotemporal aspects of Ca^{2+} signalling that will be described in Section 5.2.

2.1. *Ca^{2+} ON reactions*

In response to external stimuli, channels in the PM or endoplasmic reticulum (ER) are opened, and Ca^{2+} flows into the cytoplasm to bring about the elevation of cytosolic Ca^{2+} responsible for cell activation (Fig. 1). During these ON reactions, the cell employs a variety of both Ca^{2+} entry channels and CR channels, which are organized into different modules (Fig. 2). The entry of Ca^{2+} across the PM is carried out by many different channels whose names indicate how they are opened. For example, there are receptor OCs (ROCs, see module 1 in Fig. 2), second-messenger OCs (SMOCs, see module 2 in Fig. 2), VOCs (see module 3 in Fig. 2) and store OCs (SOCs). The last have attracted a lot of interest recently because there have been some new developments regarding the nature of the SOC and how it might be regulated.

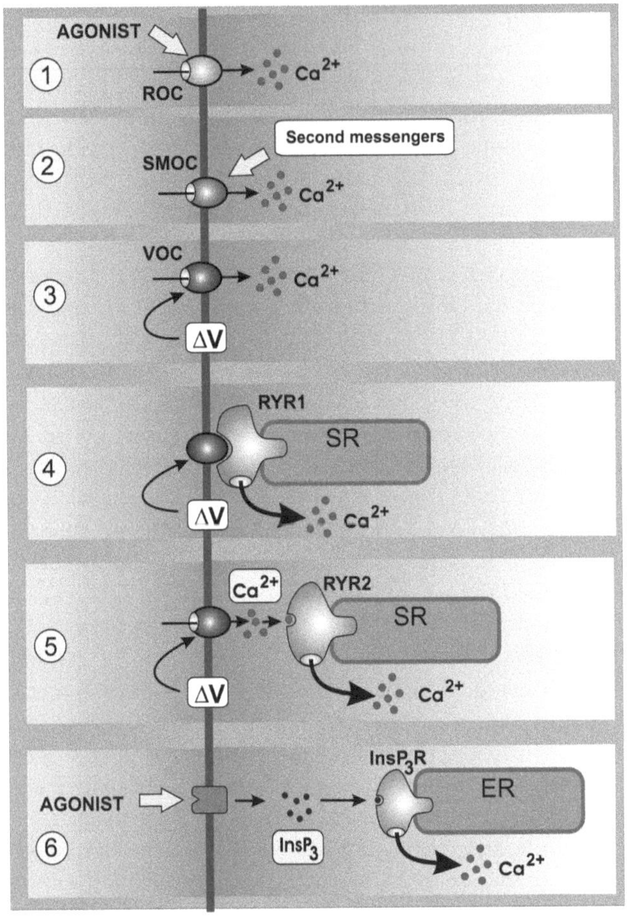

Fig. 2. Summary of some of the main modules that are mixed and matched to create different Ca^{2+} signalling systems. (1) Agonists such as the neurotransmitters acetylcholine, glutamate and ATP act directly on receptor-operated channels (ROCs) in the plasma membrane (PM) to allow external Ca^{2+} to enter the cell. (2). Second messengers such as diacylglycerol, cyclic AMP, cyclic GMP and arachidonic acid acting from the cytoplasmic side open second-messenger OCs (SMOCs) in the PM. (3) Membrane depolarization (ΔV) activates voltage OCs (VOCs) in the PM to allow a rapid influx of external Ca^{2+}. (4) Membrane depolarization (ΔV) activates the $Ca_V1.1$ L-type channel that then activates the type 1 ryanodine receptor (RYR1) in skeletal muscle through a direct conformational-coupling mechanism. (5) Membrane depolarization (ΔV) activates VOCs in the PM to allow a rapid influx of external Ca^{2+} to provide trigger Ca^{2+} that then activates the RYR2 to release Ca^{2+} stored in the sarcoplasmic reticulum (SR). This mechanism is found in cardiac muscle and neurons. (6) Agonists acting on cell-surface receptors generate inositol 1,4,5-trisphosphate ($InsP_3$) that then diffuses into the cell to activate the $InsP_3$ receptor ($InsP_3R$) to release Ca^{2+} from the endoplasmic reticulum (ER) (See Color Plate 52, p. 547).

Release of Ca^{2+} from internal stores is carried out by three main channel types: the RYRs, the inositol 1,4,5-trisphosphate receptors ($InsP_3Rs$) and a putative nicotinic acid dinucleotide phosphate (NAADP)-sensitive channel. One of the major problems in Ca^{2+} signalling has been to determine how stimuli arriving at the cell surface gain access to these internal stores. Two main mechanisms have been identified. Firstly, there is conformational-coupling through protein–protein interactions. This is a very fast mechanism that depends on a sensor in the PM interacting directly with an internal release channel. The receptor on the cell surface is the $Ca_V1.1$ L-type channel (a voltage sensor), which is coupled to the RYR1 (module 4 in Fig. 2). Information is transferred through a process of conformational-coupling. This mechanism is mainly found in skeletal muscle, but a similar mechanism may be present in certain neurons such as hypothalamic synaptic endings [5].

The second mechanism depends on the generation of diffusible second messengers. Activation of receptors or channels on the cell surface generates second messenger that then diffuses into the cell to activate release channels. One of the most significant Ca^{2+}-mobilizing messengers is Ca^{2+} itself, which is a potent activator of the two main internal release channels, the RYRs and the $InsP_3Rs$ (Fig. 3). This Ca^{2+}-induced CR (CICR) mechanism has the unique property of being autocatalytic and plays a central role in generating those Ca^{2+} signals that appear as regenerative Ca^{2+} waves. Another important messenger for releasing internal Ca^{2+} is $InsP_3$ (module 6 in Fig. 2) [1]. Other Ca^{2+}-mobilizing messengers have been described such as cyclic adenine dinucleotide phosphate [6] and NAADP [7,8].

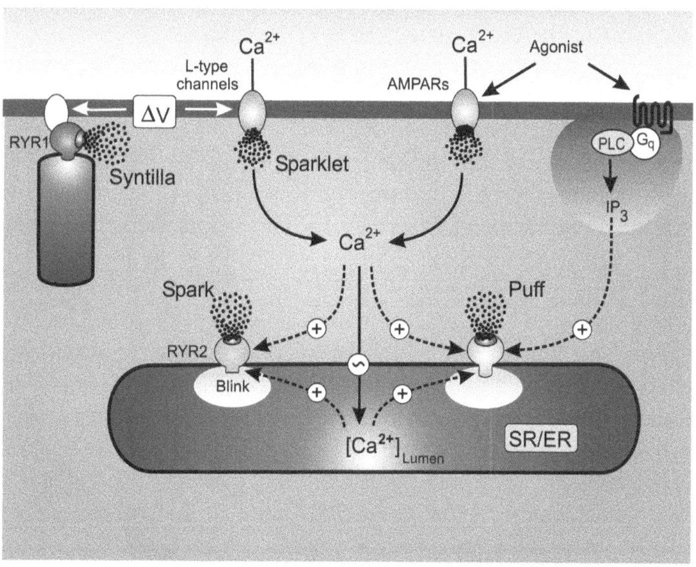

Fig. 3. The elementary events of Ca^{2+} signalling. Elementary events are the localized Ca^{2+} signals that arise from either individual or small groups of ion channels. The localized plumes of Ca^{2+} have been given different names depending on the channels that produce them. Voltage-operated channels (VOCs) in the plasma membrane (PM) produce sparklets and ryanodine receptors (RYRs) on the sarcoplasmic reticulum (SR) create sparks (and syntillas), whereas the inositol 1,4,5-trisphosphate receptors ($InsP_3Rs$) produce puffs. Channels in the PM are linked to the internal release channels through a process of Ca^{2+}-induced Ca^{2+} release (CICR). The intracellular channels have large conductances and gate so much Ca^{2+}, which result in a local depletion of Ca^{2+} within the lumen immediately below the channel. This local emptying of the endoplasmic reticulum (ER) store by RYRs has been visualized and has been called a blink.

The channels responsible for these Ca^{2+} ON reactions usually have powerful inactivation mechanisms that rapidly curtail the entry or release processes to prevent the cell being swamped with Ca^{2+}, which can result in cell stress and apoptosis. Once the ON reactions have been curtailed, the Ca^{2+} OFF reactions rapidly take over to return the activated level of Ca^{2+} back to its resting level.

2.1.1. Store-operated Ca^{2+} entry

Much of the Ca^{2+} used for signalling is released from internal stores that have a finite capacity. For signalling to continue, therefore, Ca^{2+} must be brought in from the outside to ensure that the store remains topped up. Much of this ER Ca^{2+} homeostasis depends on the sarcoplasmic–ER Ca^{2+}-ATPase (SERCA) pumps that recycle the released Ca^{2+}. However, there are always some losses to the outside resulting in store depletion, which will not only result in a decline in signalling capacity but also trigger the ER stress signalling pathways. To guard against the deleterious effects of Ca^{2+} store depletion, the cell employs SOCs that open in response to store emptying to ensure a constancy of its internal Ca^{2+} store [9]. Electrophysiological studies have revealed that there may be different types of SOCs that vary in their Ca^{2+} selectivity. The classical example of a SOC is the CR-activated Ca^{2+} (CRAC) channel found in lymphocytes, which has a very high selectivity for Ca^{2+} and a very low conductance [10]. One of the main functions of this entry mechanism, therefore, is to maintain the internal store of Ca^{2+}. However, it can also function as a source of signal Ca^{2+} especially under conditions where Ca^{2+} signalling has to be maintained over a prolonged period as occurs during the stimulation of cell proliferation.

The nature of the signal that emanates from the ER to activate SOCs is still a matter of some debate. One of the important questions to consider is whether the whole of the ER is involved in SOC activation. This seems unlikely because depletion of the ER Ca^{2+} can trigger stress signalling pathways by seriously interfering with the processes of protein synthesis and packaging. There is growing evidence that a small specialized region of the ER closely associated with the PM regulates the entry of Ca^{2+} (Fig. 4). The major unsolved problem concerns the mechanism whereby the empty store sends a message to open the SOC in the PM.

Some of the proposed mechanisms include the generation of a calcium influx factor, exocytosis of vesicles containing SOCs or a conformational-coupling mechanism whereby the release channels in the ER interact directly with the SOCs in the PM (Fig. 5). The latter mechanism resembles that found in skeletal muscle where L-type VOCs in the PM interact directly with the RYR1 channels in the SR (module 4 in Fig. 2). A similar conformational-coupling mechanism has been proposed to explain how empty ER stores can stimulate entry through SOCs in the PM [11]. In this latter case, information is flowing in the opposite direction, i.e. the output signal travels from the ER to the PM. There are a number of possible molecular components of this putative coupling mechanism (Fig. 5), which are less well defined than those of the skeletal muscle system.

One of the problems with trying to understand this entry process is that the identity of the entry channels is uncertain. An integral membrane protein called

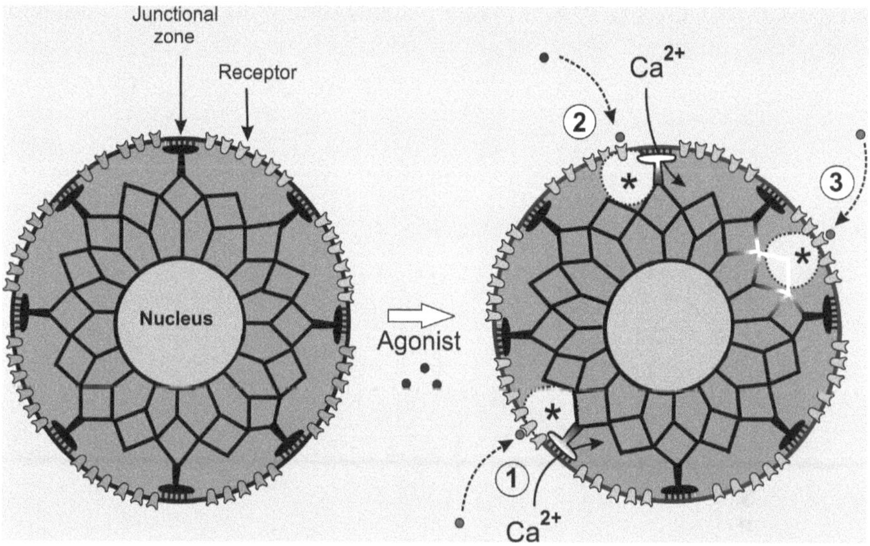

Fig. 4. The hypothesis outlined in this figure suggests that store-operated Ca^{2+} entry occurs at localized regions where there is close apposition of the endoplasmic reticulum (ER) and plasma membrane (PM). The ER is an interconnected reticular network of tubules, and short arms of this ER network come into contact with the PM to form junctional zones. The narrow gap between the ER and the PM may be the site where the signal from the empty ER is transferred to the store-operated channels (SOCs). Agonists acting on cell-surface receptors stimulate phospholipase C (PLC) to produce a microdomain of inositol 1,4,5-trisphosphate ($InsP_3$) (*), which functions to deplete Ca^{2+} in the junctional zones. The latter then sends a message to the SOCs in the PM to promote entry (i.e. regions 1 and 2). This signal might be transmitted through a protein–protein interaction as outlined in the conformational-coupling hypothesis shown in Fig. 5. In some cases, the microdomain of $InsP_3$ does not coincide with a junctional zone, and the local depletion of Ca^{2+} fails to trigger entry (see region 3).

Orai1 may be the CRAC channel [12]. There also are indications that there may be a number of other SOCs coded for by various members of the TRP ion channel family such as TRPC1, TRPC4 and TRPV6.

Another area of uncertainty concerns the nature of the protein in the ER that senses store emptying and then relays information to the SOCs. Both the $InsP_3Rs$ and the RYRs, which are known to be able to sense the luminal level of Ca^{2+}, have been implicated as signal initiators in the ER, which then convey information to the SOCs in the PM through a direct protein–protein interaction. There is some evidence to support the idea that either an $InsP_3R$ or an RYR may play a role in transmitting information from the ER to the channels in the PM (Fig. 5). Such an involvement of an RYR would mean that this release channel might play a pivotal role in both the release of internal Ca^{2+} (as occurs in skeletal muscle, see module 4 in Fig. 2) and the entry of external Ca^{2+} (Fig. 5). Indeed, there is some evidence to show that both mechanisms may coexist in skeletal muscle where the SOCs may function in long-term Ca^{2+} homeostasis to overcome fatigue during intensive exercise. The SOC may couple to the RYR1 through mitsugumin 29, which is a synaptophysin family-related protein positioned within the junctional space between the PM and the SR [13].

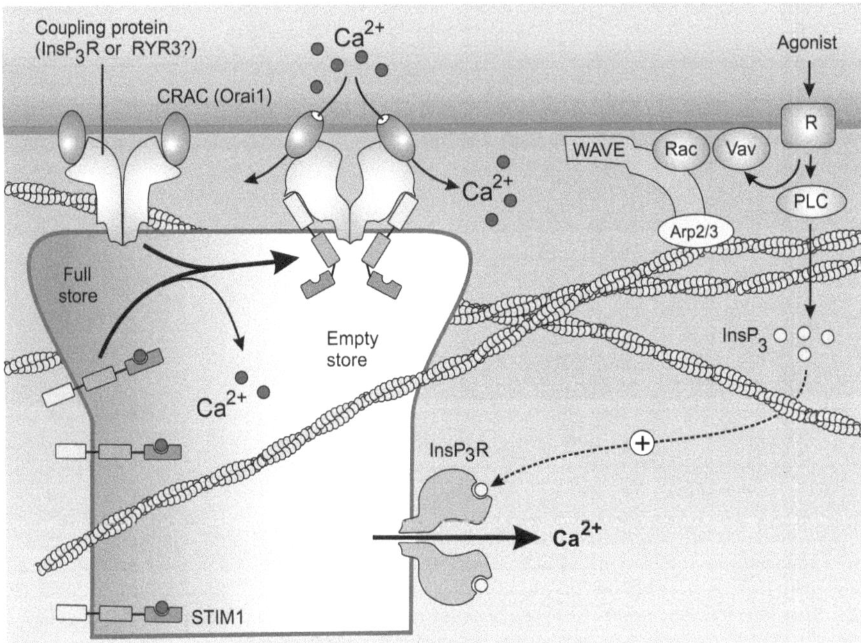

Fig. 5. A conformational-coupling hypotheses for store-operated Ca^{2+} entry as might occur in lymphocytes. Entry of Ca^{2+} depends on activation of a store-operated channel (SOC) such as the Ca^{2+} release-activated Ca^{2+} (CRAC) channel. When an agonist activates phospholipase C (PLC), the increase in inositol 1,4,5-trisphosphate ($InsP_3$) then releases Ca^{2+} from the endoplasmic reticulum (ER) and the emptying of the store opens the CRAC channel. Store emptying might be detected by the stromal interaction molecule 1 (STIM1), which has an N-terminal EF-hand domain that binds Ca^{2+} when the store is full. As the store empties, Ca^{2+} comes off STIM causing it to undergo a conformational change that is transmitted through some coupling protein [perhaps an $InsP_3$ receptor ($InsP_3R$) or a ryanodine receptor (RYR)] to open the CRAC channels. Formation of this coupling complex appears to depend on cytoskeletal remodelling driven by WAVE2 and controlled through the monomeric G proteins Vav and Rac (See Color Plate 53, p. 548).

Another candidate for the ER sensor is the stromal interaction molecule (STIM), which is located within the ER and has a single EF-hand that might function to sense Ca^{2+} (Fig. 5) [14]. If STIM is knocked out, there is a dramatic decline in store-operated entry indicating that this protein plays some direct role in the coupling mechanism. Under resting conditions when the store is full of Ca^{2+}, STIM appears to be distributed over the ER. Upon store emptying, STIM moves to the surface to aggregate at close contacts between the ER and the PM [15] and is thus ideally situated to act as a sensor for conformational coupling. However, STIM is a small molecule, and the problems remain as to how information is transferred across the 20-nm gap to activate CRAC or SOCs in the PM. In the model outlined in Fig. 5, the suggestion is that when STIM loses its Ca^{2+} and migrates towards the cell surface, it makes contact with some coupling protein (e.g. an $InsP_3R$ or an RYR) to induce the conformational change that results in channel opening.

The conformational-coupling hypothesis depends on the ER making close contact with the PM so that information can be relayed between proteins located in the two apposed membranes. Many of the components that function in store-operated entry are associated with the caveolae, which have been suggested as possible site for this entry processes. There are numerous examples of the ER making intimate contact with the caveolae as described in both smooth muscle [16] and cardiac cells [17].

Another important structural component for entry is the cytoskeleton, which may be particularly important in forming and/or maintaining the structural integrity of this complex (Fig. 5). Proteins that function in actin remodelling such as WAVE can profoundly influence store-operated Ca^{2+} entry. In the case of T cells, activation of the T-cell receptor relays information out to a number of signalling pathways. One of these is the activation of Vav, which seems to act through Rac and WAVE2 to remodel the actin cytoskeleton that has a role in regulating the entry of Ca^{2+} through the CRAC channel [18,19].

2.1.2. CICR

A process of CICR plays a central role in the way Ca^{2+} signals are generated. This positive feedback mechanism whereby Ca^{2+} triggers its own release has two important functions in cells. It enables Ca^{2+} entering across the PM to function as a messenger to release Ca^{2+} from the internal store (Fig. 3). This function of CICR was first described in cardiac cells where the $Ca_V1.2$ L-type channel provides an influx of trigger Ca^{2+} that then diffuses into the cell to activate the RYR2 (module 5 in Fig. 2). A similar interaction is particularly evident for neuronal Ca^{2+} entry and release channels.

The other main function of CICR is to set up intracellular and intercellular waves where an elevated level of Ca^{2+} in one region of the cell (the initiation site) propagates throughout the rest of the cytoplasm as a regenerative Ca^{2+} wave (Fig. 6). Waves can progress by recruiting either RYRs or $InsP_3Rs$ [20]. These Ca^{2+} waves are made up of elementary Ca^{2+} events such as the sparks and puffs produced by the RYRs and $InsP_3Rs$, respectively. It is these unitary events that are used to generate the regenerative waves that make up global Ca^{2+} signals.

3. Ca^{2+} OFF reactions

Cells use a variety of mechanisms to remove Ca^{2+} from the cytoplasm (Fig. 1). The introduction of Ca^{2+} into the cell during the Ca^{2+} ON reactions usually occurs for a relatively brief period during which there is a rapid increase in the intracellular concentration of Ca^{2+}. The increase in free Ca^{2+}, which can be measured in the cytoplasm using aequorin or fluorescent indicators, is a very small proportion of the total amount of Ca^{2+} that enters during the ON reactions. Much of this Ca^{2+} is rapidly bound by cytosolic buffers or is taken up by the mitochondria. As the $[Ca^{2+}]$ returns to its resting level, Ca^{2+} leaves the buffers and the mitochondria and is returned to the ER or is pumped out of the cell resulting in a brief Ca^{2+} transient.

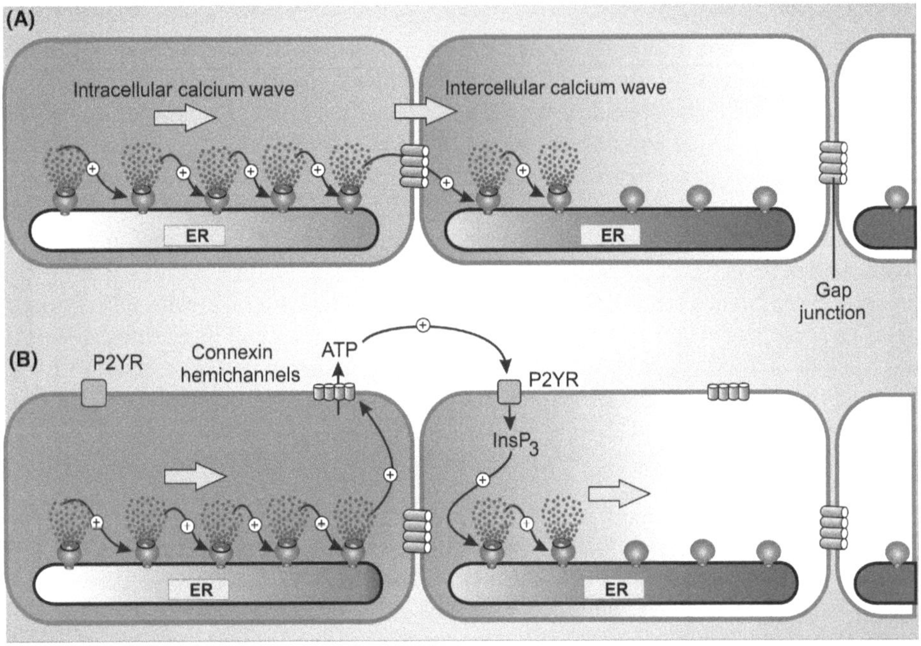

Fig. 6. Intracellular waves depend on a process of Ca^{2+}-induced Ca^{2+} release (CICR) whereby the Ca^{2+} being released from one channel diffuses across to neighbouring channels that are excited to release further Ca^{2+} thereby setting up a regenerative wave. When such waves meet a cell boundary, they can trigger waves in adjacent cells thus setting up an intercellular wave. There is still some debate about the way the wave travels between cells. (**A**) One mechanism proposes that when the intracellular wave reaches the cell boundary, some small molecular weight component, most likely to be Ca^{2+}, diffuses across the gap junction to ignite another intracellular wave in the neighbouring cell. (**B**) An alternative mechanism suggests that the intracellular wave in one cell stimulates the release of ATP through hemichannels, which then diffuses across to ignite a wave in neighbouring cells by acting on P2Y receptors to produce inositol 1,4,5-trisphosphate ($InsP_3$) (See Color Plate 54, p. 549).

The recovery process thus depends on a complex interplay between cytosolic Ca^{2+} buffers, mitochondria and Ca^{2+} pumps and exchangers on the internal stores and PM. These Ca^{2+} OFF reactions operate at different stages during the recovery phase of a typical Ca^{2+} spike. The buffers and mitochondria operate early, and the Ca^{2+} pumps and exchangers are responsible for restoring the status quo by pumping Ca^{2+} out of the cell or back into the ER. These pumps and exchangers operate at different times during the recovery process. The Na/Ca exchangers have low affinities for Ca^{2+} but have very high capacities, and this enables them to function at the beginning of the recovery process to rapidly remove large quantities of Ca^{2+}. On the contrary, the PMCA and the SERCA pumps have lower capacities, but their higher affinities mean that they can complete the recovery process and can continue to pump at lower Ca^{2+} levels thus enabling them to maintain both the internal stores and the resting level of Ca^{2+} within the cytoplasm. Some of these OFF reactions interact with each other during the recovery period, and this is particularly evident in the case of the ER/mitochondrial Ca^{2+} shuttle. When Ca^{2+} is released from the ER, Ca^{2+} is rapidly taken up by the mitochondria, and this is then released slowly back to the ER once Ca^{2+} has returned back to the resting level.

Under pathological situations, the lack of oxygen reduces the supply of energy thus compromising the function of the OFF reactions resulting in the rise of Ca^{2+} that is so damaging during stroke or cardiac ischemia. Under normal circumstances, however, the fully energized OFF reactions can rapidly reduce the pulse of Ca^{2+} introduced by the ON mechanisms thus generating a brief Ca^{2+} transient. Such brief pulses of Ca^{2+} are a characteristic feature of the Ca^{2+} signalling pathway and contribute to the spatiotemporal aspects of Ca^{2+} signalling.

4. *Ca^{2+} buffers*

Cells express a large number of Ca^{2+}-binding proteins (CaBPs), which fall into two main groups, Ca^{2+} sensors and Ca^{2+} buffers. The Ca^{2+} sensors respond to changes in intracellular Ca^{2+} by activating the downstream effectors that control cellular responses (Fig. 1). In a sense, all proteins capable of binding Ca^{2+} will act as a buffer, and this also applies to the sensors. However, the concentration of these sensors is usually rather low so they have little buffering capacity. The role of Ca^{2+} buffering is carried out by another major group of CaBPs. The major cytosolic buffers in cells are parvalbumin (PV) and calbindin D-28k (CB). The major buffers that operate within the lumen of the ER are calsequestrin and calreticulin. The latter is unusual in that it functions both as a cytosolic and as a luminal buffer. The cytosolic buffers have subtly different Ca^{2+}-binding properties and are expressed in cells in differing combinations and concentrations to create Ca^{2+} signals with kinetics that are tailored to carry out different functions. For example, neurons such as Purkinje cells express large amounts of PV and CB. As a consequence, Purkinje cells have a large endogenous Ca^{2+}-buffering capacity, e.g. their buffers bind approximately 2000 Ca^{2+} ions for each free ion. Lower capacities of 50–100:1 are found in other cells. Motor neurons have a very low-buffering capacity and consequently have large Ca^{2+} signals in both the soma and the dendrites during normal physiological responses, and this makes them particularly susceptible to neurodegeneration. Buffer concentration is one of the important parameters in determining buffer capacity. The other key parameters include affinity for Ca^{2+} and other metal ions, the kinetics of Ca^{2+} binding and release and mobility. The sole function of these Ca^{2+} buffers is to bind Ca^{2+}, and this has an important function in shaping both the spatial and the temporal properties of Ca^{2+} signals.

5. *Spatiotemporal aspects of Ca^{2+} signalling*

The use of Ca^{2+} as a universal signal for cell regulation is somewhat paradoxical because this ion can be very toxic to cells if its level remains high for a prolonged period. Such toxicity is avoided by presenting Ca^{2+} signals in a pulsatile manner

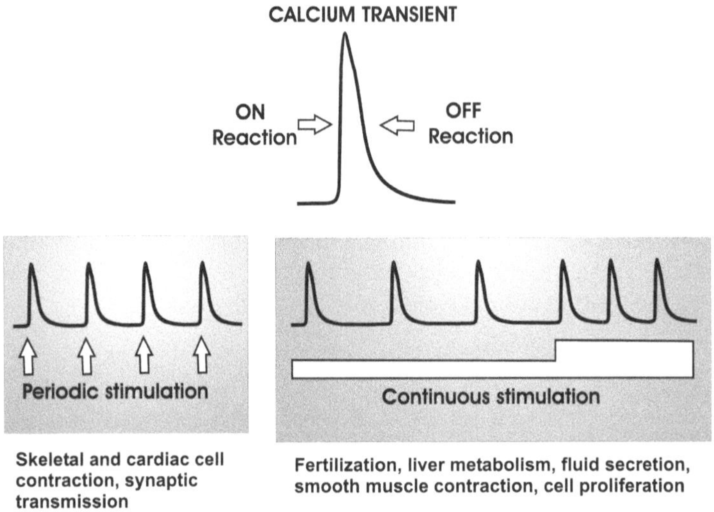

Fig. 7. In almost every example where Ca^{2+} is used as a signal, it is presented as a brief transient, which is the digital signal used to set up complex temporal patterns of Ca^{2+} signalling. In some cells, these pulses are produced on demand in that they are generated in response to periodic stimulation (arrows) as occurs in muscle contraction where a brief burst of Ca^{2+} activates the contractile machinery, which then recovers when the Ca^{2+} signal is removed. Likewise, the release of neurotransmitters from nerve terminals is triggered by brief localized pulse of Ca^{2+}. In many other tissues, which receive a continuous stimulation over a prolonged period (bar), the Ca^{2+} signal is again presented as brief spikes that are produced rhythmically to give highly regular Ca^{2+} oscillations whose frequencies vary with the level of cell stimulation.

(Fig. 7). In addition to this temporal aspect, the Ca^{2+} signal is also highly organized in space. For example, muscle contraction is activated by a global elevation in Ca^{2+}, whereas the release of neurotransmitters results from a minute punctate pulse of Ca^{2+} delivered directly to the docked vesicle by a Ca^{2+} channel tightly associated with the exocytotic machinery. In between these two extremes, there are many variations in the way the Ca^{2+} signal is presented to cells.

5.1. Temporal aspects of Ca^{2+} signalling

The temporal aspect of signalling is based on the fact that most Ca^{2+} signals appear in cells as transients (Fig. 7). There are two main ways in which this digital Ca^{2+} signal is produced. Firstly, there are 'on-demand' Ca^{2+} transients such as the signals generated in muscle cells and neurons by membrane depolarization. Secondly, transients can appear as an oscillation when cells respond to stimuli that persist for long periods. Such oscillations, which can have a wide range of frequencies, are a characteristic feature of many cellular control systems. There are two main types of oscillatory activity: membrane oscillators and cytosolic oscillators. The output from membrane oscillators usually sets up a regular train of action potentials that provide the rhythmical pacemaker activity responsible for driving cellular processes such as contraction, neuronal activity and secretion. These membrane potential oscillations can open VOCs that gate Ca^{2+} and can thus result in intracellular Ca^{2+} oscillations. Ca^{2+} oscillations can also be generated by cytosolic oscillators. The latter are of particular interest because oscillator frequency can vary with stimulus intensity, and such frequency modulation is used to encode information. There also is evidence that information carried as oscillations can greatly enhance the activation of gene transcription [21,22].

5.1.1. Membrane oscillators

Membrane oscillators are usually generated through an interplay between outward currents, usually carried by K^{+}, that induce membrane hyperpolarization that inhibits the inward currents (e.g. Na^{+} and Ca^{2+}) that cause depolarization. The entry of Ca^{2+} during the depolarizing phase is responsible for the phasic elevation of Ca^{2+} that is returned to the cytoplasm by the PMCA during the OFF reaction.

There are a number of cell types that set up pacemaking activity using such membrane oscillators. Sinoatrial node pacemaker cells establish the regular trains of action potential, which drive cardiac contraction. Thalamocortical neurons have a neuronal rhythmicity that displays oscillations at a frequency of 0.5–4 Hz using a combination of different inward and outward currents. Neurons that reside within the suprachiasmatic nucleus, which have the biological clock, also have a membrane Ca^{2+} oscillator that is responsible for generating the output signals from the biological clock.

5.1.2. Cytosolic Ca^{2+} oscillations

Cytosolic Ca^{2+} oscillations are produced by the periodic release of internal Ca^{2+} through the operation of a cytosolic oscillator [1,3]. This Ca^{2+} oscillation mechanism depends on the release of Ca^{2+} from intracellular stores. In the case of agonist-induced oscillations, the $InsP_3/Ca^{2+}$ signalling cassette is primary responsible for initiating oscillatory activity. The information encoding/decoding of Ca^{2+} oscillations depends on the ability to modulate the different parameters of these Ca^{2+} oscillations.

Such agonist-dependent cytosolic Ca^{2+} oscillations are a major feature of Ca^{2+} signalling in many cell types. Liver cells display Ca^{2+} oscillations with agonist concentration-dependent changes in frequency [23]. Ca^{2+} oscillations are evident during the acquisition of meiotic competence during development [24] and are responsible for oocyte activation during mammalian fertilization [25]. Such Ca^{2+} oscillations are also observed following artificial insemination by intracytoplasmic spermatozoa injection [26]. Smooth muscle cells surrounding cortical arterioles display spontaneous Ca^{2+} oscillations [27]. Astrocytes display spontaneous Ca^{2+} oscillations [28]. Airway epithelial cells respond to ATP by generating the Ca^{2+} oscillations that control ciliary beat frequency [29].

5.2. Spatial aspects of Ca^{2+} signalling

Components of the Ca^{2+} signalling pathways are often highly organized and localized to discrete cellular areas where Ca^{2+} microdomains are formed to regulate specific cellular processes [20,30,31]. Much attention has focused on the elementary and global aspects of Ca^{2+} signalling, which determine whether signalling occurs within highly localized regions or is spread more globally by the formation of both intracellular and intercellular Ca^{2+} waves. One of the exciting aspects of signalling microdomains is the way they are used by neurons to increase their capacity to process information. The fact that this information processing can be confined to very small volumes within the spines means that each neuron is capable of simultaneously processing large amounts of information. Such input-specific signalling is particularly relevant to the process of synaptic modifications during learning and memory. The high concentrations of neuronal Ca^{2+} buffers, such as CB, play a major role in restricting Ca^{2+} signals to individual spines, which are the smallest units of neuronal integration.

The size of these signalling microdomains will depend on a number of aspects such as the rate of signal generation, the rate of signal diffusion, the rate of signal removal by the OFF mechanisms and the degree of buffering. Buffers thus play an important role in determining the volume of these signalling microdomains, and this may have been a particularly important feature for the miniaturization of Ca^{2+} signalling within the brain.

5.2.1. Elementary aspects of Ca^{2+} signalling

The development of fluorescent indicators to visualize Ca^{2+} in real time in living cells has revealed a spatial dimension to its action that can account for both the universality and the versatility of Ca^{2+} signalling. The microdomains that have been recorded in cells are the result of elementary events produced by the brief opening of either entry channels in the PM or release channels in the ER. Elementary events are the basic building blocks of Ca^{2+} signalling. They can either perform highly localized signalling functions or can be recruited to generate global Ca^{2+} signals. Most of these elementary events are due to the brief opening of Ca^{2+} channels located either in the PM or in the ER and thus result in localized pulses of Ca^{2+} (Fig. 3). The presence of Ca^{2+} buffers helps to restrict these brief pulses to small microdomains within the cytoplasm. A number of different elementary events have now been described as outlined in the following sections.

5.2.2. Sparklet

A sparklet is formed as a result of the brief opening VOCs (Fig. 3). Sparklets have been visualized in ventricular heart cells [32]. These sparklets play a critical role in ventricular cell E–C coupling because they provide the trigger Ca^{2+} that activates the RYR2s in the junctional zone. Another function for sparklets is to control exocytosis particularly at synaptic endings where a localized pulse of Ca^{2+} is responsible for transmitter release. An elementary event, which is equivalent to a sparklet, is formed in the stereocila of hair cells upon opening of the mechanosensitive channel [33].

5.2.3. Spark

A Ca^{2+} spark is formed by the opening of a group of RYRs (Fig. 3). They were first described in cardiac cells where they are responsible for excitation–contraction coupling [34]. However, they have now been described in many other cell types where they play a variety of different functions. In ventricular cardiac cells, the spark is the unitary Ca^{2+} signal that is produced at each junctional zone. There are approximately 10 000 junctional zones in each cell, and to get a rapid contraction, the individual sparks must all be fired simultaneously. An electrical recruitment process is used for this synchronization. In atrial cells, the sparks have a different function from those in the ventricular cells. The initial sparks activated by membrane depolarization are restricted to the junctional zones at the cell surface where they provide a signal to ignite a Ca^{2+} wave that spreads into the cell by triggering a progressive series of sparks through the process of CICR. In mossy fibre presynaptic endings, spontaneous Ca^{2+} sparks can trigger transmitter release [35]. Spontaneous Ca^{2+} transients, which resemble sparks, have been recorded in cerebellar basket cell presynaptic endings [36]. Smooth muscle cell Ca^{2+} sparks function to control both contraction and relaxation. In the case of relaxation, the spark activates the large conductance (BK) channel to produce an outward current that hyperpolarizes the membrane [37].

5.2.4. Syntilla

A scintilla is an elementary event produced by RYRs and thus resembles a spark (Fig. 3). These syntillas function in hypothalamic neuronal presynaptic CR [5].

5.2.5. Blink

Blinks have been visualized within the SR of ventricular muscle cells [38]. As release channels such as the RYRs have a very high conductance, they can gate sufficient Ca^{2+} to cause a temporary depletion of Ca^{2+} within the lumen of the junctional SR (Fig. 3). These blinks are attracting considerable interest because they may play a role in the inactivation of ventricular RYR2s to curtail the cardiac Ca^{2+} transient.

5.2.6. Puff

A puff is a unitary event that results from the release of Ca^{2+} from a small group of $InsP_3Rs$ (Fig. 3). These puffs, which are very similar to the Ca^{2+} sparks, are the building blocks of the intracellular Ca^{2+} waves in cells that result in global Ca^{2+} signals. Such localized Ca^{2+} signals form microdomains of Ca^{2+} to regulate localized cellular processes such as transmitter release from the astrocyte endings that form part of the tripartite synapse [39] and transmitter release from neocortical glutamatergic presynaptic endings [40]. Release of Ca^{2+} by $InsP_3Rs$ produces the microdomains of Ca^{2+}, which are confined to individual spines in Purkinje neurons [41].

5.3. Global Ca^{2+} signals

Most of the global Ca^{2+} signals in cells are produced by the release of Ca^{2+} from internal stores. The intracellular release channels such as the $InsP_3Rs$ and RYRs can create such global signals if their release activity can be synchronized. There are two mechanisms of synchronization that is nicely illustrated by the way ventricular and atrial cardiac cells are controlled. In ventricular cells, electrical recruitment by the action potential is used to activate all the sparks in the junctional zones simultaneously. By contrast, atrial cells use a process of diffusional recruitment whereby Ca^{2+} sparks at the cell surface trigger Ca^{2+} waves that spread inwards through a process of CICR by recruiting RYRs located deeper within the cell.

There also are examples where intracellular waves can spill across into neighbouring cells to set up intercellular Ca^{2+} waves, which thus provide a mechanism for coordinating the activity of a local population of cells.

5.3.1. *Intracellular Ca^{2+} waves*

Intracellular Ca^{2+} waves can be generated by both $InsP_3Rs$ and RYRs, which are Ca^{2+}-sensitive channels that contribute to the positive feedback process of CICR responsible for forming Ca^{2+} waves (Fig. 6A). The main condition that has to be met for Ca^{2+} waves to form is that their Ca^{2+} sensitivity must be increased such that they can respond to the local Ca^{2+} spark or puff produced by their neighbours. In both cases, the level of Ca^{2+} within the lumen of the ER appears to be a critical factor. This ER loading is achieved by entry of external Ca^{2+}, and as the lumen loads up with Ca^{2+}, the $InsP_3Rs$ and RYRs gradually increase their sensitivity such that they can participate in the regenerative processes that result in a Ca^{2+} wave. In effect, this increase in Ca^{2+} sensitivity of the release channels converts the cytoplasm into an 'excitable medium' capable of spawning these regenerative waves.

These intracellular waves are an integral part of many cellular control processes. Intracellular Ca^{2+} waves provide the global Ca^{2+} signal that activates mammalian oocytes at fertilization [25]. Excitation–contraction coupling in atrial cardiac cells depends on Ca^{2+} waves that spread into the cell from the periphery [42]. Astrocyte excitability depends on an intracellular wave that spreads from the tripartite synapses down to the endfoot processes. In the pancreas, $InsP_3Rs$ in the apical region can trigger waves that then spread through the basal region through RYRs.

5.3.2. *Intercellular Ca^{2+} waves*

There are a number of instances of Ca^{2+} waves travelling from one cell to the next. Such intercellular waves may act to coordinate the activity of a local population of cells. There is still some uncertainty concerning the way in which the wave is transmitted from one cell to the next (Fig. 6). One mechanism proposes that low molecular weight components such as $InsP_3$ or Ca^{2+} spill through the gap junctions to ignite waves in neighbouring cells. In order for a cell to set up an intracellular wave, the internal release channels have to be sensitized, so it seems likely that all the cells in the population have to be in a similar state in order for an intercellular wave to pass from one cell to the next. In such a scenario, Ca^{2+} is the most likely candidate to be the stimulus that passes from one cell to the next. An alternative model proposes that the intracellular wave in one cell stimulates the release of ATP that then diffuses across to neighbouring cells where it acts on P2Y receptors to increase $InsP_3$, which then acts to trigger a new wave (mechanism B in Fig. 6).

The function for such intercellular waves has not been clearly established. Much of the work on these waves has been done on cultured cells, but there are a number of reports showing that such intercellular waves do occur between cells in situ. Intercellular waves have been described in astrocytes, but their physiological function is unclear. As they only seem to appear following intense stimulation, they may be a manifestation of some pathological change. In this respect, the astrocyte wave moves at approximately the same rate as spreading depression that appears to be linked to the onset of migraines. Intercellular waves, which have been recorded in the intact perfused liver, appear to travel in a periportal to pericentral direction [43]. This directionality has led to the suggestion that the wave might function to regulate a peristaltic contraction wave to control the flow of bile. During development, there are pan-embryonic intercellular waves that sweep around the blastoderm margin in the late gastrula of zebra fish [44]. These spatiotemporal aspects greatly enhance the versatility of Ca^{2+} signalling thus providing the flexibility to regulate so many cellular processes.

6. *Ca^{2+} signalling function*

The function of Ca^{2+} as an intracellular second messenger is carried out by a combination of Ca^{2+} sensors and Ca^{2+} effectors. The major sensors are the EF-hand proteins troponin C, calmodulin (CaM), neuronal calcium sensor proteins and the S100 proteins. These sensors are then responsible for relaying information through a range of effectors such as Ca^{2+}-sensitive K^+ channels, Ca^{2+}-sensitive Cl^- channels, Ca^{2+}–CaM-dependent protein kinases, calcineurin, phosphorylase kinase, myosin light chain kinase and Ca^{2+}-promoted Ras inactivator. The activation of these different effectors is then responsible for stimulating the large number of Ca^{2+}-sensitive cellular processes.

References

1. Berridge, M.J. (1993) *Nature* 361, 315–325.
2. Clapham, D.E. (1995) *Cell* 80, 259–268.
3. Berridge, M.J., Lipp, P. and Bootman, M.D. (2000) *Nat. Rev. Mol. Cell Biol.* 1, 11–21.
4. Berridge, M.J., Bootman, M.D. and Roderick, H.L. (2003) *Nat. Rev. Mol. Cell Biol.* 4, 517–529.
5. De Crescenzo, V., ZhuGe, R., Velázquez-Marrero, C., Lifshitz, L.M., Custer, E., Carmichael, J., Lai, F.A., Tuft, R.A., Fogarty, K.E., Lemos, J.R. and Walsh, J.V. (2004) *J. Neurosci.* 24, 1226–1235.
6. Galione, A. and White, A. (1994) *Trends Cell Biol.* 4, 431–436.
7. Patel, S., Churchill, G.C. and Galione, A. (2001) *Trends Biochem. Sci.* 26, 482–489.
8. Lee, H.C. (2003) *Curr. Biol.* 13, R186–R188.
9. Putney, J.W. (1986) *Cell Calcium* 7, 1–12.
10. Hoth, M. and Penner, R. (1992) *Nature (Lond)* 355, 353–356.
11. Berridge, M.J. (1995) *Biochem. J.* 312, 1–11.
12. Feske, S., Gwack, Y., Prakriya, M., Srikanth, S., Puppel, S.-H., Tanasa, B., Hogan, P.G., Lewis, R.S., Daly, M. and Rao, A. (2006) *Nature (Lond)* 441, 179–185.

13. Pan, Z., Yang, D., Nagaraj, R.Y., Nosek, T.A., Nishi, M., Takeshima, H., Cheng, H. and Ma, J. (2002) *Nat. Cell Biol.* 4, 379–383.
14. Roos, J., DiGregorio, P.J., Yeromin, A.V., Ohlsen, K., Lioudyno, M., Zhang, S., Safrina, O., Kozak, J.A., Wagner, S.L., Cahalan, M.D., Veliçelebi, G. and Stauderman, K.A. (2005) *J. Cell Biol.* 169, 435–445.
15. Liou, J., Kim, M.L., Heo, W.D., Jones, J.T., Myers, J.W., Ferrell, J.E. and Meyer, T. (2005) *Curr. Biol.* 15, 1235–1241.
16. Forbes, M.S., Rennels, M.L. and Nelson, E. (1979) *J. Ultrastruct. Res.* 67, 325–339.
17. Gabella, G. (1978) *J. Ultrastruct. Res.* 65, 135–147.
18. Nolz, J.C., Gomez, T.S., Zhu, P., Li, S., Medeiros, R.B., Shimizu, Y., Burkhardt, J.K., Freedman, B.D. and Billadeau, D.D. (2006) *Curr. Biol.* 16, 24–34.
19. Zipfel, P.A., Bunnell, S.C., Witherow, D.S., Gu, J.J., Chislock, E.M., Ring, C. and Pendergast, A.M. (2006) *Curr. Biol.* 16, 35–46.
20. Berridge, M.J. (1997) *J. Physiol.* 499, 291–306.
21. Dolmetsch, R.E. Xu, K. and Lewis, R.S. (1998) *Nature (Lond)* 392, 933–936.
22. Li, W.-H., Liopis, J., Whitney, M., Xlokarnik, G. and Tsien, R.Y. (1998) *Nature (Lond)* 392, 936–941.
23. Robb-Gaspers, L.D., Rutter, G.A., Burnett, P., Hajnóczky, G., Denton, R.M. and Thomas, A.P. (1998) *Biochim. Biophys. Acta* 1366, 17–32.
24. Carroll, J., Swann, K., Whittingham, D. and Whitaker, M. (1994) *Development* 120, 3507–3517
25. Miyazaki, S., Yuzaki, M., Nakada, H., Shirakawa, H., Nakanishi, S., Nakade, S. and Mikoshiba, M. (1992) *Science* 257, 251–255.
26. Nakano, Y., Shirakawa, H., Mitsuhashi, N., Kuwabara, Y. and Miyazaki, S. (2001) *Mol. Hum. Reprod.* 3, 1087–1093.
27. Filosa, J.A., Bonev, A.D. and Nelson, M.T. (2004) *Circ. Res.* 95, e73–e81.
28. Nett, W.J., Oloff, S.H. and McCarthy, K.D. (2002) *J. Neurophysiol.* 87, 528–537.
29. Zhang, L. and Sanderson, M.J. (2003) *J. Physiol.* 546, 733–749.
30. Bootman, M.D., Lipp, P. and Berridge, M.J. (2001) *J. Cell Sci.* 114, 2213–2222.
31. Rizzuto, R. and Pozzan, T. (2006) *Physiol. Rev.* 86, 369–408.
32. Wang, S.-Q., Song, L.-S., Lakatta, E.G. and Cheng, H. (2001) *Nature* 410, 592–596.
33. Lumpkin, E.A. and Hudspeth, A.J. (1998) *J. Neurosci.* 18, 6300–6318.
34. Cheng, H., Lederer, W.J. and Cannell, M.B. (1993) *Science* 262, 740–744.
35. Sharma, G. and Vijayaraghavan, S. (2003) *Neuron* 38, 929–939.
36. Conti, R., Tan, Y.P. and Llano, I. (2004) *J. Neurosci.* 24, 6946–6957.
37. Brenner, R., Peréz, G.J., Bonev, A.D., Eckman, D.M., Kosek, J., Wiler, S.W., Patterson, A.J., Nelson, M.T. and Aldrich, R.W. (2000) *Nature (Lond)* 407, 870–876.
38. Brochet, D.X.P., Yang, D., Di Maio, A.D., Lederer, W.J., Franzini-Armstrong, C. and Cheng, H. (2004) *Proc. Natl. Acad. Sci. U. S. A.* 102, 3099–3104.
39. Grosche, J., Matyash, V., Möller, T., Verkhratsky, A., Reichenbach, A. and Kettermann, H. (1999) *Nat. Neurosci.* 2, 139–143.
40. Simkus, C.R.L. and Stricker, C. (2002) *J. Physiol. (Lond)* 545, 521–535.
41. Denk, W., Sugimori, M. and Llinas, R. (1995) *Proc. Natl. Acad. Sci. U. S. A.* 92, 8279–8282.
42. Mackenzie, L., Bootman, M.D., Berridge, M.J. and Lipp, P. (2001) *J. Physiol.* 530, 417–429.
43. Robb-Gaspers, L.D. and Thomas, A.P. (1995) *J. Biol. Chem.* 270, 8102–8107.
44. Webb, S.E. and Miller, A.L. (2003) *Nat. Rev. Mol. Cell Biol.* 4, 539–551.

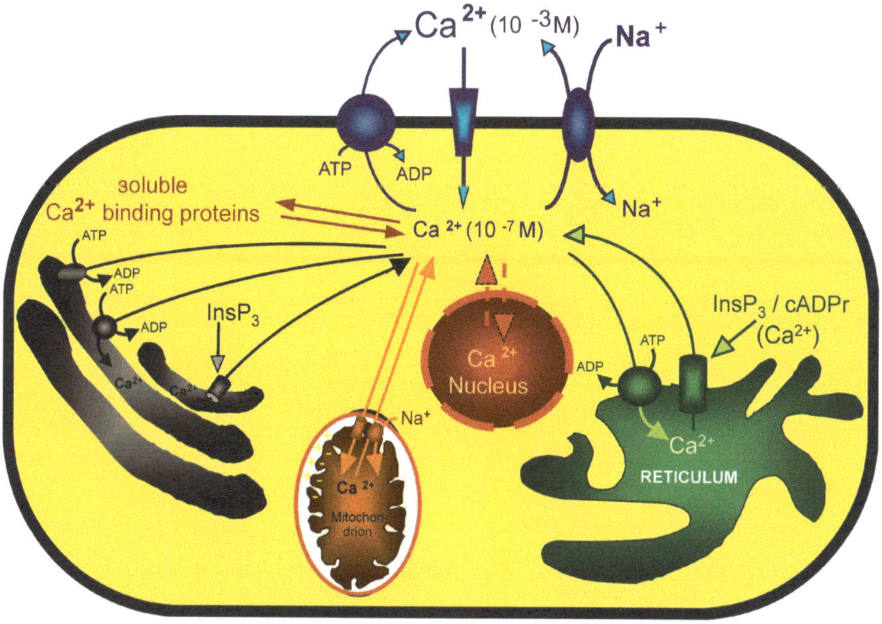

Color Plate 1. A global view of cellular Ca^{2+} homeostasis. The figure shows the systems that transport Ca^{2+} across the plasma membrane and the membranes of the organelles. The transport of Ca^{2+} in and out of the nucleus is represented by dashed arrows to indicate the present controversy over whether the traffic of Ca^{2+} across the nuclear envelope occurs continuously and passively through the pores or whether the pores are somehow gated. For simplicity, only one Ca^{2+} channel type is indicated in the plasma membrane. Ca^{2+} enters the reticulum and the Golgi through ATPases: the SERCA pump in the reticulum, and the SERCA pump plus the SPCA pump in the Golgi. It exits from these two compartments through channels activated by ligands: InsP3 and cADPr in the reticulum, InsP3 in the Golgi. The novel Ca^{2+}-linked messenger NAADP is not indicated in the figure as the membrane system on which it acts has net yet been conclusively identified. Ca^{2+} enters mitochondria through the electrophoretic uniporter, which is energized by the negative inside membrane potential maintained by the respiratory chain. It leaves them through exchangers, of which the most important and best characterized is a NCX. The figure also shows the two systems that export Ca^{2+} from cells, a high-affinity low-capacity PMCA pump and a low-affinity large-capacity NCX. Finally, it shows soluble proteins that buffer Ca^{2+} and may also decode its message (the Ca^{2+} sensor proteins) (See Figure 2, p. 6).

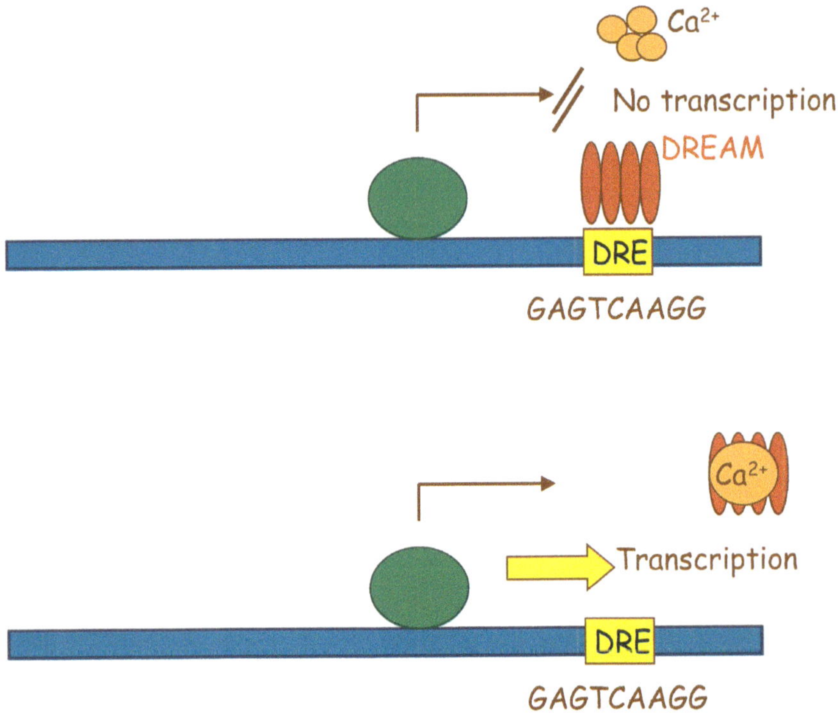

Color Plate 2. Effect of the downstream regulatory element antagonistic modulator (DREAM) on gene transcription. The DREAM tetramer binds to DNA sites (DRE sites GAGT) ideally located between the TATA box and the TS site. Each DREAM monomer contains 4 EF hands, of which only three are operational (See Figure 4, p. 11).

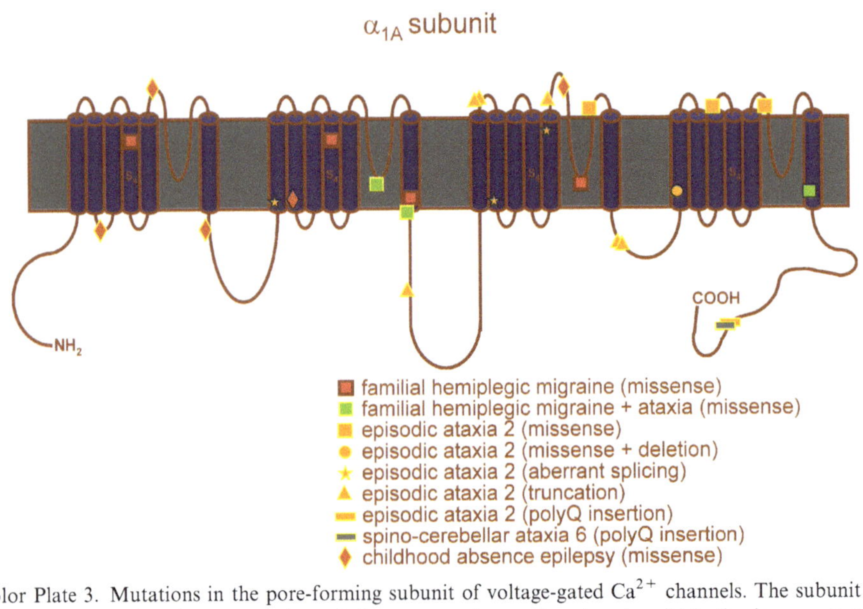

Color Plate 3. Mutations in the pore-forming subunit of voltage-gated Ca^{2+} channels. The subunit is a fourfold repeat of regions consisting of six transmembrane domains, in which the loop connecting transmembrane domains 5 and 6 on the external side of the membrane fold in to form the channel. The figure shows the disease phenotypes described so far. Adapted from Ref. 56 (See Figure 5, p. 18).

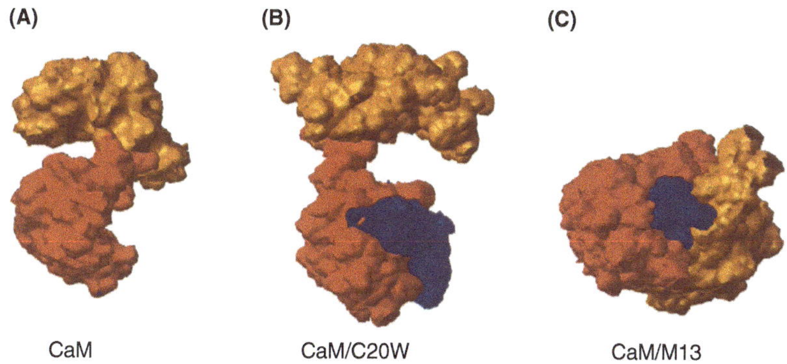

Color Plate 4. Surface representation of calcium-bound calmodulin (CaM) and its different binding modes to peptides. The N-terminal half of CaM is in orange, the C-terminal half is in red, and the peptides are in blue. The orientation of the C-terminal half of CaM is the same in all cases. **(A)** Crystal structure of calcium-bound CaM. **(B)** Solution structure of the CaM/C20W complex showing that the peptide C20W, corresponding to the N-terminal portion of the CaM-binding domain of the plasma membrane Ca^{2+}-pump, but lacking the C-terminal hydrophobic anchor residue, binds only to the C-terminal half of CaM. **(C)** The solution structure of CaM/M13 shows a compact globular complex involving both C- and N-terminal halves of CaM in binding the peptide M13 corresponding to the CaM-binding domain of MLCK. (Reprinted with permission from Elshorst et al. [20]) (See Figure 2, p. 55).

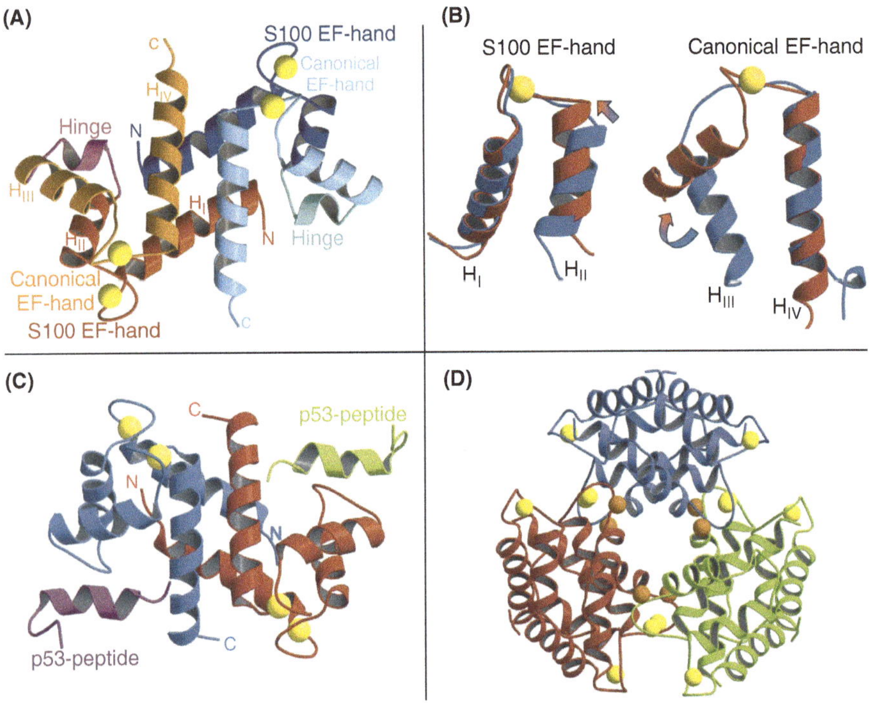

Color Plate 5. Three-dimensional structures of S100 proteins. **(A)** Schematic representation of human Ca^{2+}-loaded S100A6 showing the homodimer and bound Ca^{2+} ions (yellow). (PDB code 1K96 [171]). **(B)** Conformational change in EF-hands of S100 proteins. The S100-specific EF-hand is depicted in (left), the canonical EF-hand in (right). The Ca^{2+}-free protein is shown in blue, the Ca^{2+}-loaded form in red. Ca^{2+} ions are shown as yellow spheres. The coordinates were taken from the crystal structures of human S100A6 (PDB codes 1K8U, 1K96; [171]). **(C)** Target binding to S100 proteins. Complex of Ca^{2+}-loaded human S100B dimer with two peptides derived from p53 (PDB code 1DT7; [172]). **(D)** Hexameric structure of Ca^{2+}-loaded S100A12. Three S100A12 dimers (shown in green, blue and magenta) form a hexamer. The Ca^{2+} ions bound to the EF-hands are shown as bright yellow spheres. At the hexamer-forming interface, six additional Ca^{2+} ions are located, (orange spheres) [173]. (Reproduced from the work of Fritz and Heizmann [162], by permission of John Wiley & Sons) (See Figure 3, p. 70).

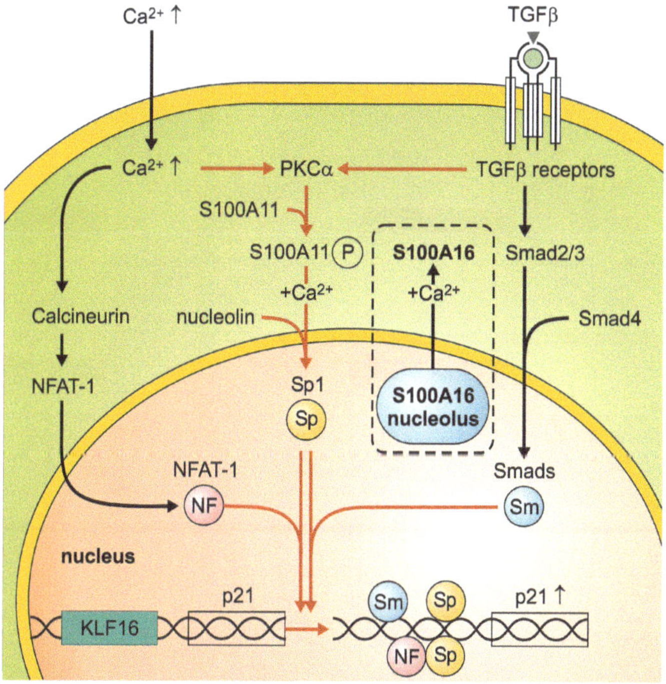

Color Plate 6. S100A11-mediated pathways in human keratinocytes. By exposure of cells to high Ca^{2+} or TGFβ, S100A11 is transferred to the nucleus, where it induces p21 through activation of Sp1/Sp3 (for details, see Ref. 199). In other cells, we detected a Ca^{2+}-dependent translocation of S100A16 from the nucleus (nucleolus) to the cytosol, appositely to S100A11 [186] (See Figure 4, p. 73).

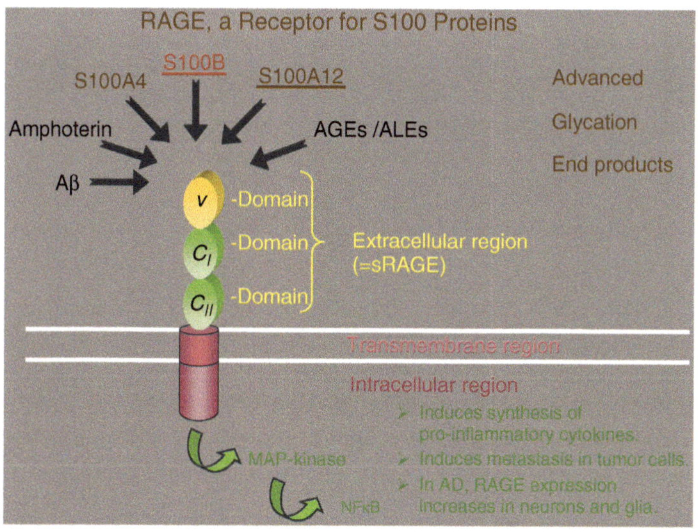

Color Plate 7. S100–RAGE signalling. RAGE is a multifunctional receptor of the immunoglobulin family binding a variety of structurally and functionally unrelated ligands. RAGE contains an extracellular domain consisting of 3Ig domains (one V- and two C-types), a transmembrane spanning region and a cytosolic tail. RAGE triggers the activation of key signalling pathways and is involved in several human diseases [167,214] (See Figure 5, p. 75).

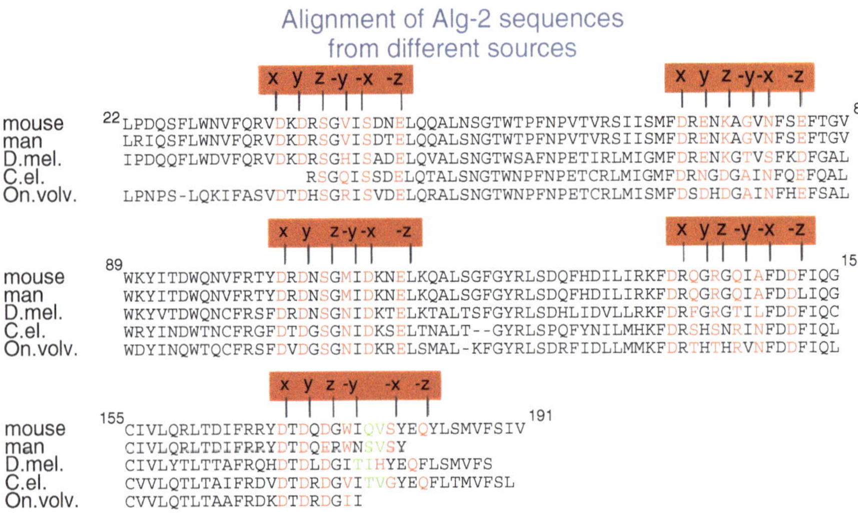

Color Plate 8. Alignment of the amino acid sequences of the penta EF-hand Ca^{2+}-binding protein ALG-2 from different organisms. The conserved amino acid residues of the five EF-hand Ca^{2+}-binding loops are indicated in red. In the fifth EF-hand domain, two amino acid residues are inserted (shown in green) between the positions –y and –x. (Reprinted from Krebs 1998, *BioMetals* 11, 375–382 with kind permission of Springer Science and Business Media) (See Figure 6, p. 79).

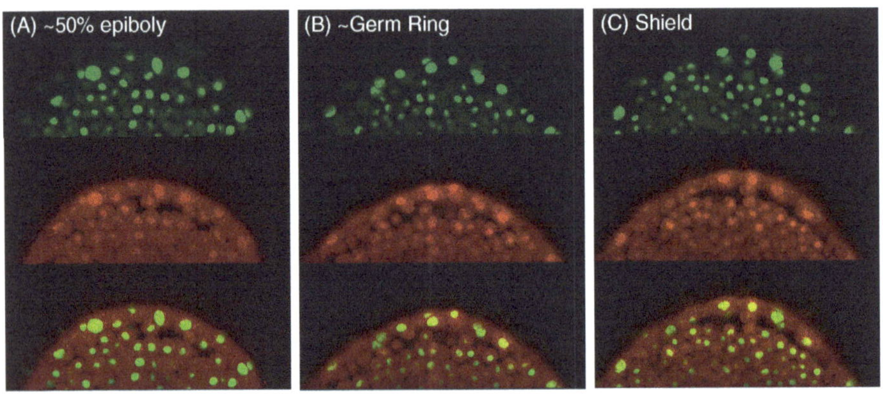

Color Plate 9. Co-localization of ALG-2 and RBM22 in the nucleus. Zebrafish embryos have been co-injected with RBM22-EGFP RNA (green fluorescent) and ALG2-mRFP RNA (red fluorescent) at the 1-cell stage. The figure shows a single confocal section at **(A)** 50% epiboly, **(B)** Germ Ring and **(C)** Shield stage of embryonal development. Merging the green and red fluorescent images provides evidence for the co-localization of the two proteins within the nucleus. For details, see Section 5.1.2.6 and Ref. 249 (See Figure 7, p. 80).

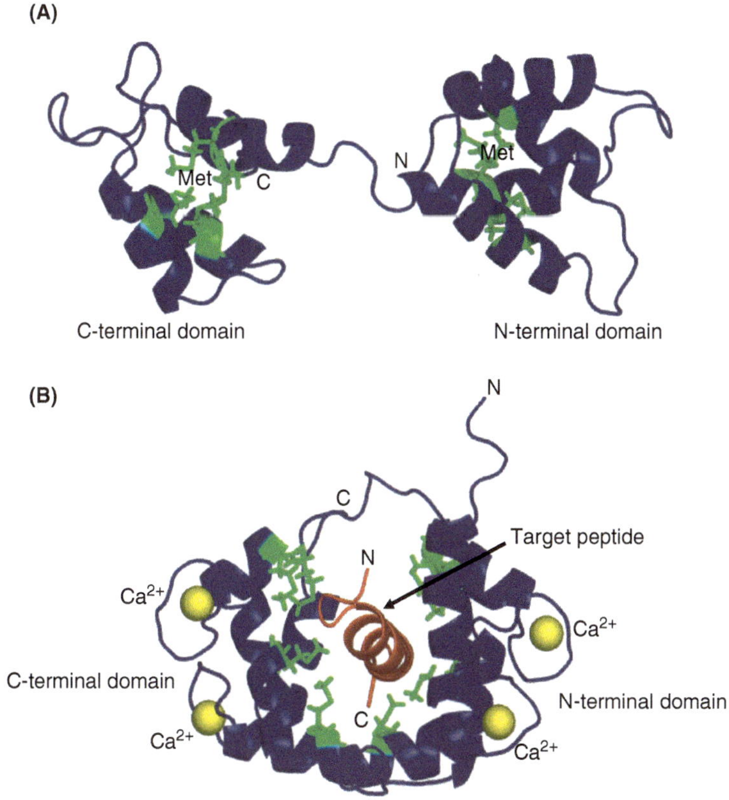

Color Plate 10. The effects of Ca^{2+} and target binding on calmodulin (CaM) structure. **(A)** Solution structure of Ca^{2+}-depleted CaM (1DMO.pdb). The blue ribbons show the backbone secondary structural elements of CaM. The green stick representations show the key Met residues (Met36, 51, 71, 72, 109, 124, 144, and 145) involved in "fine-tuning" specific target recognition for CaM. The helices making up each respective EF-hand in the N- and C-terminal domains are in a "closed" antiparallel conformation. The overall shape of apo CaM is dumbbell-like. **(B)** Solution structure of Ca^{2+}-loaded CaM complexed with a peptide from myosin light chain kinase (2BBN.pdb). Upon liganding of Ca^{2+} (yellow spheres), the inter-helical angles in each EF-hand "open" and the central helix region become much more flexible. The target peptide (red ribbon) is accommodated by a drastic shape change in CaM. N and C represent NH_2 and COOH termini, respectively (See Figure 1, p. 98).

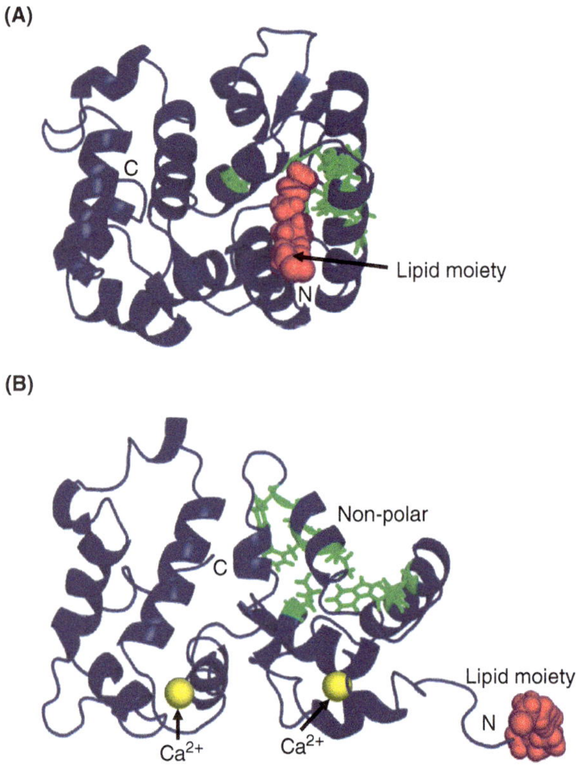

Color Plate 11. The effects of Ca^{2+} on the structure of myristoylated recoverin. **(A)** Solution structure of Ca^{2+}-depleted recoverin (1REC.pdb). The blue ribbons show the backbone secondary structural elements of recoverin. The green stick representation show conserved hydrophobic residues (3 × Phe, Tyr, Leu, Trp) providing the main scaffolding for the myristoyl moiety (magenta spheres). **(B)** Solution structure of Ca^{2+}-loaded recoverin (1JSA.pdb). Specific residues on the C-terminal portion of recoverin provide the "fine-tuning" for target specificity in the neuronal Ca^{2+} sensor (NCS) proteins. In the case of recoverin, Ca^{2+} induces a large conformational change which results in the exposure of the hydrophobics and release of the myristoyl group. Interaction of the myristoyl moiety with lipid membranes directs recoverin for the regulation of membrane-bound proteins. N and C represent NH_2 and COOH termini, respectively (See Figure 2, p. 100).

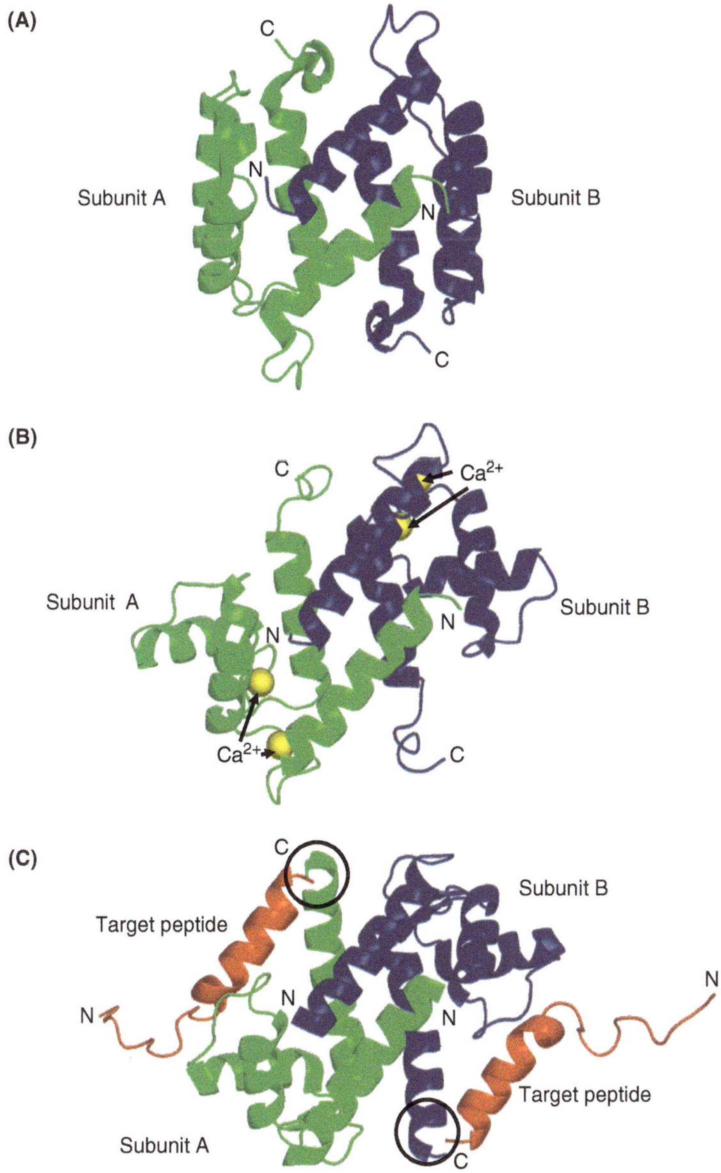

Color Plate 12. The effects of Ca^{2+} on the structure and target recognition of S100B. **(A)** Solution structure of homodimeric and Ca^{2+}-depleted S100B (1B4C.pdb). The blue ribbons represent the secondary structural elements defined by the backbone conformation of one subunit (subunit B), while the green ribbon follows the conformation of the dimeric subunits (subunit A). In total, the dimer is made up of four EF-hands where each apo motif is in the "closed" conformation. **(B)** Solution structure of Ca^{2+}-loaded S100B (1QLK.pdb). Each S100B dimer is capable of coordinating four Ca^{2+} ions (yellow spheres). Upon Ca^{2+}-loading, S100B undergoes a large conformational change with EF-hands moving to a relatively "open" conformation. **(C)** Solution structure of Ca^{2+}-loaded S100B complexed with a peptide of the N-terminal domain of NDR kinase (1PSB.pdb). Residues near the C-terminus of each subunit form an extended helical conformation upon Ca^{2+} loading (open circle), thereby accommodating and promoting target interactions. The target protein also undergoes a conformational change upon forming an S100–target protein complex. N and C represent NH_2 and COOH termini, respectively (See Figure 3, p. 104).

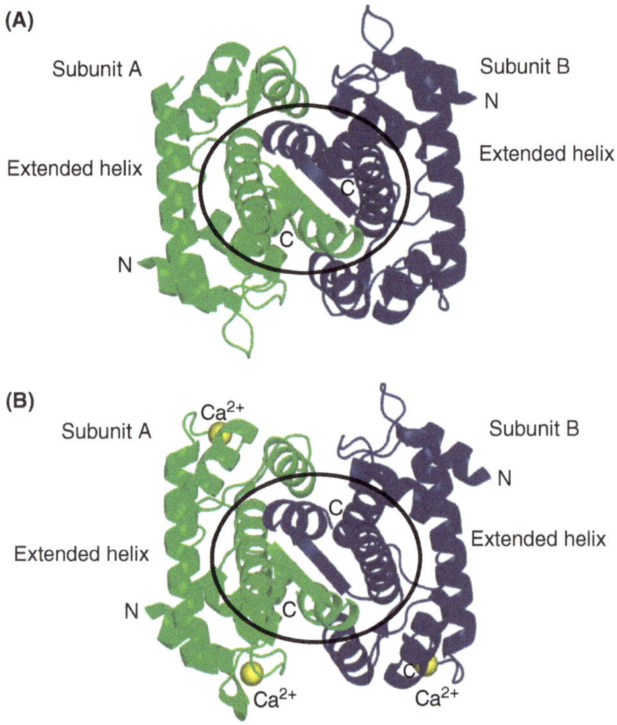

Color Plate 13. The effects of Ca^{2+} on the structure of grancalcin. **(A)** Crystal structure of N-terminally truncated (52 residues) grancalcin in the Ca^{2+}-depleted state (1F4Q.pdf). The penta-EF-hand proteins pair all five EF-hands through dimerization. The green ribbons delineate the secondary structure of one grancalcin subunit, while the blue ribbon shows the structure of the homodimerization partner. A short antiparallel β-sheet is formed from the loop region of each EF-hand motif contributed from each subunit. The open circle shows that the interface of grancalcin is made up primarily from the fifth EF-hand of each subunit. A long, "extended helix" makes up the F and E helices of two consecutive EF-hand motifs within each subunit. **(B)** Crystal structure of N-terminally truncated (52 residues) grancalcin in the Ca^{2+}-loaded state (1K94.pdb). Upon Ca^{2+} loading (yellow spheres), only minor structural changes are observed in most penta-EF-hand proteins. Nonetheless, the conformational change propagates through the entire protein to the common Gly/Pro-rich N-terminus; target interactions have also been identified at other specific locations. N and C represent NH_2 and COOH termini, respectively (See Figure 4, p. 106).

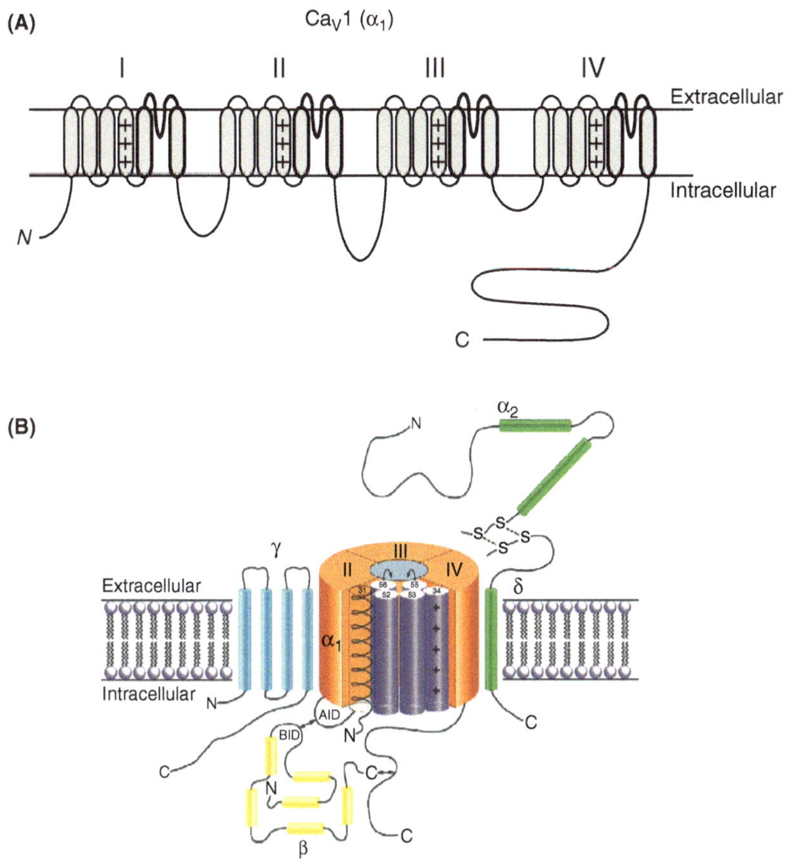

Color Plate 14. (**A**) Predicted topology of the α_1 subunit. Each one of four repeats contains six putative membrane-spanning helices (S1 through S6). Helices S4 contain positively charged amino acids critical for channel activation–deactivation gating. The P-loop (composed of S5–S6) involved in Ca^{2+} selectivity and permeation is indicated in bold. (**B**) Schematic representation of multi-heteromeric voltage-gated Ca^{2+} channels. Disulfide bonds between the α_2 and the δ subunits are indicated. AID, α_1-interacting domain, which interacts with the auxiliary subunit; BID, β-interacting domain (Reprinted with permission from Randall and Benham [13]) (See Figure 2, p. 129).

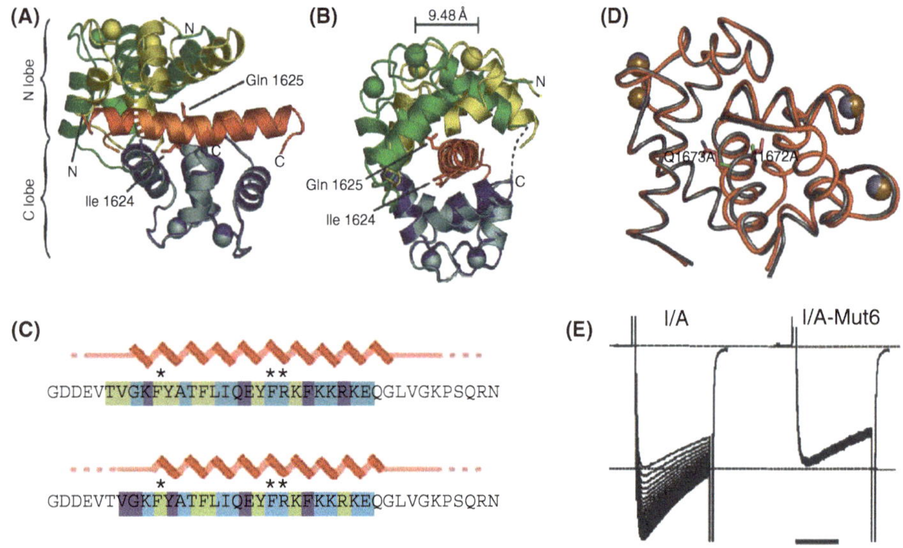

Color Plate 15. High-resolution structures of Ca^{2+}/calmodulin (CaM) binding to IQ domain. (**A**) Ribbon diagram of the complex. Green, CaM Ca^{2+}/N lobe; blue, Ca^{2+}/C lobe; red, IQ domain, with the IQ residues in stick representation. Darker shades, complex A; lighter shades, complex C. (**B**) Right angle rotation of (A). (**C**) Amino acid sequence put into the context of the three-dimensional structure. Zigzag represents α-helical regions; straight line, non-α-helical residues. IQ peptide residues contacting CaM (4 Å) are subdivided into those interacting with the Ca^{2+}/N lobe (green), Ca^{2+}/C lobe (cyan), and both lobes (purple) [82]. (**D**) A very similar structure [83] demonstrates that replacement of IQ with AA, which abolishes Ca^{2+}-dependent inactivation and enhances calcium-dependent facilitation (CDF), does not significantly alter the structure of the Ca^{2+}/CaM-IQ complex. (**E**) CDF, measured with Ca2+ ions as charge carrier in *Xenopus* oocytes, shows prominent facilitation when the isoleucine of the IQ motif is replaced by an alanine (I/A, left), but not with the additional alanine replacement of six residues upstream of the IQ motif, TVGKFY (mut6), which were implicated in interaction of Ca^{2+}/CaM-dependent protein kinase II with the $Ca_V1.2$ C-terminal domain [84]. Similar abolition of CDF was independently obtained by alanine replacement of three amino acids indicated by the asterisks [82] (See Figure 4, p. 135).

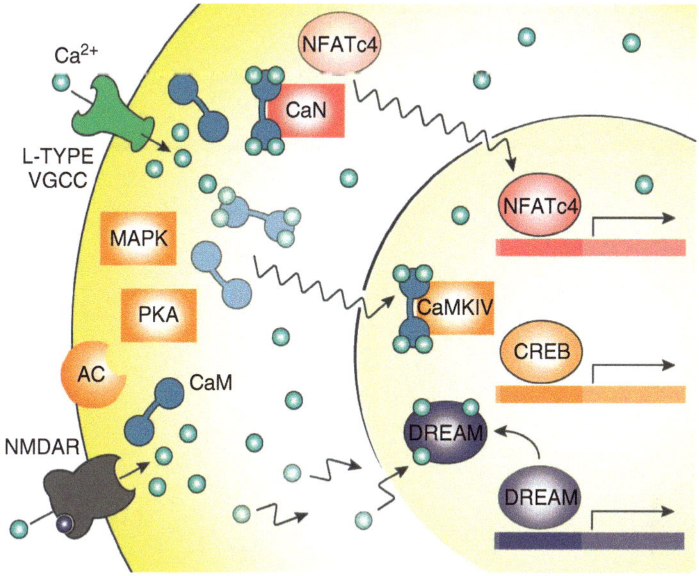

Color Plate 16. A schematic representation of the early stages (0–5 min) of activity-dependent gene expression. Following rises in intracellular Ca^{2+}, DREAM dissociates from DNA resulting in the lifting of transcriptional repression. Within seconds of Ca^{2+} entry through L-type Ca^{2+} channels and NMDA receptors, calmodulin (CaM) translocates to the nucleus, supporting CREB phosphorylation through activation of CaMKIV. Nearly as rapid, NFATc4 also undergoes translocation to the nucleus following its dephosphorylation by calcineurin (CaN). Other pathways not shown here include the MAPK/PKA pathways, which exert their influence over a longer time scale. Here, VGCC denotes voltage-gated Ca^{2+} channels, principally L-type. (Reprinted with permission from Deisseroth et al. [151]) (See Figure 6, p. 142).

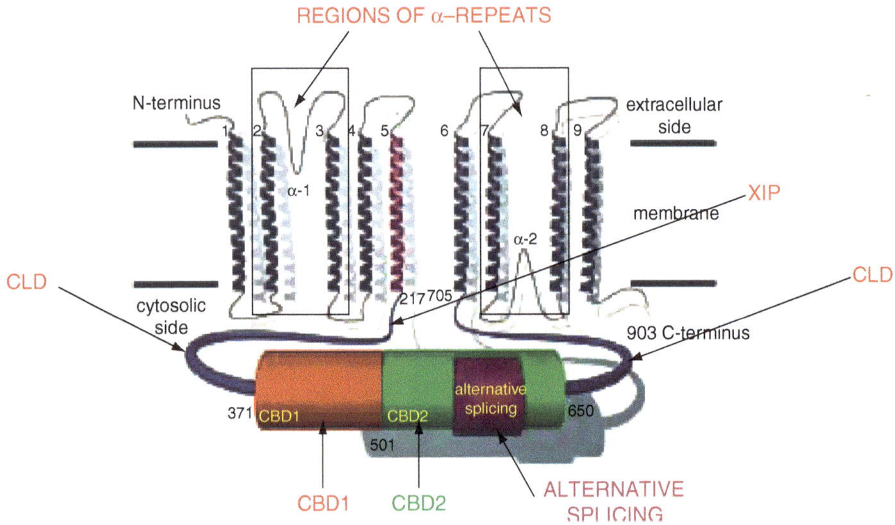

Color Plate 17. Topology model of NCX. The different regions of interest are indicated in the figure. CBD, calcium-binding domain; CLD, catenin-like domain; and XIP, inhibitory peptide of the exchanger. The figure has been adopted from Fig. 1 of Ref. [47] with permission from Elsevier (See Figure 1, p. 168).

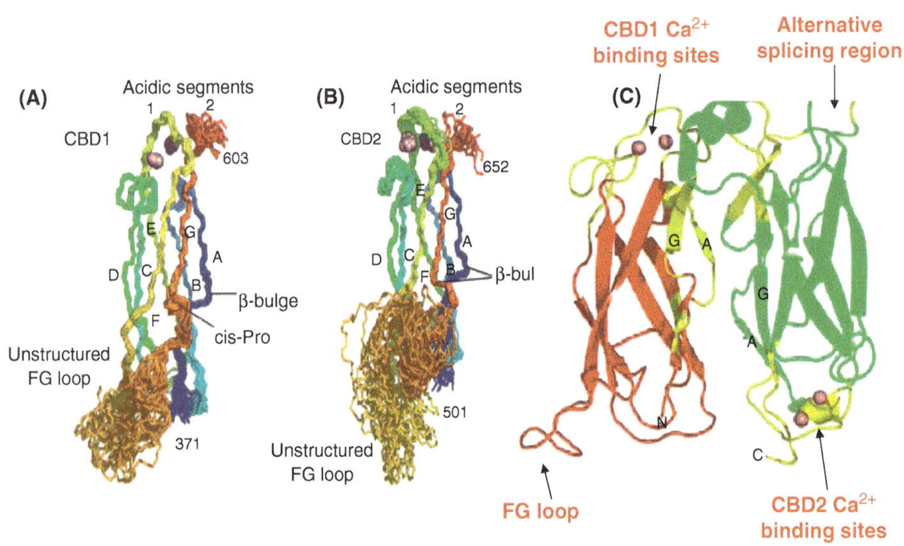

Color Plate 18. Structures of the Ca^{2+}-binding domains CBD1 (**A**) and CBD2 (**B**) obtained by an overlay of 20 nuclear magnetic resonance (NMR)-derived structures. (**C**) A possible interaction between CBD1 and CBD2. Regions of interest as discussed in the text are indicated in the figure. CBD, calcium-binding domain. The figure was adopted from Figs 2 and 5 of Ref. [47] with permission from Elsevier (See Figure 2, p. 171).

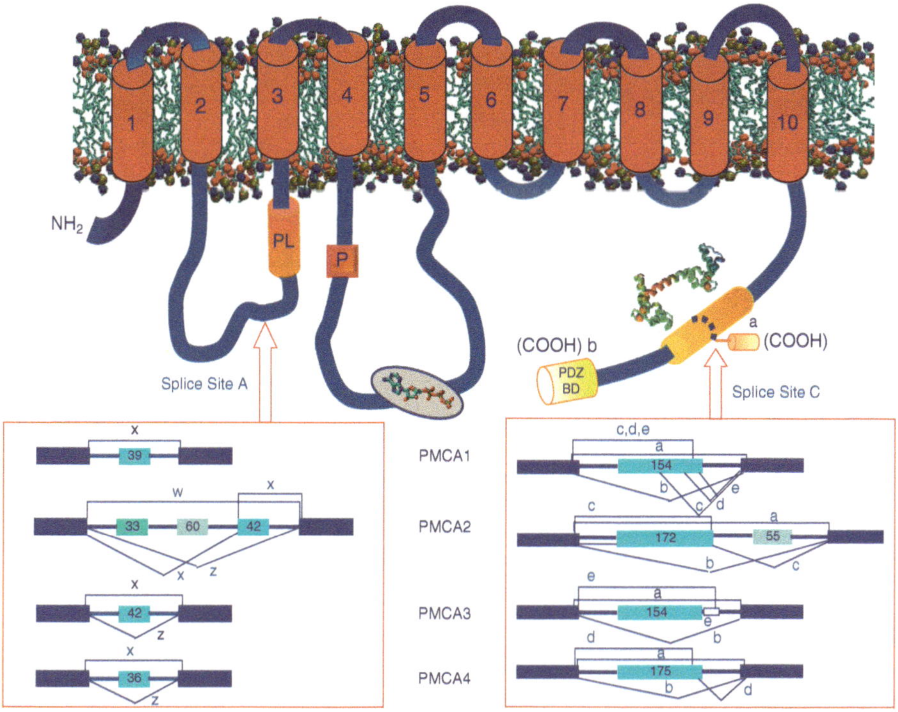

Color Plate 19. Membrane topology domains and splicing variants of the pump. The 10 putative transmembrane regions are numbered and indicated by red boxes. The acidic phospholipid (PL)-sensitive region preceding transmembrane domain 3 and the C-terminal calmodulin-binding domain (next to the 3D structure of calmodulin) is shown as orange cylinders. The P in the red square indicates the site of aspartyl-phosphate formation during pump function. The ATP-binding domain is indicated by the gray ellipse with the ATP structure inside. The putative C-terminal PDZ-binding domain (PDZ-BD) is shown as a shaded yellow box at the extreme C-terminus. The sites of alternative RNA splicing are indicated under the topology structure and are labelled as sites A and C. The exon structure of the region affected by alternative splicing is shown for each of the four plasma membrane Ca^{2+}-ATPase (PMCA) genes. Constitutively spliced exons are shown as dark blue boxes and alternatively inserted exons in light blue. The sizes of alternatively spliced exons are given as nucleotide numbers; the splicing options are indicated by connecting lines (See Figure 1, p. 187).

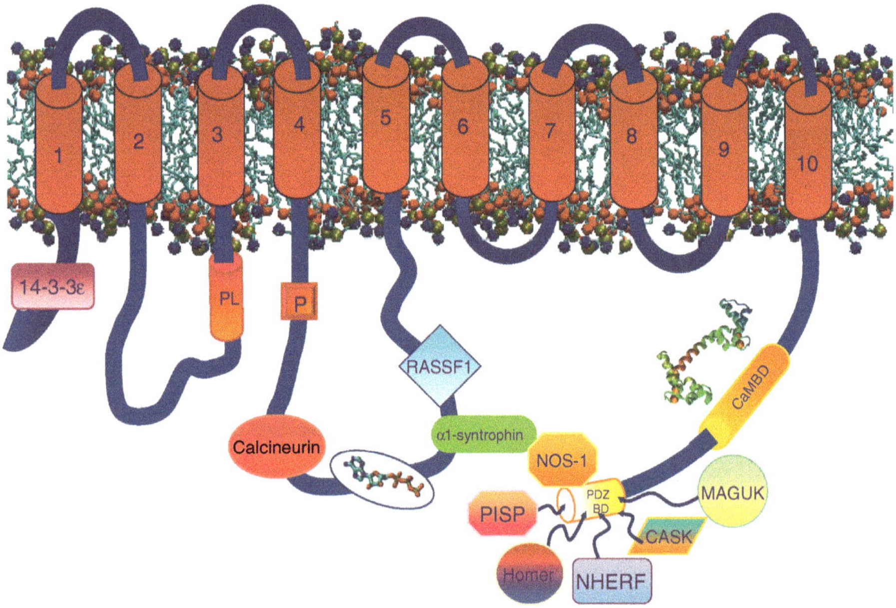

Color Plate 20. Interactors of plasma membrane Ca^{2+}-ATPase (PMCA) pumps. The cartoon shows the approximate sites of interaction of the PMCA isoforms with ligands (drawn in coloured boxes). The essential domains of the pump are designated as in the legend for Fig. 1. The 14-3-3ε protein binds to the N-terminal domain of PMCA (first 90 residues) [84]. Calcineurin binds to the amino acid region 501–575 In the main intracellular loop, the region is close to the ATP-binding domain [83], Ras-associated factor 1 (RASSF1) and α1-syntrophin interact with residues 652–840 [81,82]. As shown, most ligands interact with the C-terminal region of the pumps, interacting with the PDZ-binding domain (PDZ-BD) at the end of the C-terminal tail [nitric oxide synthase-1 (NOS-1), PMCA-interacting single-PDZ protein (PISP), Na^+/H^+ exchanger regulatory factor (NHERF), Homer, Ca^{2+}-/calmodulin-dependent serine protein kinase (CASK) and membrane-associated guanylate-kinase family (MAGUK) family] [79,81,85–88]. Note how syntrophin binds to NOS-1 to form a ternary complex with the pump [81] (See Figure 2, p. 190).

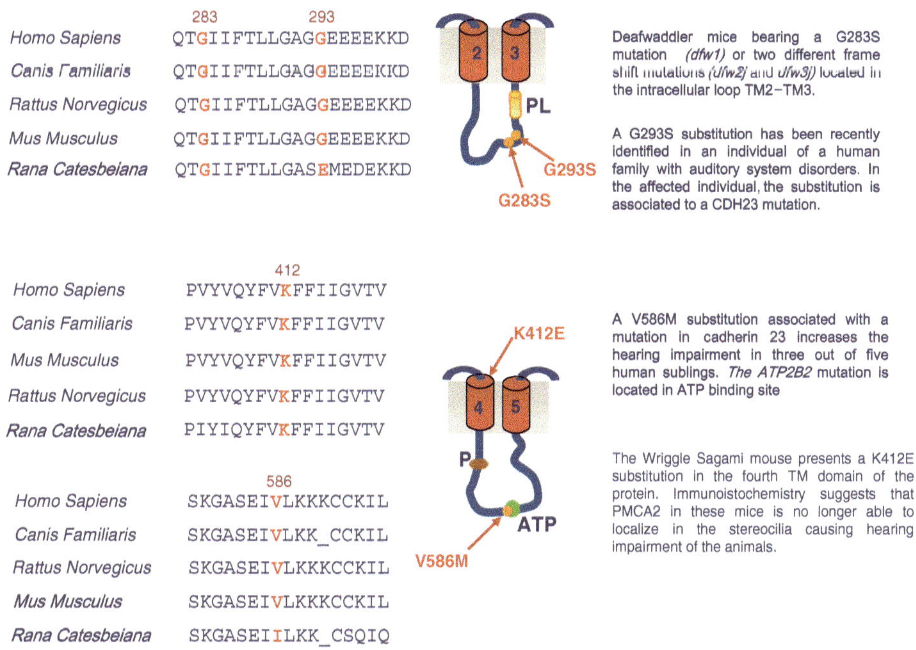

Color Plate 21. Deafness-associated mutations of plasma membrane Ca^{2+}-ATPase 2 (PMCA2). Sequence alignment showing the mutation sites in the PMCA2 genes known so far. They are given on the left in the local sequences for different species. Amino acid substitutions are highlighted in red, and the number of the residues given. The localization of the mutations is marked by arrows in the topography structures, and a summarized description of the related phenotypes is given at the side (See Figure 3, p. 192).

(A)

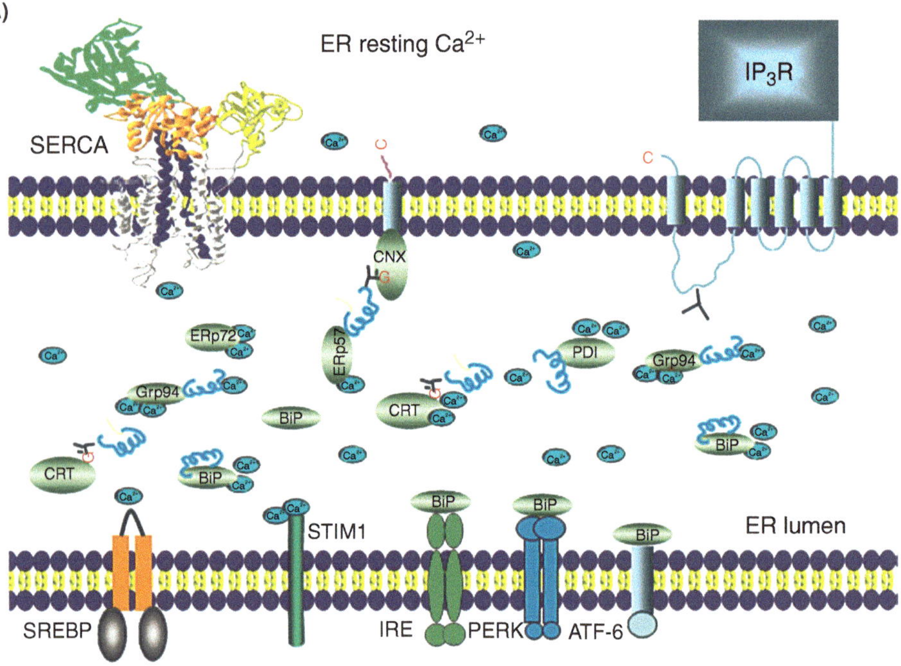

Color Plate 22. **(A)** Ca^{2+} dynamics in the lumen of the endoplasmic reticulum (ER). This diagram shows a model of events occurring in response to agonist-activated Ca^{2+} fluctuations in the lumen of the ER. These events affect numerous cellular functions. **A** represents ER Ca^{2+} resting conditions. Under these conditions, there are no major changes in the ER luminal Ca^{2+} concentration (See Figure 1(A), p. 210).

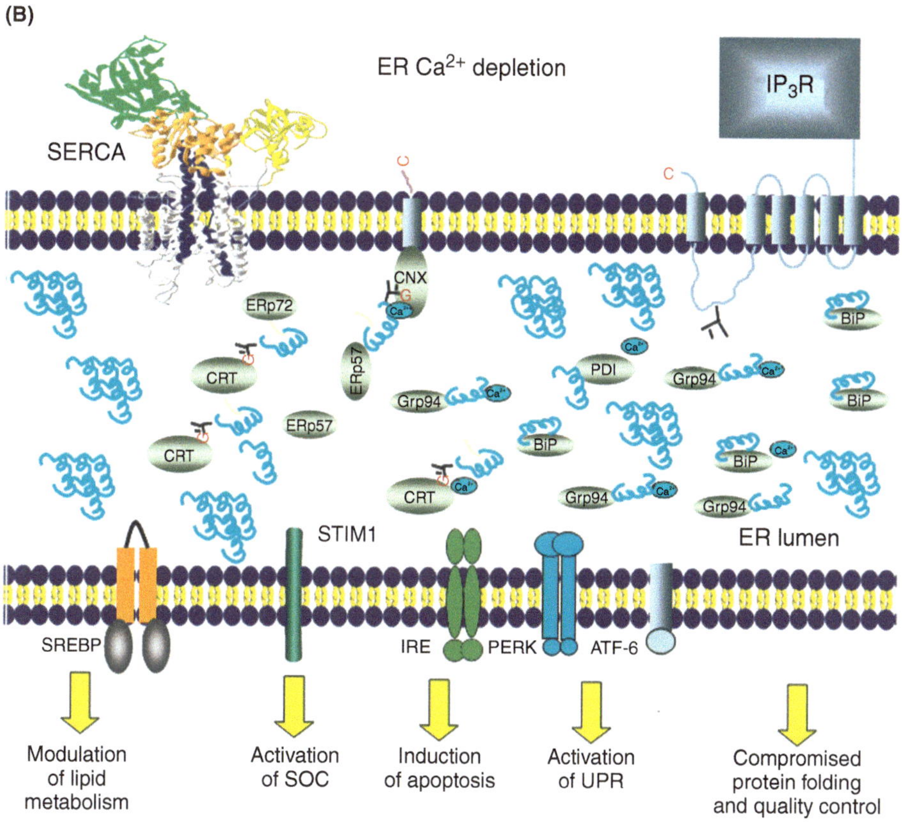

Color Plate 23. **(B)** represents ER Ca^{2+} depletion conditions, where ER luminal Ca^{2+} is below 20 μM. Under these conditions, there is activation of store-operated Ca^{2+} channel (SOC) through Ca^{2+} sensing by stromal-interacting molecule (STIM). Prolonged ER Ca^{2+} depletion leads to activation of unfolded protein response (UPR), accumulation of unfolded proteins, activation of dsRNA-activated protein kinase-like ER kinase (PERK), inositol-requiring kinase (IRE1), and activating transcription factor-6 (ATF-6), apoptosis, and cell death. Function of sterol regulatory element-binding protein (SREBP) may also be affected under ER Ca^{2+} depletion. Sustained increase in ER luminal Ca^{2+} (See Figure 1(B), p. 211).

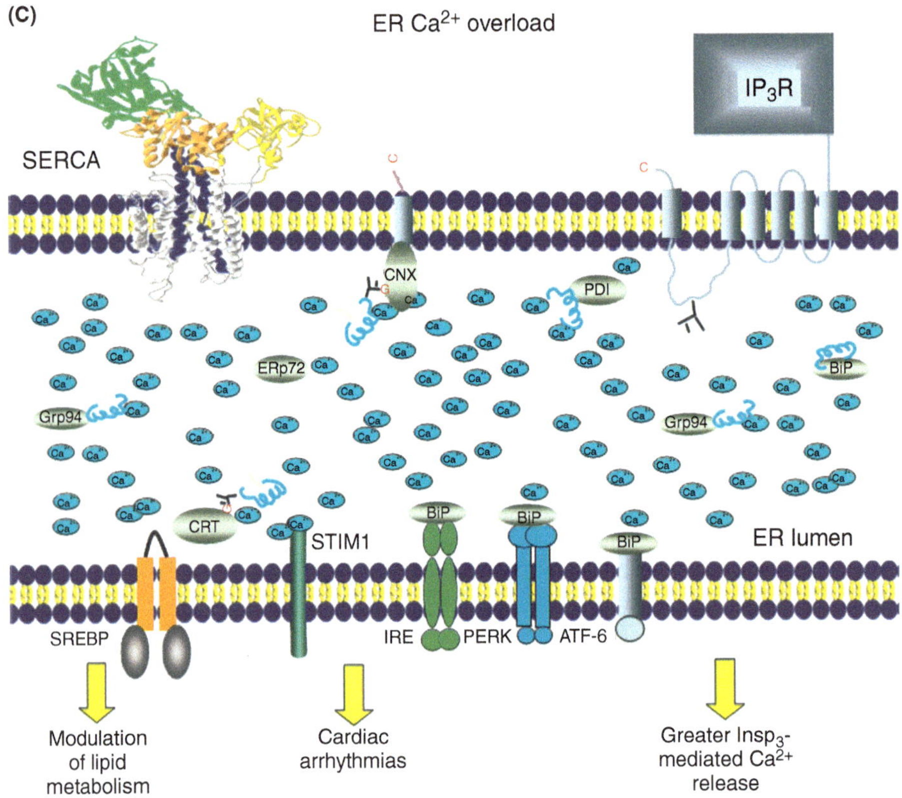

Color Plate 24. (**C**) leads to congenital problems including cardiac arrhythmias. It also affects inositol 1,4,5-trisphosphate ($InsP_3$) receptor function and may impact on lipid metabolism. CNX, calnexin; CRT, calreticulin; PDI, protein disulphide isomerase; and SERCA, sarcoplasmic/ER Ca^{2+}-ATPase (See Figure 1(C), p. 212).

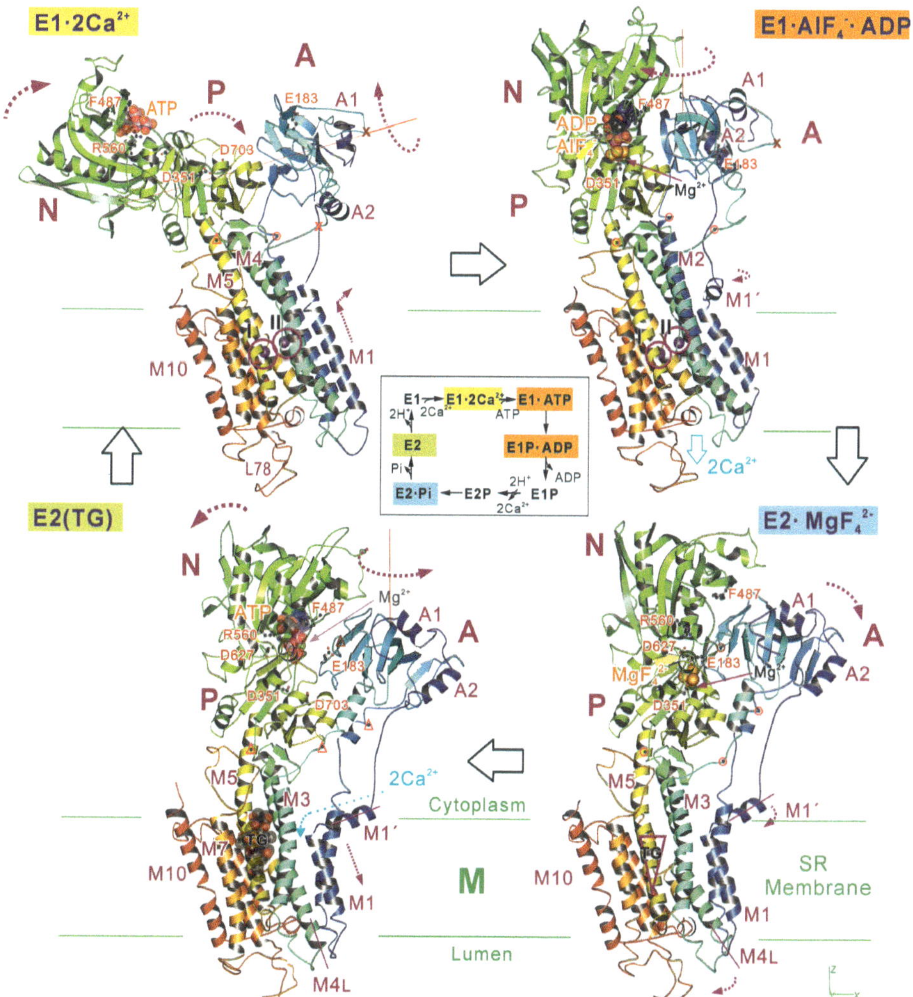

Color Plate 25. Front views [parallel to the membrane (x–y) plane] of Ca^{2+}-ATPase in four different states and a simplified reaction scheme (showing only the forward direction), in which different colours correspond to the respective structures presented here. Colours change gradually from the amino terminus (blue) to the carboxy terminus (red). Purple spheres (numbered and circled) represent bound Ca^{2+}. Three cytoplasmic domains (A, N and P), the α-helices in the A-domain (A1 and A2) and those in the transmembrane domain (M1, M4L, M5 and M10) are indicated. M1′ is an amphipathic part of the M1 helix lying on the bilayer surface. Digestion sites are shown in small circles (protected), triangles (partially protected) and crosses (not protected) with proteinase K (orange) and trypsin (T2 site, brown). Docked ATP and thapsigargin (TG) are shown in transparent space fill. Several key residues – E183 (A), F487 and R560 (N, ATP binding), D351 (phosphorylation site), D627 and D703 (P) – are shown in ball-and-stick representation. PDB accession codes are 1SU4 (E1·$2Ca^{2+}$), 1WPE (E1·AlF_x·ADP), 1WPG (E2·$MgF_4{}^{2-}$) and 1IWO (E2(TG)). Atomic co-ordinates of the aligned models are available at the author's web site (http://www.iam.u-tokyo.ac.jp/StrBiol/resource/res.html) (See Figure 1, p. 220).

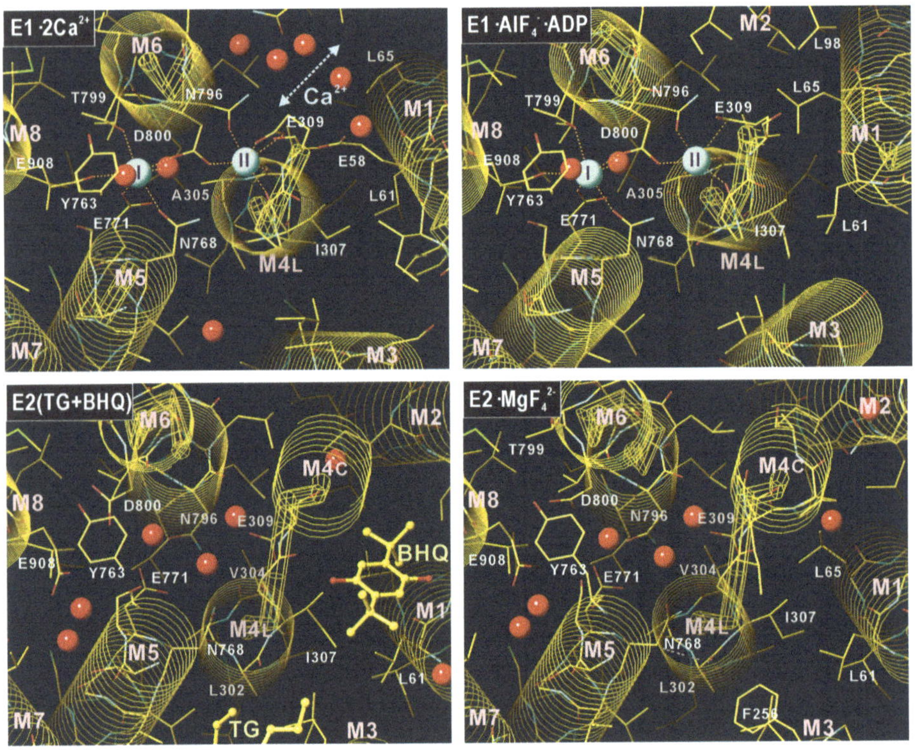

Color Plate 26. Details of the transmembrane Ca^{2+}-binding sites in four different states viewed approximately normal to the membrane. Ca^{2+} appears as cyan spheres and water molecules as red spheres. Note that site II Ca^{2+} is exchangeable with those in the cytoplasm in E1·2Ca^{2+} by conformation change of the E309 side chain but not in E1·AlF_4^-·ADP, as the space is occupied by L65 (M1) and other hydrophobic side chains. Two inhibitors, TG and 2,5-di-*tert*-butyl-1,4-dihydroxybenzene (BHQ), are shown in ball-and-stick in E2(TG + BHQ) (See Figure 2, p. 223).

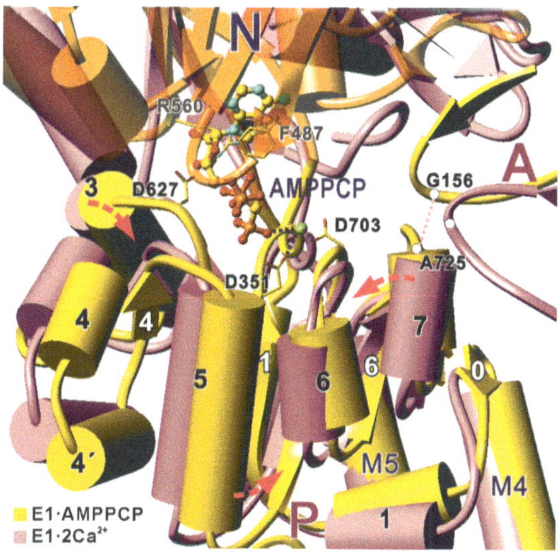

Color Plate 27. Structural changes caused by the binding of ATP and Mg^{2+} to the P-domain. AMPPCP, a non-hydrolysable analogue of ATP, is used here; the bound metal in the crystal structure (small green sphere) is most likely Ca^{2+} rather than Mg^{2+}. Two crystal structures are fitted with the 10 residues at the N-terminal end of the M5 helix. The N-domain in E1·AMPPCP is shown transparent. Note that the P-domain is bent (arrows in red broken lines) by co-ordination of the metal by the γ-phosphate, D351 and D703 (shown in stick model) and that the A-domain tilts because of the inclination of the P7 helix (See Figure 3, p. 225).

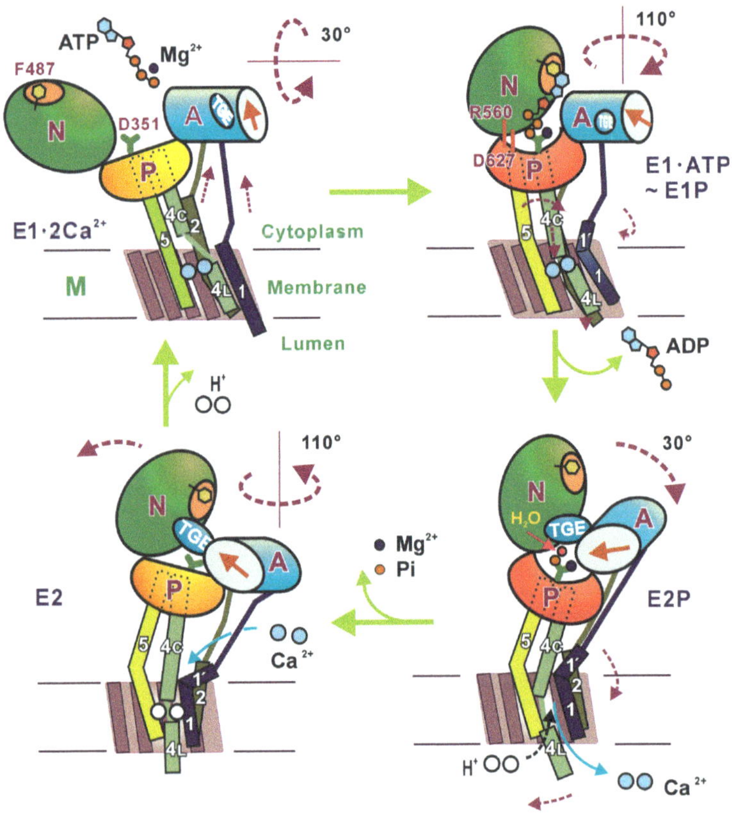

Color Plate 28. A cartoon illustrating the structural changes of the Ca^{2+}-ATPase during the reaction cycle based on the crystal structures in the five different states [14] (See Figure 4, p. 227).

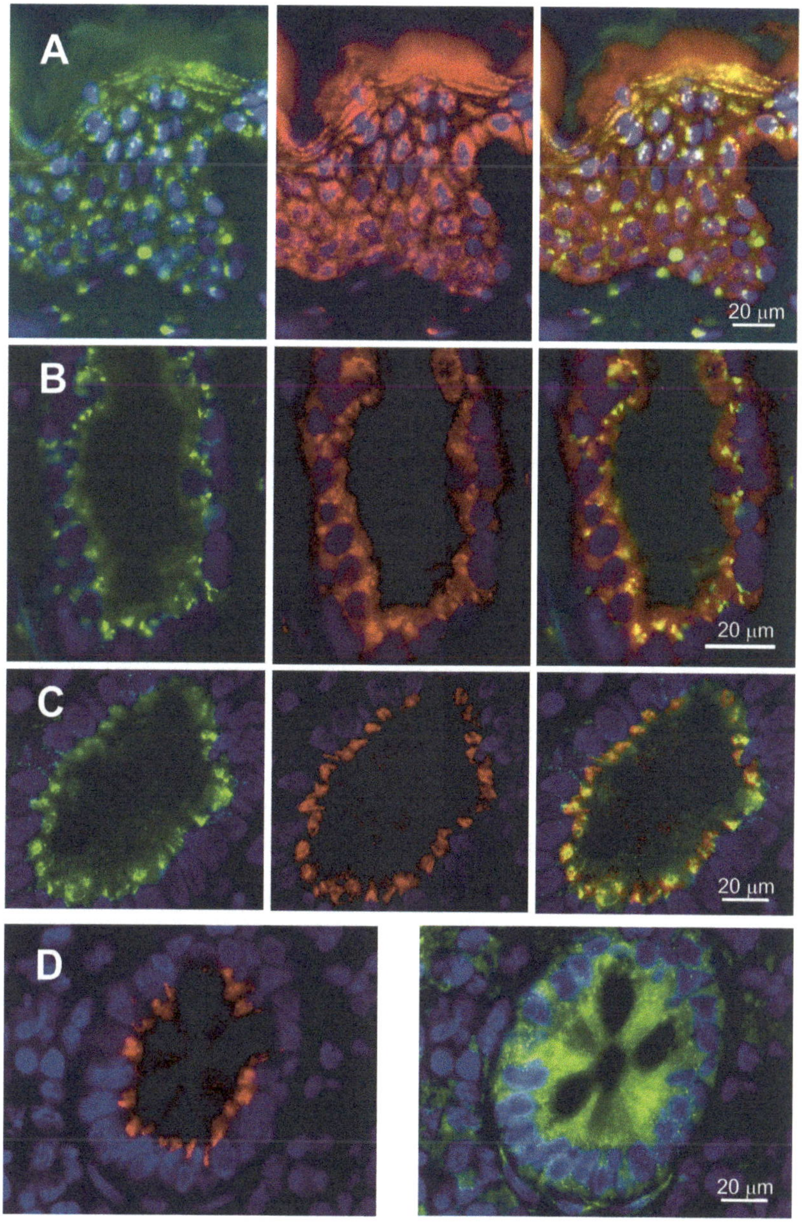

Color Plate 29. Immunocytochemical localization of secretory pathway Ca^{2+}-transport ATPase 1 (SPCA1) in human epidermis (**A**), prostate (**B**), colon (**C**), and of SPCA2 in human colon (**D**). In **A–C**, sections were triple labeled for *trans*-Golgi network 46 (TGN46) as a Golgi marker (green, left panels), for SPCA1 (red, middle panels) and with DAPI (4′,6-diamidino-2-phenylindole, dihydrochloride) to stain the nuclei (blue). The overlay is shown in the right panels. The section in **D** was stained for SPCA2 (red, left) and for SERCA2b (green, right), demonstrating the more restricted localization of SPCA2 as compared with SERCA2 (See Figure 2, p. 238).

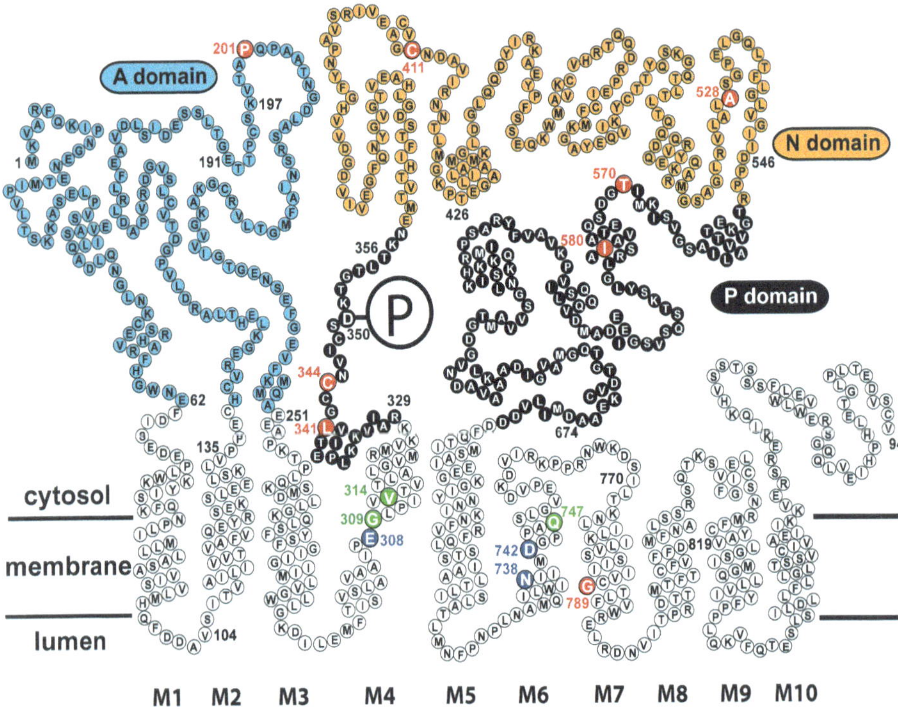

Color Plate 30. Schematic presentation of a planar model showing the primary and putative secondary structure of the human secretory pathway Ca^{2+}-transport ATPase 1d (SPCA1d) isoform highlighting various structural determinants and some of the Hailey–Hailey disease (HHD) mutants studied. The model is based on the crystal structure of sarco(endo)plasmic reticulum Ca^{2+}-transport ATPase 1a (SERCA1a) in the Ca^{2+}-bound E_1 form (reviewed in [102]) and adapted from the SERCA2b model described by Dode et al. [110]. Each amino acid residue is indicated by its single-letter code inside its corresponding circle. Stacked diagonal rows of three or four residues indicate α-helices, whereas the β-strands are shown as ladder-type residue arrangements. The loops connecting different secondary structures are shown as linear sections. The membrane portion of SPCA1d contains ten putative membrane-spanning helices, M1–M10. Critical amino acid residues involved in binding either Ca^{2+} or Mn^{2+} are shown in M4 and M6 as blue circles with white letters, whereas those involved in conferring SPCAs their well-known selectivity to Mn^{2+} binding and transport are depicted as green circles containing white letters. The phosphorylation site represented by Asp^{350} (with the phosphate group also indicated) is highlighted by inverted color. Some of the amino acids substituted in HHD patients analyzed in previous studies [8,150,151] are highlighted as red circles with white letters ($Pro^{201} \rightarrow$ Leu, $Leu^{341} \rightarrow$ Pro, $Cys^{344} \rightarrow$ Tyr, $Cys^{411} \rightarrow$ Arg, $Ala^{528} \rightarrow$ Pro, $Ile^{580} \rightarrow$ Val, $Thr^{570} \rightarrow$ Ile, and $Gly^{789} \rightarrow$ Arg). Other HHD mutations include $Gly^{309} \rightarrow$ Cys and $Asp^{742} \rightarrow$ Tyr [8], where the Gly and Asp residues are highlighted as green and blue circles, respectively, because they are structurally involved in ion recognition and binding processes. A-domain, actuator domain (azure circles); P-domain, phosphorylation domain (inverted color circles); and N-domain, nucleotide-binding domain (orange circles) (See Figure 5, p. 247).

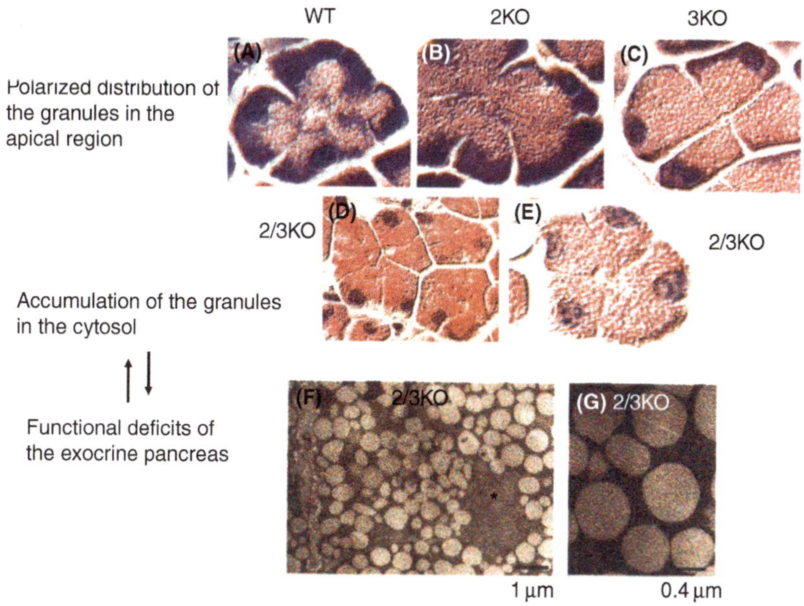

Color Plate 31. Role of IP_3 receptor type two (IP_3R2) and type 3 (IP_3R3) in exocrine secretion. Pancreatic acinar cells show polarized structure. In normal acinar cells, granules are accumulated at the apical pole. Single IP_3R deficient mice of type 2 (2KO) and type 3 (3KO) have cytoplasmic area with no secretion granules at the basal region. However, double deficient mice of type 2 and type 3 (2/3K0) show secretion granules fully in all the cytoplasm (See Figure 2, p. 271).

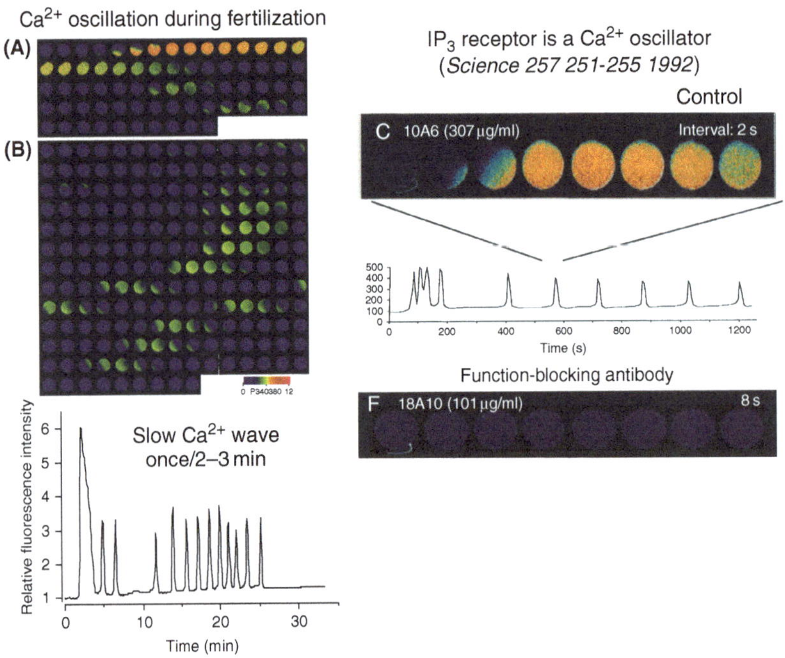

Color Plate 32. Role of IP_3 receptor type 1 (IP_3R1) in fertilization and in Ca^{2+} oscillation. Left panel: Ca^{2+} oscillation with very slow frequency (once in 2–3 min) is observed during fertilization in ascidian egg. Right panel: When function-blocking antibody (18A10) is introduced in hamster eggs, Ca^{2+} oscillation and fertilization is blocked even though the sperm is attached to the egg (left side of the bottom panel) which suggests that IP_3 and IP_3R1 is required for fertilization and Ca^{2+} oscillation (See Figure 3, p. 272).

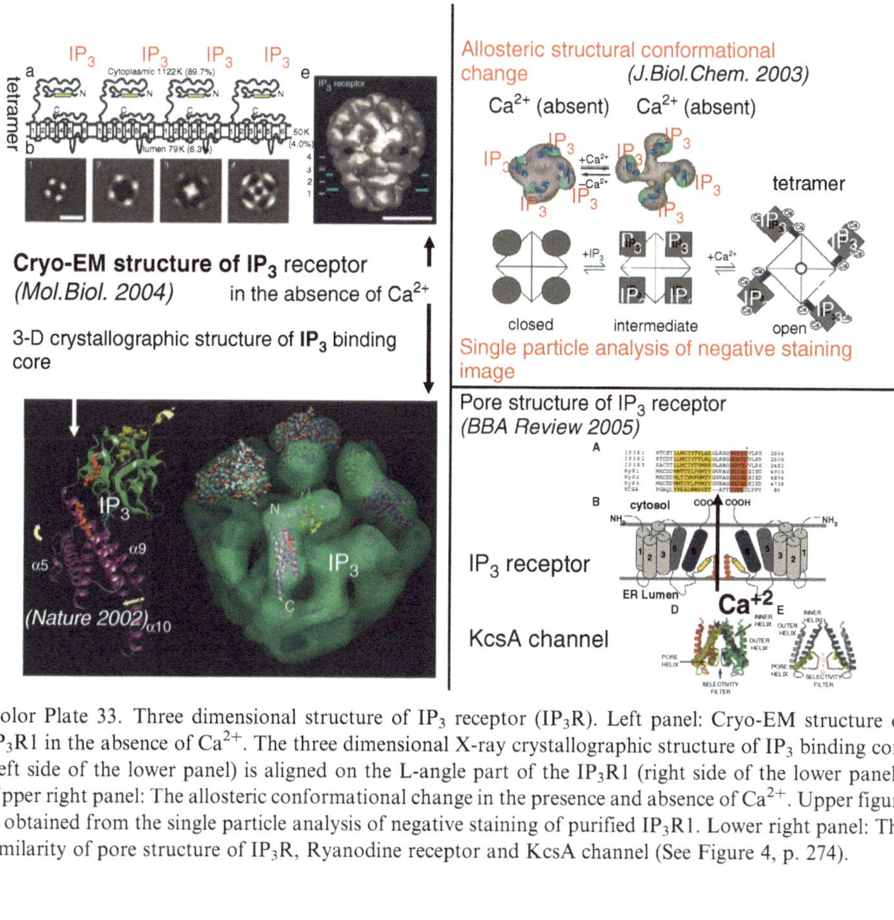

Color Plate 33. Three dimensional structure of IP_3 receptor (IP_3R). Left panel: Cryo-EM structure of IP_3R1 in the absence of Ca^{2+}. The three dimensional X-ray crystallographic structure of IP_3 binding core (left side of the lower panel) is aligned on the L-angle part of the IP_3R1 (right side of the lower panel). Upper right panel: The allosteric conformational change in the presence and absence of Ca^{2+}. Upper figure is obtained from the single particle analysis of negative staining of purified IP_3R1. Lower right panel: The similarity of pore structure of IP_3R, Ryanodine receptor and KcsA channel (See Figure 4, p. 274).

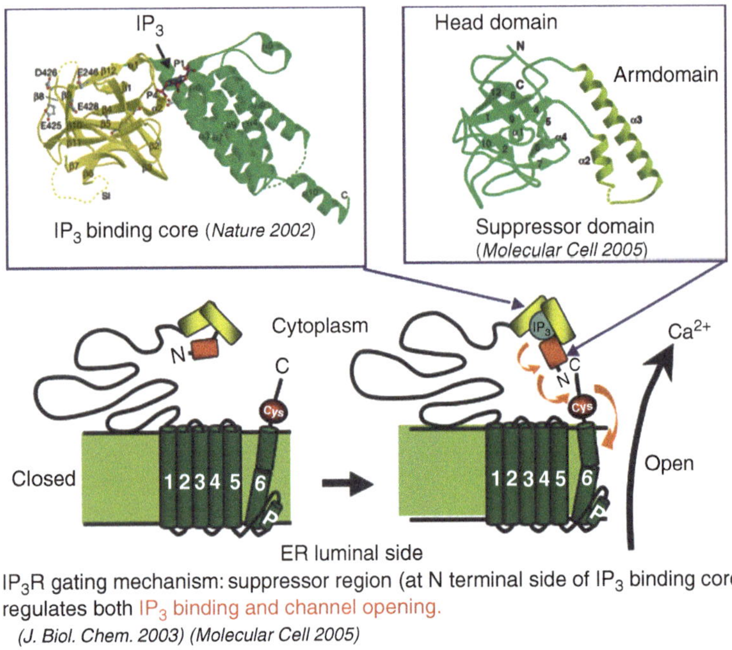

Color Plate 34. Gating mechanism of IP_3 ligated channel pore opening. IP_3 binding core and N-terminal suppressor region revealed by X-ray crystallographic analysis interact together with C-terminal portion of IP_3R to regulate both IP_3 binding and channel opening. Both cryo-EM structural analysis and X-ray crystallographic analysis provided important information to understand the gating mechanism of the IP_3R channel pore (See Figure 9, p. 281).

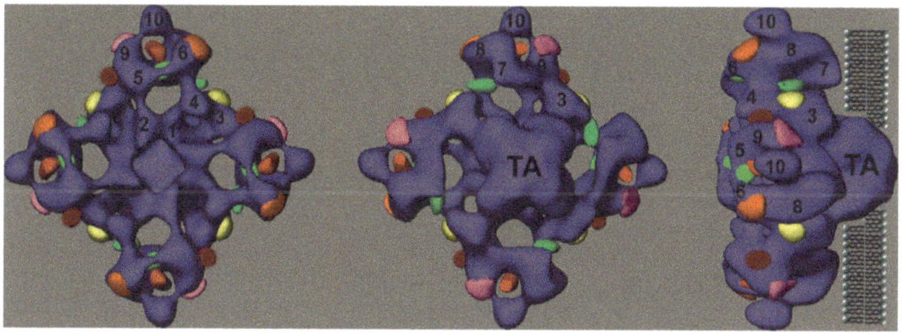

Color Plate 35. Ryanodine receptor three-dimensional structure. RyR architecture displayed as a projection view from the cytoplasm to sarcoplasmic reticulum (SR) lumen (left panel), the view from SR lumen to cytoplasm (middle panel) and side view (right panel). The numerals on the cytoplasmic assembly refer to distinguishable globular domains. TA indicates the transmembrane assembly. The location of the three divergent regions D1 (yellow), D2 (orange) and D3 (purple) as well as the N-terminal (red) and central domains (light green) is shown. The location of the binding sites for FK506-binding protein 12 (FKBP12) (brown) and calmodulin (green) is also shown. Images were kindly prepared by Dr. Zheng Liu (Wadsworth Center, Albany, NY, USA) (See Figure 1, p. 294).

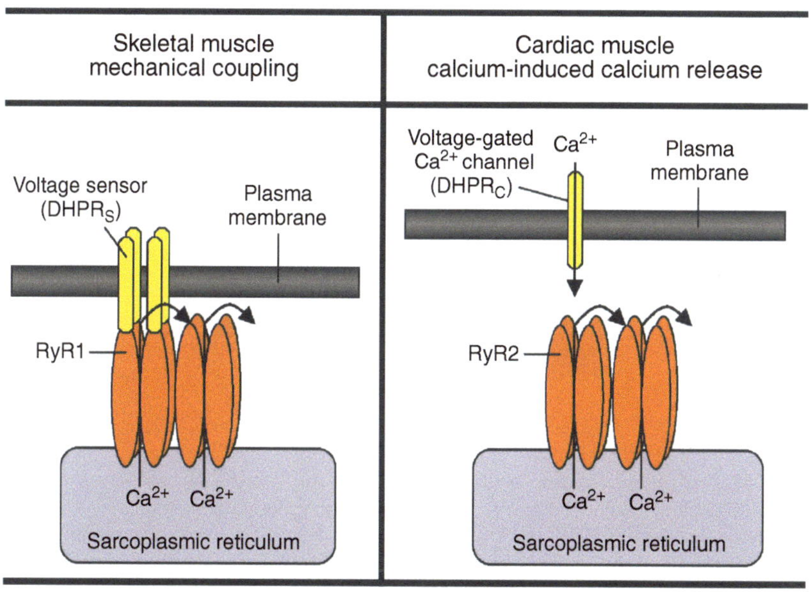

Color Plate 36. Ryanodine receptor (RyR) activation in striated muscles. In skeletal muscle dihydropyridine receptor (DHPR$_s$) channels sense the plasma membrane depolarization and transmit a conformational change to the RyR1 causing the channel to open and release Ca^{2+} from the sarcoplasmic reticulum (SR) store. Ca^{2+} influx is not required. Alternate RyR1 channels that are not coupled to DHPR$_s$ may either be activated by Ca^{2+} released from a neighbouring RyR1 or by transmission of the conformational change through physical interactions between RyR1 channels within the array. In cardiac muscle, DHPR$_c$ channels are activated by plasma membrane depolarization and allow Ca^{2+} influx into the cytoplasm. The Ca^{2+} entering the cell activates RyR2 channels allowing Ca^{2+} release from the SR store. Neighbouring RyR2 channels may in turn be activated by Ca^{2+} flowing from within the SR or by conformational changes transmitted through physical interactions between RyR2 channels within the array (See Figure 4, p. 299).

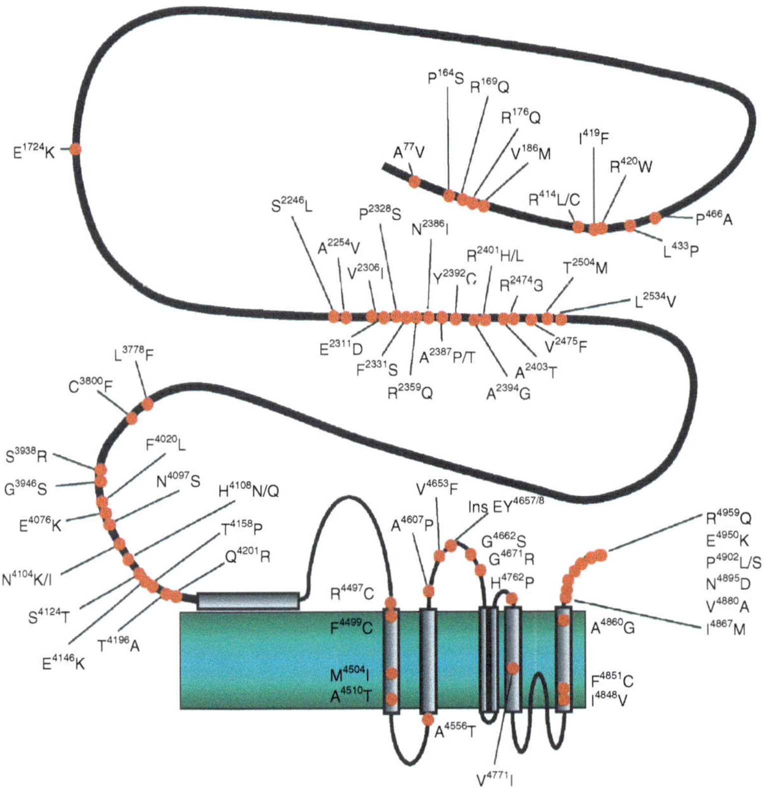

Color Plate 37. Ryanodine receptor 2 (RyR2) mutations associated with arrhythmogenic diseases. Schematic to illustrate catecholaminergic polymorphic ventricular tachycardia/arrhythmogenic right ventricular dysplasia type 2 (CPVT/ARVD2) mutations (red circles) clustered primarily in three discrete regions of the RyR2 polypeptide: the N-terminal (residues 77–466), the central (residues 2246–2534) and the C-terminal domain (residues 3778–4959). Image kindly prepared by Dr. N. Lowri Thomas (Wales Heart Research Institute, Cardiff University, UK) (See Figure 5, p. 322).

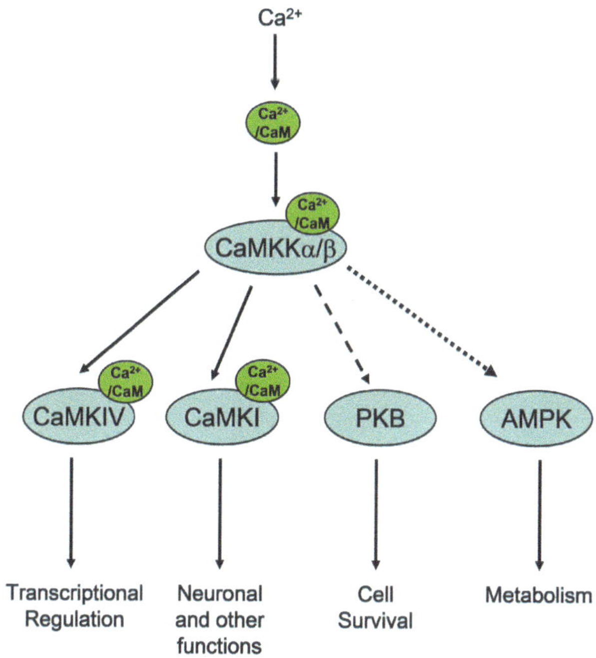

Color Plate 38. Multiple pathways are regulated by calcium/calmodulin-dependent kinase kinase (CaMKK). CaMKK is an upstream activator for CaMKI, CaMKIV, PKB (Protein Kinase B), and AMP-activated protein kinase (AMPK). CaMKKα and CaMKKβ can activate CaMKI and CaMKIV (indicated by solid arrows); CaMKKα has been shown to activate PKB (indicated by dashed arrow); CaMKKβ appears to be the major CaMKK isoform contributing to AMPK activation (indicated by dotted arrow). Ca^{2+}/CaM regulates both the upstream kinase (CaMKK) and target kinase (CaMKI or CaMKIV) of the CaMKK–CaMKI and CaMKK–CaMKIV cascades (See Figure 1, p. 350).

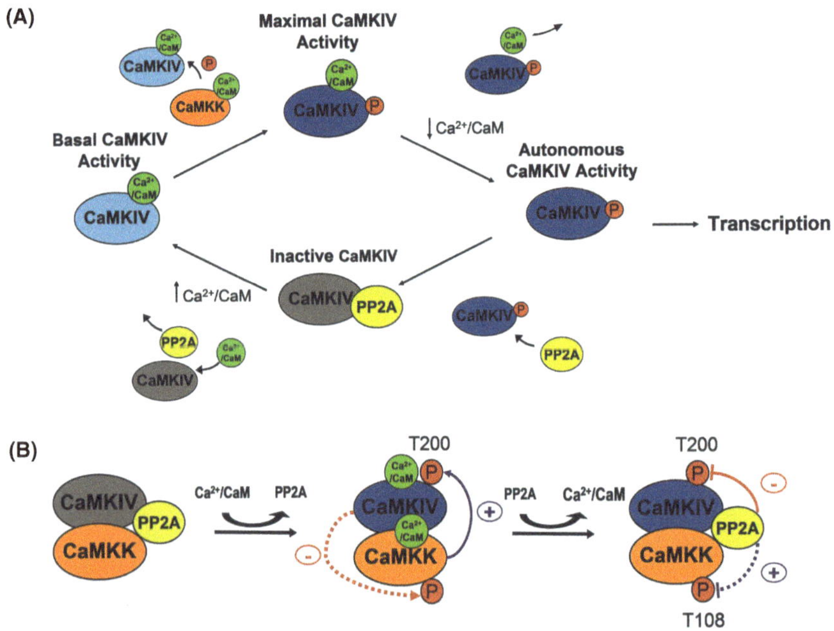

Color Plate 39. (**A**) Regulation of calcium/calmodulin-dependent kinase IV (CaMKIV) activity. CaMKIV interacts with PP2A and Ca^{2+}/CaM in a mutually exclusive manner; CaMKIV/PP2A complexes may be disrupted upon a Ca^{2+} stimulus. CaMKIV becomes basally active upon Ca^{2+}/CaM binding but only achieves autonomous activity (which is required for CaMKIV transcriptional function) after phosphorylation on residue T^{200} by CaMKK. PP2A negatively regulates CaMKIV by dephosphorylating residue T^{200} of CaMKIV. (**B**) Potential regulation of CaMKK–CaMKIV cascade components within a complex. CaMKIV stably interacts with both CaMKK and PP2A, and it is possible that a complex consisting of all three components might exist under some cellular conditions. In such a complex, CaMKIV would be readily accessible to phosphorylation by CaMKK, such that CaMKIV activation could occur almost immediately upon Ca^{2+}/CaM binding. CaMKIV can phosphorylate CaMKK in vitro, and an intriguing possibility is that activated CaMKIV might phosphorylate CaMKK on inhibitory residues such as T^{108}, thus regulating CaMKK in a negative feedback fashion. PP2A negatively regulates CaMKIV by dephosphorylating residue T^{200} and positively regulates CaMKK by dephosphorylating residue T^{108}, although it is unknown whether these two PP2A functions can occur within the same complex. Positive regulatory events are depicted in blue, whereas negative regulatory events are depicted in red. Confirmed regulatory events are represented with solid lines, whereas speculative events are represented by dotted lines (See Figure 2, p. 355).

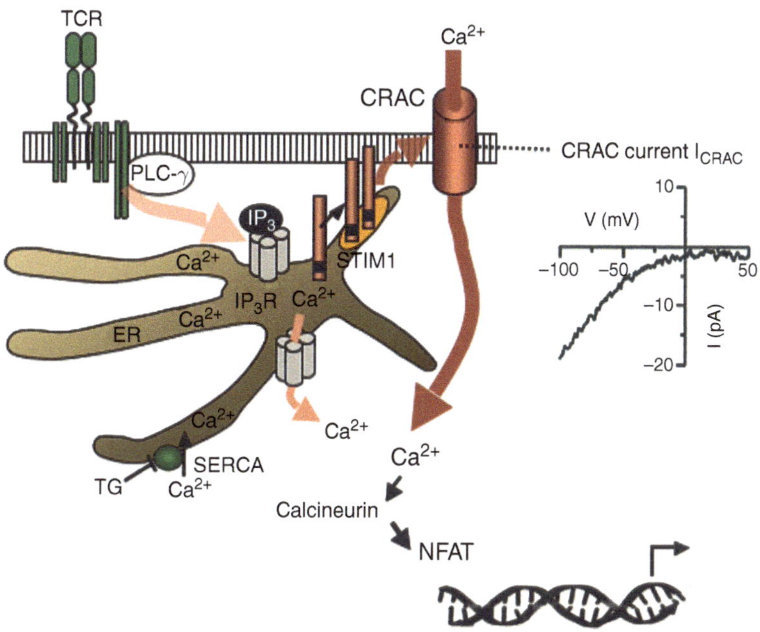

Color Plate 40. Overview of the Ca^{2+}–calcineurin–nuclear factor of activated T cells (Ca^{2+}–CN–NFAT) pathway. Stimulation of T cells through the T-cell receptor (TCR) induces activation of phospholipase C-γ (PLC-γ), which generates the second messenger inositol-1,4,5-trisphosphate (IP_3), which in turn binds to the IP_3 receptor (IP_3R), releasing Ca^{2+} from endoplasmic reticulum (ER) Ca^{2+} stores. Ca^{2+} store depletion activates an inside-out signalling pathway leading to the opening of store-operated Ca^{2+} channels like the Ca^{2+} release-activated Ca^{2+} (CRAC) channel, which gives rise to well-defined CRAC currents (I_{CRAC}). The process of CRAC channel activation is thought to involve relocation of stromal interaction molecule 1 (STIM1). Ca^{2+} influx through CRAC channels results in activation of the phosphatase CN, dephosphorylation of NFAT and gene expression. TG, thapsigargin; SERCA, sarcoplasmic–ER Ca^{2+}-ATPase (See Figure 1, p. 366).

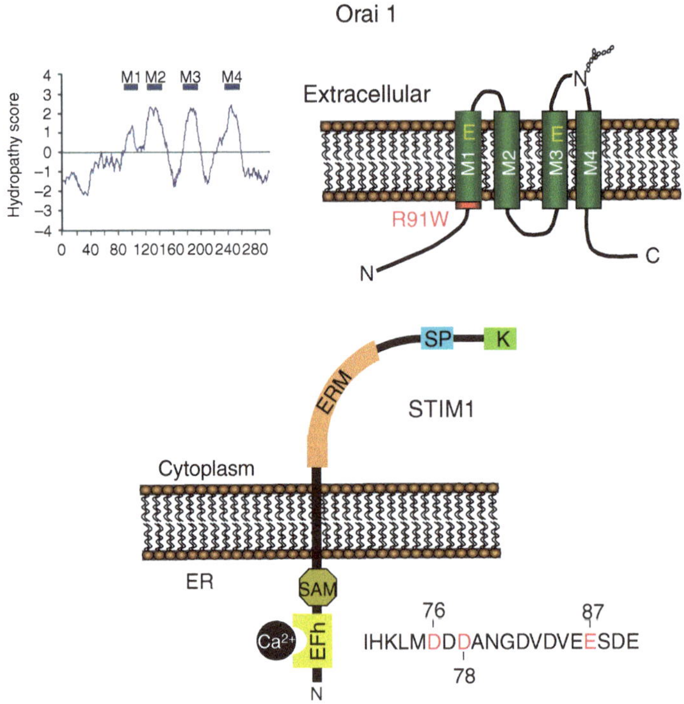

Color Plate 41. Domain architecture of Orai1 and stromal interaction molecule 1 (STIM1). Hydropathy analysis of Orai1 using the Kyte–Doolittle algorithm predicts a protein with four transmembrane segments (M1–M4) [59–61]. The predicted membrane topology of Orai1 is derived from the hydropathy plot and experiments suggesting localization of Orai1 in the plasma membrane and extension of the N- and C-termini of the protein in the cytoplasm [48,60]. The site of the R91W mutation in immunodeficient patients is indicated by a red bar [60], and two conserved glutamate (E) residues important for Ca^{2+} permeability of the Ca^{2+} release-activated Ca^{2+} (CRAC) channel are shown in yellow. STIM1 is a single-pass transmembrane protein, which is predominantly localized in the endoplasmic reticulum (ER). The N-terminus contains a Ca^{2+} binding EF-hand (EFh) motif, which senses ER Ca^{2+} levels. Several negatively charged residues critical for Ca^{2+} sensing are indicated in red (D76, D78 and E87); [28,40,239]. STIM1 also contains a sterile alpha motif (SAM), a coiled coil (CC)/ezrin–radixin–moesin (ERM) domain, a serine–proline-rich region (SP) and a lysine-rich region (K), which are likely to mediate protein–protein interactions (See Figure 2, p. 370).

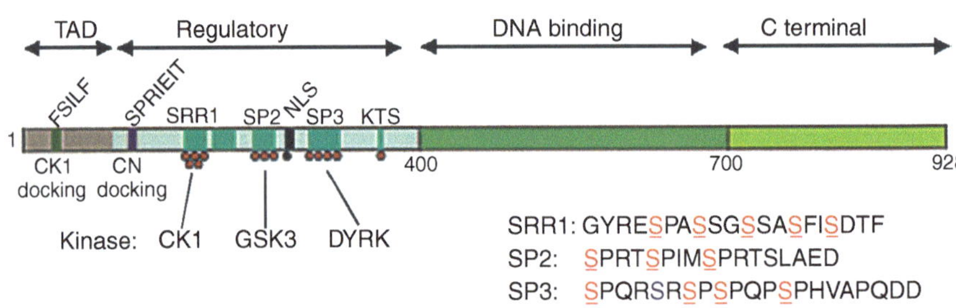

Color Plate 42. Schematic structure of the nuclear factor of activated T cells 1 (NFAT1) regulatory domain. Conserved phosphorylation sites are indicated by red circles and serine-rich motifs (SRR1, SP1, SP2, SP3 and KTS) by cyan bars. The black circle represents a conserved phosphorylation site near the nuclear localization sequence (NLS) that is not dephosphorylated by calcineurin (CN). Kinases (CK1, GSK3 and DYRK for the SRR1, SP2 and SP3 motifs, respectively) are listed below. At lower right are the sequences of the SRR1, SP2 and SP3 motifs, with the phosphorylated serines shown in red (See Figure 7, p. 381).

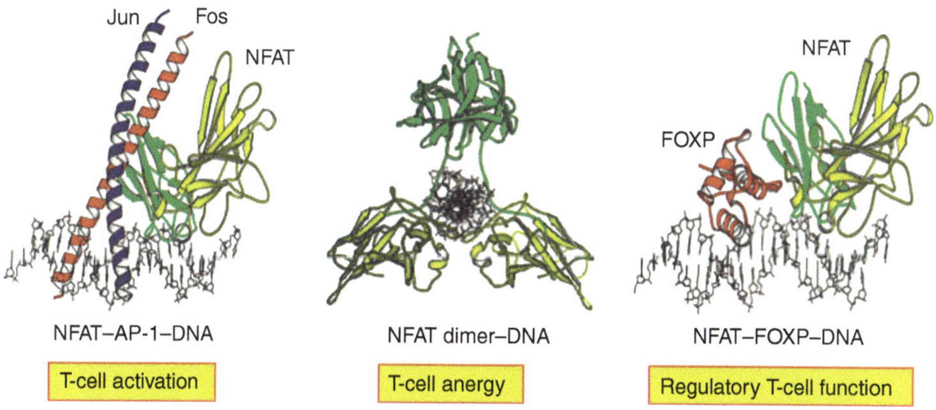

Color Plate 43. Nuclear factor of activated T cells (NFAT) can bind DNA in multiple configurations, each eliciting a distinct transcriptional programme characteristic of a distinct cellular response. Left, NFAT–DNA binding in cooperation with AP-1 (Fos-Jun) activates high-level transcription of cytokine and chemokine genes leading to productive T-cell immune responses; middle, DNA binding by NFAT in the absence of AP-1 induces T-cell anergy, in part through formation of NFAT dimers on DNA; and right, NFAT binding to NFAT–AP-1 composite DNA elements with a different transcriptional partner, FOXP3, induces 'dominant' peripheral tolerance by promoting the suppressive function of regulatory T cells (See Figure 8, p. 385).

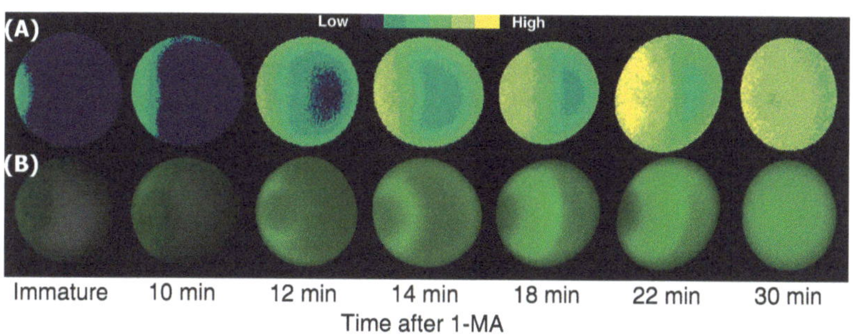

Color Plate 44. Increased sensitivity of the inositol 1,4,5-triphosphate receptors ($InsP_3Rs$) to $InsP_3$ during the starfish oocyte maturation process induced by 1-methyladenine (1-MA). (**A**) Relative fluorescence images showing the initiation of the Ca^{2+} signal in the animal hemisphere of seven oocytes at different maturation stages. Ca^{2+} release was induced by the uncaging of pre-injected caged $InsP_3$. The uncaging was induced after the injected $InsP_3$ had diffused globally to the entire oocyte. (**B**) Overlay of fluorescence and transmitted light microscopy of the same seven oocytes shown in panel **A** (See Figure 2, p. 431).

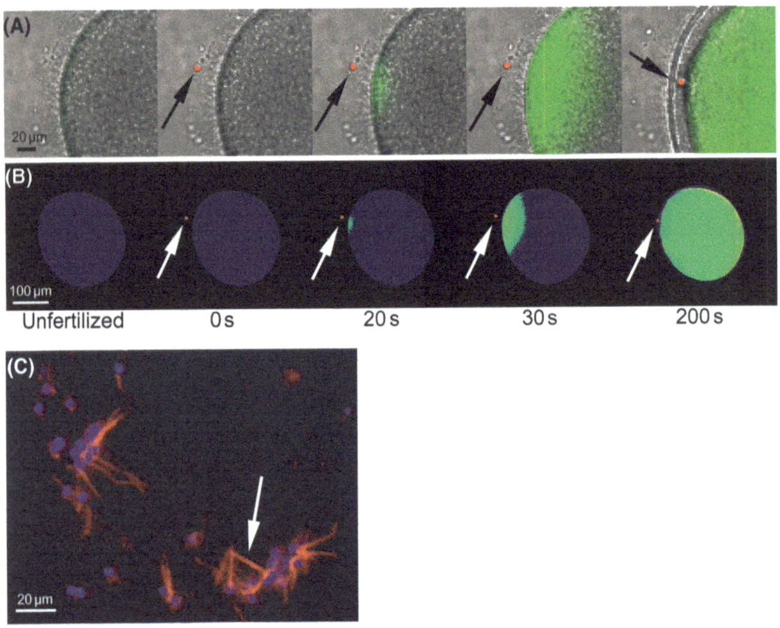

Color Plate 45. Ca^{2+} signaling induced by the sperm in starfish oocytes. Panel **A** shows an overlay of fluorescence images and transmitted light of a number of spermatozoa surrounding the oocyte. One of them (arrow), indicated in red, is going to extend the long acrosomal process that will initiate the Ca^{2+} response in the oocyte (third image from the left and panel **C**). The circumscribed Ca^{2+} elevation will spread centripetally while the spermatozoon is still at some distance from the surface of the oocyte (fourth image from the left). Eventually, the head of the spermatozoon makes contact with the oocyte at the time when the Ca^{2+} wave has become global (fifth image from the left). At this point, the fertilization membrane becomes completely elevated. At a later stage, the spermatozoon will be engulfed by the oocyte. Panel **B** shows relative fluorescence images of the same sequence of events shown in panel **A**. Panel **C** shows phalloidin-stained F-actin on the long acrosomal process of reacted spermatozoa (arrow) (See Figure 4, p. 433).

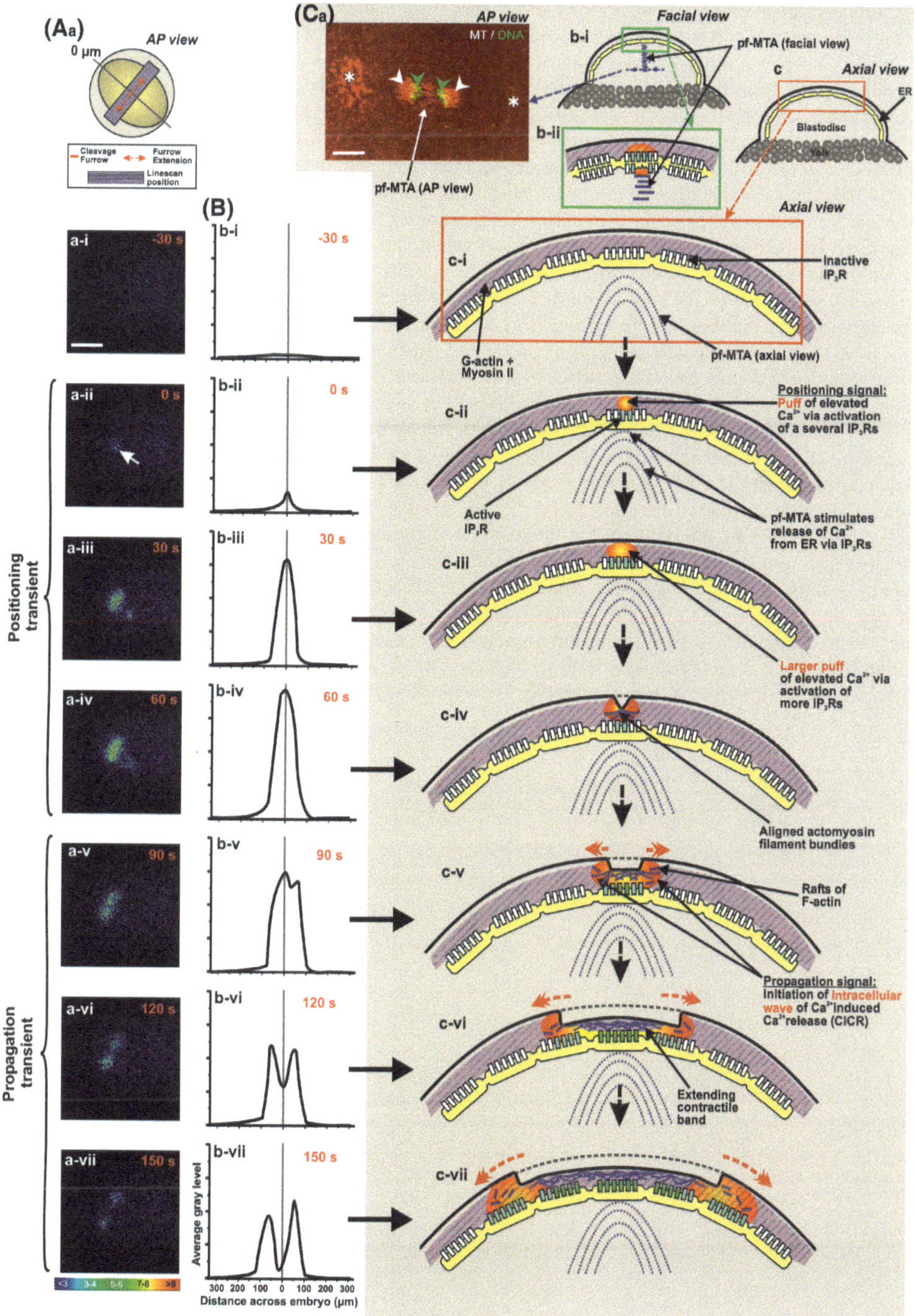

Color Plate 46. *(Continued)*

Color Plate 46 (Continued). Cleavage furrow positioning and propagation in zebrafish zygotes. **(A)** Representative animal pole (AP) view of an *f*-aequorin-loaded zebrafish embryo demonstrating the changes in intracellular free Ca^{2+} during positioning and propagation of the first cell division cycle. **(Aa)** A schematic illustration of the embryo to show the location and extension of the cleavage furrow as well as the position of a Linescan analysis (the results of which are shown in **(B)**). The luminescent images (pseudocolor panels, labeled a-i to a-vii) represent 10 s of accumulated luminescence and were obtained at the following time intervals: (a-i) 30 s before the appearance of the positioning transient; (a-ii) during the initial appearance of the positioning transient (see arrow); then (a-iii) 30 s, (a-iv) 60 s, (a-v) 90 s, (a-vi) 120 s, and (a-vii) 150 s after the initiation of the positioning transient. **(B)** Analysis of luminescence along the first cleavage furrow, before (b-i), and then during the positioning (b-ii to b-iv) and propagation (b-v to b-vii) transients. The profiles represent the average gray level of light emission from the region of the blastodisc shown in **(Aa)** and were measured using the Linescan function in the Metamorph (v.6.1) image analysis software. 0 μm represents the location on the blastodisc where the positioning signal was first detected and corresponds to the mid-point of the furrow. **(Ca)** A pre-furrowing microtubule array (pf-MTA) appears to arise from the mid-zone spindle and grows upward and outward to the blastodisc cortex where it precedes the appearance of the furrow on the surface of the blastodisc. This representative embryo was fixed during the metaphase/anaphase transition of the first division cycle and the microtubules (MTs; red) and DNA (green) labeled. In this single optical section through the blastodisc, the pf-MTA is located between the separating chromosomes (green arrowheads) and originates from the remnants of the mitotic spindle (white arrowheads). The MTOCs (asterisks) at the spindle poles are also shown. **(Cb-i)** A schematic representation of the blastodisc from a facial view to show the approximate location of the single confocal section in **(Ca)**. **(Cb-ii)** We propose that the pf-MTA transmits positional information to the blastodisc cortex, which results in the localized release of Ca^{2+} via the activation of IP_3Rs in the ER, and subsequently the correct positioning of the cytokinetic contractile apparatus. **(Cc)** Hypothetical model showing an axial view of a zebrafish blastodisc to illustrate how Ca^{2+} released via the activation of IP_3Rs in the ER might generate the furrow positioning and propagation transients during the first cell division cycle via a blip/puff/wave Ca^{2+}-signaling cascade. Scale bars represent 200 μm (in **A**) and 20 μm (in **Ca**). Reproduced with kind permission from *Zygote* [53,54] (See Figure 1, p. 454).

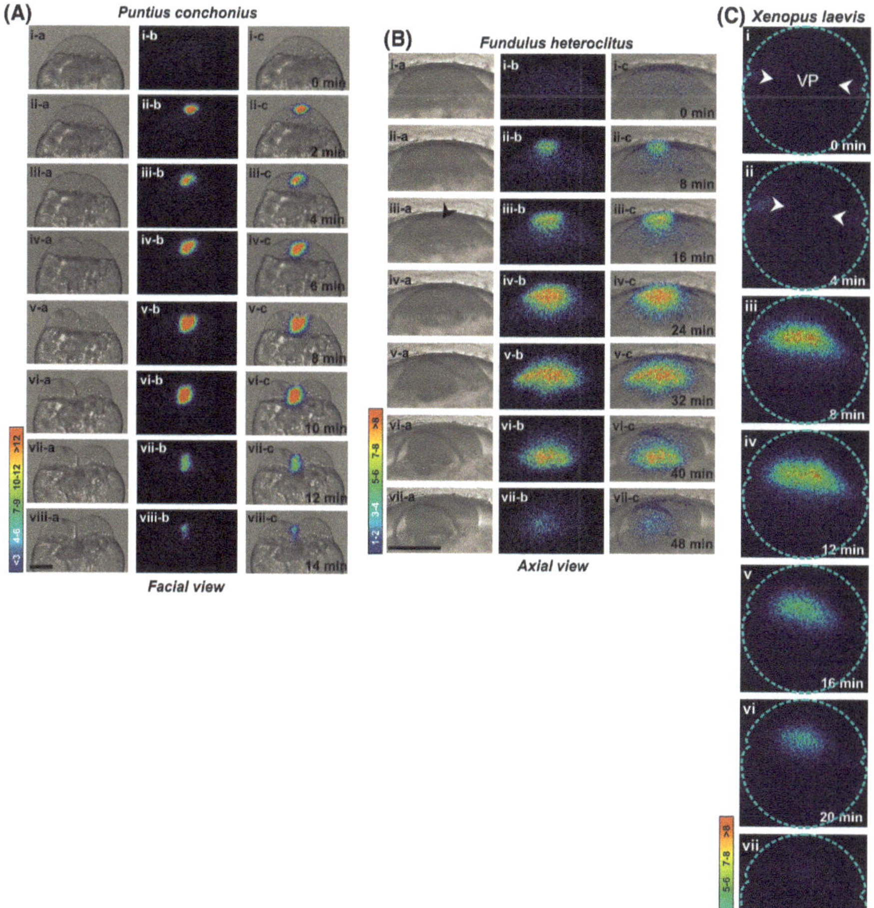

Color Plate 47. Representative examples of the changes in intracellular free Ca^{2+} that occur during the first cell division cycle in *f*-aequorin-loaded **(A)** *Puntius conchonius*, **(B)** *Fundulus heteroclitus*, and **(C)** *Xenopus laevis* embryos. In **(A)**, the luminescent images (pseudo-colored panels, labeled i-b to viii-b) represent 30 s of accumulated luminescence with a 1.5-min gap between each image, in **(B)** the luminescent images (i-b to vii-b) represent 30 s of accumulated luminescence with an 7.5-min gap between each image and in **(C)** the luminescent images (i to vii) represent 120 s of accumulated luminescence with a 2-min gap between each image. In **(A)** and **(B)**, corresponding bright-field images (in **A** labeled i-a to viii-a and in **B** labeled i-a to vii-a) were grabbed immediately after the respective luminescent images. The superimposed luminescent and bright-field images (shown in **A**, panels i-c to viii-c and in **B**, panels i-c to vii-c) show the location of the Ca^{2+} signals more clearly. Color scale indicates luminescent flux in photons/pixel. Scale bars represent 200 μm. **(A)**, **(B)**, and **(C)** are Webb et al., Chan et al., and Lee et al., unpublished results, respectively (See Figure 2, p. 456).

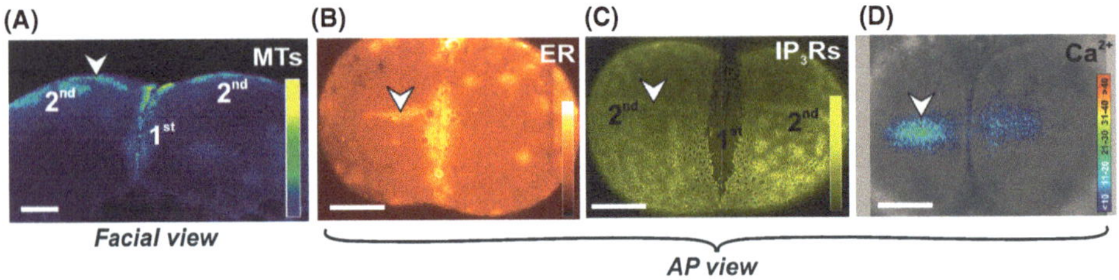

Color Plate 48. Localization of microtubules (MTs), endoplasmic reticulum (ER), IP_3 receptors (IP_3R_s), and Ca^{2+} before the first morphological appearance of the second cleavage furrow. These representative examples of embryos viewed from **(A)** a facial and **(B–D)** animal pole views show the co-localization (see arrowheads) of **(A)** the MTs, **(B)** the ER, **(C)** IP_3Rs, and **(D)** Ca^{2+} at the site of the future furrows of the second cell division cycle. **(D)** The aequorin-generated image (in pseudo-color) is superimposed on the appropriate bright-field image of the embryo (acquired just prior to the aequorin image) to show the position of the Ca^{2+} signal more clearly. This panel represents 60 s of accumulated luminescence. Color scale indicates luminescent flux in photons/pixel. Scale bars represent 100 μm. The first and second cleavage furrows are labeled in panels **(A)** and **(C)**. Reproduced with kind permission from *Zygote* [53] (See Figure 3, p. 457).

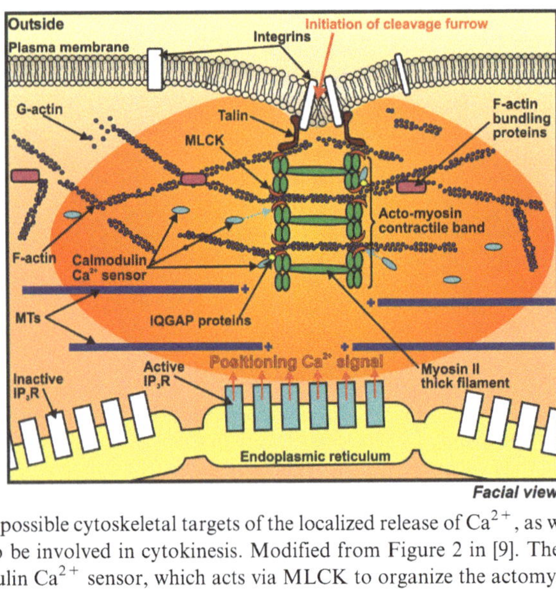

Color Plate 49. The possible cytoskeletal targets of the localized release of Ca^{2+}, as well as other molecular elements reported to be involved in cytokinesis. Modified from Figure 2 in [9]. The main Ca^{2+}-sensitive target is the calmodulin Ca^{2+} sensor, which acts via MLCK to organize the actomyosin-based contractile band (See Figure 4, p. 463).

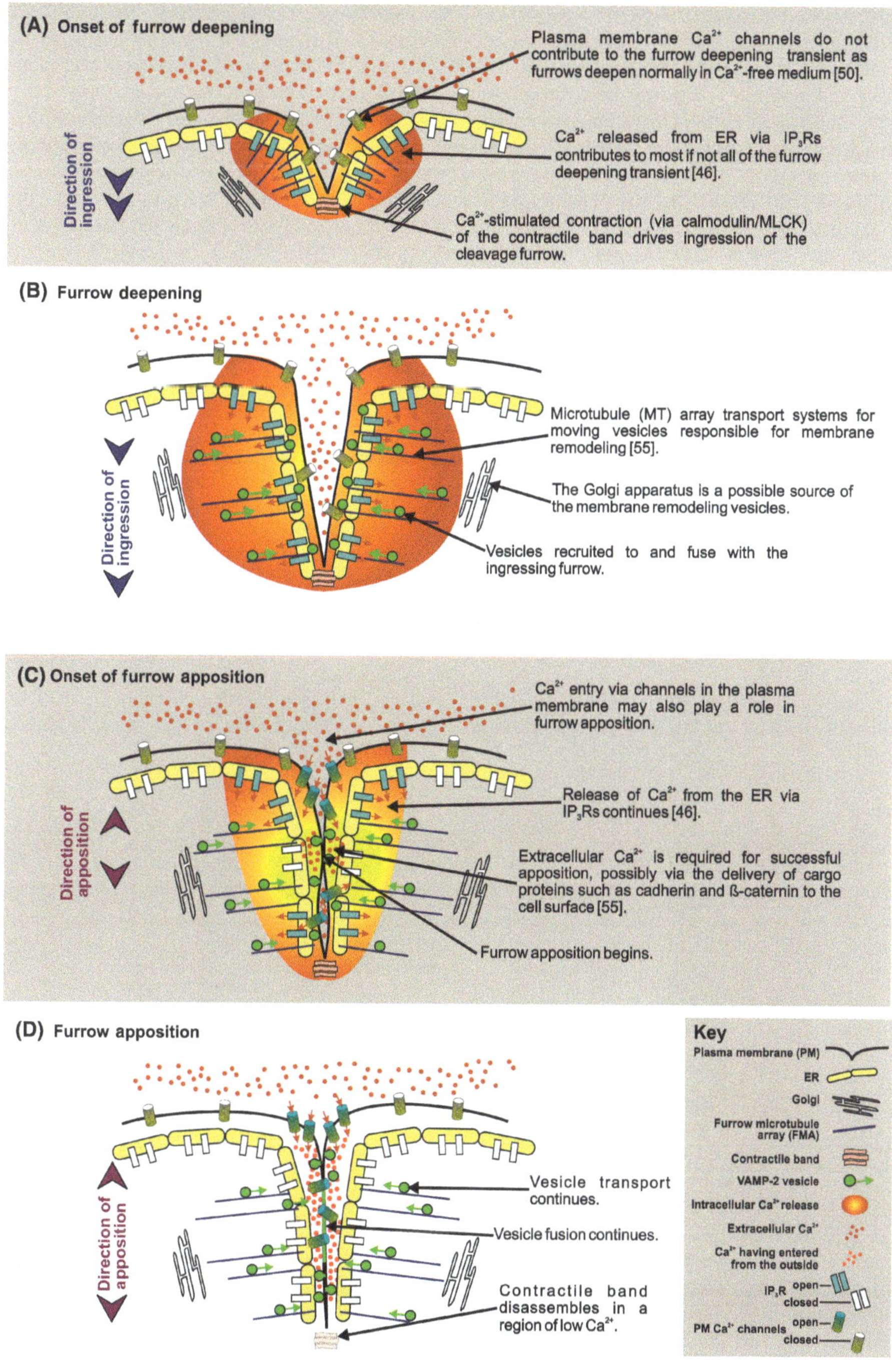

Color Plate 50. (*Continued*)

Color Plate 50 (Continued). Possible roles of Ca^{2+} signaling during cleavage furrow deepening and apposition in zebrafish zygotes. Hypothetical model showing a facial view of a zebrafish blastodisc to illustrate how Ca^{2+} released via the activation of IP_3Rs in the ER might generate the furrow deepening transient and how Ca^{2+} entering from the outside of the embryo through plasma membrane Ca^{2+} channels might generate the furrow apposition Ca^{2+} transient. The model also suggests that one of the possible downstream functions of these Ca^{2+} transients is to regulate vesicle recruitment and fusion during ingression and apposition. It is suggested that vesicle recruitment might be mediated by cognate v- and t-SNARE partners located on the vesicles and ingressing furrow membrane, respectively. It is also proposed that vesicles are transported to the deepening furrow by an array of perpendicular microtubules [55], which may have developed from the pf-MTA (See Figure 5, p. 464).

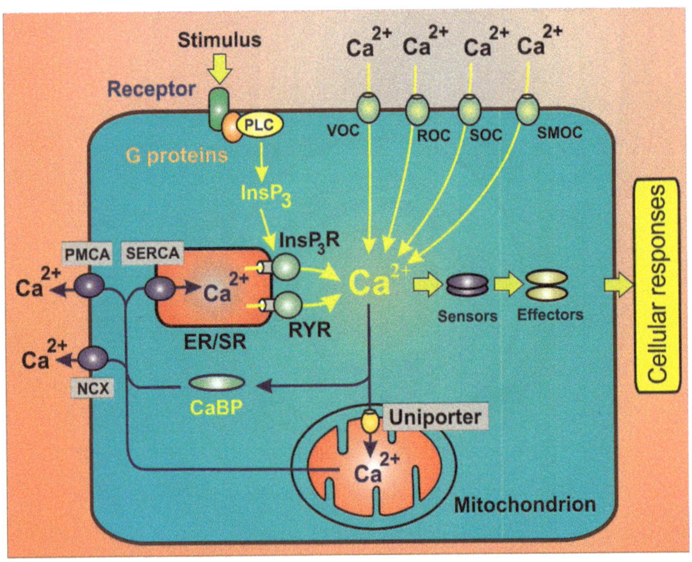

Color Plate 51. Summary of the major components of the Ca^{2+} signalling toolkit. Many of the individual components are present in multiple isoforms that further enhance the diversity of the Ca^{2+} signalling systems. The yellow arrows illustrate the ON reactions that introduce Ca^{2+} into the cell, and the blue arrows depict the OFF reactions that pump Ca^{2+} either out of the cell or back into the endoplasmic reticulum (ER). During its passage through the cytoplasm, Ca^{2+} resides temporarily on the Ca^{2+}-binding proteins (CaBPs) that function as buffers, or it can pass through the mitochondria. To carry out its signalling function, Ca^{2+} binds to sensors that then employ a range of effectors to stimulate many different cellular processes. $InsP_3R$, inositol 1,4,5-trisphosphate receptor; PLC, phospholipase C; NCX, Na^+/Ca^{2+} exchanger; PMCA, plasma membrane Ca^{2+}-ATPases; ROCs, receptor-operated channels; RYR, ryanodine receptor; SOCs, store OCs; SERCA, sarcoplasmic–ER Ca^{2+}-ATPase; SMOCs, second-messenger OCs; VOCs, voltage OCs (See Figure 1, p. 486).

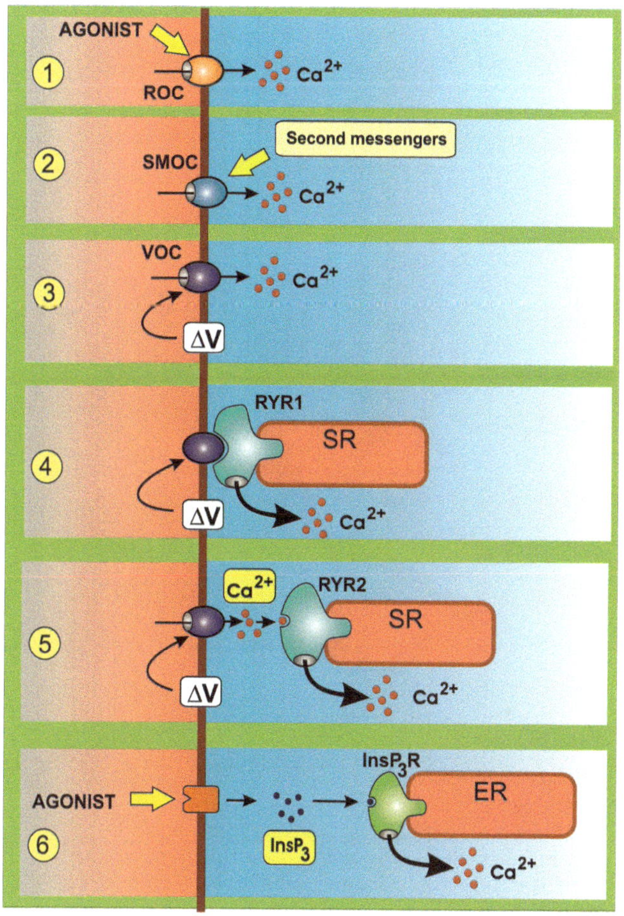

Color Plate 52. Summary of some of the main modules that are mixed and matched to create different Ca^{2+} signalling systems. (1) Agonists such as the neurotransmitters acetylcholine, glutamate and ATP act directly on receptor-operated channels (ROCs) in the plasma membrane (PM) to allow external Ca^{2+} to enter the cell. (2). Second messengers such as diacylglycerol, cyclic AMP, cyclic GMP and arachidonic acid acting from the cytoplasmic side open second-messenger OCs (SMOCs) in the PM. (3) Membrane depolarization (ΔV) activates voltage OCs (VOCs) in the PM to allow a rapid influx of external Ca^{2+}. (4) Membrane depolarization (ΔV) activates the $Ca_V1.1$ L-type channel that then activates the type 1 ryanodine receptor (RYR1) in skeletal muscle through a direct conformational-coupling mechanism. (5) Membrane depolarization (ΔV) activates VOCs in the PM to allow a rapid influx of external Ca^{2+} to provide trigger Ca^{2+} that then activates the RYR2 to release Ca^{2+} stored in the sarcoplasmic reticulum (SR). This mechanism is found in cardiac muscle and neurons. (6) Agonists acting on cell-surface receptors generate inositol 1,4,5-trisphosphate ($InsP_3$) that then diffuses into the cell to activate the $InsP_3$ receptor ($InsP_3R$) to release Ca^{2+} from the endoplasmic reticulum (ER) (See Figure 2, p. 488).

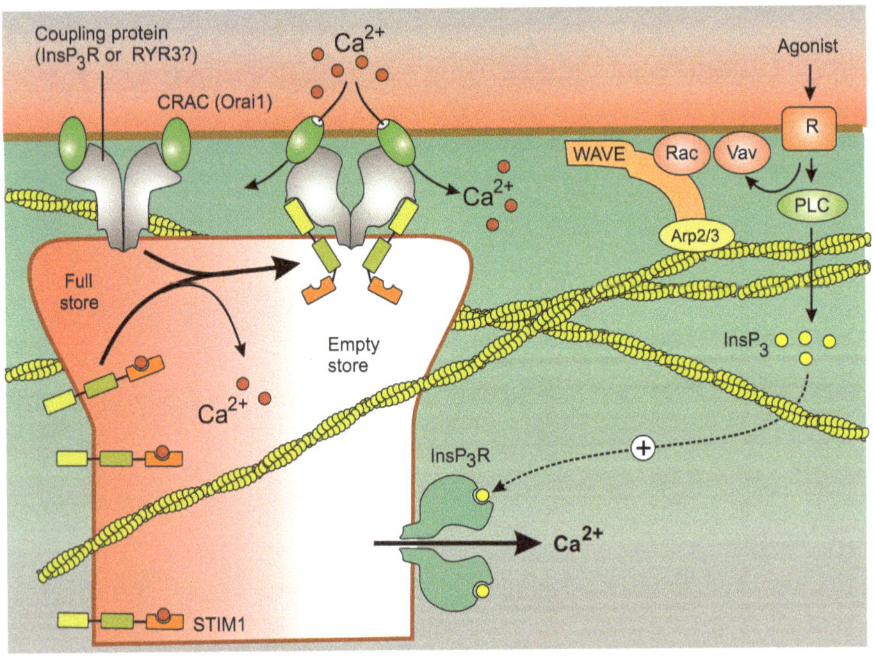

Color Plate 53. A conformational-coupling hypotheses for store-operated Ca^{2+} entry as might occur in lymphocytes. Entry of Ca^{2+} depends on activation of a store-operated channel (SOC) such as the Ca^{2+} release-activated Ca^{2+} (CRAC) channel. When an agonist activates phospholipase C (PLC), the increase in inositol 1,4,5-trisphosphate ($InsP_3$) then releases Ca^{2+} from the endoplasmic reticulum (ER) and the emptying of the store opens the CRAC channel. Store emptying might be detected by the stromal interaction molecule 1 (STIM1), which has an N-terminal EF-hand domain that binds Ca^{2+} when the store is full. As the store empties, Ca^{2+} comes off STIM causing it to undergo a conformational change that is transmitted through some coupling protein [perhaps an $InsP_3$ receptor ($InsP_3R$) or a ryanodine receptor (RYR)] to open the CRAC channels. Formation of this coupling complex appears to depend on cytoskeletal remodelling driven by WAVE2 and controlled through the monomeric G proteins Vav and Rac (See Figure 5, p. 492).

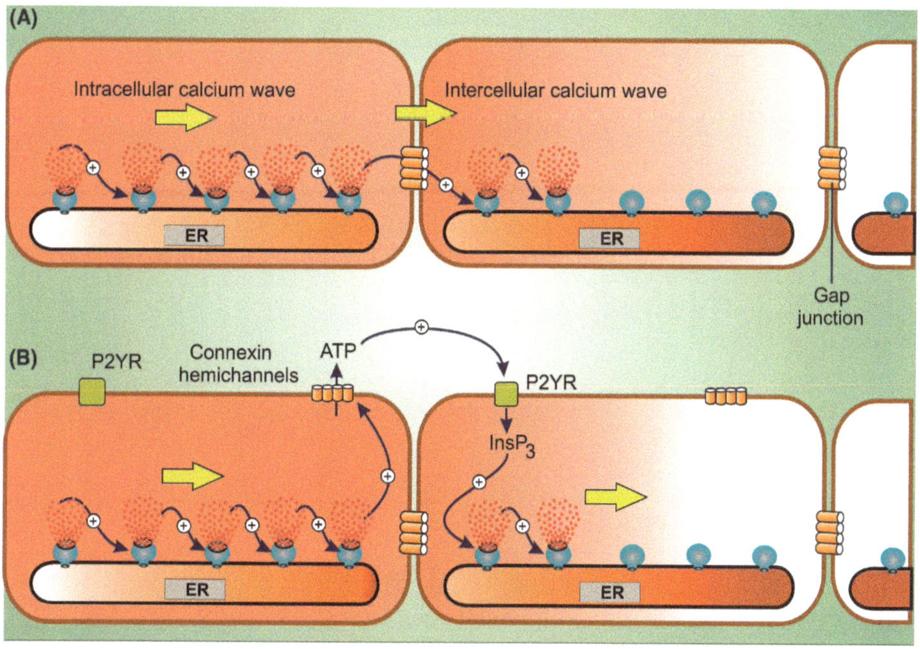

Color Plate 54. Intracellular waves depend on a process of Ca^{2+}-induced Ca^{2+} release (CICR) whereby the Ca^{2+} being released from one channel diffuses across to neighbouring channels that are excited to release further Ca^{2+} thereby setting up a regenerative wave. When such waves meet a cell boundary, they can trigger waves in adjacent cells thus setting up an intercellular wave. There is still some debate about the way the wave travels between cells. (**A**) One mechanism proposes that when the intracellular wave reaches the cell boundary, some small molecular weight component, most likely to be Ca^{2+}, diffuses across the gap junction to ignite another intracellular wave in the neighbouring cell. (**B**) An alternative mechanism suggests that the intracellular wave in one cell stimulates the release of ATP through hemichannels, which then diffuses across to ignite a wave in neighbouring cells by acting on P2Y receptors to produce inositol 1,4,5-trisphosphate ($InsP_3$) (See Figure 6, p. 494).

Index